D1359254

Compendium of Methods for the Microbiological Examination of Foods

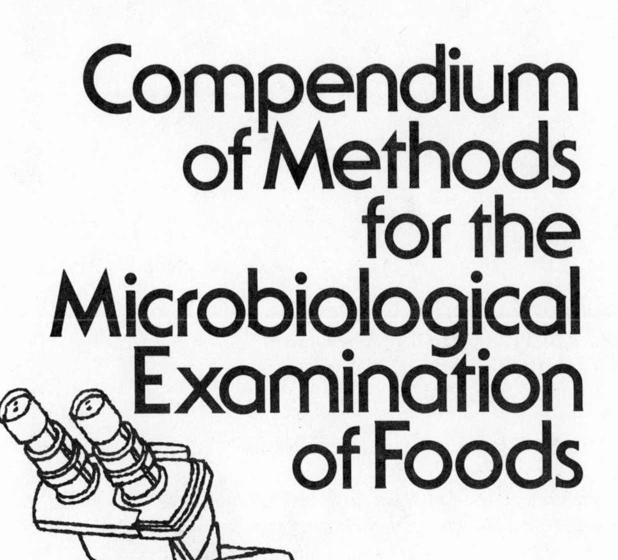

Prepared by the APHA Intersociety/Agency Committee on Microbiological Methods for Foods

Marvin L. Speck, Editor

Publisher:
American Public Health Association
1015 Eighteenth Street, NW
Washington, DC 20036

Library of Congress Catalog Number: 76-26862
International Standard Book Number: 0-87553-081-8

Library of Congress Cataloging in Publication Data

American Public Health Association. Intersociety/Agency
Committee on Microbiological Methods for Foods.
Compendium of methods for the microbiological examination of foods.

Bibliography: p.
Includes index.
1. Food—Microbiology—Technique. I. Speck,
Marvin L., 1913- II. Title
QR115.A5 1976 576'.163 76-26862
ISBN 0-87553-081-8

Printed in the United States of America
Typography: Bru-El Graphic Inc., Springfield, VA
Set in: *Baskerville, Trade Gothic*
Text and Binding: Rose Printing Co., Tallahassee, FL
Cover Design: Donya Melanson Assoc., Boston, MA

INTERSOCIETY/AGENCY COMMITTEE ON METHODS FOR THE MICROBIOLOGICAL EXAMINATION OF FOODS

MARVIN L. SPECK, Chairman
William Neal Reynolds Professor, Food Science and Microbiology, Department of Food Science, North Carolina State University, Raleigh, NC 27607
Representative: American Public Health Association

CLEVE B. DENNY
Director of Research Services, National Canners Association, 1133 20th Street, N.W., Washington, DC 20036
Representative: National Canners Association

R. PAUL ELLIOTT
Formerly Chief, Microbiology Branch, SS, APHIS, U.S. Department Agriculture; Present address: 1095 Lariat Lane, Pebble Beach, CA 93953. Resigned from IS/A Committee in 1974.

E. M. FOSTER
Director, Professor of Bacteriology, Food Research Institute, The University of Wisconsin, Madison, WI 53706
Representative: American Society of Microbiology

JOHN T. GRAIKOSKI
Supervisory Research Microbiologist, National Marine Fisheries Service, Mid-Atlantic Coastal Fisheries Center, NOAA, Milford, CT 06460
Representative: National Marine Fisheries Service

W. J. HAUSLER, JR.
Director, State Hygienic Laboratory, University of Iowa
Iowa City, IA 52240
Representative: Association of State and Territorial Public Health Laboratory Directors, Association of Food and Drug Officials of the United States

GEORGE J. HERMANN
1827 Mason Mill Road, Decatur, GA 30033
Representative: Center for Disease Control

NINO F. INSALATA
Research Manager, Microbiological Research, General Foods Corporation, Central Research, Tarrytown, NY 10591
Representative: National Environmental Health Association, Institute of Food Technologists

RALPH W. JOHNSTON
Chief, Microbiology Staff, Meat and Poultry Inspection Program, U.S.D.A., Animal and Plant Health Inspection Service, Washington, DC 20250
Representative: U.S. Department of Agriculture

Edmund M. Powers

Microbiology Division, Food Laboratory, U.S. Army Natick Laboratories, Natick, MA 01760

Representative: U.S. Department of Defense, Food Program

John H. Silliker

President, Silliker Laboratories, 1139 E. Dominquez Street, Suite 1, Carson, CA 90746

Representative: American Council of Independent Laboratories

Carl Vanderzant

Professor of Food Microbiology, Animal Science Department, Texas A & M University, College Station, TX 77840

Representative: International Association of Milk, Food and Environmental Sanitarians

PROJECT DIRECTOR

Howard L. Bodily

APHA Staff Associate for Laboratory Programs, P.O. Box 247, Midway, UT 84049

PROJECT OFFICER

Joseph C. Olson, Jr.

Division of Microbiology, Bureau of Foods, Food and Drug Administration, 200 C. Street, S.W., Washington, DC 20204

Representative: Association of Official Analytical Chemists, Food and Drug Administration

AUTHORS AND CONTRIBUTORS

JOHN A. ALFORD

Laboratory Chief, Dairy Foods Nutrition Laboratory, Nutrition Institute, USDA, Bldg 157, ARC East, Beltsville, MD 20705

R. E. ARENDS

Contech Laboratory, Pet Inc., Greenville, IL 62246

DAVID H. ASHTON

Hunt-Wesson Foods, 1645 W. Valencia Drive, Fullerton, CA 92634

EDWARD F. BAER*

Microbiologist, FB, 05, Division of Microbiology BF-124, Food and Drug Administration, 200 C Street, S.W., Washington, DC 20204

JOHN A. BAROSS

Research Associate, Department of Microbiology, Bioscience Bldg, Oregon State University, Corvallis, OR 97330

T. A. BELL

U.S. Department of Agriculture, Box 5578, Raleigh, NC 27607

REGINALD W. BENNETT

Research Microbiologist, Division of Microbiology, Food and Drug Administration, 200 C Street, S.W., Washington, DC 20204

MERLIN S. BERGDOLL

Professor-Enterotoxin, University of Wisconsin, Food Research Institute, 2115 Herrick Drive, Madison, WI 53706

A. RICHARD BRAZIS

Chief, Laboratory Development Section, Division of Microbiology, Food and Drug Administration, 1090 Tusculum Avenue, Cincinnati, OH 45226

JAMES J. BRINDA

Director, Bureau of Environmental Hygiene, City of Minneapolis, Division of Public Health, 250 S. 4th Street, Minneapolis, MN 55402

FRANK L. BRYAN

Chief, Food Borne Diseases Activity, Health Agencies, Branch Training Program, Center for Disease Control, Atlanta, GA 30333

FRANCIS F. BUSTA

Department of Food Science and Nutrition, 238 FSI Building, University of Minnesota, St. Paul, MN 55101

JEPTHA E. CAMPBELL

Chief, Microbial Biochemistry, Division of Microbiology, FDA, 1090 Tusculum Avenue, Cincinnati, OH 45226

JAMES C. CANADA

Quality Control Manager, Chemical and Microbiological Services, Gerber Products Company, 445 State Street, Fremont, MI 49412

RICHARD T. CAREY

National Supervisor, Food Technologist, Granding Branch, USDA, AMS, Poultry Division Grading Branch, Washington, DC 20250

THOMAS C. CHENG

Institute for Pathology, Lehigh University, Bethlehem, PA 18015

*Deceased

Lee R. Chugg
General Foods Corporation, Technical Center, 250 North Street, White Plains, NY 10625
D. O. Cliver
Associate Professor, University of Wisconsin, Food Research Institute, Madison, WI 53706
D. A. Corlett, Jr.
Manager, Microbiology, Del Monte Research Center, 205 N. Wiget Lane, Walnut Creek, CA 94598
Robert H. Deibel
Professor, Department of Bacteriology, 1550 Linden Drive, University of Wisconsin, Madison, WI 53706
Wallis E. Dewitt
Chief, Enteric Investigation Unit, Bacterial Diseases Branch, Center For Disease Control, Altanta, GA 30333
John M. Dryer
Manager, Microbiology, H. J. Heinz Company, P.O. Box 57, Pittsburgh, PA 15230
Charles Lee Duncan
Associate Professor, University of Wisconsin, Food Research Institute, 1925 Willow Drive, Madison, WI 53706
William V. Eisenberg
Microanalytical Branch, Division of Microbiology, HFF-127, Food and Drug Administration, 200 C Street, S.W., Washington, DC 20204
J. L. Etchells
U.S. Department of Agriculture, Box 5578, Raleigh, NC 27607
Martin S. Favero
Deputy Chief, Phoenix Laboratories, U.S. Public Health Service, CDC-Phoenix Field Station, 4402 N. Seventh Street, Phoenix, AZ 85014
James C. Feeley
Chief, Special Pathogens Unit, Epidemiology Program, Bacterial Diseases Branch, Center for Disease Control, Atlanta, GA 30333
Morris Fishbein*
Microbiologist, Food and Drug Administration, Division of Microbiology-HFF-124, 200 C Street, S.W., Washington, DC 20204
Damien Gabis
Vice-President, Silliker Laboratories, 1304 Halsted Street, Chicago Heights, IL 60411
Stanley E. Gilliland
Associate Professor, Department of Food Science, North Carolina State University, Raleigh, NC 27607
J. M. Goepfert
Associate Professor, University of Wisconsin, Food Research Institute, 2115 Herrick Drive, Madison, WI 53706
James M. Gorman
Technical Director, Food Division, Seymour Foods, Inc., P.O. Box 1220, Topeka, KS 66601

*Deceased

Rodney J. H. Gray
University of Delaware, Department of Foods and Nutrition, 206 Allison Hall, Newark, DE 19711
V. W. Greene
School of Public Health, University of Minnesota, Minneapolis, MN 55455
Philip A. Guarino
Manager, Corporate Analytical Services, 414 Light Street, Baltimore, MD 21202
Lester Hankin
Biochemist, Connecticut Agricultural Experimentation Station, New Haven, CT 06504
Stanley M. Harmon
Research Microbiologist, Division of Microbiology BF-124, Food and Drug Administration, 200 C Street, S.W., Washington, DC 20204
Paul A. Hartman
Distinguished Professor, Department of Bacteriology, Iowa State University, Ames, IA 50010
Glen L. Hayes
Senior Bacteriologist, American Can Company, 433 North Northwest Highway, Barrington, IL 60010
George R. Healy
Chief, Parasitology Section, Parasitology Branch, Center for Disease Control, Atlanta, GA 30333
John S. Hilker
Microbiological Manager, M & M / Mars, High Street, Hackettstown, NJ 07840
W. Mike Hill
Armour Foods Company, Armour Research Center, Scottsdale, AZ 85260
W. E. Hobbs
Head, Microbiological Regulatory Services, General Mills, Inc., 9000 Plymouth Avenue North, Minneapolis, MN 55427
Glen L. Hoffman
Fish Farming Experimental Station, U.S.S.W.S., P.O. Box 860, Stuttgart, AK 72160
Daniel A. Hunt
Assistant Chief, Shellfish Sanitation Branch, Division of Shellfish Sanitation, BF-230, Food and Drug Administration, 200 C Street, S.W., Washington, DC 20204
Margaret Huston
Scientific Director, Poultry and Egg Institute of America, 67 East Madison Street, Chicago, IL 60603
John J. Iandola
Associate Professor, Microbiology, Division of Biology, Kansas State University, Manhattan, KS 66506
Keith A. Ito
Head, Microbiology Section, National Canners Association, 1950-6th Street, Berkeley, CA 94710

George J. Jackson
Chief, Parasitology Laboratory, Division of Microbiology, Food and Drug Administration, 200 C Street, S.W., Washington, DC 20204

Donald A. Kautter
Assistant Chief, Division of Microbiology (BF-124), Food and Drug Administration, 200 C Street, S.W., Washington, DC 20204

John A. Koburger
Associate Professor, Department of Food Science, University of Florida, Gainsville, FL 32611

Harry E. Korab
Technical, National Soft Drinks Association, 1101 16th Street, N.W., Washington, DC 20036

Allen A. Kraft
Professor, Food Research Laboratory, Iowa State University, Ames, IA 50010

C. P. Kurtzman
Zymologist, Northern Regional Research Laboratory, Agricultural Research Service, U.S. Department of Agriculture, 1815 North University Street, Peoria, IL 61604

Louis C. LaMotte, Jr.
Licensure and Proficiency Testing Division, Bureau of Laboratories, Center for Disease Control, Atlanta, GA 30333

J. S. Lee
Department of Food Science, Oregon State University, Corvallis, OR 97331

W. H. Lee
Research Microbiologist, Division of Microbiology, Food and Drug Administration, 200 C Street, S.W., Washington, DC 20204

Harold V. Leininger
Director, Minneapolis Microbiological Facility, DHEW, PHS, Food and Drug Administration, 240 Hennepin Avenue, Minneapolis, MN 55401

J. Ralph Lichtenfels
Parasite Classification and Distribution Unit, Animal Parasitology Institute, U.S.D.A., A.R.S., B.A.R.C., East 120 Beltsville, MD 20705

John Liston
Institute for Food Science and Technology, College of Fisheries, University of Washington, Seattle, WA 98195

Joseph Lovett
Food and Drug Administration, 1090 Tusculum Avenue, Cincinnati, OH 45226

Richard K. Lynt
Division of Microbiology BF 124, Food and Drug Administration, 200 C Street, S.W., Washington, DC 20204

J. R. Matches
College of Fisheries, University of Washington, Seattle, WA 98195

Duane T. Maunder
Continental Can Company, Inc., 1350 W 76 Street, Chicago, IL 60620

J. L. MAYOU
Microbiology, R & D Laboratories, The Pillsbury Company, 311 Second Street, S.E., Minneapolis, MN 55414

IRA J. MEHLMAN
Food and Drug Administration-HFF-125, 200 C Street, S.W., Washington, DC 20204

JAMES W. MESSER
Food Microbiology Branch, Division of Microbiology, Food and Drug Administration, 1090 Tusculum Avenue, Cincinnati, OH 45226

H. DAVID MICHENER
Microbiology Research Unit, Western Regional Laboratory, U.S. Department of Agriculture, Berkeley, CA 94710

THADDEUS F. MIDURA
California State Department of Health, 2151 Berkeley Way, Berkeley, CA 94704

JOHN J. MIESCIER
N. E. Technical Service Unit, Food and Drug Administration, CBC Building S-26, Davisville, RI 02852

WILLIAM A. MOATS
Dairy Quality Investigations, Market Quality Research Division, Bldg 309, ARC East, U.S. Department of Agriculture, Beltsville, MD 20705

GEORGE K. MORRIS
Chief, Epidemiologic Services Laboratory Section Center for Disease Control, Atlanta, GA 30333

J. ORVIN MUNDT
Department of Microbiology, University of Tennessee, Knoxville, TN 37916

D. I. MURDOCK
The Coca-Cola Company, Foods Division, P.O. Box 368, Plymouth, FL 32768

M. J. NAKAMURA
Professor and Chairman, Department of Microbiology, University of Montana, Missaula, MT 59801

F. E. NELSON
Professor, 225 Agricultural Sciences, University of Arizona, Tucson, AZ 85721

Z. JOHN ORDAL
Professor, Department of Food Science, 580 Bevier Hall, University of Illinois, Urbana, IL 61801

DONALD S. ORTH
Proctor and Gamble Company, Technical Center, 6071 Center Hill Road, Winton Hill, Cincinnati, OH 45224

HENRY J. PEPPLER
Scientific Director, Universal Foods Corporation, Milwaukee, WI 53201

PAUL L. POELMA
Division of Microbiology, Bureau of Foods, No 135, Food and Drug Administration, Washington, DC 20204

DONALD J. PUSCH
Manager, Technical Section, Corporate Quality Assurance Laboratory, The Pillsbury Company, 311 Second Street, S.E., Minneapolis, MN 55414

BIBEK RAY
Visiting Assistant Professor, Food Science Department, North Carolina State University, Raleigh, NC 27607

James Redman
N.Y. State Department Environmental Conservation, SUNY, Bldg 40, Stoneybrooke, NY 11790

Joseph V. Rodricks
Chief, Biochemistry Branch, Bureau of Foods, BF-146, Food and Drug Administration, 200 C Street, S.W., Washington, DC 20204

Arnold C. Salinger
Laboratories Research Administration, Maryland Department of Health and Mental Hygiene, 201 W. Preston, Baltimore, MD 21201

Arvey C. Sanders
Food Microbiology Branch, Division of Microbiology, Bureau of Foods, Food and Drug Administration, 200 C Street, S.W., Washington, DC 20204

William E. Sandine
Professor, Department of Microbiology, Oregon State University, Corvallis, OR 97331

David C. Sands
Department Plant Pathology and Botany, Connecticut Agricultural Experiment Station, New Haven, CT 06504

Arnold E. Schulze
Division of Microbiology (HFF-127), Food and Drug Administration, 200 C Street, S.W., Washington, DC 20204

Anthony J. Sinskey
Department of Nutrition and Food Science, Massachusetts Institute of Technology, Cambridge, MA 02139

F. R. Smith
910 East Oak Street, Greenville, IL 62246

R. V. Speck
Process Microbiology, Campbell Institute for Food Research, Campbell Place, Camden, NJ 08101

D. F. Splittstoesser
Professor, Cornell University, NYS Agriculture Experimental Station, Geneva, NY 14456

Berenice M. Thomason
Analytical Bacteriology Section, Bacteriology Branch, Center for Disease Control, Atlanta, GA 30333

H. Thompson
Del Monte Corporation Research Center, 205 N. Wigot Lane, Walnut Creek, CA 94598

Paul J. Thompson
Microbiology, Gerber Products Company, 445 State Street, Fremont, MI 49412

Donald Vesley
Assistant Professor, School of Public Health, University of Minnesota, 1325 Maye, Minneapolis, MN 55455

Joy G. Wells
Chief, Salmonella-Shigella Unit, Center for Disease Control, Atlanta, GA 30333

Edmund A. Zottola
Professor, Department of Food Science and Nutrition, University of Minnesota, St. Paul, MN 55101

TABLE OF CONTENTS

GENERAL ENUMERATION PROCEDURES

INDICATOR MICROORGANISMS AND PATHOGENS; TOXIN DETECTION

MICROORGANISMS AND FOOD SAFETY: FOOD-BORNE ILLNESS

FOODS AND THE MICROORGANISMS INVOLVED IN THEIR SAFETY AND QUALITY

PREFACE

Many factors have contributed to the prevalent intensive microbiological examination of foods. Food safety has become an important responsibility of various regulatory agencies charged with maintaining food safety. The food industry also is concerned with the production of safe foods, and also must cope with problems of preventing food spoilage.

The increased dependence of consumers on food processing industries has contributed to the growth and centralization of food processing. While this situation is conducive to the specialized and expert performance of various processing operations, any lapses in the supervision and care during the processing of foods potentially may affect large quantities of food and a large number of consumers. This situation has contributed to the importance of microbiological testing of foods.

The increased microbiological surveillance of the quality of foods has resulted in the development of many useful analytical procedures. Those which allow the most accurate evaluation of food quality are the ones that should be used. At the same time, it is important that procedures used in different laboratories not be adopted indiscriminately. This is particularly important since multiple laboratories in varying geographical locations may be responsible for the examination of a given food supply. Therefore, the use of the same or equivalent procedures by analysts is essential. Otherwise, health agencies, food processors, as well as the consumer, cannot have the assurance needed for evaluating the data obtained in testing food supplies.

Following the 1971 National Conference on Food Protection sponsored by the American Public Health Association in Denver, Colorado, the need for a *Compendium of Methods for the Microbiological Examination of Foods* became evident. Different government agencies and many industries already had adopted methods for use in their laboratories. Additional publications presented methods useful in the examination of foods for foodborne pathogens. In order to consolidate useful methods in one publication, the American Public Health Association, through its Committee on Laboratory Standards and Practices, developed a proposal that a compendium of methods be developed which would include methods for the evaluation of food safety, as well as for the microbiological spoilage of foods. Subsequently, the Food and Drug Administration contracted with the American Public Health Association to develop such a compendium of methods.

The *Compendium* is to a degree a continuation of two previous publications in this field. The first edition of *Recommended Methods for the Microbiological Examination of Foods* was published in June 1958 by the Coordinating Committee on Laboratory Methods of the American Public Health Association. The committee involved in this work was chaired by Dr. Harry E. Goresline. A second edition of this manual was published in 1966 by the same organization with Dr. John M. Sharf, chairman of the subcommittee charged with the preparation of the manual. Furthermore, the APHA Committee on Laboratory Standards and Practices suggested that the procedures for shellfish examination contained in *Recommended Procedures for the Examination of Sea Water and Shellfish*, 4th Edition, 1970, be incorporated in this *Compendium*. This has been accomplished and, therefore, effectively terminates the above mentioned procedural manual. Those individuals interested in methods for the examination of sea water are referred to the 14th Edition of *Standard Methods for the Examination of Water and Wastewater*, 1975.

Some of the methods included in this compendium differ only in formats from those given in the Official Methods of Analytical Chemists (12th ed.) and in *Standard Methods for the Examination of Dairy Products* (13th ed.). These methods are used by various federal, state, and municipal regulatory agencies as official methods of analysis. Other methods were selected which have been used in the monitoring of foods for pathogens or for food spoilage types of microorganisms.

In an effort to assist analysts not familiar with different types of foods, a section has been developed in the *Compendium* for a brief description of different foods and processing operations involved in their manufacture. Emphasis in these chapters is given to the types of microorganisms likely to occur in specific foods or food commodity groups, as well as their significance in these foods. This information will allow analysts a more judicious selection of methods for the examination of certain foods.

Shortcomings in different methods are recognized widely by food microbiologists and, hopefully, research will eliminate these from current analytical methodology. While much attention is now given to correct procedures for securing representative food samples for analysis, relatively little is known about the insults to contained microorganisms in the samples during transportation to laboratories. This is particularly important where split samples must be shipped to different laboratories for comparative analyses. The development of cryoprotective additives for samples would be very helpful here. Current procedures generally ignore the stressed or injured condition that may exist in many microorganisms resulting from sublethal treatments during processing and storage of foods. Methods need to be developed that allow injury to be repaired, especially before selective cultural procedures are applied. The enterotoxigenic staphylococci present challenges in the detection of cells by cultural procedures or by the use of indicators such as thermostable nuclease; and procedures for measuring the enterotoxins in food still are adaptable to only a few specialized laboratories. Fluorescent antibody techniques for detecting salmonellae are not as adaptable as had been expected for routine examinations of foods. Procedures for the more certain identification of *Bacillus cereus*, and isolates of *Clostridium perfringens* are needed, as are selective media for detecting *Yersinia* and *Shigella* in foods. A sequel to this *Compendium* is being planned as research provides means for the updating of current methodology.

Many persons have contributed most unselfishly to the development of the Compendium. The project was conceived and developed by Dr. Howard L. Bodily. Collaboration in its implementation and development was supported by a contract from the Food and Drug Administration, U. S. Department of Health, Education, and Welfare, through the encouragement of Dr. J. C. Olson, Jr. Government agencies involved in the microbiological surveillance of foods and professional societies with a competence and interest in food microbiology were asked to appoint representatives to an Intersociety/Agency Committee to study current analytical methods for the microbiological examination of foods. Authors and contributors were solicited from microbiologists who had established competence in the different subject areas selected for the *Compendium*. The Intersociety/Agency Committee spent many hours in planning the general format and overseeing the content of the *Compendium.* The authors and contributors were especially generous and prompt in the writing, reviewing, and correcting of the materials for which they accepted responsibility. Deliberations of the Intersociety/Agency Committee during its work on the *Compendium* were facilitated and assisted by the attentive care of Mrs. H. L. Bodily to committee proceedings. Dr. Nell Hirschberg furnished exceptional dual competence as copy editor and technical reviewer. Dr. Elizabeth Robinton constructed the index for the *Compendium*. For all of the study and contributions provided by colleagues, the editor expresses his sincere appreciation.

Corrections and technical questions should be sent to the Director of Publications, APHA, 1015 Eighteenth Street, NW, Washington, DC 20036 and will be referred to the Editor or the Chairman of the IS/A Committee.

Marvin L. Speck
Editor, First Edition
Chairman, Intersociety/Agency Committee
Compendium of Methods for the
Microbiological Examination of Foods

GENERAL LABORATORY PROCEDURES

CHAPTER 1

SAMPLE COLLECTION, SHIPMENT AND PREPARATION FOR ANALYSIS

Damien A. Gabis, A. Richard Brazis, and Thaddeus F. Midura

1.1 GENERAL CONSIDERATIONS

A proper sample, its collection, transportation to the laboratory, and preparation for examination, is the first priority in the microbiological examination of any food product. The laboratory results and their interpretation are only as valid as the sample submitted for examination. Every effort must be made to insure that samples collected are representative of the lot of material under examination, that they are the proper type for the determination to be made, and are protected against extraneous contamination and improper handling, especially at temperatures that may alter the microflora present. Ample refrigeration must be provided to prevent destruction or growth of organisms in the sample. Perishable samples collected in the nonfrozen state must be refrigerated, preferably at 0 to 4.4 C, from the time of collection until receipt at the laboratory. Samples collected while frozen should be kept solidly frozen. Samples should be examined within 36 hours after sampling.

1.2 SAMPLE HISTORY AND OTHER DATA

Samples should be accompanied by a form which contains adequate identification and specific requests for microbiological studies. The following information is considered necessary: sample description, collector's name, name and address of the manufacturer, dealer and/or distributor, date, place, and time of collection. Frequently, the temperature at the time of collection is also useful to the laboratory for the interpretation of results. Further, it is often desirable that the reason for sampling be given. Samples may be collected as part of a quality control or surveillance program, as official samples to determine conformity to legal specifications, or as part of a foodborne disease investigation.

1.3 SAMPLING PROCEDURES

The objective of sampling is to obtain information about a particular lot or product by examination of a small portion removed from the bulk. The

IS/A Committee Liaison: W. J. Hausler, Jr.

number of samples to be collected, and the size of each sample must be determined with care if the laboratory results are to be meaningful. Application of statistical principles in sampling is useful so that the number of units sampled will meet the objectives of the proposed laboratory testing. Since it is not feasible to use one set of tables or instructions for all sampling situations, the reader is directed to the publication of the International Commission on Microbiological Specifications for Foods (ICMSF) on sampling plans.[3]

Consumer packages of foods should be sampled from the original unopened containers when possible. This practice eliminates contamination that might be introduced by attempting to open and sample the container contents, and allows laboratory examinations to be performed on products as they are offered to the public.

If the products are in bulk or in containers of a size impractical for submission, aseptically transfer a representative sample portion to a sterile container. The sample container should be labeled with the date, time, location, and other pertinent information necessary to identify the sample.

Sampling instruments may be presterilized in protective covers; presterilized multiple use containers, which must be cleaned and sterilized after each sampling, may be used for a set of similar samples. Such instruments include long handle dippers, sampling tubes, syringes, sampling cocks on food equipment, and tools needed to open bulk containers. A dial type metal thermometer[1] is preferred for determining the temperature of samples. The thermometer should be checked against a thermometer certified by the National Bureau of Standards. Chapter 3 contains information on equipment and its cleaning and sterilization. Sample containers should be clean, dry, sterile, leakproof, wide mouth, and of a suitable size to hold samples of the particular product under examination. Multiple use containers as well as presterilized plastic containers, including bags, are acceptable. Glass containers may break and contaminate food products so they should be avoided whenever possible. The sampling operation should be planned in advance with all the needed equipment and containers at hand. Sampling instruments should be handled as little as possible before unwrapping, and must be protected from contamination until after the sampling is completed.

In sampling bulk fluids, the amount of materials handled directly over bulk containers of food should be kept to a minimum. Before drawing a sample, mix the food mass so as to insure that the sample is as homogeneous as possible, and then rinse the collection device in the product. If adequate mixing or agitation of the bulk product is not possible, multiple samples should be drawn. Care must be taken to select sample containers that are large enough to accommodate the needed sample volume when ¾ full. Thermometers used in bulk food containers should be sanitized before use.[1] Cool samples to 0 to 4.4 C quickly if they are not already refrigerated.

Where appropriate, a temperature control sample also should be submitted. This sample will not be tested microbiologically, but allows an accurate determination of sample temperature when received at the laboratory.

It should be labeled to indicate that it is a temperature control sample and not to be analyzed microbiologically. Water collected in the same type of container as the test sample, and accompanying the samples, may be adequate for the temperature control sample. At times, it may be desirable to submit an empty sterile sample container as a control.

The sample should be sealed in its container in such a manner so that the container will not break open and introduce extraneous contamination. Pliable tape of one kind or another often is used. If the sample is to be examined for a regulatory purpose, the sample container must be sealed so that it cannot be opened without breaking the seal.

Sampling of dry or semisolid foods should be done with sterile triers, spoons, or spatulas. Sterile tongue depressors may be substituted for spatulas. Aliquots from several areas of the food under examination should be taken to insure a representative sample. Care should be taken to protect this type of sample from excess humidity.

Frozen bulk foods may be sampled with sterile corers, auger bits, and other sharp sampling instruments. A presterilized auger bit or hollow tube may be used to obtain enough material for analysis. Frozen samples should be kept frozen until arrival at the laboratory. Freezing and thawing of samples should be avoided.

In some instances samples are tested as part of a foodborne disease outbreak investigation. Such outbreaks may be the subject of legal proceedings, and laboratory personnel may be required to testify concerning the results of their examinations. If the record of the sample is incomplete, or if samples are received in nonsterile containers or in a partially decomposed state, the laboratory results may be of little or no value.

It is important to use common sense in outbreaks, and to collect all perishable leftover foods served at the suspect meal(s) as soon as possible. All of these foods should be held under suitable conditions until an analysis of the attack rate data and other facts can define more accurately the suspect food(s).[2] The original containers in which the foods were found also should be collected, identified, and submitted for examination.

It is important to consider other specimens as well as foods in outbreak situations. Human specimens may include stools, vomitus, and serum. These specimens should be collected in sterile containers with leakproof stoppers, and should be properly identified as to the patient's name, type of specimen, and date of collection.[5]

Always use aseptic techniques in obtaining samples even if the foods have been mishandled grossly. If presterilized sampling instruments or containers are not available, a statement of the method used to sterilize instruments and containers should be included.

1.4 STORAGE AND SHIPMENT OF SAMPLES

Often it is necessary to store samples prior to shipment. If this is the case, a storage area for frozen (−20 C) and for refrigerated (0 to 4.4 C) samples should be available. Refrigerated samples should not be stored for more than 36 hours. Waterproof labels should be used on sample containers to prevent the loss of labels.

Samples should be delivered to the laboratory as rapidly as possible. The time and date of arrival at the laboratory should be noted. When it is not possible to deliver samples personally to the laboratory, they should be shipped by the most economical means commensurate with the need for rapid handling and examination. The samples should be packed to prevent breakage, spillage, or change in temperature. Since laboratory examinations of food samples require some preparatory work, the laboratory should be given advance notice, if possible, of the number and types of samples to be submitted.

If the product is in a dry condition or is canned, it may not need to be refrigerated for shipment. The sample label should indicate whether refrigeration is required. Refrigerated products must be transported in an insulated shipping container with sufficient refrigerant to maintain the samples at 0 to 4.4 C until arrival at the laboratory. Water ice in plastic containers or cold packs serve well for 0 to 4.4 C shipping and should last two days under most conditions. Dry ice may be used for longer transit times if the sample is separated from the dry ice by packing material to avoid freezing. Refrigerated product samples should not be frozen since destruction of certain microorganisms will occur. Frozen samples can be kept frozen by insuring that the samples are in contact with dry ice. Such samples should be sent to the laboratory by the fastest possible means.

Samples not requiring refrigeration or freezing may be packed in a cardboard box using crumpled newspaper or other insulating material to prevent breakage. For refrigerated and frozen samples, special plastic foam containers are commercially available and should be used because of their excellent insulating properties. Mark the shipment of samples as "Perishable," "Packed in Dry Ice," "Refrigerated Biologic Material," or "Fragile," as appropriate. Label the shipment according to Federal Postal Regulations.

1.5 RESPONSIBILITY OF THE LABORATORY UPON RECEIPT OF SAMPLES

Laboratory apparatus, materials, reagents and media, which are required for the preparation and examination of food homogenates are described in Chapter 2. Such equipment, not previously mentioned, includes glass and metal thermometers, incubators, water baths, media, chemicals, reagents, distilled and dilution water, automatic sample and/or dilution container shakers, glass and aluminum blenders, stomachers, jars and tops, spatulas, spoons, dilution bottles, pipet and pipet cans, balances and weights, can openers, petri dishes, can covers, fermentation tubes, and other necessary glassware. The precautions for utilizing pH meters; colorimeter; grinding equipment; dry, moist, and ethylene oxide sterilization; as well as necessary sanitization of laboratory equipment, also are described in Chapter 2.

Temperature records of refrigerators, freezers, and incubators, should be kept on a daily basis. Automatic temperature recorders should be used periodically, particularly for water baths and other instruments which require strict temperature control. The maximum graduations of thermometers should be 1 C. For water baths and incubators with low tolerance of temperature fluctuation, thermometers should be graduated in 0.1 C intervals. When different size dilution blanks are used, the amount of diluent should not vary more than 2% from the designated volume. Balances and weights should be checked monthly to determine sensitivity, as described in APHA Standard Methods for the Examination of Dairy Products.[1] Colorimeters and pH meters should be standardized periodically. If pH test papers must be used, they must be standardized weekly against pH buffers for different hydrogen ion concentrations.

The microbial density at all working surfaces (e.g., work bench), light intensity (preferably 100 foot candles), ventilation, electric utilities, absence of insects and rodents, storage of street clothes, media and reagents, and microscopes, should be regulated as recommended in the APHA Standard Methods for the Examination of Dairy Products[1] and by other publications on laboratory control and safety. Where applicable, laboratory animals and cages, membrane or glass filters, microscopes, lamps, slides, metal and glass syringes, pipeting machines, spectrophotometers, water stills, and mortars and pestles, should be maintained in a clean, readily available and operable condition. The surfaces of laboratory benchtops should be impervious and level. The arrangement of working surfaces in each laboratory ideally provides approximately 10 linear feet of working space per person. The ceilings, walls, and floors of the laboratory should be smooth and easily cleanable. Doors and windows should be screened. The location of incubators and/or incubator rooms, refrigerators, freezers, and water baths should be easily accessible. Cabinets, drawers, or shelves should be adequate for the protection of glassware, apparatus, and other materials. It is suggested that foot operated hand washing facilities be provided, and paper towels.

On arrival at the laboratory, the sample identification form should be reviewed promptly to determine the nature of the samples, and whether each shipment contains official, investigative, potential food poisoning agents, or quality assurance samples. The microbiologist should be notified immediately upon receipt of the samples, particularly when the identification form indicates prompt examination upon arrival at the laboratory. It is essential that refrigerated and perishable samples be examined within 36 hours after collection of samples. Frozen samples should be stored in a freezer until examined. Low moisture or canned samples which are not perishable, and have been collected and shipped at ambient temperatures, may be stored at ambient temperatures or at 0 to 4.4 C. All others should be stored at 0 to 4.4 C until examined.

The temperature control sample accompanying refrigerated samples should be examined using a precooled thermometer following opening of the sample chest or package. Record the temperature of the temperature control sample as well as the date and time of arrival, and store such samples at 0 to 4.4 C until tested. Samples requiring microbiological and chemical analyses must be examined microbiologically before chemical analyses are conducted.

The physical appearance of the sample and sample container should be noted and recorded. Where applicable, examine each container for the presence of rupture. If this condition exists, there may be leakage of the sample which may result in cross contamination of the sample. The presence of the container rupture must be noted on the sample identification form. Swells or gas formation in the sample should be noted on the identification form. The sample container should be checked for potential decomposition of the container coating, and this should be noted on the identification form.

1.6 PREPARATION OF FOOD SAMPLE HOMOGENATE

All samples should be examined by appropriate microbiological procedures as promptly as possible after receipt at the laboratory. Removal of samples from their containers should be done aseptically. Frozen samples should be thawed as soon as feasible at refrigeration temperatures. Alternatively, higher temperatures may be used for a short period of time but the temperature must be low enough to prevent destruction of pathogenic microorganisms (< 45 C for ≤ 15 minutes). In thawing frozen samples, care must be taken to prevent an increase in the number of microorganisms present due to excessive exposure to growth temperatures, and the temperature must be low enough to prevent destruction of microorganisms. Frequent shaking of samples at elevated temperatures, and with a rotary motion, may aid in thawing samples. A thermostatically controlled water bath, with agitator, is recommended for the rapid thawing of samples.

Dry samples should be mixed with a sterile spoon or spatula. Prepare the initial dilution of each sample by aseptically weighing it into a sterile wide mouth (glass or polypropylene) or screw cap bottle and add sterile dilution water. For relatively insoluble proteinaceous materials such as dry milk, 1.25% sodium citrate may be used as a diluent. Liquid samples should be agitated thoroughly before weighing.

The size of the sample for analysis depends on the homogeneity of the food. Samples for analysis should be at least 10 grams, preferably 25 to 50 grams. If the sample is obviously not homogeneous (e.g., frozen food composite), remove a 50 gram sample from a representative macerated portion of the whole package or analyze each different portion of food separately, depending on the test(s) to be conducted. For samples requiring blending, weigh a representative portion of the food into a sterile tared blender jar or sterile wide mouth screw cap bottle. Then transfer the contents from the bottle to the blender when the analyses are to begin. As an alternative to the blender, the food sample may be homogenized by using a stomacher.[4] The sample is weighed into a sterile polyethylene bag and an appropriate volume of diluent is added. The bag is then placed in the stomacher and agitated for 30 to 60 seconds. The sample homogenate is then ready for dilution. Care must be taken to use bags which can withstand the pressure generated during stomaching. Sharp objects in the sample, such as bone splinters, may make pinholes in the bag and cause leakage of the suspension. The remaining portions of samples may be held under refrigeration at 0 to 4.4 C or, if the product is a dried food, it may be held at ambient temperature. These will serve as a backup sample for immediate future use if needed. High moisture foods should be frozen for long term sample retention. However, it must be understood that freezing may alter the microbial flora of the sample.

Normally, prepare a 1:10 dilution of each sample. A variety of diluents may be used depending upon the nature of the product. Those diluents most commonly used are Butterfield's phosphate buffer, phosphate buffer with or without citrate[1], and 0.1% peptone water. When analyzing for specific organisms, other diluents may be appropriate (e.g., 3% NaCl for *Vibrio parahaemolyticus*). When analyzing fatty foods or lump forming powders, wetting agents such as Tergitol Anionic-7 or Tween 80 (1%) may be used.

Samples miscible in water may be agitated in the diluent mixture by shaking horizontally or vertically 25 times, through a 1 foot arc in 7 seconds. Agitation may require a longer time to obtain a homogeneous mixture depending on the nature of the sample. Alternatively, mechanical shakers may be used where the results are comparable to hand shaking. In this case, the dilutions are agitated for 15 seconds.

For immiscible samples, add the dilution water to the blender, and blend promptly for 2 minutes at 8,000 rpm (usually low speed). Some blenders may operate at speeds lower than 8,000 rpm. It is preferable to use a higher speed for a few seconds initially. The interval between the agitation of samples and removing test aliquots should not exceed 3 minutes. It is recommended that not more than 10 minutes elapse between preparing the first 1:10 dilution, and all subsequent dilutions have been inoculated into appropriate media.

Because microorganisms in some foods may be subject to metabolic stress prior to examination of the product, it may be desirable to keep time and temperature records for the various steps in the analysis. These include holding times and temperatures of the diluted sample in the container, dilution blanks, and Petri plates prior to addition of agar.

1.7 REFERENCES

1. Standard Methods for the Examination of Dairy Products, 13th ed. 1972. American Public Health Association. Washington, D.C.
2. International Association of Milk, Food, and Environmental Sanitarians, Inc. 1966. Procedure for the Investigation of Foodborne Disease Outbreaks, 2nd ed., Shelbyville, Indiana
3. International Commission on Microbiological Specifications for Foods. 1974. Microorganisms in Foods II. Sampling for Microbiological Analysis: Principles and Specific Applications, University of Toronto Press. Toronto, Canada
4. Sharpe, A. N. and A. K. Jackson. 1972. Stomaching: a new concept in bacteriological sample preparation. Appl Microbiol **24**: 175–178
5. Collection, Handling and Shipment of Microbiological Specimens. 1973. US Department of Health, Education, and Welfare. DHEW Pub. No. (CDC) 74-8263, Atlanta, Ga.

CHAPTER 2

EQUIPMENT, MEDIA, REAGENTS, ROUTINE TESTS AND STAINS

Harold V. Leininger

2.1 EQUIPMENT

2.11 Introduction

Equipment and supplies meet the specifications described in Standard Methods for the Examination of Dairy Products unless elsewhere specified in this text. All material to be sterilized by hot air should be sterilized so that materials at the coldest part of the load are heated to not less than 170 C (338 F) for not less than one hour. (This usually requires exposure for about two hours at 170 C.) To insure sterility, do not crowd the oven, and when the oven is loaded to capacity with apparatus, preferably use longer periods or slightly higher temperatures as determined by the quality assurance program for your laboratory and equipment. Recording devices with thermocouples distributed in several locations in the load are desirable.

All media and other materials to be sterilized in the autoclave should be sterilized according to the directions given by the author, manufacturer, or described in the specific method to be employed. Specifications for the most frequently used items are listed below.

a. Autoclave:

Shall be of a size sufficient to prevent crowding of the interior and be constructed to provide uniform temperatures within the chamber up to and including the sterilizing temperature of 121 C. Additionally, it must be equipped with an accurate mercury filled thermometer or bimetallic helix dial thermometer properly located to register minimum temperatures within the sterilizing chamber (temperature recording instrument optional), pressure gauge, and properly adjusted safety valve. Optionally provide temperature recorder controller. Small 5 to 10 quart pressure cookers, equipped with a thermometer for temperature control, may be substituted for an autoclave when proper temperatures are maintained and recorded, and satisfactory results are obtained. The autoclave, especially strainers to traps or vents, should be cleaned frequently or immediately after spills or malfunction of the equipment. The laboratory should have a maintenance program designed specificially to keep the equipment in good repair

CONTRIBUTORS: Louis C. LaMotte, Jr., James W. Messer
IS/A Committee Liaison: W. J. Hausler, Jr.

and working order, with not less than annual checks by reputable repairmen. This check should include thermocouple readings to determine temperatures throughout the autoclave.

b. Balance:

Sensitive to 0.1 g with 200 g load, 2 kg capacity single pan balance preferred.

c. Colony counter:

Standard apparatus, Quebec model preferred, or one providing equivalent magnification and visibility.

d. Dilution bottles:

Capacity about 150 ml, borosilicate glass, closed with Escher type rubber stoppers or screw caps. Use friction fit liners in screw caps, as required, to make the closure leakproof. Be sure that each batch of dilution blanks is filled properly. Dilution bottles should be marked indelibly at 99 ± 1 ml graduation level. Plastic caps for bottles or tubes and plastic closures for sample containers must be treated when new to remove toxic residues. Treatment may be accomplished by autoclaving them twice while they are submerged in water, or exposing them to two successive washings in water containing a suitable detergent at 82 C (180 F).

e. Hot air sterilizing oven:

Of a size sufficient to prevent crowding of the interior. The oven should be provided with vents suitably located to insure prompt and uniform heat distribution for adequate sterilization. It should be equipped with a thermometer having a range of 0 to 220 C located to register the minimal temperature in the oven. A temperature recorder in addition to the thermometer is desirable.

f. Incubator and incubator rooms:

Must be properly constructed and controlled. Temperatures within the incubator must not vary more than ± 1 C from the setting. Determine temperature variations within the incubator or incubator room (filled to maximum capacity) that are consistent with good laboratory practice. Incubators must be loaded so that there will be not less than 1 inch of space between adjacent stacks of plates, and between walls and stacks. Plates must reach incubation temperature within 2 hours.

Incubator rooms must be well insulated, equipped with properly distributed heating units, and have forced air circulation which does not interfere with good humidity control. Humidity should be such that plates do not have a tendency to collect moisture which encourages spreaders, but should be low enough that agar plates do not lose more than 15% moisture in 48 hours.

Keep incubators in rooms where temperatures are within the range of 16 to 27 C (60.8 to 80.6 F). Laboratory personnel are responsible (1) for recording temperatures in the top and bottom of the incubator, or incubator room, twice daily—in the morning after the incubator has been closed all night, and just before the end of the day's work; and (2) for adjusting the thermoregulator. Place thermometers on top and bottom shelves, and in between as needed. Thermometer bulbs must be submerged in water within small, tightly closed vials, since air temperature is an unreliable index of actual agar temperatures.

Optionally use automatic devices of predetermined accuracy for controlling and recording temperatures in incubators and/or incubator rooms. Check the accuracy of these devices periodically by comparing readings with those from standard thermometers.

When standard thermometers are used, minimum and maximum registering thermometers may be placed in each incubator to indicate undetected gross temperature deviations and the degree of such deviations. Do not depend on readings from these special thermometers for daily records of temperatures. Minimum and maximum registering thermometers are unnecessary when automatic devices of predetermined accuracy for controlling and recording temperatures are in continuous and proper operation.

g. Mechanical shaker:

As prescribed in the 13th edition of Standard Methods for the Examination of Dairy Products may be used. Mechanical shakers must shake the container 25 complete up and down movements of about 1 foot in 7 seconds.

h. Metal syringe:

For the rapid and convenient transfer of 0.01 ml quantities of liquid foods having a semiautomatic spring actuated plunger and a stainless steel piston sliding in a close fitting stainless steel measuring tube, and with an adjustable setscrew to permit presetting the instrument for repeated accurate measurements of 0.01 ml quantities (Figure 1). The syringe is available from the Applied Research Institute, 90 Brighton Avenue, Perth Amboy, N.J. 08861; or from Dairy Research Products, Inc., Box 288, Yarmouth, Me.

The accuracy of the syringe adjustment should be determined before use and periodically thereafter as necessary. Where certification is required before use, appropriate regulatory agencies should verify the accuracy of the setting before applying the certification symbol over the setscrew. This provides the necessary assurance that the setscrew cannot be readjusted accidentally or otherwise without visible evidence thereof.

NOTE: The mechanical principle of the syringe makes possible positive measurement of 0.01 ml portions without necessitating manual adjustment of the milk column. It also makes possible the accurate measurement of test portions of cold heavy fluid creams, an operation impossible with 0.01 ml pipets. Artifacts (metallic oxide particles) may be encountered with metal syringes.

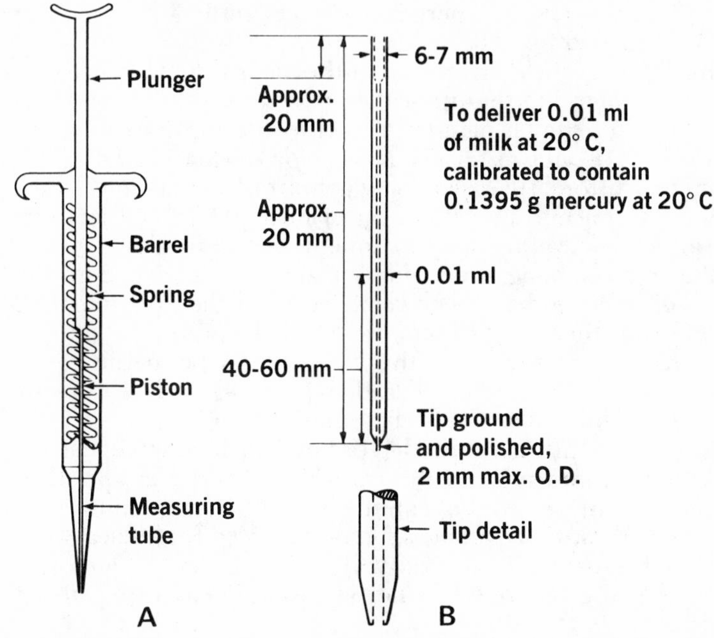

FIGURE 1. Transfer Instruments for 0.01 ml Amounts. A—Syringe; B—Pipet

To determine the weight (and indirectly the volume) of milk delivered by a 0.01 ml syringe or pipet, first determine the specific gravity of a homogenized milk sample. Withdraw a representative test charge, wiping the exterior of the tip of the instrument as usual. Weigh the charged instrument on an analytical balance. Expel the charge as usual near the 1 cm^2 area and, if a syringe is being used, spread the charge evenly over the area with the tip of the piston. Weigh the discharged instrument. The difference in weight between the instrument charged and discharged should average (over 5 to 10 weighings) approximately 0.0103 g when measuring milk.

Do not allow residues to dry on the instrument. Immediately after use, disassemble and clean the measuring tube and piston. Do not interchange parts from one syringe with parts of another. Handle parts carefully when they are disassembled. When assembled, the instrument is reasonably sturdy, but should not be dropped or handled carelessly. Unless it is necessary, do not unscrew the barrel at the finger grip joint. To disassemble, unscrew the tip assembly from the barrel. If necessary to remove the spring, reassemble it with the larger end toward the tip. Clean all residues from the piston and lower part of the plunger by wiping exposed portions with dry paper tissue or cloth. Use soapless detergents and fat solvents somewhat sparingly, but as needed (metal is not as resistant to corrosion as glass). Keep in mind that, unless the syringe is properly maintained, it can contribute both bacterial cells and artifacts (metallic oxide particles).

Ordinarily do not submerge the barrel and finger grip in detergents or solvents. With forced suction or pressure from a closefitting rubber bulb applied at the delivery end of the measuring tube, clean the interior by submerging the tip of the opposite end in soapless detergent and then, in a similar manner, rinse the interior in clean water before air drying. Do not use acids, alkalies or ordinary soap solutions for cleaning (customarily used for glassware). Use chemical cleaners in very weak concentrations only, at or near pH 7.0. Observe the extra precaution of cleaning new instruments before first use.

Before assembling the syringe, make certain that both detergent and milk residues have been removed from the instrument. To reassemble, slide the spring (if removed) over the piston and plunger, carefully insert the piston into the measuring tube at the countersunk opening, and tighten the threaded joint. Always be sure threaded joints are tightened to identical positions.

All syringes, officially branded or not, which, after repeated use, are tested and found not to conform with deliveries as specified above, should be returned to the manufacturer with instructions to recondition them. Ordinarily syringes should be tested for accuracy biennially, or more frequently if needed. WARNING: Except in emergencies, do not loosen the setscrew for the purpose of readjusting the plunger stroke.

i. Compound microscope:

Binocular preferred, with 1.8 mm oil immersion objective, substage actuated by a rack and pinion carrying an Abbé condenser having a numerical aperture (NA) of 1.25 or higher, and equipped with an iris diaphragm, a flat mirror if the light is not an integral part of the microscope assembly, or mounted on the base, a mechanical stage, and oculars as follows: single for monocular or paired for binocular, 10X magnification recommended (12.5 to 15X permitted). For both enumerative and grading purposes, the ocular(s) should be calibrated to provide a microscopic factor of 300,000 to 600,000. Wide field oculars with high eyepoints are advantageous, especially for those who wear eyeglasses.

j. Ocular disc subdivided into quadrants:

To use the disc place it in the ocular so that it rests on the edge of the reducing diaphragm. Face the etched surface of the disc according to directions provided by the microscope manufacturer. One disc only is required for a binocular microscope. Before using the disc, determine the microscopic factor corresponding to the area to be examined.

k. Microscope lamp:

Necessary where the light is not a part of or attached to the microscope frame. Advanced or professional type, with a light source equivalent to 100 watt illumination, having a reflector, a condensing lens system, and an iris diaphragm to provide Köhler type illumination. Any simple lamp system of proper intensity that provides critical illumination is satisfactory.

Modern "on base" or attached illuminators equipped with an aspheric condenser lens, reflector, and 110 volt, 15 watt clear bulb are acceptable if light quality and intensity are adequate. Choose a type specifically designed for a monocular or binocular scope.

The ordinary separate substage illuminator lacking a reflector and an aspheric condenser lens is not acceptable for use in estimating bacterial numbers.

l. Mechanical state:

Regular, as supplied with the microscope; or a special stage for slides 2 × 4.5 inches.

m. Stage micrometer slide:

Ruled in 0.1 and 0.01 mm divisions.

n. Pipet:

For bacteriological use only, calibrated to deliver 0.01 ml quantities of milk according to APHA specifications. Straight, thick walled capillary tubing with a bore of such diameter that a single graduation mark is 40 to 60 mm from the pipet tip. The tip is blunt and formed so as to discharge the milk cleanly. Calibrated to contain 0.1395 g (approximately 140 mg) of mercury, it will discharge 0.01 ml of milk at 20 C (68 F).

NOTE: When not in use, keep pipets submerged in, and the bore filled with, suitable soapless detergent or strong cleaning solution. When preparing for use, rinse the bore and exterior thoroughly in clean water until free of the detergent or cleaning solution.

o. Petri dishes or plates:

Bottoms of at least 85 mm inside diameter and 12 mm deep, with interior and exterior surfaces of bottoms free from bubbles, scratches or other defects. Petri dishes may be made of glass or of plastic, and should have flat bottoms.

p. Petri dish containers:

Of stainless steel or aluminum, with covers, preferred. Char resistant sacks of high quality sulfite pulp kraft paper may be substituted. Single service Petri dishes may be stored in their original containers.

q. Pipets, glass and plastic:

Nontoxic, straight wall, tips ground or fire polished (glass), calibrated for bacteriological use, conforming to APHA specifications. Pipets having graduations other than those listed in the APHA specifications may be used provided they meet all the other APHA specifications. Volume delivered in 4 seconds maximum, last drop of undiluted milk blown out, or last drop of diluted milk touched out, 1 ml, tolerance ± 0.025 ml. To allow for residual milk and milk dilutions on walls and in the tip of the glass pipet under the specific technic hereinafter described for rapid transfers, such pipets shall be graduated to contain 1.075 ml of water at 20 C (68 F). In the case of styrene plastic, the pipet should be calibrated to contain 1.055 ml of water at 20 C. Use only pipets with unbroken tips and having graduations distinctly marked so as to contrast sharply with the color of the fluid and dilutions being employed. Discard pipets which are damaged in any way.

r. Pipet containers:

Stainless steel or aluminum preferred. Copper containers shall not be used. Char resistant, high quality sulfite pulp kraft paper may be substituted.

s. Refrigerator:

To cool and maintain the temperature of samples between 0 C and 4.4 C (32 to 40 F) until tested, and for storage of prepared media when desired.

t. Storage space:

Cabinets, drawers or shelves should be free of dust and insects, and adequate for the protection from breakage of glassware, apparatus, and other equipment.

u. Thermometers:

Of appropriate range, mercury filled (or having a distinctively colored fluid with a freezing point lower than −1 C; or an adjustable type, with a graduation interval not to exceed 1 C unless otherwise specified, accuracy checked at least every 2 years with a thermometer certified by the National Bureau of Standards (or one of equivalent accuracy). Where a record is desired of temperatures in refrigerators, autoclaves, hot air ovens, or incubators, automatic temperature recording instruments may be used.

There are two general types of mercury thermometers in use in laboratories: those calibrated for total immersion, and those that are designed to be partially immersed. Partial immersion thermometers have a line all the way around the stem of the thermometer at the point to which they should be immersed. If this line is not indicated, then the thermometer is designed for total immersion. As examples, a partial immersion thermometer should be used in a water bath because only part of the thermometer is immersed in the medium (water, in this case), the temperature of which is being measured. Conversely, a thermometer in an incubator or refrigerator is immersed totally in the warm or cold environment, and should be of the complete immersion type.

The easiest way to check the calibration of laboratory thermometers is to put them in a water bath, either partially or totally immersed in the water, according to the way they will be used in the laboratory. Also place in the water bath a thermometer certified by the National Bureau of Standards. Most, but not all, of these thermometers are calibrated for total immersion, so that they typically are immersed totally in the water of the water bath. Vigorous stirring of the water in the bath is essential to insure uniform temperature during thermometer calibration.

Check the calibration by comparing the temperature reading on the certified thermometer with that of the laboratory thermometer at or very near the temperature the thermometer will be used to measure (e.g., an incubator thermometer should be checked at 32 C because this is the temperature of interest). If the thermometer is to be used for several purposes, it should be checked at three different temperatures at least. If there is a difference in the temperature reading of the laboratory thermometer and the certified thermometers, after the reading of the certified thermometer has been corrected as indicated by the certificate, attach a tag to the laboratory thermometer to show the amount of correction that should be applied to get an accurate measurement of temperature.

v. Tally:

A mechanical counting device or counter that marks colonies as the operator counts them on the plates.

w. Water bath:

Of appropriate size, and thermostatically controlled, for holding melted media at 44 to 46 C (111.2 to 114.8 F).

Water baths should be covered whenever possible, and the water level maintained sufficiently high to protect the temperature of the agar surface in the flasks or other containers.

CAUTION: Specifications for incubating certain tests are called for throughout the text. In those instances, water baths meeting the requirements indicated by the test being conducted must be employed.

x. Work area:

Table or other rigid support, level, with ample surface, in a clean, well lighted (at least 50 and preferably 100 foot candles at each working surface), and well ventilated room reasonably free from dust and drafts. The microbial density of air (bacteria, yeasts and molds) in plating areas, as determined during plating by exposure of poured plates, should not exceed 15 colonies per plate during a 15 minute exposure.

2.2 CULTURE MEDIA

2.21 Introduction

This chapter contains the formulae for all culture media mentioned in the text of the various chapters, and the directions for making and using media. The names of the various media are those by which they are known generally in the literature with some few exceptions where we have used the title given by the author in the chapter wherein the medium is used. In the text, the recommended medium is referred to by name and is listed alphabetically in Chapter 2.

In most cases, media recommended here can be obtained in the dehydrated form. The equivalent product of any manufacturer may be used. Catalogs of Bioquest Division of Becton Dickinson and Co., P.O. Box 243, Cockeysville, Maryland 21030, and Difco Laboratories, Detroit, Michigan 48201, contain products with their formulation. Since it is difficult to define equivalency in objective terms, the user must test products in his own laboratory as part of his routine laboratory quality assurance program to determine that the medium functions as expected. It is suggested that dehydrated media be employed unless the author specifically recommends that the user prepare the media from individual ingredients. In such cases, the laboratory should test the medium with sufficient cultures to determine that the performance of his preparation is satisfactory. Also, we have supplied formulation for each medium believing that this information will be of interest to those engaged in research. Only in rare instances are the names of any manufacturer given and then only to provide a source of the material mentioned in the formulation, especially when that material is available from a single company or commonly is not employed in the routine laboratory.

The reader is invited to refer to Chapter 4 of Standard Methods for the Examination of Dairy Products, 13th Edition, for other details concerning culture media and preparation.

Determine the hydrogen ion concentration of culture media at 25 C (77 F) electrometrically or colorimetrically, and record the reaction in terms of pH. This temperature specification is used in the commercial preparation of media and should be adhered to in the laboratory determination of pH prior to use. Determinations made at 45 C (113 F) to take advantage of the fluid state are not accurate, having been shown to differ significantly from those obtained at the recommended temperature.

Experience has indicated that demineralizers for demineralizing water are particularly difficult to maintain, and distilled water is recommended for all media. A test for the suitability of water, based on growth of *Enterobacter aerogenes* in a minimal medium, is described in the 13th Edition of Standard Methods for the Examination of Dairy Products. Water used for the preparation of dilution blanks is particularly critical. Residual chlorine or chloramines may be found in distilled water prepared from chlorinated water supplies. The presence of these compounds always should be ruled out. The quality of the water in this respect should be verified by a suitable quantitative procedure such as starch iodide titration. If chlorine compounds are found in the distilled water, they should be neutralized by the addition of an equivalent amount of sodium thiosulfate or sodium sulfite. Following this step, redistill the water. In this way these compounds can be eliminated from distilled water used in the preparation of dilution water.

2.22 Individual Media

ACETATE AGAR

Sodium chloride	5.0 g
Magnesium sulfate	0.1 g
Monoammonium phosphate	1.0 g
Dipotassium phosphate	1.0 g
Sodium acetate	2.0 g
Bromthymol blue	0.08 g
Agar	20.0 g
Distilled water	1.0 liter

Mix ingredients in distilled water and heat gently to dissolve. Dispense 7 ml portions into 16 × 150 mm tubes.

Sterilize at 121 C for 15 minutes, and slant the tubes so as to obtain a 1 inch butt and a 1.5 inch slant, pH 6.8 ± 0.2.

ACID PRODUCTS TEST BROTH

Yeast extract	7.5 g
Peptone	10.0 g
Invert sugar*	10.0 g
Distilled water	1.0 liter

*Invert sugar, Catalog #758C, may be obtained from ICN Nutritional Biochemicals 26201 Miles Rd., Cleveland, Ohio 44128.

To prepare the broth, suspend yeast extract and peptone in distilled water. Slowly stir in the invert sugar and bring the mixture to a boil. Continue heating until all ingredients are dissolved. Cool the broth and adjust to pH 4.0 ± 0.2 using tartaric acid 25% w/v. The broth should be distributed in 500 ml Erlenmeyer flasks with screw type closures or equivalent. Total volume of broth per flask is 300 ml. Sterilize at 121 for 15 minutes.

AGAR MEDIUM FOR DIFFERENTIAL ENUMERATION OF LACTIC STREPTOCOCCI

Tryptone	5.0 g
Yeast extract	5.0 g
Casamino acids	2.5 g
L-arginine-hydrochloride	5.0 g
Dipotassium phosphate	1.25 g
Calcium citrate	10.0 g
Carboxymethyl cellulose*	15.0 g
Agar	15.0 g

*DuPont CMC Grade P-754 at the rate of 0.6% may be substituted for CMC Cekol MV, Uddeholm, Sweden.

Suspend 15 g of agar in 500 ml distilled water and steam until dissolved.

In another glass container suspend 10 g of calcium citrate and 15 g of CMC in 500 ml of distilled water. Stir while heating until a homogeneous, white, turbid suspension is formed. Mix the two portions in a stainless steel vessel containing the required amounts of tryptone, yeast extract, dipotassium phosphate, casamino acids, and arginine. Cover the mixture and steam for 15 minutes. After steaming adjust to pH 5.6 ± 0.2 with 6N hydrochloric acid. Dispense agar in bottles in 100 ml quantities and sterilize at 121 C for 15 minutes.

Just before pouring plates, add 5 ml sterile, reconstituted nonfat milk (11% solids), 10 ml of sterile 3% (w/v) calcium carbonate in distilled water and 2 ml of sterile 0.1% bromcresol purple in distilled water to 100 ml of sterile agar (melted and tempered to 55 C), and mix to obtain homogeneity. After the addition, pH should be 5.9 ± 0.2. Pour the mixture into previously chilled sterile Petri dishes to obtain a layer 4 to 5 mm thick. After solidification, dry plates for 18 to 24 hours in an incubator at 37 C.

ALKALINE PEPTONE WATER

Peptone	10.0 g
Sodium chloride	10.0 g
Distilled water	1.0 liter

Mix ingredients in distilled water and adjust to pH 8.4 to 8.6. Dispense 7 ml aliquots into screw cap vials. Sterilize 10 minutes at 121 C.

ANAEROBIC EGG AGAR

Fresh eggs (antibiotic free)	3.0
Yeast extract	5.0 g
Tryptone	5.0 g
Proteose peptone	20.0 g
Sodium chloride	5.0 g
Agar	20.0 g
Distilled water	1.0 liter

Wash eggs with a stiff brush and drain. Soak in 70% alcohol 10 to 15 minutes; remove and allow eggs to air dry. Crack eggs aseptically; separate and discard the whites. Add the yolks to an equal volume of sterile saline and mix thoroughly.

Combine the remainder of the ingredients, dissolve, adjust to pH 7.0, dispense, and sterilize at 121 C for 15 minutes. Melt 1 liter of the sterile agar, cool to 45 to 50 C, add 80 ml of the egg yolk emulsion, mix thoroughly, and pour plates immediately.

ANDERSEN'S PORK PEA AGAR

Pork infusion	800.0 ml
Pea infusion	200.0 ml
Peptone	5.0 g
Tryptone	1.6 g
Dipotassium phosphate	1.25 g
Soluble starch	1.0 g
Sodium thioglycollate	0.5 g
Agar	16.0 g
Distilled water	1.0 liter

Prepare the pork infusion by adding 1 lb of fresh lean ground pork to 1 liter of distilled water and steam in flowing steam for 1 hour. Filter out the meat through cheese cloth. Chill to solidify and remove the remaining fat, centrifuge to remove remaining solids, and use infusion in the above formula.

Prepare the pea infusion by blending 1 lb of good quality fresh green peas or frozen peas with 450 ml of water and steam for 1 hour in flowing steam. Remove solids by centrifugation and clarify with celite. Use filtrate in above formula.

Mix the pork infusion, pea infusion and other ingredients and adjust to pH 7.2 ± 0.2. Autoclave 5 minutes at 121 C. Clarify while hot by adding 25 g of celite and filtering through Whatman No. 4 filter paper with suction. Tube and store at 4 C. As needed, thaw and autoclave at 121 C for 12 minutes.

Prepare a 5% solution of sodium bicarbonate and sterilize by filtration. Keep refrigerated.

Place 0.4 ml of bicarbonate into each plate, mix the molten agar thoroughly with the ingredients, and allow to solidify. When solid, pour a layer of thioglycollate agar over the surface and allow to harden.

APT AGAR

(All Purpose Medium with Tween 80)

APT agar is prepared by adding 15.0 grams of agar to the basal formula below. Bring to a boil to melt the agar. Dispense and autoclave as below.

APT AGAR + BCP

Prepare agar as indicated. Add 2 ml of bromcresol purple, 1.6% in alcohol, per liter of medium prior to dispensing.

APT BROTH

Trypticase, peptone, or tryptone	10.0 g
Yeast extract	7.5 g
Dipotassium phosphate	5.0 g
Sodium chloride	5.0 g
Sodium citrate	5.0 g
Sodium carbonate	1.25 g
Thiamine	0.0001 g
Dextrose	10.0 g
Tween 80	0.2 g
Magnesium sulfate	0.8 g
Manganese chloride	0.14 g
Ferrous sulfate	0.04 g
Distilled water	1.0 liter

Dissolve ingredients in distilled water and dispense into flasks or bottles. Sterilize for 15 minutes at 121 C. Longer sterilization is not recommended. The final pH should be 6.7 ± 0.2.

APT BROTH (FOR CULTIVATION OF *LACTOBACILLUS*)

Ingredient	Amount
Yeast extract	7.5 g
Tryptone	12.5 g
Dextrose	10.0 g
Sodium citrate	5.0 g
Thiamine hydrochloride	0.0001 g
Sodium chloride	5.0 g
Dipotassium phosphate	5.0 g
Manganese chloride	0.14 g
Magnesium sulfate	0.8 g
Ferrous sulfate	0.04 g
Polysorbate 80	0.2 g
Distilled water q.s.	1.0 liter

Dissolve ingredients in distilled water, dispense and sterilize at 121 C for 15 minutes.

BAIRD-PARKER MEDIUM

Basal medium

Ingredient	Amount
Tryptone	10.0 g
Beef extract	5.0 g
Yeast extract	1.0 g
Sodium pyruvate	10.0 g
Glycine	12.0 g
Lithium chloride · $6H_2O$	5.0 g
Agar	20.0 g

Suspend ingredients in 950 ml distilled water. Heat to boiling to dissolve completely. Dispense 95 ml portions in screw capped bottles. Autoclave 15 minutes at 121 C. Final pH 6.8 to 7.2 at 25 C.

Egg Enrichment

a. Soak eggs in aqueous mercuric chloride 1:1000 for not less than 1 minute. Rinse in sterile water and dry with a sterile cloth.
b. Aseptically crack eggs, and separate whites and yolks.
c. Blend yolk and sterile physiological saline solution (3 + 7 v/v) in high speed sterile blender for 5 seconds.
d. Mix 50 ml blended egg yolk to 10 ml of filter sterilized 1% potassium tellurite. Mix and store at 2 to 8 C.

Enrichment—Bacto EY tellurite enrichment

Complete medium

Add 5 ml prewarmed (45 to 50 C) enrichment to 95 ml melted basal medium, which has been adjusted to 45 to 50 C. Mix well (avoiding bubbles), and pour 15 to 18 ml into sterile 15 × 100 mm Petri dishes. Plates of complete medium should be stored at 2 to 8 C for no longer than 48 hours before use. Medium should be densely opaque; do not use nonopaque plates. Plates should be dried before use (a) in a convection type oven or incubator for 30 minutes at 50 C, with lids removed and agar surface downward; (b) in a forced air oven or incubator for 2 hours at 50 C, with lids on and agar surface upward; (c) in an incubator for 4 hours at 35 C with lids on and agar surface upward; or (d) on laboratory bench for 16 to 18 hours at room temperature, with lids on and agar surface upward.

BASE LAYER MEDIUM (Lipolysis)

Fat	50.0 g
Victoria Blue B 1:1500 solution*	200.0 ml
Agar	15.0 g

*Victoria Blue solution may be omitted when tributyrin is used.

Fat may consist of tributyrin, corn oil, soybean oil, any available cooking oil, lard tallow, triglycerides which do not contain antioxidant or other inhibitory substances.

Dissolve the fat in petroleum ether 5 to 10 g/100 ml and pass it through a column of activated alumina to remove FFA (free fatty acids). Evaporate the petroleum ether from the fat on a steam table and under a stream of nitrogen. (If substances are known to be free of FFA, the purification step may be omitted.)

Sterilize the fat in the autoclave for 30 minutes at 121 C. Sterilize the Victoria Blue by filtration. Dissolve the agar in 800 ml distilled water and sterilize for 15 minutes at 121 C. Cool the above sterile ingredients to 50 C and mix in a warm sterile blender for 1 minute. Pour 3 to 4 ml in the bottom of sterile Petri dishes. (Petri dishes may require a volume larger because of the difficulty of wetting the surface with the fat containing medium.) Plates may be used immediately or stored 3 to 4 days.

BASAL MEDIUM (PROTEOLYSIS)

See Soft Agar Gelatin Overlay (Basal Medium)

BEEF HEART INFUSION MEDIUM

Trim the fat from a beef heart and grind or mince 500 g of material. Add 1,000 ml of distilled water and allow to stand overnight in a refrigerator with occasional stirring.

Remove from refrigerator and boil over a free flame for 15 minutes.

Separate the tissue from the liquid by passing through two layers of cheese cloth and save both portions.

To the liquid portion add 10 g peptone and 5 g of NaCl and restore to original volume.

Adjust to pH 7.6 ± 0.2 with 1N NaOH for anaerobic cultivation (7.0 to 7.2 for aerobic cultivation) and boil for 15 to 20 minutes and filter through paper.

If medium is to be used immediately: Place a 2 cm column of tissue in a test tube and fill the tube half full with the broth. Sterilize for 20 minutes at 121 C.

If medium is to be stored: Place the liquid in screw capped bottles and sterilize for 15 minutes at 121 C. Rapidly dry the tissue in a thin layer in an incubator with forced circulation and store in a sealed jar. When medium is needed, prepare tubes as above and sterilize as for the freshly made tubes.

Use: Check tubes for sterility by incubation for 24 hours at 37 C. This medium may be used for the cultivation of pathogenic and other anaerobes and is suitable for carrying stock cultures and demonstrating spore formation. It is not suitable for the butyric-butyl group of thermophilic anaerobes.

BEEF HEART INFUSION BROTH
(From dehydrated commercial preparation)

Beef heart, infusion from	500.0 g
Tryptose	10.0 g
Sodium chloride	5.0 g

To rehydrate the medium, dissolve 25 grams of Beef Heart Infusion broth in 1000 ml distilled water, distribute in tubes and sterilize in the autoclave for 15 minutes at 121 C. The final reaction of the medium should be adjusted to pH 7.4.

For best results, the medium should be freshly prepared. If it is not used the same day as sterilized, heat in boiling water or flowing steam for several minutes to remove absorbed oxygen and cool quickly without agitation, just prior to inoculation.

BILE ESCULIN AGAR

Beef extract	3.0 g
Peptone	5.0 g
Bile salts (oxgall)	40.0 g
Ferric citrate	0.5 g
Esculin	1.0 g
Agar	15.0 g
Distilled water	1.0 liter

Dissolve beef extract, peptone, and agar in 400 ml of distilled water. Separately dissolve the ferric citrate in 100 ml of water. Dissolve the oxgall in 400 ml of water. Combine the solutions, mix and heat to 100 C for 10 minutes. Then autoclave at 121 C for 15 minutes and temper to 50 C. Aseptically add 100 ml of a 1% esculin solution that has been filter sterilized. (The 1% esculin solution should be warmed to facilitate complete solution prior to filtration.) Dispense in screw capped tubes and slant to cool. Optionally add 50 ml of horse serum to the agar concurrently with the esculin.

BISMUTH SULFITE AGAR

Peptone	10.0 g
Beef extract	5.0 g
Dextrose	5.0 g
Disodium phosphate	4.0 g
Ferrous sulfate	0.3 g
Bismuth ammonium citrate	1.85 g
Sodium sulfite	6.15 g
Agar	20.0 g
Brilliant green	0.025 g
Distilled water	1.0 liter

Dissolve ingredients in distilled water by boiling approximately 1 minute. Adjust to pH 7.7 ± 0.2, cool to 45 to 50 C, suspending precipitate with gentle agitation, and pour plates without sterilizing medium. Let plates dry with covers partially open. Caution: Plates lose selectivity after 72 hours.

BISMUTH SULFITE SALT BROTH*

Solution A

Peptone	10.0 g
Potassium chloride	0.7 g
Sodium chloride	25.0 g
Magnesium chloride · $6H_2O$	5.0 g
Distilled water	950.0 ml

Dissolve ingredients in distilled water. Using ten percent aqueous sodium carbonate, adjust pH to 9.1. Sterilize 15 minutes at 121 C. Cool to room temperature.

Solution B

Sodium sulfite	20.0 g
Distilled water	100.0 ml

Boil to dissolve

Solution C

Ammonium bismuth citrate	0.1 g
Distilled water	100.0 ml

Boil to dissolve

Solution D

Mannitol	20.0 g
Distilled water	100.0 ml

Boil to dissolve

Mix solution B and C and boil for 1 minute, then add solution D. Mix well before using. The solution should be white and turbid.

Prepare complete Bismuth Sulfite Broth by aseptically adding 100 ml of the "Bismuth Sulfite Mannitol" mixture and 1 ml of 95 percent ethanol to solution A. Mix well. Dispense 10 ml aseptically into sterile culture tubes without additional heating of the medium.

*This is a complex mixture. If dehydrated medium is not employed and the laboratory prepares its medium from separate ingredients, the mixture should be tested carefully against cultures with known characteristics to determine the suitability of the medium.

BLOOD AGAR BASE

Infusion from beef heart	500.0 g
Tryptose or Thiotone	10.0 g
Sodium chloride	5.0 g
Agar	15.0 g
Distilled water	1.0 liter

Suspend 40 grams of commercial heart infusion agar in distilled water and bring to a boil. Dispense in tubes or flasks and autoclave for 15 minutes at 121 C. Final reaction should be 7.3 ± 0.2.

When pouring blood agar plates melt medium and cool to 45 to 50 C and add aseptically 5 percent of sterile defibrinated sheep blood. Mix thoroughly avoiding incorporation of air bubbles and distribute in sterile tubes or pour into Petri plates in a thick layer.

BRAIN HEART INFUSION AGAR

Brain Heart Infusion Agar may be prepared by adding 15.0 g of agar to the basal formula below. Boil to melt agar. Dispense into bottles or flasks and autoclave as below.

BRAIN HEART INFUSION BROTH

Calf brain, infusion from	200.0 g
Beef heart, infusion from	250.0 g
Proteose peptone or Polypeptone	10.0 g
Dextrose	2.0 g
Sodium chloride	5.0 g
Disodium phosphate	2.5 g
Distilled water	1.0 liter

Dissolve 37 grams of commercial dehydrated ingredients in distilled water by bringing to a boil and make up to 1 liter. Dispense into tubes and autoclave for 15 minutes at 121 C. Final reaction should be pH 7.4 ± 0.2.

BRILLIANT GREEN AGAR
(For *Salmonella*)

Yeast Extract	3.0 g
Proteose peptone No. 3 or Polypeptone	10.0 g
Sodium chloride	5.0 g
Lactose	10.0 g
Sucrose	10.0 g
Phenol red	0.08 g
Brilliant green 0.25% solution	5.0 ml
Agar	20.0 g
Distilled water	1.0 liter

Dissolve ingredients by boiling in distilled water. Autoclave 121 C for 12 minutes. Cool by 45 to 50 C and pour 20 ml portions into 15 × 100 mm Petri dishes. Let dry for about 2 hours with covers partially open, then close plates. Final pH 6.9 ± 0.2.

BRILLIANT GREEN LACTOSE BILE AGAR

Peptone	8.25 g
Lactose	1.9 g
Oxgall	0.00295 g
Sodium sulfite	0.205 g
Ferric chloride	0.0295 g
Monopotassium phosphate	0.0153 g
Erioglaucine	0.0649 g
Basic fuchsin	0.0776 g
Brilliant green	0.0000295 g
Agar	10.15 g
Distilled water	1.0 liter

Dissolve ingredients in distilled water by bringing to a boil. Dispense into tubes or flasks and autoclave for 15 minutes at 121 C. Final reaction should be approximately pH 6.9. The medium is sensitive to light and should be stored in the dark.

BRILLIANT GREEN LACTOSE BILE BROTH, 2%

Peptone	10.0 g
Lactose	10.0 g
Oxgall	20.0 g
Brilliant green	0.0133 g
Distilled water	1.0 liter

Dissolve ingredients by suspending in distilled water and heating gently. Dispense into tubes. Add inverted fermentation vial. Autoclave 15 minutes at 121 C. Final pH 7.2 ± 0.2. Autoclave temperature must drop slowly below 75 C before the door is opened.

BROMCRESOL PURPLE CARBOHYDRATE BROTH

Peptone	10.0 g
Beef extract (optional)	3.0 g
Sodium chloride	5.0 g
Bromcresol purple	0.04 g
Distilled water	1.0 liter

Dissolve 10 g glucose, 5 g adonitol, 5 g arabinose, 5 g mannitol, 5 g maltose, 5 g sucrose, 5 g lactose, 5 g sorbitol, 5 g cellobiose, 5 g salicin or 5 g trehalose, in distilled water as required in title of test. Adjust pH to 7.0 ± 0.2. Dispense 8 ml aliquots to 16 × 150 mm tubes containing inverted 12 × 75 mm tubes. Autoclave 10 minutes at 121 C. Final pH 6.8 to 7.0. Allow autoclave temperature to drop slowly.

BUFFERED GLUCOSE BROTH

See MR-VP Medium.

CARBOHYDRATE FERMENTATION MEDIA

Fermentation Broth Base

Peptone	10.0 g
Meat extract	3.0 g
Sodium chloride	5.0 g
Andrade's indicator	10.0 ml
Distilled water	1.0 liter

Dissolve ingredients in distilled water and adjust pH to 7.1 to 7.2. Dispense in tubes with inverted insert tubes and sterilize at 121 C. for 15 minutes. Autoclave temperature must drop slowly before the door is opened. (See exceptions noted below.)

Glucose, lactose, sucrose, and mannitol are employed in a final concentration of 1%. Other carbohydrates such as dulcitol, salicin, etc. may be used in a final concentration of 0.5%. Glucose, mannitol, dulcitol, salicin, adonitol, and inositol may be added to the basal medium prior to sterilization. Medium containing neutral glycerol should be sterilized at 121 C for 10 minutes. Disaccharides such as lactose, sucrose, and cellobiose (10% solution in distilled water, neutral pH) should be sterilized by filtration or at 121 C for 10 minutes and added to previously sterilized basal medium. Arabinose, rhamnose, and xylose also should be sterilized separately. If basal medium is tubed in 3.0 ml amounts, add 0.3 ml of sterile aqueous carbohydrate solution, i.e., one-tenth the volume.

CARY AND BLAIR TRANSPORT MEDIUM

Sodium thioglycollate	1.5 g
Disodium phosphate	1.1 g
Sodium chloride	5.0 g

These ingredients are added to 990 ml of demineralized or distilled water in the order given. Add: Agar 5.0 g

Heat medium only until clear. After mixture has cooled to 50 C, add calcium chloride, 1% solution, 9.0 ml

Adjust pH to 8.4. Dispense in 7 ml quantities in 9 ml screw capped vials (previously rinsed and sterilized). Steam vials for 15 minutes, cool, and tighten caps.

CASEIN SOY PEPTONE AGAR

See Trypticase or Tryptic Soy Agar.

CHOPPED LIVER BROTH

Ground beef liver	500.0 g
Soluble starch	1.0 g
Peptone	10.0 g
Dipotassium phosphate	1.0 g
Distilled water	1.0 liter

Add finely ground beef liver to the distilled water and boil for 1 hour. Adjust broth to pH 7.0 ± 0.2 and boil another 10 minutes. Press through cheese cloth and make broth to 1 liter with distilled water. Add peptone and dipotassium phosphate and adjust to pH 7.0 ± 0.2. Place liver particles from the pressed cake in the bottom of culture tubes (about 1 cm deep), cover with 8 to 10 ml of broth. Sterilize for 20 minutes at 121 C. Before use, exhaust for 20 minutes in flowing steam.

CHRISTENSEN'S UREA AGAR

Basal Ingredients:

Peptone	1.0 g
Sodium chloride	5.0 g
Glucose	1.0 g
Monobasic potassium phosphate	2.0 g
Phenol red (6 ml of 1:500 solution)	0.012 g

Urea concentrate:

Urea	20.0 g
Distilled water	100.0 ml

Adjust to pH 6.8 to 6.9. Filter sterilize.

Dissolve 15 g agar in 900 ml distilled water, add basal ingredients, and sterilize at 121 C for 15 minutes. Cool to 50 to 55 C, then add 100 ml urea concentrate. Mix and distribute in sterile tubes. The medium is slanted with a deep butt. This medium may be employed in fluid form (without the agar) if desired.

Alternate Method. Stuart et al., 1945, modification of the highly buffered medium of Rustigian and Stuart, 1941.

Yeast extract	0.1 g
Monobasic potassium phosphate	0.091 g
Dibasic sodium phosphate	0.095 g
Urea	20.0 g
Phenol red	0.01 g
Distilled water	1.0 liter

Mix ingredients in the distilled water. This medium is filter sterilized and tubed in sterile tubes in 3 ml amounts. The basal medium (without urea) may be prepared in 900 ml of distilled water and sterilized at 121 C for 15 minutes. After cooling, 100 ml of 20% sterile urea solution are added and the medium dispensed in sterile tubes in 3 ml amounts.

Inoculation: Three loopsful (2 mm loop) from an agar slant culture are inoculated into a tube of medium and the tube is shaken to suspend the bacteria.

Incubation: Tests are incubated in a water bath at 37 C, and the results are read after 10 minutes, 60 minutes, and 2 hours.

CITRATE AGAR (Christensen)

Sodium citrate	3.0 g
Glucose	0.2 g
Yeast extract	0.5 g
Cysteine monohydrochloride	0.1 g
Ferric ammonium citrate	0.4 g
Monopotassium phosphate	1.0 g
Sodium chloride	5.0 g
Sodium thiosulfate	0.08 g
Phenol red	0.012 g
Agar	15.0 g
Distilled water	1.0 liter

Mix ingredients in the distilled water. Tube and sterilize at 121 C for 15 minutes and slant (1 inch butt, 1.5 inch slant).

The ferric ammonium citrate and sodium thiosulfate may be omitted from the formula, if desired, since they do not affect the value of the medium as an indicator of citrate utilization.

CITRATE AGAR, SIMMONS'

Sodium chloride	5.0 g
Magnesium sulfate	0.2 g
Monoammonium phosphate	1.0 g
Dipotassium phosphate	1.0 g
Sodium citrate	2.0 g
Agar	15.0 g
Distilled water	1.0 liter

Mix ingredients in the distilled water. Add 40 ml of 1:500 bromthymol blue indicator solution. Sterilize at 121 C for 15 minutes, and slant the tubes so as to obtain a 1 inch butt and a 1.5 inch slant.

COOKED MEAT MEDIUM (CM)

Beef heart	454.0 g
Proteose peptone or peptone	20.0 g
Dextrose	2.0 g
Sodium chloride	5.0 g

Suspend 12.5 g of dry commercial preparation in 100 ml of distilled water at room temperature, mix thoroughly to wet particles and allow the mixture to soak for 15 minutes. Shake to maintain an even suspension and distribute into tubes.

Alternately: Distribute 1.25 g of commercial preparation into test tubes and add 10 ml distilled water to each tube. Mix thoroughly to wet particles and let soak for 15 minutes.

Sterilize at 121 C for 15 minutes. Final pH 7.2 ± 0.2. Autoclave temperature must cool slowly to less than 75 C before the door is opened.

Before use, exhaust in flowing steam for 20 minutes.

CRYSTAL VIOLET TETRAZOLIUM AGAR

To Plate count agar, add

Crystal violet	1 ppm
2,3,5 triphenyl tetrazolium chloride	50 ppm

Dissolve plate count agar, add crystal violet and sterilize at 121 C for 15 minutes. Prepare solution of 2,3,5 triphenyl tetrazolium chloride (TTC) and sterilize by filtration. Add sterile TTC aseptically to melted and cooled (44 to 46 C) agar prior to pouring plates.

Prepare 0.1% crystal violet solution by dissolving 0.1 g crystal violet in 100 ml ethyl alcohol. Add 1.0 ml of the 0.1% crystal violet solution per liter to obtain 1 ppm in final medium. Prepare 1% solution of TTC. Add 5 ml of the 1% TTC solution per liter to obtain 50 ppm.

DECARBOXYLASE TEST MEDIA
FALKOW METHOD
(Basal Medium)

Peptone	5.0 g
Yeast extract	3.0 g
Glucose	1.0 g
Bromcresol purple (1.6% solution)	1.0 g
Distilled water	1.0 liter

Mix ingredients in distilled water. Adjust pH to 6.7 to 6.8 if necessary (medium with ornithine requires adjustment).

The basal medium is divided into four parts and treated in the same manner as that given above for the Moeller method, except that only 0.5% of the L-amino acid is added. After the amino acids are added to three of the four portions of basal medium, the media are tubed in small (13 × 100 screw capped) tubes and sterilized at 121 C for 10 minutes. The remaining portion of basal medium, without amino acid, serves as the control.

Inoculate lightly from a young agar slant culture. Use an oil seal and inoculate a control tube with each culture under investigation. Incubate at 37 C, examine daily for 4 days. This medium first becomes yellow because of acid production. Later, if decarboxylation occurs, the medium becomes alkaline (purple). The control tubes remain acid (yellow).

WARNING: This method is not recommended for *Klebsiella* and *Enterobacter*.

DECARBOXYLASE TEST MEDIA*
BASAL FOR USE WITH LYSINE, ARGININE, ORNITHINE

Moeller Method (1954, 1955)

Basal Medium

Peptone (Orthana special)**	5.0 g
Beef extract	5.0 g
Bromcresol purple (1.6%)	0.625 ml
Cresol red (0.2%)	2.5 ml
Glucose	0.5 g
Pyridoxal	5.0 mg
Distilled water	1.0 liter

Adjust pH to 6 or 6.5

*When used for *Vibrio parahaemolyticus*, 3% sodium chloride must be added.

**Kemish Fabrik, Copenhagen, Denmark.

The basal medium is divided into four equal portions, one of which is tubed without the addition of any amino acid. These tubes of basal medium are used for control purposes. To one of the remaining portions of basal medium is added 1% of L-lysine dihydrochloride; to the second, 1% of L-arginine monohydrochloride; and to the third portion, 1% of L-ornithine dihydrochloride. If DL amino acids are used, they should be incorporated into the medium in 2% concentration, since the microorganisms apparently are active against the L forms only. The pH of the fraction to which ornithine is added should be readjusted after the addition and prior to sterilization. The amino acid medium may be tubed in 3 or 4 ml amounts in small (13 × 100 mm) screw capped tubes and sterilized at 121 C for 10 minutes. A small amount of floccular precipitate may be seen in the ornithine medium. This does not interfere with its use.

Inoculation: Inoculate lightly from a young agar slant culture. After inoculation add a layer (about 10 mm in thickness) of sterile mineral (paraffin) oil* to each tube including the control. A control tube always should be inoculated with each culture under investigation. Incubate at 37 C; examine daily for 4 days. Positive reactions are indicated by alkalinization of the medium with a color change from yellow to violet. Weakly positive reactions may be bluish gray.

*The mineral oil (heavy) should be sterilized in an autoclave at 121 C for 30 minutes to one hour, depending on the size of the containers. Large screw capped test tubes, about half full of oil, are convenient. The cloudiness seen after sterilization disappears in a short time.

DESOXYCHOLATE CITRATE AGAR (Leifson, 1935 modified)

Meat infusion (pork, beef, or beef heart)	330.0 g
Proteose No. 3 or Thiotone	10.0 g
Lactose	10.0 g
Sodium citrate	20.0 g
Sodium desoxycholate	5.0 g
Ferric ammonium citrate	2.0 g
Agar	13.5 g
Neutral red	0.02 g
Distilled water	1.0 liter
(pH 7.3–7.5)	

Mix ingredients in distilled water. This medium may be boiled for a minute or two to dissolve the ingredients, but it should not be subjected to excessive heat. NOTE: It should not be autoclaved.

The original formulation given by Leifson (1935) employed 1 liter of infusion made from fresh pork adjusted to pH 7.5. To this the other ingredients were added in the order given (and as described under desoxycholate agar, above). Twenty-five g of sodium citrate (11 H_2O) were used, and the neutral red was added after the final adjustment of the pH (2 ml of 1% solution per liter).

DESOXYCHOLATE LACTOSE AGAR

Peptone, polypeptone, or tryptose	10.0 g
Lactose	10.0 g
Sodium chloride	5.0 g
Sodium citrate	2.0 g
Sodium desoxycholate	0.5 g
Neutral red	0.03 g
Agar	15.0 g
Distilled water	1.0 liter

Dissolve ingredients in distilled water by bringing to a boil and adjust to pH 7.1 ± 0.2. There is no need to sterilize further unless medium is to be stored. Overheating will inhibit solidification of agar.

DEXTROSE TRYPTONE AGAR

See Dextrose Tryptone Bromcresol Purple Agar.

DEXTROSE TRYPTONE BROMCRESOL PURPLE AGAR

Tryptone or trypticase	10.0 g
Dextrose	5.0 g
Bromcresol purple 2% ethanol solution	2.0 ml
Agar	15.0 g
Distilled water	1.0 liter

Mix ingredients in distilled water. Heat to boiling to dissolve ingredients. Dispense into tubes or flasks and autoclave for 15 minutes at 121 C. Final reaction should be pH 6.7 ± 0.2.

Dextrose tryptone bromcresol purple broth may be made using the same formula above by withholding the agar.

DIFFERENTIAL BROTH FOR LACTIC STREPTOCOCCI

Tryptone	5.0 g
Yeast extract	5.0 g
Dipotassium phosphate	1.0 g
Arginine	5.0 g
Sodium citrate	20.0 g
Bromcresol purple	0.02 g
Distilled water q.s.	1.0 liter

Suspend ingredients in approximately 800 ml distilled water, add 35 ml of 11% reconstituted skim milk and bring volume to one liter. Steam the medium 15 minutes, cool to 25 C, and adjust pH to 6.2. Dispense 7 ml quantities into 10 × 126 mm screw cap tubes containing Durham fermentation tubes. Sterilize at 121 C for 15 minutes. Final pH 6.2 ± 0.2. Do not open the door until the temperature has dropped below 75 C.

DIFFERENTIAL REINFORCED CLOSTRIDIAL MEDIUM (DRCM)

Peptone	10.0 g
Beef extract	10.0 g
Hydrated sodium acetate	5.0 g
Yeast extract	1.5 g
Soluble starch	1.0 g
Glucose	1.0 g
L-cysteine hydrochloride	0.5 g
Distilled water	1.0 liter

Add peptone, beef extract, sodium acetate and yeast extract to 800 ml of distilled water. Add the starch to the remaining 200 ml of water by making a cold paste in a little of the water, boil the rest of the water and stir it into the paste. Combine the two solutions, mix and steam for 30 minutes to complete the dissolving of ingredients. After steaming adjust to pH 7.1 to 7.2 with 10N NaOH. Filter through filter paper, dispense in 25 ml quantities into screw capped bottles, and sterilize by autoclaving at 121 C for 15 minutes. Immediately before using, add sodium sulfite and ferric citrate to give final concentrations of 0.04% and 0.07% respectively. Prepare a 4% solution of anhydrous sodium sulfite and a 7% solution of ferric citrate (scales). The ferric citrate must be heated to dissolve. Filter both solutions to sterilize and store at 3 to 5 C until needed. After steaming the basal medium to drive off dissolved oxygen, cool it quickly, and add the sulfite and citrate solutions immediately before using it.

DIHYDROLASE BROTH

Yeast extract	3.0 g
Sodium chloride	30.0 g
Glucose	1.0 g
Bromcresol purple (1.6% solution)	1.0 ml
Decarboxylase basal broth	1.0 liter

Adjust pH to 6.7 to 6.8. Divide basal broth into 3 portions. Add 0.5% L-arginine to first portion, add 0.5% L-lysine to second portion; third portion is the basal broth control.

Dispense 3 ml into 13 mm × 100 mm screw cap tubes. Sterilize fo 10 minutes at 121 C.

DNase TEST MEDIUM (Jeffries et al, 1957 Martin and Ewing, 1967)

Tryptose*	20.0 g
Deoxyribonucleic acid	2.0 g
Sodium chloride	5.0 g
Agar	15.0 g
Distilled water	1.0 liter

*Phytone 5 g and trypticase 15 g may be substituted for the tryptose.

Mix ingredients in distilled water. Heat mixture to boiling temperature with frequent agitation (swirling). Sterilize at 121 C for 15 minutes.

DNase TEST MEDIUM (Alternate method: Smith, Hancock, and Rhoden, 1969)

DNase test medium	1000.0 ml
Methyl green 0.5% solution	10.0 ml

Dissolve 500 mg of methyl green (Fischer M295) in 100 ml of distilled water. Extract repeatedly with chloroform employing approximately 100 ml each time until chloroform layer becomes colorless. This procedure removes methyl violet, which is an impurity in the methyl green. About six extractions should suffice.

Add one ml of methyl green solution per 100 ml of melted DNase test medium and sterilize at 121 C for 15 min. Plates may be kept for several days (about a week), if refrigerated. If desired, the dye solution can be filter sterilized and added aseptically.

Plates should be inoculated (see above), incubated overnight, and examined by transmitted light against a white background. A zone of clearing around the colony indicates a positive reaction.

EC BROTH

Pancreatic digest of casein	20.0 g
Bile salt mixture or Bile Salts No. 3	1.5 g
Lactose	5.0 g
Dipotassium phosphate	4.0 g
Potassium phosphate	1.5 g
Sodium chloride	5.0 g
Distilled water	1.0 liter

Dissolve ingredients in distilled water, and distribute 8 ml portions into 16 × 150 mm test tubes containing inverted 10 × 75 fermentation tube. Autoclave 15 minutes at 121 C. Final pH, 6.9 ± 0.2. Do not open autoclave door until temperature has dropped below 75 C.

ENTERIC ENRICHMENT (EE) BROTH

Peptone	10.0 g
Dextrose	5.0 g
Disodium phosphate	8.0 g
Monopotassium phosphate	2.0 g
Oxbile	20.0 g
Brilliant green, certified	0.015 g
Distilled water	1.0 liter

Dissolve ingredients in distilled water with stirring. Dispense 30 ml portions into 100 ml Erlenmeyer flasks. Autoclave 5 minutes at 121 C or 30 minutes at 100 C. Final pH should be 7.2 ± 0.2.

EOSIN METHYLENE BLUE AGAR (LEVINE)

Peptone	10.0 g
Lactose	10.0 g
Dipotassium phosphate	2.0 g
Agar	15.0 g
Eosin Y	0.4 g
Methylene blue	0.065 g
Distilled water	1.0 liter

Mix ingredients in distilled water. Boil to dissolve peptone, phosphate, and agar in 1 liter of water. Dispense in 100 or 200 ml portions, and autoclave 15 minutes at 121 C. Cool to 45 to 50 C, swirl contents, and pour plates. Final pH should be 7.1 ± 0.1.

EUGON AGAR

Tryptose or Trypticase	15.0 g
Soytone or Phytone	5.0 g
Glucose	5.0 g
L-cystine	0.2 g
Sodium chloride	4.0 g
Sodium sulfite	0.2 g
Sodium citrate	1.0 g
Agar	15.0 g
Distilled water	1.0 liter

Dissolve ingredients in distilled water with gentle heating. Sterilize medium for 15 minutes at 121 C. Final reaction of the medium will be pH 7.0 ± 0.2. Swirl flask just prior to pouring medium.

FLUID THIOGLYCOLLATE MEDIUM

Pancreatic digest of casein USP	15.0 g
L-cystine	0.5 g
Dextrose	5.0 g
Yeast extract	5.0 g
Sodium chloride	2.5 g
Sodium thioglycollate	0.5 g
Agar	0.75 g
Resazurin	0.001 g
Distilled water	1.0 liter

Heat ingredients in distilled water to dissolve and adjust to final pH 7.1 ± 0.2. Dispense in test tubes, filling them to at least two-thirds of tube capacity. Sterilize at 121 C for 15 minutes. Just before use, heat to boiling, or expose in flowing steam for 10 minutes to remove dissolved oxygen. Cool rapidly to incubation temperature.

FPA MEDIUM (FLUORESCENT PECTOLYTIC AGAR)

Proteose peptone No. 3	20.0 g
Dipotassium phosphate	1.5 g
Magnesium sulfate • $7H_2O$	0.73 g
Pectin	5.0 g
Agar	15.0 g
Distilled water	1.0 liter

Dissolve ingredients in distilled water by boiling. Adjust to pH 7.1 and sterilize at 121 C for 15 minutes.

ANTIBIOTICS SOLUTION

Penicillin G	75,000 units
Novobiocin	45.0 mg
Cycloheximide	75.0 mg

Add 1 ml ethanol to above mixture of antibiotics and allow to stand 30 minutes. Dilute with 9 ml sterile distilled water and add 1 ml of this diluted antibiotic solution to 100 ml of sterile and cooled basal medium before pouring plate. After the plates are inoculated by a spread plate technique and incubated, the fluorescent bacteria are detected under long wavelength ultraviolet light. Plates are then treated with the polysaccharide precipitant as described under MP medium.

GEL-PHOSPHATE BUFFER

Gelatin	2.0 g
Disodium phosphate	4.0 g
Distilled water	1.0 liter

Dissolve gelatin and phosphate in distilled water with gentle heat. Sterilize at 121 C for 20 minutes. Final pH should be 6.2.

GELATIN AGAR

Gelatin	30.0 g
Agar	15.0 g
Sodium chloride	10.0 g
Trypticase	10.0 g
Distilled water	1.0 liter

Melt all ingredients, adjust pH to 7.2 ± 0.2, sterilize for 15 minutes at 121 C.

GLUCOSE BROTH

Trypticase or Tryptone	10.0 g
Dextrose	5.0 g
Sodium chloride	5.0 g
Distilled water	1.0 liter

Preparation: Dissolve ingredients in distilled water. Heat gently to obtain solution. Adjust to pH 7.3. Dispense and insert inverted fermentation vials, if desired. Sterilize at not over 118 C for 15 minutes. Do not open autoclave until the temperature has dropped below 75 C.

GLUCOSE SALT TEEPOL BROTH (GSTB)

Beef extract	3.0 g
Peptone	10.0 g
Sodium chloride	30.0 g
Glucose	5.0 g
Methyl violet	0.002 g
Teepol*	4.0 ml
Distilled water	1.0 liter

*Teepol is available through Shell Chemical Company, Industrial Chemical Division, 823 Commerce Avenue, Oak Brook, Illinois. Catalog listing: Teepol 610.

Dissolve ingredients in distilled water and dispense in 10 ml portions. If larger quantities of sample are to be examined (25 g), use screw capped jars containing 225 ml of broth. Sterilize at 121 C for 15 minutes. Final pH should be 9.4 ± 0.2.

GLUCOSE TRYPTONE AGAR

Tryptone	10.0 g
Glucose	5.0 g
Agar	15.0 g
Bromcresol purple	0.04 g
Distilled water	1.0 liter

Heat to dissolve ingredients in distilled water and autoclave 15 minutes at 121 C. Final pH 6.7 ± 0.2.

GN BROTH (Hajna, 1955)

Glucose	1.0 g
D-mannitol	2.0 g
Sodium citrate	5.0 g
Sodium desoxycholate	0.5 g
Dipotassium phosphate	4.0 g
Monopotassium phosphate	1.5 g
Sodium chloride	5.0 g
Tryptose	20.0 g
Distilled water	1.0 liter

Dissolve ingredients in distilled water by heat. Dispense in tubes in convenient amounts and sterilize at 116 C for 15 minutes. Final pH is 7.0 ± 0.2. Avoid excessive heating.

GUM TRAGACANTH-ARABIC

Gum tragacanth	2.0 g
Gum arabic	1.0 g
Distilled water	100.0 ml

Dissolve ingredients in distilled water, heating slightly. Sterilize at 121 C for 15 minutes.

HALOPHILIC AGAR

Casamino acids	10.0 g
Yeast extract	10.0 g
Proteose peptone	5.0 g
Trisodium citrate	3.0 g
Potassium chloride	2.0 g
Magnesium sulfate·$7H_2O$	25.0 g
Sodium chloride	250.0 g
Agar	20.0 g
Distilled water	1.0 liter

Dissolve ingredients in distilled water and sterilize at 121 C for 15 minutes. Final pH should should be 7.2.

HALOPHILIC BROTH

Prepare as for Halophilic Agar except omit agar.

HEKTOEN ENTERIC AGAR

Proteose peptone	12.0 g
Yeast extract	3.0 g
Lactose	12.0 g
Sucrose	12.0 g
Salicin	2.0 g
Bile complex	9.0 g
Sodium chloride	5.0 g
Sodium thiosulfate	5.0 g
Ferric ammonium citrate	1.5 g
Bromthymol blue	0.065 g
Acid fuchsin	0.1 g
Agar	14.0
Distilled water	1.0 liter

Suspend 75.7 grams in distilled water. Boil with frequent stirring. Do not overheat or autoclave. When completely in solution, cool to 55 to 60 C and distribute into plates. Allow plates to solidify with lids ajar to provide a dry surface for inoculation. Plates may be refrigerated for future use. Final pH 7.5 ± 0.2.

HUGH-LEIFSON GLUCOSE BROTH (LISTON-BAROSS) (HLGB)

Peptone	2.0 g
Yeast extract	0.5 g
Sodium chloride	30.0 g
Glucose	10.0 g
Bromcresol purple	0.015 g
Agar	3.0 g
Distilled water	1.0 liter

Dissolve ingredients in distilled water. Adjust the pH to 7.4 ± 0.2 before autoclaving at 121 C for 15 minutes.

HYA AGAR

Beef extract	1.0 g
Proteose peptone #3	10.0 g
Glucose	2.5 g
Galactose	2.5 g
Lactose	5.0 g
Agar	15.0 g
Distilled water	1.0 liter

Dissolve ingredients in distilled water. Adjust medium to pH 6.8 ± 0.2 before sterilization. Autoclave in 90 ml aliquots at 121 C for 20 minutes. Add sugars as cold, sterilized, 10% solutions (Millipore filtered—0.2μ), just prior to plating samples.

INDOLE-NITRITE BROTH

Trypticase or Tryptone	20.0 g
Disodium phosphate	2.0 g
Potassium nitrate	1.0 g
Glucose	1.0 g
Agar	1.0 g
Distilled water	1.0 liter

Dissolve ingredients in distilled water by heating with stirring. Autoclave 15 minutes at 121 C. Final pH should be 7.2.

KF STREPTOCOCCUS AGAR

Proteose peptone #3 or Polypeptone	10.0 g
Yeast extract	10.0 g
Sodium chloride	5.0 g
Sodium glycerophosphate	10.0 g
Maltose	20.0 g
Lactose	1.0 g
Sodium azide	0.4 g
Bromcresol purple	0.015 g
Agar	20.0 g
Distilled water	1.0 liter

Dissolve ingredients in distilled water by boiling, and dispense in 100 ml portions. Autoclave at 121 C for 10 minutes. When ready to use, cool to 50 C and add 1 ml of 1% solution TTC (Triphenyl tetrazolium chloride 1%) per 100 ml. Final pH should be 7.2. Do not overheat this medium.

KF STREPTOCOCCUS BROTH

Proteose peptone #3 or Polypeptose	10.0 g
Yeast extract	10.0 g
Sodium chloride	5.0 g
Maltose	20.0 g
Sodium glycerophosphate	10.0 g
Lactose	1.0 g
Sodium azide	0.4 g
Bromcresol purple*	0.015 g
Distilled water	1.0 liter

*Phenol red (0.018 g) may be substituted for the bromcresol purple.

Dissolve ingredients in distilled water, and dispense in 10 ml portions. Autoclave at 10 minutes at 121 C. Final pH, 7.2 ± 0.1.

KG AGAR

Preparation A

Peptone	1.0 g
Yeast extract	0.5 g
Phenol red	0.025 g
Agar	18.0 g
Distilled water	900.0 ml

Dissolve ingredients in distilled water and adjust to pH 6.8. Autoclave basal media at 121 C for 20 minutes, cool to 50 C, and add 100 ml preparation B, and 1 ml of preparation C. Mix well, pour into Petri dishes, allow to solidify, and store in a manner to eliminate excess surface moisture. Plates may be stored at 4 C for up to 7 days.

Preparation B—Colbecks Egg Yolk Broth or Concentrated Egg Yolk Emulsion

1. Colbecks Egg Yolk Broth

Obtainable from Difco Laboratories, Detroit, Mich., this egg yolk emulsion has been successfully used in the formulation of KG agar where it is added to the sterilized, cooled basal agar medium at a rate of 100 ml per 900 ml base. No studies on the efficacy of incorporating this form of emulsion into MYP agar have been reported, thus it is advocated that the investigator employing MYP agar use the alternate emulsion.

2. Concentrated Egg Yolk Emulsion

This product is manufactured by Oxoid Ltd, London, England, and is marketed in the U.S.A. by Difco Laboratories, Detroit, Michigan, and by Flow Laboratory, Rockville, Md. 20852. It is added to the sterilized and cooled agar base of both the KG and MYP agar at a rate of 100 ml emulsion to 900 ml base.

Preparation C—Polymyxin B Sulfate

This selective agent is obtainable in sterile powdered form (500,000 units, i.e., 50 mgm per vial) from Pfizer Inc., New York, N.Y., and from Difco laboratories as Sterile Antimicrobic Vial P. To use, add aseptically by syringe 5 ml sterile distilled water. Mix to dissolve powder and remove by syringe. Add 1 ml per liter to final medium.

KOSER'S CITRATE BROTH

Sodium ammonium phosphate	1.5 g
Monopotassium phosphate	1.0 g
Magnesium sulfate	0.2 g
Sodium citrate	3.0 g
Distilled water	1.0 liter

Dissolve ingredients in distilled water and dispense in 10 ml portions in test tubes, and autoclave 15 minutes at 121 C. Final pH, 6.7 ± 0.2.

KRANEP AGAR

a. Basal medium:

Peptone	10.0 g
Sodium chloride	3.0 g
Disodium phosphate • 12 H_2O	2.0 g
Agar	25.0 g
Distilled water	1.0 liter

Suspend ingredients in 1 liter of distilled water and heat to boiling with frequent agitation to dissolve ingredients. Dispense into flasks, 900 ml per flask, and autoclave 15 minutes at 121 C.

b. Egg emulsion

Immerse fresh eggs in alcohol, drain off excess and carefully flame to sterilize exterior. Break eggs aseptically and separate yolks and whites. Mix equal parts egg yolk and sterile physiological saline using a mechanical blender at low speed (available as 50% Egg Yolk Emulsion, Difco Laboratories, Detroit, Michigan).

c. Complete medium

Cool melted basal medium to 50 C then add

Potassium thiocyanate	25.5 g
D-mannitol	5.1 g
Lithium chloride	5.1 g
Sodium pyruvate	8.2 g

Heat 20 minutes at 110 C. Cool to 50 C and add filter sterilized solutions at 50 C as follows:

Sodium azide (0.5% solution)	10.0 ml
Actidione (Cycloheximide—0.41% solution)	10.0 ml
Egg emulsion (b)	100.0 ml

Dispense into plates 15 to 20 ml per plate. Incubate inoculated plates 48 hr at 37 C, then 24 hours at room temperature (23 to 25 C). Final pH 6.8 to 7.0.

LACTOBACILLUS MRS BROTH

See MRS Broth.

LACTOBACILLUS SELECTION MEDIUM
(Rogosa Medium)

Pancreatic digest of casein	10.0 g
Yeast extract	5.0 g
Monopotassium phosphate	6.0 g
Ammonium citrate	2.0 g
Dextrose	20.0 g
Sorbitan monooleate	1.0 g
Sodium acetate hydrate	25.0 g
Magnesium sulfate	0.575 g
Manganese sulfate	0.120 g
Ferrous sulfate	0.034 g
Agar	15.0 g
Distilled water	1.0 liter

Mix ingredients thoroughly in distilled water, and boil for two minutes. Use without autoclaving. Cool to 45 C, add 1.32 ml of glacial acetic acid and pour as for plate counts.

LBS Modified:

Adjust pH to 5.4 ± 0.2 using acetic acid. Add .0075% sterile brilliant green prior to cooling. Mix, cool, and pour as above.

LACTIC AGAR

Tryptone	20.0 g
Yeast extract	5.0 g
Gelatin	2.5 g
Glucose	5.0 g
Lactose	5.0 g
Sucrose	5.0 g
Sodium chloride	4.0 g
Sodium acetate	1.5 g
Ascorbic acid	0.5 g
Agar	15.0 g
Distilled water	1.0 liter

Dissolve ingredients in distilled water by gentle heating. Sterilize at 121 C for 15 minutes.

LACTOSE BROTH

Beef extract	3.0 g
Peptone	5.0 g
Lactose	5.0 g
Distilled water	1.0 liter

Heat slightly to dissolve ingredients in distilled water and dispense into fermentation tubes. Autoclave for 15 minutes at 121 C. pH should be between 6.8 and 7.0, preferably 6.9 ± 0.2. Allow temperature in autoclave to drop slowly below 75 C before opening.

LACTOSE GELATIN MEDIUM

Lactose	10.0 g
Sodium phosphate (dibasic)	5.0 g
Phenol red	0.05 g
Gelatin	120.0 g
Distilled water	1.0 liter

Dissolve ingredients in distilled water. Adjust the pH to 7.5 prior to adding the lactose and phenol red. Dispense ten ml amounts of the medium into 15 × 150 mm screw capped tubes. Autoclave at 121 C for 15 minutes. Just before use, heat to boiling, or expose to flowing steam for 10 minutes to remove dissolved oxygen. Cool rapidly to incubation temperature.

LACTOSE GELATIN MEDIUM
(For *Cl. perfringens*)

Tryptose	15.0 g
Yeast extract	10.0 g
Lactose	10.0 g
Disodium phosphate	5.0 g
Phenol red	0.05 g
Gelatin	120.0 g
Distilled water	1.0 liter

Suspend ingredients in distilled water and dissolve by heating gently. Final pH 7.5 ± 0.2. Dispense 10 ml in 16 mm × 150 mm screw capped test tubes. Sterilize at 121 C for 15 minutes. Heat to exhaust oxygen prior to use.

LAURYL SULFATE TRYPTOSE BROTH

Tryptose or Trypticase (pancreatic digest of casein)	20.0 g
Lactose	5.0 g
Dipotassium phosphate	2.75 g
Monopotassium phosphate	2.75 g
Sodium chloride	5.0 g
Sodium lauryl sulfate	0.1 g
Distilled water	1.0 liter

Dissolve ingredients in distilled water and dispense 10 ml portions in 20 × 150 mm test tubes containing inverted 10 × 75 mm fermentation tubes. Autoclave 15 minutes at 121 C. Final pH 6.8 ± 0.2.

LEE'S AGAR

Tryptone	10.0 g
Yeast extract	10.0 g
Lactose	5.0 g
Sucrose	5.0 g
Calcium carbonate	3.0 g
Dipotassium phosphate	0.5 g
Bromcresol purple*	0.02 g
Agar	18.0 g
Distilled water to make	1.0 liter

*Add 1 ml of a sterile (121 C for 15 minutes) 0.2% solution per 100 ml of medium just before pouring plates.

Dissolve ingredients in distilled water with gentle heating. Adjust medium to pH 7.0 ± 0.2 before sterilization for 20 minutes at 121 C.

Carefully mix the melted medium to evenly suspend calcium carbonate and pour medium into previously chilled sterile Petri dishes to obtain a layer 4 to 5 mm thick. After solidification, dry plates in a 30 C incubator for 18 to 24 hours.

LITMUS MILK

Skim milk powder	100.0 g
Litmus	0.75 g
Distilled water	1.0 liter

Suspend ingredients in distilled water and dissolve. Dispense in suitable containers. Sterilize 15 minutes at 121 C.

LIVER BROTH

Fresh beef liver	500.0 g
Distilled water	1.0 liter
Tryptone	10.0 g
Soluble starch	1.0 g
Dipotassium phosphate	1.0 g

Remove the fat from 1 pound of fresh beef liver, grind, mix with 1000 ml of distilled water, and boil slowly for 1 hour. Adjust the pH to 7.6 and remove the liver particles by straining through cheese cloth. Make the volume of the broth back to 1000 ml with distilled water and add the tryptone, dipotassium phosphate, and soluble starch, and refilter. Dispense 15 ml of the broth into 20 × 150 mm tubes and add liver particles previously removed to a depth of one inch in each tube. Autoclave 20 minutes at 121 C.

LIVER INFUSION—SORBIC ACID AGAR

Make liver broth as indicated in liver broth medium. To 1 liter of liver broth add:

Sorbic acid	120.0 g
Tryptone or trypticase	10.0 g
Agar	20 g

Dissolve ingredients by bringing to a boil. Adjust pH to 5.5 ± 0.2. Sterilize for 15 minutes at 121 C.

LIVER VEAL AGAR

Liver, infusion from	50.0 g
Veal, infusion from	500.0 g
Proteose peptone	20.0 g
Neopeptone	1.3 g
Tryptone	1.3 g
Dextrose	5.0 g
Soluble starch	10.0 g
Isoelectric casein	2.0 g
Sodium chloride	5.0 g
Sodium nitrate	2.0 g
Gelatin	20.0 g
Agar	15.0 g
Distilled water	1.0 liter

Mix ingredients in distilled water. Sterilize at 121 C for 15 minutes. Final pH should be 7.3 ± 0.2.

Alternately: Suspend 97 grams of dehydrated commercial product in 1 liter of distilled water. Heat to dissolve ingredients. Sterilize at 121 C for 15 minutes.

LIVER VEAL EGG AGAR

Fresh eggs (antibiotic free)	3.0
Liver veal agar	1.0 liter

Wash eggs with a stiff brush and drain. Soak eggs in 0.1% mercuric chloride solution for 1 hour. Pour off mercuric chloride solution and replace with 70% ethyl alcohol. Soak in 70% ethyl alcohol for 30 minutes. Crack the eggs aseptically and discard the whites. Remove the yolk with a 50 ml Luer-lok syringe. Place in a sterile container and add an equal volume of sterile saline (0.85% sodium chloride). Mix thoroughly. To each 500 ml of melted liver veal agar at 50 C, add 40 ml of the egg yolk-saline suspension. Mix thoroughly and pour plates. Dry plates at room temperature for 2 days or at 35 C for 24 hours. Discard contaminated plates and store sterile plates in the refrigerator.

LYSINE IRON AGAR (Edwards and Fife)

Peptone	5.0 g
Yeast extract	3.0 g
Glucose	1.0 g
L-lysine	10.0 g
Ferric ammonium citrate	0.5 g
Sodium thiosulfate	0.04 g
Bromcresol purple	0.02 g
Agar	15.0 g
Distilled water	1.0 liter

Dissolve ingredients in distilled water and adjust to pH 6.7 ± 0.2. Dispense in 4 ml amounts in 100 × 13 mm tubes and sterilize at 121 C for 12 minutes. Slant tubes so as to obtain a deep butt and a short slant.

M 16 MEDIUM

See Rogosa SL Agar.

MacCONKEY AGAR

Peptone or Gelysate	17.0 g
Proteose peptone No. 3 or Polypeptone	3.0 g
Lactose	10.0 g
Bile salts, purified	1.5 g
Sodium chloride	5.0 g
Agar	13.5 g
Neutral red	0.03 g
Crystal violet	0.001 g
Distilled water	1.0 liter

Dissolve ingredients in distilled water by heating with stirring. Autoclave 15 minutes at 121 C. Final pH should be 7.1 ± 0.2.

MacCONKEY BROTH

Peptone or Gelysate	20.0 g
Lactose	10.0 g
Oxgall	5.0 g
Bromcresol purple	0.01 g
Distilled water	1.0 liter

Dissolve ingredients in distilled water by stirring. Dispense 225 ml portions to 500 ml Erlenmeyer flasks. Autoclave 15 minutes at 121 C. Final pH should be 7.3 ± 0.2. Medium may be dispensed into suitable size tubes and autoclaved in same manner.

MALONATE BROTH

Yeast extract	1.0 g
Ammonium sulfate	2.0 g
Dipotassium phosphate	0.6 g
Monopotassium phosphate	0.4 g
Sodium chloride	2.0 g
Sodium malonate	3.0 g
Glucose	0.25 g
Bromthymol blue	0.025 g
Distilled water	1.0 liter

Dissolve ingredients in distilled water by heating if necessary. Dispense into test tubes and autoclave for 15 minutes at 121 C. Final pH 6.7 ± 0.1.

MALT AGAR

Malt extract	30.0 g
Agar	15.0 g
Distilled water	1.0 liter

Dissolve ingredients in 1 liter of distilled water with occasional agitation and boil gently for one minute. Dispense into suitable containers and sterilize at 121 C for 15 minutes. Final pH is 5.5 ± 0.2. Do not over-autoclave this medium.

MALT AGAR (ACIDIFIED)

Malt agar acidified with 10% sterile tartaric acid to pH 3.5 ± 0.2. Prepare acid solution by weighing 10 g of tartaric acid into beaker and bringing up to 100 ml with distilled water. Dissolve and sterilize at 121 C for 15 minutes. Acidify the sterile and tempered medium with a predetermined quantity of acid solution immediately before pouring plates. Do not attempt to reheat medium once acid has been added. Determine accuracy of adjusted pH by pouring an aliquot of the medium into a small beaker, cooling to incubation temperature, and placing a recently standardized pH electrode directly into the solidified medium.

MALT AGAR (WITH ANTIBIOTIC)
Solution A

Prepare Malt Agar.

Solution B

Add 500 mg each of chlortetracycline HCl and chloramphenicol to 100 ml sterile buffered distilled water and mix. (Not all material dissolves.

Therefore, the suspension must be evenly dispersed prior to pipetting into the medium.)

To prepare mixture:

Melt medium (solution A above), temper to 45 ± 1 C and add 2 ml of antibiotic solution per 100 ml medium.

METHYLENE BLUE AGAR

Glucose	10.0 g
Agar	20.0 g
Methylene blue	0.1 g
Distilled water	1.0 liter

Dissolve ingredients in distilled water by heating. Add 10 ml of 0.1N sodium hydroxide. Sterlize at 121 C for 15 minutes.

This agar is used as a seal for culture tubes. Atmospheric oxygen will gradually cause the surface to turn blue and progress downward. As long as the bottom of the seal is colorless, anaerobic conditions are being maintained.

MOTILITY GELATIN INFUSION (GI) MEDIUM

Beef Heart Infusion broth	1.0 liter
Gelatin	53.4 g
Agar	3.0 g

Dissolve by heating with stirring. Tube and autoclave 15 minutes at 121 C. Final pH should be 7.2. Cool tubes in cool running water.

MOTILITY-NITRATE MEDIUM
(Not for *Clostridium perfringens*)

Beef extract	3.0 g
Peptone	5.0 g
Sodium nitrate	1.0 g
Glycerol	5.0 ml
Galactose	5.0 g
Agar	3.0 g
Distilled water	995.0 ml

Dissolve ingredients by boiling in distilled water, tube and sterilize at 121 C for 15 minutes. Cool quickly in cold running water. Final pH 7.0 ± 0.2.

MOTILITY-NITRATE MEDIUM (BUFFERED)
(For *Clostridium perfringens*)

Beef extract	3.0 g
Peptone	5.0 g
Potassium nitrate	5.0 g
Disodium phosphate	2.5 g
Agar	3.0 g
Galactose	5.0 g
Glycerol	5.0 g
Distilled water	1.0 liter

Dissolve ingredients by boiling in distilled water. Adjust to pH 7.4 and dispense 9 ml portions into 125 x 16 mm tubes. Sterilize 15 minutes at 121 C. Cool quickly in cool running water.

MOTILITY TEST MEDIUM

Beef extract	3.0 g
Gelysate peptone	10.0 g
Sodium chloride	5.0 g
Agar	4.0 g
Distilled water	1.0 liter

Dissolve ingredients by gentle boiling in distilled water. Adjust reaction to pH 7.4 and tube, about 8 ml per tube, and sterilize at 121 C for 15 minutes. Cool quickly in cold running water.

MP-7 (MINERAL PECTIN 7) MEDIUM

Basal medium:

Pectin (citrus or apple)	5.0 g
Monopotassium phosphate	4.0 g
Disodium phosphate	6.0 g
Yeast extract	1.0 g
Ammonium sulfate	2.0 g
Agar	15.0 g

Mix dry ingredients prior to placing in liquid for better dispersion, then add 500 ml distilled water. Add the 500 ml of the mineral solutions below, bring the volume of the mixture up to 1 liter. Adjust pH to 7.2 ± 0.2. Dissolve ingredients with gentle heating and sterilize at 121 C for 15 minutes.

Prepare separate mineral solutions as below and add as directed:

Ferrous sulfate	0.2 g plus 200 ml distilled water
Magnesium sulfate	40.0 g plus 200 ml distilled water
Calcium chloride	0.2 g plus 200 ml distilled water
Boric acid	0.002 g plus 200 ml distilled water
Manganese sulfate	0.002 g plus 200 ml distilled water
Zinc sulfate	0.014 g plus 200 ml distilled water
Cupric sulfate	0.010 g plus 200 ml distilled water
Molybdenum trioxide	0.002 g plus 200 ml distilled water

Mix 1 ml each of the above salts in 492 ml of distilled water producing 500 ml of mineral solution for addition to the basal medium.

After colonies have grown (preferably on spread plates), pour polysaccharide precipitant, described in reagents section, over the surface of the plate taking care not to dislodge the colonies. Zones of pectin hydrolysis will appear quickly, usually within a few minutes and can best be viewed against a black background. The reagent precipitates intact pectin, and pectolytic colonies are seen surrounded by a halo in an otherwise opaque medium.

MP-5 (MINERAL PECTIN 5) MEDIUM

MP-7 medium double strength (without agar) 500 ml, adjust pH to 5 to 6 with 1N hydrochloric acid. Sterilize. Cool to 48 C.

Add 500 ml sterile 3% agar solution tempered to 48 C and mix. Pour plates immediately to prevent hydrolysis.

MRS BROTH (de Man, Rogosa and Sharpe)

Proteose peptone No. 3	10.0 g
Beef extract	10.0 g
Yeast extract	5.0 g
Glucose	20.0 g
Tween 80	1.0 g
Dipotassium phosphate	2.0 g
Sodium acetate trihydrate	5.0 g
Triammonium citrate	2.0 g
Magnesium sulfate • $7H_2O$	0.2 g
Manganese sulfate • $4H_2O$.	0.05 g
Distilled water	1.0 liter

Dissolve ingredients in distilled water. Adjust medium to pH 6.2 to 6.6 before sterilization for 15 minutes at 121 C. The pH after sterilization should be between 6.0 and 6.5. MRS agar is prepared by adding 15 g agar per liter of medium.

MR-VP MEDIUM
(See VP medium for *B. cereus*)

Peptone	7.0 g
Glucose	5.0 g
Dipotassium phosphate	5.0 g
Distilled water	1.0 liter

Dissolve ingredients in distilled water and autoclave 15 minutes at 121 C. Final pH is 6.9 ± 0.2.

MUCATE BROTH

Peptone	10.0 g
Mucic acid	10.0 g
Bromthymol blue	0.024 g
Distilled water	1.0 liter

Dissolve ingredients in distilled water by stirring and adding small increments of 5N NaOH to final pH of 7.4 ± 0.2. Autoclave 10 minutes at 121 C.

MUCATE CONTROL BROTH

Peptone	10.0 g
Bromthymol blue	0.024 g
Distilled water	1.0 liter

Dissolve ingredients in distilled water with stirring. Autoclave 10 minutes at 121 C. Final pH should be 7.4 ± 0.2.

MUELLER HINTON AGAR

Beef, infusion from	300.0 g
Casamino acids, technical	17.5 g
Starch	1.5 g
Agar	17.0 g
Distilled water	1.0 liter

To rehydrate the agar, suspend 38 grams of Mueller Hinton Agar in cold distilled water and heat to boiling to dissolve the agar completely. Distribute in tubes or flasks and sterilize for 15 minutes at 121 C. The final reaction of the agar should be pH 7.4.

MUELLER HINTON BROTH

Beef, infusion from	300.0 g
Casamino acids or acidicase peptone	17.5 g
Soluble starch	1.5 g

Dissolve ingredients in distilled water to reach a volume of 1.0 liter. Sterilize in the autoclave 15 minutes at 121 C. Final pH 7.4 ± 0.2.

MY-40 AGAR (Malt, Yeast Extract 40% Sucrose)

Malt extract	20.0 g
Yeast extract	5.0 g
Agar	20.0 g
Sucrose	400.0 g
Distilled water	1.0 liter

Dissolve ingredients in distilled water heating gently. Sterilize 20 minutes at 121 C. pH is not adjusted. Do not overheat.

MYP AGAR (Mannitol Yolk Polymyxin)

Preparation A—	Ingredient	Amount
	Meat extract	1.0 g
	Peptone	10.0 g
	D-mannitol	10.0 g
	Sodium chloride	10.0 g
	Phenol red	0.025 g
	Agar	15.0 g
	Distilled water	900.0 ml

Mix ingredients in distilled water. Adjust to pH 7.1 ± 0.2, sterilize at 121 C for 20 minutes, cool to 49 ± 1 C and add 100 ml of preparation B and 1 ml of preparation C. Mix well, pour into Petri dishes, allow to solidify and store in a manner to eliminate excess surface moisture. Plates may be stored at 4 C for 7 days.

Preparation B—Colbecks Egg Yolk Broth or Concentrated Egg Yolk Emulsion.
Concentrated Egg Yolk Emulsion. This product is manufactured by Oxoid Ltd, London, England. Available from Difco Laboratories, Detroit, Michigan, as Egg York Enrichment 50%. Add 100 ml per 900 ml basal medium.

Preparation C—Polymyxin B Sulfate. This selective agent is obtainable in sterile powdered form (500,000 units, i.e., 50 mg per vial) from Pfizer Inc., New York, N.Y., and from Difco Laboratories, Detroit, Michigan, as Antimicrobic Vial P. To use, add aseptically, by syringe, 5 ml sterile distilled water. Mix to dissolve powder. Add 1 ml by syringe to a liter of final medium.

MYCOLOGICAL (MYCOPHIL) AGAR

Phytone or Soytone (papaic digest of soya meal)	10.0 g
Dextrose	10.0 g
Agar	18.0 g
Distilled water	1.0 liter

Dissolve ingredients in distilled water with heat and autoclave 12 minutes at 118 C (12 lb steam pressure for 10 minutes).

For yeast and mold counts of carbonated beverages, sugar, and other similar materials adjust the pH to 4.5 to 4.7 by adding up to 15 ml of sterile 10 per cent lactic acid to each liter of melted medium prior to plating. Do not reheat after acidification.

MYCOPHIL AGAR + ANTIBIOTIC

Preparation of antibiotic solution: Add 500 mg each of chlortetracycline HCl and chloramphenicol to 100 ml sterile phosphate buffered distilled water and mix. (Not all material dissolves, therefore the suspension must be evenly dispersed before pipetting into the medium). Two ml of this solution is added per 100 ml of tempered agar giving a final concentration in the medium of 100 mg/l of each of the antibiotics. After swirling, the medium is ready for use.

NITRATE BROTH

Beef extract	3.0 g
Peptone	5.0 g
Potassium nitrate	1.0 g
Distilled water	1.0 liter

Dissolve ingredients in distilled water. Distribute in tubes and sterilize for 15 minutes at 121 C. The final pH is 7.0.

NUTRIENT AGAR

Beef extract	3.0 g
Peptone	5.0 g
Agar	15.0 g
Distilled water	1.0 liter

Suspend ingredients in distilled water and melt agar by gentle boiling. Dispense into suitable flasks or bottles and sterilize 15 minutes at 121 C. Final pH 7.3. (If this medium is used as a blood agar base, add 8 g of sodium chloride per liter to make the medium isotonic so that red cells will not rupture, and adjust pH to 7.3 ± 0.2.)

NUTRIENT AGAR WITH MANGANESE

Preparation A:
Prepare nutrient agar as indicated.

Preparation B:
Dissolve 3.08 g Manganese sulfate in 100 ml distilled water. Add 1.0 ml of Preparation B to nutrient agar (Preparation A) and sterilize 15 minutes at 121 C.

NUTRIENT BROTH

Beef extract	3.0 g
Peptone	5.0 g
Distilled water	1.0 liter

Heat to dissolve, dispense into tubes or flasks, and autoclave 15 minutes at 121 C. Final pH, 6.7 ± 0.2.

NUTRIENT GELATIN

Beef extract	3.0 g
Peptone	5.0 g
Gelatin	120.0 g
Distilled water	1.0 liter

Dissolve ingredients in distilled water with gentle heating. Dispense in test tubes. Sterilize at 121 C for 12 minutes. Final pH 6.8 ± 0.2. Commercial dehydrated medium is satisfactory unless otherwise directed. Inoculate by stabbing the medium with a wire using inoculum from an agar slant culture. Incubate at 20 to 22 C for 30 days.

NUTRITIVE CASEINATE AGAR

Isoelectric casein	3.0 g
Peptonized milk, dry	7.0 g
Agar	12.0 g
Bromcresol purple	0.04 g
Distilled water	1.0 liter

Dissolve ingredients in distilled water. Heat to boiling to dissolve ingredients and dispense in tubes or flasks. Autoclave 15 minutes at 121 C. Final reaction should be pH 6.5 ± 0.2.

OF BASAL MEDIUM

Tryptone	2.0 g
Sodium chloride	5.0 g
Dipotassium phosphate	0.3 g
Agar	2.0 g
Bromthymol blue	0.08 g
Distilled water	1.0 liter

Add ingredients to cold distilled water and heat to boiling to dissolve completely. Dispense in 100 ml amounts and sterilize in the autoclave for 15 minutes at 15 pounds pressure (121 C). To 100 ml of sterile basal medium, aseptically add 10 ml of a sterile 10% solution of carbohydrate; incubate as directed.

OF Basal Medium is used for differentiating fermentative from oxidative metabolism of carbohydrates by various gram negative bacteria. Hugh and Leifson showed that when a gram negative organism is inoculated into two tubes of a suitable medium containing a carbohydrate and the medium in one tube is covered with petrolatum to exclude oxygen while the medium in the second tube is uncovered, reactions of differential value may be observed. Fermentative organisms will produce an acid reaction in both the covered and uncovered media. Oxidative organisms will produce an acid reaction in the uncovered medium and yield slight to no growth without change in reaction in the covered medium. Organisms which are not classified either as oxidative or fermentative yield no change in the covered medium, and an alkaline reaction in the uncovered medium.

ORANGE SERUM AGAR

Tryptone or trypticase	10.0 g
Yeast extract	3.0 g
Dextrose	4.0 g
Dipotassium phosphate	2.5 g
Agar	17.0 g
Cysteine	0.001 g
Orange serum	200.0 ml
Distilled water	800.0 ml

Dissolve ingredients in distilled water. Prepare orange serum by heating 1 liter of freshly extracted orange juice or reconstituted frozen orange juice concentrate to approximately 93 C (200 F). Add 30 g of filter aid and mix thoroughly. Filter under suction through a Buchner funnel using coarse filter paper precoated with filter aid. Discard the first few ml of filtered serum.

Sterilize 15 minutes at 121 C. After sterilization, the pH should be about 5.5. Do not over autoclave.

This medium is excellent for cultivating acid tolerant organisms, including *B. coagulans, B. thermoacidurans*, the lactobacilli, and supports vigorous growth of butyric acid forming anaerobes in shake tubes. However, it is not satisfactory for counting the butyric anaerobes due to the large amount of gas produced.

ORANGE SERUM BROTH (For Cultivation of Acid Tolerant Microorganisms)

Prepare as Orange Serum Agar except omit the agar.

PE-2 MEDIUM

Yeast extract	3.0 g
Peptone	20.0 g
Bromcresol purple 1% ethanol solution	4.0 ml
Distilled water	1.0 liter

Dissolve ingredients in distilled water by heating if necessary and dispense 18 to 20 ml portions into 18 × 150 mm screw cap test tubes. Add 8 to 10 untreated Alaska seed peas (Rogers Bros. Company, Seed Division, P.O. Box 2188, Idaho Falls, ID. 83401, Catalog No. 423, or Northrup King Seed Company, 1500 N.E. Jackson Street, Minneapolis, MN. 55413). Allow ingredients to stand 1 hour to effect hydration. Sterilize 15 minutes at 121 C.

PEPTONE WATER DILUENT 0.1%

Peptone	1.0 g
Distilled water	1.0 liter

Dissolve peptone in distilled water. Adjust to pH 6.8 ± 0.2. Prepare dilution blanks with this solution, dispensing a sufficient quantity to allow for loss during autoclaving. Autoclave at 121 C for 15 minutes.

PEPTONE WATER 0.5%

Peptone or Gelysate	5.0 g
Distilled water	1.0 liter

Dissolve peptone in distilled water. Dispense in amounts to provide 99 ± 2.0 or 9 ± 0.2 ml after sterilizing at 121 C for 15 minutes. Final pH 6.8 ± 0.2.

PHENOL RED CARBOHYDRATE BROTH

Trypticase or Proteose peptone No. 3	10.0 g
Sodium chloride	5.0 g
Beef extract	1.0 g
Phenol red	7.2 ml of a .25% solution
Carbohydrate as specified	*
Distilled water	1.0 liter
* Dulcitol—if used	5.0 g
Lactose—if used	10.0 g
Sucrose—if used	10.0 g

Dissolve ingredients by heating with gentle agitation until dissolved. Autoclave 10 minutes at 118 C. Final pH should be 7.3 ± 0.1.

PHENYLALANINE AGAR (Ewing et al., 1957)

Yeast extract	3.0 g
DL-phenylalanine	2.0 g
(or L-phenylalanine)	(1.0 g)
Disodium phosphate	1.0 g
Sodium chloride	5.0 g
Agar	12.0 g
Distilled water	1.0 liter

Dissolve ingredients in distilled water. Tube and sterilize at 121 C for 10 minutes and allow to solidify in a slanted position (long slant).

Test for deamination of phenylalanine to phenylpyruvic acid. Test reagent: 0.5 M (about 13% w/v) solution of ferric chloride.

Inoculation:

Inoculate the slant of the PA agar with a fairly heavy inoculum from an agar slant culture.

Incubation:
4 hours or, if desired, 18 to 24 hours at 37 C. Following incubation, 4 or 5 drops of ferric chloride reagent are allowed to run down over the growth on the slant. If phenylpyruvic acid has been formed, a green color develops in the syneresis fluid and in the medium. pH 7.3 ± 0.2.

PHYTONE YEAST EXTRACT AGAR WITH STREPTOMYCIN AND CHLORAMPHENICOL

Phytone or Soytone	10.0 g
Yeast extract	5.0 g
Dextrose	40.0 g
Streptomycin	0.03 g
Chloramphenicol	0.05 g
Agar	17.0 g
Distilled water	1.0 liter

Suspend ingredients in distilled water and dissolve by boiling gently. Dispense into suitable containers and sterilize by autoclaving 118 C for 15 minutes. Final pH should be 6.6 ± 0.2.

PLATE COUNT AGAR (Standard Methods Agar)

Tryptone (Pancreatic Digest of Casein USP) or Trypticase	5.0 g
Yeast extract	2.5 g
Glucose	1.0 g
Agar	15.0 g
Distilled water	1.0 liter

Dissolve ingredients in distilled water by boiling, and adjust to pH 7.1 ± 0.1. Dispense into tubes or flasks and autoclave 15 minutes at 121 C. Final reaction should be pH 7.0 ± 0.1.

To make plate count agar with bromcresol purple, add 0.04 g BCP per liter of medium.

PLATE COUNT AGAR + CYCLOHEXIMIDE

Prepare basal PCA. Autoclave at 121 C for 15 minutes. Cool PCA to 50 C.

Reconstitute Cycloheximide in distilled water; sterilize using membrane filter. Add 100 micrograms/ml to agar just prior to pouring plates.

PLATE COUNT AGAR WITH ANTIBIOTIC

Rehydrate the basal medium according to directions and sterilize. Temper medium to 45 ± 1 C, and add 2 ml of antibiotic solution per 100 ml medium as described under Mycophil Agar + Antibiotic.

POLYPECTATE GEL MEDIUM

Sodium polypectate	70.0 g
Peptone	5.0 g
Dipotassium phosphate	5.0 g
Monopotassium phosphate	1.0 g
Calcium chloride·$2H_2O$	0.6 g
Distilled water	1.0 liter

Heat 500 ml of distilled water and place all ingredients except polypectate in blender with the heated water to dissolve. Add the polypectate last, in small amounts per addition, with slow stirring to diminish occluded air. Finally adjust to pH 7.0 and add remainder of distilled water. Sterilize by autoclaving at 121 C for 15 minutes.

POLYPEPTONE YEAST EXTRACT (PY) MEDIUM
(*C. perfringens* Carbohydrate Fermentation)

Polypeptone	20.0 g
Yeast extract	5.0 g
Sodium chloride	5.0 g
Salicin or raffinose	10.0 g
Distilled water	1.0 liter

Add salicin or raffinose as indicated in Chapter 35. Dissolve ingredients in distilled water. Adjust pH to 6.9 to 7.0. Dispense 9 ml portions into 125 × 16 screw capped tubes and sterilize 15 minutes at 121 C.

POTASSIUM CYANIDE (KCN) BROTH

Basal broth:

Proteose peptone No. 3 or Polypeptone	3.0 g
Disodium phosphate	5.64 g
Monopotassium phosphate	0.225 g
Sodium chloride	5.0 g
Distilled water	1.0 liter

Dissolve ingredients in distilled water with stirring. Autoclave 100 ml portions 15 minutes at 121 C. Final pH should be 7.6. Prepare 0.5% potassium cyanide by weighing 0.5 g into 100 ml sterile distilled water. Using pipet filler transfer 1.5 ml cold potassium cyanide solution to 100 ml basal broth (precooled). DO NOT PIPETTE BY MOUTH. Mix. Distribute 1 ml portions to sterile 13 × 100 mm tubes and stopper immediately with No. 2 corks impregnated with paraffin. (Prepare corks by boiling in paraffin for 5 minutes.) Store medium at 5 to 10 C. Storage life is two weeks. **Exercise caution because potassium cyanide is lethal.**

POTATO DEXTROSE AGAR (ACIDIFIED)

Infusion from white potatoes	200.0 ml
Dextrose	20.0 g
Agar	15.0 g
Distilled water	1.0 liter

Suspend 39 grams of commercial dehydrated ingredients in distilled water. Heat mixture to boiling to dissolve the ingredients. Distribute into tubes or flasks, and autoclave 15 minutes at 121 C (15 lb pressure). When used as plating medium for yeasts and molds, melt in flowing steam or boiling water, cool, and acidify to pH 3.5 with sterile 10 percent tartartic acid solution. (For use in the cultivation of yeasts and molds, adjust to the desired pH if different from pH 3.5.) Mix thoroughly and pour into plates. To preserve solidifying properties of the agar do not heat medium after the addition of tartartic acid. For preparation of Potato Dextrose Agar with Antibiotic, add antibiotics as described under Mycophil Agar with Antibiotic.

PURPLE CARBOHYDRATE BROTH

Proteose peptone No. 3 or Gelysate	10.0 g
Sodium chloride	5.0 g
Beef extract	1.0 g
Bromcresol purple	0.015 to 0.02 g
Distilled water	1.0 liter
Carbohydrate as specified	*

* Dulcitol—if used	5.0 g
Lactose—if used	10.0 g
Sucrose—if used	10.0 g

Dissolve ingredients in distilled water by heating with gentle agitation. Autoclave 10 minutes at 118 C. Final pH 6.8 ± 0.1.

RAPPAPORT BROTH MODIFIED (MAGNESIUM CHLORIDE-MALACHITE GREEN-CARBENICILLIN MEDIUM)

Solution A:

Tryptone	10.0 g
Distilled water	1.0 liter

Solution B:

Disodium hydrogen phosphate	9.5 g
Distilled water	1.0 liter

Solution C:

Magnesium chloride·$6H_2O$	40.0 g
Distilled water	100.0 ml

Sterilize separately 121 C for 15 minutes.

Solution D:

Malachite green	0.2 g
Distilled water	100.0 ml

Not sterilized.

Solution E:

Carbenicillin* (filter sterilized)	1.0 mg/ml

*Available as Carbenicillin Indanyl Sodium available from Roerig, 235 East 42nd Street, New York, N. Y. 10017 under the name Geocillin.

To make 250 ml of this medium, mix 155 ml of solution A and 40 ml of solution B, and sterilize at 121 C for 15 minutes. Cool to 50 C, then add 53 ml of sterile solution C. Add 1.6 ml solution D and 0.6 ml of solution E.

REDDY'S DIFFERENTIAL BROTH FOR SEPARATING LACTIC STREPTOCOCCI

See Differential Broth for Lactic Streptococci.

REDDY'S DIFFERENTIAL MEDIUM FOR ENUMERATION OF LACTIC STREPTOCOCCI

Tryptone	5.0 g
Yeast extract	5.0 g
Casamino acids	2.5 g
L-arginine-hydrochloride	5.0 g
Dipotassium phosphate	1.25 g
Calcium citrate	10.0 g
Carboxy methyl cellulose*	15.0 g
Agar	15.0 g

*DuPont CMC grade P-754 at the rate of 0.6% may be substituted.

Dissolve agar by suspending 15 g of agar in 500 ml distilled water and heat until dissolved. In another glass beaker mix 10 g of calcium citrate and 15 g of CMC in 500 ml of distilled water. Heat with continuous stirring until a homogeneous, white, turbid suspension is formed. Mix the two portions in a stainless steel vessel containing the required amounts of tryptone, yeast extract, dipotassium phosphate, casamino acids, and arginine. Cover mixture and heat gently for 15 minutes. Adjust to pH 5.6 ± 0.2 with 6N HCl after heating. Sterilize at 121 C for 15 minutes. Just before pouring plates, add 5 ml sterile, reconstituted nonfat milk (11% solids), 10 ml of sterile 3% (w/v) calcium carbonate in distilled water, and 2 ml of sterile 0.1% bromcresol purple in distilled water, to 100 ml of sterile agar (melted and tempered to 55 C). Carefully mix the melted medium and pour into previously chilled sterile Petri dishes to obtain a layer 4 to 5 mm thick. After solidification of medium, dry plates for 18 to 24 hours in an incubator at 37 C.

ROGOSA, MITCHELL, AND WISEMAN MEDIUM

Prepare same as Lactobacillus Selection Medium.
Final pH should be approximately 5.4. Sterilize 15 minutes at 121 C.

ROGOSA SL AGAR (M-16 MEDIUM)
M 16 agar (A Modification of Rogosa SL)

Beef extract	5.0 g
Yeast extract	2.5 g
Ascorbic acid	0.5 g
Phytone or Soytone	5.0 g
Polypeptone or Tryptose	5.0 g
Sodium acetate trihydrate	3.0 g
Lactose or dextrose*	5.0 g
Agar	10.0 g
Distilled water q.s.	1.0 liter

*Added from a sterile stock solution after medium is autoclaved and cooled to 50 C.

Dissolve ingredients in distilled water. Adjust medium to pH 7.2 ± 0.2 with 2N sodium hydroxide before autoclaving.

ROPE SPORE MEDIUM (AMOS AND KENT-JONES MEDIUM)

Peptone	10.0 g
Beef extract	5.4 g
Sodium chloride	9.0 g
Distilled water	1.0 liter

Dissolve ingredients in distilled water by boiling. Adjust pH to 7.2. Dispense in 5 ml amounts into tubes and sterilize 15 minutes at 121 C.

SABOURAUD DEXTROSE AGAR
(For Cultivation of Yeasts and Molds)

Dextrose	40.0 g
Peptone	10.0 g
Agar	15.0 g
Distilled water	to 1.0 liter

Dissolve ingredients in distilled water and heat to boiling; dispense in flasks; sterilize in autoclave at 121 C for 15 minutes; pH 5.6 ± 0.2. Do not over autoclave.

SABOURAUD'S DEXTROSE BROTH (SD)

Polypeptone	10.0 g
Dextrose	40.0 g
Distilled water	1.0 liter

Dissolve ingredients in distilled water completely, and dispense 40 ml portions in prescription bottles. Final pH should be 5.8. Autoclave at 121 C for 15 minutes. Do not exceed 121 C nor autoclave more than 15 minutes.

SABOURAUD'S DEXTROSE BROTH (Modified) and SABOURAUD LIQUID BROTH MODIFIED

Pancreatic digest of casein (Trypticase or Tryptone)	5.0 g
Peptic digest of animal tissue (Thiotone or Proteose peptone No. 3)	5.0 g
Dextrose	20.0 g

Dissolve ingredients in a liter of distilled water. Dispense and sterilize at 121 C for 15 or 20 minutes. The autoclave temperature should be reached in ten minutes. Care should be taken not to heat this broth unduly. Because of the high carbohydrate content, excessive heating darkens the medium and it becomes less efficient.

SABOURAUD'S DEXTROSE BROTH
(For Osmophilic Microorganisms)

Neopeptone	20.0 g
Dextrose	80.0 g

Dissolve ingredients in 1.0 liter of distilled water. Adjust pH to 6.0 ± 0.2. Autoclave at 121 C for 15 minutes. After sterilization, cool to 25 to 30 C and aseptically add 20,000 units penicillin potassium G, 40,000 micrograms streptomycin sulfate. Final pH should be 6.0.

NOTE: Industrial experience with this medium has indicated that satisfactory results are obtained by acidifying the medium after sterilization to pH 4.7 to 4.8 with sterile 10% lactic acid rather than using antibiotics.

SALINE AGAR BASE

Purified agar	15.0 g
Sodium chloride	8.5 g
Distilled water	1.0 liter

Dissolve sodium chloride in water and adjust to pH 7.0 ± 0.2. Add agar and heat to dissolve the agar, dispense in 100 ml volumes, and autoclave for 15 minutes at 121 C.

SEA WATER (SYNTHETIC)

Sodium chloride	24.0 g
Potassium chloride	0.7 g
Magnesium chloride $\cdot 6H_2O$	5.3 g
Magnesium sulfate $\cdot 7H_2O$	7.0 g
Distilled water	1.0 liter

Suspend ingredients in distilled water. Using 1N sodium hydroxide, adjust so that final pH is 7.0 ± 0.2. Sterilize 15 minutes at 121 C.

SEA WATER AGAR (SWA)

Yeast extract	5.0g
Peptone	5.0 g
Beef extract	3.0 g
Agar	15.0 g
Sea water (synthetic)	1.0 liter

Dissolve ingredients in the synthetic sea water and sterilize at 121 C for 15 minutes. Final pH should be 7.2 ± 0.2.

SELENITE-CYSTINE BROTH

Tryptone	5.0 g
Lactose	4.0 g
Disodium phosphate	10.0 g
Sodium selenite	4.0 g
Cystine	0.01 g
Distilled water	1.0 liter

Dissolve ingredients in distilled water by heating with frequent agitation. Do not autoclave. Heat 10 minutes in flowing steam. Final pH 7.0 ± 0.2. Medium is not sterile. Use the same day as prepared.

SELENITE CYSTINE BROTH (NORTH AND BARTRAM)

Polypeptone or Tryptose	5.0 g
(or use 4 grams tryptone in place of polypeptone)	
Lactose	4.0 g
Sodium acid selenite	4.0 g
Disodium phosphate	5.5 g
Monopotassium phosphate	4.5 g
L-cystine (1% in sodium hydroxide)	1.0 ml
Distilled water	1.0 liter

1% cystine solution in sodium hydroxide is prepared by adding 1.0 g L-cystine to 15 ml 1N NaOH and diluting to 100 ml with sterile distilled water.

SELENITE F BROTH

Peptone component	5.0 g
Lactose	4.0 g
Sodium phosphate, anhydrous*	10.0 g
Sodium acid selenite	4.0 g
Distilled water	1.0 liter

*The ratio of monosodium phosphate to disodium phosphate must be determined with each lot of sodium acid selenite. The total phosphate concentration should be 1%, adjusted so that the solution containing the sodium acid selenite is buffered at pH 7.0. Uninoculated media exhibiting large amounts of the red precipitate of oxidized selenite should not be used.

Dissolve the dry ingredients in distilled water with gentle heat. Dispense in tubes to a depth of at least 2 inches and sterilize by exposure to flowing steam for not more than 30 minutes. Do not autoclave. Tubed media may be stored in the refrigerator. Final reaction should be pH 7.0

Alternatively, a proportionate amount of the dry ingredients may be added to a sewage effluent, water sample, or other large specimen and dissolved without heat for the enrichment of suspected enteric pathogens.

SIM AGAR

Beef extract	3.0 g
Peptone	30.0 g
Peptonized iron	2.0 g
Sodium thiosulfate	0.025 g
Agar	3.0 g
Distilled water	1.0 liter

Dissolve ingredients in distilled water and heat to boiling. Adjust pH to 7.3 ± 0.2. Distribute the medium to about 3 inch depth in test tubes and sterilize for 15 minutes at 121 C. Cool quickly in running water.

SIMMON'S CITRATE AGAR

Magnesium sulfate	0.2 g
Monoammonium phosphate	1.0 g
Dipotassium phosphate	1.0 g
Sodium citrate	2.0 g
Bromthymol blue	0.08 g
Sodium chloride	5.0 g
Agar (wash vigorously for 3 days)	20.0 g
Distilled water	1.0 liter

Dissolve ingredients in distilled water with gentle heating, adjust to pH 6.8 ± 0.2, and tube. Sterilize 15 minutes at 121 C. Slant tubes with a 2 to 3 cm deep butt for cooling.

SKIM MILK AGAR

Standard Methods agar	1.0 liter
Reconstituted Skim Milk (10% solids)	100.0 ml

Melt Standard Methods agar. Cool to 50 C and add 100 ml sterile skim milk, mix well and pour into Petri dishes.

SOFT-AGAR-GELATIN OVERLAY* (BASAL MEDIUM)

Peptone	5.0 g
Beef extract	3.0 g
Sodium chloride	5.0 g
Manganese sulfate	0.05 g
Agar	15.0 g
Distilled water	1.0 liter

Dissolve ingredients in distilled water by gentle heating. Adjust to pH 7.0. Sterilize 15 minutes at 121 C.

*The soft overlay is the same as the Basal Medium except that 0.8% agar and 1.5% gelatin are used.

SODIUM LACTATE AGAR

Trypticase	10.0 g
Yeast extract	10.0 g
Sodium lactate	10.0 g
Dipotassium phosphate	0.25 g
Agar	15.0 g
Distilled water q.s.	1.0 liter

Dissolve ingredients in distilled water with gentle heating. Adjust medium to pH 7.0 ± 0.2 before sterilization for 20 minutes at 121 C.

SPORULATION BROTH (*C. perfringens*)

Polypeptone or Tryptose	15.0 g
Yeast extract	3.0 g
Soluble starch	3.0 g
Magnesium sulfate	0.1 g
Sodium thioglycollate	1.0 g
Disodium phosphate	11.0 g
Distilled water	1.0 liter

Suspend ingredients in distilled water and dissolve by boiling gently. Final pH 7.8 ± 0.2. Dispense 20 ml into 20 mm × 150 mm screw cap test tubes. Sterilize 121 C for 15 minutes. Before use exhaust in flowing steam for 20 minutes.

SS AGAR

Beef extract	5.0 g
Proteose peptone or Polypeptone	5.0 g
Lactose	10.0 g
Bile salts	8.5 g
Sodium citrate	8.5 g
Sodium thiosulfate	8.5 g
Ferric citrate	1.0 g
Brilliant green	0.00033 g
Neutral red	0.025 g
Agar	13.5 g
Distilled water	1.0 liter

Dissolve ingredients in distilled water by bringing to a boil. Final reaction should be approximately pH 7.0 ± 0.2.

Do not sterilize in the autoclave.

As soon as all ingredients are in solution, cool until the flask can be handled, and pour about 20 ml of medium into each Petri plate. The plates should be in a draftless area of low contamination. After pouring, partially remove the covers of the dishes to allow vapor to escape and dry the surface of the agar for 2 hours. When used for streaking, agar surface should appear dry in order to obtain well isolated colonies.

STANDARD METHODS AGAR

See Plate Count Agar.

STANDARD METHODS AGAR + BROMCRESOL PURPLE

Add 1 ml of 1.6% alcohol solution of bromcresol purple per liter of agar prior to sterilizing.

STANDARD METHODS AGAR WITH CASEINATE

Pancreatic digest of casein (Tryptone or Trypticase)	5.0 g
Yeast extract	2.5 g
Glucose	1.0 g
Agar	15.0 g
Sodium caseinate	10.0 g
Trisodium citrate hydrated	1.0 liter of a 0.015 molar solution
Calcium chloride	20 ml of a 1.0 molar solution

Prepare a .015 molar solution of trisodium citrate by placing 4.41 g of trisodium citrate containing 2 molecules of water into a volumetric flask, bring to 1 liter volume with distilled water and mix thoroughly. Dissolve the pancreatic digest of casein, yeast extract, glucose, and agar in 500 ml of the 0.015 molar trisodium citrate with gentle heating.

Dissolve the sodium caseinate in the other 500 ml of 0.015 molar trisodium citrate. Mix the two trisodium citrate solutions and autoclave 15 minutes at 121 C.

Prepare a 1 molar solution of calcium chloride solution by adding 218.99 grams of calcium chloride hexahydrate to a volumetric flask; bring to 1 liter with distilled water, and mix thoroughly. Sterilize 15 minutes at 121 C.

Prior to pouring plates add 20 ml of the sterile 1 molar calcium chloride solution to the molten mixture containing all other ingredients, mix and pour 2 mm thick layers in Petri dishes.

STAPHYLOCOCCUS MEDIUM NO. 110

Yeast extract	2.5 g
Tryptone	10.0 g
Gelatin	30.0 g
Lactose	2.0 g
D-mannitol	10.0 g
Sodium chloride	75.0 g
Dipotassium phosphate	5.0 g
Agar	15.0 g

Suspend ingredients in distilled water and heat to boiling with frequent agitation to dissolve ingredients. Adjust volume to 1000 ml with distilled water. Dispense into appropriate containers and autoclave 15 minutes at 121 C. Final pH 7.0 ± 0.2.

SULFITE AGAR
(For the Detection of Thermophilic Anaerobes Producing H_2S)

Tryptone or Trypticase	10.0 g
Sodium sulfite (anhydrous)	1.0 g
Agar	20.0 g
Distilled water	1.0 liter

Dissolve ingredients in distilled water, dispense into tubes in about 15 ml amounts, and into each tube place an iron nail or a small clean strip of iron or base plate. No adjustment of reaction is necessary. Autoclave 20 minutes at 121 C. Tubes should be used within a week after making.

As an alternate for the iron strip or nail, 10 ml of a 5% solution of iron citrate may be substituted in the sulfite medium formula. It is necessary to heat the citrate solution to completely dissolve ferric citrate scales or pearls.

SULFITE-POLYMYXIN-SULFADIAZINE (SPS) AGAR

Tryptone	15.0 g
Yeast extract	10.0 g
Ferric citrate	0.5 g
Agar	15.0 g
Distilled water	1.0 liter

Dissolve ingredients in distilled water by heating to boiling for 1 to 2 minutes. Autoclave at 121 C for 15 minutes. Final pH should be 7.0 ± 0.2.

To each liter of sterile medium, add the following filter sterilized solution:

10% sodium sulfite ($Na_2SO_3 \bullet 7H_2O$) solution	5.0 ml
0.12% polymyxin B sulfate solution	10.0 ml
Sodium sulfadiazine solution containing 12 mg/ml	10.0 ml

TELLURITE POLYMYXIN EGG YOLK AGAR

Basal medium:

Tryptone	10.0 g
Yeast extract	5.0 g
Mannitol	5.0 g
Sodium chloride	20.0 g
Lithium chloride	2.0 g
Agar	18.0 g
Distilled water	1.0 liter

Suspend ingredients in distilled water and heat to boiling with frequent agitation to dissolve ingredients. Adjust to 900 ml with distilled water. Adjust pH to 7.2 ± 0.2. Dispense into bottles or flasks and autoclave 15 minutes at 121 C.

Enrichment:

Egg yolk emulsion (30% v/v) in physiological saline. Prepared by soaking fresh eggs for about 1 minute in a 1:1000 dilution of saturated mercuric chloride solution. Crack eggs aseptically and separate yolks and whites. Suspend egg yolk in 0.85% sodium chloride solution (30% v/v) and blend in high speed blender for about 5 seconds.

Complete medium:

Cool melted basal medium to 50 to 55 C in water bath. Add 100 ml enrichment to 900 ml basal medium, followed by 0.4 ml of a 1% filter sterilized solution of polymyxin B to a final concentration of 40 μg/ml, and 10 ml of a sterile 1% solution of potassium tellurite. Final pH 7.2 ± 0.1.

TERGITOL 7 AGAR

Proteose Peptone No. 3 or Polypeptone	5.0 g
Yeast extract	3.0 g
Lactose	10.0 g
Tergitol-7 (sodium heptadecyl sulfate)	0.1 ml
Bromthymol blue	0.025 g
Agar	15.0 g
Distilled water	1.0 liter

Suspend ingredients in distilled water and dissolve by gentle boiling. Dispense into suitable containers and sterilize 15 minutes at 121 C.

Before using, cool to 45 C, add 0.3 ml of one percent triphenyl tetrazolium chloride and pour into Petri plates. pH 6.9 ± 0.2.

TETRATHIONATE BROTH
(Muller, 1923: modified by Kauffmann, 1935)

Basal medium:

Polypeptone or proteose peptone	5.0 g
Bile salts	1.0 g
Calcium carbonate	10.0 g
Sodium thiosulfate	30.0 g
Distilled water	1.0 liter

Iodine solution:

Iodine	6.0 g
Potassium iodide	5.0 g
Distilled water	20.0 ml

Heat the ingredients of the basal medium in distilled water to boiling temperature, cool to less than 45 C, add 2 ml of iodine solution to each 100 ml of base. Add 1 ml of 1:1000 solution of brilliant green per 100 ml of base medium as recommended in Kauffmann's Combined Enrichment Medium. The basal medium, with or without added brilliant green, may be tubed, sterilized at 121 C for 15 minutes, and stored. In this case, iodine solution is added (0.2 ml per 10 ml of medium) prior to use.

Sulfathiazole (0.125 mg per ml of medium) may be added to prevent excessive growth of *Proteus* (Galton et al., 1952).

THERMOACIDURANS AGAR

Yeast extract	5.0 g
Proteose peptone	5.0 g
Dextrose	5.0 g
Dipotassium phosphate	4.0 g
Agar	20.0 g
Distilled water	1.0 liter

Suspend ingredients in distilled water and dissolve agar by boiling. Distribute into flasks or bottles. Autoclave 15 minutes at 121 C. Final pH should be 5.0.

THIOGLYCOLLATE OVERLAY AGAR

Agar	20.0 g
Sodium thioglycollate	0.5 g
Distilled water	1.0 liter

Add ingredients to distilled water and bring to a boil. Dispense into tubes or flasks, and autoclave 15 minutes at 121 C.

THIOGLYCOLLATE AGAR*

Pancreatic digest of casein USP	15.0 g
L-cystine	0.5 g
Dextrose	5.0 g
Yeast extract	5.0 g
Sodium chloride	2.5 g
Sodium thioglycollate	0.5 g
Resazurin	0.001 g
Agar	20.0 g
Distilled water	1.0 liter

Suspend ingredients in distilled water and heat to boiling to dissolve completely. Distribute approximately 16 ml quantities to 20 × 150 mm screw capped tubes. Add to each tube with head down, an acid cleaned 6 d nail. Sterilize at 121 C for 15 minutes. Final reaction should be 7.0 to 7.1. This medium should be used within one week of preparation.

*This medium is the same formulation as Fluid Thioglycollate Medium except for the addition of 20 grams of agar.

THIOSULFATE-CITRATE-BILE SALTS-SUCROSE AGAR (TCBS)

Yeast extract	5.0 g
Polypeptone or Proteose Peptone No. 3	10.0 g
Sucrose	20.0 g
Sodium thiosulfate ($5H_2O$)	10.0 g
Sodium citrate ($2H_2O$)	10.0 g
Sodium cholate	3.0 g
Oxgall	5.0 g
Sodium chloride	10.0 g
Ferric citrate	1.0 g
Bromthymol blue	0.04 g
Thymol blue	0.04 g
Agar	15.0 g
Distilled water	1.0 liter

Dissolve ingredients in distilled water by bringing to boil. Final pH should be 8.6 ± 0.2. This medium should not be autoclaved.

TRIPLE SUGAR IRON AGAR

Polypeptone	20.0 g
Lactose	10.0 g
Sucrose	10.0 g
Glucose	1.0 g
Sodium chloride	5.0 g
Ferrous ammonium sulfate • $6H_2O$	0.2 g
Sodium thiosulfate	0.2 g
Phenol red	0.025 g
Agar	13.0 g
Distilled water	1.0 liter

Add ingredients to distilled water and bring to a boil. Distribute in tubes using enough medium to obtain a deep butt. Autoclave 15 minutes at

121 C. Remove from autoclave and slant to obtain a deep butt. Final reaction should be approximately pH 7.4 ± 0.1.

or

Beef extract	3.0 g
Yeast extract	3.0 g
Peptone	15.0 g
Proteose peptone	5.0 g
Lactose	10.0 g
Sucrose	10.0 g
Dextrose	1.0 g
Ferrous sulfate	0.2 g
Sodium chloride	5.0 g
Sodium thiosulfate	0.3 g
Agar	12.0 g
Phenol Red	0.024 g
Distilled water	1.0 liter

Add ingredients to distilled water and bring to a boil. Distribute in tubes using enough medium to obtain a deep butt. Autoclave 15 minutes at 121 C. Remove from autoclave and slant to obtain a deep butt. Final pH 7.4 ± 0.2.

TRYPTICASE (OR TRYPTIC) SOY AGAR

Trypticase or Tryptone	15.0 g
Phytone or Soytone	5.0 g
Sodium chloride	5.0 g
Agar	15.0 g

Suspend ingredients in 1.0 liter of distilled water, mixing thoroughly. Heat with frequent agitation, and boil for about 1 minute to dissolve completely. Autoclave 15 minutes at 121 C. Final pH, 7.3 ± 0.2.

TRYPTICASE (WITH GLUCOSE) OR TRYPTIC SOY AGAR

Trypticase or Tryptone (pancreatic digest of casein)	17.0 g
Phytone or Soytone (papain digest of soymeal)	3.0 g
Sodium chloride	5.0 g
Dipotassium phosphate	2.5 g
Glucose	2.5 g
Agar	15.0 g
Distilled water	1.0 liter

Dissolve ingredients in distilled water; warm slightly if necessary to complete solution. Dispense into tubes or bottles and sterilize by autoclaving 15 minutes at 121 C. Final reaction should be pH 7.3.

TRYPTICASE SOY-TRYPTOSE BROTH

Tryptose	10.4 g
Trypticase or Tryptone	8.5 g
Sodium chloride	5.0 g
Phytone or Soytone	1.5 g
Dextrose	1.8 g
Dipotassium phosphate	1.25 g
Yeast extract	3.0 g
Distilled water	1.0 liter

Dissolve ingredients in distilled water. Sterilize at 121 C for 15 minutes. Final pH 7.2 ± 0.2.

TRYPTICASE PEPTONE GLUCOSE YEAST EXTRACT BROTH (TPGY) and TPGY WITH TRYPSIN

Trypticase or Tryptone	50.0 g
Peptone	5.0 g
Yeast extract	20.0 g
Glucose	4.0 g
Sodium thioglycollate	1.0 g
Distilled water	1.0 liter

Dissolve solid ingredients in distilled water, adjust to pH 7.0, if necessary, and dispense in volumes appropriate for use. Sterilize the dispensed medium at 121 C for 8 minutes (15 minutes for large volumes), and refrigerate until used.

If trypsin is to be added, prepare a 1.5% aqueous solution of trypsin. Sterilize by filtration through a Millipore (or comparable) 0.45 mμ filter and refrigerate until needed. After steaming to drive off oxygen and cooling, add the trypsin to the TPGY immediately before inoculating.

TRYPTONE GLUCOSE EXTRACT AGAR

Beef extract	3.0 g
Tryptone	5.0 g
Dextrose	1.0 g
Agar	15.0 g
Distilled water	1.0 liter

Dissolve ingredients in distilled water by boiling gently. Adjust to pH 7.0 ± 0.2. Distribute in tubes or flasks. Sterilize 15 minutes at 121 C.

TRYPTONE GLUCOSE YEAST BROTH

Tryptone or trypticase	10.0 g
Dextrose	5.0 g
Dipotassium phosphate	1.25 g
Yeast extract	1.0 g
Distilled water	1.0 liter

Dissolve ingredients in distilled water and adjust to pH 6.8 ± 0.2, tube in 8-10 ml amounts and sterilize 20 minutes at 121 C.

Unless freshly sterilized, tubes should be exhausted in flowing steam for at least 20 minutes before using.

For anaerobes after inoculation, stratify tubes with melted tryptone agar.

If medium is to be incubated at 55 C, the stratifying layer should be solidified by cooling and then tubes preheated in a water bath to 55 C before being placed in the incubator.

TRYPTONE (TRYPTICASE) SOY BROTH

Tryptone or Trypticase	17.0 g
Phytone or Soytone	3.0 g
Sodium chloride	5.0 g
Dipotassium phosphate	2.5 g
Dextrose	2.5 g
Distilled water	1.0 liter

Dissolve ingredients in distilled water; warm slightly if necessary to complete solution. Dispense into tubes or bottles, and sterilize by autoclaving 15 minutes at 121 C. Final reaction should be pH 7.3 ± 0.2.

TRYPTOPHANE BROTH

Tryptone	10.0 g
Distilled water	1.0 liter

Dissolve with stirring. Autoclave 15 minutes at 121 C. Final pH should be 6.9 ± 0.2.

TRYPTOSE-SULFITE CYCLOSERINE (TSC) AGAR

Tryptose	15.0 g
Soytone	5.0 g
Yeast extract	5.0 g
Sodium bisulfite (meta)	1.0 g
Ferric ammonium citrate	1.0 g
Agar	20.0 g
Distilled water	1.0 liter

Dissolve ingredients in distilled water and adjust the pH to 7.6 ± 0.2 and autoclave for 10 minutes at 121 C. To each liter of autoclaved medium, add 10 ml of a 4.0% filter sterilized solution of D-cycloserine to give a final concentration of approximately 400 μg per ml. Cool the medium to 50 C and add 40 ml of a sterile 50% egg yolk in saline emulsion per 500 ml of medium, with the exception of that used to overlay the plates. Egg Yolk enrichment 50% may be obtained from Difco Laboratories, Detroit, Michigan. Dispense the medium in standard Petri dishes for surface plating. Air dry at room temperature for 24 hours or until the surface of the agar is somewhat dry prior to use. Prepare plates fresh each time they are to be used.

EGG YOLK-FREE TRYPTOSE-SULFITE-CYCLOSERINE (EY-FREE TSC) AGAR

This medium is identical to tryptose-sulfite-cycloserine (TSC) agar except for the omission of the egg yolk. The medium is used for pour plates rather than for surface plating.

UREA BROTH (Rapid Test)

Yeast extract	0.1 g
Monopotassium phosphate	0.091 g
Disodium phosphate	0.095 g
Urea	20.0 g
Phenol red	0.01 g
Distilled water	1.0 liter

Dissolve ingredients in distilled water, sterilize by filtration and dispense under aseptic conditions in 1 ml amounts into sterile 13 × 100 mm tubes. Final reaction should be pH 6.8 ± 0.2. Do not autoclave or steam sterilize this medium.

"V-8" MEDIUM FOR LACTOBACILLI

"V-8" vegetable juice*	500.0 ml
Tryptose	10.0 g
Lactose	5.0 g
Beef extract	3.0 g
Agar	15.0 g
Bromcresol green	0.1 g

*V-8 is a trademark for a blend of 8 vegetable juices. It may be purchased in retail food outlets.

Filter the "V-8" juice through 2 layers of cheese cloth. Dissolve ingredients in sufficient water to make a total of 1.0 liter. Heat gently to dissolve. Adjust pH to 5.7 ± 0.2. Sterilize 121 C for 15 minutes.

VEAL INFUSION AGAR

Veal, infusion from	500.0 g
Proteose peptone No. 3 or Polypeptone	10.0 g
Sodium chloride (NaCl)	5.0 g
Agar	15.0 g
Distilled water q.s.	1.0 liter

Suspend 40 grams of commercial dehydrated medium in distilled water. Dissolve by heating with stirring. Dispense 7 ml portions to 16 × 150 mm tubes. Autoclave 15 minutes at 121 C. After sterilization, incline tubes to give 5 cm slant. Final pH should be 7.4 ± 0.1.

VEAL INFUSION BROTH

Veal, infusion from	500.0 g
Proteose peptone No. 3	10.0 g
Distilled water	1.0 liter

Suspend 22 grams of dehydrated commercial product in distilled water. Boil gently and dissolve with stirring. Autoclave 15 minutes at 121 C. Final pH should be 7.3 ± 0.2.

VIOLET RED BILE AGAR (VRBA)

Yeast extract	3.0 g
Peptone or Gelysate	7.0 g
Sodium chloride	5.0 g
Bile salts or Bile Salts No. 3	1.5 g
Lactose	10.0 g
Neutral red	0.03 g
Crystal violet	0.002 g
Agar	15.0 g
Distilled water	1.0 liter

Suspend the ingredients in distilled water and allow to stand for a few minutes. Mix thoroughly and adjust to pH 7.4 ± 0.2. Heat with agitation and boil for 2 minutes. Do not sterilize.

Prior to use, cool to 45 C and use as a plating medium. After solidification, add a cover layer above the agar of approximately 3 to 4 ml to prevent surface growth and spreading of colonies.

VOGEL AND JOHNSON AGAR (TELLURITE GLYCINE RED AGAR BASE)

Basal medium:

Trypticase or Tryptose (pancreatic digest of casein)	10.0 g
Yeast extract	5.0 g
D-mannitol	10.0 g
Dipotassium phosphate	5.0 g
Lithium chloride • $6H_2O$	5.0 g
Glycine	10.0 g
Agar	16.0 g
Phenol red	0.025 g
Distilled water	1.0 liter

Suspend ingredients in distilled water, heat to boiling with frequent agitation to dissolve ingredients. Dispense into bottles or flasks, 98 ml per container, and autoclave 15 minutes at 121 C. Final pH 7.2 ± 0.2.

Complete medium:

Cool molten basal medium to 45 to 50 C. Add 2 ml sterile 1% potassium tellurite solution to each 98 ml container and mix thoroughly just prior to pouring. Pour 15 to 18 ml into sterile Petri dishes.

V-P BROTH (MODIFIED FOR *B. cereus*)
Smith, Gordon, and Clark

Proteose peptone	7.0 g
Glucose	5.0 g
NaCl	5.0 g
Distilled water	1.0 liter

Dissolve ingredients in distilled water. Dispense 5 ml in 18 mm test tubes. Sterilize 15 minutes at 121 C.

NOTE: The medium is a modified medium and must be formulated in the laboratory. See the modified V-P test reagents for use with this medium.

WAGATSUMA AGAR

Peptone	10.0 g
Yeast extract	3.0 g
Dipotassium phosphate	5.0 g
Sodium chloride	70.0 g
Mannitol	10.0 g
Crystal violet	0.001 g
Agar	15.0 g
Distilled water	1.0 liter

Suspend ingredients in distilled water and dissolve ingredients by boiling gently. Adjust to pH 8.0. Do not autoclave. Wash rabbit erythrocytes three times in physiological saline and reconstitute to original blood volume. Add 2 ml of washed erythrocytes to 100 ml of agar cooled to 50 C just prior to pouring.

WORT AGAR

Malt extract	15.0 g
Peptone	0.78 g
Maltose	12.7 g
Dextrin	2.75 g
Glycerol	2.35 g
Dipotassium phosphate	1.0 g
Ammonium chloride	1.0 g
Agar	15.0 g
Distilled water	1.0 liter

Dissolve ingredients in distilled water by heating gently. Sterilize 15 minutes at 121 C. pH 4.8 ± 0.2. Do not over autoclave.

XL AGAR BASE

Yeast extract	3.0 g
L-lysine	5.0 g
Xylose	3.5 g
Lactose	7.5 g
Sucrose	7.5 g
Sodium chloride	5.0 g
Phenol red	0.08 g
Agar	13.5 g
Distilled water	1.0 liter

Heat mixture in distilled water to boiling temperature to dissolve the ingredients. Sterilize at 121 C for 15 minutes, and then cool to 55 to 60 C. Aseptically add 20 ml of sterile solution containing:

Sodium thiosulfate	34.0 g
Ferric ammonium citrate	4.0 g
Distilled water	100.0 ml

Mix well to obtain a uniform suspension.

XLD AGAR

The completed XLD medium is prepared by adding 25 ml of 10% sterile solution of sodium desoxycholate per liter of the above mentioned (cooled) medium. Mix well and adjust to pH 6.9.

YEAST NITROGEN BASE AGAR

Solution A:

Agar	40.0 g
Dextrose	40.0 g
Distilled water	1.0 liter

Suspend ingredients in distilled water. Sterilize 12 minutes at 121 C.

Solution B:

N-base dehydrated*	13.0 g
Distilled water	1.0 liter

*N-base dehydrated may be obtained from Difco Laboratories, Detroit, Michigan.

Suspend N-base in distilled water. Sterilize 12 minutes at 121 C.

Cool solution A and B to 50 C and mix the contents of the two containers. Add 3 ml of 5% sterile tartaric acid/100 ml to the complete mixture just prior to pouring plates. Pour plates using 20 to 30 ml per plate. Allow to solidify, invert plates, incubate 24 hours at 30 C, and observe for contaminating colonies. Store sterile plates inverted in the refrigerator until used. Spread 0.01 ml of suitable dilutions on the surface with a platinum loop. Incubate 3 to 5 days at 28 to 32 C prior to counting.

YESAIR'S PORK INFUSION AGAR

Lean pork	450.0 g
Peptone	5.0 g
Tryptone	1.5 g
Dipotassium phosphate	1.25 g
Soluble starch	1.0 g
Glucose	1.0 g
Agar	15.0 g
Distilled water	1.0 liter

Remove as much fat from the pork as possible and grind the meat. To 450 grams of the defatted, ground pork add 1.0 liter of distilled water, and boil for 1 hour. Filter through a double layer of cheese cloth to remove the meat. Cool in the refrigerator and skim off any fat that comes to the surface and make the volume back to 1 liter with additional distilled water. Add the remaining ingredients and adjust the pH to 7.4 to 7.6. Bring to a boil and add the agar and continue heating until dissolved. Distribute in flasks, keep at 50 to 55 C overnight in slanting position to permit settling; decant the clear portion into tubes or flasks. Autoclave 30 minutes at 121 C; the final pH should be between 6.8 and 7.

YXT AGAR

Yeast extract	4.0 g
Glucose	4.0 g
Malt extract	10.0 g
Agar	15.0 g
Distilled water	1.0 liter

Dissolve ingredients in distilled water, heating if necessary. Sterilize at 121 C for 15 minutes.

YXT AGAR WITH TETRACYCLINE

Yeast extract	4.0 g
Glucose	4.0 g
Malt extract	10.0 g
Agar	15.0 g
Distilled water	1.0 liter

Mix ingredients in distilled water. Dispense in suitable flasks, sterilize at 121 C for 15 minutes.

Tetracycline should be reconstituted with distilled water and sterilized using a membrane filter. Cool medium to 50 C and add 10 micrograms tetracycline/ml just prior to pouring plates.

2.3 REAGENTS AND INDICATORS

This section contains the formulae for preparation of all reagents and indicators used throughout this book, with the exception of a few of which the preparation is relatively complex. The preparation of those reagents or indicators has been thoroughly described in the appropriate chapters in which they are needed. Reagents should be prepared using chemicals of the highest purity only and with double distilled water. The reagents may be heat sterilized or sterilized by membrane filtration when necessary.

ACETIC ACID, 5N

Glacial acetic acid	287.0 ml
Distilled water	1.0 liter (final volume)

ANDRADE'S INDICATOR

Distilled water	100.0 ml
Acid fuchsin	0.5 g
Sodium hydroxide (1.0 N)	16.0 ml

The fuchsin is dissolved in the distilled water and the sodium hydroxide is added. If, after several hours, the fuchsin is not sufficiently decolorized, add an additional 1 or 2 ml of alkali. The dye content of different samples of acid fuchsin varies quite widely, and the amount of alkali which should be used with any particular sample usually is specified on the label. The reagent improves somewhat on aging, and should be prepared in sufficiently large amount to last for several years. The indicator is used in the amount of 10 ml per liter of medium.

Other indicators such as bromthymol blue or bromcresol purple may be used instead of Andrade's. However, the investigator prefers Andrade's indicator for several reasons (a) It is colorless and early fermentation within insert tubes is observed easily; (b) It is not reduced easily; (c) Reversion from an acid to an alkaline condition is more clear cut than is usually the case with other indicators; (d) It is inexpensive.

BARIUM CHLORIDE, 1% AQUEOUS SOLUTION

Barium chloride	10.0 g
Distilled water	1.0 liter

BROMCRESOL PURPLE INDICATOR

Carefully weigh 0.1 g of bromcresol purple indicator and suspend in 20 ml 0.05 N NaOH. Heat gently with constant stirring. Add 5 ml of 0.05N NaOH. This constitutes the stock solution.

To make up the indicator solution dilute 1 ml of the stock solution with 9 ml of distilled water.

BUTTERFIELD'S BUFFERED PHOSPHATE DILUENT

Stock solution:

Monopotassium hydrogen phosphate	34.0 g
Distilled water	500.0 ml

Adjust to pH 7.2 with about 175 ml/N sodium hydroxide solution; dilute to one liter and store.

Diluent:

Dilute 1.25 ml stock solution to 1 liter with distilled water. Prepare dilution blanks in suitable containers. Sterilize at 121 C for 15 minutes.

CARBONATE-BICARBONATE BUFFER

Solution A:

Sodium carbonate	5.3 g
Distilled water q.s.	100.00 ml

Solution B:

Sodium bicarbonate	4.2 g
Distilled water q.s.	100.0 ml

Theoretically, a pH of 9.0 should result from mixing 4.4 ml of Solution A with 100 ml of Solution B. In practice, it is sometimes necessary to add as much as 17 ml of Solution A to 100 ml of Solution B. The pH should be checked on a pH meter.

CATALASE TEST

Flood plates with 3.5% hydrogen peroxide solution. Observe for bubble formation using a hand lens or wide field binocular microscope. Colonies exhibiting no evidence of gas formation are catalase negative. Transfer a loopful of colony to a slide, mix with 2 to 5% hydrogen peroxide and observe as above.

COAGULASE PLASMA

A. Desiccated coagulase plasma (rabbit) with EDTA: Reconstitute according to manufacturer's directions.

B. If plasma containing EDTA is not available, reconstitute desiccated rabbit plasma and add Na_2H_2 EDTA to a final concentration of 0.1% in the reconstituted plasma.

CYTOCHROME OXIDASE REAGENT

N,N,N,N-tetramethyl-p-phenylenediamine	5.0 g
Distilled water	1.0 liter

Store in dark glass bottle at 5 to 10 C. Storage life is 14 days. To perform the test add 0.3 ml to 18 hour blood agar base slant. A positive reaction is the development of blue color within 1 minute.

DECARBOXYLASE TEST

See Decarboxylase Test Medium.

DICHLOROFLUORESCEIN INDICATOR

Dichlorofluorescein	0.5 g
95 per cent ethyl alcohol	100.0 ml

FORMALINIZED SALINE SOLUTION

Sodium chloride	5.0 g
Formaldehyde (36%)	5.0 ml
Distilled water	1.0 liter

GLYCEROL FORMALDEHYDE SOLUTION

Glycerol (95%)	50 ml
Formalin (37% Formaldehyde)	50 ml

Mix the two solutions and make up to 200 ml with distilled water.

HEPES BUFFERED SALT SOLUTION

N-2 Hydroxyethyl Piperazine N′-2-Ethane sulfonic acid (HEPES)	6.0 g
Sodium chloride	11.7 g
Distilled water	500.0 ml

Dissolve ingredients in the water and adjust to pH 8.0 with 5N sodium hydroxide. Store in refrigerator.

INDOLE INDICATOR STRIPS
(Also see Kovacs' Reagent)

Prepare a saturated solution of oxalic acid. Soak filter paper in this solution and dry.

Cut paper into strips 1/2 inch wide and 2 to 3 inches long.

Inoculate SIM agar and insert an oxalic acid indicator strip into the tube. Fold the end of the strip over the edge of the tube and insert the cap (or plug) so that the strip is suspended in the tube. Take care that the strip does not contact the medium or become moist. The strip turns pink during incubation when the organism produces indole.

IODINE 2% IN 70% ALCOHOL

Iodine	2.0 g
70% Ethyl alcohol in water	100.0 ml

KIRKPATRICK'S FIXATIVE

Absolute ethanol	60.0 ml
Chloroform	30.0 ml
Formalin solution	10.0 ml

KOVACS' REAGENT

p-Dimethylaminobenzaldehyde	5.0 g
Amyl alcohol	75.0 ml
Hydrochloric acid (concentrated)	25.0 ml

Dissolve p-dimethylaminobenzaldehyde in the amyl alcohol, and then slowly add the hydrochloric acid. To test for indole, add 0.2 to 0.3 ml of reagent to 5 ml of a 48 hour culture of bacteria in tryptone broth. A dark red color in the surface layer constitutes a positive test for indole.

LECITHOVITELLIN SOLUTION

Mix an egg yolk thoroughly with 250 ml of physiological saline solution and clarify by centrifuging for 20 minutes at 14,000 × g at 4 C. Sterilize the supernatant by Seitz filtration and store in the refrigerator.

LUGOL'S SOLUTION
(Not for Gram stain)

Potassium iodide	10.0 g
Iodine	5.0 g
Distilled water	100.0 ml

Mix potassium iodide with about 30 ml distilled water and dissolve. Then add iodine, bring volume to 100 ml, and dissolve the iodine.

MERCURIC IODIDE SOLUTION

For preparation of suspensions for slide agglutination tests
Stock solution:

Mercuric iodide	1.0 g
Potassium iodide	4.0 g
Distilled water	100.0 ml

Working solution (1:1000):

Stock solution	10.0 ml
0.5 or 0.85% sodium chloride solution	90.0 ml
Formalin	0.05 ml

METHYL RED INDICATOR

Methyl red	0.10 g
Alcohol, 95% (ethanol)	300.0 ml

Dissolve methyl red in 300 ml of alcohol, and make up to 500 ml with distilled water. Incubate test cultures 5 days at 30 C. Alternatively incubate at 37 C for 48 hours. Add 5 or 6 drops of reagent to cultures. Do not perform tests on cultures incubated less than 48 hours. If equivocal results are obtained, repeat tests on cultures incubated 4 or 5 days. Duplicate tests should be incubated at 22 to 25 C.

MOUNTING FLUID

Glycerol	90.0 ml
Carbonate-bicarbonate buffer, pH 9	10.0 ml

NITRATE REDUCTION
(Not for *Clostridium perfringens*)*

Method 1

Solution A:

Sulfanilic acid	8.0 g
5N acetic acid	1.0 liter

Solution B:

Alpha-naphthol	5.0 g
5N acetic acid	1.0 liter

*Alpha-naphthylamine has been dropped as a reagent and is considered a carcinogen.

To 3 ml of an 18 to 24 hour culture in indole-nitrite broth add two drops of sulfanilic acid reagent and two drops of alpha-naphthol reagent. An orange color indicates that nitrate has been reduced to nitrite. If the reaction is negative, one must examine for residual nitrate since conceivably the nitrite may have been reduced to another state. Add a few grains of powdered zinc. If an orange color does not develop, nitrate has been reduced. Perform tests on uninoculated medium (control).

Method 2

Solution A:

Sulfanilic acid	0.5 g
Glacial acetic acid	30.0 ml
Distilled water	120.0 ml

Solution B:

5 amino-2-naphthalene sulfonic acid (Cleve's acid)	0.2 g
Glacial acetic acid	30.0 ml
Distilled water	120.0 ml

Add 0.1 ml Solution A followed by 0.1 ml reagent B to a culture incubated 24 to 48 hours at 35 C in nitrate broth. Development of a red color in 1 minute indicates reduction of nitrate to nitrite.

NITRATE REDUCTION
(For *Clostridium perfringens*)

Inoculate suspected *Clostridium perfringens* cultures into 9 ml of Buffered Motility Nitrate Medium and incubate 24 hours at 35 C. Add 0.5 ml of sulfanilic acid (Solution A) and 4 drops of alpha-naphthol (Solution B) to 9 ml of the Buffered Motility Nitrate medium. A positive test is indicated by an orange color developing as the reagents diffuse into the semisolid medium. This takes approximately 15 minutes.

NITRATE TEST STRIPS

1. Into a sterile, dry, screw capped tube 20 × 110 mm, place 0.5 ml distilled water.
2. Using growth from a 24 hour culture in trypticase or tryptic soy broth, prepare a dense suspension in the test tube without wetting the upper portion of the tube.
3. Place a nitrite test strip with forceps into the bottom of the tube with the arrow of the strip pointing down.

4. Tighten the cap and incubate at 37 C for 2 hours. Gently agitate the tube after 1 hour and after 2 hours without wetting the middle portion of the strip.
5. After incubation, carefully tilt the tube back and forth several times completely wetting the strip and leaving it in a horizontal position.
6. Nitrate reduction is indicated by a blue color in the top of the strip.

ONPG TEST

Test for Beta-D-galactosidase activity (LeMinor and Ben Hamida, 1962; Lubin and Ewing, 1964; Costin, 1966).

Reagent A: 1.0 M monosodium phosphate·H_2O solution, pH 7.

Dissolve 6.9 g $NaH_2PO_4 \cdot H_2O$ in approximately 45 ml distilled water. Add approximately 3 ml of 30% (w/v) sodium hydroxide solution and adjust to pH 7. Bring volume to 50 ml with distilled water and store in refrigerator (about 4 C).

Reagent B: .0133 M o-nitrophenyl-B-D-galactopyranoside (ONPG) in 0.25 M monosodium phosphate solution, pH 7.

Dissolve 80 mg ONPG in 15 ml distilled water at 37 C. Add 5 ml of 1.0 M monosodium phosphate solution (see (a) above). Solution should be colorless and should be stored in refrigerator. Prior to its use an appropriate portion (sufficient for the number of tests to be done) of the buffered 0.0133 M ONPG solution should be warmed to 37 C.

Procedure:

Cultures to be tested are inoculated onto triple sugar iron agar (TSI) slants and incubated for 18 hours at 37 C (or at another appropriate temperature if required). Nutrient agar slants containing 1.0% lactose also may be used. A large loopful of growth from each culture is emulsified in 0.25 ml of physiological saline solution (heavy suspension).

One drop of toluene is added to each tube and the tubes shaken well (to aid liberation of the enzyme). Following this the tubes are allowed to stand for about five minutes at 37 C. Add 0.25 ml of buffered 0.0133 molar ONPG solution to each suspension to be tested, shake the tubes well, and incubate in a waterbath at 37 C. Readings should be made after 30 minutes, 1 hour, and 24 hours of incubation. Positive results are indicated by the development of a yellow color.

ONPG ALTERNATE METHOD

ONPG Disks: disks impregnated with ortho-nitrophenyl-B-D-galactopyranoside. Perform test as follows:

Emulsify a loopful of growth from an 18 hour TSI (or BAB) slant in 0.2 ml sterile 0.85% sodium chloride in sterile 13 × 100 mm tubes. Add disk and mix.

Incubate 4 to 6 hours at 35 C. Positive reaction is a yellow color.

OXIDASE TEST

Para-amino dimethylphenylenediamine monohydrochloride	1.0 g
Distilled water	100.0 ml

Suspend ingredient in distilled water and dissolve. Prepare this preparation fresh, and do not store.

Drop several drops of the reagent on suspect colonies. If the organisms produce oxidase, the colonies will turn pink, red, then black, in that order. Discard plates after 20 minutes.

PHENOLPHTHALEIN INDICATOR 0.1%

Dissolve 0.1 g phenolphthalein in 100 ml 50% ethyl alcohol.

PHENOLPHTHALEIN SOLUTION 1% ALCOHOLIC

Phenolphthalein	1.0 g
Ethyl alcohol	100.0 ml

Dissolve by agitating.

PHENYLALANINE DEAMINASE

See Phenylalanine Medium.

PHOSPHATE BUFFERED SALINE (PBS), pH 7.6, 0.01 M

Concentrated (10X) Stock Solution (pH is not 7.6 at this point):

Disodium phosphate anhydrous, reagent grade	12.35 g
Monosodium phosphate·H_2O reagent grade	1.80 g
Sodium chloride, reagent grade	85.0 g
Distilled water to a final volume of	1.0 liter

The salts dissolve much more readily at 37 C than at room temperature.

Working Solution—PBS, pH 7.6, 0.01 M:

Concentrated stock solution	100.0 ml
Distilled water q.s.	1.0 liter

POLYSACCHARIDE PRECIPITANT

1% acqueous solution of hexadecyltrimethylammoniumbromide. (This solution may be autoclaved).

OR

7 normal hydrochloric acid prepared by diluting 548.3 ml of the 37% hydrochloric acid solution to 1 liter.

POTASSIUM HYDROXIDE 40% + CREATINE

Potassium hydroxide	40.0 g
Distilled water q.s.	100.0 ml
Creatine (reagent grade)	0.3 g

Stir until dissolved. Store not over 3 days at 2 to 8 C (40 F) nor more than 21 days at −17.8 C (0 F).

SALINE SOLUTION, 0.5% AQUEOUS SOLUTION

Sodium chloride	5.0 g
Distilled water	1.0 liter

Autoclave 15 minutes at 121 C. Final pH should be 7.0.

SILVER NITRATE REAGENT

Silver nitrate	29.063 g
Distilled water	1.0 liter

This makes a 0.171N silver nitrate solution. Dissolve silver nitrate, place in a brown bottle, standardize and store in refrigerator.

When 1 ml of sample containing chloride is titrated, each ml of silver nitrate solution used is equal to 1 g of sodium chloride per 100 ml of sample. Use dichlorofluorescein indicator. It changes to a light pink salmon color at the end point.

SODIUM BICARBONATE (for neutralization)

Sodium bicarbonate	100.0 g
Distilled water	1.0 liter

Sterilize by filtration.

SODIUM CHLORIDE, PHYSIOLOGICAL 0.85% NaCl

Sodium chloride	8.5 g
Distilled water	1.0 liter

Dispense into flasks, bottles, or tubes and autoclave at 121 C for 15 minutes.

SODIUM CHLORIDE 3%

Sodium chloride	30.0 g
Distilled water	1.0 liter

Dispense into flasks, bottles, or tubes and autoclave at 121 C for 15 minutes.

STABILIZER SOLUTION

Cellulose Gum*	2.5 g
Formalin 40% (37% formaldehyde by weight)	10.0 ml
Distilled water	500.0 ml

*Na carboxymethyl cellulose (Cellulose Gum CMC-7H3SF, Hercules, Inc., Cellulose and Protein Products Department, Wilmington, DE 19899)

Place boiling distilled water in high speed blender. With blender running, add the cellulose and formalin. Blend for about 1 minute.

ALTERNATE STABILIZER SOLUTION

Pectin	3.0 to 5.0 g
or Algin	1.0 g
Formalin 40% (37% formaldehyde by weight)	2.0 ml
Distilled water	100.0 ml

Add pectin and formalin directly to distilled water while agitating in high speed blender. Treat solution with vacuum or heat to remove air bubbles.

(If blender is not available, mix dry stabilizer with a small quantity of alcohol to facilitate incorporation with water.)

Adjust to pH 7.0 to 7.5.

SULFURIC ACID, 1% AQUEOUS SOLUTION

Sulfuric acid (95%)	5.7 ml
Distilled water	1.0 liter

UREASE REAGENT RAPID

Solution A:

Urea	4.0 g
Ethyl alcohol, 95%	4.0 ml
Distilled water	8.0 ml

Do not heat. Store in refrigerator. If urea precipitates out, warm in the hand or in warm water until it dissolves.

Solution B: Stock Buffer Solution

Dipotassium phosphate	0.1 gm
Monopotassium phosphate	0.1 gm
Sodium chloride	0.5 gm
Phenol red (0.2 per cent)	1.0 ml
Distilled water	100.0 ml

Sterilize this solution in the autoclave.

Procedure:

Dilute 1 part of Solution A with 19 parts of Solution B. This constitutes the test solution.

The mixture must not be heated and need not be sterilized by filtration. A fresh mixture should be made from the stock solutions each time it is used.

A marked red coloration of the solution is a positive reaction for bacterial urease production.

VOGES-PROSKAUER (V-P) TEST REAGENTS

Solution A:

α-Naphthol	5.0 g
Absolute ethanol	100.0 ml

Solution B:

Potassium hydroxide	40.0 g
Distilled water q.s.	100.0 ml

Perform Voges-Proskauer (V-P) test at room temperature by transferring 1 ml of 48 hour culture to test tube and adding 0.6 ml of α-naphthol (Solution 1) and 0.2 ml of 40% potassium hydroxide (Solution 2); shake after addition of each solution. To intensify and speed reactions, add a few crystals of creatine to test medium. Read results 4 hours after adding reagents. Positive V-P test is the development of an eosin pink color.

V-P TEST (MODIFIED FOR *B. cereus*)

Potassium hydroxide	40.0 g
Distilled water	100.0 ml

For test: Incubate purified cultures 48 hours at 37 C in modified V-P medium (Smith, Gordon, Clark). Add 5 ml of above solution plus 0.5 to 1 mg creatine to 5 ml of incubated V-P Broth Modified (Smith, Gordon, Clark). The appearance of a red color within 30 to 60 minutes indicates the presence of acetylmethyl carbinol, i.e., a positive test. Note the medium is a modification of the usual medium and must be formulated in the laboratory.

2.4 STAINS

The staining procedures including reagents and their preparation are adapted from Staining Procedures, (1973), Third Ed., George Clark, ed., Williams and Wilkins Co., Baltimore, Md.

Dyes employed should be from batches certified by the Biological Stain Commission or dyes of equal purity. (No batch is approved by the Commission unless it meets chemical and physical tests and has been found to produce satisfactory results in the procedures for which it is normally used.)

Control organisms should be employed frequently, preferably with every batch of slides stained to assure the analyst that the completed preparations and technique employed are producing appropriate results.

CRYSTAL VIOLET STAIN (0.5—1% SOLUTION)

Crystal violet (90% dye content)	0.5 to 1.0 g
Distilled water	100.0 ml

Dissolve crystal violet in distilled water. Filter through coarse filter paper. Prepare smear, air dry, heat fix, and stain 20 to 30 seconds; rinse with tap water, air dry, and examine.

ERYTHROSIN STAIN

Stock Solution:

Erythrosin B, certified	1.0 g
Distilled water	100.0 ml

Dissolve dye in distilled water and store in the refrigerator.

Buffer Solution:
Buffer solution is made by dissolving equal parts of 0.2M disodium phosphate and 0.2M monosodium phosphate in distilled water.

To Use:
Make a 1:5,000 concentration of the stain by diluting 1 ml of the stock solution with 50 ml of the buffer solution.
To 1 ml liquid sample add 1 ml of the 1:5,000 erythrosin solution in a small serum tube. Shake to obtain an even suspension of organisms and transfer a drop of solution to the hemacytometer with a 3 mm platinum loop.

GRAM STAIN HUCKER*

Hucker Crystal Violet

Solution A:

Crystal violet (85% dye content)	2.0 g
Ethyl alcohol 95%	20.0 ml

Solution B:

Ammonium oxalate monohydrate	0.2 g
Distilled water	20.0 ml

Mix equal parts of Solutions A and B. (Sometimes it is found that the crystal violet is so concentrated that gram negative organisms do not properly decolorize. To avoid this difficulty, the crystal violet solution may be diluted as much as tenfold prior to mixing with equal parts of Solution B.)

a. Lugol's solution, Gram's modification: Dissolve 1.0 g iodine crystals and 2.0 g potassium iodide in 300.0 ml distilled water.
b. Counterstain: Dissolve 2.5 g safranin dye in 100.0 ml 95 percent ethyl alcohol. Add 10.0 ml of the alcoholic solution of safranin to 100.0 ml distilled water.
c. Ethyl alcohol, 95 percent

Procedure: Stain the heat fixed smear for 1 minute with the ammonium oxalate-crystal violet solution. Wash the slide in water; immerse in Lugol's solution for 1 minute.
Wash the stained slide in water; blot dry. Decolorize with ethyl alcohol for 30 seconds, using gentle agitation. Wash with water. Blot and cover with counterstain for 10 seconds, then wash, dry, and examine as usual.
Cells which decolorize and accept the safranin stain are gram negative. Cells which do not decolorize but retain the crystal violet stain are gram positive.

*Preferably stain vigorously growing 24 hour cultures from nutrient or other agar free of added carbohydrates. Use positive and negative culture controls.

GRAM'S STAIN, KOPELOFF AND BEERMAN'S MODIFICATION*

1. 1 per cent aqueous methyl violet . . . . 30.0 ml
   5 per cent sodium bicarbonate solution . . . . 8.0 ml
2. Iodine . . . . 1.0 g
   Potassium iodide . . . . 2.0 g
   5 per cent sodium bicarbonate solution . . . . 60.0 ml
   Distilled water . . . . 240.0 ml
3. Counterstain
   a. Safranin, saturated alcoholic solution
      of safranin . . . . 10.0 ml
      Distilled water . . . . 90.0 ml
   b. Basic fuchsin
      Basic fuchsin . . . . 0.05 g
      Distilled water . . . . 100.0 ml
4. Staining Procedure
   a. Flood heat fixed slide with methyl violet solution and allow to remain 5 minutes.
   b. Wash with water.
   c. Flood with iodine solution and leave 2 minutes.
   d. Drain and decolorize with acetone.
   e. Decoloriation is very rapid and must be carefully controlled.
   f. Wash in gently running water.
   g. Counterstain for 30 seconds to 1 minute with safranin or basic fuchsin.
   h. Wash, blot, and dry.

*Preferably stain vigorously growing 24 hour cultures from nutrient or other agar free of added carbohydrates. Use positive and negative culture controls.

GRAY'S DOUBLE DYE STAIN

1. 1 per cent aqueous methylene blue . . . . 50.0 ml
   Methanol . . . . 50.0 ml
   Add dye to the methanol.
2. 1 per cent aqueous basic fuchsin . . . . 25.0 ml
   Methanol . . . . 25.0 ml
   Add dye to the methanol.

Mix the two solutions before use. Stain slide for a few seconds with the second stain, dry and examine. Bacteria are stained blue and the background pink.

If the slide needs defatting, use xylene, and wash off with methyl alcohol before staining.

LOEFFLER'S METHYLENE BLUE SOLUTION
(Staining of Colonies on Membrane Filter Pad)

Loeffler's methylene blue solution

Solution A:

Methylene Blue (90% dye content) . . . . 0.2 g
Ethanol 95% . . . . 60.0 ml

Solution B:

0.1N KOH . . . . 2.0 ml
Distilled water . . . . 200.0 ml

Mix solutions A and B.

Glycerol—formaldehyde solution

Solution A:

Glycerol 95% . . . . 50.0 ml

Solution B:

Formaldehyde 37% . . . . 50.0 ml

Mix Solutions A and B and make up to 200 ml with distilled water.

NEWMAN LAMBERT STAIN—Levowitz and Weber

Methylene Blue Chloride . . . . 0.6 g
Ethanol 95% . . . . 52.0 ml

(Stir to dissolve.)

Tetrachlorethane (technical) . . . . 44.0 ml

Place in a closed container and refrigerate at 4.4 to 7.2 C for 12 to 24 hours. Bring to room temperature.

Glacial acetic acid . . . . 4.0 ml

Filter through Whatman No. 2 filter paper. Store in a tightly stoppered container in a cool dark area, but do not refrigerate.

Preparation of Slides:

Clean glass slides are required since the rather thick smears produced from dairy products may peel from the slide during the staining procedure. If possible, use new slides. If old slides must be used, first clean with Bon-Ami, rinse in distilled water, immerse in dichromate-sulfuric acid cleaning solution overnight, rinse in distilled water, heat in an oven at 200 C for several hrs. with the glass surfaces exposed to the hot air, cool, and then store in 95% ethanol until required. If a quantitative determination of the numbers of organisms is required, a measured quantity (0.01 ml) of the dairy product (or a known dilution of it) is evenly spread over a 1 cm^2 area of the glass slide. A calibrated microscope is required and the counts should be made as described in the 13th edition of Standard Methods for the Examination of Dairy Products.

Some dairy products should be diluted in 1.25% sodium citrate to aid the solution of casein.

Procedure:

1. The smears should be dried rapidly in a small hot air oven or on a thermostatically controlled hot plate at 45 to 50 C; the temperature should not rise over 51 C. Cool to room temperature.
2. Stain and defat for 2 minutes with the methylene blue ethanol tetrachlorethane solution. Use a closed staining dish which is capable of preventing the evaporation of the solution. The slide should be covered completely by the solution and it should be discarded at the first sign of any formation of precipitate.
3. Remove the slide from the staining dish and drain off the excess stain by resting the long edge on the surface of absorbent paper. Dry thoroughly, using forced air from a blower or fan.
4. Rinse the dried stained slides by passing through 3 changes of tap water at 38 to 43 C.
5. Dry rapidly and thoroughly, using forced air from a blower or fan, and examine.

Results:

Normal bacterial cells are stained heavily; the background material should be evenly and lightly stained. Plasmolyzed bacterial cells should stain with various degrees of intensity depending on the degree of plasmolysis. Leucocytes should be stained with the cytoplasm slightly darker than the background and the nuclear areas more deeply stained which allows their differentiation into types.

NORTH'S ANILINE OIL—METHYLENE BLUE STAIN

To prepare stain, mix 3 ml of aniline oil and 10 ml of 95 per cent ethyl alcohol. Add 1.5 ml of concentrated hydrochloric acid slowly with constant agitation. Slowly add 30 ml of a saturated alcoholic methylene blue solution and then add distilled water to make 100 ml. Filter and keep in tightly stoppered bottle.

SPORE STAIN

A Modification of the Bartholomew and Mittwer Spore Stain Which Omits the Heat Step

Preparation of Solutions:

A. A saturated solution of malachite green chloride is prepared by dissolving 10 g of malachite green with 100 ml of distilled water. Filter to remove any undissolved dye.

B. A 0.25% aqueous solution of safranin O is prepared. This is the same safranin solution as used for the counterstain in the Gram stain.

Procedure:

1. Prepare an air dried smear of the organisms and heat fix by flaming gently.
2. Cool the slide and flood it with Solution A.

3. Rinse with water.
4. Flood the slide with Solution B and let stand for 15 seconds.
5. Rinse, dry and examine the slide.

Spores stain green; vegetative cells are red.

SPORE STAIN
(Schaeffer and Fulton Modification of the Wirtz Spore Stain)

Solution A:

Malachite green	5.0 g
Distilled water	100.0 ml

Solution B:

Safranin O	0.5 g
Distilled water	100.0 ml

Procedure:

1. Prepare a smear of the organisms. Air dry. Heat fix by passing through the flame of a Bunsen burner.
2. Flood the smear with Solution A and heat to steaming 3 to 4 times within 30 seconds, then cool.
3. Wash in tap water for 30 seconds.
4. Flood with Solution B for 3 seconds.
5. Wash lightly, blot dry, and examine.

Results: Spores should stain green; the vegetative cells red.

SPORE STAIN (SCHAEFFER–FULTON)

Solution A:

Malachite green	10.0 g
Distilled water	100.0 ml

Filter to remove undissolved dye.

Solution B:

Safranin O	0.25 g
Distilled water	100.0 ml

Procedure:

1. Prepare a smear of the organism and air dry.
2. Heat fix by passing the slide through the flame of a Bunsen burner about 20 times. Each passage should be similar to that used for normal heat fixation. Let the slide cool before proceeding.
3. Flood the slide with Solution A and let stand for 10 minutes at room temperature.
4. Rinse with tap water.
5. Flood the slide with Solution B and let stand for 15 seconds.
6. Rinse very lightly in tap water, blot dry, and examine.

Results: The spores should be stained green; the vegetative cells red.

WOOLFAST PINK RL STAIN

Preparation A:

Woolfast Pink RL*	1.0 g

Preparation B:

Trichloroacetic acid	5.0 ml
Acetic acid	1.0 ml
Ethanol	25.0 ml
Distilled water	75.0 ml

Final Solution:

Dissolve Preparation A in 100 ml of Preparation B.

*American Hoechst Corporation
1515 Broadway
New York, New York 10036

WOLFORD'S STAIN

Solution A:

Basic fuchsin	1.0 g
Ethyl alcohol, 95 per cent	100.0 ml

Solution B:

North's aniline oil methylene blue stain.

Make dilute stain for making direct microscopic counts by mixing 2 ml of solution A with 10 ml of solution B and 88 ml distilled water. Filter before use.

The dilute stain solution may be kept under refrigeration for 3 to 4 weeks, but should be discarded after that time.

ZIEHL-NEELSEN'S CARBOL-FUCHSIN STAIN

Solution A:

Basic fuchsin 90% dye content	0.3 g
Ethyl alcohol 95%	10.0 ml

Solution B:

Phenol crystals C.P.	5.0 g
Distilled water	95.0 ml

Dissolve the basic fuchsin in alcohol.

Dissolve the phenol in distilled water.

Combine the two solutions. Filter through coarse filter paper. Resultant solution should be removed from light and refiltered as necessary to remove crystal formation.

This is an excellent simple stain for primary staining of a food product. Prepare smear, air dry and heat fix 15 to 20 seconds. Rinse with tap water, air dry, and examine.

CHAPTER 3

SAMPLING, EQUIPMENT, SUPPLIES, AND ENVIRONMENT

Damien A. Gabis, Donald Vesley, and Martin S. Favero

3.1 GENERAL CONSIDERATIONS

Surfaces of equipment and containers which come in contact with food products should be practically free from microorganisms. In addition, the food plant environment, including the air and water supply, should not be a source of contamination. Contamination of food products, if it occurs following a final biocidal treatment, has direct public health and keeping quality significance. Many foods do not receive biocidal treatments, and it is especially important that the equipment and environment for such products be microbiologically controlled.

The following tests have been adapted from, and are consistent with, the latest edition of Standard Methods for the Examination of Dairy Products.[1] These standardized tests are equally applicable to nondairy food plants. The reader is referred to that publication for additional, more detailed descriptions of relevant procedures.

3.2 TESTS FOR SANITIZATION OF EQUIPMENT

3.21 Rinse Solution Method

The sanitary condition of containers and equipment can be determined by repeated flushing of a measured volume of sterile, buffered sanitizer neutralizing solution or nutrient broth over the product contact surface, and then determining the bacterial population of the rinse solution by plating or membrane filter techniques.[1]

Large equipment assemblies may be checked by circulating water (chlorinated, then neutralized) through the equipment, and determining the numbers of microorganisms by membrane filter method.[1] Sanitation of assemblies may also be checked by sampling the food product progressively at points of access along processing lines.

The equipment and supplies needed for this procedure are pipets, Petri plates, plate count agar, stock phosphate buffer solution, and sterile buffered rinse solution. The stock buffer solution is prepared by dissolving 34 g of potassium dihydrogen phosphate (KH_2PO_4) in 500 ml of distilled water, adjust to pH 7.2 with 1 N NaOH solution, and make up to 1 liter with

IS/A Committee Liaison: W. J. Hausler, Jr.

distilled water. The stock buffer may be sterilized at 121C for 15 minutes and stored in the refrigerator. To prepare the buffered rinse solution add 1.25 ml of stock phosphate buffer solution, 5 ml of 10% aqueous sodium thiosulfate ($Na_2S_2O_3 \cdot 5H_2O$), 4 g Asolectin (Associated Concentrates, Woodside, N. Y. 11377), and 10 g of Tween-20 or Tween-80 to distilled water and make up to 1 liter. Dispense in screw capped vials made up to contain, after sterilization, 20 ml, 100 ml or other needed volumes. Asolectin is hygroscopic and should be stored in a dessicator. Weigh the powder and rapidly dissolve by heating over boiling water.

To prepare the 10% sodium thiosulfate solution dissolve 100 g of $Na_2S_2O_3 \cdot 5H_2O$ in distilled water and make up to 1 liter; filter; store in the refrigerator or dark place, or in an amber bottle. Nutrient broth may be applied as a rinse also since it effectively neutralizes residual active chlorine or quaternary ammonium compounds,[19] and may be used instead of buffered rinse solution. Sterile hypodermic syringes and needles will also be needed as well as 70% ethanol.

Firm walled, capped and uncapped items may be rinsed by grasping the container firmly with its long axis in a horizontal position. Shake vigorously 10 times, each shake being a "to-and-return" thrust of about 8 inches. Turn the container 90 degrees and repeat the horizontal shaking treatment; turn the container 90 degrees twice more and repeat the horizontal shaking. Swirl the container vigorously 20 times in a small circle with the long axis in a vertical position, then invert and repeat.

For flexible walled items, arrange the collapsed liner on a smooth, clean, firm horizontal surface as flat as its construction permits. With the hands or a roller, "work" the rinse solution back and forth 10 times, contacting all interior surfaces completely. Lift the liner and hang with the fill tube down to permit the rinse solution to collect there.

Rinse solutions from container samples are analyzed by distributing 10 ml of the rinse solution equally among three sterile Petri dishes; seed two other Petri dishes with 1 ml portions. Pour 15 to 18 ml of Plate Count Agar (Chapter 2) into the 10 ml portions and 10 to 12 ml into the 1 ml portions. Incubate the plates at 32 C for 48 ± 3 hours and count the colonies. If 20 ml of rinse solution was used then the total number of colonies on the three plates which received the 10 ml rinse, multiplied by 2, or the total number of colonies on the two plates which received 1 ml each, multiplied by 10 yields the number of colonies per container.

When the plates which received 1 ml each show more than 300 colonies, report the counts as more than 6,000 per container. When 100 ml of rinse solution are used, membrane filter procedures should be followed for the analysis.[1] Membrane filter procedures also may be used with the 20 ml samples.

The presence and concentration of coliforms are determined by dividing 5 ml of rinse solution between two plates and pouring with suitable medium. Yeasts, molds, proteolytic bacteria, and other specific microorganisms may be determined similarly by the use of appropriate differential media, and incubation temperatures and times.

Rinse solutions obtained from cleaned-in-place (CIP) systems or processing assemblies require the use of membrane filter procedures[1] for analysis of both rinse and control samples. Average the yields of the rinse samples taken at the beginning, middle and end of drainage and substract the yield (if any) of the control samples: Calculate the ratio of sample volume to rinse volume and multiply by the corrected yield to obtain an indication of the numbers of organisms present in the entire system. The presence of specific types of organisms may be determined by employing appropriate differential media and incubation temperatures.

3.22 Surface Contact Methods

Microbiological examination of surfaces requires selection of the proper method. Swab procedures and RODAC (Replicate Organism Direct Agar Contact plates) are usually the methods of choice. Swab technics should be used for areas such as cracks, corners or crevices, i.e., areas of such dimension that the swab will be more effective in recovering organisms from them. The RODAC procedure is better utilized on flat surfaces such as walls, floors, ceilings and equipment surfaces. Selection of the proper technic is essential for meaningful results.

Two swab contact methods are given. The first consists of brushing a sterile swab moistened in an appropriate solution over a measured surface area. The swab is then rinsed thoroughly in a measured amount of sterile solution and the liquid examined microbiologically. Select areas for swabbing which appear to be unclean or difficult to clean. Equipment surfaces which are regular, smooth, ground, polished, and large or accessible, are easier to clean than those which are irregular, rough seamed, ridged, small, rounded on small radii, or less conveniently reached. If such latter areas are not properly cleaned, residues will accumulate and will not be sanitized when chemical solutions are applied.

The general supplies needed for this procedure are given in Chapter 2. Specific supplies required for the swab contact method include small screw-capped vials (3 to 4 inches long) prepared to contain 5 ml of rinse solution after autoclaving. In addition, sterile nonabsorbent cotton swabs are needed. These swabs should be twisted firmly to approximately 3/16 inch in diameter by ¾ inch long over one end of a wood applicator stick 5 to 6 inches long. They should be packaged in individual or multiple convenient protective containers with the swab heads away from the closure. Dacron and rayon, or soluble calcium alginate swabs may be used.[1]

To sample equipment surfaces, open the sterile swab container, grasp the end of a stick, being careful not to touch any portion which might be inserted into the vial, and remove the swab aseptically. Open a vial of buffered rinse solution, moisten the swab head, and press out excess solution against the interior of the wall of the vial with a rotating motion. Hold the swab handle to make a 30 degree angle contact with the surface. Rub the swab head slowly and thoroughly over approximately 8 square inches of surface. Rub the swab three times over this surface, reversing direction between successive strokes. Return the swab head to the solution vial, rinse briefly in the solution, then press out the excess. Swab four more 8 square inch areas of the surface being sampled, as above, rinsing the swab in the solution after each swabbing, and removing the excess.

After the fifth area has been swabbed, position the swab head in the vial, and break or cut it with sterile scissors or other device,[7] leaving the swab head in the vial. Replace the screw cap, put the vial in a waterproof container packed in cracked ice or other suitable refrigerant, and deliver to the laboratory. When sampling utensils, use one swab for each group of four utensils and sample as for equipment surfaces.

At the laboratory remove the vial from refrigerated storage. Shake vigorously, making 50 complete cycles of 6 inches in 10 seconds, striking the palm of the other hand at the end of each cycle. Groups of vials may be put into appropriate racks and shaken manually, or mechanical shakers may be employed for a time interval which treats cotton swab heads similarly. Plate 1 and 0.1 ml portions of rinse solution.

Plate additional decimal dilutions, if deemed necessary. Pour plates with plate count agar (Chapter 2). Incubate plates, count colonies, and then calculate the number of colonies recovered from 8 square inches (equivalent to 1 ml of rinse). In the case of utensils, report the count as the residual bacterial count per utensil examined. For example, if four utensils are swabbed, and if 1 ml of the 5 ml of the diluent is plated which results in a count of 100 colonies, then the count per utensil will be 125. When other than "total counts" are sought, plate with appropriate differential media and incubate as required.

Food processing equipment and environment surfaces may be sampled using the cellulose sponge swab technic.[14] Cellulose sponges are cut into 2 × 2 × 2 inch pieces, placed in individual Kraft paper bags, and steam sterilized at 121 C for 20 minutes. The sterile sponge is removed from the bag using sterilized or sanitized crucible tongs and moistened with 10 ml of buffered rinse solution or 10% solution of Tergitol Anionic-7 (Union Carbide, Chicago). It may be desirable, depending on the organism being sought, to replace the Tergitol Anionic-7 with another sampling fluid (0.1% peptone water or buffered rinse solution) because of the possible deleterious effects of Tergitol on the gram positive bacteria.

Instead of using crucible tongs, the hands may be covered with sterile or sanitized rubber or plastic gloves to handle the sponge. When sampling for *Salmonella*, it is sufficient to thoroughly wash the hands with soap, followed by sanitization in a solution of 200 ppm available chlorine.

The surface is sampled by vigorously rubbing the sponge over the designated area until the soil is removed. An area of several square meters may be efficiently swabbed. If the surface is flat, the rinse solution may be applied directly to the surface and then taken up into the sponge by the rubbing action. After sampling, the sponge is introduced into a sterile plastic bag, and transported under refrigeration to the laboratory.

Because of the large areas that can be sampled with the cellulose sponge, this technic is particularly useful in detecting *Salmonella* on food plant equipment and in the environment. For *Salmonella* analyses, the sponge is introduced directly into lactose broth which is then tested by conventional technics (Chapter 27).

The sponge sample may be subjected to a variety of analytical procedures, such as the total plate count, coliforms, yeasts and molds, staphylococci, etc. For quantitative analyses, 50 to 100 ml of diluent is added to the bag containing the sponge. The sponge is then vigorously massaged with the diluent for 1 minute or more. For quantitative analysis, aliquots of the diluent are then removed from the bag and plated in the desired medium, followed by appropriate incubation time and temperature. The numbers of microorganisms per unit surface area are calculated on the basis of the area swabbed, the amount of diluent used, and the size of the aliquot plated. For example, if 50 colonies are obtained from a 1 ml aliquot derived from a sponge in 100 ml of diluent which swabbed 1 square meter, the total plate count/square meter will be 5,000.

In addition to the swab methods, the RODAC plate method (agar contact) provides a simple agar contact tool for sampling surfaces.[4, 11, 18] It is of value in estimating the sanitary quality of surfaces encountered in dairy operations, food processing plants, and hospitals. Collaborative studies have indicated that the mean percent recovery of aerosolized *Bacillus subtilis* spores from stainless steel surfaces was slightly less than 50%, both for the swab (47%) and for the RODAC plate (41%) methods.[4] However, results from the RODAC plate method were more reproducible than those from the swab technic. The RODAC plate method is recommended particularly when quantitative data is sought from flat, impervious surfaces. It should not be used for pervious, creviced or irregular surfaces.

Disposable plastic RODAC plates (Falcon Plastics, Los Angeles) may be purchased prefilled with test medium, or they may be filled in the laboratory. They should be prepared with 15.5 to 16.3 ml of plate count agar (Chapter 2) for quantitative sampling. If qualitative data are sought, differential media may be used, or, preferably, colonies should be transferred to the differential media from the original broad spectrum medium.

If sampling is to be carried out on surfaces previously subjected to chemical germicide treatment, appropriate neutralizers should be incorporated into the medium. The most commonly used neutralizers are 0.5% polysorbate (Tween) 80 plus 0.07% soy lecithin. The polysorbate 80 is used to neutralize some substituted phenolic disinfectants, and the soy lecithin is used to neutralize quaternary ammonium compounds, but the efficacy of the neutralizers for agar contact sampling has not been demonstrated definitively.

A recent development in this area is the availability of Dey-Engley medium (Difco, Inc., Detroit) which incorporates a variety of ingredients capable of neutralizing any of the germicidal chemicals likely to be encountered on surfaces.[9] Following preparation of the plates, they should be incubated at 32 C for 18 to 24 hours as a sterility check. They should be used within the succeeding 12 hours unless appropriately wrapped and refrigerated. In addition, the agar in the RODAC plate should have a slightly convex surface with the center of the meniscus rising above the rims of the plate. This is necessary to make proper contact of the agar with the surface to be sampled.

When sampling, remove the plastic cover and carefully press the agar surface to the surface being sampled. Make certain that the entire agar meniscus contacts the surface, using a rolling uniform pressure on the back of the plate to effect contact. Replace the cover and incubate in an inverted position for 24 to 48 hours. When heavy contamination is present, the 24 hour incubation period should be observed to prevent overgrowth. When contamination is light, 48 hours should be allowed for the growth of colonies. Colonies should be counted using a Bactronic or Quebec colony counter, and should be recorded as number of colonies per RODAC plate, or number of colonies per square inch.

Product contact surfaces selected for RODAC sampling will necessarily be flat and impervious, and thus relatively easy to clean and disinfect. Other environmental surfaces, such as floors, more likely will be contaminated heavily. In either event, a sufficient number of spots should be sampled to obtain representative data. Randomization of site selection may permit additional comparisons and inferences to be made. Ideally the RODAC plate method should be used on previously cleaned and sanitized surfaces. Samples taken from heavily contaminated areas will result in overgrowth on the plates. Accurate colony counts can be made on plates with 200 or fewer colonies only.

3.3 STANDARD TEST FOR WATER SUPPLIES

To be practically and economically suitable for use, the water used in food plants must, in addition to complying with sanitary requirements, be free from microorganisms that could initiate spoilage. Spoilage of refrigerated milk and milk products frequently has been caused by inoculation with waterborne organisms (primarily psychrotrophs).[20] This contamination has occurred either directly, through product contact with the water itself, or indirectly, by microbes metabolizing nutrient residues on incompletely cleaned equipment surfaces. Since the microorganisms of water usually are destroyed by chlorination or heat, many plants treat all their water to keep spoilage at a minimum.

The procedures for sampling and for testing water supplies to determine conformance with coliform concentration by both standard, most probable number (MPN), and membrane filter methods, are completely detailed in the latest edition of Standard Methods for the Examination of Water and Wastewater.[2] Although limitations for total bacterial counts are not included in drinking water specifications, information on the numbers of organisms developing at various incubation temperatures is helpful both in classifying untreated waters and in assaying the efficiency of treatment procedures. The membrane filter technic is particularly useful in determining the levels of organisms in supplies containing such low numbers that the usual plating procedures do not yield meaningful results.

3.4 SCREENING TESTS FOR AIR

Bacteria, molds, and yeasts become airborne from numerous sources. The two most important sources of such contamination are (1) people, especially by shedding, and (2) various dusts. Other sources are drains, insects, rodents, and products. The specific food product and its desired microbiological quality dictate the quality of the air supply required. Sterile air in laminar flow, within practical limits, is recommended for the reduction of airborne contamination in aseptic packaging areas for sterilized products and products expected to have a long shelf life. Microbial contaminants in air exist as aerosol particles which vary in size from one to 50 microns. These particles may consist of a single unattached organism, or clumps composed of a number of organisms adhering to a dust particle, or free-floating and surrounded by a film of dried organic or inorganic material. Bacterial and fungal spores are generally more numerous as air contaminants than are vegetative cells. Vegetative bacterial cells are more important than spores since these types include the primary etiologic agents of communicable diseases. However, many vegetative cells ordinarily do not survive long in the air unless the relative humidity, temperature, and other factors are favorable, and unless the organism is surrounded by a protective cover. The objectives of an air sampling program should be defined before initiation of the sampling. For example, is information required on (1) the number of viable organisms suspended in a sample of air, (2) the number of particle bearing viable organisms, (3) the size distribution of the particles, or (4) information relating all of these factors? Viable bacteria and fungi in the air may be determined by several methods. These include impingement in liquids, impaction on solid surfaces, filtration, sedimentation, centrifugation, electrostatic precipitation, and thermal precipitation. The methods most commonly employed are sedimentation, impaction onto agars, and liquid impingement.

The results of sedimentation sampling are influenced by particle size, and by the velocity and direction of air currents in relation to the agar plates. Therefore, microorganisms that do not settle onto agar from the air are not included in the count obtained.

In the impaction methods, organisms from a specific volume of air are deposited on agar and on coated or dry surfaces. Under satisfactory conditions, the agar method is relatively efficient, but it has limitations because too high a count causes excessive colony growth and counting difficulties. The air temperature must be above freezing. Too long an exposure to air flow during sampling causes dehydration of the agar surface, which reduces adhesion of particles and subsequent colony growth. However, this problem may be overcome by the "O.E.D." method described by May.[13] Oxyethylenedocosyl ether plus a spreading agent is flooded onto the dried agar plate surfaces; the excess is poured off leaving a protective film which prevents dehydration without interfering with the collection of viable microbes, thus prolonging potential sampling time. Coated and dry surfaces may not retain all the organisms.

Liquid impingement consists of depositing organisms from the air into a liquid, such as buffered distilled water or nutrient broth. The method is not suited to small numbers of airborne organisms. Not all the organisms are retained, and some may be destroyed by high impingement velocity. With extended sampling, air movement tends to reduce the volume of liquid.

The viability and number of microorganisms in the air may be influenced by temperature, humidity and light rays. Air currents, the number of microorganisms on various surfaces, ambient air conditions surrounding the area, the number or activity of persons in the area, and sanitary conditions also affect the viable counts. These factors should be recorded during the period in which the air is sampled.

Prepare agar plates using the equipment, supplies, and procedures described for the Aerobic Plate Count, for psychrotrophic counts, or for yeast and mold counts. Pour 15 ml portions of sterilized tempered agar into sterile Petri plates avoiding contamination. Replace the cover and allow the agar to solidify on a level surface. Hold plates at 2 to 7 C to prevent dehydration of the agar. These plates are used for the sedimentation method of estimating microorganisms in the air.

When using the sedimentation method, place a new paper towel or parchment paper at the location where the air is to be sampled and set the Petri plate horizontally on a flat surface. Expose the agar medium by removing the plate cover and, without inversion, place it on the paper beside the bottom of the Petri plate. After exposure of the agar or other medium for 15 minutes to 2 hours, replace the cover and incubate. Care must be taken that the agar plate does not become dehydrated with extended exposure time. This can be controlled partially by adding more medium than usual to the plates, and using the "O.E.D." method.[13]

Following incubation, colonies are counted, and results expressed as the number per square foot per minute, since a standard Petri plate bottom area is approximately $^1/_{15}$ square feet. Plate count agar and yeast and mold agar (Chapter 2) generally are used. Other selective media are used occasionally, but not all species for which the selective agar was designed will grow on it after their recovery from the air. For example, some aerosolized *Escherichia coli* cells will not grow on desoxycholate agar, but will produce colonies on plate count agar.

Several types of commercial impaction samplers are available. Two common designs are the single-stage slit and the multi-stage sieve sampler. The single-stage slit samplers have the advantage of being more versatile, while the multi-stage sieve samplers permit the determination of various particle sizes in the air. In order to obtain optimum results with any air sampler, attention must be given to correct air flow rate, proper positioning of the agar plate in the sampler, the volume of air sampled, and the concentration of viable microorganisms. Particular attention must be given to the Petri plate and volume of agar required for the proper operation of each sampler.

One of the most common types of single-stage slit is the Fort Detrick Slit Sampler (Reynier and Sons, Chicago, Illinois). The sampler is cylindrical with a slit tube, agar plate, and interval timer. The slit tube is threaded into the top of the sampler and is easily adjusted to the proper level (2 to 3 mm) above the agar surface as determined by a height indicator. The slit width is set at 0.15 mm by an adjustable plate on the bottom of the slit tube. The agar containing plate is rotated by means of a clock mechanism. The rate of rotation depends on the timer (30 minutes to 12 hours). Air is drawn through the slit, impacting the suspended particles on the rotating agar surface. After sampling, the plate is incubated as required, and the colonies counted. The optimum flow rate for this sampler is 28.3 liters/minute. The flow rate may be varied by adjustment of the slit width and vacuum supply.

The Casella type single-stage slit sampler is also commonly used (C.A. Brinkman and Co., Long Island, N.Y. and Eliott Engineering Co., Decatur, Ga.). This type of sampler consists of an airtight chamber enclosing a rotating platform holding an agar plate. Air enters through a slit (0.33 mm wide) at the top of the chamber. Particles in the air stream are impacted onto the agar directly below the slit opening. The air leaves the chamber through an opening in the wall or base. The distance between the slit and agar surface is controlled by raising or lowering the rotating platform with an arm and screw adjustment. The rotating platform can be operated at three different rotation speeds (0.5, 2, and 5 rpm) by an electrically driven mechanism. A small opening at the bottom of the airtight chamber connected to a manometer measures the air flow through the slit in the Casella sampler.

The most common multi-stage sieve sampler used is the Andersen sampler (2000 Inc., 5899 South State Street, Salt Lake City, Utah 84107). This type of sampler is a cascaded arrangement of sieve type samplers having smaller diameter holes in each succeeding plate. The six-stage sampler consists of sieves with holes of the following diameters: 1.81 mm, 0.91 mm, 0.71 mm, 0.53 mm, 0.343 mm and 0.25 mm. The distance of the agar collecting surface from the holes of each sieve is controlled by utilizing a special Petri dish containing 27 ml of medium. Air is drawn successively through each of the sieves at increasing velocity so that larger particles are impacted in the first stages and the smaller particles, depending on their sizes and inertia, are impacted on the last stages. After sampling the plates are removed and incubated. The optimum flow rate is 28.3 liters/minute.

Liquid impingers may be used to determine microbial air contamination. Several commercial liquid impingers are available. The basic technic is common to almost all types. The impinger is connected to the vacuum pump by means of adequate tubing. Adjust the airflow rate through the impinger as specified for the unit. Place buffered distilled water with antifoam into the impinger and sterilize. Place the sampling probe in the proper location for procurement of the air sample. Connect the impinger to the suction tube and conduct sampling for 30 minutes, or for a specific time in accordance with the expected number of microorganisms in the air. Prepare dilutions, plate and incubate in accordance with enumeration procedures. Count the colonies and record as the number per cubic foot of air.

It is important that the manufacturer's directions be followed for each sampler, and that the limitations of each be understood. The methods given above for sampling of air supplies is by no means comprehensive. The laboratory worker is directed to the reference Public Health Monograph 60 for a detailed discussion of this subject.[17]

3.5 REFERENCES

1. Standard Methods for the Examination of Dairy Products. 1972. 13th Ed. American Public Health Association. Washington, D.C.

2. Standard Methods for the Examination of Water and Wastewater, 13th ed. 1971. American Public Health Association—American Water Works Association—Water Pollution Control Federation. Washington, D.C.
3. ANONYMOUS. 1969. Standard methods for measuring and counting particulate contamination of surfaces. In Book of A.S.T.M. Standards, Part 8. American Society for Testing and Materials. Philadelphia, Pa. pp 511–514.
4. ANGELOTTI, R., J. L. WILSON, W. LITSKY, and W. G. WALTER, 1964. Comparative evaluation of the cotton swab and RODAC methods for the recovery of *Bacillus subtilis* spore contamination from stainless steel surfaces. Health Lab Sci **1**:289–296
5. BATCHELOR, H. W. 1960. Aerosol samplers. Adv. Applied Microbiol **2**:31–64
6. BUCHBINDER, L., T. C. BUCK, JR., P. M. PHELPS, R. V. STONE, and W. D. TIEDMAN. 1947. Investigations of the swab rinse technic for examining eating and drinking utensils. Amer J Pub Health **37**:373–378
7. BUCK, T. C., JR. and E. KAPLAN. A sterile cutting device for swab vial outfits utilizing wood applicators. J. Milk Technol **7**:141–142
8. DIMMICK, R. L., and A. B. AKERS. 1969. An Introduction to Experimental Aerobiology. John Wiley & Sons, Inc., New York, N.Y.
9. ENGLEY, F. B. and B. P. DEY. 1970. A universal neutralizing medium for antimicrobial chemicals. Chemical Specialities Manufacturer's Association, Proceedings of the 56th Meeting
10. GREENE, V. W., D. VESLEY, R. G.BOND, and G. S. MICHAELSEN. 1962. Microbiological contamination of hospital air. I. Quantitative studies. Applied Microbiol. **10**:561–566
11. HALL, L. B. and M. J. HARTNETT. 1964. Measurement of bacterial contamination of surfaces in hospitals. Pub Health Rep **79**:1021–1024
12. HARPER, W. J. 1969. Growth of psychrophilic bacteria in commercial fluid milk products. Proc 29th Ann Meeting, Institute of Food Technologists. Chicago, Ill.
13. MAY, K. R. 1969. Prolongation of microbiological air sampling by a monolyer on agar gel. Applied Microbiol **18**:513
14. SILLIKER, J. H. and D. A. GABIS. 1975. A cellulose sponge sampling technic for surfaces. J Milk Food Technol. **38**:504
15. SING, E. L., P. R. ELLIKER, L. J. CHRISTENSEN, and W. E. SANDINE. 1967. Effective testing procedures for evaluating plant sanitation. J Milk Food Technol **30**:103–111
16. TIEDMAN, W. D., Chairman. 1948. Technic for the bacteriological examination of food utensils. Committee report. Amer J Pub Health Yearbook 1947–48. Part 2, pp 68–70
17. US Department of Health, Education & Welfare. 1959. Sampling Microbiological Aerosols. Publ Health Monogr 60, PHS Pub No. 686. US Govt Ptg Off, Washington, D.C.
18. WALTER, W. G. and J. POTTER. 1963. Bacteriological field studies on eating utensils and flat surfaces. J. Environ Health **26**:187–197
19. WEBER G. R. and L. A. BLACK. 1948. Inhibitors for neutralizing the germicidal action of quaternary ammonium compounds. Soap Sanitary Chemicals **24**:137–142
20. WITTER, L. D. 1961. Psychrophilic bacteria—A review. J. Dairy Sci **44**:983–1015

GENERAL ENUMERATION PROCEDURES

CHAPTER 4

AEROBIC PLATE COUNT

S. E. Gilliland, F. F. Busta, James J. Brinda, and Jeptha E. Campbell

4.1 INTRODUCTION

The development of agar media in the late 1800s opened the way for the development of methods for enumerating microorganisms by colony counts. These methods have been the most used procedures for determining populations of viable microorganisms. Enumeration procedures are based on the assumption that the microbial cells present in a sample mixed with an agar medium each form visible, separated colonies. The Aerobic Plate Count (APC), however, does not necessarily measure the actual total number of viable microorganisms per gram of analyzed sample, since bacterial cells occur singly and in pairs, chains, clusters, or clumps. The counts obtained by these methods should not be reported as viable cell counts, but as colony counts per unit, or colony forming units per unit. When determining colony counts on a food product, it must be remembered that all types of microorganisms will not grow on a single agar medium incubated under one set of conditions. This is due to special requirements with regard to nutrients, oxygen, incubation temperature, damaged cells, or other factors. A more thorough representation of the total bacterial count may be obtained by plating the sample on more than one nonselective medium, and incubating under more than one condition, i.e., temperature, aeration, etc.

Colony count methodology can provide a useful tool for estimating microbial populations in foods. The optimum medium and conditions for determining the colony count may vary from one food to another. However, once the optimum procedure for a given food is determined, it can be very useful for routine microbial analysis of the food. Since minor variations in procedures can alter the results obtained with the colony count, the competency and accuracy of the analyst are very important.

4.2 TREATMENT OF SAMPLE (Chapter 1)

4.21 Examination of Sample (Chapter 1)

IS/A Committee Liaison: W. J. Hausler, Jr.

4.22 Other Tests on Same Sample

If additional tests are to be performed on the sample, first remove portions for microbiological analysis.

4.3 REAGENTS AND MEDIA

4.31 Plating Media

Nonselective media such as standard methods agar (plate count agar) or tryptone glucose yeast extract agar (Chapter 2) should be used for determining the Aerobic Plate Count. Examples of other nonselective media which may prove useful for some food products are: trypticase soy agar, APT agar, and MRS agar (Chapter 2).

4.32 Diluents

Phosphate buffered dilution water prepared according to Butterfield[9] is a suitable diluent for most foods. One-tenth percent peptone dilution water is also a suitable diluent. The peptone water provides more protection for bacteria.[15,19,20] Distilled water should be used for preparing diluents, and should be of the quality described in Chapter 2.

Fill dilution bottles with phosphate buffered water or peptone water, so that after sterilization each will contain 99 ml ± 2 ml, or other desired amount. Autoclave the bottles at 121 C for 15 minutes. When more or less than 99 ml amounts are required in bottles or tubes, apply proportionally smaller deviation tolerances for the amounts in the dilution blanks. Optionally use correctly calibrated automatic water measuring devices. When bulk sterilized diluent is used, measure water directly into sterile dilution bottles or other dilution container and use promptly.

4.4 PRECAUTIONS AND LIMITATIONS OF METHODS

The Aerobic Plate Count provides an estimate of the number of viable microorganisms in food according to the medium employed and the time and temperature of incubation. Microbial cells often occur as clumps or groups in foods. While shaking of samples and dilutions tends to uniformly distribute the clumps of bacteria, it may not completely disrupt them. Mixing the initial dilution in a mechanical blender may provide better breakage of the clumps. However, this does not insure that the microorganisms will be distributed as single cells in the dilutions. Consequently, each colony that appears on the agar plates can arise from a clump of cells or from a single cell, and are often referred to as colony forming units (CFU).

The accuracy of the APC method also may be limited by the failure of some microorganisms to form visible colonies on the agar medium. This failure can result from nutritional deficiencies of the medium, unfavorable oxygen tension, unfavorable incubation temperature, or cell injury. Length of incubation time and temperature may also be a factor. The presence of inhibitory substances on glassware or in diluents may affect adversely some bacteria so that they will not form colonies. Other factors which may influence the accuracy of the colony count include: improper sterilization and protection of sterilized diluents, media, and equipment, inaccurate measurement of samples and dilutions, improper distribution of the sample in or on the agar medium, errors in counting colonies and in computing counts.

While there are some inherent limitations in enumerating microorganisms by the APC method, many of the errors can be minimized if the analyst follows directions carefully and exercises extreme care in making all measurements. Consistently accurate and meaningful results can be obtained from the routine examination of a given type of food only if each sample of that food is analyzed with the same methods or procedures. This includes sampling procedures, sample preparation, preparation of dilutions, plating medium, incubation conditions, and counting procedures.

4.5 PROCEDURE

4.51 Sample Preparation

All Petri plates should be labeled with the sample number, dilution, date, and any other desired information prior to preparing the dilutions.

Frozen samples should be thawed in the container in which they arrived at the laboratory. In most cases suitable thawing can be achieved by placing the samples in a refrigerator at 0 to 4.4 C for not more than 18 hours just prior to plating.[13] An alternate procedure for obtaining test portions of frozen foods (particularly from larger samples) is to use an electric drill combined with a funnel.[1] The sterile auger bit is inserted through a sterile plastic funnel (which has been cut off so the hole is just slightly larger than the bit) held against the frozen sample (Figure 1). The frozen shavings are conveyed to the surface and collected in the funnel. The shavings then can be placed in a sterile sample container. For large solid food samples (frozen or unfrozen), test portions should be taken aseptically from several areas using sterile knives and forceps, then mixed as a composite, so that a more representative test portion can be evaluated. If the sample is not too large and consistency permits, the best way to obtain a representative test portion is to macerate the entire sample by blending for 2 minutes in a sterile blender cup (caution: be sure sufficient sample is present to cover the blender blades; avoid excessive localized heating due to excessive blending times).

Liquid or semiliquid samples can be mixed by rapidly inverting the sample container 25 times. Where practical, shake 25 times in 7 seconds over a one foot arc. The interval between mixing and removing the test portion should not exceed 3 minutes.[15]

Prior to opening any sample container, the exterior immediately surrounding the area from which the sample is to be removed should be cleaned to remove any material that might contaminate the sample. The area may be swabbed with 70% ethanol to further prevent inadvertent contamination.

4.52 Diluting Samples

FIGURE 1. Funnel Collection Apparatus (Adams and Busta. 1970. Appl Microbiol. 19:878)

4.521 Selecting dilutions

For the most accurate colony count, the dilution(s) should be selected so that the total number of colonies in a plate will be between 30 and 300.[8] If the count is expected to be in the range of 3.0×10^3 to 3.0×10^5 per ml or g, prepare plates containing 1 : 100 and 1 : 1000 dilutions. Figure 2 shows a schematic drawing of examples for preparing dilutions.

4.522 Measuring portions of sample for initial dilution

a. Liquid products

Test portions of nonviscous liquid products (i.e., viscosity not greater than milk) may be measured volumetrically using a sterile pipet. Do not insert the pipet more than 2.5 cm below the surface of the sample. The pipet

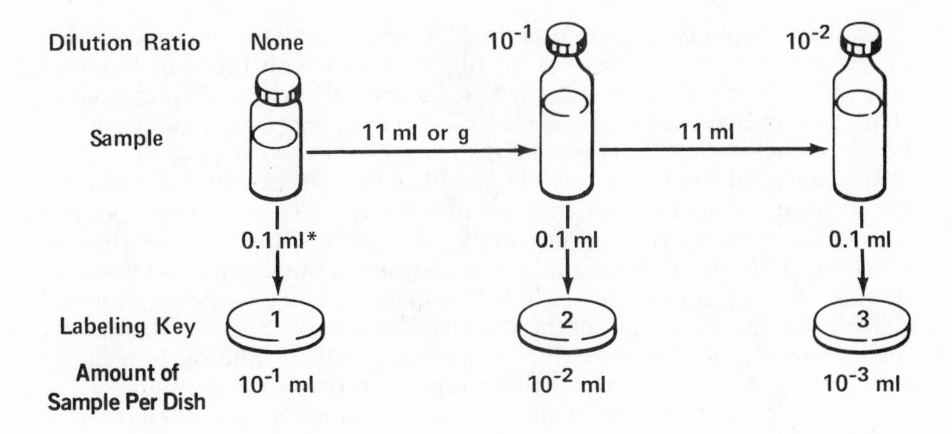

* If a nonviscous liquid (4.522)

FIGURE 2. Preparation of Dilutions from a Nonviscous Liquid Food Sample

should be emptied into the diluent (phosphate buffered distilled water or 0.1% peptone water) by letting the column drain from the graduation mark to the rest point of the liquid in the tip of the pipet within 2 to 4 seconds. Promptly and gently blow out the last drop.[15] Do not rinse the pipet in the dilution water. For viscous liquid products, the test portion for the initial dilution should be aseptically weighed (11 ± 0.1 g) into a sterile 99 ml dilution blank (or 10 ± 0.1 g into 90 ml or 50 ± 0.1 g into 450 ml). This provides a 1 : 10 dilution.

The initial dilution, usually 1 : 100 or 1 : 10, should be mixed by shaking the container, making 25 up-and-down (or back-and-forth) movements of about 1 foot in 7 seconds.[15] Optionally, a mechanical shaker may be used to shake the dilution blanks for 15 seconds.[2]

b. Solid or semisolid foods

Fifty g (± 0.1 g) of a representative test portion of solid or semisolid food should be weighed aseptically (using sterile forceps or spatulas) into a tared sterile blender cup.[13, 15] Add 450 ml of sterile diluent to the blender cup. Blending for 2 minutes at low speed (approximately 8000 rpm) should be sufficient to disperse adequately the material.[13] The blending time required may vary depending on the type of food.[15] All subsequent dilutions from the initial dilution should be prepared within 15 minutes.[13]

Optionally, if the entire food sample is less than 50 g, weigh to the nearest 0.1 g a portion approximating one-half of the sample into a sterile tared blender cup. Add sufficient sterile diluent to make a 1 : 10 dilution (i.e., add an amount of diluent equal to 9 times the weight of the test portion in the blender cup). The total volume in the blender cup must cover the blades completely. Proceed as in the previous paragraph.

Caution should be exercised in the blending step to avoid or prevent excessive heating. The amount of heating may vary with foods of different consistencies and may be expected to increase if blending times greater than 2 minutes are required. Chilled diluent (i.e., tempered in an ice-water bath) may be employed to decrease the chances of excessive heating.

In some solid food products the microbial flora is restricted primarily to the surface area. More accurate enumerations of these microorganisms may be obtained by rinsing the sample with sterile diluent rather than by blending. This can be accomplished by placing the sample in a suitable sterile container (plastic bag or sealable bottle) and adding a volume of sterile diluent equal to the weight of the sample. The container is then shaken in a manner similar to that used for preparing an initial dilution of a liquid food sample. Each ml of "rinse" thus prepared represents 1 g of sample.

Use sterile pipets for initial and subsequent transfers from each container. If the pipet does become contaminated before completing the transfers, replace it with a sterile pipet. Use a separate sterile pipet for transfers from each different dilution. Dilution blanks should be adjusted to 15 to 25 C.

Caution. Do not prepare or dispense dilutions, or pour plates in direct sunlight. When removing sterile pipets from the container, do not drag tips over the exposed exteriors of the pipets remaining in the case because exposed ends of such pipets are subject to handling and other contamination. Do not wipe or drag the pipet across the lips and necks of vials or dilution bottles. Do not insert pipets more than 2.5 cm below the surface of the sample or dilution. Draw test portions above the pipet graduation, then raise the pipet tip above the liquid level, and adjust to the mark by allowing the lower side of the pipet tip to contact the inside of the containers (Figure 3) in such a manner that drainage is complete and excess liquid does not adhere[11] when pipets are removed from sample or dilution bottles. Ordinarily, do not subject sterile pipets to flaming.

When measuring diluted samples of a food, hold the pipet at an angle of about 45 degrees with the tip touching the inside bottom of the petri dish or the inside neck of the dilution bottle. Lift the cover of the Petri dish just high enough to insert the pipet. Allow 2 to 4 seconds for the diluted food to drain from the 1 ml graduation mark to the rest point in the tip of the pipet; then, holding the pipet in a vertical position, touch the tip once against a dry spot on the plate. Do not blow out. When 0.1 ml quantities are measured, hold the pipet as directed and let the diluted sample drain from the 1.1 ml graduation point down to the 1.0 ml mark. Do not retouch the pipet to the plate.

After depositing test portions in each series of plates, pour the medium. Duplicate plates should be prepared for each dilution plated.

4.53 Plating

4.531 Melting medium

Melt the required amount of medium quickly in boiling water or by exposure to flowing steam in a partially closed container, but avoid prolonged exposure to unnecessarily high temperatures during and after melting. If the medium is melted in two or more batches, use all of each batch in order of melting, provided that the contents in separate containers remain fully melted. Melted standard methods agar that contains precipitate must be discarded. Do not resterilize the plating medium.

Cool the melted medium promptly to approximately 45 C and hold in a water bath between 44 C and 46 C until used. Set a thermometer into water or medium in a separate container similar to that used for medium; this temperature control medium must have been exposed to the same heating and cooling as the medium. Do not depend upon the sense of touch to indicate the proper temperature of the medium when pouring agar.

4.532 Pouring and mixing agar

Select the number of samples to be plated in any one series so that not more than 20 minutes (and preferably 10 minutes) elapse between diluting the first sample and pouring the last plate in the series.[7, 18] Should a continuous plating operation be conducted by a team, plan the work so that the time between the initial measurement of a test portion into the diluent or

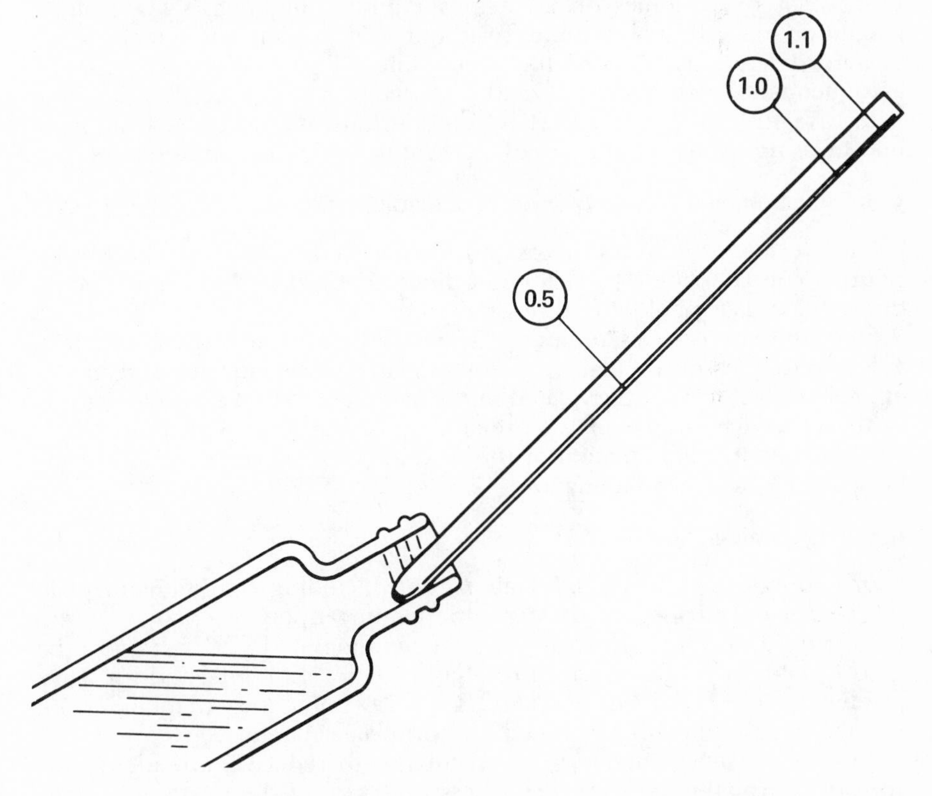

FIGURE 3. Volumetric Measurement of Sample or Diluted Sample
NOTE: Lower Side of the Pipet Touches the Inside Surface of the Container

directly into a dish and the pouring of the last plate for that sample is not more than 20 minutes. Introduce 10 to 12 ml of liquified medium at 44 to 46 C into each plate by gently lifting the cover of the Petri dish just high enough to pour the medium. Carefully avoid spilling the medium on the outside of the container or on the inside of the plate lid when pouring. As each plate is poured, thoroughly mix the medium with the test portions in the Petri dish, taking care not to splash the mixture over the edge, by rotating the dish first in one direction and then in the opposite direction, by tilting and rotating the dish, or by using mechanical rotators. Having thus spread the mixture evenly over the bottom of the plate, allow it to solidify (usual time, 10 minutes) on a level surface. After solidification, invert the plates to prevent spreaders, and promptly place them in the incubator. Note: To obtain countable plates for foods having low plate counts, low dilutions must be used. For some foods this results in the presence of a considerable amount of food particles in the plate which makes it difficult to distinguish the colonies easily for accurate counting. This problem often can be overcome by adding 1 ml of 0.5% (w/v) 2, 3, 5, triphenyltetrazolium chloride (TTC) per 100 ml of melted agar medium just prior to pouring the plates. Most bacteria form red colonies on an agar medium containing TTC. Counts should be made initially with and without TTC to determine if the TTC has any deleterious effect on the count. The TTC should be prepared as an aqueous solution and sterilized by passage through a sterile Seitz filter or equivalent sterilizing filter. The solution must be protected from light and must not be exposed to excessive heat to avoid decomposition.

4.54 Recommended Sterility Controls of Medium, Dilutions and Equipment

The sterility of dilution waters and plating medium may be checked by pouring control plates for each lot of dilution blanks and medium, and, if desired, for each lot of Petri dishes and pipets.

Since the number of samples in a series may vary considerably, prepare at least one agar control plate for samples plated in the morning and another agar control for samples plated in the afternoon. If several colonies appear on the agar control plates, prepare additional control plates to determine potential contamination of the medium, water blanks, plates, or pipets, as well as air contamination.

4.55 Incubation

Procedures for the Aerobic Plate Count, including conditions of plate incubation for various foods, differ depending upon the nature of the food and the type of microbial flora to be enumerated.[6, 15, 23, 27, 31] In the AOAC method,[5] plates are incubated at 35 C for 48 hours, and for dairy products[3], plates are incubated at 32 C for 48 ± 3 hours. Adaptations for specific commodities are presented in appropriate chapters.

Avoid excessive humidity in the incubator to reduce the tendency for spreader formation, but prevent excessive drying of the medium by controlling ventilation and air circulation. Agar in plates should not lose weight by more than 15% during 48 hours of incubation.

4.56 Counting Colonies on Plates and Recording

4.561 Selecting and counting colonies

Count colonies with the aid of magnification under uniform and properly controlled artificial illumination. (Use of a hand tally is optional.) Routinely use a colony counter[4, 26] equipped with a guide plate ruled in square centimeters. Plates should be examined in subdued light. Avoid mistaking particles of undissolved medium, sample or precipitated matter in plates for pinpoint colonies. Examine doubtful objects carefully, using higher magnification, where required, to distinguish colonies from foreign matter. A stereo microscope may be useful for this examination. Schedules of the laboratory analyst should be arranged to prevent eye fatigue and the inaccuracies that inevitably result from eyestrain.

Count all colonies on selected plates promptly after the incubation period. If impossible to count at once, the plates may be stored, after the required incubation, at approximately 0 to 4.4 C for a period of no more than 24 hours. This should be avoided as routine practice. When counting colonies on plates from individual samples, observe the directions given for spreaders as well as those listed below. Also record on the report for each lot of samples the results of sterility tests on control materials (dilution blanks, agar, etc.) used when pouring plates.

4.562 One plate with 30 to 300 colonies

Select a plate with 30 to 300 colonies unless excluded by 4.569. Count all colonies on the selected plate, including those of pinpoint size, record the dilution used, and report the total colonies as a basis for the Aerobic Plate Count. See Table 1, Sample Nos. 1001 and 1004.

4.563 Duplicate plates

Count plates with 30 to 300 colonies and average the counts to obtain the Aerobic Plate Count (Table 1, Sample No. 1011). If more than one plate of a given dilution is prepared, but only one plate yields 30 to 300 colonies, count both plates unless excluded by 4.569a or 4.569b. Include these counts in the arithmetic average, and compute the count per milliliter by multiplying the average number of colonies by the dilution used. See Table 1, Sample Nos. 1012 and 1013. When duplicate plates from consecutive decimal dilutions are counted, compute the count per milliliter for each dilution and proceed as in 4.564 and 4.567. See Table 1, Sample Nos. 1013, 1014 and 1015.

4.564 Consecutive dilutions (30 to 300 colonies)

If plates from two consecutive decimal dilutions yield 30 to 300 colonies each, compute the count per milliliter for each dilution by multiplying the number of colonies per plate by the dilution used, and report the arithmetic average as the Aerobic Plate Count per milliliter, unless the higher computed count is more than twice the lower one, in which case report the lower computed count as the Aerobic Plate Count per milliliter or per gram, as applicable. See Table 1, Sample Nos. 1002, 1003, 1014 and 1015.

Table 1: Examples for Computing Aerobic Plate Counts per Milliliter or per Gram

Sample No.	Colonies/Dilution 1:100	1:1000	Count Ratio[a]	Aerobic Plate Count	Rule Applicable under 56[b]
Common application where 2 plates, one from each of 2 decimal dilutions, are poured					
1001	234	28		23,000	4.562
1002	293	41	1.4	35,000	4.564
1003	140	32	2.3	14,000	4.564
1004	Spr[c]	31	—	31,000	4.562
1005	0	0	—	< 100 Est	4.567
1006	TNTC	7150	—	> 6,500,000 Est	4.568
1007	18	2	—	1800 Est	4.566
1008	Spr[c]	Spr[c]	—	Spr[c]	4.569c
1009	325	25	—	33,000 Est	4.565
1010	32	265	—	LA	4.569c
Procedure where two or more plates per dilution are poured					
1011	175	16	—	19,000	4.563
	208	17	—		
1012	322	23	—	30,000	4.563
	278	29	—		
1013	281	40	—	33,000	4.563
	378	24	—		
1014	138	42	2.4	15,000	4.563
	162	30	—		
1015	274	35	1.4	30,000	4.563
	230	Spr[c]	—		

[a]Count ratio is the ratio of the greater to the lesser plate count, as applied to plates from consecutive dilutions having between 30 and 300 colonies.

[b]All counts should be made in accordance with Section 4.56 as well as any other rules listed or given in the text.

[c]Spreader and adjoining area of repressed growth covering more than one-half of plate.

4.565 No plate with 30 to 300 colonies

If there is no plate with 30 to 300 colonies and one or more plates have more than 300 colonies, use plate(s) having a count nearest 300 colonies and count plate(s) as in 4.564 and 4.568. Report as the Estimated Aerobic Plate Count. See Table 1, Sample Nos. 1006 and 1009.

4.566 All plates with fewer than 30 colonies

If plates from all dilutions yield fewer than 30 colonies each, record the actual number of colonies on the lowest dilution (unless excluded by 4.569a or 4.569c) and report the count as the Estimated Aerobic Plate Count per milliliter or per gram, as applicable. See Table 1, Sample No. 1007.

4.567 Plates with no colonies

If plates from all dilutions of any sample have no colonies and inhibitory substances have not been detected, report the count as less than (<) one times the corresponding lowest dilution. For example, if no colonies appear on the 1 : 100 dilution, report the count as "less than 100 (< 100) Estimated Aerobic Plate Count" per milliliter or per gram, as applicable. See Table 1, Sample No. 1005.

4.568 Crowded plates (more than 300 colonies)

If the number of colonies per plate exceeds 300, count the colonies in those portions of the plate that are representative of colony distribution, and calculate the Estimated Aerobic Plate Count from these data. If there are fewer than 10 colonies per square centimeter, count the colonies in 13 squares, selecting, if representative, 7 consecutive squares horizontally across the plate and 6 consecutive squares at right angles, being careful not to count a square more than once. (The sum of the colonies in 13 representative square centimeters multiplied by 5 yields the estimated colonies per plate when the area of the plate is 65 cm^2).

When there are more than 10 colonies per square centimeter, count the colonies in 4 such representative squares. Multiply the average number found per square centimeter by the appropriate factor to determine the estimated colonies per plate. Because the average inside diameter of the bottom of a standard glass Petri dish is 91 mm, normally multiply the average number of colonies per square centimeter by 65. If pressed glass dishes or single use plastic dishes with either larger or smaller diameters are used, the laboratory should determine the actual diameter and use a correction factor for multiplication. Generally, the area of a standard plastic petri dish is approximately 56 cm^2 and, therefore, the appropriate factor is 56.

Do not report counts on crowded plates from the highest dilution as "too numerous to count" (TNTC). Where bacterial counts on crowded plates are greater than 100 colonies per square centimeter, report as greater than (>) 6,500 times the highest dilution plated. See Table 1, Sample No. 1006. Report counts as "Estimated Aerobic Plate Count" per milliliter or per gram, as applicable.

When all colonies on a plate are accurately counted and the number exceeds 300, report as "Estimated Aerobic Plate Count" per milliliter or per gram, as applicable. See Table 1, Sample No. 1009.

4.569 Spreaders

a. Distribution

If spreaders occur on the plate(s) selected (see 4.569b below), count colonies on representative portions thereof (4.568) only when colonies are well distributed in spreader free areas, and the area covered by spreader(s) including the total repressed growth area, if any, does not exceed one-half of the plate area. Where the repressed growth area alone exceeds one-quarter of the total area, report the test as directed in 4.569c.

b. Counting spreading colonies

When the counting of spreading colonies cannot be avoided, count each of three distinct types as one source. The first type is a chain of colonies, not too distinctly separated, that appears to be caused by disintegration of a bacterial clump when the Petri dish is rotated to mix the agar with the test material. If only one such chain exists, count it as a single colony. If one or more chains appear to originate from separate sources, count each source as one colony. Do not count each individual growth in such chain(s) as separate colonies.

The second type of spreading colonies is that which develops in a film of water between the agar and the bottom of the dish. The third type is that which forms in a film of water at the edge or over the surface of the agar. These two latter types develop largely because of an accumulation of moisture at the point from which the spreader originates. When dilution water is distributed uniformly throughout the medium, bacteria rarely develop into spreading colonies. Any laboratory with 5% of plates more than one-fourth covered by spreaders should take immediate steps to eliminate this trouble. Note the conditions for reporting results as laboratory accidents when spreader(s) covers more than one-half of the plate (4.569c).

c. Spreader growth, laboratory accidents, or bacterial growth inhibitors.

If all plates prepared from the samples have excessive spreader growth (4.569b) or are known to be contaminated or are otherwise unsatisfactory, report as "Spreaders" (Spr) or "Laboratory Accident" (LA). See Table 1, Sample Nos. 1008 and 1010. Inhibitory substances in a sample may be responsible for the lack of colony formation. The analyst may be inclined to suspect the presence of inhibitory substances in the sample under examination when plates show no growth, or show proportionately less growth in lower dilutions; such developments cannot, however, always be interpreted as evidence of inhibition.

4.57 Computing and Recording Counts

To compute the Aerobic Plate Count, multiply the total number of colonies or the average number (if duplicate plates of the same dilution) per plate by the reciprocal of the dilution used. Record the dilutions used, and the number of colonies counted or estimated on each plate.

When colonies on duplicate plates and/or consecutive dilutions are counted and the results are averaged prior to recording, round off counts to two significant figures only at the time of conversion to the Aerobic Plate Count. See Table 1, Sample No. 1013.

Avoid creating fictitious ideas of precision and accuracy when computing Aerobic Plate Counts by recording only the first two left-hand digits. Raise the second digit to the next highest number only when the third digit from the left is 5, 6, 7, 8 or 9; use zeroes for each successive digit toward the right from the second digit. (See examples in Table 1.)

Record the results of sterility control tests on materials. Record the temperature of the incubator during the incubation of agar plates.

4.58 Reporting and Interpreting Counts

Report counts as the Aerobic Plate Count or the Estimated Aerobic Plate Count, per milliliter or per gram, as applicable.

The matter of interpreting microbial colony counts is beyond the scope of this manual. Local regulations should be consulted regarding standards or limits.

When using Aerobic Plate Counts to compare the microbial quality of several samples of food, it is important that counts be obtainable by the same procedure. Furthermore, extreme care must be exercised in following the procedures if the results are to be meaningful.

Inaccuracies in counting plates due to carelessness, impaired vision, or failure to recognize colonies, can lead to erroneous results. Laboratory workers who cannot duplicate their own counts on the same plate within 5%, and the counts of other analysts within 10%, should discover the cause(s) and correct such disagreements.[11, 12]

This method is adapted from the Aerobic Plate Count method as specified in the AOAC Official Methods of Analysis, 11th Edition, 1970, and the Standard Plate Count method as specified in the APHA Standard Methods for the Examination of Diary Products, 13th Edition, 1972. With the exception of the conditions specified for the incubation of plates in the various chapters, this method is in substantive agreement with the methods described in the AOAC and APHA publications.

4.6 SPECIAL METHODS

4.61 Surface or Spread Plate Method

Methods of plating designed to produce all surface colonies on agar plates have certain advantages over the pour plate method.[28] Use of translucent media is not essential with a surface or spread plate, but is necessary with the pour plate to facilitate location of colonies. The colonial morphology can be observed better for surface colonies, improving the ability of the analyst to distinguish between different colony types. Organisms are not exposed to the heat of the melted agar medium. Such exposure may result in lower counts in some cases.[24, 28, 30]

In some studies, the use of a spread plate method resulted in higher counts than were observed with normal pour plate methods.[10, 22, 24] On the other hand, since relatively small volumes (0.1 to 0.5 ml) of sample must be used, the method may lack accuracy for samples containing few microorganisms.[28]

4.611 Equipment and supplies

In addition to the equipment and supplies needed for the pour plate method, glass spreaders (glass rods shaped like hockey sticks or rakes) are needed to spread the samples. Suitable "hockey sticks" can be made by fire polishing both ends of 20 cm long (3.5 mm diameter) glass rods and bending them at right angles approximately 3 cm from the end.[28] Care must be taken to assure that the 3 cm portion is kept straight.

4.612 Regents and media

The reagents and media are the same as for the pour plate method (4.3).

4.613 Procedure

a. Sample preparation (See section 4.51)

b. Prepoured agar plates

The desired nonselective agar should be melted and tempered as in section 4.531. Pour approximately 15 ml of the agar into sterile Petri plates. To facilitate uniform spreading, the surface of the agar should be dried by holding the plates at 50 C for 1.5 to 2 hours.[28] Plates may also be dried suitably at a lower temperature (25 to 35 C) for longer periods (18 to 24 hr).

c. Preparation of dilutions

Ten-fold serial dilutions should be prepared following general procedures in section 4.521, 4.522 and 4.523 except that 11 ml or g of sample or dilution is transferred to subsequent 99 ml dilution blanks (see Figure 4). Other suitable sample sizes and diluent amounts can be used to give 1:10 dilutions.

d. Measuring dilutions onto plates

Using a sterile pipet (graduated into 0.1 ml divisions) place 0.1 ml of the desired dilution onto the surface of the agar medium.

e. Spreading the dilution

Using a sterile hockey stick, spread the 0.1 ml sample as quickly and carefully as possible on the surface of the agar medium. A separate "hockey stick" should be used for each plate.[28] Allow the plates to dry at least 15 minutes.

f. Incubation (See section 4.55)

g. Counting colonies (See section 4.56)

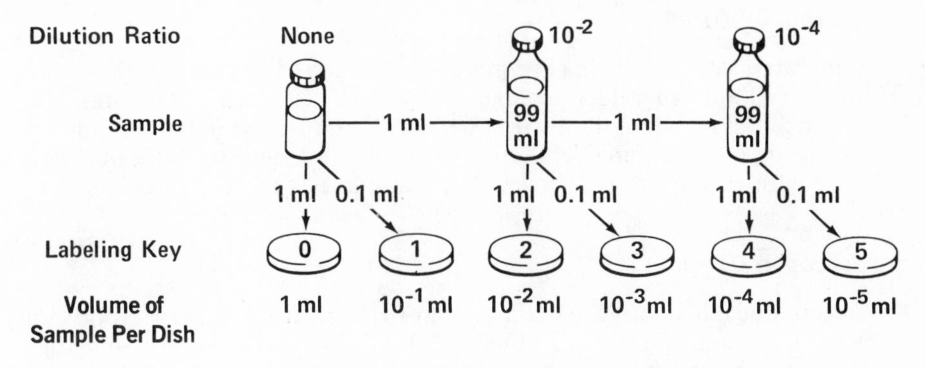

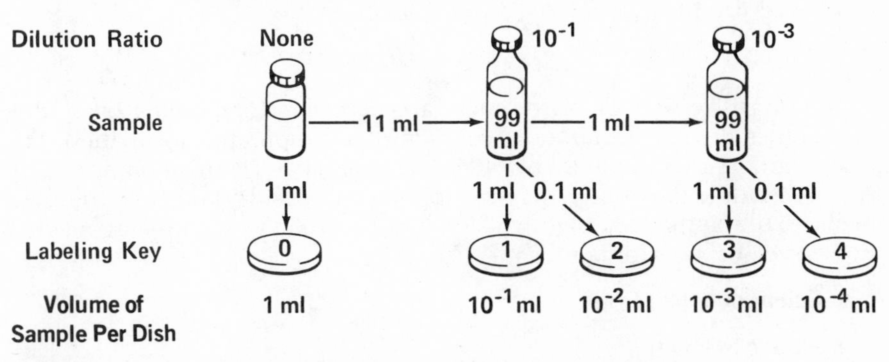

FIGURE 4. Preparation of Dilutions for Spread Plate Method

4.62 Oval Tube or Bottle Culture Method[16, 21]

The method described here appears to be useful only for nonviscous liquid foods (4.522) having counts greater than 3,000/ml or for viscous or solid foods (4.522) having counts greater than 30,000/g.

4.621 Equipment and supplies

a. Oval tube

Round neck modified form test tubes, plugged with cotton, or capped with stainless steel closures (*Bellco No. 2005-00020) and having bulb dimensions, of 100 × 17 × 27 mm, with ¾ inch neck (Bellco No. 2042-17027*) are recommended.

---

*Bellco Glass, Inc., 340 Edruda Road, Vineland, N.J. 08360

b. Water bath (Chapter 2)

c. Loop, 0.001 ml

Calibrated 0.001 ml loops are made of B & S #26 gauge platinum rhodium wire. They are welded in a true circle loop (I.D. 1.45 ± 0.06 mm) and attached to a 3 inch length of wire (Scientific Products N2075-2**) which is mounted in a suitable holder. Loops may be reshaped by being fitted over a No. 54 twist drill. (Loops of proper size should not fit over a No. 53 twist drill.)

d. Loop, 0.01 ml

Acceptable 0.01 loops are made similar to those in Standard Methods for the Examination of Dairy Products, 13th edition C.071 except that they are calibrated to contain 0.01 ml, and made of B & S #19 gauge platinum-rhodium wire, (ID 4 mm ± 0.03 mm) (Scientific Products N2075-1). (Loop of proper size fits over a No. 22 but not over a No. 21 twist drill.)

e. Racks

Noncorrosive wire or plastic racks are necessary for holding oval tubes and bottles during sterilization, cooling, inoculation, and incubation. For oval tubes the racks should hold 40 or 50 tubes (1 × 1¼ in. openings). The top shelf should have one-half inch supports on one side (long dimension) to facilitate tilting the rack slightly when it is placed on its side during incubation. (Size of spaces for bottles 1 × 1⅝ in.)

f. Colony counter (4.56)

g. Plating medium

See section 4.31. Dispense 4 ml of melted medium in each oval tube, or 3.5 ml in a 1 oz bottle. Plug each oval tube with cotton (or a stainless steel cap), or apply a screw cap to the bottle and sterilize.

4.622 Procedure

Samples should be prepared as in section 4.51. An initial dilution (1:10) should be prepared for viscous or solid foods (4.522b). Flame sterilize the 0.01 ml or 0.001 ml standard loop and allow to cool. Transfer a loopful of food (liquid) or dilution to an oval tube (or a bottle) containing sterile melted agar (tempered to 44 to 46 C), being careful to dip the standard loop only 2 to 3 mm below the surface of the sample or dilution in an area as free from foam as possible. Hold the plane of the loop in a vertical position when withdrawing it from the sample and move the loop back and forth through the agar to insure removal of all material from the loop.

Replace the closure, mix agar and diluted food thoroughly by swinging the tube back and forth rapidly through a small arc for a period of 5 seconds (about 25 complete excursions), and lay the oval tube flat on the table, or, preferably, slant it slightly. After the agar has solidified, incubate the oval

**Scientific Products, 1430 Waukegan Rd., McGaw Park, Ill. 60085

tubes horizontally in a wire rack (with the agar adhering to the upper side of the container), at the appropriate incubation time and temperature.

Count colonies by placing the tube or bottle on a colony counter. Record colony counts from tubes or bottles having 100 or fewer colonies, compute the number per milliliter or gram and report the results as "Oval Tube Colony Count per milliliter or gram;" or as "Bottle Culture Colony Count per milliliter or gram." If oval tube or bottle cultures yield fewer than 30 colonies each, record the actual number of colonies and report the count as "Estimated Oval Tube Colony Count per milliliter or gram," or as "Estimated Bottle Colony Count per milliliter or gram." If tubes or bottles from dilutions of any sample contain no colonies, and inhibitory substances have not been detected, report the count as less than one times the corresponding lowest dilution. If all tubes or bottles have more than 100 colonies, results should be reported as greater than 100 times the corresponding dilution.

When computing counts, remember that the amount of sample transferred by the 0.01 ml loop is equivalent to a 100 fold dilution (1000 fold dilution for the 0.001 ml loop).

4.63 Plate Loop Method[29]

The usefulness of the plate loop method is similar to that of the previous method (4.62) in that accurate data cannot be expected for nonviscous liquid foods (4.522b) having counts less than 3,000/ml nor for viscous and solid foods (4.522b) having counts less than 30,000/g.

4.631 Equipment and supplies

The total equipment assembly necessary for this method cannot be purchased as a unit, but can be made up from the following components:

a. Loop, 0.01 or 0.001 ml

(See section 4.621d & 4.621c). Make an approximate 30 degree bend about 3 to 4 mm from the loop, with the loop opening toward the hub. Kink the opposite end of the wire in several places.

b. Luer-Lok hypodermic needle

A 13 gauge needle should be sawed off 24 to 36 mm from the point where the barrel enters the hub. Insert the kinked end of the wire shank of the loop into the sawed off needle to a point where the bend is about 12 to 14 mm from the end of the needle.

c. Cornwall continuous pipetting outfit

A 2 ml capacity Becton, Dickinson & Co. No. 3052* pipetting assembly (consisting of a metal pipetting holder, a Cornwall Luer-Lok syringe, and a filling outfit) is suitable. Adjust to deliver 1.0 ml. The length of the rubber tubing should be long enough for adequate mobility.

---

*Becton, Dickinson & Co., Rutherford, N.J. 07070

The apparatus (Figure 5) and other parts may be sterilized by autoclaving at 121 C for 15 minutes, or sanitized by submerging the completely disassembled unit in boiling water for 10 minutes.

d. Dilution blanks (4.32)

e. Petri dishes (Chapter 2)

f. Media (4.31)

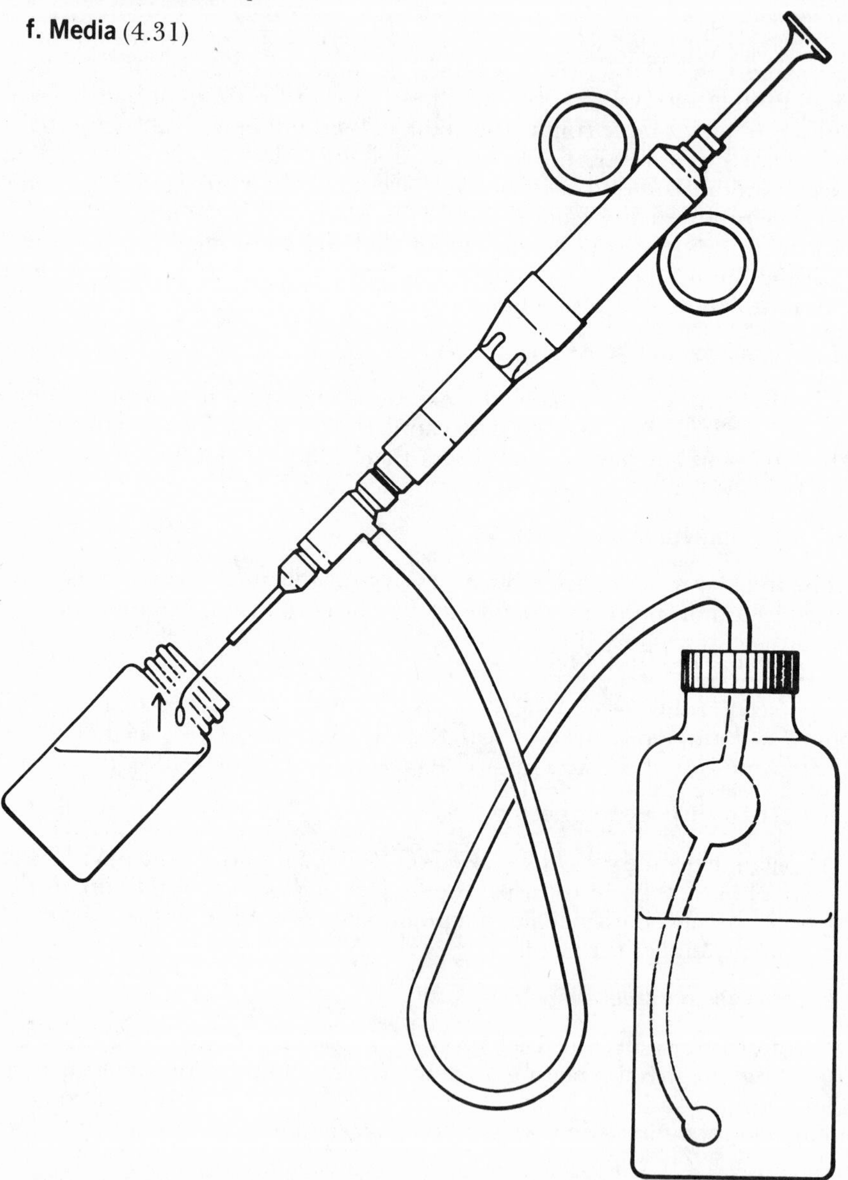

FIGURE 5. Plate Loop Apparatus, Showing Vertical Removal of Loop

4.632 Procedure

a. Assemble the sterilized measuring and transfer instrument. Aseptically place the end of the rubber supply tube attached to the syringe in a bottle of sterile diluent. Depress the syringe plunger rapidly several times to pump diluent into the glass syringe (which previously has been calibrated to deliver one ml with each depression of the plunger).

b. Prepare samples as in section 4.51. For viscous or solid foods an initial dilution (1:10) should be prepared (4.522b).

c. Before initial transfer is made in examining a series of samples, briefly flame the loop (preferably in a clean, high temperature gas flame) and allow to cool 15 seconds or more. Carefully dip the loop into the sample or dilution (avoiding foam) as far as the bend in the shank. (The bend serves as a graduation mark and also permits vertical removal of the loop.) To measure the sample, insert the loop vertically into the sample three times, moving the loop with a uniform up-and-down movement over a distance of about an inch. Avoid water droplets that rinse off the loop. Each downward movement should be at a rate of about 55 to 60 beats per minute. A metronome may be used to establish uniform timing. (Wide mouth sample bottles and good illumination facilitate the inoculation of plates.)

d. Raise the cover of a sterile Petri dish, insert loop, and depress the plunger, causing sterile dilution water to flow across the charged loop, and thus washing a measured volume (0.01 ml or 0.001 ml) of sample into the dish.

Caution: Do not depress the plunger so rapidly that water fails to follow the shank and flow across the loop.

e. Normally the residue remaining on the loop after discharging the sample is not significant. However, small imperfections in the welding of the loop or in smoothness of the metal surface may lead to incomplete rinsing. This possibility should be determined for each loop by running a series of control plates. If the loop is determined to be free-rinsing, no flaming between samples is necessary.

f. The speed of removal of the loop from the surface of the sample affects the accuracy of the measurement: Removing the loop slowly causes less than the desired amount to adhere; jerking the loop out rapidly causes more than the desired amount to adhere.

g. Pour plates with 12 to 15 ml of agar; incubate at the appropriate incubation time and temperature.

h. Computing and reporting of counts. Refer to section 4.56 for guidelines governing counting of plates and computing of counts. If a 0.01 ml loop was used, the amount of sample or dilution placed in the Petri dish was equivalent to a 100 fold dilution, i.e., if 0.01 ml of a 1 : 10 dilution of food was plated, the final dilution would be 1 : 1000. (The final dilution would be 1 : 10,000 if a 0.001 ml loop were used.) The reproducibility of these methods is often less than the level desired for enumeration. Report results as "Plate Loop Colony Count per milliliter or gram."

4.64 Drop Plate Method[25]

This method of enumerating microorganisms is similar in principle to the spread plate method (4.61) except that glass hockey sticks are not used to spread the diluted sample on the agar surface. The sample may be diluted as in sections 4.521 and 4.522. Measurement of the diluted samples onto the surface of prepoured plates (4.613) is done by adding a predetermined number of drops from a specially calibrated pipet.[25, 28] The drops are allowed to spread and dry over an area (usually 1.5 to 3 cm in diameter) of the agar surface. The plates are incubated at the desired temperature and time. The colonies are counted (4.56), and the computation of the colony count is based on the number of drops per plate, the number of drops per milliliter, and the dilution factor. The method is not recommended for food samples having counts of less than 3000 per gram. For details the reader is referred to the literature.[25, 28]

4.65 Membrane Filter Method

For certain foods or food ingredients the ability to test relatively large samples will improve the accuracy of quantitative microbiological analyses. Large volumes of liquid foods or solutions of dry foods which can be dissolved and passed through a bacteriological membrane filter (pore size 0.45 μm) may be analyzed for microbial content by the membrane filter method. The method is especially useful for samples which contain low numbers of bacteria.

4.651 Equipment and supplies

The filtration apparatus and supplies listed in this section are those manufactured by Millipore Corporation, Bedford, Mass. 01730. However, comparable equipment manufactured by other companies may be used.

a. Membrane filters

Hydrosol Analysis (Type HA 0.45 μm pores) filters having 47 mm diameter with a grid marked on the pad are recommended (Millipore number HAWG 047AG or HAWG 047 SO).

b. Nutrient pads

The Millipore filters (HAWG 047AO and HAWG 047SO) are packed in containers which also contain cellulosic pads (47 mm diameter). These pads may be used as nutrient pads.

c. Filter holders

Millipore Corporation manufacturers several types of filter holders made of Pyrex glass, stainless steel, or plastic, any of which may be employed.

d. Filtering flask

Two flasks (1 liter capacity) are needed: one to receive the filtrate and the other as an in-line trap.

e. Vacuum tubing

23 inch length with an internal diameter of 0.25 inch is needed.

f. Vacuum source

A small laboratory vacuum pump or water aspirator system is required for the filtration.

g. Forceps

Smooth tipped forceps are necessary for handling aseptically the membrane filters.

h. Petri plates

Specially designed plastic Petri plates are available from Millipore Corporation (PD 1004700).

i. Volumetric measuring devices

Ten ml borosilicate glass serological pipets are needed along with other desired pipets. For measuring larger samples, graduated cylinders are recommended.

j. Incubator (Chapter 2)

k. Stereoscopic microscope and illuminator

Low power magnification is necessary for counting colonies on membrane filters. This may be achieved with Millipore Corporation's stereoscopic dissecting microscope (XX7550500) equipped with an illuminator (XX7550510), or with comparable equipment.

4.652 Media

For total counts, nonselective broth (such as trypticase soy broth or brain heart infusion broth, Chapter 2) may be utilized. Selective and differential media are available for use in enumerating selected groups of bacteria. Diluents are described in section 4.32.

4.653 Procedure

a. Aseptically assemble the membrane filter apparatus (following manufacturer's directions) and connect to the vacuum system.

b. The sample or sample solution should be shaken vigorously to insure an even dispersal of the microorganisms.

c. The desired volume is introduced into the funnel with a sterile 10 ml pipet or graduated cylinder. (For sample volumes less than 10 ml, aseptically pour approximately 20 ml of sterile diluent in the funnel prior to adding the sample.) If a graduated cylinder is used, rinse the cylinder with approximately 50 ml of sterile diluent and add the rinse to the funnel.

d. Apply vacuum to the filtering apparatus and allow the liquid to pass completely through the filter into the flask. Do not turn off vacuum.

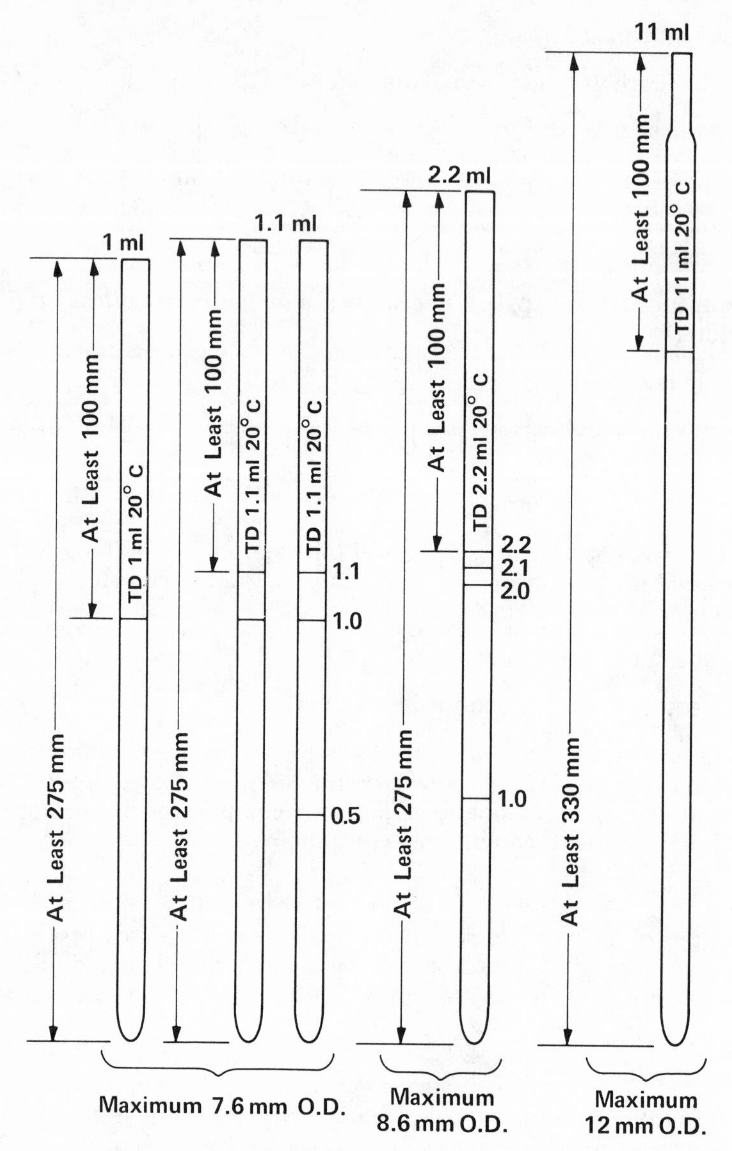

All pipets are calibrated to contain indicated amount of water 20° C.

Accuracy

1.0 and 1.1 ml	± 0.025 ml	From any graduation to tip
2.2 ml	± 0.040 ml	
11 ml	± 0.200 ml	

FIGURE 6. Bacteriological Transfer Pipets

e. Rinse inside of funnel with sterile diluent. (Use an amount at least equal to the volume of liquid just filtered.) After rinse has passed completely through the filter, turn off the vacuum.

f. Carefully and aseptically disassemble the part of the apparatus containing the filter. Using sterile smooth tipped forceps remove the filter and carefully place it (avoiding air bubbles) onto the surface of the chosen saturated nutrient pad or agar medium.

g. Incubation (4.55)

h. Counting colonies

Colonies should be counted with the aid of low power magnification. An acceptable range of colonies per filter is 20 to 200.

i. Computing counts

Membrane filter colony counts per milliliter or gram can be computed based on the amount of sample filtered.

4.66 Automated Plating Technique

Gilchrist *et al*[14] have described an automated device for determining the number of colony forming units in liquids. A known volume of sample is dispensed onto a rotating agar plate in an Archimedes spiral. The amount of sample decreases as the spiral moves out toward the edge of the plate. A modified counting grid, which relates the area of the plate to sample volume, is used to count the colonies on an appropriate area of the plate. With this information the colony count for the sample can be computed. Such a method offers an important advantage in the use of an electronic colony counter if the sensing device of the counter is designed to travel in the same Archimedes spiral as the automatic plating machine. The combination of automatic plating and electronic colony counting can provide means for more efficiently and accurately analyzing foods for microbial content.

4.67 Enumeration of Anaerobes

Methods for handling samples and incubation conditions for determining total anaerobic colony counts are outlined in Chapter 19 for anaerobic spore formers except that steps to eliminate vegetative cells are omitted. If the analyst is concerned with enumerating strict anaerobes, the methodology developed at the Anaerobe Laboratory,[17] Virginia Polytechnic Institute and State University, Blacksburg, Va. 24061, may be of help.

4.7 REFERENCES

1. Adams, D. M. and F. F. Busta. 1970. Simple method for collection of samples from a frozen food. Appl. Microbiol. **19**:878
2. Standard Methods for the Examination of Dairy Products, 1960. 11th ed., American Public Health Association. New York, N.Y.

3. Standard Methods for the Examination of Dairy Products, 1972. 13th ed., American Public Health Association. Washington, D. C.
4. ARCHAMBAULT, J., J. CUROT, and M. H. MCCRADY. 1937. The need of uniformity of conditions for counting plates (with suggestions for a standard colony counter). Amer. J. Pub. Hlth. **27**:809–812
5. Official Methods of Analysis, 1970. 11th ed., Association of Official Analytical Chemists. Washington, D. C.
6. BABEL, F. J., E. B. COLLINS, J. C. OLSON, I. I. PETERS, G. H. WATROUS, and M. L. SPECK. 1955. The standard plate count of milk as affected by the temperature of incubation. J. Dairy Sci. **38**:493–503
7. BERRY, J. M., DEBORAH A. MCNEILL, and L. D. WITTER. 1969. Effect of delays in pour plating on bacterial counts. J. Dairy Sci. **52**:1456–1457
8. BREED, R. S. and W. D. DOTTERRER. 1916. The Number of Colonies Allowable on Satisfactory Agar Plates. Tech. Bull. 53. New York Agri. Exper. Sta., Geneva, N.Y.
9. BUTTERFIELD, C. T. 1932. The selection of a dilution water for bacteriological examinations. J. Bacteriol. **23**:355–368
10. CLARK, D. S. 1967. Comparison of pour and surface plate methods for determination of bacterial counts. Can. J. Microbiol. **13**:1409–1412
11. COURTNEY, J. L. 1956. The relationship of average standard plate count ratios to employee proficiency in plating dairy products. J. Milk Food Technol. **19**:336–344
12. DONNELLY, C. B., E. K. HARRIS, L. A. BLACK, and K. H. LEWIS. 1960. Statistical analysis of standard plate counts of milk samples split with state laboratories. J. Milk Food Technol. **23**:315–319
13. Bacteriological Analytical Manual for Foods. 1972. 3rd ed., Food and Drug Administration. Washington, D.C.
14. GILCHRIST, J. E., J. E. CAMPBELL, C. B. DONNELLY, J. T. PEELER, and J. M. DELANEY. 1973. Spiral plate method for bacterial determination. Appl. Microbiol. **25**:244–252
15. HARTMAN, P. A. and D. V. HUNTSBERGER. 1961. Influence of subtle differences in plating procedure on bacterial counts of prepared frozen foods. Appl. Microbiol. **9**:32–38
16. HEINEMANN, B. and M. R. ROHR. 1953. A bottle agar method for bacterial estimates. J. Milk Food Technol. **16**:133–135
17. HOLDEMAN, L. V. and W. E. C. MOORE. Anaerobe Laboratory Manual, VPI Anacrobe Laboratory, Virginia Polytechnic Institute and State University. Blacksburg, Va.
18. HUHTANEN, C. N., A. R. BRAZIS, W. L. ARLEDGE, E. W. COOK, C. B. DONNELLY, R. E. GINN, J. J. JEZESKI, D. PUSCH, H. E. RANDOLPH, and E. L. SING. 1972. Effects of time of holding dilutions on counts of bacteria from raw milk. J. Milk Food Technol. **35**:126–130
19. JAYNE-WILLIAMS, D. J. 1963. Report of a discussion on the effect of the diluent on the recovery of bacteria. J. Appl. Bacteriol. **26**:398–404
20. KING, W. L. and A. HURST. 1963. A note on the survival of some bacteria in different diluents. J. Appl. Bacteriol. **26**:504–506
21. MEYERS, R. P. and J. A. PENCE. 1941. A simplified procedure for the laboratory examination of raw milk supplies. J. Milk Technol. **4**:18–25
22. PUNCH, J. D. and J. C. OLSON JR. 1964. Comparison between standard methods procedure and a surface plate method for estimating psychrophilic bacteria in milk. J. Milk Food Technol. **27**:43–47
23. RANDOLPH, H. E., B. K. CHAKRABORTY, O. HAMPTON, and D. L. BOGART. 1973. Effect of plate incubation temperature on bacterial counts of grade A raw milk. J. Milk Food Technol. **36**:152–154
24. RAY, B. and M. L. SPECK. 1973. Discrepancies in the enumeration of *Escherichia coli*. Appl. Microbiol. **25**: 494–498
25. REED, R. W. and G. B. REED. 1948. "Drop Plate" method of counting viable bacteria. Can. J. Res. **26E**:317–326
26. RICHARDS, O. W. and P. C. HEIJN. 1945. An improved darkfield Quebec colony counter. J. Milk Technol. **8**:253–256
27. SHARF, J. M. 1966. Recommended Methods for the Microbiological Examination of Foods. 2nd ed. American Public Health Assoc., New York, N.Y.
28. THATCHER, F. S. and D. S. CLARK. 1968. Microorganisms in Foods: Their Significance and Methods of Enumeration. 1st ed. Univ. Toronto Press, Toronto, Canada

29. Thompson, D. I., C. B. Donnelly, and L. A. Black. 1960. A plate loop method for determining viable counts of raw milk. J. Milk Food Technol. **23**:167–171
30. Vanderzant, C. and A. W. Matthys. 1965. Effect of temperature of the plating medium on the viable count of psychrophilic bacteria. J. Milk Food Technol. **28**:383–388
31. Yale, M. W. and C. S. Pederson. 1936. Optimum temperature of incubation for standard methods of milk analysis as influenced by the medium. Amer. J. Pub. Hlth. **26**:344–349

CHAPTER 5

DIRECT MICROSCOPIC COUNT

D. J. Pusch, F. F. Busta, W. A. Moats, and A. E. Schulze

5.1 INTRODUCTION

The direct observation and enumeration of microorganisms in a known volume of food material is the quantitative principle on which the direct microscopic count is based. This principle applies in films, counting chambers, or membranes.

The direct microscopic counting of microbial cells in a food film on a slide is an adaptation of the procedure developed by Breed,[3, 4] which has been used extensively for the enumeration of bacteria in milk.[1] A known quantity of food product is spread evenly over a prescribed area on a microscope slide. The film is dried, fixed, defatted if necessary, stained, and individual and/or clumps of microorganisms in a given number of fields are counted using a compound microscope. The number of fields counted can be translated into the amount of product examined and a count per gram or milliliter thus can be calculated.

The advantages of this procedure are that a) it is rapid, b) films may be stained and read later, c) it requires a minimum amount of equipment, d) morphological and/or Gram stain types may be identified, and e) the slides may be kept for further reference. The disadvantages are that a) it is suited only for foods which contain large numbers of microorganisms, b) only a small quantity (0.01 or 0.001 ml) is examined, thus limiting precision, c) debris may make identification of microorganisms difficult, and d) analyst fatigue reduces precision of the test.

In foods such as yeast, fermented foods, and processed fruits and vegetables where large numbers of organisms may be expected, a counting chamber may be used to enumerate bacteria, yeast, and mold spores or hyphae fragments.[2] A counting chamber is a glass slide so constructed that when the chamber is charged with a food suspension and a cover slip placed over it, a definite volume of the suspension can be related to a specific area of the microscopic field. Alternatively, the slide may be ruled to define volumes and the number of organisms per ruled area is counted microscopically. The organisms may be stained or counted with dark field illumination.

CONTRIBUTOR: William V. Eisenberg
IS/A Committee Liaison: W. J. Hausler, Jr.

5.2 TREATMENT OF SAMPLE (Chapter 1)

5.21 Examination of Sample (Chapter 1)

5.22 Other Tests on Same Sample

If other microbiological or chemical tests are to be performed on the sample, the sample portion for the direct microscopic procedure should be removed first aseptically, or the portions for the other procedures should be removed aseptically.

5.3 FOOD FILMS

5.31 Special Equipment and Reagents

5.311 Microscopic slides

Clean glass slides, 1 × 3 inches or 2 × 3 inches, plain or with margins etched to delineate a circular 1 cm^2 area. Slides delineated with a 1 cm^2 circle are acceptable and are available from Bellco Glass Inc., Vineland, N.J. 08360—Cat. No. 5638-01930. If plain slides are used, a guide which delineates the 1 cm^2 (2 cm^2 for egg products) may be used under the slide.

5.312 Transfer instruments

A metal syringe[1, 17] calibrated to deliver 0.01 ml, a pipet calibrated to deliver 0.01 ml according to APHA specifications,[1] or a platinum alloy loop (4 mm I.D., made of Bond S gauge No. 19 wire) that delivers 0.01 ml may be used.

5.313 Film preparation

A bent point dissecting needle of 30 gauge wire may be used for spreading the sample on a slide. A level, dust free surface, preferably with a thermostatic control, should be used to dry films at 40 to 45 C. The staining apparatus consists of a tray or jar for holding the stain and a rack for submerging the slides. Stains and solvents should be stored in jars or bottles with tight fitting covers. Forceps may be used for dipping and holding slides.

5.314 Film examination

A compound microscope and lamp calibrated with a stage micrometer should be used. Immersion oil should have a refractive index of 1.51 to 1.52 at 20 C.

5.315 Stains

Many stains have been suggested for direct microscopic examination of foods. The following are among those more commonly used: Gram stain, Levowitz-Weber modification of the Newman-Lampert stain,[11] Crystal violet,[20] North's analine oil methylene blue stain,[18] Gray's double dye stain,[10] Wolford's stain,[20] and Erythrosin.[20]

5.316 Standardization of equipment

Extreme care must be observed in standardization and adjustment of the microscope and its light source. The microscopic field area determines the quantity of product being examined and is therefore fundamental to the calculation of the microbial population in a given unit volume. The illumination must be adjusted to secure maximum brightness without producing excessive glare.

The microscope should be calibrated by the Breed direct count method.[1] Microscopes used for this procedure should have microscopic factors (MF) between 300,000 and 600,000. The average number of microorganisms seen in a microscopic field of known area is multiplied by the MF to calculate the number of organisms in 1 ml of the suspension used to prepare the film. Conversely, the reciprocal of the MF is that portion of 1 ml of material seen in one microscopic field. Dilutions made prior to making the films must, of course, be considered in calculating the microbial density in the original food product.

Field diameters of 0.206 to 0.146 mm will give the appropriate microscopic factor of 300,000 to 600,000. The microscopic factor may be specified at the time the microscope is purchased if it is to be used routinely for direct microscopic enumeration.

To determine the microscopic factor, adjust the illuminator to provide maximum optical resolution. Place a stage micrometer ruled in 0.1 and 0.01 mm divisions on the stage of the microscope, and with the 1.8 mm oil immersion objective proceed as follows:

a. Measure the field diameter in millimeters to the third decimal place, e.g., 0.146.
b. To determine the area of the field, square the radius (r = 1/2 field diameter) and multiply by π, e.g., ($r^2 = 0.005329$) $\times$ 3.1416 = area of field in square millimeters.
c. To convert the area of one field in mm^2 to cm^2, divide field area in mm^2 by 100.
d. To determine the number of such fields in 1 cm^2, divide by the area (in cm^2) of one field.
e. Since only 0.01 ml of sample is spread over the 1 cm^2 area, multiply the number of fields by 100 to determine the number of fields per ml of dilution used to make the film. The value thus obtained is the microscopic factor and can be calculated by use of the following condensed formula:

$$\text{M.F.} = \frac{10{,}000}{3.1416 \times r^2}$$

The microscopic factor should be expressed to two significant numbers (e.g., 480,000). This factor, when multiplied by the average number of organisms per field, will represent an estimation of the total number of organisms present in 1 ml of the suspension used to prepare the film. As the microscopic factor increases, the area of the film in each field examined decreases; thus more fields must be examined to observe approximately the same total area of film.

For proper illumination adjustment, when an "in base" or "on base" illuminator is used, the condenser should be adjusted to give maximum brightness of the field. Use the condenser diaphragm only to reduce glare. To obtain maximum resolution when a separate lamp is used, preferably adjust both illumination source and microscope condenser properly to obtain Kohler type illumination.[1]

5.32 Precautions and Limitation of Methods

As with other methods of enumerating microorganisms in food and food ingredients, the direct microscopic method will give only an estimate of the actual number present. A basic limitation of this method is that it does not lend itself to foods which contain relatively low microbial populations. A population of about 10^5/ml is minimal for statistical significance. Also it is not possible generally to distinguish accurately between viable and nonviable cells using this procedure, and consequently direct microscopic counts may exceed the corresponding viable colony count by a factor of two or more.

5.321 Sources of error

Some of the factors which may lead to erroneous direct microscopic counts are the following:

a. failure to obtain a representative sample;
b. failure of some microorganisms to be released from a solid food mass during the blending operation;
c. inaccuracies in measuring 0.01 ml portions because the pipet and the syringe were designed for examining milk;
d. incorrect preparation and staining of slides;
e. uneven spreading or failure to allow film to dry in a level position may cause irregular distribution of microorganisms;
f. failure to count a sufficient number of fields;
g. errors in observations and calculations;
h. poor microscopic technique such as inadequate or excessive illumination or improper focusing;
i. analyst fatigue.

5.33 Procedure

5.331 Sample preparation

Samples must be drawn, shipped, and stored properly so that the measurement is a quantitative and qualitative representative of the microflora present in the food sampled. Initial general sample preparation follows the procedures in Chapter 4.51. To assure that the sample is homogeneous and contains an even distribution of the microflora, proper shaking, mechanical blending, or mixing is essential before removing the aliquot for the direct microscopic count. To avoid debris in the suspension, allow the food particles to settle for a few minutes, or strain the material through cheese cloth before transferring the test portion.

5.332 Preparation of films

With a clean 0.01 ml syringe, dip the tip of the measuring tube not more than 1 cm below the surface of a well mixed test sample and repeatedly rinse the tube by drawing in and expelling portions. Holding the tip beneath the surface, fully release the plunger and withdraw a test portion. With a clean paper tissue, remove excess test portion from the exterior of the tip. Holding the syringe in a nearly vertical position, expel the test portion into the center of the film area, touch off the tip of the instrument, and with the tip of the piston, spread exactly over a circular 1 cm^2 area. Do not release the plunger until after the syringe is removed from the film. A bent point needle also may be used to spread the film.

Alternatively, with a clean 0.01 ml pipet, rinse traces of residual water from bore by drawing in and expelling the test portion. Withdraw portion of well mixed test sample slightly above the graduation mark. Wipe the exterior of the pipet with a clean paper tissue and by absorbing the test portion at the tip, adjust the length of the column to the 0.01 graduation mark. Place the tip near the center of the film area and expel the 0.01 ml test portion. Apply a slight amount of air pressure on the pipet as it is withdrawn from the film area. With a bent point dissecting needle, spread the sample over the circular 1 cm^2 area.

Also alternatively, a 0.01 ml calibrated loop can be adapted to this procedure. With a clean loop, withdraw the loopful vertically from the sample suspension. Transfer and spread the 0.01 ml portion over the 1 cm^2 area.

Prepared films should be dried without delay on a level surface at 40 to 45 C. Films should dry within 5 minutes, but too rapid drying can cause cracking or peeling.

Films of food high in fat should be defatted by rinsing the slide in xylol and washing off in methyl alcohol before staining. Levowitz-Weber stain does not require a defatting step.[1]

Specific stains may be selected to identify certain microorganisms in specific food products or commodities.

5.333 Microscopic examination

When examining stained films for screening purposes, preliminary examination of the film should be made using the high dry objective. A drop of immersion oil is placed on the film and, using the oil immersion objective, an estimate is made of the clumps of microorganisms present in 1 ml of the test portion. Clumps are counted separately if any cell or group of cells of the same morphological type is separated by a distance equal to or greater than twice the smallest diameter of the two cells nearest each other. Count cells of different morphology, or which are stained differently, as separate units regardless of their proximity to other cells.

To examine a representative portion of the film, select a starting field midway on any side and 2 or 3 fields in from the edge. Count separate fields in a series across the film. Then start midway at the top or bottom of the film and count a series of separate fields in a line perpendicular to the first series. If more fields are to be examined, repeat the selection of fields about 2 mm away from the first series. Depending on the cell density in the film, and precision desired, from 10 to 100 fields usually are counted.

The methods for determining the number of fields suggested in Standard Methods for the Examination of Dairy Products[1] also may be useful. If the purpose is to determine whether or not the count is above or below a given standard rather than to obtain an absolute or specific number, a sequential counting procedure may be used.[14, 15] With this method, the maximum number of fields are counted to determine if the sample meets a specific standard.

5.34 Interpretation of Data

The results obtained with this procedure are only estimates of the total microbial levels present in the food product examined. They may represent viable or nonviable or both types of cells. The estimates obtained in direct microscopic counting, therefore, must be interpreted in this manner.

The effect of staining methods for the counting of bacteria in milk has been reviewed by Moats[12, 13] and by Olson and Black.[19] Correlation between the standard plate count and direct microscopic procedures for milk also have been compared.[7] A rapid graphical method has been developed for estimating the precision of direct microscopic counting data, applicable to both dried films and counting chambers.[5] Although these reports deal primarily with milk, the information given is applicable to direct microscopic enumeration in many other commodities.

5.341 Computing and recording microscopic count

To estimate the microbial population in the food product, it is necessary to determine the number of organisms per unit volume, and to multiply this number by the reciprocal of the dilutions used to make the film.

In computing counts for the film method, the average number of microorganisms per field is multiplied by the microscopic factor (MF) and by the reciprocal of the dilution used.

$$\begin{matrix}\text{Average number of} \\ \text{organisms per field}\end{matrix} \times \text{MF} \times \begin{matrix}\text{Reciprocal of} \\ \text{dilution used}\end{matrix} = \begin{matrix}\text{Direct microsopic} \\ \text{count per ml or g}\end{matrix}$$

If a constant number of fields is counted, a working factor (WF), which is the MF divided by the number of fields counted, may be used to calculate the count.

$$\text{WF} \times \begin{matrix}\text{Total number of} \\ \text{organisms counted in} \\ \text{a given number of} \\ \text{fields}\end{matrix} \times \begin{matrix}\text{Reciprocal of} \\ \text{dilution used}\end{matrix} = \begin{matrix}\text{Direct microscopic} \\ \text{count per g or ml}\end{matrix}$$

5.342 Reporting and interpreting counts

The results should be reported as the direct microscopic count (DMC) per g or ml.

5.35 Special Methods

5.351 Egg products

The following special procedures are suggested for liquid and frozen eggs and for dried eggs.[2]

a. Liquid and Frozen Eggs (AOAC Method)

Place 0.01 ml of undiluted well mixed egg on a clean slide spreading it evenly over a 2 cm^2 area. A circular area 1.6 cm in diameter will give the 2 cm^2 area. Dry the film on a level surface at 35 to 40 C. Defat by immersing in xylene 1 minute or less. Fix the smear by immersing in ethanol (95%) 1 minute or less and stain in North's aniline oil methylene blue stain for 10 to 20 minutes.[18] Wash slides by repeated immersion in distilled water and dry thoroughly. Count the microorganisms and report as the number of bacteria per gram of egg material. **CAUTION:** Since the egg material was spread over 2 cm^2 on the slide, the microscope factor must be doubled when making the calculations.

b. Dried Eggs

Place 0.01 ml of a 1:10 or 1:100 dilution of dried eggs on a clean, dry microscope slide and spread over 2 cm^2. A 0.1 normal LiOH solution may be used as the diluent if necessary to obtain a good solution of dried eggs. Dry, stain, and count the film as for liquid and frozen eggs. **CAUTION:** When making calculations, in addition to doubling the microscope factor, the analyst must multiply by the dilution factor.

5.352 Membrane filter method

Microorganisms can be collected on appropriately sized membrane filters. With proper treatment and use of solvents, stains, and immersion oil, the microorganisms can be enumerated by direct microscopic examination.[16]

5.4 COUNTING FUNGI

5.41 Howard Mold Count

The Howard Mold Count Method is applicable to a variety of processed fruit and vegetable products. It is considered an indicator of the presence of raw material which has not been properly sorted or trimmed in the factory to remove rot. The method depends on a count of microscope fields containing fungus filaments in a standardized counting chamber.[2]

5.411 Special equipment and stabilizer solution

a. Blender, High Speed

The term "high speed blender" designates a mixer with 4 canted, sharp edged, stainless steel blades rotating at the bottom of a 4 lobe jar at 10,000 to 12,000 rpm. A Waring blender or equivalent meets these requirements. Use a one liter 4 lobe jar fitted with a 4 blade assembly, 2 blades tilted upward about 30° with diameter of 60 mm, and 2 blades tilted downward about 25° with diameter of 55 mm. Operate at 3000 to 3500 rpm using a variable transformer. To measure speed, attach a one hole No. 8 rubber stopper to the square rotor shaft and insert a tachometer.

b. Cyclone

The laboratory cyclone or pulper consists of a cylindrical perforated metal screen in which revolves a paddle which forces soft material from food products out through openings in the screen.

Tough materials such as seeds, skins, and stems are moved along and out the opening in the end of the cylinder. Use a ¼ horsepower, 110v, 1725 rpm electric motor as a power source. The screen is 22 gauge material, 400 holes/square inch, each 0.027 inch diameter. The screen is 2.5 inches inside diameter and length of effective screen is 3 inches. The paddle has 2 fins, each $^{25}/_{32}$ inches wide, set alternately and extending 1 $^{3}/_{16}$ inches from the center of the shaft. The pulper is fed thru a hopper which leads into a basin 3.5 inches long and 2.5 inches inside diameter. The portion of the paddle with fins inserted at 30° angle forces material from the basin into the screening compartment. The cyclone is so constructed that the waste opening may be closed as needed. Sieved material is caught in the shield and delivered thru a spout to the container. The machine may be disassembled readily for washing.

(Blueprints available from Division of Microbiology, Food and Drug Administration, Washington, D.C. 20204.)

c. Howard Mold Counting Slide

A Howard Mold Counting chamber is a glass slide of one piece construction with a flat plane circle about 19 mm in diameter or rectangle 20 × 15 mm surrounded by moat and flanked on each side by shoulders 0.1 mm higher than the plane surface. The cover glass is supported on the shoulders and leaves a depth of 0.1 mm between the underside of the cover glass and the plane surface. The central plane, shoulders, and cover glass have optically worked surfaces. To facilitate calibration of the microscope, newer slides are engraved with a circle 1.382 mm in diameter or with two fine parallel lines 1.382 mm apart.

d. Compound Microscope

For mold counting, the microscope must have a standardized field of view of 1.382 mm diameter at 90 to 125X and be equipped with a glass disk that fits into one microscope eyepiece, ruled into squares, each side of which is equal to $^{1}/_{6}$ of diameter of field.

e. Stabilizer Solutions

A 0.5% Na carboxymethylcellulose solution is preferred (Cellulose Gum CMC-7H3SF, Hercules Inc., Cellulose and Protein Products Dept., Wilmington, DE 19899). Place 500 ml of boiling water in high-speed blender. With blender running, add 2.5 g cellulose gum and 10 ml formaldehyde solution (ca 37%), and blend for about 1 minute. Alternatives are 3 to 5% pectin or 1% algin. Treat the solution with vacuum or heat to remove air bubbles. Adjust to pH 7.0 to 7.5. If blender is not available, mix dry stabilizer with alcohol to facilitate incorporation with water.

5.412 Limitations

The Howard Mold Count procedure shows the presence of rot due to mold, but it is not designed to differentiate species of molds. Fungus fragments encountered on a mold count slide are not readily identifiable as to genus and species.

5.413 Procedure

The general mold count method is represented by the procedure for tomato juice. The unmodified juice is placed directly on the Howard Mold Counting Chamber without preliminary treatment except mixing.

Clean the Howard slide, so that Newton's rings are produced between slide and cover glass. Remove the cover glass and, with a knife blade or scalpel, place a portion of well mixed sample upon the central disk; with the same instrument, spread the sample evenly over the disk and place the cover glass on the slide to give uniform distribution. Use only enough sample to bring the material to the edge of the disk. Discard any mount showing uneven distribution or absence of Newton's rings, or liquid that has been drawn across the moat and between the cover glass and the shoulder.

Place the slide under the microscope and examine with such adjustment that each field or view covers 1.5 mm^2. (This area, which is essential, frequently may be obtained by so adjusting the draw tube that the diameter of the field becomes 1.382 mm. When such an adjustment is not possible, make an accessory ocular diaphragm with an aperture accurately cut to the necessary size. The diameter of the field of view can be determined by use of a stage micrometer. When the instrument is properly adjusted, the volume of liquid examined per field is 0.15 mm^3.) Use a magnification of 90 to 125X. In those instances where identifying characteristics of mold filaments are not clearly discernible in the standard field, use a magnification of about 200X (18 mm objective) to confirm identity of mold filaments previously observed in the standard field.

From each of 2 or more mounts examine 25 or more fields taken in such manner as to be representative of all sections of the mount. Observe each field, noting presence or absence of mold filaments and recording results as positive when the aggregate length of 3 filaments present exceeds $1/6$ of the diameter of the field. Calculate the proportion of positive fields from the results of the examination of all observed fields and report as the percent of fields containing mold filaments.

a. Sample Preparation for Specific Products

Some products on which mold counting is to be performed require preliminary treatment before the material is placed on the counting chamber, as indicated under the individual items below.

1. Ade Concentrate

   Make mold count on well mixed sample without diluting.

2. Apple Butter (AOAC Method)
   Make mold count after first diluting 50 ml well mixed sample with 50 ml stabilizer solution.
3. Apple Pomace
   Weigh 200 g of pomace in a tared 1500 ml beaker and add 1000 ml of water. Heat to boiling and simmer 2 hours. Add sufficient water to bring mixture to the original volume. Pulp through laboratory cyclone, and mix thoroughly. Weigh 50 g of 3% pectin or other stabilizer solution in a 250 ml beaker and add 50 g of the pulp. Mix thoroughly and make mold count.
4. Blackberries, Raspberries, and Other Drupelet Berries Frozen With or Without Sugar (AOAC Method)
   Pulp berries through a cyclone and mix thoroughly. Mix 25 g pulp with 50 ml stabilizer solution. Make mold count.
5. Blackberries, Raspberries, and Other Drupelet Berries Frozen in Syrup, Canned in Syrup or Water (AOAC Method)
   Drain berries 2 minutes on No. 20 sieve. Pulp, dilute, and make mold count.
6. Blueberries, Canned
   Pulp through a laboratory cyclone. Mix the pulp thoroughly. Weigh out 50 g 3% pectin or other stabilizer solution and add 50 g of the pulp. Mix thoroughly and make a mold count. Before placing the coverglass in position, check the pulp on the mold count slide under the widefield microscope and remove any seeds or seed fragments.
7. Blueberries, Frozen with no Added Sugar
   Pulp without draining, through a laboratory cyclone; dilute with stabilizer solution, and make mold count.
8. Blueberries, Frozen with Sugar or Syrup
   Berries frozen with sugar or syrup may not yield a satisfactory pulp for mold counting without cooking. Boil the berries for 5 minutes or until tender. Pulp the sample while hot, without draining; dilute with stabilizer solution and make a mold count.
9. Capsicums, Ground (Red Pepper, Chili Powder, etc.) (AOAC Method)
   Weigh 10 g thoroughly mixed sample of ground capsicum and transfer to high speed blender. Add 200 ml 1% sodium hydroxide solution in 3 or 4 successive portions, stirring after each addition, washing down with final portion any material that may stick to walls of blender. Agitate mixture in blender 1 minute. With rubber policeman, rub down into mixture any material sticking to walls and repeat blending 2 minutes longer. Add 2 or 3 drops capryl alcohol to break foam. Mix 100 g of this mixture with 50 g stabilizer solution and count with Howard mold counting slide.

   Occasionally blended mixture contains particles of seed tissue that make it difficult to obtain Newton's rings in preparing slide for mold counting. Clamp devised for holding cover slip in place to obviate this difficulty consists of metal plate with circular opening, 2.5 cm diameter, in center of plate; 2 clips attached to edge of plate hold cover slip in position when slide is placed on plate.

10. Citrus juice (AOAC Method)

Pour contents of can into beaker and mix thoroughly by pouring back and forth between beaker and can 12 or more times. After mixing, transfer 50 ml juice to graduated 50 ml conical bottom centrifuge tube. Centrifuge 10 minutes at 2200 rpm in a centrifuge with arm length of 5 ¼ inches (13.34 cm), or any other centrifuge giving equivalent centrifugal force as computed by following formula: $N_1^2r_1 = N_2^2r_2$, where N = rpm, and r = radius of centrifuge arm. Check speed with a tachometer since rheostat does not necessarily indicate speed in rpm.

Let the centrifuge come to a complete stop before removing tubes, and read volume of sediment in centrifuge tube. Remove the tube and decant the supernate without disturbing the sediment. Add water to the tube to bring the level of contents to the 10 ml mark and then add 5 ml stabilizer solution. Thoroughly mix the sediment, water, and stabilizer solution and pour into a small beaker. Mix by pouring back and forth between the beaker and tube 6 or more times. Stir the mixture thoroughly in the beaker and proceed as in the general method. In addition to checking microscopic fields, indicate those fields containing *Geotrichum candidum.*

11. Citrus Juice Concentrate

Make mold count on well mixed sample without diluting.

12. Cranberry Sauce (AOAC Method)

Strained: Immerse an unopened can of sauce in boiling water bath 30 to 45 minutes to facilitate breaking the gel. Remove the can from the bath and open carefully to avoid loss of sauce through sudden release of pressure. Transfer the contents into beaker (one liter for No. 2 can). Stir sauce to break the gel. Slow speed mechanical mixer (350 to 450 rpm) may be used. Thoroughly mix 50 g of the stirred sauce with 50 g stabilizer solution. Make mold count of the mixture.

13. Cranberry Sauce (AOAC Method)

Whole (Seeds and Skins Included): Pulp contents of the container. If considerably greater than 1 lb (500 g), such as No. 10 can, remove well mixed aliquot of 1 lb through a cyclone to remove skins and seeds, and prepare homogeneous pulp. Mix 50 g of this pulp with 50 g stabilizer solution. Make a mold count.

14. Garlic Powder (AOAC First Action Method)

Proceed as with Ground Capsicums.

15. Grape Pulp (Heat Processed)

Weigh 100 g portion of well mixed pulp in a Waring blender or similar mixer. Add 100 ml of 0.5% sodium hydroxide solution and mix for 3 minutes. If necessary, break the foam by adding 2 or 3 drops of capryl alcohol and stirring. Add 20 g of blended pulp to 20 g of a stabilizer solution. Mix thoroughly and make mold count.

16. Grapes

Stemmed and Crushed (not Heat Processed). Pulp sample in the laboratory pulper and mix thoroughly. Make mold count on the undiluted pulp.

17. Infant Food, Pureed (AOAC Method)

Add about 0.2 g sodium hydroxide to approximately 6 g product and stir thoroughly until sodium hydroxide is dissolved. Make mold count.

18. Jams and Preserves

Blackberry, Raspberry and Other Drupelet Berries. Pulp the entire sample, if 2 lbs or less, through a laboratory cyclone. If sample is over 2 lbs, pulp a representative portion of about 2 lb and mix. Weigh 50 g of 3% pectin or other stabilizer solution in a 250 ml beaker. Add 50 g of the jam or preserves and 10 drops of capryl alcohol and mix thoroughly. Apply suction and shake gently until most of the air bubbles are removed (5 to 10 minutes). Mix thoroughly and make mold count.

19. Jams and Preserves

Strawberry and Other Fruit. Pulp and mix sample as directed in 5.413a18, above. Add approximately 100 g of pulped material to a 250 or 500 ml suction flask. Heat the flask gently under vacuum until most of the air bubbles are removed from the jam or preserves.

20. Pineapple Juice (AOAC Method)

Proceed as for citrus juice, but after decanting the supernate in centrifuge tube add 0.5 ml concentrated hydrochloric acid to dissolve the oxalate crystals.

21. Pineapple Juice Concentrate

Make mold count on well mixed sample without diluting.

22. Pineapple, Crushed

Drain the pineapple on a No. 14 sieve of suitable diameter for 2 minutes. Make mold count on the drained juice as in 5.413a20.

23. Pineapple, Sliced, Chunk, and Tidbit

Drain the pineapple on a No. 8 sieve of suitable diameter for 2 minutes. Make mold count on the drained juice as in 5.413a20.

24. Strawberries, Frozen (AOAC Method)

Pulp thawed berries through a cyclone and mix thoroughly. Pour juice through the cyclone last. If necessary remove air bubbles with suction or by mixing about 100 g pulp with 3 to 5 drops capryl alcohol. Again mix thoroughly and make mold counts.

25. Tomato Catsup (AOAC Method)

Place 50 ml stabilizer solution in a 100 ml graduate, add 50 ml well mixed sample by displacement, and mix thoroughly. Make mold count.

26. Tomatoes, Canned (AOAC Method)

Drain contents of can 2 minutes on No. 2 sieve. For containers of less than 3 lb net weight, use 8 inch diameter sieve; for containers of 3 lb or greater net weight, use 12 inch sieve. Examine drained tomatoes and record number and size of any rotten portions present. Pass the drained tomatoes through a laboratory cyclone. Make mold counts on both drained juice and pulped tomatoes.

27. Tomato Juice (AOAC Method)

    Make mold count on undiluted sample.

28. Tomato Powder, Dehydrated (AOAC First Action Method)

    Weigh 11 g of thoroughly mixed sample into a high speed blender, containing 150 ml water to produce a mixture equivalent to tomato juice, or use 17 g tomato powder and 150 ml water to produce a mixture equivalent to tomato puree. Blend 30 seconds and rub down any material adhering to walls with a rubber policeman. Rinse walls with 50 ml water and blend 1 minute. Add 2 drops capryl alcohol to break foam, and count mold.

29. Tomato Puree and Paste (AOAC Method)

    Add water to make a mixture having a tomato soluble solids content such that the refractive index of the filtered or centrifuged liquid portion is 1.3448 to 1.3454 at 20 C (1.3442 to 1.3448 at 25 C). Make mold count.

30. Tomato Sauce (AOAC First Action Method)

    Make mold count on undiluted sample.

31. Tomato Sauce in Pork and Beans, Spaghetti, Ravioli, Chili con Carne, Tamales, etc. (AOAC Method)

    Place unopened can in hot water and heat until the contents are thoroughly warmed. Open the can and transfer the contents onto a No. 6 sieve. Drain until the major portion of the liquid passes through. (With some products, sauce runs through at once, but in the case of some beans and spaghetti more than 10 minutes may be required.) Mix the sauce thoroughly, place 10 ml in a centrifuge tube, and proceed as with tomato soup.

    Use care in counting products containing meat so as not to confuse mold filaments and muscle fibers that superficially resemble each other; muscle fibers are usually much thicker and striations are often visible.

32. Tomato Sauce Packing Medium on Fish (AOAC Method)

    Place the unopened can in hot water (90 to 95 C) until contents are thoroughly warmed. Open the can and drain the contents on No. 6 sieve until the major portion of sauce and oil passes through. Mix liquid, place up to 50 ml in a 50 ml centrifuge tube, and centrifuge as in 5.413a33. Record volume of lower oil free sauce layer, and discard oil and part of mold free aqueous layer. Add water to bring to recorded volume, mix, and count mold, removing bits of fish tissue from slide if necessary, before counting.

33. Tomato Soup (AOAC Method)

    Place the unopened can in hot water and heat until the contents are thoroughly warmed; then open. Transfer 10 ml of thoroughly mixed soup to a 50 ml centrifuge tube and add 3 ml sodium hydroxide solution (1 + 1). If starch is absent, omit the sodium hydroxide. Stir until the starch dissolves and tissues clear. Add enough water to fill the tube, and centrifuge. (Time required to centrifuge the samples varies greatly. With centrifuge arm length of 5 ¼ inches and speed of approximately 1600 rpm, about 20 minutes are required for an average sample. In heavy soups, gelatinizing of much starch sometimes interferes with the proper settling out of solids during centrifuging. If liquid remains cloudy, it may be necessary to discard the sample and start again by adding 3 ml sodium hydroxide solution to only 5 ml of soup.) When the supernate is clear, pour off; if not entirely clear, check supernate for mold before discarding. Add enough water to the residue in tube to bring it to the original volume of soup; mix, and count mold.

5.414 Interpretation of data

There is a sufficiently close relation between the mold count and the amount of decayed tissue in a fruit or vegetable product that the mold count may be used as an index for determining whether good manufacturing practice has been followed.

5.42 Rot Fragment Count

The Rot Fragment Count is a microscopic count of particles of fruit tissue with attached mold hyphae. It is used for comminuted tomato products (5.422a) and apple butter (5.422b). Rot fragments are derived from decayed tissue which was not eliminated by sorting and trimming the raw fruit used in preparing the product.

5.421 Special equipment and solutions

a. Pipet for tissue transfer

Use a 1 ml measuring pipet (bore 3.0 ± 0.5 mm) with the tip cut off at the 1.0 ml mark.

b. Rot fragment counting plate and cover preparation

The glass plate measures 55 × 100 mm, 1.5 to 4.0 mm thick with a cover 50 × 85 mm, about 1.5 mm thick. Carefully paint on a coat of material resistant to hydrofluoric acid over the entire surface, avoiding pinholes. Asphaltum varnish makes excellent coating material; paraffin wax may also be used. Carefully scribe through the coating crosswise parallel lines, 4.5 mm apart with a 15 mm space at each end. If asphaltum varnish is used, the lines may be scribed with a new steel wheel glass cutter.

Place the coated scribed slides face down over HF in a polyethylene container. Determine proper acid fume exposure by trial and error. Following etching, remove the coating by placing the slide in water containing detergent. If the coating is not easily scrubbed off, use benzene or toluene for cleanup.

Counting plates may also be prepared from plastic, as follows: Use clear plastic plate 55 × 100 mm, 4 to 6 mm thick, with a glass cover 50 × 85 mm, approximately 2 mm thick. With a sharp needle, carefully scribe parallel lines 4.5 mm apart across the plate leaving 15 mm spaces at each end. Several slides can be made at one time by using a strip of plastic 100 mm wide and any multiple of 55 mm for length plus allowance for each cut of 2 to 3 mm. Plastic slides are easier to make and are also unbreakable. However, they require more careful handling to avoid scratching.

Fasten ½ of a square cover slip (about 22 mm on side and about 0.25 mm thick) at each end of the counting plate to raise the cover plate above ruled plate. See Figure 1.

c. Widefield stereoscopic microscope

The microscope should have the following minimal specifications: Binocular body with inclined oculars; sliding or revolving nosepiece to accommodate 3 objectives; 3 parfocal objectives 1X, 3X, and 6 or 7.5X; paired 10X or 15X widefield oculars; and capable of illumination by transmitted or reflected light.

In counting rot fragments, remove the mirror and metal contrast plate and replace with box type substage lamp fitted with a daylight or blue groundglass filter and a 15 watt bulb. Place lamp so that the center of glass filter is directly below objective and within 2 cm of the glass microscope plate. In place of a substage lamp, a substage mirror with a ground glass surface may be used.

d. Crystal violet solution

Dissolve 10 g dye (Color Index 42555) in 100 ml alcohol and filter.

e. Stabilizer solution

See 5.411e.

5.422 Procedure

a. Comminuted tomato products (AOAC Method)

Weight 10 g of juice or 5 g of catsup or sauce, and transfer with 100 ml water to a 400 ml beaker. In the case of puree or paste, add water to

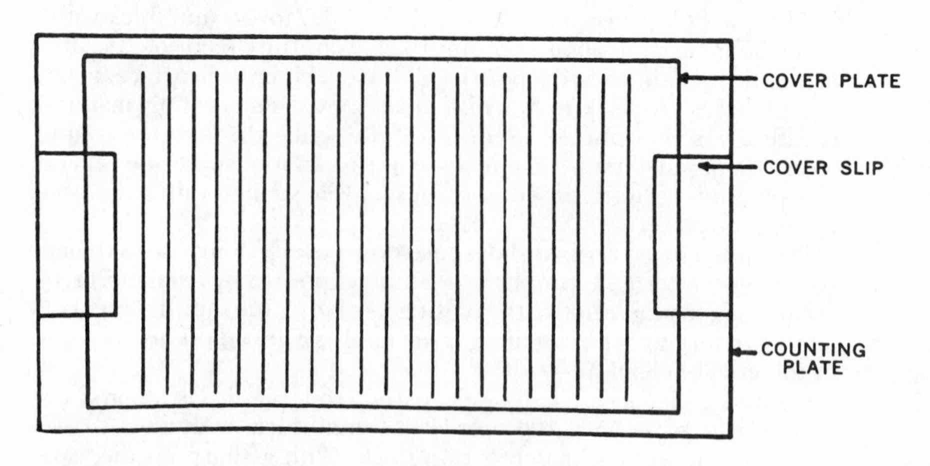

FIGURE 1. Rot Fragment Counting Slide

make a mixture having a tomato soluble solids content such that the refractive index of the filtered or centrifuged liquid portion is 1.3448 to 1.3454 at 20 C (1.3442 to 1.3448 at 25 C). Use 5 g of the mixture for a sample.

Add 10 drops crystal violet solution, stir, and let stain 3 minutes. Add 200 ml of water and spread evenly over surface of a 3 inch No. 60 sieve. Rinse the beaker with 200 ml of water and pour water evenly over the tomato tissue on the sieve. Tilt sieve to about a 30° angle and wash tissue to the lower edge with water. Let the tissue drain, and transfer about 12 × 3 cm to the bottom of a graduated tube with a spatula. Transfer the remaining tissue by washing down with water from a dropper and immediately taking up the tissue in the wash water before it has run through the sieve. Bring the volume of water and tissue up to 10 ml with water. Add stabilizer solution to bring the volume to 20 ml and mix well. Pipet two separate 0.5 ml portions and spread evenly over two counting slides using the pipet. Let the material flow slowly onto the slide, spreading uniformly on the center of the slide to cover an area about 6 × 2 cm. Touch the lower end of the pipet to the slide several times to ensure complete removal of the material. Blow out the last drop if necessary. Examine each slide at 30 to 45X, using transmitted light. Count the number of rot fragments on each of two slides, add results, and multiply by 2 (for 10 g sample) or 4 (for 5 g sample) to obtain the number of rot fragments per g of juice, catsup, sauce, or diluted puree or paste. A rot fragment is defined as a particle of tomato cellular material with one or more mold filaments attached. Some particles may appear to be almost solid masses of mold.

b. Apple butter

Weigh 1.0 g of apple butter into a 50 ml beaker. Add about 20 ml of water and 0.5 to 1.0 ml of saturated aqueous gentian violet solution and stain the product 3 minutes. Decant through a No. 100 sieve, rinse the contents of the sieve with 100 to 200 ml of water and transfer the retained pulp to a graduated cylinder. Transfer is most readily accomplished by rinsing the pulp to one edge of the sieve, scraping out most of the pulp with a spatula, and then rinsing the sieve with water from pipet or wash bottle. Do not transfer more than 10 ml of water to the tube during this operation. Fill to the 10 ml mark with water and then fill with the stabilizer solution to the 20 ml mark and mix. Measure out two separate 0.5 ml portions and spread over two counting slides. Examine each slide using a widefield microscope at 30 to 45X and transmitted light. Count the number of rot fragments per slide. Calculate and report number per g of product.

5.423 Interpretation of data

In addition to showing whether decayed raw material has been used in a product, the rot fragment count can be used in conjunction with the mold count to indicate the degree of product comminution that has been produced in the process by factory equipment. The ratio of rot fragment count to mold count decreases with a greater degree of product comminution. The ratio can be used in interpreting the count in terms of actual rotten tissue in product[8, 9].

5.43 *Geotrichum* Count

Geotrichum candidum, frequently called machinery mold, may grow on moist surfaces in fruit and vegetable processing plants and bottling works. Mycelial fragments then may enter bottled drinks and fruit and vegetable products from the factory equipment. The amount of *G. candidum* in a product is determined by a microscopic count of fragments of mycelium with the characteristic feathery branching habit of the species.

5.431 Special equipment and solutions

See 5.421.

5.432 Procedure

a. Fruits, vegetables, and juices, canned

This procedure is applicable to products where mold is not masked by large amounts of tissues.[2] Determine the net weight of the can contents. Drain contents for 3 minutes on 8 inch No. 9 sieve in a pan. Remove the fruit from the sieve with a spoon and discard. Wash the can and sieve with about 300 ml of water from a wash bottle, saving the liquid and washings. Quantitatively transfer the combined liquid and washings onto a 5 inch No. 16 sieve resting in 2 liter beaker. Wash the residue on the sieve with about 50 ml of water, and discard the residue. Quantitatively transfer the combined liquid and washings onto a 5 inch No. 230 sieve, tilted at about a 30° angle, and discard the liquid and washings. Wash the tissue to the lower edge of the sieve with water.

With a wash bottle and spatula, transfer the residue from the sieve to a 50 ml graduated thick walled centrifuge tube with a minimum volume of water.

1. For volumes of 10 ml or less, dilute to 10 ml. Add 1 drop of crystal violet staining solution and mix thoroughly. Add 10 ml of stabilizer solution to bring total volume to 20 ml. Proceed with pipetting (5.432a4).
2. For volumes between 10 ml and 30 ml, dilute to 40 ml. Add 3 drops of crystal violet solution. Mix well. Centrifuge about 6 minutes at about 2200 rpm using centrifuge with arm length of 5 ¼ inches or equivalent (see 5.413a10).

   Decant and discard the supernate. Bring the volume of the sediment in the centrifuge tube to the nearest 5 ml graduation by adding water. Note the combined volume of sediment and water, and add an equal volume of stabilizer solution. Mix thoroughly but gently. Record the total volume of mixture in the centrifuge tube (V). Proceed with pipetting (5.432a4).

3. For volumes greater than 30 ml, transfer to a 100 ml glass stoppered graduate. Dilute to 100 ml and mix well. Quickly pour off two 25 ml aliquots into separate centrifuge tubes and proceed as in (5.432a2). Keep final volumes equal; V = sum of volume in both tubes.
4. Pipet four separate 0.5 ml portions, using a pipet, and apply as a streak about 4 cm long to each of 4 rot fragment counting slides (8 in case of (5.432a3)). With transmitted diffused light, examine each slide at 30 to 45X using a stereoscopic microscope. Count recognizable *Geotrichum* mycelial fragments (usually 3 or more characteristic hyphal branches) on each of four slides. Record the total from all slides.
5. Calculate mycelial fragments/500 g of product as follows: $N = (S \times (V/2) \times 500)/W$, where N = mycelial fragments/500 g; S = total mycelial fragments on four slides; V = ml total volume of mixture in centrifuge tube; and W = net weight product sample in g. (For volume greater than 30 ml, (5.432a3), S = total on 8 slides, and V = 50.)

b. Cream style corn, canned

Weigh 250 g of the product into a 600 ml beaker and bring the volume up to 600 ml with distilled water. Quantitatively transfer contents of the beaker onto an 8 inch No. 8 sieve in a suitable pan and drain for about 1 minute. Evenly distribute the material on the sieve with a spoon and vigorously wash with about 300 ml of distilled water from a wash bottle. Remove the residue from the sieve with a spoon and discard. Wash the beaker and the sieve with about 300 ml of water from a wash bottle, saving the liquid and washings. Quantitatively transfer the combined liquid and washings from the pan onto a 5 inch No. 16 sieve resting in a 2000 ml beaker. Wash the residue on the sieve with about 100 ml of water and discard the residue. Add 50 g of sodium hydroxide pellets to the beaker and bring the volume to 100 ml with water. While constantly stirring, bring the contents of the beaker to a boil. Quantitatively transfer the combined liquid and washings onto an 8 inch No. 230 sieve, tilted at a 30° angle, and discard sieved liquid material. Wash the residue to the lower edge of the sieve with water. With a spatula and wash bottle, transfer the residue from the sieve to a 100 ml graduated cylinder keeping the volume below 50 ml. Add 20 drops of crystal violet staining solution, and stir for about 2 minutes with a spatula. Bring the volume to 50 ml, add 50 ml of stabilizer solution, and mix thoroughly. Pipet four separate 0.5 ml portions, and proceed as in 5.432a4.

c. Whole kernel corn, frozen

Determine the net weight of the container. Allow the product to thaw on a No. 8 sieve in a suitable pan. Using a wash bottle, wash the product with a quantity of water equal to the net weight. Drain 3 minutes and proceed as in method for Canned Fruits, Vegetables, and Juices. See 5.432a1.

d. Soft drinks, carbonated and noncarbonated (AOAC First Action Method)

Determine the net volume, V, of the container and transfer the contents to a 3 inch No. 230 sieve. Transfer the residue from the sieve to a 50 ml graduated centrifuge tube, and proceed as in the method for Canned Fruits, Vegetables and Juices. See 5.432a1. Express the results as mycelial fragments/350 ml (12 fluid ounces) = 3500 S/V, where S = total number of mycelial fragments on the four slides.

5.433 Interpretation of data

The presence in a product of *Geotrichum* fragments characterized microscopically by their typical feathery appearance demonstrates growth of the mold on food contact equipment,[6] and may indicate insanitary conditions due to inadequate cleanup.

5.5 COUNTING CHAMBERS

Methods utilizing hemocytometers and similar counting chambers should be reserved for estimations of populations at relatively high densities in systems with minimal interference from food particles.

5.51 Special Equipment

Two chambers representative of those available are the bright line counting chamber, American Optical Cat. No. 1492 and the Petroff-Hauser bacteria counting chamber, Scientific Products Cat. No. C8400-1.

Standard bacteriological loops (3 mm) or capillary tubes may be used for charging chambers.

5.52 Procedure

To transfer a portion of the sample or diluted sample to charge the counting chamber, dip the loop in a well mixed sample and transfer to a clean counting chamber with a cover slip in place. Touch the charged loop to the edge of the cover slip and allow the area between the cover slip and the counting surface to fill with the test suspension by capillary action. A capillary tube may be used as an alternative for the loop.

After the hemocytometer chamber has been charged and the cell motion is minimal, the microorganisms in 25 to 50 squares are counted at magnifications ranging from 300 to 600X. The depth of a bright line hemocytometer cell is 0.1 mm and the rulings cover an area of 1 mm^2. The depth of the counting cell in the Petroff-Hauser chamber is only 0.02 mm and each square is 2.5×10^{-4} cm^2 representing a volume of 5×10^{-8} ml for each square observed. Therefore, a microbial density in excess of 10^7/ml is essential.

In computing counts, it is again necessary to determine the number of organisms per unit volume and multiply by the reciprocal of the dilution used.

$$\text{Number of organisms per unit volume} \times \text{Reciprocal of dilution made} = \text{Direct microscopic count per g or ml}$$

5.6 REFERENCES

1. Standard Methods for the Examination of Dairy Products. 1972. 13th ed. American Public Health Association. Washington, D.C.
2. Official Methods of Analysis. 1975. 12th ed. Association of Official Analytical Chemists. Washington, D.C.
3. Breed, R. S. 1911. The determination of bacteria in milk by direct microscopic examination. Zentral Bl. f. Bakt. II. Abt. **30**:337–340
4. Breed, R. S. and J. D. Drew. 1916. Counting Bacteria by Means of the Microscope. Tech. Bull. 49. New York Agr. Exp. Sta., Albany, N.Y.
5. Cassell, E. A. 1965. Rapid graphical method for estimating the precision of direct microscopic counting data. Appl. Microbiol. **13**:493–496
6. Cichowicz, S. M. and W. V. Eisenberg. 1974. Collaborative study of the determination of *Geotrichum* mold in selected canned fruits and vegetables. J. Assoc. Off. Anal. Chem. **57**:957–960
7. Dabbah, R. and W. A. Moats. 1966. Correlation between standard plate count and four direct microscopic count procedures for milk. J. Milk Food Technol. **29**:366–371
8. Eisenberg, W. V. 1968. Mold counts of tomato products as influenced by different degrees of product comminution. Association of Food and Drug Officials of the United States. Quarterly Bulletin **32**:173–179
9. Eisenberg, W. V., M. Parran, Jr., A. E. Schulze, and R. G. Douglas. 1969. Effect of comminution on mold counts of tomato products. J. Assoc. Off. Anal. Chem. **52**:749–752
10. Gray, P. H. H. 1943. Two stain method for direct bacteria count. J. Milk Food Technol. **2**:76
11. Levowitz, D. and M. Weber. 1956. An effective "single solution" stain. J. Milk Food Technol. **19**:121–127
12. Moats, W. A. 1964. Staining of bacteria in milk for direct microscopic examination—A review. J. Milk Food Technol. **27**:308–310
13. Moats, W. A. 1972. A comparison of three staining methods for direct microscopic counting of bacteria in milk. J. Milk Food Technol. **35**:496–498
14. Morgan, M. E. and P. MacLeod. 1953. A more critical sequential procedure for microscopic grading of milk. J. Milk Food Technol. **16**:228–231
15. Morgan, M. E., P. MacLeod, E. O. Anderson, and C. I. Bliss. 1952. An improved procedure for microscopic grading of milk intended for pasteurization. J. Milk Food Technol. **15**:3–7
16. Mulvany, J. G. 1969. Membrane filter techniques in microbiology. *In* Methods in Microbiology. Vol. 1. J. R. Norris and D. W. Ribbons (ed). Academic Press, New York, N.Y. p. 243
17. Newman, R. W. 1952. The Smith 0.01 ml syringe in the microscopic grading of milk. J. Milk Food Technol. **15**:101–103
18. North, W. R. 1945. Aniline oil-methylene blue stain for the direct microscopic count of bacteria in dried milk and dried eggs. J. Assoc. Off. Anal. Chem. **28**:424–426
19. Olson, J. C., Jr. and L. A. Black. 1951. A comparative study of stains proposed for the direct microscopic examination of milk. J. Milk Food Technol. **14**:49–52
20. Sharf, J. M. 1966. Recommended Methods for the Microbiological Examination of Foods. 2nd ed. American Public Health Association, New York, N.Y.

CHAPTER 6

MPN—MOST PROBABLE NUMBER

Joseph Mayou

6.1 INTRODUCTION

The Most Probable Number (MPN) technique utilizes a "multiple dilution to extinction" approach that has proved valuable in estimating populations of microorganisms, especially in situations where extremely low densities are encountered, where particular foods may complicate other enumeration methods, or where special media or cultural systems may be desired. For many food products and food ingredients, the particulate structure of the sample, or other characteristics, makes it difficult to use colony count procedures when the microbial density of the test organism is less than 10 microorganisms per gram. In these cases the Most Probable Number (MPN) method is very useful in estimating microbial numbers, especially in extremely low populations.[4] A well known example of this application is the enumeration of coliforms in water, dairy products, and other foods.[1, 2, 5, 8, 9] Here special cultural conditions are applied sequentially for selective or differential purposes.

In naturally inoculated or accidentally contaminated foods, the distribution of microorganisms on or in the product may be homogeneous throughout the product, may be on the surface of a solid food item, or may be at isolated points of contamination in specific or random locations throughout the food. Values listed in the MPN tables are calculated on the assumption that the microorganism sought is homogeneously distributed in the sample and the homogenate. In food samples and slurries, fats and insoluble food particles prevent the homogeneity assumed. For this reason, MPNs derived from food samples are somewhat less accurate and reproducible than those derived from water samples where homogeneity is more easily attainable.

In general, the MPN procedure results in a value that is considerably less accurate than plating procedures. For example, in the enumeration of a pure culture of *Enterobacter aerogenes,* a nonselective plating procedure is preferable to the MPN. If the same organism is to be enumerated in samples such as river water, where other bacterial species predominate, the MPN procedure may be the method of choice. Accuracy sometimes must be sacrificed because of the lack of good differential plating procedures, or

IS/A Committee Liaison: Ralph W. Johnston

the need to test large volumes of sample slurries heavily laden with particulate material and mixed bacterial species.

Depending upon the sampling and information sought, the sample may be mixed or mascerated to bring about the homogeneous distribution of microorganisms. As this homogeneous sample is divided into subsamples by serial dilution and distribution of aliquots, some of the subsamples eventually contain such small amounts of sample that they will contain none of the microorganisms sought. The presence or absence of microbial cells in given subsamples can be used to estimate statistically the population in the original sample.[4] The MPN method is based on this subdividing of the sample and therefore may be described as the "multiple tube dilution to extinction" method. The accuracy of a single test depends on the number of tubes used per dilution. The most satisfactory information is obtained when all of the tubes with the larger portions show growth, and the tubes with the smaller portions show no growth.

6.2 TREATMENT OF SAMPLE

Procurement, transport, examination of sample, etc., should be done in accordance with procedures in Chapter 1.

6.21 Examination of Sample

6.211 Temperature, moisture, container

Refer to chapters on specific microorganisms.

6.212 Collection, storage, shipment, preparation, examination times, and identities.

Chapter 1

6.213 Other tests on same samples

Remove portions for microbiological analysis first if other tests are to be performed.

6.3 EQUIPMENT AND SUPPLIES

6.31 General

Most of the equipment and supplies for MPN determinations are used in the determination of colony counts. Therefore, for general equipment, see Chapter 2.

6.32 Special Equipment Required

a. Flasks or tubes which can be sterilized and which are large enough to hold the amount of media needed (i.e., 100 gram samples will need flasks capable of holding at least one liter of medium plus the 100 g of sample).

b. Inoculating loop (3 mm i.d.)

c. Test tube racks.
d. Commercially available closures for tubes and flasks or nonabsorbent cotton for plugging tubes.
e. Bunsen burner or other system for flaming and sterilizing loops and needles.
f. Microscope with 900x magnification.
g. Microscope slides.

6.4 REAGENTS AND MEDIA

6.41 General

6.411 Liquid media

a. Nonselective media

1. Tryptone glucose yeast broth
2. Beef heart infusion broth
3. A.P.T. broth
4. Trypticase soy broth
5. Nutrient broth

b. Selective media

Refer to chapters for specific microorganisms or commodities.

6.412 Dilution medium Chapter 2.

6.42 Special Media

Refer to chapters for specific microorganisms.

6.5 RECOMMENDED CONTROLS

6.51 Sterility Controls on Media

It is recommended that one set of tubes from each batch of medium prepared should be used as an uninoculated control. If, for example, the 5 tube MPN method is being used, then a set of 5 tubes should be incubated as uninoculated controls to assure that the medium was properly sterilized.

6.52 Other Controls

Temperatures in the incubators should be controlled (Chapter 2).

6.6 PRECAUTIONS AND LIMITATIONS OF METHODS

6.61 Sources of Error

This method is used most frequently when low microbial densities are expected. Therefore, the use of large sample sizes is common. If the sample contains inhibitory substances, or the sample itself is inhibitory (i.e., sodium chloride), growth in the tubes with high concentrations of sample may be inhibited.

6.62 Special Considerations or Special Methods

The MPN method of enumeration is used frequently for determining the numbers of specific types of microorganisms. In this case, media selective for the specific types of microorganisms are used. Types of bacteria that are enumerated include salmonellae, staphylococci, coliforms, fecal coliforms, *E. coli,* and others. For more specific information, refer to the section relating to the specific microorganisms or specific commodities.

6.7 PROCEDURE

6.71 Preparation of Samples and Dilutions

The manipulation and dilution of samples for the MPN method is essentially identical to the procedure used in colony counts (Chapter 4).

6.72 Introduction of Sample or Diluted Sample into Liquid Medium

The ratio of sample volume to medium volume must be considered. Basically, one part of sample to ten parts of medium should be maintained. Thus, if a 10 g sample is used, it should be dispersed in 100 ml of medium. When the microorganisms in liquids or diluted samples are enumerated, the strength of the medium can be adjusted so that the concentration of medium after the addition of the sample is equal to single strength medium. If a double strength medium is advisable and permissible, an equal amount of sample or diluted sample may be added.

6.73 Incubation

6.731 General

Incubate tubes for total MPN at 32 C ± 1 for 48 hours ± 3 hours. This temperature and time is identical to those used to determine colony counts. Incubation may be in an air incubator or in a water bath.

6.732 Special

When using the MPN method for specific microorganisms, very specific and critically controlled incubation temperatures may be required. Refer to chapters on the specific microorganisms to be enumerated, or specific commodities to be evaluated.

6.74 Detection of Positive Tubes

6.741 Turbidity

When using samples that do not cloud the medium in the tubes, the development of turbidity after incubation can be used to detect growth (positive tubes). However in many cases the sample is visible so that turbidity cannot be determined. When this is true, other methods for the detection of positive tubes must be employed.

6.742 Metabolic end products

a. Detection of gas production

Gases produced by developing microorganisms can be captured and observed. This can be accomplished with gas traps or inverted vials that are placed in the medium in the growth vessels prior to sterilization. A positive reaction is recorded when gas bubbles are observed in traps or inverted vials at the end of the incubation period. Other methods can be used to capture and observe the gases produced. These include overlay with vaspar or agar, and use of a respirometer or similar device. Obviously this is useful only when the microorganisms to be enumerated are known to produce gases under the conditions of the test.

b. Detection of acid or base

Acid or base production can be determined after incubation by measuring the pH or titratable acidity in each tube or by using a medium containing a pH indicating dye which gives a color change with changes in pH. Detecting positive tubes by this method requires that the microorganisms being enumerated produce acid or base from a defined substrate.

c. Detection with reduction methods

Electron acceptors such as resazurin, methylene blue, or 2,3,5 triphenyltetrazolium chloride can be incorporated into the medium. Reduction of any of these compounds by microbial action also serves as an indicator of growth.

6.743 Direct examination

a. Microscopic method

Microscopic examination of stained smears of the contents of tubes may be a useful technique. This can be done by placing a loopful of the medium from the tube on a slide, drying, heat fixing, staining with a simple crystal violet stain, rinsing, and drying. The slide is examined for microorganisms using oil emersion at 900x magnification. If the original sample contains high numbers of inactivated microorganisms, this method is not applicable.

6.75 Confirmation of Positive Tests

To confirm growth in questionable tubes, a loopful of the medium from the tube should be transferred to a nonselective medium and incubated for an appropriate additional period of time. Growth in this medium confirms the presence of viable microorganisms in the tube. To confirm a questionable growth reaction (such as acid) in tubes of selective media with heavy food turbidity, a loopful of the medium from the incubated tube should be transferred to a tube of the identical medium and similarly incubated. Growth reactions in such subcultures can be readily observed since they are free of the color, turbidity, etc., contributed by the food under test.

6.8 INTERPRETATION OF DATA

6.81 Calculations of MPN

When the multiple tube method is used, results may be reported as "the most probable number of microorganisms per gram". Tables 2 and 3 show the most probable numbers of microorganisms corresponding to the frequency of positive tubes, secured from a variety of multiple portion inoculation systems, beginning with 1 ml test portions of a 1 : 10 food dilution (0.1 g portion). Since more than one organism may be responsible for each positive test, the calculation of the most probable number of organisms involves a logarithmic function when test portions are related decimally.

To obtain the MPN, determine the number of positive tubes from each of three selected dilutions. Refer to the appropriate MPN table and record the MPN.

Table 1: Examples and Solutions of Results that Can Occur When More Than Three Different Portions Are Used

Example	0.1g	0.01g	0.001g	0.0001g
a)	*5/5	**5/5**	**2/5**	**0/5**
b)	**5/5**	**4/5**	**2/5**	*0/5
c)	**0/5**	**1/5**	**0/5**	*0/5
d)	**5/5**	**3/5**	**1/5**	**1/5**
e)	**5/5**	**3/5**	**2/5**	*0/5

*When all tubes are positive, the smallest three volumes tested are chosen, and a "greater than" sign (>) is used to indicate that the most probable number is greater than the one given.

When more than three different volumes are employed, use results from only three consecutive volumes (Table 1). Use the results of the smallest volumes for which the results are 5/5, and the results obtained with the next two smaller volumes. Significant results in the following examples are in boldface type; asterisked figures should not be utilized in the computations. In Example c where none of the volumes tested was 5/5 the first three volumes should be used so as to place positive results in the middle volume. If a positive test occurs in a volume smaller than the three chosen according to rule (Example d), add the value appearing above the line for that result to the value above the line for the smallest chosen volume, for a result to read as in Example e.

Often it is necessary to calculate the MPN from initial volumes different from those listed in Tables 2 and 3. If the greatest portion used for the table reference is 0.01g rather than 0.1g, multiply the MPN listed in the table by 10. Thus, results of a 5 tube MPN determination showing 3 positive 0.01g portions, 2 positive 0.001g portions, and 1 positive 0.0001g portion (3-2-1)

Table 2: Most Probable Number (MPN) per 1g of Sample and 95% Confidence Limits (6)

Using Five Tubes with 0.1, 0.01, and 0.001g Portions

Number of Positive Tubes			MPN	Limit MPN	
0.1	0.01	0.001	/g	Lower	Upper
0*	0*	0*	<2	<0.5	—
0	0	1	2	<0.5	7
0	0	2	4	<0.5	11
0	1	0	2	<0.5	7
0	1	1	4	<0.5	11
0	1	2	6	<0.5	15
0	2	0	4	<0.5	11
0	2	1	6	<0.5	15
0	3	0	6	<0.5	15
1	0	0	2	<0.5	7
1	0	1	4	<0.5	11
1	0	2	6	<0.5	15
1	0	3	8	1	19
1	1	0	4	<0.5	11
1	1	1	6	<0.5	15
1	1	2	8	1	19
1	2	0	6	<0.5	13
1	2	1	8	1	19
1	2	2	10	2	23
1	1	0	8	1	19
1	3	1	10	2	23
1	4	0	11	2	23
2	0	0	5	<0.5	13
2	0	1	7	1	17
2	0	2	9	2	21
2	0	3	12	3	28
2	1	0	7	1	17
2	1	1	9	2	21
2	1	2	12	3	28
2	2	0	9	2	21
2	2	1	12	3	28
2	2	2	14	4	34
2	3	0	12	3	28
2	3	1	14	4	34
2	4	0	15	4	37
3	0	0	8	1	19
3	0	1	11	2	25
3	0	2	13	3	31
3	1	0	11	2	25

Table 2: Most Probable Number (MPN) per 1g of Sample and 95% Confidence Limits (6)

Using Five Tubes with 0.1, 0.01, and 0.001g Portions					
Number of Positive Tubes			MPN /g	Limit MPN	
0.1	0.01	0.001		Lower	Upper
3	1	1	14	4	34
3	1	2	17	5	46
3	1	3	20	6	60
3	2	0	14	4	34
3	2	1	17	5	46
3	2	2	20	6	60
3	3	0	17	5	46
3	3	1	21	7	63
3	4	0	21	7	63
3	4	1	24	8	72
3	5	0	25	8	73
4	0	0	13	3	31
4	0	1	17	5	46
4	0	2	21	7	63
4	0	3	25	8	75
4	1	0	17	5	46
4	1	1	21	7	63
4	1	2	26	9	78
4	2	0	22	7	67
4	2	1	26	9	78
4	2	2	32	11	91
4	3	0	27	9	80
4	3	1	33	11	93
4	3	2	39	13	126
4	4	0	34	12	95
4	4	1	40	14	108
4	5	0	41	14	110
4	5	1	48	16	124
5	0	0	23	7	70
5	0	1	31	11	89
5	0	2	43	15	114
5	0	3	58	19	144
5	0	4	76	24	180
5	1	0	33	11	93
5	1	1	46	16	120
5	1	2	64	21	134
5	1	3	84	26	197
5	2	0	49	17	126
5	2	1	70	23	168

Table 2: Most Probable Number (MPN) per 1g of Sample and 95% Confidence Limits (6)

Using Five Tubes with 0.1, 0.01, and 0.001g Portions					
Number of Positive Tubes			MPN	Limit MPN	
0.1	0.01	0.001	/g	Lower	Upper
5	2	2	94	28	219
5	2	3	120	33	281
5	2	4	148	38	366
5	2	5	177	44	515
5	3	0	79	23	187
5	3	1	109	31	251
5	3	2	141	37	343
5	3	3	175	44	503
5	3	4	212	53	669
5	3	5	253	77	788
5	4	0	130	35	300
5	4	1	172	41	484
5	4	2	221	57	698
5	4	3	278	90	849
5	4	4	345	117	999
5	4	5	436	145	1161
5	5	0	240	68	734
5	5	1	348	118	1005
5	5	2	542	180	1405
5	5	3	920	210	3000
5	5	4	1600	350	5300
5	5	5	>1600	800	—

*Number of positive tubes in five tested.

are read from Table 2 as 17, and multiplied by 10 to arrive at 170 as the actual MPN/g for the sample. Similarly, if the greatest portion used for the table reference is 10.0g rather than 0.1g, divide the MPN derived from the table by 100. Thus, results of a 3 tube MPN determination for salmonellae showing 3 positive 10 gram portions, 1 positive 1g portion, and no positive 0.1g portions (3-1-0), are read from Table 3 as 43 and divided by 100 to arrive at 0.43 as the actual MPN/g for the sample.

An alternative approach to obtain the MPN per gram of food utilizes the following formula.[9]

$$\frac{\text{MPN from Table x dilution factor of the middle tube}}{100} = \text{MPN/g}$$

Table 3: Most Probable Number (MPN) per 1g of Sample and 95% Confidence Limits

Using Three Tubes with 0.1, 0.01 and 0.001g Portions

Number of Positive			MPN	Limit MPN	
	Tubes		/g	Lower	Upper
0.1	0.01	0.001			
0	0	0	<3		
0	0	1	3	0.5	9
0	1	0	3	<0.5	13
1	0	0	4	<0.5	20
1	0	1	7	1	21
1	1	0	7	1	23
1	1	1	11	3	36
1	2	0	11	3	36
2	0	0	9	1	36
2	0	1	14	3	37
2	1	0	15	3	44
2	1	1	20	7	89
2	2	0	21	4	47
2	2	1	28	10	150
3	0	0	23	4	120
3	0	1	39	7	130
3	0	2	64	15	380
3	1	0	43	7	210
3	1	1	75	14	230
3	1	2	120	30	380
3	2	0	93	15	380
3	2	1	150	30	440
3	2	2	210	35	470
3	3	0	240	36	1,300
3	3	1	460	71	2,400
3	3	2	1,100	150	4,800
3	3	3	>2,400		

6.811 Statistical interpretation

All of the statistical considerations are embodied in the confidence limits given in the two tables. The strict interpretation of these limits is that if one always asserts that the true density of organisms lies between these limits, then he will be right in 95% of such assertions.

Recently, a statistical treatment and computer program has been reported in which any number of multiples may be used for any number of samples and dilutions, and the program rapidly will produce MPNs, averages, and confidence limits at any level of significance.[7] Although procedures for such manipulations were available previously,[3] this computer program procedure greatly increases the ease of using odd number multiples and averaging several MPN values.

6.82 Reporting and Interpreting Counts

Microbiological counts should be recorded as "number of microorganisms per quantity of sample by MPN method", e.g., coliform MPN/ g = 10. Include with the report of microbiological counts by the MPN method the number of tubes used in each dilution, i.e., 5 tube MPN or 3 tube MPN and the method used.

6.9 REFERENCES

1. Standard Methods for the Examination of Water and Wastewater. 13th Ed. American Public Health Association, New York, N.Y.
2. Official Methods of Analysis. 11th Ed. 1970. Association of Official Analytical Chemists, Washington, D.C.
3. HALVORSON, H. O. 1958. Some Elements of Statistics for Microbiologists. The Technical University of Norway, Trondheim, Norway
4. HALVORSON, H. O. and N. R. ZIEGLER. 1935. Application of statistics to problems in bacteriology. 4. Experimental comparison of the dilution method, the plate count, and the direct count for the determination of bacterial populations. J. Bacteriol. **29**:609–634
5. HAUSLER, W. J., Jr. 1972. Standard Methods for the Examination of Dairy Products. 13th Ed. APHA, New York, N.Y.
6. HOSKINS, J. K. 1933. The most probable number of *B. coli* in water analysis. J. Amer. Water Works Assoc. **25**:867–877
7. PARNOW, R. J. 1972. Computer program estimates bacterial densities by means of the most probable numbers. Food Technol. **26**:56–62
8. SHARF, J. M. 1966. Recommended Methods for the Microbiological Examination of Foods. APHA, New York, N.Y.
9. THATCHER, F. S. and D. S. CLARK. 1968. Microorganisms in Foods: Their Significance and Methods of Enumeration. Univ. Toronto Press, Toronto, Canada

CHAPTER 7

DETECTION AND ENUMERATION OF INJURED MICROORGANISMS

Z. J. Ordal, J. J. Iandola, B. Ray, and A. G. Sinskey

7.1 INTRODUCTION

Foods which have been heated, refrigerated, frozen, dried or irradiated, and other foods which have a high osmotic pressure (high ionic strength) or low water activity, as well as equipment surfaces which have been sanitized, may contain viable bacteria that are physiologically deficient because of the environmental stress to which the cells have been subjected.

Ideally, the method for enumerating bacteria should detect environmentally stressed cells as well as normal cells. However, until recently the accepted methodology for the enumeration of the microbial flora in foods has failed to consider the altered physiological state of the microorganisms that may be present. It is now recognized that stressed, but viable cells, in the appropriate environment, can repair themselves or recover from the environmental stress and become a potential public health hazard. Therefore, it is desirable to have knowledge of the history of the food product and the effect of the processing stress in order to choose an effective enumeration procedure.

Selective agents commonly added to media used for the detection and enumeration of index organisms, or selected foodborne pathogens, are generally inhibitory to the repair or recovery of stressed bacteria.[6, 15] When such media must be used, the stressed bacteria in the sample must be permitted to recover or repair themselves prior to being added to the selective medium.[18]

Under currently recommended procedures, the use of such selective media results in failure to detect stressed but viable cells, and hence does not yield an accurate estimation of the bacterial content of the particular food product.[7]

Caution should be exerted to avoid stress to microorganisms during the enumeration process. When pour plating procedures are used, the bacteria present may be subject to the heat stress caused by the molten agar.[13, 21, 26] Klein et al[15] reported that heterotrophic organisms in water samples were susceptible to the transient stress of agar at 45 C used in the standard methods pour plate procedure. Significant increases in numbers of cells recovered from water samples were noted when surface plating techniques were used.

IS/A Committee Liaison: M. L. Speck

Current research has indicated that regardless of the types of stress imposed upon a bacterial population (a) cellular injury is rapidly repaired when injured cells are incubated in an appropriate repair medium at an appropriate temperature[16]; (b) the process of repair precedes cell multiplication; and (c) the completely repaired cells respond normally to the selective agents present in the medium used for their detection. It is therefore desirable to allow stressed cells to repair any damage before they are enumerated by customary procedures on selective media. The procedures that promote maximum repair must be evaluated to minimize or eliminate multiplication of competing organisms.

The value of a period of repair before attempting to quantitate a foodborne pathogen was recognized by North.[19] He used lactose as a nonselective or preenrichment broth prior to enrichment in selenite cystine or tetrathionate broth for detecting small numbers of salmonellae in dried egg products. The success of this method was attributed correctly to restoration of a large number of cells to a state of active growth after a period of dormancy or trauma caused by the processing of egg products. Nevertheless, the 24 hour preincubation period made it difficult to distinguish repair from survivor growth. Recognizing this problem, Montford and Thatcher[18] recommended that the preincubation period be reduced to 5 hours, minimizing the growth of competing organisms present in the sample. Currently recognized procedures for the detection of salmonellae in dried egg products, frozen pasteurized liquid eggs, and frozen processed foods call for incubation of the sample in lactose broth or equivalent for 24 to 48 hours at 35 to 37 C before proceeding with selective enrichment.[4]

Bissonnette et al [5] found that *E. coli* and *S. faecalis* in water samples from stream environments could not grow on selective medium, but retained their colony forming ability on a nutritionally rich, nonselective medium. Their results suggested that a preenrichment period in a rich medium to allow repair of injury results in enhanced recovery of sanitary indicator organisms from stressful aquatic environments.

A preenrichment in trypticase soy broth may be necessary for the recovery of stressed *Staphylococcus aureus* cells present in processed food products.[12] Various selective agents, including high levels of potassium tellurite, used in media to select for pathogenic staphylococci, are inhibitory to the growth of stressed cells.[10, 14] In addition, trypticase soy broth + 10% NaCl, used in an MPN procedure to quantitate *S. aureus* from foods, is lethal to thermally stressed cells.[3, 10] A preenrichment period was necessary to obtain reasonable estimates of numbers of viable heat stressed *S. aureus* cells present in a food slurry using an MPN procedure in trypticase soy broth + 10% NaCl.[3] However, preliminary results indicated that use of a preenrichment period did not increase numbers of staphylococci recovered from contaminated commercial cheese and noodle samples.[3] Several enumeration procedures incorporating a preenrichment period to facilitate recovery of stressed staphylococci have been proposed.

Warseck et al [27] using information derived from studies by Ray and Speck,[21, 22] have reported on a method for the repair and enumeration of injured coliforms in frozen foods. It was found that injured cells repaired rapidly without multiplying if incubated in trypticase soy broth for one hour at 25 C from foods frozen at −20 C. Preincubation in trypticase soy broth resulted in a 20 fold increase in coliform count in some samples.

Speck and associates[24, 25] used a preenrichment period on trypticase soy agar (TSA) for enumerating coliforms injured by freezing. Surface or pour plated TSA plates were incubated for one to two hours at 25 C before overlaying with violet red bile agar (VRBA). Hartman et al[11] surface plated food samples containing injured coliforms on VRBA made without the selective bile salts and dyes. The agar plates were overlaid with VRBA containing double the usual concentrations of bile salts, neutral red, and crystal violet. The time between plating of samples and overlaying the plates was not critical.

Several investigators, as cited by Adams,[1] have shown that the recovery of severely heated *Clostridium perfringens* spores can be improved greatly if the enumeration medium is supplemented with lysozyme. Alderton et al [2] also have shown that the lysozyme in the recovery medium can increase the recovery of heated spores of *C. botulinum* type E and type A.

7.2 TREATMENT OF SAMPLE.

Chapter 1

7.3 SPECIAL EQUIPMENT AND SUPPLIES.

Chapter 2

7.4 SPECIAL REAGENTS AND MEDIA.

Chapter 2

7.5 PREPARATION OF STANDARDS

None required.

7.6 RECOMMENDED CONTROLS

The detection and enumeration of all microorganisms require that suitable uninoculated control media be included in the incubation. However, for the detection of stressed microorganisms, a rich noninhibitory medium that is suitable for the growth of both normal and environmentally stressed organisms, must be used to indicate the total viable population present and to eliminate any false changes in the population due to multiplication of cells in the sample.

Current investigations show that the preincubation time needed to facilitate repair of stressed cells varies with the type of food product in question, the number of contaminating microorganisms present, and the degree to which these microorganisms are injured. Therefore, another valuable control is the selection of plating times after preincubation of the sample in a nonselective medium. Ideally, several platings should be made during the preincubation period on both a nonselective medium and on a selective plating medium. The plating time used to report the number of organisms present in the sample should indicate that recovery of injured cells is complete, and that cell multiplication has not yet begun.

7.7 PRECAUTIONS AND LIMITATIONS OF METHODS

Several limiting constraints must be considered when attempting to assay stressed organisms. These arise as a result of the processing of the food product, or as a consequence of the handling of the cells subsequent to the debilitating treatment.

The first of these to be considered is that the injured microorganisms may be unable to multiply on the selective media used for their enumeration. This may be due to too high a concentration of the selective agent in the medium, or to an extreme degree of stress experienced during the processing treatment. A second limitation concerns the toxicity of the selective agent to injured bacteria. Unreal decreases in the number of organisms present can occur when the selective agent is partially lethal at the onset of recovery.

The problem of multiplication during recovery has been alluded to in the section regarding controls. Nonetheless, when repair is done in liquid media, care must be exercised in distinguishing between recovery and multiplication. A consequence of undetected multiplication is a false high estimate of not only the injured population, but of the total viable numbers as well. To circumvent this problem, monitoring of the sample (during recovery) should be maintained by plating both on a rich noninhibitory medium as well as on a selective medium.[3, 22, 25] To avoid such problems, repair on solid media has been proposed.[11, 22, 23]

Low natural populations create an additional problem when chemically processed foods are in question.[24] In these situations the agents responsible for the stress may be carried over to the incubation mixture for recovery, or to the noninhibitory medium used to estimate the total number of cells present. These materials may cause apparent decreases in the viable number and hence errors in the interpretation of data.[24]

7.8 PROCEDURE

7.81 Staphylococci

a. Aseptically weigh a 50 g sample in a sterile beaker.

b. Transfer to a sterile blender jar or a stomacher bag and add 450 ml of room temperature trypticase soy broth (this is a 1 : 10 dilution). Blend for one minute at high speed, or stomacher for 2 to 4 minutes.

c. Incubate the blender jar or stomacher bag at 35 to 37 C for 2 hours.

d. Make additional decimal dilutions using 0.1% peptone water as required.

e. Surface plate a 0.1 or 0.5 ml sample per plate in duplicate from each dilution on Baird-Parker agar. If samples contain low levels of organisms and plating more of the sample is desired, the number of plates per sample may be increased.

f. Incubate the plates at 35 C for 36 to 48 hours before counting colonies.

g. Samples, especially with low levels of staphylococci, also can be enumerated after the recovery treatment by the Most Probable Number (MPN) technique using trypticase soy broth containing 10% NaCl.

7.82 Salmonellae[4]

See Chapter 25 for details. *Salmonella* in processed foods should be enumerated using a preenrichment procedure. Lactose broth is the medium most commonly used.

7.83 Coliforms

7.831 Liquid media repair method[27]

This procedure has been used experimentally for frozen foods.

a. Thoroughly mix a 50 g sample with 450 ml trypticase soy broth.
b. Incubate food slurry at 25 C for one hour.
c. Make appropriate additional decimal dilutions in 0.1% peptone water.
d. Samples in 1.0 ml portions are pour plated in duplicate with violet red bile agar from appropriate dilutions. After 10 to 15 minutes at room temperature, layer an additional 4 ml of violet red bile agar tempered to 45 C over the surface.
e. Incubate plates at 35 C for 20 to 24 hours before counting colonies.

7.832 Solid media repair method[24, 25]

This method has been tested with commercial semipreserved foods.

a. Thoroughly mix a 50 g sample with 450 ml of 0.1% peptone water.
b. Make appropriate decimal dilutions with 0.1% peptone water.
c. Samples in 0.5 ml or 1.0 ml portions are pour plated in duplicate with 5 ml trypticase soy agar. Sample in 0.5 ml portions can be used per slate if surface plating method is used; in such case prepoured TSA plates with well dried surfaces are used. Preliminary studies indicate that plate count agar (PCA) may replace the TSA.
d. Incubate the plates at 35 C for 2 hours to effect repair. This incubation period is necessary when surface plating method is used; one hour is adequate for the pour-plating procedure.
e. Overlay the plates with 10 ml of violet red bile agar tempered to 45 C.
f. Incubate the plates at 35 C for 24 hours.
g. Count red to pink colonies that are at least 0.5 mm in diameter and report as coliform count per gram.
h. Confirm representative colonies on brilliant green lactose bile broth for gas production within 48 hours at 35 C.

7.84 Membrane Filter Method

Goff et al [9] have used membrane filters experimentally for recovering and enumerating stressed cultures. Dilutions of stressed cell samples were filtered through Millipore filters (type HAWG, 0.45 μm pore size; Millipore Corp., Bedford, Mass.). The membrane was placed on trypticase soy agar or a filter pad soaked with trypticase soy broth and incubated at 37 C for 1 to 4 hours to allow injured cells to recover. The membrane filter was then transferred to the surface of a selective medium and incubated for 48 to 72 hours at 37 C to isolate staphylococci. Claydon has applied this technique to pasteurized milk.[8]

This method has the advantage that the repair treatment can be adjusted to allow even the most severely damaged microorganisms to repair without danger of overgrowth by the others. Theoretically, this procedure can be made applicable to any bacteria by altering either the repair medium and medium used for isolation, the temperature of incubation, or the oxygen tension during incubation.

One drawback to this procedure is that food samples containing very low numbers of microorganisms cannot be enumerated due to the clogging of the filter with food particles present in low dilutions. This problem can be minimized by prefiltering the suspension before passing it through the Millipore filter.

7.9 INTERPRETATION OF DATA

The number of organisms per gram of food is equal to the number of colonies, either on the surface of the agar plate or on the surface of the filter membrane, multiplied by the dilution factor used to achieve countable plates.

If MPN techniques are used, standard tables relative to the number of tubes used (e.g., three tube or five tube) should be consulted.

7.10 REFERENCES

1. Adams, D. M. 1974. Requirement for and sensitivity to lysozyme by *Clostridium perfringens* spores heated at ultrahigh temperatures. Appl. Microbiol. **27**:797–801
2. Alderton, G., J. K. Chen, and K. A. Ito. 1974. Effect of lysozyme on the recovery of heated *Clostridium botulinum* spores. Appl. Microbiol. **27**:613–615
3. Allen, M. G., Z. J. Ordal, and J. C. Olson, Jr. 1975. Use of preincubation period in recovering stressed cells of *Staphylococcus aureus* from foods. J. Food Sci. Submitted for publication
4. Bacteriological Analytical Manual for Foods. 1972. Food and Drug Administration, Washington, D.C.
5. Bissonnette, G. K., J. J. Jezeski, G. A. McFeters, and D. G. Stuart. 1975. Influence of environmental stress on enumeration of indicator bacteria from natural waters. Appl. Microbiol. **29**:186–194
6. Busta, F. F. and J. J. Jezeski. 1963. Effect of sodium chloride concentration in an agar medium on growth of heat-shocked *Staphylococcus aureus*. Appl. Microbiol. **11**:404–407
7. Clark, C. W., L. D. Witter, and Z. J. Ordal. 1968. Thermal injury and recovery of *Streptococcus faecalis*. Appl. Microbiol. **16**:1764–1769
8. Claydon, T. J. 1975. A membrane-filter technique to test for the significance of sublethally injured bacteria in retail pasteurized milk. J. Milk Food Technol. **38**:87–88
9. Goff, J. H., T. J. Claydon, and J. J. Iandola. 1972. Revival and subsequent isolation of heat-injured bacteria by a membrane filter technique. Appl. Microbiol. **23**:857–862
10. Gray, R. J. H., M. A. Gaske, and Z. J. Ordal. 1974. Enumeration of thermally stressed *Staphylococcus aureus* MF 31. J. Food Sci. **39**:844–846
11. Hartman, P. A., P. S. Hartman, and W. W. Lanz. 1975. Violet red bile 2 agar for stressed coliforms. Appl. Microbiol. **29**:537–539
12. Heidelbaugh, N. D., D. B. Rowley, E. M. Powers, C. T. Bourland, and J. L. McQueen. 1973. Microbiological testing of skylab foods. Appl. Microbiol. **25**:55–61
13. Huhtanen, C. N., A. R. Brazis, W. L. Arledge, C. B. Donnelly, R. E. Ginn, H. E. Randolph, and E. J. Koch. 1975. Temperature equilibration times of plate count agar and a

14. IANDOLA, J. J. and Z. J. ORDAL. 1966. Repair of thermal injury of *Staphylococcus aureus*. J. Bacteriol. **91**:134–142
15. KLEIN, D. A. and S. WU. 1974. Stress: a factor to be considered in heterotrophic microorganism enumeration from aquatic environments. Appl. Microbiol. **27**:429–431
16. LAWTON, W. C. and F. E. NELSON. 1955. Influence of sublethal treatment with heat or chlorine on the growth of psychrophilic bacteria. J. Dairy Sci. **38**:380–386
17. MAXCY, R. B. 1970. Non-lethal injury and limitations of recovery of coliform organisms on selective media. J. Milk Food Technol. **33**:445–448
18. MONTFORD, J. and F. S. THATCHER. 1961. Comparison of four methods of isolating salmonellae from foods and elaboration of a preferred procedure. J. Food Sci. **26:**510–517
19. NORTH, W. R. 1961. Lactose pre-enrichment method for isolation of salmonellae from dried egg albumen. Appl. Microbiol. **9**:188–195
20. ORDAL, Z. J. 1970. Current developments in detection of microorganisms in foods: Influence of environmental factors on detection methods. J. Milk Food Technol. **33**:1–5
21. RAY, B. and M. L. SPECK. 1973. Discrepancies in the enumeration of *Escherichia coli*. Appl. Microbiol. **25**:494–498
22. RAY, B. and M. L. SPECK. 1973. Enumeration of *Escherichia coli* in frozen samples after recovery from injury. Appl. Microbiol. **25**:499–503
23. SHARPE, A. N. and A. K. JACKSON. 1972. Stomaching: A new concept of bacteriological sample preparation. Appl. Microbiol. **24**:175–178
24. SPECK, M. L. and B. RAY. 1975. The effect of freezing and storage on microorganisms in frozen foods. J. Food Sci. Submitted for publication
25. SPECK, M. L., B. RAY, and R. B. READ, JR. 1975. Repair and enumeration of injured coliforms by a plating procedure. Appl. Microbiol. **29**:549–550
26. VANDERZANT, C. and A. W. MATTHYS. 1965. Effect of temperature of the plating medium on the viable count of psychrophilic bacteria. J. Milk Food Technol. **28**:383–388
27. WARSECK, M., B. RAY, and M. L. SPECK. 1973. Repair and enumeration of injured coliforms in frozen foods. Appl. Microbiol. **26**:919–924

MICROORGANISMS INVOLVED IN PROCESSING AND SPOILAGE OF FOODS

CHAPTER 8

PSYCHROTROPHIC MICROORGANISMS

S. E. Gilliland, H. D. Michener, and A. A. Kraft

8.1 INTRODUCTION

8.11 Definition

Microorganisms which grow in foods at refrigeration temperatures have usually been called psychrophilic. Although this term implies optimum growth at low temperatures, relatively few of the psychrophilic microorganisms isolated from foods have optimum growth temperatures below 20 C.[15]

It has been suggested that the term "psychrotrophic" be applied to those organisms able to grow relatively rapidly at commercial refrigeration temperatures without reference to optimum temperature for growth.[9] Species of *Pseudomonas, Achromobacter, Flavobacterium* and *Alcaligenes* are often included among the psychrotrophic bacteria, although some of these genera may no longer be recognized. Certain molds (*Geotrichum, Botrytis*) also are able to grow in refrigerated foods.

8.12 Significance in Foods

Many psychrotrophic bacteria when present in large numbers can cause a variety of off flavors as well as physical defects in foods. Their growth rate is highly temperature dependent, and becomes increasingly slower as the temperature is reduced. Therefore, shelf life or the rate of quality loss, and subsequent spoilage of a refrigerated food is also highly temperature dependent.[4,15]

The enumeration of psychrotrophic bacteria in foods that are to be stored refrigerated (0 to 10 C) is important, because their presence (particularly in large numbers) indicates a high potential for spoilage during extended storage. The extended period of refrigerated storage of both raw and processed foods, by centralization of food handling and processing facilities, has increased the importance of psychrotrophic bacteria. Raw foods held under refrigeration prior to processing, as well as nonsterile heat processed foods that rely on refrigeration for shelf life, are subject to quality loss and possible spoilage by psychrotrophic bacteria.

IS/A Committee Liaison: Carl Vanderzant

Most psychrotrophic bacteria are destroyed by a mild heat treatment such as pasteurization. However, some heat resistant types, such as some species of *Bacillus* and *Clostridium*, may survive.[3,5,6,11,13] In most cases the presence of psychrotrophic bacteria in heat processed foods implies post-processing contamination.

The presence of psychrotrophic bacteria is important also in frozen foods such as chicken or turkeys when they are thawed (and sometimes reprocessed) for retail sale as refrigerated food. In foods which are kept frozen until they reach the consumer, large numbers of psychrotrophic bacteria may indicate a history of unsanitary handling. However, they will not cause spoilage unless such foods are subjected to temperature abuse.

Although psychrotrophic bacteria will not grow in frozen foods, they can grow and cause spoilage if the food is allowed to thaw partially, and is subsequently held at too high a temperature (i.e., unfrozen but still refrigerated). Some microbial species can grow slowly at temperatures as low as –5 to –12 C. Quality losses because of microbial activities may occur in such foods after prolonged storage periods.[7,8]

8.2 GENERAL METHODS

Psychrotrophic microorganisms usually are enumerated by plating (pour or spread plate method) food samples on nonselective media, with incubation of plates at refrigeration temperatures such as 7 C for 10 days.

In some refrigerated foods, a majority of the psychrotrophic bacteria responsible for quality losses and subsequent spoilage are gram negative rods. To detect postpasteurization contamination of pasteurized milks, plating media have been proposed (such as crystal violet tetrazolium agar) which are intended to inhibit gram positive bacteria and detect gram negative species. Plates with these selective media are incubated at a temperature higher than 7 C, and consequently colonies can be counted sooner. However, in foods where gram positive species are a significant part of the psychrotrophic population (cured meats for example), this approach is not advisable.

A "Keeping Quality Test", in which the agar plate count of a food is determined prior to, and following, preliminary incubation of the food (5 to 7 days at commercial refrigeration temperature), can provide valuable information about the potential for the development of a psychrotrophic flora in a refrigerated food.[1]

8.3 TREATMENT OF SAMPLE

Collection and treatment of samples for the psychrotrophic count are essentially the same as for the agar plate count (Chapter 4). However, refrigerated storage of samples should be minimized because psychrotrophic bacteria will grow under these conditions. Microbial counts after storage may no longer reflect the count at the time of sampling.

8.4 EQUIPMENT AND MEDIA

Refer to Chapter 4 for standard equipment and supplies needed. A low temperature incubator should maintain a temperature of 7 C ± 1 C.

8.41 Media

8.411 Nonselective agar: media such as standard methods agar (Chapter 2) or trypticase soy agar (Chapter 2) can be employed for the psychrotrophic count.

8.412 Selective agar: Crystal violet tetrazolium agar (CVT).[10, 12] (Chapter 2)

8.5 PRECAUTIONS AND LIMITATIONS

8.51 Plate Incubation Temperature

In addition to the precautions and limitations listed for the plate count method (Chapter 4), it is important to consider the plate incubation temperature for the psychrotrophic count as compared to the temperature at which the food is actually stored. If the temperatures differ considerably, the counts may not reflect the microorganisms which will grow under refrigerated storage of the food. Also, microorganisms which grow on laboratory media at low temperatures do not necessarily grow in foods at these temperatures. The reverse may also be true in some cases.

8.52 Pour Plate versus Spread Plate Technique

Psychrotrophic bacteria are easily injured or killed when plates are poured with agar held above 44 to 46 C.[10, 16] Even in that range some cells will die; therefore, as an alternative, a surface or spread plate method can be employed (Chapter 4).[14]

8.53 Injured Cells

Psychrotrophic bacteria may have received sublethal injury if the food has been heated or frozen (Chapter 7). This may extend their lag phase so that the recommended incubation time is insufficient to detect them, or this may make them more fastidious so that they do not grow on the routinely used plating media.[14] In either case they could eventually grow and cause deterioration in refrigerated food, although undetected by the usual analytical methods.

8.54 Blending of Sample

Standard plating techniques in which an electric blender is used to mix sample and diluent usually give reproducible counts with bacteria and yeasts, but not necessarily with molds. The more vigorous the blending, the more the mycelium is cut into small pieces, each of which may develop into a colony. The number of mycelium fragments depends not only on blending time, but also on speed, sharpness of the blades, and volume. Details on enumeration of molds and yeasts are presented in Chapter 16.

In some blenders and particularly when blending time exceeds about 2 minutes, localized heating and injury or death of sensitive psychrotrophic bacteria must be considered. When longer blending times are required, blend the sample for 2 minutes, turn the blender off for 2 minutes to allow cooling, then blend again for 2 minutes if necessary.

8.6 PROCEDURES

8.61 Plate Count Method: Nonselective Media

The agar plate count method using nonselective media such as standard methods agar (Chapter 2) or trypticase soy agar (Chapter 2) is recommended. Prepare dilutions, pour plates, or spread plates as described for the agar plate count (Chapter 4). Incubate plates at 7 C ± 1 C for 10 days.[1]

With the spread plate method the incubation time can be reduced by 2 to 3 days.[2,14] In another method the plates are incubated first at 17 C for 16 hours and then at 7 C for 3 days. Visible colonies do not develop during the first 16 hours but development at 7 C is much more rapid thereafter.[14]

Count colonies and compute counts as described in Chapter 4. Report counts as psychrotrophic plate count per milliliter, gram or cm^2 as applicable.

8.62 Plate Count Method: Selective Media

Prepare dilutions and plates as described for the agar plate count (Chapter 4) except that crystal violet tetrazolium (CVT) agar (Chapter 2) is used. Incubate plates for 48 hours at 30 C or 5 days at 22 C. Count red colonies and compute count as for plates with nonselective media.

8.63 Keeping Quality Test

Determine the agar plate count (Chapter 4) on a fresh sample of the product and again after storage (5 to 7 days) of a sample at normal commercial refrigeration temperatures for the product. Incubate plates for 48 hours at 30 C or 5 days at 22 C.

8.7 INTERPRETATION

When psychrotrophic bacteria are recovered from a food by plating, with incubation of plates at an arbitrarily selected low temperature, it means only that the organisms appearing on the plates may grow in the food at refrigeration temperatures. Since temperatures in this range often are close to the minimum growth temperature for these organisms, a slight reduction in temperature may suppress growth, thus changing the level and composition of the emergent microbial population.

The selection of a temperature for plate incubation such as 7 C is arbitrary and the analyst may want to select a temperature equal to that at which the refrigerated food is normally stored. This approach, however, does not allow comparison of counts between various types of samples. Also, the choice of 7 C, or any other incubation temperature in this range, is arbitrary, because there is no sharp dividing line for growth of psychrotrophic and mesophilic bacteria.

The presence of large numbers of psychrotrophic bacteria in refrigerated foods such as dairy products, meat, poultry, and seafood may reflect growth of the initial population during storage and/or massive contamination at some point prior to or during refrigerated storage. Whether or not flavor or physical defects will appear depends largely on the level and biochemical characteristics of the microbial species. Even if no defect appears, the bacterial counts of such foods may exceed legal levels.

Psychrotrophic plate counts on freshly processed products (to be stored under refrigeration) should be interpreted with caution, particularly if the food is to be stored for extended periods. A few cells capable of growing at low temperatures may result in large populations (10^6 to 10^8/ml or g) in days or weeks, particularly if the food is held at marginal refrigeration temperatures.

The "Keeping Quality Test" has the advantage that the food is preincubated at temperatures similar to those used in commercial samples, and aids in interpretation of counts as described above.

The use of special selective media which allow growth of gram negative spoilage bacteria and suppress gram positive organisms to evaluate 'low temperature spoilage bacteria" in foods has certain limitations. The selective agents are frequently inhibitory to some gram negative species, particularly when the bacteria are sublethally injured. However, quality losses and subsequent spoilage during refrigerated storage of some major food commodities are not confined to gram negative species only.

8.8 REFERENCES

1. Standard Methods for the Examination of Dairy Products. 13th ed. 1972. American Public Health Association, New York, N.Y.
2. Baumann, D. P. and G. W. Reinbold. 1963. Enumeration of psychrophilic microorganisms. A review. J. Milk Food Technol. **26**:81–86
3. Bhadsavle, C. H., T. E. Shehata, and E. B. Collins. 1972. Isolation and identification of psychrophilic species of *Clostridium* from milk. Appl. Microbiol. **24**:699–702
4. Elliott, R. P. and H. D. Michener. 1965. Factors affecting the growth of psychrophilic microorganisms in foods—A review. U. S. Dept. Agr. Tech. Bull. No. 1320, 110 pp
5. Grosskopf, J. C. and W. J. Harper. 1969. Role of psychrophilic sporeformers in long life milk. J. Dairy Sci. **52**:897 (Abstr)
6. Larkin, J. M. and J. L. Stokes. 1966. Isolation of psychrophilic species of *Bacillus*. J. Bacteriol. **91**:1667–1671
7. Michener, H. D. and R. P. Elliot. 1964. Minimum growth temperatures for food-poisoning, fecal indicator, and psychrophilic microorganisms. Advances in Food Research, Vol. 13, Academic Press, New York, N.Y. pp. 349–396
8. Michener, H. D. and R. P. Elliott. 1969. Microbiological conditions affecting frozen food quality, p. 43–84. In W. B. Van Arsdel, M. J. Copley, and R. L. Olson (ed.), Quality and Stability of Frozen Foods. Wiley and Sons, New York, N.Y.
9. Mossel, D. A. A. and H. Zwart. 1960. The rapid tentative recognition of psychrotrophic types among *Enterobacteriaceae* isolated from foods. J. Appl. Bacteriol. **23**:185–188

10. OLSON, H. C. 1963. Selective plating technique for detecting contamination in pasteurized milk. J. Dairy Sci. **46**:362. (Abstr)
11. OLSON, H. C. 1963. Spoilage of milk by thermoduric psychrophiles. J. Dairy Sci. **46**:362 (Abstr)
12. OLSON, H. C. 1971. Bacteriological testing of milk for regulatory purposes—usefulness of current procedures and recommendations for change. V. Pasteurized milk. J. Milk Food Technol. **34**:279–281
13. SHEHATA, T. E. and E. B. COLLINS. 1971. Isolation and identification of psychrophilic species of *Bacillus* from milk. Appl. Microbiol. **21**:466–469
14. THOMAS, S. B. 1969. Methods of assessing the psychrotrophic bacterial content of milk. J. Appl. Bacteriol. **32**:269–296
15. TOMPKIN, R. B. 1973. Refrigeration temperature as an environmental factor influencing the microbial quality of food—A review. Food Technol. **27**:54–58
16. VANDERZANT, C. and A. W. MATTHYS. 1965. Effect of temperature of the plating medium on the viable count of psychrophilic bacteria. J. Milk Food Technol. **28**:383–388

CHAPTER 9

THERMODURIC MICROORGANISMS

F. E. Nelson

9.1 INTRODUCTION

Thermoduric organisms can be defined as those which will survive some significant measure of heat treatment. The survival may constitute only a small percentage of the initial population, or most of the cells may survive. Differentiation between thermophilic and thermoduric organisms commonly is on the basis that the former not only survive the heat treatment, but also grow at the elevated temperatures used. Bacteria are the principal thermoduric organisms encountered in foods, but molds such as *Byssochlamys fulva* (which will be treated separately in Chapter 17) and an occasional strain of *Aspergillus* or *Penicillium* qualify as thermoduric under some circumstances. Thermoduric bacteria have received the greatest attention in connection with milk and milk products, where the term commonly is applied to organisms which remain viable after pasteurization.[3,5,11]

The genera *Micrococcus, Streptococcus* (primarily the enterococci), *Microbacterium, Arthrobacter, Lactobacillus, Bacillus,* and *Clostridium* (the last requiring special anaerobic technics for recovery) are recognized usually as containing some species which will qualify as thermoduric. Other genera may be represented occasionally when very high initial counts result in a few cells surviving a heat treatment which ordinarily would result in a complete kill. Because of the extremely diverse characteristics of the organisms involved, few generalizations concerning their actions on foods can be made. They may range from no detectable change, through considerable acid production to extensive proteolysis and/or lipolysis. Thermoduric organisms usually do not grow to any significant degree at good refrigeration temperatures, but psychrotrophic *Bacillus* sp. have been encountered,[4, 9] and enterococci may grow slowly.

The extent of survival after heat treatment depends upon many factors. Even within a species, the sensitivity to heat can vary markedly between strains. Rapidly growing organisms usually are more sensitive than are those that have reached a growth plateau. Large populations increase considerably the probability that an occasional cell, or a few cells, will survive a given heat treatment, since the death curve usually is dominantly logarith-

IS/A Committee Liaison: Carl Vanderzant

mic. The character of the food being heated, particularly the pH, and the presence of protective materials such as sugars, fats, and proteins, influence survival to a considerable degree.

9.2 TREATMENT OF SAMPLE

Samples shall be taken with the same precautions against contamination and growth as used for samples for "total" counts (Chapter 1). Swab, rinse, and similar samples shall be taken as outlined in Chapter 3.

9.3 EQUIPMENT

9.31 Low Temperature

For the usual low temperature, long holding time procedure, such as recommended in Standard Methods for the Examination of Dairy Products (SMEDP)[1]:

a. Test tubes, sterile: screw capped 20 × 125 mm; 18 mm cap with rubber or plastic liner.
b. Pipets, sterile: graduated, 5 ml, 10 ml, or 11 ml delivery.
c. Thermometers: The thermometers used in the water bath and in the pilot tube with sample should cover the critical range of temperature during heat treatment, should have divisions of 0.1 C (0.2 F), and should be checked at least biennially with an NBS certified thermometer.
d. Water bath: electrically heated, thermostatically controlled (at 62.8 ± 0.5 C {145 ± 0.9 F} for milk work), equipped with stirrer and thermometer. The water volume must be enough to absorb the cooling effect of tubes placed in the bath without a drop in temperature of more than 0.5 C (0.9 F).
e. Metal or wire rack for holding test tubes.

9.32 Higher Temperatures

For heating to higher temperatures for shorter periods of time, special equipment and materials adequate for the particular conditions involved must be employed.[2,10]

9.4 MEDIA

Either standard methods agar (Chapter 2) or, alternatively, trypticase soy agar (Chapter 2) may be used for pouring the plates for enumeration. Incubation of plates shall be at 32 C (30 C, if this is authorized in other sections) for 48 ± 3 hours (unless 72 hours incubation is generally accepted).

9.5 PRECAUTIONS

The conditions used for recovery of sublethally heat stressed cells (Chapter 7) markedly influence apparent survival, since the stressed cells are, to a considerable degree, more demanding in their requirements for initiation of growth (recovery from stress as employed by some microbiologists) than are unstressed cells.[12,13,14] Although many factors involved remain to be determined, presence of some complex nutrients, absence or low levels of selective agents such as NaCl, sodium azide and some antibiotics, pH level, and time and temperature of incubation during growth for enumeration, all affect the level of apparent survival, in some instances very markedly. Standardization of conditions for enumeration is necessary in order to obtain comparable results in several laboratories, especially when organisms stressed by heat or by other factors are to be enumerated.

9.6 PROCEDURE

Liquid samples shall be mixed thoroughly to provide homogeneity. Solid and semisolid samples shall be prepared in an initial 1:10 dilution by homogenization with a mechanical blender (Chapter 1), using phosphate-buffered distilled water (SMEDP).[1]

A 5 ml quantity of the initial liquid sample or of the 1:10 dilution is transferred aseptically to a sterile test tube, using precautions to avoid contamination of the tube above the level filled by the sample. Material deposited on the upper portions of the tube may lead to false results, as it may not be subjected to the same degree of heat as the bulk of the material, and it also may dry somewhat, and dry heat is considerably less microbicidal than is moist heat. While a group of tubes is being prepared for heating, all should be held in an ice bath, both to retard microbial growth, and to have all tubes at a uniform temperature for standardization of subsequent heating rate. A pilot tube containing sample, with thermometer inserted therein, shall be used to monitor temperatures at all stages. A rack of tubes is placed in the bath after the bath temperature has stabilized at the desired level. The tubes may be immersed completely (tightly closed) or they may be immersed so that the water line is approximately 4 cm above sample level. Timing begins when the pilot material is within 0.5 C (0.9 F) of the treatment temperature. At the end of the holding period (30 minutes for milk), cool the samples below 10 C (50 F) by immersion in an ice water bath. Determine the thermoduric count by the plate count method (Chapter 4) or, for nonregulatory purposes, by one of the simplified viable count procedures (Chapter 4). The results are reported as thermoduric count per unit of sample (milliliter, gram, cm^2, etc.).

9.7 INTERPRETATION

The thermoduric count has been used in the milk industry primarily as a test of the care employed in utensil sanitation and as a means of detecting sources of organisms responsible for high counts in pasteurized products. Regulatory agencies seldom incorporate this test in their control procedures for producer milk, but are involved indirectly by the enforcement of the requirement that the standard plate count on Grade A pasteurized milk shall not exceed 20,000 per ml. Levels of thermoduric organisms tolerated in producer milk vary in different situations, but counts of

<1000/ml are achievable, and finding of populations appreciably above this ordinarily would result in attempts to find the cause and correct it.[3,5]

The significance of thermoduric microorganisms in nondairy foods is not well established. However, similar considerations hold for pasteurized egg products as those given for the interpretation of thermoduric counts on pasteurized milk. Also thermoduric coryneform bacteria, such as *Microbacterium* sp. in heated sausage, may be a cause of spoilage of these products. Some types of enterococci have a degree of resistance to heating, but this group ordinarily is determined on the basis of tests other than thermal resistance, even in dairy products.[8] The thermoduric character of *Streptococcus faecium* frequently permits this organism to survive the pasteurization treatment given canned hams, and thus cause sour odor and flavor, as well as poor color retention after removal from the cans.[7]

Spoilage of refrigerated previously cooked foods usually is the result of postheating contamination, but a number of the thermoduric types can grow slowly and produce a defect, particularly when contamination with more definitely psychrotrophic organisms following heating was avoided or minimized.

Thermoduric bacteria frequently have low reducing ability in the dye reduction tests, and thermoduric and/or psychrotrophic bacteria constitute a high proportion of the microflora of manufacturing grade bulk tank milk when the reduction tests agree poorly with the plate count.[6]

Foodborne illness has not been associated with the presence of thermoduric organisms in foods, but enterococci (Chapter 30), which may be thermoduric, have been considered a possible cause of foodborne illness, even though the evidence is not unequivocal.[7]

9.8 REFERENCES

1. Standard Methods for the Examination of Dairy Products, 13th ed. 1972. W. J. Hausler, ed. American Public Health Assoc., Washington, D. C.
2. Dickerson, R. W., Jr., and R. B. Read, Jr. 1968. Instrument for study of microbial thermal inactivation. Appl. Microbiol. **16**:991–997
3. Foster, E. M., F. E. Nelson, M. L. Speck, R. N. Doetsch, and J. C. Olson, Jr. 1957. Dairy Microbiology. Prentice-Hall, Inc., Englewood Cliffs, N. J.
4. Grosskopf, J. C. and W. J. Harper. 1969. Role of psychrophilic sporeformers in long life milk. J. Dairy Sci. **52**:897 (Abstr.)
5. Hammer, B. W. and F. J. Babel. 1957. Dairy Bacteriology, 4th ed. John Wiley and Sons, Inc., New York, N.Y.
6. La Grange, W. S. and F. E. Nelson. 1965. Evaluation of dye reduction tests for manufacturing-grade bulk-tank milk. J. Dairy Sci. **48**:1129–1133
7. Niven, C. F., Jr. 1963. Microbial indexes of food quality: Fecal streptococci, pp 119–131 In L. W. Slanetz et al (ed.), Microbiological Quality of Foods. Academic Press, Inc., New York, N. Y.
8. Shannon, E. L., G. W. Reinbold, and W. S. Clark, Jr. 1970. Heat resistance of enterococci. J. Milk Food Technol. **33**:192–196
9. Shehata, T. E. and E. B. Collins. 1971. Isolation and identification of psychrophilic species of *Bacillus* from milk. Appl. Microbiol. **21**:466–469
10. Stroup, W. H., R. W. Dickerson, Jr., and R. B. Read, Jr. 1969. Two-phase slug flow heat exchanger for microbial thermal inactivation research. Appl. Microbiol. **18**:889–892

11. Thomas, S. B., R. G. Druce, G. J. Peters, and D. G. Griffiths. 1967. Incidence and significance of thermoduric bacteria in farm milk supplies: A reappraisal and review. J. Appl. Bacteriol. **30**:265–298
12. Thomas, W. R., G. W. Reinbold, and F. E. Nelson. 1963. Effect of temperature and time of plate incubation on the enumeration of pasteurization-resistant bacteria in milk. J. Milk Food Technol. **26**:357–363
13. Thomas, W. R., G. W. Reinbold, and F. E. Nelson. 1966. Effect of pH of plating medium on enumeration of pasteurization-resistant bacteria in milk. J. Milk Food Technol. **29**:156–160
14. Thomas, W. R., G. W. Reinbold, and F. E. Nelson. 1966. Effect of the type of bacteriological peptone in the plating medium upon the enumeration of pasteurization-resistant bacteria in milk. J. Milk Food Technol. **29**:182–186

CHAPTER 10

LIPOLYTIC MICROORGANISMS

J. A. Alford

10.1 INTRODUCTION

10.11 Hydrolytic and Oxidative Lipolysis

Many foods contain significant amounts of fat, and these fats are susceptible to hydrolysis and oxidation which lead to changes in flavor. Although many of the problems of fat breakdown are nonmicrobial in origin, numerous bacteria, yeasts and molds are capable of causing both hydrolytic and oxidative deterioration. Lipolytic counts usually are not performed on a routine basis. Food manufacturers and processors usually enumerate lipolytic types only when a problem occurs.

In the enumeration of lipolytic microorganisms, the microbiologist usually is concerned only with detection of hydrolytic activity, since detection of lipid oxidizing microorganisms does not lend itself to routine agar plate counts. Microbial oxidation of lipids has been investigated,[9] but additional studies are needed to determine chemical reactions indicative of important oxidative changes, and to develop methods for the routine detection of microorganisms bringing them about. Although only fatty acids of low molecular weight are sufficiently volatile to contribute directly to flavor changes, any free fatty acid (FFA) released by hydrolysis is more susceptible to oxidation than is a fatty acid esterified in a triglyceride. Thus, measurement of fat hydrolysis gives some indication of the potential for oxidative changes as well.

10.12 Foods Involved

The foods most often involved in problems of lipolysis are cream, butter, margarine, salad dressings, and other high fat products. Animal carcasses, seeds, and other natural fat containing materials may develop high lipolytic counts under some storage conditions (e.g., high humidity). Lipolytic and oxidative changes in meats and other fat containing foods are often associated with quality loss or spoilage. Desirable flavors in many cheeses and some other fermented foods also are associated with changes in the fat.

IS/A Committee Liaison: Carl Vanderzant

10.13 Microorganisms

The genera *Pseudomonas, Achromobacter* and *Staphylococcus* among the bacteria, *Rhizopus, Geotrichum, Aspergillus,* and *Penicillium* among the molds, and the yeast genera *Candida, Rhodotorula,* and *Hansenula* contain many lipolytic species.[5, 8]

10.14 Lipase Stability

Lipases are generally quite stable and may be active over a long period of time particularly at low temperatures, including frozen storage.[1]

10.2 TREATMENT OF SAMPLE

Samples are prepared and dilutions made as outlined in Chapters 1 and 3.

10.3 SPECIAL REAGENTS AND MEDIA

10.31 Reagents

a. Victoria Blue B (National Aniline Company, New York). If the Victoria Blue B is not available, Victoria Blue without a subdesignation may be used, although variations in color changes can be expected.

b. Petroleum ether. ACS specifications. bp 30 to 60 C.

c. Activated alumina. Glass columns diameter 2 to 3 cm × length 10 to 15 cm are packed with 25 g activated alumina which has been prewashed with petroleum ether. A column of this size will remove the FFA from 50 g fat containing 2 to 3% FFA.

10.32 Fat Substrates

a. Tributyrin: Reagent grade, from which the FFA have been removed as follows: Dissolve tributyrin in petroleum ether (5 to 10 g/100 ml) and pass through a column of activated alumina. Remove the petroleum ether from the purified triglyceride by evaporating it on a steam table under a stream of nitrogen. Note: If the tributyrin is known to contain only negligible amounts of FFA, the purification step may be omitted.

b. Corn oil, soy bean oil, and other fats which contain a good percentage of unsaturated fatty acids. Any fresh, commercially available cooking oil is suitable provided it does not contain antioxidants or other materials at an inhibitory level. FFA should be removed as described for tributyrin.

c. Solid fats such as lard, tallow, etc. also may be used, particularly when examining a food where only the solid fat is present. FFA should be removed as above and the fat melted before emulsification.

d. Other triglycerides such as triolein may be used, if cognizance is taken of the problems of emulsification, mono- and di-glyceride contamination and lipase specificity noted in 10.5.

10.33 Media

10.331 Base layer medium (Chapter 2)

10.332 Nutrient overlay medium (Chapter 2)

The following media are listed in order of preference

Nutrient agar
Standard methods agar
Casein soy peptone agar

10.333 Single layer medium

a. Base layer medium as in 10.331.

b. Nutrient overlay media prepared and sterilized as in 10.332 except that the ingredients per liter are contained in 800 ml.

To 20 ml of melted and mixed base layer medium, add 80 ml of the concentrated nutrient overlay medium, mix in blender an additional 5 seconds before use.

10.4 RECOMMENDED CONTROLS

Prepare test plates with each new batch of fat substrate and dye. Streak them with *Geotrichum candidum*, another lipolytic microorganism such as *Pseudomonas fragi*, and a nonlipolytic organism. Incubate plates at 20 to 25 C for 3 to 4 days. Interpretation: Tributyrin will be hydrolyzed very slowly, if at all, by *G. candidum.* Inhibitory substances in the oils or other fats are indicated by weak zones of hydrolysis by one or both lipolytic microorganisms.

10.5 PRECAUTIONS AND LIMITATIONS

10.51 Temperature

The temperature of incubation is important since lipase production is sometimes inhibited in cultures incubated at the upper limits of their growth range.[3, 8] Incubation should be at least 5 C below maximum growth temperature.

10.52 Substrate

True lipases attack only insoluble substrates, therefore, avoid soluble substrates such as simple esters, monoglycerides, Tweens, etc., which may make substrate preparation or detection of hydrolysis easier. Tributyrin is the simplest triglyceride occurring in natural fats and oils. Although a true fat, it is hydrolyzed by some microorganisms that will not hydrolyze triglycerides or fats containing only longer chain fatty acids. However, for screening purposes to enumerate lipolytic microorganisms of potential importance in foods, it is the substrate of choice,[3, 6, 8] unless *Geotrichum* species are present (10.53). As noted in 10.32, other substrates may give more meaningful information when examining a specific product (e.g., lard as substrate for lipolytics from pork). However, Section 10.5, and particularly 10.55, should be considered carefully.

10.53 Lipase Specificity

Most lipases have a broad specificity with respect to substrate, although the lipase from *G. candidum* and some *Candida* species hydrolyzes only unsaturated fatty acids from the triglycerides.[2, 5, 7] These species will appear nonlipolytic on tributyrin, but readily hydrolyze fats containing oleic, linoleic, and other unsaturated fatty acids.[4, 5] Additional research is needed to determine whether other microorganisms produce lipases with distinctive characteristics.

10.54 Growth Medium

Lipase production by some microorganisms is limited by the presence of a readily fermentable carbohydrate,[3, 8] while others require it for growth.[5, 10] The amount included in a medium should be limited to the amount required for reasonable growth. If the microorganism will grow on nutrient agar, it is the medium of choice; if not, media such as standard methods agar or casein soy peptone agar which contain limited amounts of carbohydrate may be used.

10.55 Emulsification

Good emulsification of the substrate is essential since the lipase activity occurs at the fat-water interface. Although vigorous shaking often will give a satisfactory medium, most consistent results are obtained by blending or homogenizing to produce small globules. Stabilizing gums such as gum acacia may be used, but emulsifiers such as Tweens, mono- and di-glycerides, should not be used because of their possible use as a substrate.

10.56 Multiple Plating

Tributyrin is the most widely used substrate because of its ease of handling and ease of hydrolysis. However, as indicated above, this may lead to both false positives and false negatives. Use of two substrates, at least for spot checks, is desirable.

10.57 Indicator Dyes

Victoria Blue, Spirit Blue, Nile Blue Sulfate, Night Blue, and other dyes have been used as indicators of fat hydrolysis. However, toxicity to one or more microorganisms has been reported for all of them.[4, 5, 8] Victoria Blue has as little toxicity as any of them, and is the dye of choice at present.

10.6 PROCEDURE

10.61 Double Layer Method

Pour base layer as described in 10.331

Prepare dilutions of the product and plate in the same manner described for Standard Plate Count (Chapter 4) except that the dilution is placed on surface of base layer and 10 to 12 ml of nutrient overlay is added and mixed within 2 to 3 minutes.

10.62 Single Layer Method

Double layered agar plates give the best detection of weakly lipolytic bacteria and should be used, particularly when added carbohydrate is necessary for good growth.[4, 5, 6, 10] However, in many cases regular plating procedures with a fat containing medium will give satisfactory counts.

Prepare dilutions of the product and plate as described under standard plate count method (Chapter 4) except that plates are poured with 12 to 15 ml of the single layer medium (10.333).

10.63 Incubation

Incubate plates at 20 to 25 C for 3 days if tributyrin is the fat substrate, and 4 to 7 days for other fats.

10.64 Counting Colonies and Reporting

On tributyrin agar without Victoria Blue B lipolytic colonies are indicated by a transparent zone surrounding the colony on an opaque background. On media containing the dye a dark blue zone surrounds lipolytic colonies on an opaque, light blue background. Indistinct zones sometimes occur because of weak lipolysis, acid production, etc. (see 10.5). Use of a stereoscopic microscope to detect these weak zones may be helpful in some instances. However, the investigator should remember that careful substrate preparation and incubation will do more to minimize these problems than trying to interpret their meaning. Count colonies and report as lipolytic count (substrate) per milliliter (or per gram).

10.7 REFERENCES

1. ALFORD, J. A. and D. A. PIERCE. 1961. Lipolytic activity of microorganisms at low and intermediate temperatures. III. Activity of microbial lipases at temperatures below 0 C. J. Food Sci. **26**:518–524
2. ALFORD, J. A., D. A. PIERCE, and F. G. SUGGS. 1964. Activity of microbial lipases on natural fats and synthetic triglycerides. J. Lipid Res. **5**:390–394
3. ALFORD, J. A., J. L. SMITH, and H. D. LILLY. 1971. Relationship of microbial activity to changes in lipids of foods. J. Appl. Bacteriol. **34**:133–146
4. ALFORD, J. A. and E. E. STEINLE. 1967. A double layered plate method for the detection of microbial lipolysis. J. Appl. Bacteriol. **30**:488–494
5. BOURS, J. and D. A. A. MOSSEL. 1973. A comparison of methods for the determination of lipolytic properties of yeasts mainly isolated from margarine, moulds and bacteria. Arch. Lebensm. Hyg. **24**:197–203
6. FRYER, T. F., R. C. LAWRENCE, and B. REITER. 1967. Methods for isolation and enumeration of lipolytic organisms. J. Dairy Sci. **50**:477–484
7. JENSEN, R. G., J. SAMPUGNA, J. G. QUINN, D. L. CARPENTER, T. A. MARKS, and J. A. ALFORD. 1965. Specificity of a lipase from *Geotrichum candidum* for *cis*-octadecenoic acid. J. Am. Oil Chem. Soc. **42**:1029–1032
8. LAWRENCE, R. C. 1967. Microbial lipases and related esterases. Dairy Sci. Abst. **29**:1–8, 59–70
9. SMITH, J. L. and J. A. ALFORD. 1969. Action of microorganisms on the peroxides and carbonyls of fresh lard. J. Food Sci. **34**:75–78
10. UMEMOTO, Y. 1969. A method for the detection of weak lipolysis of dairy lactic acid bacteria on double-layered agar plates. Agr. Biol. Chem. (Japan) **33**:1651–1653

CHAPTER 11

PROTEOLYTIC MICROORGANISMS

J. S. Lee

11.1 INTRODUCTION

Protein hydrolysis by microorganisms in foods may produce a variety of odor and flavor defects. Some of the common psychrotrophic spoilage bacteria are strongly proteolytic and cause undesirable changes in dairy, meat, poultry, and seafood products particularly when high populations are reached after extended, refrigerated storage. On the other hand, microbial proteolytic activity may be desirable in certain foods, such as in the ripening of cheese where it contributes to the development of flavor, body, and texture. Opinions differ about the usefulness of proteolytic counts to evaluate quality losses of refrigerated dairy, meat, poultry, and fishery products.[4,6,7]

In some foods the level of proteolytic microorganisms may be of value to project refrigerated storage life and to assess processing methods.[6,7]

Proteolytic species are common among the genera *Bacillus, Clostridium, Pseudomonas,* and *Proteus*. Microorganisms that carry out protein hydrolysis and acid fermentation are called acid proteolytic, for example, *Streptococcus faecalis* var. *liquefaciens* and *Micrococcus caseolyticus*.[2]

11.2 GENERAL METHODS

11.21 Milk Agar Medium

The hydrolysis of casein in an opaque skim milk agar often is used to determine proteolysis by microorganisms on or in agar plates. Colonies of proteolytic bacteria will be surrounded by a clear zone as a result of the conversion of casein into soluble nitrogenous compounds. However, clear zones on milk agar can be produced by bacteria that produce acid from fermentable carbohydrates in the medium.[3] The clear zone on common milk agar medium reflects only the more complete breakdown of casein, since the early stages of proteolysis cannot be detected against the opaque background.

An improved skim milk agar medium has been developed[7] by adding sodium caseinate, trisodium citrate and calcium chloride to standard meth-

IS/A Committee Liaison: Carl Vanderzant

ods agar. Its greater sensitivity is related to the detection of the early stages of casein breakdown, the formation of a zone of precipitation (insoluble paracaseins) in a transparent medium. This medium is well buffered, which reduces the occurrence of false positive zones caused by acid production. No significant differences were found between total counts of raw milks on standard methods medium and the caseinate standard methods medium. This latter medium then can be used for the simultaneous determination of total and proteolytic counts.

11.22 Gelatin Agar Medium

Numerous methods have been used to detect gelatin hydrolysis by microorganisms: gelatin liquefaction (gelatin stab), and the detection of hydrolyzed gelatin in agar media with or without chemical protein precipitants. Pitt and Dey[8] developed a gelatin agar medium for the detection of gelatinase without the use of a protein precipitant.

A double layer gelatin medium with a soft agar gelatin overlay is available which detects both weakly and strongly proteolytic bacteria in a direct plating procedure.[6] Advantages claimed for the medium are: 1) rapid diffusion of proteolytic enzymes through the soft overlay, creating large zones of clearing, 2) reduction in swarming, and 3) more rapid colony development than with pour plates. If enumeration of proteolytic bacteria is desired from dairy or seafoods, incorporation of skim milk or fish juice in the overlay will detect proteolysis without application of chemical precipitant. This will be advantageous if further purification or identification of colonies is needed.

A double layer plating technique also can be applied when one or more components of the growth medium are not compatible with the protein used to check hydrolysis. A two layer plate has been developed[9] with a lower indicator layer of milk agar and an upper layer of marine agar to detect proteolytic marine bacteria. Samples are placed on the upper layer by the spread plate method. No chemical precipitant is required to detect proteolysis.

11.3 TREATMENT OF SAMPLE

Prepare samples and appropriate dilutions as described for the agar plate count (Chapter 4).

11.4 MEDIA

11.41 Skim Milk Agar[1] (Chapter 2)

11.42 Standard Methods Caseinate Agar[7] (Chapter 2)

11.43 Soft-Agar-Gelatin-Overlay Medium[6] (Chapter 2)

11.5 PROCEDURES

11.51 Skim Milk Agar Method (Chapter 2)

Prepare pour or streak plates. Incubate plates at 21 C for 72 hours or use other plate incubation conditions recommended for total counts of the food concerned. Flood plates with 1% HCl or 10% acetic acid solution for 1 minute.[1] Pour off excess acid solution. Count colonies surrounded by clear zones produced by proteolysis.

11.52 Standard Methods Caseinate Agar Method[7] (Chapter 2)

Place appropriate dilutions of sample in 0.1 ml quantities on the surface of plates and distribute evenly by spreading with a sterile bent glass rod. Allow the plates to dry for 15 minutes. Incubate plates for 24 to 72 hours at 30 C. To enumerate proteolytic psychrotrophic bacteria, incubate plates at 7 C for 10 days. Count colonies which form white or off white precipitate around the colony. (Organisms which are strongly proteolytic can further breakdown the precipitate to soluble components with the formation of an inner transparent zone).

11.53 Soft-Agar-Gelatin-Overlay Method[6]

Place appropriate dilutions of sample in 0.1 ml quantities on the surface of the basal medium and distribute uniformly by spreading with a sterile bent glass rod. Pour 2.5 ml of melted gelatin overlay medium and distribute evenly over surface of agar. Incubate plates at 20 C for 3 days. Flood surface of plate with 10 ml of 5% acetic acid for 15 minutes. Count colonies with clear zones.

The standard methods caseinate agar method and the soft-agar-gelatin-overlay method have not yet received collaborative evaluation.

11.54 Reporting

Report as **proteolytic count per milliliter or gram** as applicable.

11.6 PRECAUTIONS

The most serious disadvantage associated with the milk agar method is the need to flood the medium with a protein precipitant to confirm that the zones of clearing are caused by proteolysis and not from acid formed by fermentation of carbohydrates. After treatment with protein precipitant, colonies on the plates cannot be used for further analysis unless a replicate plate has been prepared.

Simultaneous determination of total and proteolytic counts on the same plate is difficult if the ratio of the two counts vary widely. The zones of clearing may not be distinct unless the colonies are well separated. This requires the counting of plates that contain fewer colonies, with resulting loss of accuracy. In addition, weakly proteolytic bacteria may not be detectable unless the plate incubation period is extended, sometimes beyond the optimum length for the total count. This difficulty may be further compounded if a mixture of bacteria with widely varying proteolytic activities is present on the same plate.

11.7 INTERPRETATION

Ideally the proteolytic activity of a microorganism should be measured against the specific protein(s) of the food being examined. The temperature of plate incubation should reflect that of the food during the time that microbial proteolytic activity is expected, or has taken place. Convenience and the need for standardization has limited the media mainly to those containing gelatin or casein (skim milk). Gelatin is an incomplete protein and the ability to liquify gelatin, or the lack of it, may not be correlated to the specific proteolytic potential being measured.[10] Ability to hydrolyze casein, on the other hand, is more closely related to the ability to hydrolyze animal protein.[5] The level of proteolytic bacteria and/or the proportion in terms of the total microbial flora can be useful in some foods[6,7] to indicate quality (potential refrigerated shelf life). In addition, proteolytic activity is one of the most readily demonstrable characteristics and is used widely in the description and taxonomic classification of bacteria.

11.8 REFERENCES

1. Standard Methods for the Examination of Dairy Products. 13th ed. 1972. American Public Health Association, Inc., New York, N.Y. p. 148
2. Frazier, W. C. 1967. Food Microbiology, 2nd ed. McGraw-Hill, New York, N.Y. p. 59
3. Frazier, W. C. and P. Rupp. 1928. Studies on the proteolytic bacteria of milk. I. A medium for the direct isolation of caseolytic milk bacteria. J. Bacteriol. **16**:57–64
4. Jay, J. M. 1972. Mechanism and detection of microbial spoilage in meats at low temperatures: A status report. J. Milk Food Technol. **35**:467–471
5. Kazanas, N. 1968. Proteolytic activity of microorganisms isolated from fresh water fish. Appl. Microbiol. **16**:128–132
6. Levin, R. E. 1968. Detection and incidence of specific species of spoilage bacteria on fish. I. Methodology. Appl. Microbiol. **16**:1734–1737
7. Martley, F. G., S. R. Jayashankar, and R. C. Lawrence. 1970. An improved agar medium for the detection of proteolytic organisms in total bacterial counts. J. Appl. Bacteriol. **33**:363–370
8. Pitt, T. L. and D. Dey. 1970. A method for the detection of gelatinase production by bacteria. J. Appl. Bacteriol. **33**:687–691
9. Sizemore, R. K. and L. H. Stevenson. 1970. Method for the isolation of proteolytic marine bacteria. Appl. Microbiol. **20**:991–992
10. Manual of Microbiological Methods. 1957. Society of American Bacteriologists. McGraw-Hill, New York, N.Y. p. 55

CHAPTER 12

HALOPHILIC MICROORGANISMS

J. A. Baross

12.1 INTRODUCTION

Microorganisms which require certain minimal concentrations of salt (NaCl) for growth are called halophilic. In general, the requirement for salt is not an exclusive need for NaCl because many halophiles require low levels of K^+, Mg^{++}, and other cations and anions in addition to NaCl.[4, 9, 15, 18] Furthermore, for some bacteria, the apparent requirement for NaCl is not specific and other salts and sugars can be substituted. The levels of salt required by microorganisms vary greatly. Therefore, the microbial types associated with a particular salted food depend on the concentration and type of salt and the type of food.

The most practical classification of halophilic microorganisms is based on the level of salt required.[4, 10, 16] Slight halophiles grow optimally in media containing 2 to 5% salt, moderate halophiles in media containing 5 to 20% salt, and extreme halophiles in media containing 20 to 30% salt. Additionally, there are many halotolerant microorganisms which grow without added salt as well as in salt concentrations exceeding 5%. Some halotolerant microorganisms are involved in the spoilage of salted foods, whereas others, such as *Staphylococcus aureus*, are human pathogens. Yeasts and molds, and other osmophilic microorganisms are involved in the spoilage of foods with low a_w values, including salted foods.

Among salted foods, low salted foods (1 to 7% salt) are more susceptible to microbiological spoilage and are also likely to become contaminated with human pathogens more than are foods containing high levels of salt. This is particularly true of untreated "fresh" seafoods.[23, 24] Heavily brined foods do not spoil easily unless maintained at elevated temperatures.[23]

A compilation of the types of halophilic, spoilage, and pathogenic microorganisms associated with various salted foods is shown in Table 1.

12.2 DILUENTS AND MEDIA

12.21 Diluents

a. **Phosphate buffer with salt** (SMEDP-NaCl). (Chapter 2). Add required amount of NaCl prior to sterilization.

IS/A Committee Liaison: Carl Vanderzant

b. **Synthetic sea water** (SW)[3] (Chapter 2)

12.22 Media

a. **Trypticase soy agar with added salt** (TSA-NaCl) (Chapter 2). Add required amount of NaCl prior to sterilization.

b. **Sea water agar** (SWA) (Modified MacLeod[19]) (Chapter 2)

c. **Halophilic agar** (HA)[9, 10] (Chapter 2)

d. **Halophilic broth** (HB) (Chapter 2)

Halophilic medium (HA) without agar is used as a diluent and as an enrichment medium for the isolation of extremely halophilic bacteria. Sterilize medium by autoclaving at 121 C for 15 minutes.

12.3 SLIGHTLY HALOPHILIC MICROORGANISMS

12.31 General

Most of the slightly halophilic bacteria originate from marine environments. Marine psychrotrophic bacteria of the genera *Pseudomonas, Moraxella, Acinetobacter,* and *Flavobacterium* contribute to the spoilage of marine fish and shellfish. Similar gram negative rods of terrestrial origin and other gram negative and gram positive psychrotrophic organisms are frequently involved.[22] Some of these organisms have a complex ionic requirement and may require Mg^{++} and K^{+} in addition to NaCl for growth and proteolytic activity, whereas the salt requirement for other slightly halophilic bacteria is osmotic. Growth of many of these marine bacteria is inhibited if the NaCl concentration of the growth medium is lower than 0.5% or higher than 5%. Dilution of the seafood sample with distilled water frequently causes lysis of many representative spoilage bacteria. Holding marine foods for sustained time periods at temperatures exceeding 25 C will reduce the numbers of psychrotrophic microorganisms significantly. To identify the degree of initial microbial contamination of marine origin, prechill media, reagents, and sampling equipment to 5 C. Avoid washing food sample with distilled water, and do not use diluents containing less than 3% NaCl.

The microbial flora associated with salted meats and vegetables is variable, and dependent on many factors including the type of food, the presence of other salts or organic preservatives, the concentration of salt, and the storage conditions (temperature, packaging conditions).[10, 12, 20] In general, microorganisms involved in the spoilage of low salted meats and vegetables (1 to 7% NaCl) can be enumerated without the use of special media, provided the diluents and plating media are supplemented with NaCl equivalent to the concentration in the food sample. Occasionally, however, spoilage caused by clostridia occurs particularly in vacuum packaged

Table 1: Halophilic, Spoilage and Pathogenic Microorganisms Associated with Various Salted Foods and Seafoods

Food type	% Salt associated with food	Halophilic types	Spoilage microorganisms	Pathogens[a]	References
I. Marine fish (Teleosts)	2 to 4 (similar to seawater)	Slight and halotolerant types: *Pseudomonas* *Moraxella* *Acinetobacter* and other gram negative bacteria	Proteolytic: *Pseudomonas, Moraxella* and *Acinetobacter*	*Vibrio parahaemolyticus* *Clostridium botulinum* E *Clostridium perfringens* *Staphylococcus aureus* *Salmonella* and other pathogenic *Enterobacteriaceae, Erysipelothrix insidiosa*	11, 17, 22, 23, 24
II. Molluscan shellfish	1 to 4 (similar to seawater)	Same as for fish—but greater percentage of *Vibrio*	Same as for fish at early stages. Lactic-acid bacteria & yeasts in later stages of spoilage.	Same as for fish plus human enteric viruses and *Gonyaulax*	5, 6, 11, 17
III. Crustaceans	1 to 4 (similar to seawater)	Same as for fish—but greater percentage of *Vibrio*	Same as for fish. Also yeasts and chitinoclastic microorganisms	Same as for fish	5, 6, 11, 17, 25
IV. Brined meats (Ham, bacon, beef, prepared meats and sausage)	1 to 7 brine[b]	Halotolerant molds, yeasts & gram positive bacteria	Molds, *Lactobacillus, Micrococcus*, and *Vibrio* in bacon. *Clostridium* in packaged salted meats	*Clostridium botulinum* *Clostridium perfringens* *Staphylococcus aureus* Pathogenic *Enterobacteriaceae* in meats containing low salt	7, 8, 12, 14, 21

V. Salted vegetables	1 to 15	Moderate and halotolerant molds, yeasts & gram positive bacteria	Lactic acid bacteria, yeasts & molds, *Bacillus*, *Enterobacteriaceae* in foods with low salt content. *Clostridium* in packaged foods.	Dependent on level of salt in food. Pathogenic members of *Enterobacteriaceae* in low salt foods. *Staphylococcus aureus* in highly salted food.	12, 20
VI. Salted fish (A) Light salt	1 to 10	Slight and halotolerant types	*Pseudomonas* in lightly salted fish. *Clostridium* in packaged fish. *Micrococcus* in fish containing 5 to 10% salt	Dependent on salt concentration, same as for salted vegetables. *Vibrio parahaemolyticus* may occur when salt is 1 to 7%	13, 22, 23, 24
(B) Heavy salt	75 to 80 brine (10 to 15 salt in interior of fish)	Moderately & extremely halophilic types	*Halobacterium*, *Halococcus* (Cause condition called "pink" in fish). Also some members of the *Micrococcaceae*	*Staphylococcus aureus*	13, 22, 23, 24

[a] Refer to specific section for the isolation of specific pathogens.

[b] Brine concentration is $\frac{\text{g salt}}{\text{g salt} + \text{g water}} \times 100$.

foods (refer to specific section for the isolation and enumeration of anaerobes).

12.32 Sampling

Low salted foods to be analyzed for psychrotrophic spoilage microorganisms should be tested without delay, and definitely within 24 hours; otherwise, growth will have occurred. Samples should be maintained at 5 C until tested.

12.33 Procedures

12.331 Fish (Teleosts).

Remove skin samples with a sterile cork borer (d = 1.6 cm). Collect six pieces, three from the ventral side and three from the dorsal side of the fish. Take flesh samples from just below the skin by peeling back the skin with a sterile scalpel and forceps and dissecting the flesh. Weigh skin or flesh in a sterile container. Add 10 g of skin to 90 ml phosphate buffer (3% NaCl) or artificial sea water (12.21b) with either sterile sand or glass beads (10 g). Mix sample and diluent thoroughly by shaking vigorously for 1 minute. Prepare dilutions (to 10^{-8}) in phosphate buffer (3% NaCl) or sea water (12.21b). Place 0.1 ml aliquots on either trypticase soy agar plus 3% NaCl (12.22a) or sea water agar (12.22b) with a spread plate technique. Incubate plates at 7 C for 10 days for the enumeration of psychrotrophic spoilage bacteria. Report results as counts per gram of skin or flesh.

For fish flesh samples, fillets, or small whole fish (less than 6 inches) cut samples aseptically into slices (2.5 cm^2). Add 50 g of flesh to 450 ml of diluent (3% NaCl) in a sterile blender jar and blend for 2 minutes. If the sample is not sufficiently homogenized, let the sample stand for 2 minutes before blending for an additional 2 minutes. Prepare serial dilutions and plates as described above (12.331). Report results as counts per gram.

The detection and enumeration of the slightly halophilic pathogen *Vibrio parahaemolyticus* is described in detail in Chapter 29.

12.332 Molluscan shellfish

Collect and prepare samples for microbiological analysis as described in Chapter 41.[1] Prepare initial 1 : 1 dilution by blending 100 g of shellfish meat with 100 ml sterile diluent (12.21a-3% NaCl). Plate samples as described in 12.331. Report results as counts per gram sample at 7 C or 25 C. It is advisable to prepare two sets of plates, one to be incubated at 7 C and the other at 25 C because shellfish generally reside in near shore environments which are subject to wide fluctuations in temperature and, therefore, harbor both a psychrotrophic and a mesophilic bacterial flora.

12.333 Crustaceans

Collect and prepare samples as described in Chapter 40. Prepare dilutions as for fish flesh. Plate samples are described in 12.331.

12.334 Lightly salted meats and vegetables (1 to 7% NaCl).

Prepare dilutions by blending 50 g samples with 450 ml diluent (12.21a) with added NaCl equivalent to salt concentration of food sample. Plate samples on trypticase soy agar supplemented with NaCl equivalent to NaCl concentration of food. If food sample usually is not refrigerated, use 1 ml aliquot for pour plates and incubate plates for 4 days at 25 C. For refrigerated foods such as bacon, plate 0.1 ml aliquots with the spread plate technique. Incubate plates for 10 days at 7 C. Report results as counts per gram.

12.34 Interpretation

In general, fish contaminated with greater than 10^8 psychrotrophic bacteria/cm^2 of skin or per gram of muscle is considered spoiled.[23, 24] The most common seafood spoilage bacteria are *Pseudomonas* species which are psychrotrophic (Chapter 8) and actively proteolytic (Chapter 11). In particular, *P. putrefaciens, P. fragi*, and related species are known to produce the principal chemical changes associated with fish spoilage.[11] These species comprise less than 4% of the bacterial flora associated with fresh fish, and 20% and occasionally as high as 80% of the flora of spoiled fish.[2, 11] In contrast, the dominant spoilage organisms frequently isolated from shellfish belong to the *Moraxella-Acinetobacter* group.[11] Chemical and organoleptic tests (odor particularly associated with gills) are used in conjunction with bacterial counts to assess the extent of spoilage. High total volatile nitrogen (TVN) and trimethylamine (TMA) values are frequently used as an indication of bacterial activity.[25]

There are no established microbiological standards for low salted meats and vegetables.

12.4 MODERATELY HALOPHILIC MICROORGANISMS

12.41 General

Most of the moderately halophilic bacteria involved in the spoilage of salted foods (5 to 20% NaCl) are gram positive species of the *Bacillaceae* and *Micrococcaceae*. *Micrococcus halodenitrificans* and other *Micrococcus* sp. have a specific requirement for NaCl.[15, 16] This is also true for a moderately halophilic *Achromobacter* species isolated from salted herring.[16] In contrast, the requirement for salt by many moderately halophilic *Bacillus* species is not specific for NaCl, and many other Na^+ and K^+ salts can be substituted. Salted foods, which can spoil because of moderately halophilic microorganisms, commonly harbor high numbers of halotolerant gram positive bacteria, yeasts and molds. (Consult Table 1 for references on isolation and enumeration procedures of osmophilic and halotolerant pathogens. In addition consult Chapters 12 and 13 of this compendium.)

12.42 Procedure

Moderately salted foods are sampled for spoilage microorganisms as described in 12.32. Prepare 1:10 dilution by mixing 50 g of food in 450 ml sterile buffer (12.21a) with added NaCl equivalent to the salt concentration of the food sample. Plating procedures are described in 12.334. For the isolation and enumeration of specific bacterial types consult Table 1 and related chapters.

12.43 Interpretation

There are no microbiological standards for foods containing 5 to 20% NaCl. Use organoleptic tests (odor and visual evidence of spoilage, such as slime or gas formation) in conjunction with total bacterial counts to determine the extent of spoilage of these foods. *Staphylococcus aureus* and *Clostridium perfringens* can grow in some moderately salted foods.

12.5 EXTREMELY HALOPHILIC MICROORGANISMS

12.51 General

The extreme halophiles are normally found in aquatic environments of unusually high salt concentrations and in solar evaporated sea salts. These microorganisms require a minimum of 15% NaCl for growth but will grow optimally in media containing 20 to 30% NaCl.[4,16] They are principally species of the genera *Halobacterium* and *Halococcus* which produce bright red or pink pigments, grow very slowly even under optimal conditions, and are readily lysed when exposed to low salt concentrations (less than 10%). Extremely halophilic bacteria have been incriminated in the spoilage of fish, bacon, and hides preserved in sea salts. Severe contamination of foods with *Halobacterium* or *Halococcus* will generally result in a pink discoloration on the outer surface of the sample, accompanied by decomposition and putrefaction.

12.52 Procedure

For isolation, transfer surface slime from salted fish or bacon to an HA agar plate (12.223) using a cotton or alginate swab. For the quantitative enumeration of *Halobacterium* and *Halococcus* from food samples, blend 50 g of sample with 450 ml HB. Place 0.1 ml aliquots of each dilution (to 10^{-6}) on halophilic agar (12.22c) plates with a spread plate technique. For the detection of extremely halophilic bacteria from solar sea salts or brine solutions (70 to 80%), prepare serial dilutions in halophilic broth (12.22d) of up to 10^{-6}. Prepare plates on halophilic agar (12.22c). Incubate plates at 33 to 35 C in a humid incubator for 5 to 12 days. The presence of pink or red colonies is indicative of extreme halophiles. Report results as halophilic counts per ml or g. Alternately, for samples suspected of containing low levels of halophilic bacteria (less than 10 per g or ml), place 10 ml or 10 g sample into 90 ml HB broth and incubate at 35 C for up to 12 days. Then streak from broth onto halophilic agar (12.22c) and incubate plates as described. Report the presence or absence of extreme halophiles.

12.53 Interpretation

There are no microbiological standards for heavily brined foods and usually only an organoleptic observation (presence of red or pink slime and putrefaction) is performed to check for spoilage. Since *Halobacterium* and *Halococcus* are normally present in sea salts, food spoilage by these organisms can be prevented by sterilizing the salt prior to use for curing. Extremely halophilic bacteria will not grow on foods stored at temperatures below 7 C.

Extreme halophiles are not pathogenic to man and any incidence of food poisoning associated with heavily salted foods is usually caused by *Staphylococcus aureus* (refer to Chapter 31 for the isolation and enumeration of *S. aureus*).

12.6 HALOTOLERANT MICROORGANISMS

12.61 General

Microorganisms capable of growing in NaCl concentrations exceeding 5%, as well as in media containing no NaCl are called halotolerant. Most halotolerant bacteria are gram positive and are species of the *Micrococcaceae*, the *Bacillaceae*, and some of the corynebacteria. A few gram negative species will also grow in foods cured with 10% salt. Many human pathogens, such as *Staphylococcus aureus* and *Clostridium perfringens* and some strains of *C. botulinum* can be responsible for food poisoning outbreaks involving salted foods.[21]

Most halotolerant microorganisms in foods are isolated when tested for slight or moderate halophiles. Specific media, however, are required for the isolation of halotolerant molds and yeasts from food samples (refer to Chapter 13 for the isolation and enumeration of osmophiles).

12.62 Procedure

Refer to sections 12.3 and 12.4.

12.63 Interpretation

Refer to sections 12.3 and 12.4.

12.7 REFERENCES

1. Recommended Procedures for the Examination of Sea Water and Shellfish. 4th ed. 1970. American Public Health Association, Inc., New York, N.Y.
2. CHAI, T., C. CHEN, A. ROSEN, and R. E. LEVIN. 1968. Detection and incidence of specific species of spoilage bacteria on fish. II. Relative incidence of *Pseudomonas putrefaciens* and fluorescent pseudomonads on haddock fillets. Appl. Microbiol. **16**:1738–1741
3. COLWELL, R. R. and R. Y. MORITA. 1964. Reisolation and emendation of description of *Vibrio marinus* (Russell) Ford. J. Bacteriol. **88**:831–837
4. EIMHJELLEN, K. 1965. Isolation of extremely halophilic bacteria, p. 126–138. *In* M. Schlegel (ed.), Anreicherungskultur und Mutantenauslese. Gustav Fisher Verlag, Stuttgart, Germany
5. EKLUND, M. W., J. SPINELLI, P. MIYAUCHI, and H. GRONINGER. 1965. Characteristics of yeasts isolated from Pacific crab meat. Appl. Microbiol. **13**:985–990
6. FIEGER, E. A. and A. F. NOVAK. 1961. Microbiology of shellfish deterioration, p. 561–611. In G. BORGSTROM (ed.) Fish as Food, Vol. I. Academic Press, New York, N.Y.
7. GARDNER, G. A. 1973. A selective medium for enumerating salt requiring *Vibrio* spp. from Wiltshire bacon and curing brines. J. Appl. Bacteriol. **36**:329–333

8. GARDNER, G. A. and A. G. KITCHELL. 1973. The microbiological examination of cured meats, p. 11–20. In R. G. BOARD, and D. W. LOVELOCK (ed.), Sampling—Microbiological Monitoring of Environments. Academic Press, New York, N.Y.
9. GIBBONS, N. E. 1957. The effect of salt concentration on the biochemical reactions of some halophilic bacteria. Can. J. Microbiol. **3**:249–255
10. GIBBONS, N. E. 1969. Isolation, growth and requirements of halophilic bacteria, p. 169–183. In J. R. NORRIS and D. W. RIBBONS. (ed), Methods in Microbiology. Academic Press. New York, N.Y. Vol. 3B
11. HERBERT, R. A., M. S. HENDRIE, D. M. GIBSON, and J. M. SHEWAN. 1971. Bacteria active in the spoilage of certain seafoods. J. Appl. Bacteriol. **34**:41–50
12. HARRIGAN, W. F. and M. E. MCCANCE. 1966. Laboratory Methods in Microbiology. Academic Press, London, England
13. HENNESSEY, J. P. 1971. Salted and dried groundfish products, p. 114–116. In R. KREUZER (ed.), Fish Inspection and Quality Control. Fishing News (Books) Limited, London, England
14. INGRAM, M. and R. H. DAINTY. 1971. Changes caused by microbes in spoilage of meats. J. Appl. Bacteriol. **34**:21–39
15. KUSHNER, D. J. 1968. Halophilic bacteria, p. 73–99. *In* W. W. UMBREIT and D. PERLMAN (ed.), Advances in Applied Microbiology. Vol. 10. Academic Press, New York, N.Y.
16. LARSEN, H. 1962. Halophilism, p. 297–342. In I. G. GUNSALUS and R. Y. STANIER (ed.), The Bacteria, a Treatise on Structure and Function. Vol. IV, The Physiology of Growth. Academic Press, New York, N.Y.
17. LISTON, J., J. R. MATCHES, and J. BAROSS. 1971. Survival and growth of pathogenic bacteria in seafoods, p. 246–249. In R. KREUZER (ed.), Fish Inspection and Quality Control, Fishing News (Books) Limited, London, England
18. MACLEOD, R. A. 1965. The question of the existence of specific marine bacteria. Bacteriol. Rev. **29**:9–23
19. MACLEOD, R. A. and E. ONOFREY. 1957. Nutrition and metabolism of marine bacteria. VI. Quantitative requirements for halides, magnesium, calcium, and iron. Can. J. Microbiol. **3**:753–759
20. RIEMANN, H. 1969. Food processing and preservation effects, p. 489–541. *In* H. RIEMANN (ed.), Food Borne Infection and Intoxications, Chapter 12. Academic Press, New York, N.Y.
21. RIEMANN, H., W. H. LEE, and C. GENIGEORGIS. 1972. Control of *Clostridium botulinum* and *Staphylococcus aureus* in semi-preserved meat products. J. Milk Food Technol. **35**:514–523
22. SHAW, B. G. and J. M. SHEWAN. 1966. Psychrophilic spoilage bacteria of fish. J. Appl. Bacteriol. **31**:89–96
23. SHEWAN, J. M. 1971. The microbiology of fish and fishery products—a progress report. J. Appl. Bacteriol. **34**:299–315
24. SHEWAN, J. M. and G. HOBBS. 1967. The bacteriology of fish spoilage and preservation, p. 169–208. In D. J. D. HOCKENHULL (ed.), Progress in Industrial Microbiology, Vol. 6. Chemical Rubber Co. Press. Cleveland, Ohio
25. VANDERZANT, C., B. F. COBB, and C. A. THOMPSON, Jr. 1973. Microbial flora, chemical characteristics and shelf life of four species of pond-reared shrimp. J. Milk Food Technol. **35**:443–446

CHAPTER 13

OSMOPHILIC MICROORGANISMS

Edmund A. Zottola

13.1 INTRODUCTION

Osmophilic microorganisms most commonly encountered in the food industry are yeasts. They can grow in highly concentrated sugar solutions.[2] They are frequently the cause of spoilage of honey, chocolate candy with soft centers, jams, molasses, corn syrup, concentrated fruit juices, and other similar products.[4, 9, 10] Walker and Ayres,[9] in their review on yeasts as spoilage organisms, differentiate between osmophilic yeasts, those which can grow in high sugar concentrations, and osmoduric yeasts, those which tolerate but do not grow in high sugar concentrations. Yeasts which can grow in, or tolerate, high salt concentrations may also be termed osmophilic or osmoduric, but the mechanisms involved are different. For purposes of this discussion the definition for osmophiles of Walker and Ayres[9] given above will be used. Excluded are those yeasts which tolerate high salt concentration, and which are more correctly called halophiles (Chapter 12).[1, 2] Almost all of the known osmophilic yeasts are species of *Saccharomyces: Saccharomyces rouxii, S. rouxii* var. *polymorphus*, and *S. mellis*. For a more complete discussion of yeasts as spoilage organisms, the recent review by Walker and Ayres[9] should be consulted.

Enumeration procedures for osmophilic yeasts are similar to those used for other yeasts. Improved recovery of osmophilic yeasts has been reported on media which imitate the composition of the food under examination, or contain high sugar concentrations.[2, 3] MY-40 agar[11] and modifications of this medium[6, 8] meet the above requirements.

13.2 SAMPLE COLLECTION[1]

Carefully cleansed, rinsed, dried, and sterilized sampling containers should be used. Care must be taken to select samples representative of the lot. Take precautions to handle samples in such a manner as to prevent contamination or change prior to examination (Chapter 1).

Liquid sugar received in tank trucks should be sampled from the truck under aseptic conditions. Likewise, the product in storage tanks should be

IS/A Committee Liaison: Carl Vanderzant

examined at periodic intervals. Some food processors receive sugar in bulk form either in Air Slide Cars or by tank truck, then liquefy the product on their own premises, and store it in large tanks. Products in these tanks should be checked for yeast contamination.

13.3 EQUIPMENT

13.31 Membrane Filter Apparatus (Chapter 4)

13.311 Millipore membrane filter equipment

a. Hydrosol stainless filter holder
b. Pyrex filter flask, 1 liter
c. Vacuum pressure pump
d. Stainless steel forceps
e. Plastic Petri dishes
f. Filters, Type HA white grid-marked, and absorbent pads, presterilized
g. Microscope illuminator, variable intensity, transformer powered

13.312 Other equipment

a. Erlenmeyer flasks, screw cap, 500 ml
b. Stereoscopic microscope: 1× and 2× objectives with 10× widefield eye pieces

13.313 Sterilization

Wrap top and bottom sections separately with Kraft paper or aluminum foil.

Autoclave 15 minutes at 121 C.

Note: Filter may be rinsed with 90 to 100 ml of sterile water (99 ml dilution bottle) when several samples from the same batch are run.

13.4 MEDIA AND STAINS

13.41 Media

13.411 Potato dextrose agar[1] (Chapter 2)

13.412 MY-40 agar[6, 8, 11] (Chapter 2)

13.413 Modified Sabouraud's dextrose broth[3, 6] (Chapter 2)

Note: Industrial experience with this medium has indicated that satisfactory results are obtained by acidifying the medium after sterilization to pH 4.7 to 4.8 with sterile 10% lactic acid rather than utilizing antibiotics.[6]

13.42 Stains[3]

13.421 Loefflers methylene blue solution (Chapter 2)

13.422 Glycerol formaldehyde solution (Chapter 2)

13.5 PRECAUTIONS

If difficulty is encountered in recovering osmophilic organisms from high sugar foods with a medium of low sugar content, add sufficient sugar to the medium to duplicate the sugar content of the food under examination.[3, 6, 11]

When examining the liquid syrups, standardize sample to approximately 20° Brix.

With the membrane filter technique, use a stainless steel filter holder because liquid sugar is difficult to clean from the glass grid of the Pyrex filter holder.[6]

13.6 PROCEDURES

13.61 Plate Count Method[1, 7]

13.611 Dry samples

Prepare appropriate dilutions with sterile, phosphate buffered dilution water as diluent. Add 15 to 20 ml of MY-40 agar, allow plates to solidify and incubate at 30 C for 48 hours.[6, 8]

Alternately use acidified potato dextrose agar and incubate plates at 30 C for 4 to 5 days.[7]

13.612 Liquid samples[1, 7]

Prepare appropriate dilutions, taking into account the water content of sample, so that dilution is based on the solids in sample. Proceed as in 13.611. Report the number of osmophilic yeasts per gram.

13.62 Membrane Filter Technique[3, 6]

13.621 Preparation of sugar sample

a. Weigh 75 g of the 67-70° Brix liquid sugar sample aseptically into a sterile tared 500 ml Erlenmeyer flask (volumetric methods are not satisfactory for measuring concentrated sugar solutions).

b. Add approximately 175 g of sterile water (most easily accomplished by precalibrating the flasks with a mark at the appropriate level). This results in a solution of approximately 20° Brix.

13.622 Preparation of filter assembly and filtration of a sample

a. Remove Kraft paper from bottom assembly and mount on filter flask.

b. Cover filter holder with sterile Petri dish.

c. Place a sterile type HA filter grid side up under Petri dish on sterile stainless steel filter holder. Use sterile forceps (liquid sugar is difficult to clean from glass grid of the Pyrex filter holder).

d. Remove paper from bottom of funnel and assemble filter unit.

Note: Keep sterile paper on top of funnel, except when adding sample.

e. Connect the filter flask to a vacuum source and apply suction. Check assembly for leakage by filtering approximately 100 ml of sterile water.

f. Filter 20° Brix sample under aseptic conditions. Filtration requires less than 5 minutes.

g. Rinse walls of sample flask with about 100 ml of sterile distilled water divided into 2 portions (use for example 99 ml sterile diluent). Likewise, flush sides of funnel with 2 portions of sterile distilled water. Allow each aliquot to be pulled through filter before the next is added.

h. Open stopcock on Erlenmeyer flask to release vacuum. Shut off pump.

Note: To extend the life of the vacuum pump, do not connect the filter flask directly to the vacuum pump. Connect the sidearm of the filter flask (with the filter holder) with tubing to a glass tube placed in a rubber stopper in the mouth of a second filter flask. The glass tube should extend about 2.5 cm below the sidearm of this flask. Install a stopcock in the rubber stopper also. Connect the sidearm of the second filter flask to the vacuum pump. Break vacuum by opening stopcock rather than by cutting pump off. Let pump run a few minutes to expel moisture in pump chamber.

13.623 Cultivation

a. Place a sterile absorbent pad in bottom of a plastic Petri dish.

b. Pipette 1.8 to 2.0 ml of modified Sabouraud's dextrose broth onto pad. Pad should be thoroughly saturated so that the filter will be uniformly wet.

Caution: Excess liquid can result in contamination of medium by growth from upper side.

c. Remove filter from filter base with sterile forceps and place, grid side up, on absorbent pad saturated with broth. Filter should be carefully rolled onto surface of pad to avoid entrapment of air bubbles.

d. Incubate dish in an inverted position for 28 hours at 30 C.

13.624 Staining

a. Transfer filter (colony side up) from Petri dish to an absorbent pad in another dish saturated with Loeffler's methylene blue solution. Stain for 2 minutes.

b. Fix colonies for 4 minutes by transferring disc to another absorbent pad with glycerol-formaldehyde solution.

c. Carefully transfer disc containing fixed colonies to original washed and dried Petri dish. Place in oven at 80 C for 10 minutes or until dried (pad changes from dark to pale blue when dry).

13.625 Counting

a. Count colonies under widefield microscope.

b. Report results as a number of osmophilic yeasts per gram of original sample. If syrup with 67° Brix is used, results can be converted to organisms per gram of dry solids by multiplying the number per ml by the factor 1.1.[3]

13.626 Alternate procedure to staining

Incubate dish (13.623d) for 48 hours at 30 C and count colonies.

Additional procedures are available for the examination of liquid sugar and syrups for yeasts and molds.[5]

13.7 INTERPRETATION

The interpretation of results is difficult since in some products osmophilic yeasts may have little if any significance. However, their presence in high sugar content products may result in spoilage. Numbers of osmophiles in a liquid sugar sample in excess of 10/gram usually indicate contamination of equipment used to handle the product. In many instances such equipment has not been designed or engineered for complete and adequate cleaning. This type of equipment can be a constant source of osmophilic spoilage organisms.[1, 3, 6, 7]

13.8 REFERENCES

1. Recommended Methods for the Microbiological Examination of Foods, 2nd ed., 1966. American Public Health Association. New York, N. Y.
2. ANAND, J. C. and A. D. BROWN. 1968. Growth rate patterns of the so-called osmophilic and non-osmophilic yeasts in solutions of polyethylene glycol. J. Gen. Microbiol. **52**:205–212
3. COLEMAN, M. C. and C. R. BENDER. 1957. Microbiological examination of liquid sugar using molecular membrane filters. Food Technol. **11**:398–403
4. COOK, A. H. 1958. The Chemistry and Biology of Yeasts. Academic Press, Inc., New York, N.Y.
5. Millipore Corporation. 1966. Techniques for Microbiological Analysis ADM-40, p 16–17. Millipore Corp., Bedford, Mass.
6. MURDOCK, D. I. 1966. Examination of Liquid Sugar Transported by Tank Truck or Rail Car. Minute Maid Company Technical Services. Plymouth, Fla.
7. National Canners Association. 1968. Laboratory Manual for Food Canners and Processors. Volume 1. Microbiology and Processing. The AVI Publishing Company, Westport, Conn.
8. SNYDER, W. L. 1969. Microbiological Study of Storage of Liquid Sugar. The Coca-Cola Company Branch Chemists' Meeting, San Francisco, Cal.
9. WALKER, H. W. and J. C. AYRES. 1970. Yeasts as spoilage organisms. In A. H. ROSE and J. S. HARRISON (ed.), The Yeasts. Vol. 3. Academic Press, Inc., New York, N.Y.
10. WHITE, J. 1954. Yeast Technology. John Wiley and Sons, Inc., New York, N.Y.
11. WICKERHAM, L. J. 1951. Taxonomy of Yeasts. Tech. Bull. No. 1029, U.S.D.A. Washington, D. C.

CHAPTER 14

PECTINOLYTIC MICROORGANISMS

Lester Hankin and David C. Sands

14.1 INTRODUCTION

The pectic substances are polymers of axial-axial (1–4) linked D-galacturonic acid units containing L-rhamnose-rich regions, with side chains composed mainly of arabinose, galactose, and xylose. The carboxyl groups are partially methylated, and the secondary hydroxyl groups may be acetylated. The molecular weights range from 30,000 to 120,000. For a detailed discussion of the chemistry of pectic materials the reader is referred to the book by Kertesz[13] and the review by Doesburg.[3]

Calcium hydrates of pectin are a part of the middle lamellar structure in the cell wall and provide rigidity to fruits and vegetables. For commercial use pectin usually is prepared from citrus fruits or apples, and is added to foods as a gelling agent, or to provide body to the product. It is used in pharmaceutical preparations, and is added to a variety of commercial products including confections, bakery goods, dairy products, cosmetics, and soaps.[13] Pectinolytic microorganisms can cause important changes, (notably softening) in the properties of stored fruits and vegetables, as well as processed products, by alterations of the pectin molecule.

14.11 Pectinolytic Enzymes

Pectic materials, when present either naturally or when added to foods, can be degraded by three different enzymes. (1) Pectate lyase or pectate transeliminase, [Poly (1,4-α-D-galacturonide)lyase, E.C.4.2.2.2],[10] eliminates Δ-4,5,-D galacturonate residues from pectate, thus bringing about depolymerization. (2) Polygalacturonase, pectin depolymerase or pectinase, [Poly (1,4,-α-D-galacturonide)glycanohydrolase, E.C.3.2.1.15][10] hydrolyzes 1,4-α-D-galactosiduronic linkages in pectate and other galacturonans.[10] Both of these enzymes exist as *endo* types, i.e., they cleave randomly between any units of the polymer. *Exo* forms also exist which cleave only at the terminal end of the molecule.[20, 24] (3) Pectinesterase, pectin demethoxylase, pectin methoxylase, pectin methylesterase (Pectin pectyl-hydrolase, E.C.3.1.1.11),[10] catalyze the hydrolysis of the methyl ester, yielding methanol and pectate, and thus make the molecule more susceptible to attack by polygalacturonase and/or pectate lyase.[10, 20, 24]

IS/A Committee Liaison: Carl Vanderzant

The first two enzymes destroy the integrity of pectic materials in plants. Thus the tissues soften and ultimately may collapse.[5, 23] In fruit products, the destruction of the pectin can cause subsequent loss of gelling power when the fruit is used in processing, and in products to which pectin is added the texture of the product may be changed.

14.12 Sources of Pectinolytic Enzymes

Most pectin degrading organisms are associated with raw agricultural products and with soil. Up to 10% of the organisms in soil have been shown to be pectinolytic.[7] These include, but are not limited to, species of *Achromobacter, Aeromonas, Arthrobacter, Agrobacterium, Enterobacter, Bacillus, Clostridium, Erwinia, Flavobacterium, Pseudomonas, Xanthomonos,*[20, 24] and many yeasts and molds, protozoa, and nematodes.[2] Many of these organisms are plant pathogens.[2, 4] The discussion which follows refers to aerobic procedures. However, the detection of anaerobic pectinolytic bacteria also has been described.[16, 17] The media described in sections 14.32 and 14.33 have been used successfully under anaerobic conditions.

14.2 METHODS AND PRECAUTIONS

14.21 Polypectate Gel Medium

Historically the most commonly used medium for the detection of organisms producing pectinolytic enzymes is the semisolid gel medium originally described by Wieringa.[25] This medium has been modified many times[14, 16, 22] and one version[14] is presented in 14.31. The organisms growing on the gel medium degrade polypectate and make depressions in the medium. This gel medium must be prepared and poured carefully to avoid air bubbles. Since the pH of the medium is 7.0, pectate lyase will be detected.[20, 24]

Many fungi produce polygalacturonases, enzymes which have more acid optima. With media adjusted to pH 7.0, fungal polygalacturonases may be missed. Lowering the pH to 6.0 or less to detect polygalacturonase activity interferes with the gelling properties of the medium,[22] hence the concentration of polypectate in the medium must be increased. At pH 7.0 the fungi may also be overgrown by bacteria. Antibiotics may be added to suppress bacterial growth, but caution is suggested because certain antibiotics inhibit enzyme production in some fungi. Inhibitors, however, have been used in the gel medium to suppress bacterial growth.[14] The use of specific inhibitors such as antibiotics or crystal violet is cautioned against, since exclusion of sensitive strains, even within the same genus, is often a problem.

14.22 Medium To Detect Pectate Lyase (MP-7 Medium) (See 14.32)

Hankin et al[6] developed a solid agar medium with mineral salts, pectin, and yeast extract for the detection of pectinolytic bacteria. It is a modification of a medium described by Jayasankar and Graham[11] and can be used for both pour and spread plates. Pectinolytic activity is determined, after growth of the organisms, by flooding the plate with hexadecyltrimethylammonium bromide which precipitates intact pectin. Within 5 minutes pectinolytic colonies are surrounded by a clear zone against an otherwise opaque medium. Plates are viewed best against a dark background. Colonies remain viable for at least 5 minutes and can be removed for subsequent studies. If HCl is used as a precipitant, the cells may be killed.

The pectic enzyme produced by organisms and detected on agar plates is dependent on the pH of the medium as well as on the calcium level.[20, 24] To obtain pectate lyase production, Voragen[24] has shown that at low pH values (*ca* 5.0) less calcium is needed ($<$0.075M) than at higher pH values (6.0 to 7.0). Thus, even though the level of calcium in the MP-7 medium is sub-optimal, the pH of the MP-7 medium is 7.4, and the pectic enzyme activity is probably pectate lyase. Pectate lyase activity is not observed below pH 6.0, nor is the polygalacturonase active above this pH. However, a polygalacturonase produced at high pH has been reported by Rombouts.[20]

14.23 Medium To Detect Polygalacturonase (MP-5 Medium) (See 14.33)

Polygalacturonase production can be determined on a modified MP-7 medium, adjusted to pH values from 5.0 to 6.0. This medium has been used successfully to grow bacteria which produce polygalacturonase. Calcium and phosphate at the levels used in this medium did not inhibit polygalacturonase production.

14.24 Selective Medium For Fluorescent Pectinolytic Pseudomonads (FPA medium) (See 14.34)

A more selective medium designed specifically to detect fluorescent pectinolytic pseudomonads has been described by Sands, Hankin, and Zucker.[21] This pectin medium contains antibiotics to curtail growth of unwanted organisms. With this medium both the fluorescent capabilities, as well as the pectinolytic ability of the pseudomonads, can be ascertained. After growth, the plate is flooded with a precipitant (section 14.36), and pectinolytic activity determined.

This medium which demonstrates pectate lyase production was especially designed for detecting *P. marginalis* and *P. fluorescens* species producing soft rot on plants. Lowering the pH of the medium from 7.0 to 6.0 interferes with fluorescent pigment production in the medium. This medium has the distinct advantage that pseudomonads can be enumerated or isolated from natural sources, including insect vectors, where they appear in low numbers. *P. phaseolicola* reportedly produces a pectin liquifying enzyme at pH 4.9 to 5.1.[8] Thus, the activity of this pseudomonad will go undetected in this selective medium.

14.25 Pectinolytic Enzyme Activity Assay

Under some circumstances it may be both necessary and important to assay the pectinolytic enzyme activities produced by microorganisms. The MP-7 medium (with agar deleted) can be used for this purpose.[6, 26, 27, 28] For pectate lyase activity the liquid medium is adjusted to pH 7.3 and enzyme activity measured on culture filtrates by the method of Albersheim and Killias[1] as detailed by Zucker and Hankin.[26] This method is based on the absorbance at 235 nm of the galacturonides formed by enzymatic degradation of the pectic materials. Culture filtrates or other test solutions are mixed with a polypectate substrate buffered at pH 8.8 with tricine. The increase in absorbance is followed spectrophotometrically. In the production of the lyase enzyme, some organisms do not produce a high constitutive level of the enzyme. A mixture of pectin together with an extract of potato is a good inducer of enzyme production for some bacteria.[28] Such an induction situation, i.e., mixtures of pectin and vegetable extract material, can be found in some prepared foods. Also, catabolite repression by glucose should be avoided. This situation exists for some of the pectate lyase producing organisms, and some food products may contain sufficient levels of glucose to cause repression. This may, in fact, limit lyase production by the organism, but not its growth.

For polygalacturonase activity, MP-7 medium containing pectin can be adjusted to pH 5.0 or 6.0. Measurement of enzyme activity can be made as described by Patil and Dimond,[18] or by Phaff.[19] The regulation of polygalacturonase production and catabolite repression has been studied.[12, 18] An assay for pectinesterase is given by Langcake et al.[15] This enzyme removes methoxyl groups from pectin, and thus leaves free carboxyl groups (pectic acid) which can be titrated with alkali, and allows enzyme activity to be determined.

The levels of calcium in any medium used to detect pectinolytic activity is of paramount importance. High levels of calcium (0.075M) stimulate, or are needed for, lyase production, whereas these high levels may limit polygalacturonase production.[24] Induction factors probably differ with the organism under study. Certain components should be avoided in media since they inhibit pectic enzyme activity. These include phenolic compounds, indoleacetic acid, fatty acids, and end products of polygalacturonase action.[2] Some of these compounds may occur in foods, and thus may act as natural inhibitors of pectinolytic action. In addition, production of the enzyme itself can be inhibited by plant extracts.[28]

14.3 MEDIA AND REAGENTS

14.31 Polypectate Gel Medium (Chapter 2)

14.32 Medium to Detect Pectate Lyase (MP-7 Medium) (Chapter 2)

14.33 Medium to Detect Polygalacturonase (MP-5 Medium) (Chapter 2)

14.34 Selective Medium For Fluorescent Pectinolytic Pseudomonads (FPA Medium) (Chapter 2)

14.341 Antibiotics solutions

Penicillin G	75,000 units
Novobiocin	45 mg
Cycloheximide	75 mg

Add 1 ml ethanol to above mixture of antibiotics, and allow to stand 30 minutes. Dilute with 9 ml sterile H_2O and add 1 ml of the diluted antibiotic solution to 100 ml of sterile and cooled basal medium before pouring plates. After the plates are inoculated by a spread plate technique and incubated, the fluorescent bacteria can be detected under long wavelength ultraviolet light (366 nm). The plates are then treated with the polysaccharide precipitant as described under MP-7 medium.

14.35 Pectinolytic Enzyme Activity Assay (14.25)

14.36 Polysaccharide Precipitant

1% solution of hexadecyltrimethylammonium bromide in water. This solution may be autoclaved if colonies are to be isolated from flooded plates. Some investigators have used 6 to 8 N HCl as the precipitant (see section 14.22).

14.4 PROCEDURES

14.41 Sample Preparation

Prepare a homogeneous suspension and appropriate dilutions of the food (Chapter 1). In food containing a large amount of glucose, sufficient dilution may be necessary to avoid possible catabolite repression of pectic enzyme synthesis.[9, 26, 27, 28] Also in high sugar foods such as jams and jellies, care should be exercised in dilution of the sample to prevent osmotic shock to the cells, by the use of an appropriate buffer or isotonic solution.

14.42 Preparation and Incubation of Plates

Place not more than 0.25 ml quantities of appropriate dilutions (so as to obtain 20 to 30 colonies) on the surface of prepoured plates. Distribute by a spread plate technique. Incubation of plates inoculated with a variety of plant materials has been carried out at 30 C for 48 hours. Incubation of plates inoculated with foods probably should be made at the temperature at which the food is stored. For example, *Erwinia carotovora* grows at 37 C but does not produce pectate lyase (Sands and Hankin, unpublished data). Some pectinolytic bacteria grow at 37 C but do not produce pectate lyase until the temperature is 32 C or below.

14.43 Counting and Reporting

Colonies which cause depressions in polypectate gel medium or are surrounded by a clear zone after flooding MP-7, MP-5, or FPA plates with a pectin precipitant, are pectinolytic. Fluorescent bacteria are detected on FPA plates with long wavelength U.V. light before adding pectin precipitant.

14.5 INTERPRETATION

Media, such as described in this chapter, on which pectinolytic microorganisms can be detected, enumerated, and isolated, are useful to characterize microbial degradation of fruits, vegetables, and processed foods, and to screen for plant pathogens or soil organisms that can degrade pectin. Clearly, the varieties of organisms that can degrade pectin are diverse. Thus, the most suitable medium to use depends on the type of organism expected as well as the type of food to be examined.

14.6 REFERENCES

1. Albersheim, P. and U. Killias. 1962. Studies relating to the purification and properties of pectin transeliminase. Arch. Biochem. Biophys. **97**:107–115
2. Bateman, D. F. and R. L. Millar. 1966. Pectic enzymes in tissue degradation. Ann. Rev. Phytopath. **4**:119–146
3. Doesburg, J. J. 1973. The Pectic Substances. Phytochemistry, Vol. I; The Process and Products of Photosynthesis. L. P. Miller ed., Van Nostrand Reinhold Co., New York, N.Y.
4. Dowson, W. J. 1957. Plant Diseases Due to Bacteria, 2nd ed. Cambridge University Press, Cambridge, England.
5. Etchells, J. L., T. A. Bell, R. J. Monroe, P. M. Masley, and A. L. Demain. 1958. Populations and softening enzyme activity of filamentous fungi on flowers, ovaries, and fruit of pickling cucumbers. Appl. Microbiol. **6**:427–440
6. Hankin, L., M. Zucker, and D. C. Sands. 1971. Improved solid medium for the detection and enumeration of pectolytic bacteria. Appl. Microbiol. **22**:205–209
7. Hankin, L., D. C. Sands, and D. E. Hill. 1974. Relation of land use to some degradative enzymatic activities of soil bacteria. Soil Science **118**:38–44
8. Hildebrand, D. C. 1971. Pectate and pectin gels of *Pseudomonas* sp. and other bacterial plant pathogens. Phytopathol. **61**:1430–1436
9. Hsu, E. J. and R. H. Vaughn. 1969. Production and catabolite repression of the constitutive polygalacturonic acid *trans*-eliminase of *Aeromonas liquefaciens*. J. Bacteriol. **98**:172–181
10. International Union of Biochemistry. 1973. Recommendations (1972) of the Intl. Union of Pure and Appl. Chem., and the Intl. Union of Biochem. on the Nomenclature and Classification of Enzymes. Elsevier Publ. Co. New York, N.Y.
11. Jayasankar, N. P. and P. H. Graham. 1970. An agar plate method for screening and enumerating pectinolytic microorganisms. Can. J. Microbiol. **16**:1023
12. Keen, N. T. and J. C. Horton. 1965. Sugar repression of endopolygalacturonase and cellulase synthesis during pathogenesis by *Pyrenochaeta terrestris* as a resistance mechanism in onion pink root. Phytopathol. **55**:1063–1064
13. Kertesz, Z. I. 1951. The Pectic Substances. Interscience Publishers, Inc. New York, N.Y.
14. King, A. D., Jr. and R. H. Vaughn. 1961. Media for detecting pectolytic gram-negative bacteria associated with the softening of cucumbers, olives, and other plant tissues. J. Food Sci. **26**:635–643
15. Langcake, P., P. M. Bratt, and R. B. Drysdale. 1973. Pectinmethylesterase in *Fusarium*-infected susceptible tomato plants. Physiol. Plant Pathol. **3**:101–106
16. Ng, H., and R. H. Vaughn. 1963. *Clostridium rubrum* sp. N and other pectinolytic clostridia from soil. J. Bacteriol. **85**:1104–1113
17. Lund, B. M. 1972. Isolation of pectolytic *Clostridia* from potatoes. J. Appl. Bacteriol. **35**:609–614
18. Patil, S. S. and A. E. Dimond. 1968. Repression of polygalacturonase synthesis in *Fusarium oxysporum* f. sp. *lycopersici* by sugars and its effect on symptom reduction in infected tomato plants. Phytopathol. **58**:676–682
19. Phaff, H. J. 1966. α-1,4-polygalacturonide glycanohydrolase (endopolygalacturonase) from *Saccharomyces fragilis*. In Methods in Enzymology, Ed. by E. F. Neufeld and V. Ginsburg, **8**:636–641
20. Rombouts, F. M. 1972. Occurrence Properties of Bacterial Pectate Lyases. Agric. Res. Rep. (Versl. landbouwk. Onderz.) 779. Centre for Agric. Publ. and Doc. Wageningen, Neth.
21. Sands, D. C., L. Hankin, and M. Zucker. 1972. A selective medium for pectolytic fluorescent Pseudomonads. Phytopathol. **62**:998–1000

22. VAUGHN, R. H., G. D. BALATSOURAS, G. K. YORK II, and C. W. NAGEL. 1957. Media for detection of pectinolytic microorganisms associated with softening of cucumbers, olives, and other plant tissues. Food Res. **22**:597–603
23. VAUGHN, R. H., A. D. KING, C. W. NAGEL, H. NG, R. E. LEVIN, J. D. MACMILLAN, and G. K. YORK II. 1969. Gram negative bacteria associated with sloughing, a softening of California ripe olives. J. Food Sci. **34**:224–227
24. VORAGEN, A. G. J. 1972. Characterization of Pectin Lyases on Pectins and Methyl Oligogalacturonates. Agric. Res. Rep. (Versl. landbouwk. Onderz.) 780. Centre for Agric. Publ. and Doc., Wageningen, Neth.,
25. WIERINGA, K. T. 1949. A Method for Isolating and Counting Pectolytic Microbes. Intl. Congr. Microbiol. Proc., 4th Congr. 1947, p. 482
26. ZUCKER, M. and L. HANKIN. 1970. Regulation of pectate lyase synthesis in *Pseudomonas fluorescens* and *Erwinia carotovora*. J. Bacteriol. **104**:13–18
27. ZUCKER, M. and L. HANKIN. 1971. Inducible pectate lyase synthesis and phytopathogenicity of *Pseudomonas fluorescens*. Can. J. Microbiol. **17**:1313–1318
28. ZUCKER, M., L. HANKIN, and D. C. SANDS. 1972. Factors governing pectate lyase synthesis in soft rot and non-soft rot bacteria. Physiol. Plant Pathol. **2**:59–67

CHAPTER 15

ACID PRODUCING MICROORGANISMS

W. E. Sandine, W. M. Hill, and H. Thompson

15.1 INTRODUCTION

A variety of acid producing bacteria is found in nature, in the soil, on raw agricultural products and in certain processed foods. One of the most important groups of acid producing bacteria in the food industry is that of the lactic acid bacteria. Members of this group are gram positive, nonsporulating cocci or rods, dividing in one plane only (with the exception of pediococci), catalase negative, usually nonmotile, obligate fermenters, producing mainly lactic acid and sometimes also volatile acids and CO_2. They are subdivided into the genera *Streptococcus, Leuconostoc, Pediococcus,* and *Lactobacillus.* The homofermentative species produce lactic acid from available sugars, while the heterofermentative types produce, in addition to lactic acid, mainly acetic acid, ethanol, CO_2, and other compounds in trace amounts. Lactic acid bacteria are widespread in nature and are best known for their activities in major foods such as dairy, meat, and vegetable products.

Many sporeforming species belonging to the genera *Bacillus* and *Clostridium* also are important acid producers. Their role in quality deterioration and subsequent spoilage of foods, particularly canned foods, is discussed in detail in other parts (18 to 23) of this text.

Propionibacterium and *Acetobacter* species are other well known acid producing bacteria. These organisms play important roles in several industrial processes. For instance, certain *Propionibacterium* species are important in the development of the characteristic flavor and eye production in Swiss-type cheeses. Members of the genus *Acetobacter* are best known for their use in the manufacture of vinegar.

Most enteric bacteria are able to carry out a mixed acid type, or butylene glycol fermentation. The significance in foods of species of *Escherichia, Enterobacter, Salmonella,* and *Shigella,* and methods for their enumeration are presented in chapters 24, 25, 26 and 27.

IS/A Committee Liaison: Carl Vanderzant

15.2 GENERAL METHODS

15.21 Titration and pH

Titratable acidity expressed as lactic or acetic acid may be used as an indirect measure of bacterial growth in brines and liquid foods. However, since the acid available to be neutralized with standard alkali depends on the buffering capacity of the liquid, pH is a more accurate indicator.

15.22 Indicator Media

Complex media containing an indicator such as bromcresol purple may be used to enumerate different acid producing types of bacteria present in food products. For organisms producing considerable acid, bromcresol green or bromphenol blue may be used and for moderate acid producers, phenol red. Acid produced and accumulating around colonies will change the color of the indicator in the medium and identify those colonies to be counted.

15.23 Special Media

Numerous complex media are available which support growth of various acid producing bacteria. A few can be employed for the qualitative and/or quantitative differentiation of certain species, for example the lactic streptococci and propionibacteria. However, most of these media are nonselective and other microorganisms may grow on them. Some of these media can only be employed in special situations where one or only a limited number of species is present. In other instances, a numerical estimation may depend on (a) recognition of colony characteristics such as size, shape, color, and biochemical reactions in the medium (for example, acid production or arginine hydrolysis), or (b) identification by picking individual colonies from plates and placing them into different test media.

15.3 REAGENTS AND MEDIA

15.31 Reagents

a. 0.1 N NaOH (Chapter 2)

b. Phenolphthalein solution

Unless otherwise specified, phenolphthalein used as an indicator is 1% alcoholic solution.

c. Bromcresol purple solution

The dye may be added directly to the medium (0.02 to 0.04 g/l) before autoclaving, or prepared as a stock solution. In the latter case, about 1 ml of a 1.6% alcoholic solution is added per liter of medium.[15] This indicator has a pH range of 5.4 (yellow) to 7.0 (purple). Stock solutions of bromcresol green (yellow 3.8 to 5.4 blue) or bromphenol blue (yellow 3.1 to 4.7 blue) may be similarly prepared and used. Caution: Some lots of indicator may be more inhibitory to bacterial growth than others. In addition, some indicators are more inhibitory to a particular bacterial species than to others.

15.32 Media

a. **Plate count agar** (Standard Methods Agar) (Chapter 2) containing bromcresol purple (PCA-BCP)

b. **Lactic agar**[5] (Chapter 2)

c. **MRS broth**[4] (Chapter 2)

d. **RMW agar**[14] (Chapter 2)

e. **M 16 agar**[9] (Chapter 2)

f. **Eugon agar**[17] (Chapter 2)

g. **APT agar**[6] (Chapter 2)

h. **Differential broth for lactic streptococci**[12] (Chapter 2)

i. **Agar medium for differential enumeration of lactic streptococci**[13] (Chapter 2)

j. **Lee's agar**[8] (Chapter 2)

k. **HYA agar**[11] (Chapter 2)

l. **Sodium lactate agar**[10] (Chapter 2)

m. **Chopped liver broth**[3] (Chapter 2)

15.4 EQUIPMENT

To enumerate acid producers, materials commonly used in bacteriological analyses are required (Chapters 2 and 4). To determine titratable acidity of milk and milk products, consult "Techniques of Dairy Plant Testing." E. F. Goss, Iowa State Univ. Press, Ames, Iowa.

15.5 PROCEDURES

15.51 Acid Producing Bacteria

Various media are used to determine acid producers in foods (Chapters 2 and 15). Following is a description of methods and media frequently employed to determine the major types and activities of acid producers in foods.

15.52 Thermophilic Flat Sour Bacteria (Chapter 21)

15.53 Total Acid Producer Count

Prepare pour plates (Chapter 4) of samples with plate count agar containing bromcresol purple (PCA–BCP) or with dextrose tryptone agar (Chapter 2). For conditions of plate incubation, consult Chapter 4. After incubation of plates, count colonies with yellow halo and report as acid-forming organisms per g of product.

15.54 Lactic Acid Producing Bacteria

The large number of broth and agar plating media proposed for lactic acid bacteria, particularly for streptococci and/or lactobacilli, is indicative of the difficulties encountered in growing some strains of these groups of organisms. The choice of medium is governed to some extent by the particular strains under study, and therefore by product or habitat. In the opinion of the authors, the media listed below may have merit in the support of colony development of lactic acid bacteria. These media are not highly selective, some such as eugon agar are not selective at all. Hence, organisms other than lactic acid bacteria may develop on these media and produce acid.

a. Lactic agar[5] (Chapter 2)

This medium was developed to support colony development of lactic streptococci and lactobacilli. Prepare pour plates of samples as described in Chapter 4 above using lactic agar. After incubation of plates prepare Gram stains of individual colonies, examine these microscopically and test for catalase reaction. Gram positive, catalase negative cocci or rods may be tentatively considered as lactic acid bacteria. If further identification is needed consult "Identification of the Lactic Acid Bacteria" by M. E. Sharpe, T. F. Fryer and D. C. Smith. In B. M. Gibbs and F. A. Skinner (ed.), Identification Methods for Microbiologists. Vol. 1A. Academic Press Inc., New York.

b. MRS agar (Chapter 2)

MRS broth was developed by de Man, Rogosa and Sharp[4] to support growth of various lactobacilli, particularly from dairy origin. MRS broth with added agar may be employed in preparing pour plates of samples as described in 15.54a using MRS agar instead of lactic agar. Identification of individual colonies may be carried out as described in 15.54a.

c. RMW agar[14] (Rogosa SL agar) (Chapter 2)

This medium is a selective medium for the cultivation of oral and fecal lactobacilli. For the preparation of agar plates, proceed as in 15.54a, but substitute RMW agar for lactic agar.

d. M 16 agar (Chapter 2)

This medium was developed by Lowrie and Pearce[9] to support growth of lactic streptococci used in Cheddar cheese manufacture in New Zealand. To prepare agar plates proceed as in 15.54a but substitute M 16 agar for lactic agar.

e. Eugon agar (Chapter 2)

This medium is reported to support surface growth of lactic acid bacteria.[17] To prepare agar plates, proceed as in 15.54a.

f. APT agar (Chapter 2)

APT agar was developed for the cultivation and enumeration of heterofermentative lactic acid bacteria of discolored cured meat products.[6] To prepare plates, proceed as in 15.54a.

15.55 Differential Media for Lactic Acid Bacteria

In this section methods are presented for the qualitative and quantitative differentiation of some lactic acid bacteria employed in the dairy industry. These media are not selective media and must be used only in pure culture studies or as recommended in this section. The broth and agar media for lactic streptococci can be used to identify members of lactic group streptococci when commercial starter cultures or products are first plated on a general purpose agar, or on media more efficient for the detection of lactic acid bacteria, such as lactic agar or eugon agar.

a. Separation of lactic streptococci—*S. lactis, S. cremoris,* and *S. lactis* var. *diacetylactis (S. diacetilactis)*

The most common microorganisms in starter cultures used in dairy products are streptococci. *S. lactis, S. cremoris,* and *S. lactis* var. *diacetylactis* belong to this group. Separation of these species can be made by biochemical tests, major criteria being arginine hydrolysis and tests for diacetyl and acetoin. Reddy et al.[12] developed a broth medium (15.32h) that differentiates these species. In addition to direct inoculation with a loopful of an active (pure) milk culture, the broth is also suitable for qualitative differentiation of individual colonies of lactic streptococci developing on agar plates containing dilutions of a commercial starter culture. After inoculation, close test tube caps tightly to prevent escape of liberated NH_3 and CO_2. Incubate tubes at 30 C for 24 to 72 hours and observe indicator color reactions and CO_2 accumulation at 24 hour intervals. *S. cremoris* produces a deep yellow color (acid) in the broth. *S. lactis* initially turns the broth yellow (acid), but later the violet hue returns due to liberation of NH_3 from arginine. *S. lactis* var. *diacetylactis* yields a violet color, and produces copious amounts of CO_2 (from citrate) in the fermentation tubes within 48 hours. *S. lactis* var. *diacetylactis* produces a more intense purple than *S. lactis. Leuconostoc* starter strains cause no appreciable color change in the violet differential broth, and only minute amounts of gas are observed with some *L. dextranicum* strains. Arginine hydrolysis in the differential broth can be further checked by testing a portion of the broth with Nessler's reagent on a porcelain spot plate. A deep red precipitate indicates arginine hydrolysis.

b. Differential enumeration of lactic streptococci

Reddy et al.[13] described an agar plating medium (15.32i) that can be used for the qualitative and quantitative differentiation of mixture of *S. cremoris, S. lactis,* and *S. lactis* var. *diacetylactis* strains. This medium contains arginine and calcium citrate as specific substrates, diffusible

(K_2HPO_4) and undiffusible ($CaCO_3$) buffer systems, and bromcresol purple as the pH indicator. Milk is added to provide carbohydrate (lactose) and growth stimulating factors. Production of acid from lactose causes yellow bacterial colonies. Subsequent arginine utilization by *S. lactis* and *S. lactis* var. *diacetylactis* liberates NH_3, and results in a localized pH change back to neutrality, and a return of the purple indicator hue. *S. lactis* var. *diacetylactis* utilizes suspended calcium citrate and after 6 days of incubation, the citrate degrading colonies exhibit clear zones against a turbid background. The buffering capacity of $CaCO_3$ limits the effects of acid and NH_3 production around individual colonies. From a practical standpoint this medium can be used (a) to study associative growth relationships in starter mixtures of these species, (b) to verify the composition of mixed starter cultures, and (c) to screen single strains for compatibility in mixed cultures.

Prepare decimal dilutions of the culture with sterile phosphate buffer (Chapter 2) by spreading 0.1 ml quantities of the dilution evenly over the surface of agar plates with a sterile bent glass rod. Incubate plates in a candle oats jar (15.56b) at 32 C and examine plates after 36 to 40 hours and after day 6 of incubation.

Counting. After 36 to 40 hours, count all colonies and then the yellow *S. cremoris* colonies separately. Return plates to the candle oats jar for an additional 4 days. After this period, expose the plates to the air for 1 hour. First, determine the total count, and then count all colonies showing zones of clearing of the turbid suspension of calcium citrate (*S. lactis* var. *diacetylactis*). Subtract the *S. cremoris* (after 36 to 40 hours) and *S. lactis* var. *diacetylactis* counts from the total count to obtain the *S. lactis* population in the mixture. Slow arginine hydrolyzing or nonhydrolyzing strains of *S. lactis* var. *diacetylactis* in cultural mixtures sometimes produce yellow colonies after 36 to 40 hours similar to *S. cremoris*. In such an instance, mark the yellow colonies (after 36 to 40 hours) with an indelible felt pen. When the final count is taken, count the marked colonies which show clearing as *S. lactis* var. *diacetylactis* and subtract their number from the original yellow colony count to obtain the accurate value for *S. cremoris*.

Maximum differential efficiency is obtained only when the counts on the individual plates do not exceed 250 colonies, and when fresh medium is employed. This medium is not selective and must be used in pure culture studies only.

c. Ratio of *Streptococcus thermophilus* and *Lactobacillus bulgaricus* in yogurt

Yogurt is a fermented milk product in which *S. thermophilus* and *L. bulgaricus* are the essential microbial species and are active in a symbiotic relationship. To obtain optimum consistency, flavor, and odor, many investigators claim that the two species should be present in about equal numbers in the culture. Dominance by either species can cause defects. Because of the emphasis on the maintenance of a balance between coccus and rod, there is a need for techniques to determine the relative proportions of *S. thermophilus* and *L. bulgaricus* when grown together in milk cultures. A microscopic examination to determine the ratio of coccus to rod is inadequate because dead cells cannot be distinguished from viable ones by this technique.

An agar medium (Lee's agar) for differential enumeration of yogurt starter bacteria has been described by Lee et al.[8] This medium contains sucrose, which most *L. bulgaricus* strains will not ferment, but *S. thermophilus* will, and lactose which both species utilize. With a suitable combination of sucrose and lactose the rate of acid production by *S. thermophilus* is enhanced and that of *L. bulgaricus* restricted. Sufficient lactose is provided to obtain adequate colony formation of *L. bulgaricus* on the agar. Directions for the preparation of the agar plates should be followed very carefully (15.32j).

Dilute culture with sterile phosphate buffer (Chapter 2) to 1×10^{-6} and spread 0.1 ml samples of the dilutions over the agar surface with a sterile bent glass rod. Incubate plates for 48 hours at 37 C in a CO_2 incubator. *S. thermophilus* will produce yellow colonies and *L. bulgaricus* will yield white colonies. For satisfactory differentiation, the total number of colonies on the plates should not exceed 250. A preponderance of either species in a mixture does not allow distinction of colony types on this medium. This is because differentiation on this medium is based on acid-producing activity and restriction of acid diffusion within a small area and its detection with a pH indicator. Obviously, this is not a selective medium, and many other microorganisms can be expected to grow on it. In addition, some strains of *L. bulgaricus* can form yellow colonies indistinguishable from those of *S. thermophilus*. This difficulty may be obviated by the use of pretested strains in the culture mixtures.

Porubcan and Sellars[11] described a medium (HYA agar) on which *L. bulgaricus* grows as diffuse, low mass colonies (2 to 10 mm in diameter) and *S. thermophilus* as discrete, high mass colonies (1 to 3 mm in diameter). Differentiation is achieved in this medium by adding an appropriate sugar or sugar mixture to the molten agar base before plating. The limitation of this method, particularly for personnel with limited training, is that differentiation is based on colony morphology.

15.56 Propionic Acid Producing Bacteria

Isolation and enumeration of propionibacteria has been difficult because of their tendency toward anaerobiosis. Ordinary plating procedures generally are of little value in enumerating them because they do not readily grow under conventional plating conditions.

a. Sodium lactate agar

Vedamuthu and Reinbold,[16] and Malik et al[10] described the use of sodium lactate agar, with subsequent incubation of plates in the candle oats jar* for the enumeration and characterization of *Propionibacte-*

*The candle oats jar is a dessicator with moistened oats in the dessicant chamber. A candle is place on the platform in the chamber and lit just before closing of the lid. The atmosphere is one of high humidity and increased CO_2.

rium. Prepare pour plates with sodium lactate agar. Incubate plates in a candle oats jar at 32 C for 6 days. For interpretation of count, consult "pouch method".

b. Pouch method

Higher counts of *Propionibacterium* species generally are achieved by this method[7] which was first described for use in enumeration of clostridia by Bladel and Greenberg.[2] Other advantages of the pouch method are its simplicity as compared to the candle oats jar incubation method, ease with which colonies can be counted and isolated, reduced incubation time, and better use of incubator space. Pouches are prepared by sealing two sheets of plastic film for a few seconds with a pouch shaped sealing iron previously heated to 190 C. Details of the aluminum sealing iron, films, and methods of preparing pouches are described by Bladel and Greenberg[2] and Hettinga et al.[7]

Pipet 1 ml aliquots of dilutions of sample through the neck into the pouch. Add 30 ml of tempered (60 C) sodium lactate agar (2% agar) and mix sample and medium by gently working the mixture between the fingers. The pouches are then placed in a "holder" during solidification of the medium. The agar seal in the neck of the pouch will exclude oxygen sufficiently so that sealing the top is not necessary. Incubate pouches at 32 C for 4 days or longer. With longer incubation, gas will accumulate, separating the film from the agar and enhancing spreading action of growth. Colonies formed in the medium can be counted by placing the pouch on a Quebec counter.

Except when pure cultures are used, not all colonies that grow on sodium lactate agar are propionibacteria. To be certain, individual colonies must be picked and identified. If the technician is thoroughly familiar with propionibacteria, tentative identification can be made on the basis of colony characteristics (size, shape, color). If further identification cannot be made, report count as "count on sodium lactate agar" and not as count of propionibacteria.

15.57 Butyric Acid Producing Bacteria

If acid, the food product should be neutralized with an excess of sterile calcium carbonate. Heat a 50 to 100 g sample in a water bath at 80 C for 20 minutes to kill vegetative cells. Prepare decimal dilutions, and inoculate previously heated and cooled tubes of chopped liver broth.[3] Seal with melted vaspar (vasoline, 50 g; agar, 2 g; Tween 80, 1 ml; H_2O, 50 ml), and incubate 7 days at 32 C. Record positive tubes daily as evidenced by production of gas and a strong butyric acid odor.

15.58 pH and Titratable Acidity

An indirect measure of the growth of acid producing bacteria in fermented foods is carried out frequently for quality control purposes. This may be done by determining the acid produced in the product by pH or titratable acidity measurements.

a. pH

A number of commercially manufactured pH meters are available and acidity measurement can be made readily by following the meter manual instructions. For specific instructions with individual commodities, consult Chapters 40 to 53.

b. Titratable acidity

Measurement of acid present may be accomplished by titration of a suitable quantity of sample with 0.1 N NaOH to the phenolphthalein end point (pH 8.3). The percent acid present as lactic acid is calculated from the following formula:

$$\% \text{ lactic acid} = \frac{(\text{ml NaOH})(\text{N NaOH})(\text{Milliequivalent weight of lactic acid})(100)}{\text{weight of sample in g}}$$

Milliequivalent weight of lactic acid = 90/1000 = 0.09
Report acidity as % lactic acid by weight. Specific instructions for individual foods are found in the latest edition of "Official Methods of Analysis of the Association of Official Analytical Chemists", for milk and milk products in "Techniques of Dairy Plant Testing," E. F. Goss, Iowa State Univ. Press, Ames, Iowa.

15.6 REFERENCES

1. Standard Methods for the Examination of Dairy Products. 13th ed. 1972. American Public Health Association, New York, N.Y.
2. Bladel, B. O. and R. A. Greenberg. 1965. Pouch method for the isolation and enumeration of clostridia. Appl. Microbiol. **13**:281–285
3. Cameron, E. J. 1936. Report on culture media for non-acid products. J. Assoc. Off. Agr. Chem. **19**:433–438
4. DeMan, J. C., M. Rogosa, and M. Elisabeth Sharpe. 1960. A medium for the cultivation of lactobacilli. J. Appl. Bacteriol. **23**:130–135
5. Elliker, P. R., A. W. Anderson, and G. Hannesson. 1956. An agar culture medium for lactic acid streptococci and lactobacilli. J. Dairy Sci. **39**:1611
6. Evans, J. B. and C. F. Niven, Jr. 1951. Nutrition of the heterofermentative lactobacilli that cause greening of cured meat products. J. Bacteriol. **62**:599–603
7. Hettinga, D. H., E. R. Vedamuthu, and G. W. Reinbold. 1968. Pouch method for isolating and enumerating propionibacteria. J. Dairy Sci. **51**:1707–1709
8. Lee, S. Y., E. R. Vedamuthu, C. J. Washam, and G. W. Reinbold. 1974. An agar medium for the differential enumeration of yogurt starter bacteria. J. Milk Food Technol. **37**:272–276
9. Lowrie, R. J. and L. E. Pearce. 1971. The plating efficiency of bacteriophages of lactic streptococci. New Zealand J. Dairy Sci. Technol. **6**:166–171
10. Malik, A. C., G. W. Reinbold, and E. R. Vedamuthu. 1968. An evaluation of the taxonomy of *Propionibacterium*. Can. J. Microbiol. **14**:1185–1191
11. Porubcan, R. S. and R. L. Sellars. 1973. Agar medium for differentiation of *Lactobacillus bulgaricus* from *Streptococcus thermophilus*. J. Dairy Sci. **56**:634
12. Reddy, M. S., E. R. Vedamuthu, and G. W. Reinbold. 1971. A differential broth for separating the lactic streptococci. J. Milk Food Technol. **34**:43–45

13. Reddy, M. S., E. R. Vedamuthu, C. J. Washam, and G. W. Reinbold. 1972. Agar medium for differential enumeration of lactic streptococci. Appl. Microbiol. **24**:947–952
14. Rogosa, M., J. A. Mitchell, and R. F. Wiseman. 1951. A selective medium for the isolation and enumeration of oral and fecal lactobacilli. J. Bacteriol. **62**:132–133
15. Society of American Bacteriologists. 1957. Manual of Microbiological Methods. McGraw-Hill, New York, N.Y.
16. Vedamuthu, E. R. and G. W. Reinbold. 1967. The use of candle oats jar incubation for the enumeration, characterization and taxonomic study of propionibacteria. Milchwissen schaft **22**:428–431
17. Vera, H. D. 1947. The ability of peptones to support surface growth of lactobacilli. J. Bacteriol. **54**:14

CHAPTER 16

YEASTS AND MOLDS

J. A. Koburger

16.1 INTRODUCTION

Yeasts and molds are widely distributed in the environment, and may be found as part of the normal flora of a food product, on inadequately sanitized equipment, or as airborne contaminants. Although certain yeasts and molds are useful in the manufacture of various foods, such as mold ripened cheese and bread, they also can be responsible for spoilage of many types of foods. Because of their slow growth and poor competitive ability, yeasts and molds often manifest themselves on or in foods in which conditions are less favorable to bacterial growth. These are foods of low pH, low moisture, high salt or sugar content, or under conditions such as low storage temperature, the presence of antibiotics, or following exposure of food to irradiation. From a practical standpoint they can become a problem in certain foods, and must be considered separately in a microbiological analysis.

Yeasts and molds can utilize such substrates as pectins and other carbohydrates, organic acids, proteins, and lipids. While yeasts are not generally considered proteolytic, recent studies have shown that some are capable of hydrolyzing a wide range of proteinaceous materials.[1, 11]

Additionally, yeasts and/or molds can cause problems through (a) synthesis of toxic metabolites, (b) resistance to heat,[19] freezing,[8] antibiotics, or irradiation[18] and (c) their ability to alter otherwise unfavorable substrates allowing for the outgrowth of pathogenic bacteria.[5] They may cause off odors, off flavors, and discoloration of food surfaces.[18]

Enumeration of yeasts and molds in foods generally provides sufficient information for routine control procedures. If identification is needed, excellent references are available.[2, 14]

16.2 GENERAL METHODS

The classical method for the enumeration of yeasts and molds uses an acidified medium to inhibit bacteria.[22] The literature does not clearly explain the specific effect of acidulants,[10] nor does there appear to be any consensus as to the best medium for enumeration.[20] Experience indicates

IS/A Committee Liaison: Carl Vanderzant

that potato dextrose agar (Chapter 2), acidified with tartaric acid to pH 3.5 following sterilization, is satisfactory. It is necessary to incubate plates at a temperature lower than that used for a total plate count,[13, 23] usually within the range of 20 to 25 C for 5 days. Occasionally high concentrations of sugar or other materials are added to the medium to aid in the isolation and enumeration of certain types.[6]

Acidified media present certain shortcomings,[9, 16, 17, 21] such as spreading mold colonies, occasional growth of bacteria, precipitation of food particles, and inability of some yeasts and molds to grow at the low pH of the medium. Other agents which have been used to control bacterial growth include dyes,[15] antibiotics,[4, 15] and various chemicals.[3, 7] Results with these media have shown increased recovery of yeasts and molds, better control of bacteria, less difficulty with the precipitation of food particles, and more rapid growth resulting in larger colonies. While these procedures are commonly used in areas other than food microbiology, no attempt has been made to evaluate fully their efficacy with foods. The antibiotic method outlined in 16.42 and 16.52 has been tested in several studies.[9, 10, 11, 12, 13] It is efficient and inexpensive; the medium is easy to prepare and use; and the method lends itself to physiological differentiation of the colonies.[11]

To determine mold or yeast counts of foods by microscopic examination, consult the latest edition of "Official Methods of Analysis of the Association of Official Analytical Chemists", Washington, D. C.

16.3 TREATMENT OF SAMPLE

Preparation of the sample is the same as for the aerobic plate count (Chapter 4).

16.4 MEDIA

16.41 Acidified Method

Medium: Acidify potato dextrose (Chapter 2) or malt agar (Chapter 2) with sterile 10% tartaric acid to pH 3.5 ± 0.1. Acidify the sterile and tempered medium with a predetermined quantity of acid solution immediately before pouring plates. Do not reheat medium once acid has been added. Determine accuracy of adjusted pH by pouring an aliquot of the medium into a small beaker, cool it to plate incubation temperature, and place the electrodes directly into the solidified medium.

16.42 Antibiotic Method

a. Medium: Use plate count (Chapter 2), mycophil (Chapter 2), or malt agar (Chapter 2); pH adjustment is not necessary. Rehydrate the medium according to directions and sterilize. Temper medium to 45 ± 1 C, and add 2 ml of antibiotic solution (16.42b) per 100 ml medium. After mixing, the medium is ready for use.

b. Preparation of antibiotic solution: Add 500 mg each of chlortetracycline HCl and chloramphenicol (Calbiochem., LaJolla, CA) to 100 ml sterile phosphate buffered distilled water and mix. Not all material dissolves, therefore the suspension must be evenly dispersed before pipetting into the medium.

16.5 PROCEDURES

16.51 Acidified Method

Prepare pour plates (Chapter 4) with 15 to 20 ml agar tempered to 45 C ± 1 C, allow to solidify, invert plates and incubate at 20 to 25 C for 5 days. If excessive mold growth develops on the plates, count first after 3 days and then again after 5 days. Counting plates from the reverse side is sometimes helpful if they are overgrown with molds.

16.52 Antibiotic Method

The procedure is the same as for the Acidified Method (16.51).

16.6 REPORTING

Report as yeast and mold count per gram, or milliliter of sample, as applicable.

16.7 PRECAUTIONS

Because of the sensitivity to high temperatures of many yeasts and molds in foods, do not incubate plates at 32 C. If controlled incubation is not available at a lower temperature, use room temperature and report as such, but note range of temperature.

When using the acidified method, at least 10 colonies should be picked periodically from a countable plate and gram stained to ensure that acid tolerant bacteria are not growing. Generally, yeasts and the asexual spores of molds will be gram positive. The vegetative growth of molds will appear gram negative.

Under alkaline conditions (pH 8.0 or above) the effectiveness of the antibiotics is diminished.[12] If highly alkaline foods are being analyzed, care should be taken to ensure that the pH of the medium is not raised above 8.0, allowing for growth of bacteria. Preferably the medium should be between pH 5.0 and 6.0; however, no problems have been encountered if the medium is maintained between pH 4.0 and 7.0. The antibiotic suspension can be stored at 5 C for up to 8 weeks without loss of activity. No difficulty has been experienced due to contaminated antibiotics. However, the stock suspension should be checked by preparing a control plate without sample.

While other inhibitors, notably rose bengal, have been used for yeasts and mold counts, they are not sufficiently inhibitory to bacteria for use in routine analyses.[21] If other inhibitors are used, check the plates for recovery of yeasts and molds for bacterial growth. The methods described in 16.5 are general procedures, and modifications may be needed when working with unusual food materials.

Precautions for blending of the sample should be observed when using plating techniques for mold counts (Chapter 8.54).

16.8 INTERPRETATION

Interpretation of results requires background data on the expected levels of yeasts and molds to be found in a food, predominant groups present, and knowledge of the potential for growth of these microorganisms in the final product. It is during the collection of these background data, and the subsequent application of this information to a control program, that the selective enumeration of yeasts and molds can be used to maximum advantage. Additional information on the use of mold and yeast counts in various foods is presented in Chapters 40 to 53.

16.9 REFERENCES

1. AHEARN, D. G., S. P. MEYERS, and R. A. NICHOLS. 1968. Extracellular proteinases of yeasts and yeastlike fungi. Appl. Microbiol. **16**:1370–1374
2. BARNETT, H. L. and B. B. HUNTER. 1972. Illustrated Genera of Imperfect Fungi. 3rd ed. Burgess Publishing Co., Minneapolis, Minn.
3. BEECH, F. W. and J. G. CARR. 1955. A survey of inhibitory compounds for the separation of yeasts and bacteria in apple juices and ciders. J. Gen. Microbiol **12**:85–94
4. COOKE, W. B. 1954. The use of antibiotics in media for the isolation of fungi from polluted water. Antibiotics and Chemotherapy **4**:657–662
5. DUITSCHAEVER, C. L. and D. M. IRVINE. 1971. A case study: Effect of mold on growth of coagulase-positive staphylococci in cheddar cheese, J. Milk Food Technol. **34**:583
6. FABIAN, F. W. and M. C. WETHINGTON. 1950. Spoilage in salad and French dressing due to yeasts. Food Res. **15**:135–137
7. HERTZ, M. R. and M. LEVINE. 1942. A fungistatic medium for enumeration of yeasts. Food Res. **7**:430–441
8. JONES, A. H. and F. W. FABIAN. 1952. The viability of microorganisms isolated from fruits and vegetables when frozen in different menstrua. Mich. State Coll. Tech. Bull. 229, p. 42
9. KOBURGER, J. A. 1970. Fungi in foods. I. Effect of inhibitor and incubation temperature on enumeration. J. Milk Food Technol. **33**:433–434
10. KOBURGER, J. A. 1971. Fungi in foods. II. Some observations on acidulants used to adjust media pH for yeast and mold counts. J. Milk Food Technol. **34**:475–477
11. KOBURGER, J. A. 1972. Fungi in foods. III. The enumeration of lipolytic and proteolytic organisms. J. Milk Food Technol. **35**:117–118
12. KOBURGER, J. A. 1972. Fungi in foods. IV. Effect of plating medium pH on counts. J. Milk Food Technol. **35**:659–660
13. KOBURGER, J. A. 1973. Fungi in foods. V. Response of natural populations to incubation temperatures between 12 and 32 C. J. Milk Food Technol. **36**:434–435
14. LODDER, J. The Yeasts. A Taxonomic Study. North-Holland Publishing Co., Amsterdam, Holland p. 1385
15. MARTIN, J. P. 1950. Use of acid, rose bengal and streptomycin in the plate method for estimating soil fungi. Soil Sci. **69**:215–232
16. MOSSEL, D. A. A., M. VISSER, and W. H. J. MENGERINK. 1962. A comparison of media for the enumeration of moulds and yeasts in foods and beverages. Lab Practice **11**:109–112
17. OVERCAST, W. W. and D. J. WEAKLEY. 1969. An aureomycin-rose bengal agar for enumeration of yeast and mold in cottage cheese. J. Milk Food Technol. **32**:442–445

18. ROSE, A. H. and J. S. HARRISON. 1970. The Yeasts. 3. Yeast Technology. Academic Press, New York, NY. p. 590
19. SCHARDING, J. H., D. C. KELLEY, J. E. COOK, and D. H. KROPF. 1973. Experimental survival studies of *Sporothrix schenckii* in a meat product. J. Milk Food Technol. **36**:311–314
20. SHADWICK, G. W., Jr. 1938. A study of comparative methods and media used in microbiological examination of creamery butter. 1. Yeast and mold counts. Food Res. **3**:287–298
21. SKIDMORE, N. C. AND J. A. KOBURGER. 1966. Evaluation of Cooke's Rose Bengal Agar for the enumeration of fungi in foods. W. Va. Acad. Sci. Proc. **38**:63–68
22. WHITE, A. H. and E. G. HOOD. 1931. A study of methods for determining numbers of moulds and yeasts in butter. 1. The relation of the pH of the medium. J. Dairy Sci. **14**:463–476
23. WHITE, A. H. and E. G. HOOD. 1931. A study of methods for determining numbers of moulds and yeasts in butter. 2. The influence of temperature and time of incubation. J. Dairy Sci. **14**:494–507

CHAPTER 17

ENUMERATION OF HEAT RESISTANT MOLD (*BYSSOCHLAMYS*)

D. F. Splittstoesser

17.1 INTRODUCTION

In recent years spoilage of thermally processed fruits and fruit products by the mold *Byssochlamys* has been recognized as a problem in the United States.[6] Spoilage of canned fruit is often made manifest by a complete breakdown of texture due to the mold's pectinolytic enzymes, while growth in juices or juice drinks results in off flavors and the presence of objectionable mycelial masses. Byssochlamic acid, a metabolite of the mold, has exhibited a low level of toxicity in animal feeding studies.[5]

Byssochlamys presents a problem because its ascospores are able to survive the hot-fill temperatures, 100 C and lower, commonly used in the processing of acidic foods. Increasing the severity of the thermal process is not always a solution, because this may cause a marked reduction in product quality. Often the best method for preventing spoilage is to prevent mold growth from developing on equipment or lug boxes. Because the heat resistant forms of the mold (asci) usually do not form in less than two weeks, sanitation is effective in controlling *B. fulva.*

17.11 Distribution

Byssochlamys has been isolated from orchards, vineyards, and growing fields, as well as from containers used to transport fruit following the harvest.[2, 10] Vegetation in contact with soil often yields some of the highest spore populations. The number of ascospores on sound fruit generally is low, less than one per gram.[4, 10]

17.2 GENERAL METHODS

17.21 Samples

Because of the low incidence of *Byssochlamys* ascospores on many foods, detection often depends upon the use of relatively large samples. For example, 100 g or more have been cultured for the enumeration of spores on sound grapes, apples, and other fruits.[10] Centrifugation may be used to concentrate the spores and asci present in fruit juices and other liquid

IS/A Committee Liaison: Carl Vanderzant

foods. The speed and length of centrifugation will be influenced by sample size, viscosity, and specific gravity.

Food samples can be stored frozen prior to culturing for *Byssochlamys*; the ascospores survive repeated freezing and thawing.

17.22 Enumeration Principles

The procedure for *Byssochlamys* depends upon first heating the sample to activate dormant spores[11] and to destroy aciduric microorganisms. The material is then cultured on an acidified agar medium that permits growth of only surviving molds: *Byssochlamys* and, rarely, sclerotia formers such as *Penicillium* and *Aspergillus* species.[10, 13] *Byssochlamys* is recognized by the structure of its conidiophores, and by the fact that the asci are not enclosed in a peridium (Figure 1).[1, 12] The conidiophores are like penicillium, but differ in that the elongated phialides terminate in slender spore bearing tips which tend to bend away from the main axis. The globose asci which contain 8 ascospores have dimensions of 8.5 to 10.5 × 7.5 to 9.5 μm. The heat resistant species can be distinguished by the color of their colonies on acidified agar medium after 14 days of incubation: those of *B. nivea* are predominantly white, while those of *B. fulva* are dull yellow, fulvous, in color. The colonies on agar start as small mycelial zones and spread rapidly to cover the entire plate.

An antibiotic containing medium has been developed also to select the molds surviving an initial heat treatment[7] (Chapter 16). The method does not appear to be advantageous over acidified agar, however, because *Byssochlamys* spores germinate and grow very well in media of pH 3.0 and lower.[8]

Most probable number procedures using a broth medium have the advantage that relatively large samples can be cultured. The procedure has a serious drawback, however, in that an extended incubation, as long as 22 days, is required before all potentially positive broths exhibit growth.[9]

The enumeration of ascospores may be hampered by the spreading of colonies. This prevents accurate counts when the density is over 10 to 15 per plate. The addition of 8.3 μg of rose bengal dye per ml of nonacidified, pH 5.6, potato dextrose agar restricts colony growth and thus permits as many as 200 per plate to be readily counted. This concentration of dye does not reduce the viable recovery of spores subjected to thermal stress.

Water is a suitable dilution medium for *Byssochlamys* ascospores. Their resistance to various toxic substances[3] obviates the need for special diluents such as peptone water or phosphate buffer.

17.3 PROCEDURE FOR FRUIT[9]

Add approximately 100 g of fruit to a tared, sterile, screw cap blender jar. Reweigh the jar to determine the weight of added sample, then add 100 ml of sterile water. Blend for 5 minutes or until the mixture appears homogeneous. Place the jar (enclosed in a polyethylene bag to safeguard against the leakage of water through the bottom bushing) in a 70 C water bath. A 2 hour hold in the water bath usually assures that the homogenate

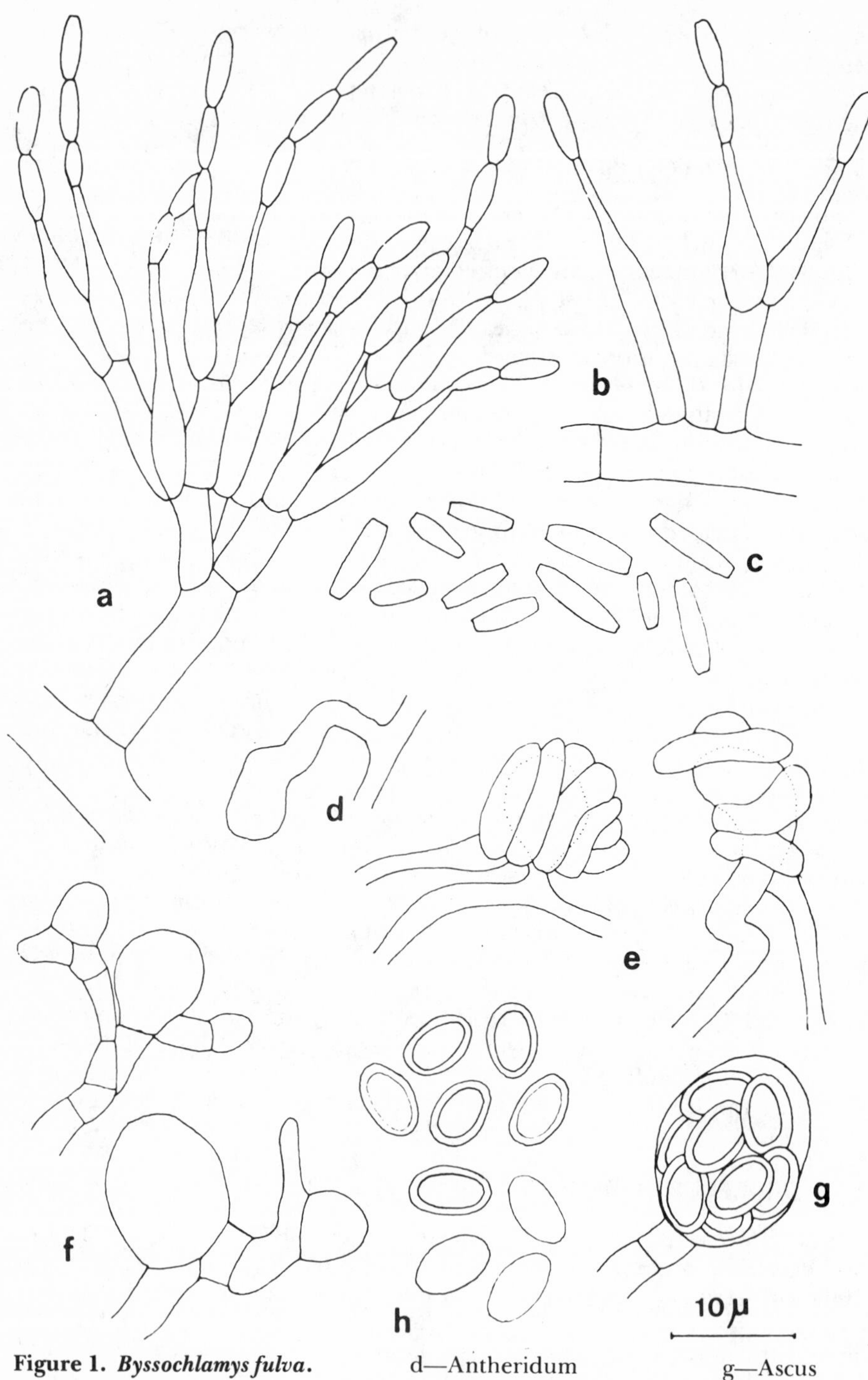

Figure 1. ***Byssochlamys fulva*.**
a, b—Sporing structures
c—Conidia
d—Antheridum
e—Asconatal initials
f—Ascus production
g—Ascus
h—Ascopores

will be at the equilibrium temperature for at least 1 hour, the minimum treatment time. After heating, distribute the entire homogenate into Petri dishes, about 10 ml per plate. Add equal volumes of potato dextrose agar, acidified to pH 3.5 with 10% tartaric acid, to the plates and mix. To assure adequate solidification, increase the agar content of the medium to 3%. Count colonies after incubation for 5 days at 32 C.

17.4 PRECAUTIONS

17.41 Air Contamination

The large number of Petri plates, 20 or more, that may be poured for each sample increases the opportunity for chance contamination by air-borne mold spores. Often the contaminants are recognizable species that do not possess special heat resistance. The development of a *Byssochlamys* colony on one or even several plates, however, presents a problem of interpretation, since one can never be sure of its source. This is particularly true in laboratories where *Byssochlamys* is undergoing study. When available, hoods or sterile rooms should be used when samples are cultured.

17.42 Recognizing *Byssochlamys*

Some strains may be identified incorrectly as *Paecilomyces*, the imperfect state, because the few asci that are produced cannot be detected under the microscope. It is especially difficult to see the asci when large numbers of conidia are formed. Asci, which apparently are more dense than conidia and hyphae, may be found at the bottom of a tube after homogenates have been allowed to settle. They also can be concentrated by brief centrifugation at a low speed.

An estimate of the number of asci present in a homogenate can be made by plating the material after heating 1 hour at 70 C.

17.5 INTERPRETATION

The presence of *Byssochlamys* ascospores is not uncommon on fruit when received at the processing plant. The average spore counts generally are low, however, under 10 per 100 g.[10] This level of contamination on raw fruit usually does not present a spoilage problem to the processor because a majority of the spores are removed by various processing steps such as washing and filtering. The remaining spores usually are destroyed by the heat process. A count of 10 spores per 100 g may indicate a serious problem, on the other hand, if this level of contamination is found in the product at a stage just prior to the retort or heat exchanger.

17.6 REFERENCES

1. Brown, A. H. S. and G. Smith. 1957. The genus *Paecilomyces* Bainier and its perfect stage *Byssochlamys* Westling. Trans. Brit. Mycol. Soc. **40**:17–89
2. Hull, R. 1933–34. Investigation of the control of spoilage of processed fruit by *Byssochlamys fulva*. Annual Report of the Fruit and Vegetable Preservation Research Station, University of Bristol, Campden, England p. 64–73
3. Ito, K. A., M. L. Seeger, and W. H. Lee. 1972. The destruction of *Byssochlamys fulva* asci by low concentrations of gaseous methyl bromide and by aqueous solutions of chlorine, an iodophor and peracetic acid. J. Appl. Bacteriol. **35**:479–483
4. King, A. D., Jr., H. D. Michener, and K. A. Ito. 1969. Control of *Byssochlamys* and related heat-resistant fungi in grape products. Appl. Microbiol. **18**:166–173
5. King, A. D., Jr., A. N. Booth, A. E. Stafford, and A. C. Waiss, Jr. 1972. *Byssochlamys fulva*, metabolite toxicity in laboratory animals. J. Food Sci. **37**:86–89
6. Maunder, D. T. 1969. Spoilage problems caused by molds of the *Byssochlamys-Paecilomyces* group. pp 12–16. *In Byssochlamys* Seminar Abstracts, Research Circular number 20, New York State Agr. Exp. Sta., Geneva, N.Y.
7. Put, H. M. C. 1964. A selective method for cultivating heat resistant moulds, particularly those of the Genus *Byssochlamys*, and their presence in Dutch soil. J. Appl. Bacteriol. **27**:59–64
8. Splittstoesser, D. F., M. C. Cadwell, and M. Martin. 1969. Ascospore production by *Byssochlamys fulva*. J. Food Sci. **34**:248–250
9. Splittstoesser, D. F., F. R. Kuss, and W. Harrison. 1970. Enumeration of *Byssochlamys* and other heat-resistant molds. Appl. Microbiol. **20**:393–397
10. Splittstoesser, D. F., F. R. Kuss, W. Harrison, and D. B. Prest. 1971. Incidence of heat-resistant molds in eastern orchards and vineyards. Appl. Microbiol. **21**:335–337
11. Splittstoesser, D. F., M. Wilkison, and W. Harrison. 1972. Heat activation of *Byssochlamys fulva* ascospores. J. Milk and Food Technol. **35**:399–401
12. Stolk, A. C. and R. A. Samson. 1971. Studies on *Talaromyces* and related genera. I. *Hamigera* gen. nov. and *Byssochlamys*. Persoonia **6**:341–357
13. Williams, C. C., E. J. Cameron, and O. B. Williams. 1941. A facultatively anaerobic mold of unusual heat resistance. Food Res. **6**:69–73

CHAPTER 18

MESOPHILIC SPOREFORMING AEROBES

Paul J. Thompson

18.1 INTRODUCTION

18.11 Classification—Taxonomy

The aerobic to facultative sporeforming bacteria are placed in the genus *Bacillus*. Members of the group are rod shaped, but may appear as spindles, clubs or wedges when bulged by endospores. They are catalase positive, a characteristic useful for distinguishing them from air tolerant *Clostridium* species. All of the bacilli have the ability to produce endospores under aerobic conditions. Temperature for growth is not a criterion for separation of species in the *Bacillus*, except for *B. stearothermophilus*.[1] Overlapping temperature ranges for growth of *Bacillus* strains necessitate an arbitrary distinction between mesophiles and thermophiles. The mesophilic, aerobic, sporeforming bacteria are considered here as all strains of *Bacillus* species that grow at 35 C but not at 55 C, a delimitation compatible (but not identical) with views of others.[2, 3, 7]

18.12 Occurrence in Canned Foods

Spoilage in canned, low acid (pH > 4.6) foods that is caused by aerobic mesophilic sporeformers is usually of the "flat sour" type; infrequently, loss of vacuum, or swelling, occurs. Inadequate heat processing is commonly responsible since spores of mesophilic bacteria are moderately resistant to moist heat. The D_{250} values are of the order of 0.1 minutes or less.[8] Pure cultures of mesophilic *Bacillus* species have been recovered from warehouse spoilage of low acid food. The apparent cause was leakage during cooling. This was indicated by container seam inspection. Possibly, the concentration of chlorine in the cooling water killed vegetative cells but not spores, which were then drawn into containers during cooling. Mesophilic sporeformers, as defined above, do not cause spoilage in properly processed acid foods (pH 4.1 to 4.6). A report of spoilage in commercially canned fruits and fruit products due to *B. macerans* or *B. polymyxa* left open the question of whether under processing or container leakage (post process) was at fault.[9]

On occasion, aciduric, mesophilic bacilli may be cultured from high acid foods (pH < 4.1) that received a "hot fill" process, provided the plating me-

CONTRIBUTOR: Keith Ito
IS/A Committee Liaison: Cleve B. Denny

dium is near neutral pH. The food is not spoiled and the significance of such a finding is nil, since germination of spores and vegetative growth in the food is prevented by the acid content. The product is commercially sterile (see Chapter 52).

18.13 Temperature and pH Requirements

The bulk of *Bacillus* species grow throughout a temperature range of 20 to 40 C in media of pH 5.0 to 9.0. An incubation temperature range of 32 to 35 C is favorable for the growth of mesophilic sporeformers important to the food microbiologist. All strains studied by Gordon et al grew in media adjusted to pH 6.8 or 7.0.[3]

18.14 Sources of Organisms

The aerobic mesophilic sporeformers are widely distributed in nature in soil, dust, and decomposing organic materials.[1] Therefore, it is not surprising that they are isolated readily from many foods and ingredients such as starches, dried fruits, vegetables, cereal grains, dried milk, and spices. Food handling equipment, properly designed to eliminate niches wherein bacteria may multiply, creates no problems of buildup in sporeforming cell populations during food processing. Equipment with pits, crevices, dead ends, open seams, and square corners provides opportunities for bacterial buildup that may lead to deterioration of food, even as it is being processed, especially when a temporary shutdown extends the dwell time. The terminal heating stage of low acid food canning provides lethality sufficient to inactivate spores of mesophilic bacteria. However, canned, cured meats have been reported to display spoilage due to mesophilic *Bacillus* species.[5] Handing equipment surfaces that contact seams or lids on jars following the heat process should be kept clean, or as nearly free of organisms as possible.

18.2 SPECIAL EQUIPMENT AND SUPPLIES

Thermostatically controlled water bath, with stirrer, adjusted to 45 C;

Weighted rings that fit 500 ml Erlenmeyer flasks to prevent bobbing and possible upset in bath;

Thermostatically controlled water or oil bath, with stirrer, equilibrated at 80 C;

Incubators rated for operation throughout 30 to 35 C temperature range with uniformity of temperature of ± 1 C at the extremes of the range.

18.3 SAMPLING

Ingredients need not be sampled routinely for aerobic, mesophilic spores except in special cases, e.g., spices or gelatin to be used in pasteurized canned ham or ham products which receive mild heat treatment. The commonly used aerobic plate count, a portion of which is contributed by mesophilic spores, indicates overall "bacteriological quality" of ingredients without resort to a separate count of mesophilic spores. A similar index of good manufacturing procedures holds for food processing lines and equipment where swab or surface contact platings indicate the adequacy of cleaning and other sanitation practices.

Freshly canned, finished product is not likely to contain mesophilic spores or vegetative cells in numbers detectable by usual plating procedures.[6] Consequently, about twenty-four cans of each code lot should be placed in incubators adjusted to 35 C where they are held, usually for 14 days. If "flat sour" spoilage is detected by a lowered pH and/or liquefaction in a given sample set held at 35 C, the procedures listed in Chapter 53 should be followed.

18.4 PROCEDURE

Twenty-two grams of ingredient or food material are weighed in a sterile, tared plastic cup, or other suitable container, and transferred to 198 ml of sterile 0.1% peptone water in a blender jar. Dispersion is accomplished by blending at high speed for 1.5 minutes. If a stomacher is used, dispersion requires about 30 seconds.

Tryptone glucose extract (TGE) agar (Difco) or trypticase glucose agar (BBL) (Chapter 2) is prepared, 100 ml per 500 ml Erlenmeyer flask. One additional flask of medium is prepared to serve as a sterility control. Sterilization at 121 C moist heat for 15 minutes is followed by cooling to 45 C in a water bath. Volumes of blended food are pipetted into a set of three flasks of TGE agar while they are held in the bath: 10 ml into the first, 1 ml into the second, and 0.1 ml into the third. Flasks are agitated gently to disperse the blended material throughout the medium.

Flasks are transferred without delay to a stirred water or oil bath adjusted to 80 C and held there for 30 minutes. Flasks are agitated gently occasionally to assist heat distribution. Cooling is done in cold tap water taking care that the temperature does not fall to a point where agar congeals. Flasks are transferred to the 45 C bath following the rapid cooling step and held there for a period not to exceed 10 minutes. The 100 ml volume in each flask, representing test material and sterility control, is poured into a set of 5 plates in approximately equal volumes, i.e., about 20 ml per plate. When agar has solidified, plates are inverted in the 35 C incubator and allowed to stand 48 hours.

Counts are made of surface and subsurface colonies. The sum of colonies on the set of 5 plates poured from TGE agar, containing 10 ml of blended food, represents the number of aerobic, mesophilic spores per gram. Similarly, the number of colonies in sets of plates receiving 1 and 0.1 ml of blended food are equal to 0.1 and 0.01 of the number of spores per gram and must be multiplied by 10 and 100 respectively to get the count per gram. The number of spores which can be detected by this method ranges from 30 to 150,000 spores per gram.

18.5 PRECAUTIONS AND LIMITATIONS OF PROCEDURE

The dilutions of sample material selected for plating in the given procedure permit enumeration of mesophilic spores in foods which contain relatively few spores, and also in materials such as spices where spore populations are often elevated. However, spore crops prepared in a laboratory will require greater dilution prior to the 80 C heating step. Certain strains of *Bacillus* that are considered thermophilic because they grow at 65 C are also capable of growth (but not spore germination) at 35 C.[4] Streak plates incubated at 55 C and 65 C will yield growth if the food material samples contain spores of thermophiles. A separate enumeration as described in Chapter 21 must be conducted if their numbers are to be subtracted from the mesophilic spore count.

18.6 INTERPRETATION OF RESULTS

Elevated aerobic, mesophilic spore populations in ingredients to be used in low acid products that undergo a mild heat process, such as canned pasteurized luncheon meats, or canned cured hams, may cause spoilage if mishandling occurs in marketing channels or by consumers. Otherwise, the mesophilic spore count is of little significance to the canner. The presence of aerobic, mesophilic spores in foods preserved by freezing or drying is innocuous, provided mishandling does not occur that causes buildup in populations of *Bacillus cereus*. (Chapter 33).

When aerobic, mesophilic sporeformers are recovered in pure culture from a large percentage of apparently sound containers of low acid canned food in a given spoiled pack, underprocessing should be considered, but cannot be assumed (Chapter 53). Container microleakage (post process) is a likely alternative. Time temperature records of processing and container closure records of low acid canned foods must be kept under Section 128b.8 of the Code of Federal Regulations, and they must be available for inspection according to Part 90.20(h). Time temperature records relating to the spoiled lot should be reviewed, but evidence should be sought also for rough handling immediately after heating, for faulty containers, or for inadequate chlorine residual in the cooling water.

18.7 REFERENCES

1. Buchanan, R. E. and N. E. Gibbons, ed. 1974. Bergey's Manual of Determinative Bacteriology, 8th Ed. Williams and Wilkins Co., Baltimore, Md.
2. Cameron, E. J. and J. R. Esty. 1926. The examination of spoiled canned foods. II Classification of flat sour spoilage organisms from nonacid foods. Jour. Infect. Dis. **39**:89–105
3. Gordon, R. E., W. C. Haynes, and C. Hor-Nay Pang. 1973. The Genus *Bacillus*. USDA Agriculture Handbook No. 427
4. Harris, O. and M. L. Fields. 1972. A study of thermophilic, aerobic, sporeforming bacteria isolated from soil and water. Can. J. Microbiol. **18**:917–923
5. Jensen, L. B. 1954. Microbiology of Meats, 3rd Ed., Garrard Press, Champaign, Ill.
6. National Canners Association. 1968. Laboratory Manual for Food Canners and Processors. Vol 1. Microbiology and Processing, 3rd Ed., AVI Publishing Co., Inc., Westport, Conn.
7. Richmond, B. and M. L. Fields. 1966. Distribution of thermophilic, aerobic sporeforming bacteria in food ingredients. Appl. Microbiol. **14**:623–626
8. Stumbo, C. R. 1973. Thermobacteriology in Food Processing. 2nd Ed. Academic Press, New York, N.Y.
9. Vaughn, R. H., I. H. Kreulevitch, and W. A. Mercer. 1952. Spoilage of canned foods caused by the *Bacillus macerans-polymyxa* group of bacteria. Food Res. **17**:560–570

CHAPTER 19

MESOPHILIC SPOREFORMING ANAEROBES

G. L. Hays and R. K. Lynt

19.1 INTRODUCTION

The mesophilic sporeforming anaerobes of greatest interest in foods fall into 2 main groups. One group consists of (1) *C. sporogenes*, and other relatively heat resistant putrefactive anaerobes, and (2) the proteolytic and the nonproteolytic strains of *C. botulinum*. The other group consists of *C. perfringens* and a variety of other similar clostridia, such as the butyric anaerobes, that are relatively nonresistant to heat. Only the putrefactive anaerobes (including *C. sporogenes*) will be considered here, since the enumeration of *C. perfringens* in foods is treated in Chapter 35, and *C. botulinum* is discussed in Chapter 34.

All mesophilic sporeforming anaerobes which blacken and digest brain or meat media with a putrid odor will be grouped together as putrefactive anaerobes in this discussion. These organisms are capable of decomposing proteins, peptides or amino acids anaerobically, with resulting foul smelling sulfur containing products, such as hydrogen sulfide, methyl and ethyl sulfide, and mercaptans. Ammonia and the amines, putrescine and cadaverine, are usually produced, along with indole and skatole as well as carbon dioxide and hydrogen.[8, 10]

The putrefactive anaerobes are rather large, gram positive rods measuring from 0.4 to 1.0 by 3.0 to 15.0 microns. All are motile except *Clostridium putrefaciens*.[5] With the exception of *C. putrefaciens*, the putrefactive anaerobes will grow within an approximate temperature range of 10 to 50 C.[10] *C. putrefaciens* will grow from 0 C. to about 30 C.[5] This growth range covers the temperature of the normal storage of processed canned foods, including the refrigerated storage of cured meats.

The spores of putrefactive anaerobes germinate and grow readily in low acid foods having pH values of 6.0 or higher, although inconsistent swell spoilage has been found in inoculated packs of low acid foods in the range of pH 5.5 to 5.6.[15] Cameron and Esty noted that putrefactive anaerobes may show abnormal development resulting in nongaseous spoilage in low acid products in the range of pH 4.6 to 5.0.[6] Thus it may be generalized that putrefactive anaerobes may cause spoilage in any canned food having a pH of 4.8 or above (the lowest pH at which *C. botulinum* spores

CONTRIBUTORS: Keith Ito
IS/A Committee Liaison: Cleve B. Denny

have been observed to grow out), if the spores are not destroyed by the thermal process to which the canned foods have been subjected. It follows that there is no necessity for testing for putrefactive anaerobes in products with a pH of less than 4.6.

Putrefactive anaerobes are distributed widely in nature (soil), and they may be normal contaminants of vegetables at the time of harvest.[5, 10] Some species commonly are found in the intestinal tracts and excreta of animals and, hence, may become contaminants of milk and meat.[10] Consequently, a number of species of putrefactive anaerobes have been isolated from spoilage in low acid canned foods with an original pH of above 4.8, but which had been underprocessed. Organisms that have been isolated from such spoilage are *C. sporogenes, C. bifermentans, C. putrefaciens, C. histolyticum, C. botulinum,* types A and B, and the closely related nontoxic organism identified as P.A. 3679.[8, 9, 10, 16] Two other putrefactive anaerobes, *C. lentoputrescens* and *C. botulinum* type F have caused spoilage in foods but not in canned foods.

Because of their public health significance, the toxin producing *C. botulinum* types A and B (Chapter 34) are often considered the most important putrefactive anaerobes. However, the spores of many of the nontoxic species are much more heat resistant than are the botulinum spores and the destruction of the nontoxic spores during the thermal process of low acid canned foods is of the utmost importance in the prevention of economic losses due to inadequate heat processing. The relative thermal resistance of *C. botulinum* types A and B spores and spores of P.A. 3679 may be illustrated by their respective D_{250} values. The D_{250} values of *C. botulinum* spores range from 0.10 to 0.20 minutes while those of P.A. 3679 spores are 0.5 to 1.5 minutes.[16]

Putrefactive anaerobic spores are distributed widely in nature, and may be normal contaminants of vegetables, meat, milk, and other food ingredients. The lack of anaerobic conditions in most equipment, except for the dead ends of pipes and possibly porous surfaces, is not conducive to the growth and sporulation of putrefactive anaerobes; thus, areas of heavy buildup seldom occur in food plants. Other sources include spices, cereals and cereal products, dried eggs, dried milk and milk products, dried vegetables (onions and garlic),[3] and rendered tallow.

It is sometimes desirable to have an estimate of the total number of mesophilic sporeforming anaerobes in a food material or ingredient. However, there is probably no way in which such an estimate can be completely representative of the total number present, and methods in use inevitably favor one group of anaerobes or the other of the total possible flora. For example, if an estimate of the proteolytic clostridia is desired, it is usual to heat for maximum germination of the spores, but such treatment is destructive of the spores of most nonproteolytic clostridia, such as those of *Clostridium botulinum* type E. (Chapter 34). If on the other hand, alcohol treatment is used to destroy other contaminating nonsporulating organisms, a more accurate estimate of the nonproteolytic clostridia will be obtained, but this does not provide for maximal germination and outgrowth of proteolytic strains. Thus, in either case, the total count will be in error by the failure of some clostridia to grow. Furthermore, neither method will destroy the spores of aerobic bacilli (Chapter 18), many of which are facultative with respect to anaerobiosis and will be counted inevitably along with the strict anaerobes. The same, of course, is also true with respect to both clostridia and facultative aerobic sporeformers (Chapter 21) which are also facultative thermophiles,[1] so that some aerobes and some facultative thermophiles also may be counted by any method designed for the enumeration of mesophilic anaerobes. An additional complication arises, moreover, if the total microflora capable of surviving either treatment contains bacteriocin or antibiotic producing organisms, and, at the same time, strains sensitive to these agents.[12]

Given these limitations, therefore, it is probably best to choose a procedure most likely to give a reasonable estimate of the groups expected to be present in a particular food, since it is likely that only when spore stocks of pure cultures are being counted that estimates can be made with any real degree of certainty.

There are several general methods used for counting mesophilic anaerobes. MPN methods (Chapter 6) seem to be the most widely used, and generally give more reproducible results with higher counts on pure cultures than other methods. However, colony counts in Prickett tubes and anaerobic Petri dishes also have had considerable use. Methods which incorporate the inoculum into melted agar with the production of deep colonies, as in pour plates or Prickett tubes, often have the handicap that gas production concomittant with the development of colonies splits the agar and renders accurate counting of well developed colonies impossible, unless a very short incubation period is used. Such counts are, therefore, subject to the possible further inaccuracy that slow growing colonies may not be counted at all. Counts made by spreading the inoculum over the surface of a solid agar medium usually show too much variation between replicates to be reliable, although the use of prereduced agar may solve this problem.

19.2 EQUIPMENT AND SUPPLIES

a. Culture tubes, either clip top or screw cap, for liquid media
b. Pricket tubes
c. Petri dishes
d. Veillon tubes
e. Serological pipettes
f. Quebec colony counter or similar device.
g. Anaerobic jar (Gaspak, Case, or similar jar)
h. Incubator set at 30 to 35 C
i. Transfer loops

19.3 SAMPLING (Chapter 3)

Where putrefactive spoilage has occurred, or if it is desirable to monitor a canning line to determine the degree of putrefactive anaerobe contamination of low acid foods, this can be accomplished best by collecting canned samples at the closing machine before the heat process and culturing for putrefactive anaerobes. If putrefactive anaerobic spores are found in relatively high numbers in these samples, then it may be advantageous to collect product samples at key points along the canning line. Samples should be collected from the washed or blanched product, the filler, and finally from closed, but unprocessed, samples. Formulated product samples should be collected from mixing tanks, filler bowls, and from closed, but unprocessed, cans.

19.4 PREPARATION OF THE SAMPLE

19.41 Sugar

Weigh 20 grams of granulated sugar or 30 grams of liquid sugar (approximately 67° Brix) into sterile 250 ml. Erlenmeyer flasks and add sterile water to make 100 mg. samples. Shake the flasks until the sugar is in solution. The samples are then heated rapidly to boiling and held at boiling for 5 minutes before chilling rapidly to room temperature.[14]

19.42 Starch, Flour, and Other Cereal Products

Weigh 11 grams of the product into a sterile 250 ml Erlenmeyer flask and add 99 ml. of sterile water. Shake until the sample is in suspension. Do not heat before culturing.[3]

19.43 Dehydrated Vegetables

Weigh 10 grams of dehydrated vegetable into a sterile 250 ml. Erlenmeyer flask and add 190 ml. of sterile water. Rehydrate for 30 minutes at refrigerator temperature. Then shake vigorously for 2 to 3 minutes. Heat the sample for 10 minutes at 5 psi (108 C) and cool to room temperature. Samples may be heated for 20 to 30 minutes at 100 C and then cooled.[3]

19.44 Rendered Tallow

Place 11 grams in a sterile 250 ml. Erlenmeyer flask and add 99 ml. of sterile water. Heat to boiling and hold for 5 minutes. Sample must be held at 45 to 50 C. and shaken vigorously just prior to culturing.[2]

19.45 Spices

The weight of the spice sample to be examined will depend upon the type of spice to be examined. Ten grams of whole spice, 2 g of bulky spice, or 1 g of ground spice are weighed into a sterile 250 ml. Erlenmeyer flask; sterile water is added to make 100 ml. Shake vigorously and boil for 5 minutes. Allow coarse particles to settle out before culturing or making dilutions for culturing.[3]

19.46 Dried Eggs

Eleven grams of dried eggs are weighed into dilution bottles containing 99 ml. of sterile water or saline, and glass beads. The samples are shaken vigorously until all lumps of dried eggs have been dispersed. Do not heat sample before culturing.[2, 3]

19.47 Dried Milk or Milk Products

Eleven grams of dried milk are dissolved in 99 ml. of sterile water in a sterile 250 ml. Erlenmeyer flask. Shake vigorously and after the dried milk is in solution, heat the sample to boiling and hold for 5 minutes before cooling.[2]

19.48 Equipment and Filled Can (Line) Samples Before Heat Processing

All samples from equipment or line samples should be examined immediately after collection at the cannery. However, if they must be transported to a laboratory for examination, the samples should be heated and cooled before being transported to the laboratory under refrigeration. These heated samples should be cultured the same day that they are collected.

Swab samples from equipment, and scraping or splinters from wooden or other porous material that have been placed in 10 ml. water blanks are heated for 20 to 30 minutes at 100 C and cooled.

The line samples collected in 202 diameter cans are weighed and an equal weight of sterile water is added. This will approximate the product brine ratio of most canned vegetables. The samples are shaken thoroughly while holding the end of the can firmly in place. Half fill 25 × 200 mm screw cap tubes with the liquid. Heat in steam for 20 to 30 minutes at 100 C. Samples of formulated products usually are diluted with an equal weight of sterile water to facilitate handling. After shaking, half fill 25 × 200 mm screw cap tubes with the diluted product and either heated at 100 C for 20 to 30 minutes or pipette portions of the unheated sample into anaerobic medium of choice and then heat as above.

The unprocessed cans collected at the closing machine are opened aseptically and portions of brine are poured into sterile 25 × 200 mm screw cap tubes, or portions of formulated product are pipetted into medium of choice; these samples are heated for 20 to 30 minutes at 100 C.

If an autoclave is available at the cannery where the samples are collected, the above samples may be heated for 10 minutes at 5 psi (108.4 C) instead of at 100 C.

19.5 CULTURAL PROCEDURES

The prepared samples of heated sugar, dehydrated vegetables, rendered tallow, spices, and dried milk or milk products are cultured as follows: 20 ml. portions of these heated ingredients are divided equally among 6 tubes of liver infusion broth which just have been heated to exclude the air, beef heart infusion broth or PE-2 medium.[7] The infusion cultures are stratified with sterile 2% agar before incubation. The PE-2 cultures are not stratified. These cultures are incubated at 30 to 35 C for 72 hours. Although most of these organisms will show growth within 2 or 3 days in those tubes in which it can be expected, it is sometimes desirable to incubate for seven days since some spores may be slow in germinating and growing out. The presence or absence of putrefactive anaerobic spores may be detected by this method. However, if it is necessary to obtain a putrefactive anaerobic spore count of any particular ingredient, this may be accomplished by culturing multiple tubes of serial dilutions of the ingredient, and determining the spore count from MPN tables (Chapter 6). Serial dilutions may be made in distilled water, saline, or culture medium.

Twenty ml portions of the unheated starch, flour or cereal products and dried egg samples are distributed between 6 tubes of liver broth, beef heart infusion broth, or PE-2 medium. These cultures are heated at 100 C for 20 minutes. The cultures must be agitated several times during this heating period. After heating, the infusion tubes must be stratified with sterile 2% agar. The cultures are incubated for 72 hours at 30 to 35 C. and examined for the growth of putrefactive anaerobes. If a putrefactive spore count is required, decimal dilutions may be cultured, heated, and incubated as above. The spore counts are obtained from MPN tables (Chapter 6).

If counts of putrefactive spores are desired, decimal dilutions of the heated equipment swab samples are cultured in freshly exhausted liver, beef heart infusion, or PE-2 medium. However, 1.0 ml and 0.1 ml portions of heated swab, scraping, or line samples usually are cultured to determine the presence of putrefactive anaerobic spores.[2, 3, 14]

19.51 Medium

Although several media for enumerating mesophilic anaerobic spores are available, experience in a number of laboratories has shown that Andersen's pork pea medium[4] is the medium of choice for recovering spores that have received a high heat treatment such as in thermal death time determinations. Therefore, it is suggested for use in pure culture work. This medium has the advantage of being clear so that growth readily may be detected visually. If the medium is not freshly sterilized, it is important to steam it at 100 C for 10 to 15 minutes to drive off dissolved oxygen shortly before inoculating. It is also important to cool the tubes quickly by partial immersion in tempered water to avoid reabsorption of oxygen. Fifteen grams of agar are used in the medium if it is to be used as a solid medium.

19.52 Enumeration by the Prickett Tube Method

Prickett tubes are deep, flattened oval tubes with a cylindrical neck.[13] The tubes are filled nearly to the neck with agar medium and autoclaved to sterilize. If tubes of sterile agar are stored between preparation and inoculation, melt the agar, and steam in flowing steam or boiling water 10 to 15 minutes to drive off dissolved oxygen just before use. Cool quickly to 42 to 45 C and inoculate with serial dilutions of the material being examined after preparing as described in Chapter 4. Distribute the inoculum uniformly throughout the medium by swirling the melted agar, taking care not to reincorporate fresh oxygen. Cool and solidify the agar quickly by immersing in cold water and stratify a layer of thioglycolate agar, methylene blue agar, or sterile vaseline over the surface to fill the neck of the tube partially to aid in maintaining anaerobic conditions during the incubation period. At the end of the incubation period, count the colonies and average the counts for replicate tubes. Calculate the number of viable spores per gram from the count and dilution factor. Counts may be made with the aid of a Quebec colony counter or similar grid. Counts should be made before gas generated by growth of the colonies has caused splitting and blowing of the agar. Record the temperature and time of the incubation period, normally 48 to 72 hours at 30 to 35 C.

19.53 Isolation of Colonies

If colonies are to be isolated from the tubes used for counts, Veillon tubes[17] may be substituted for Prickett tubes and the counts made in a similar manner. After colonies have been counted, the Veillon tubes (which are made of cylindrical glass tubing plugged at one end with a rubber stopper and at the other with cotton) are aseptically opened at both ends, the agar pushed out onto a sterile surface, and material fished from isolated colonies with a loop.

19.54 Enumeration by the Plate Count Method

Prepare samples and dilutions (Chapter 4) and inoculate duplicate plates of agar medium with 0.1 ml of each of the appropriate dilutions of the food material, using a sterile 1.0 ml serological pipette. Spread the inoculum uniformly over the surface of the agar with a sterile, bent glass rod. It is important that the surface of the agar medium be relatively dry to prevent spreading growth which frequently results from excessive moisture. Freshly poured plates, with their lids on, should be left out at room temperature overnight before being inoculated. Older plates, after storage under refrigeration, should be allowed to stand on the work bench at least an hour before inoculating.

Incubate the inoculated plates at 30 to 35 C for 2 days in an anaerobic atmosphere such as that produced by the BBL Gaspak system or a Case jar, evacuated and refilled with nitrogen. For the latter, the jar should be flushed 3 times with nitrogen by drawing a vacuum and flushing before finally refilling with nitrogen the 4th time.

Count all colonies, multiply by the dilution factor, and report as anaerobic sporeformers per gram of food if the samples were heated.

Pour plates can be prepared by placing the inoculum in the Petri dish and pouring with Andersen's agar.[4] This is overlaid with a heavy layer of thioglycollate agar (Chapter 2) and the plates incubated at 32 C for 5 to 6 days.[3] Count and report as for surface inoculated plates.

Plating is the least reliable method of counting mesophilic sporeforming anaerobes. It is not uncommon to find wide variations in the numbers of colonies developing on replicate plates of the same dilution, and when pour plates are used, the production of gas bubbles by subsurface colonies frequently causes separation of the agar, thereby complicating their enumeration.

19.6 PRECAUTIONS AND LIMITATION OF PROCEDURES

It is not possible with our present knowledge to make an unqualified statement of the significance of numbers or presence of putrefactive anaerobic spores in ingredients or at any step in a canning operation. It must be remembered also that heat processes for low acid commercially canned foods are designed to destroy an average load of putrefactive anaerobic spores. Thus to be significant the spore count would have to be extremely large or consist of a population of extremely resistant spores.

19.7 INTERPRETATION OF RESULTS

There are no standards for putrefactive anaerobic spores in ingredients as there are for thermophilic spores in sugars.[14] If putrefactive spoilage should develop in processed products, it is indicative of contaminated ingredient(s) or improperly sanitized processing equipment. Remedial measures must be taken immediately.

When putrefactive spoilage develops in incubated samples and there has been no change in product preparation or canning procedure from lots with no spoilage, the thermal process should be increased, if product quality is not impaired, until remedial action has eliminated the source of the spoilage spores. Finally, if putrefactive spoilage has developed in some of the incubated commercially canned samples, and it is determined that it was not caused by leakage, the processed product lot in the warehouse should be sorted by low vacuum detection, and incubated at 27 to 35 C for successive 30 day periods until no spoilage results. All normal cans should be reprocessed. Finally, the spoiled product should be examined for *C. botulinum* (Chapter 34), and the lot destroyed if toxin is detected.

19.8 REFERENCES

1. ALLEN, M. B. 1953. The themophilic aerobic sporeforming bacteria. Bact. Rev. **17**:125–173
2. American Can Company, 1972. Microbiology of Canned Foods. American Can Company. Barrington, Ill.
3. Recommended Methods for the Microbiological Examination of Foods. 1966. American Public Health Association, Inc. New York, N.Y.
4. ANDERSEN, A. A. 1951. A rapid plate method of counting spores of *Clostridium botulinum*. J. Bacteriol. **62**:425–432
5. BREED, R. S., E. G. D. MURRAY, and N. R. SMITH. 1957. Bergey's Manual of Determinative Bacteriology. 6th ed. The Williams and Wilkins Co., Baltimore, Md.
6. CAMERON, E. J. and J. R. ESTY. 1940. Comments on the microbiology of spoilage in canned foods. Food Research **5**:549–557
7. FOLINAZZO, J. F. and V. S. TROY. 1954. A simple medium for the growth and isolation of spoilage organisms from canned foods. Food Technol. **8**:280–281
8. FRAZIER, W. C. 1967. Food Microbiology. 2nd ed. McGraw-Hill Book Co. New York, N.Y.

9. Gross, C. E., E. Vinton, and C. R. Stumbo. 1946. Bacteriological studies relating to thermal processing of canned meat. V. Characteristics of putrefactive anaerobes used in themal resistance studies. Food Research **11**:405–410
10. Hersom, A. C. and E. D. Hulland. 1964. Canned Foods, An Introduction To Their Microbiology. (Baumgartner), 5th ed. Chemical Publishing Co., Inc. New York, N.Y.
11. Johnston, R., S. Harmon, and D. Kautter. 1964. Method to facilitate the isolation of *Clostridium botulinum* type E. J. Bacteriol. **88**:1521–1522
12. Kautter, D. A., S. M. Harmon, R. K. Lynt, Jr., and T. Lilly, Jr. 1966. Antagonistic effect on *Clostridium botulinum* type E by organisms resembling it. Appl. Microbiol. **14**:616–622
13. Miller, N. J., O. W. Garrett, and P. S. Prickett. 1939. Anaerobic technique—A modified deep agar shake. Food Research **4**:447–451
14. National Canners Association Research Laboratories, 1968. Laboratory Manual For Food Canners And Processors. Vol. 1. Microbiology and Processing. The AVI Publishing Co., Inc. Westport, Conn.
15. Sognefest, P., G. L. Hays, E. Wheaton, and H. A. Benjamin, 1948. Effect of pH on thermal process requirements of canned foods. Food Research **13**:400–416
16. Stumbo, C. R. 1973. Thermobacteriology in Food Processing, 2nd ed. Academic Press, New York, N. Y. and London, Eng.
17. Veillon, R. 1922. Sur quelques microbes thermophiles strictement anaerobies. Ann. Inst. Pasteur. **36**:422–438

CHAPTER 20

ACIDURIC FLAT SOUR SPOREFORMERS

R. V. Speck

20.1 INTRODUCTION

In 1933, during the course of an investigation of off flavor in commercially canned tomato juice, Berry isolated and described a new type of spoilage organism.[2] He established the organism to be a spore forming bacterium of soil origin. The taste of spoiled tomato juice has been described as "medicinal", "phenolic", and "fruity", and is usually accompanied by a reduction of from 0.3 to 0.5 in pH.[8] Spoiled cans remain flat; hence the term "flat sour".

20.11 Classification

The organism responsible for "flat sour" spoilage of tomato products was named *Bacillus thermoacidurans* by Berry.[2] From comparative cultural studies, Smith et al[7] concluded that *B. thermoacidurans* was identical with *Bacillus coagulans* of Hammer.[4] From their careful studies of the two species, Becker and Pederson[1] stated: "There is no justification for considering *B. thermoacidurans* as a species distinct from *B. coagulans*, and the latter name has priority."

B. coagulans is a nonpathogenic, motile, spore forming aerobe having as many as ten flagella per cell. The Gram stain reaction is usually positive, although a few variable strains have been observed.

20.12 Occurrence

B. coagulans is a common soil organism. It has been isolated from canned tomato products, particularly tomato juice, cream, evaporated milk, cheese, and silage. Spoilage and subsequent curd formation of evaporated milk is frequently caused by the organism. Hammer's original studies on the coagulation of evaporated milk led to the naming of the organism.[4] With regard to tomato products, *B. coagulans* has been isolated from canned whole tomatoes, tomato juice, tomato puree, tomato soup, and tomato vegetable juice mixes.

20.13 Temperature and pH Requirements

Becker and Pederson reported that *B. coagulans* was not obligately thermophilic.[1] They were successful in growing the organism at temper-

CONTRIBUTORS: Keith Ito
IS/A Committee Liaison: Cleve B. Denny

atures as low as 18 C. Optimum growth in artificial medium occurred between 37 and 45 C. In 1949, Gordon and Smith reported that 53 of 73 cultures studied grew at 28 C; 73 at 33 C, 37 C, and 45 C; 72 at 50 C; 66 at 55 C; 23 at 60 C; and none at 65 C.[3]

Packers of tomato products have observed spoilage development at temperatures of 21 to 38 C. The organism will not grow in tomato products with normal pH at 55 C. Berry indicated that a temperature of 37 C appeared optimum for the production of off flavor.[2]

B. coagulans grows well in artificial media at pH values between 5.0 and 7.0. Pederson and Becker showed that many cultures in their vegetative form could grow at values as low as pH 4.02.[6] In artificial media heat resistant spores were incapable of germinating and producing growth below pH 5.0.

20.14 Sources

B. coagulans can be isolated from many field soils. It is common to field grown tomatoes as well as to milk, cream, and condensed milk. Spores of the organism have been isolated from chip board separators frequently used in the packaging of empty cans. It has been found in the sweepings of rail cars previously used to ship grain products.

The organism has been found to multiply in tomato washing equipment where there is insufficient makeup water and the water temperature may reach 27 to 32 C. Spores have been isolated from empty cans and empty can washers, tomato product lines, conveyor belts, and filled can runways.

20.2 SPECIAL EQUIPMENT AND SUPPLIES

When analyzing sugar, starch, flour, nonfat dry milk, and cream, reference is made to the use of Erlenmeyer flasks marked for 100 ml. The flasks should be of 250 ml capacity and permanently marked for 100 ml. The newer type graduated flasks should be checked for accuracy.

20.3 SAMPLING

20.31 Ingredients

Since the adequacy of a canned food process is, among other things, related to the bacterial spore load of the product to be processed, it is often advantageous to determine the load of flat sour spores in the unprocessed product as well as in the product ingredients. Pinpointing the ingredient contributing most to the total spore load may prove very beneficial to a packer. Depending on the canned tomato product being manufactured, spore analysis on ingredients such as raw tomatoes, fresh tomato pulp, puree, concentrated or evaporated milk, cream, and nonfat dry milk may be considered. Sampling of dairy products is conducted in accordance with the procedures given in 3.13 to 3.222 of Standard Methods for the Examination of Dairy Products, 13th edition.[5]

20.32 Equipment and Systems

Periodic sampling of tomato wash water, material from conveyor belts, pipe lines, and tanks may reveal potential problem areas. A preseasonal bacteriological survey of pipe lines, valves and valve bonnets, storage tanks, heaters, and other equipment surfaces normally contacting product, may indicate where cleaning and sanitation need to be stressed. Sampling of such equipment may be carried out using swabbing or membrane filter techniques.

20.33 Product in Process

Canners of tomatoes and tomato products, may find it advantageous to monitor certain phases of manufacture for flat sour spores. On analysis, finished tomato products, such as juice, puree, samples of chopped tomatoes before and after the hot break, or extracted tomatoes, may indicate potential foci of spore buildup. In the canning of whole tomatoes, sampling should include juice accumulated during peeling, and puree made from sound tomatoes, both of which are used frequently as the liquid portion of canned tomatoes. Aciduric flat sour spore counts of tomato products prior to processing may assist in preventing spoilage where the product is not presterilized.

20.34 Finished Product

Since most processes for tomato products, either completely eliminate flat sour spores or drastically reduce their number, the advantage of making spore counts of the finished product must be seriously questioned. Where spore numbers have been reduced substantially, the probability is small of recovering spores by subculturing 1 ml in replicates up to five or more. No attempt should be made to plate more than 1 ml quantities of tomato juice because of its inhibitory action. Experience has shown that incubation of the canned finished product is much more meaningful.

20.4 PROCEDURE—METHODS OF ANALYSIS

20.41 Raw Tomatoes, Fresh Tomato Pulp, Puree, Concentrated Milk, Tomato Washer Water

20.411 Sample preparation

Extract the juice from raw whole or chopped tomatoes by pressing the sample in a sterile colander or sieve. The sample also may be prepared using a sterile blender jar and suitable blender. Transfer 10 ml of the expressed juice to a 20 × 150 mm screw cap tube for shocking, and tighten the rubber lined caps securely. Samples of tomato puree, and products of similar consistency are handled more conveniently using 25 × 150 mm tubes.

Completely immerse the tubes containing the samples in a water bath adjusted to 88 to 90 C. Using an extra tube fitted with a slotted rubber stopper, check the rising temperature of a similar sample of ingredient or product. After the temperature in the "control" tube reaches 88 C, the timing of the shock treatment begins, and should last 5 minutes. Normally 10 ml of water, or product of a similar consistency in a 20 × 150 mm screw cap tube requires approximately 3 minutes to reach 88 C. Cool the sample tubes immediately after the shock treatment in cold water, keeping the screw cap tops well above the surface of the water. Samples of ingredient or product which, in their preparation or manufacture, have had a recent heat treatment of 82 C, or higher, need no further shock treatment and can be plated directly. Place a control tube containing sterile tomato juice inoculated with spores known to be *B. coagulans* in the water bath with the tubes of samples to be shocked.

20.412 Cultural procedures

Transfer 1 ml of the shocked sample or decimal volume thereof into each of four Petri dishes. Add to each of two plates 18 to 20 ml of dextrose tryptone agar (Chapter 2) and add to each of two plates 18 to 20 ml of thermoacidurans agar (Chapter 2) tempered to 44 to 46 C. After solidification, invert the plates and promptly incubate at 55 C ± 1 C for 48 hours ± 3 hours. Surface colonies on dextrose tryptone agar resulting from the germination, and growth of spores of *B. coagulans* will appear slightly moist, usually slightly convex, and of a pale yellow color. Subsurface colonies on this medium are compact with fluffy edges. They are slightly yellow to orange in color and usually 1 mm or greater in diameter. Both surface and subsurface colonies will be surrounded by a yellow zone due to acid formation. In 48 hours, plates may turn completely yellow. *B. stearothermophilus* will also grow on dextrose tryptone agar giving pinhead size colonies, usually brown in color, and are of no consequence since they will not spoil tomato juice. Suspicious colonies should be transferred to litmus milk (Chapter 2) and will show coagulation if *B. coagulans, B. stearothermophilus* will not grow on the thermoacidurans agar, and therefore counts on the latter acid medium may have more significance. Typical colonies on the latter agar are large and white to cream colored.

20.42 Nonfat Dry Milk

20.421 Sample preparation

Weigh 10 g of the sample into a sterile 250 ml Erlenmeyer flask marked to 100 ml. Add N/50 sodium hydroxide to the mark and shake to completely dissolve the sample. Heat 10 minutes at 5 psi (108.4 C) steam pressure, then cool immediately.

20.422 Cultural procedures

Transfer 2 ml of the solution to each of 10 sterile Petri dishes. Add to each dish 18 to 20 ml of dextrose tryptone agar tempered to 44 to 46 C. After solidification, invert the plates and promptly incubate at 55 C ± 1 C for 48 hours ± 3 hours. Count the typical acid flat sour colonies previously described and report on the basis of 10 g of sample.

20.43 Cream

20.431 Sample preparation

Mix 2 g of gum tragacanth and 1 g of gum arabic in 100 ml of water in an Erlenmeyer flask. Sterilize in the autoclave for 20 minutes at 121 C. Transfer 20 ml of sample to a sterile 250 ml Erlenmeyer flask marked for 100 ml. Add the sterilized gum mixture to the 100 ml mark and carefully shake, using a sterile rubber stopper. Loosen stopper and autoclave for 5 minutes at 5 psi (108.4 C).

20.432 Cultural procedures

Due to the viscosity of the mixture, first pour five Petri plates with dextrose tryptone agar, then immediately transfer 2 ml of the cream emulsion and swirl in the usual manner. After solidification, invert the plates and promptly incubate at 55 C ± 1 C for 48 hours ± 3 hours. Count the typical acid flat sour colonies and report on the basis of 1 ml of sample.

20.5 PRECAUTIONS AND LIMITATIONS OF THE PROCEDURE

When plating samples of acid products such as tomato juice and tomato puree, the melted medium should be poured directly on the sample in the Petri dish. A minimum of 18 ml of medium per plate should be used per 1 ml of tomato product. Pouring should be followed immediately by gentle swirling to assure adequate dispersion of the sample in the melted medium. In addition to providing a uniform distribution of any surviving spores, the precautions will also assure a uniform color to the poured plate (due to the acid product plus the indicator of the medium). Precautions also must be taken to prevent drying out or splitting of agar in plates during incubation at 55 C. This can be accomplished by placing the plates in appropriate canisters or by providing additional moisture in the air within the incubator.

20.6 INTERPRETATION OF RESULTS

True acid "flat sour" (*B. coagulans*) spores have significance in canned products in the pH range of 4.1 to 5.0. Surviving spores of *B. stearothermophilus* will not grow in this pH range and, therefore, care must be used in distinguishing between the two organisms (20.412). Presence of *B. coagulans* can lead to economic spoilage while presence of *B. stearothermophilus* spores at this pH range are of no consequence.

Some representatives of industry show concern when counts in excess of 5 spores of *B. coagulans* per gram are encountered in the ingredients used in canned foods in the pH range of 4.1 to 5.0.

20.7 REFERENCES

1. BECKER, M. E. and C. S. PEDERSON. 1950. The physiological characters of *Bacillus coagulans (Bacillus thermoacidurans)*. J. Bact. **59**:717–725
2. BERRY, R. N. 1933. Some new heat resistant, acid tolerant organisms causing spoilage in tomato juice. J. Bact. **25**:72–73
3. GORDON, R. E. and N. R. SMITH. 1949. Aerobic spore forming bacteria capable of growth at high temperatures. J. Bact. **58**:327–341
4. HAMMER, B. W. 1915. Bacteriological Studies on the Coagulation of Evaporated Milk. Iowa Agr. Expt. Sta., Res. Bull. 19, Iowa City, Ia.
5. HAUSLER, W. J., JR., ed. 1973. Standard Methods for the Examination of Dairy Products, 13 ed. American Public Health Association, Washington, D.C.
6. PEDERSON, C. S. and M. E. BECKER. 1949. Flat Sour Spoilage of Tomato Juice. Tech. Bull. No. 287, N.Y. State Agricultural Experiment Station, Cornell University, Geneva, N.Y.
7. SMITH, N. R., R. E. GORDON, and F. E. CLARK. 1946. Aerobic Mesophilic Sporeforming Bacteria. U.S.D.A. Misc. Publ. No. 559. Washington, D.C.
8. STERN, R. M., C. P. HAGARTY, and O. B. WILLIAMS. 1942. Detection of *Bacillus thermoacidurans* (Berry) in tomato juice, and successful cultivation of the organism in the laboratory. Food Res. **7**:186–191

CHAPTER 21

THERMOPHILIC FLAT SOUR SPOREFORMERS

Duane T. Maunder

21.1 INTRODUCTION

Thermophilic "flat sour" bacteria are sporeforming facultative aerobes capable of spoiling certain low acid canned foods. They characteristically ferment carbohydrates with the production of lower fatty acids which "sour" the product but do not produce sufficient, if any, gas to change the normal "flat" appearance of the ends of the can. In practice, the flat sour bacteria are considered obligate thermophiles. However, it is possible to grow some cultures at temperatures as low as 36 to 38 C, especially if the inoculated organisms are in the vegetative state, and if proper environmental conditions are imposed. The group's upper temperature limit for growth is about 75 C. In canned low acid foods, thermophilic flat sour spoilage seldom occurs if holding temperatures below about 43 C are maintained. *Bacillus stearothermophilus* is the typical species responsible for thermophilic flat sour spoilage of such foods.[4, 9, 11]

Low acid canned foods, particularly those having a pH no lower than about 5.3, may undergo thermophilic flat sour spoilage if they contain viable spores capable of germinating and growing out in the product, and provided the foods are held at a temperature above about 43 C for a sufficient length of time.[2]

Because of the exceptionally high thermal resistance of flat sour spores, their presence in some cans of any given lot of commercially sterile low acid canned foods may be considered normal. Flat sour spoilage can develop only if the product is held in the thermophilic growth range during warehouse storage or distribution. Thus, inadequate cooling subsequent to thermal processing is a major contributor to the development of flat sour spoilage. Localized heating of sections of stacks of canned foods placed in too close proximity to heaters is another.

Another flat sour bacterium, *Bacillus coagulans*, is a facultative thermophile of particular importance in the spoilage of acid foods (Chapter 20), although it has also been implicated in spoilage of products such as evaporated milk and certain canned meats and fish.

Thermophilic bacteriological studies are sometimes complicated by some of the mesophilic aerobic sporeformers, because some of these are facul-

CONTRIBUTOR: Keith Ito
IS/A Committee Liaison: Cleve B. Denny

tative with respect to termperature to the extent that they may grow at temperatures up to 45 to 55 C.[9] Similarly, members of the genus *Thermoactinomyces*[1] may be encountered, especially in ingredient examination procedures where only mild heat shock treatment is used. Comparatively low heat resistance characteristics of all of these organisms preclude heat process survival in most shelf stable, low acid canned foods. Thus, they rarely cause problems unless processing is markedly inadequate, in which case mesophiles of even greater heat resistance, if present, may be expected to survive (Chapter 19).

The bacterial spores enter canneries in soil, on raw foods, and in ingredients, e.g., spices, sugar, starch, flour, etc.[6, 7, 10] Populations may increase at any point where a proper environment exists. For example, an item of food handling equipment in a canning line which is operated within the thermophilic growth range (about 43 to 75 C) may serve as a focal point for the buildup of an excessive flat sour spore population.

21.2 EQUIPMENT AND SUPPLIES

Equipment and supplies are needed in accordance with specifications in Chapter 2. The following additional apparatus is recommended.

21.21 Autoclave

For heat shocking food samples in addition to normal laboratory sterilization purposes.

21.22 Bacti-Disc Cutter (Figure 3 in Chapter 53)

For opening canned food samples aseptically. It may be purchased from Wilkens-Anderson Company, 4525 West Division Street, Chicago, Illinois 60651, No. 4768 or from Marmora Machine Company, 1956 North Latrobe Avenue, Chicago, Illinois 60639, as a bacteriological can opener.

21.23 Balance

For weighing ingredient samples, media, etc., 0.01 gm sensitivity.

21.24 Crystal Violet Solution

Approximately 0.5% aqueous, or other dye formulation suitable for a simple stain.

21.25 Glassware

Dilution Bottles: 6 to 8 oz, for dilution blanks containing 99 ± 2 ml 0.1% peptone in water.

Flasks: 250 ml and 300 ml Erlenmeyer (for analysis of thermophilic spores in ingredients).

Petri dishes: Sterile; either glass or plastic may be used.

Pipets: sterile, 1 ml and 10 ml Mohr pipets; sample pipets made from straight wall borosilicate tubing (7 to 8 mm ID × 35 to 40 cm).

21.26 Incubator

Of sufficient capacity to accommodate more than the number of samples to be in incubation at any given time. Incubator rooms are preferable for large numbers of canned samples. Temperature controlled to about 55 C.

21.27 Media

Dextrose tryptone agar.

Nutrient broth or dextrose tryptone broth (if needed for nutrient supplementation, see 21.42).

21.28 Microscope

With good light source, to provide magnification of about 1,000×.

21.29 pH Meter, Electrometric

pH Color Comparator with brom cresol purple and methyl red reagents and standards may be substituted.

21.210 Sample containers

Cans and end units, jars and caps, or pouches, as used in the commercial operation. Cans, metal beakers, or heavy plastic bags may be used to collect line samples.

21.211 Special water bath (starch)

The Department of Defense has drawn specifications for this bath for boiling water.[3]

21.212 Swabs

Sterile six inch cotton or alginate swabs.

21.3 SAMPLING

21.31 Ingredients

Approximately one-half pound samples from each of five bags, drums or boxes of a shipment or lot of dry sugar, starch, flour, or similar ingredients, should be collected and sealed in cans, jars, plastic bags, or other appropriate containers and taken to the laboratory.[6] Samples of ingredients used in lesser proportions in the finished product, e.g., spices, may be taken in appropriately smaller amounts. In any case, samples should be reasonably representative of the entire lot or shipment in question. Liquid sugar: collect five 6 to 8 oz samples from a tank or truck when it is being filled or emptied.[5]

21.32 Equipment and Product in Process

Only those units held at temperatures within the thermophilic growth range are of direct concern. Scrapings or swab samples of food contacting surfaces or of wet surfaces positioned directly over food materials (from which drippage may gain access to the food materials) may be cultured. Collect the samples in sterile tubes. Examination of food samples taken before and after passage through a particular piece of equipment, e.g., a blancher, a filler, etc., will reveal whether a significant buildup of flat sour spores is occurring in that item of equipment. Multiple samples of a volume equivalent to the volume of the container being packed generally are taken. Unused clean metal cans and covers are convenient sample containers. (Do not use glass containers to collect samples because of the danger of breakage or of being dropped into the equipment.) Solid materials in the presence of excess liquid may be collected in a sieve or similar device; this permits draining excess liquid. Chill the samples thoroughly without delay to preclude activity of thermophilic bacteria prior to the laboratory examination. Immersion of the sample container in cold tap water is usually adequate for this purpose.

21.33 Finished Product

The method of sampling is dependent on the object of the examination. When spore contamination levels during production are the concern, take processed containers representing (a) the start of operations at the beginning of the shift, (b) before midshift shutdown, (c) at start up after midshift shutdown, and (d) at the end of the shift. Samples of each time period should consist of at least ten containers (0.95 probability of finding one positive can if 25% of the production contains viable flat sour spores).[6] Incubate at 55 C for 5 to 7 days.

When known or suspected undercooling, or storage at temperatures above about 40 C, is a reason for concern, take containers at random from the production lot in question. Record the location from which each container was taken, i.e., position on a pallet, and location of the pallet. The larger the samples examined, the greater will be the probability of detecting flat sour spoilage if it exists. (The probability of detecting at least one spoiled container in the sample when the real spoilage level is 1% is about 95% with a 300 unit sample, 89% with a 200 unit sample and 62% with a 100 unit sample).[6] A slight loss of vacuum or sloppy consistency of product often allows separation of flat sour products without destruction of normal product cans.[6]

21.4 PROCEDURE

Thermophilic flat sour spores possess greater heat resistance than most other organisms encountered in foods. This characteristic is advantageous to the examination of foods, ingredients, etc., because, by controlled heat treatment of samples (heat shock), it is possible to eliminate all organisms except the spores with which we are concerned. Further, heat shock, or activation, is necessary to induce germination of the maximum number of spores in a population of many species, including the flat sours.[2, 10] Because the most heat resistant spores are generally the ones of concern in food canning operations, a heat shock favoring recovery of such spores is preferable. Unless otherwise specified, i.e., in a standard procedure, 30 minutes at 100 C or 10 minutes at 109 C, followed by rapid cooling, should be used.

21.41 Sample Preparation and Examination

Sugar and Starch: The National Canners Association has suggested a method and standard for determining thermophilic flat sour spore contamination of sugar and starch to be used in low acid canned foods.[3, 6, 10]

21.411 Sugar (AOAC)[5]

Place 20 gm of dry sugar in a sterile 250 ml Erlenmeyer flask marked to indicate a volume of 100 ml. Add sterile water to the 100 ml mark. Agitate thoroughly to dissolve the sugar. (Liquid sugar is examined by the same procedure, with this difference: a volume of liquid sugar calculated to be equivalent, based on degree Brix, to 20 g dry sugar is added to the 250 ml flask and diluted with water to 100 ml,[5] see 22.42).

Bring the prepared sample rapidly to a boil and continue boiling for five minutes, then water cool immediately. Pipette 2 ml of the heated sugar solution into each of five Petri plates. Add dextrose tryptone agar (Chapter 2), swirl gently to distribute the inoculum, allow to solidify. Incubate the inverted plates at 50 to 55 C for 48 to 72 hours.

21.412 Starch

Place 20 g of starch in a dry, sterile, 250 ml Erlenmeyer flask and add sterile cold water to the 100 ml mark, with intermittent shaking. Shake well to obtain a uniform suspension of the starch in water. Pipette 10 ml of the suspension into a 300 ml flask containing 100 ml of sterile dextrose tryptone agar at a temperature of 55 to 60 C. Use large bore pipets; keep the starch suspension under constant agitation during the pipetting operation. After the starch has been added to the agar, shake the flask in boiling water for a period of three minutes to thicken the starch. Then place the flask in the autoclave and heat at 5 lb pressure (108.4 C) for ten minutes. After autoclaving, the flask should be gently agitated while cooling as rapidly as possible. Violent agitation will incorporate air bubbles into the medium which subsequently may interfere with the reading of the plates. When the agar starch mixture is cooled to the proper point, distribute the entire mixture about equally into five plates and allow to harden. Then stratify with a thin layer of sterile plain 2% agar in water and allow to harden. This prevents possible "spreader" interference. Incubate the inverted plates at 50 to 55 C and count in 48 and 72 hours.

21.413 Other ingredients

The NCA procedures for sugar and starch may be applied to other ingredients used in low acid canned foods.[6] Modifications may be necessary because of physical or chemical characteristics of a particular ingredient, e.g., use of smaller sample sizes or plating smaller volumes of suspension in more than five plates because of colony particle size interference during counting.

21.414 Calculation of counts

Flat sour colonies are round, 2 to 5 mm in diameter, show a dark, opaque center, and usually are surrounded by a yellow halo in a field of purple. The yellow color (acid) of the indicator may be missing when low acid producing strains are present, or where alkaline reversion has occurred. Subsurface colonies are compact and biconvex to pinpoint in shape. If the analyst is unfamiliar with subsurface colonies of flat sour bacteria, it is advisable to streak subsurface colonies on dextrose tryptone agar to confirm surface colony morphology.

The combined count of typical flat sour colonies from the five plates represents the number of flat sour spores in 2 gm of the original sample (20 g sample diluted to 100 ml; 10 ml of this dilution plated). Multiply this count by five to express results in terms of number of spores per 10 g of sample.

The total thermophilic spore count is made by counting every colony on each of the five plates, then calculating in terms of number of spores per 10 g of sample.

21.42 Equipment and Product in Process

The source of excessive flat sour spores in a canning operation may best be determined by "line samples" taken as described above (21.32). Use quantities of sample equivalent to the amount of the material included in a container of finished product; prepare several replicates (5 to 10); after closing, warm the containers to the initial temperature of the commercial process, subject them to the normal commercial thermal process, incubate at 55 C for five to seven days; open and determine growth of flat sour bacteria by pH measurement, supplemented by microscopic examination of a direct smear if necessary. Many line samples are nutritionally complete so that plain water may be added to fill the container; however, some, i.e., formulated product components, may be lacking in essential nutrients, in which case nutrient broth or glucose broth (Chapter 2), for example, should be added instead of water. If in doubt, inoculate a control sample with *B. stearothermophilus* spores, heat shock, and incubate at 55 C for 48 to 72 hours to determine whether the sample material will support spore germination and outgrowth.

Another procedure is to make agar plate counts on serial dilutions of each heat shocked line sample. Use dextrose tryptone agar and incubate at 55 C for 48 to 72 hours.

Swabs, or scrapings from equipment surfaces should be shaken in a known volume of diluent, and the suspension heat shocked and plated on dextrose tryptone agar and incubated at 55 C for 48 to 72 hours. Resulting counts are related back to the approximate surface area.

21.43 Finished Product

Incubate containers of processed product at 55 C for five to seven days, open and examine for flat sour spoilage.[6] Comparison of pH of incubated samples and normal unincubated controls will usually be sufficient to show the presence of flat sour spoilage. If results are not clear, confirm presence or absence of spoilage flora by direct microscopic examination of smears of the product from both incubated and unincubated control containers. The bacteriological condition of products whose physical characteristics provide confusing artifacts when seen in a stained smear can be examined best in a wet mount with phase optics.

Samples collected from a warehouse, when undercooling or storing at elevated temperatures is known or suspected, are examined as above, but without preliminary incubation at 55 C.

21.5 PRECAUTIONS

Samples, other than finished products, must be handled so that opportunities for spore germination or spore production will not be provided between the collection of the samples and the start of examination procedures.

Before making a positive judgement on a sample based on pH or microscopic examination of direct smears, be sure that these characteristics are known for "normal" control products. Controls should be from the same production code as the suspect samples. However, if such are not available, a product from the same manufacturer and bearing the next closest production code must be used. This is particularly important where formulated products are concerned, although it is not necessarily confined to such products.

21.6 INTERPRETATION OF RESULTS

21.61 Ingredients

NCA standards for thermophilic flat sour spores in sugar or starch for canners' use[6] state: "For the five samples examined there shall be a maximum of not more than 75 spores and an average of not more than 50 spores per 10 g of sugar (or starch)."

The total thermophilic spore count standard is: "For the five samples examined there shall be a maximum of not more than 150 spores and an average of not more than 125 spores per 10 g of sugar (or starch)."[6]

The sugar starch standard may be used as a guide for evaluating other ingredients, keeping in mind the proportion of the other ingredients in the finished product relative to the quantity of sugar or starch which may be used in a product.[3, 6, 10]

The presence of thermophilic flat sour spores in ingredients for other than thermal processed low acid foods is probably of no significance provided those foods are not held within the thermophilic growth range for many hours. The flat sour bacteria have no public health significance.[8]

21.62 Equipment and Product in Process

Canned and processed line samples usually indicate a point or points at which a spore buildup has occurred. A high percentage of positive samples taken from one point in the line, when a low level or no positive samples were found prior to this point, shows a spore buildup in this piece of equipment. The time of day yielding positive samples may indicate whether the buildup is due to operating temperatures within the thermophilic growth range, or whether inadequate cleanup and sanitization procedures were used prior to sampling. The former condition is suggested when a majority of positive samples occur in those taken after the line has been in operation for several hours. The latter condition is suggested when samples at the startup of the line are predominately positive, and those taken later are all or mostly negative.

Plate count data for line samples may be meaningful, especially when taken over a long period of time. They can show trends regarding buildups, inadequate cleanups, etc. Because counts are made in a laboratory medium which may be a better spore germination and growth medium than certain specific food products themselves, and because the sample does not receive the equivalent of the commercial thermal process, results may reflect greater than the actual potential surviving spore load in the finished product; however, it can indicate buildup situations which are undesirable.

Further, the presence and number of the facultative thermophilic aerobic sporeformers may be determined best by plate count procedures, provided a low heat shock treatment is used.

21.63 Finished Products

Dormant thermophilic spores are of no concern in commercially sterile canned foods destined for storage and distribution where temperatures will not exceed about 43 C. However, some canned foods are destined for exposure to temperatures above 43 C during part or most of their shelf life, i.e., those shipped to tropical locales, and those intended for hot vend service. To be considered commercially sterile, these specialized foods must not contain thermophilic spores capable of germination and outgrowth in the product.

Thus, examination for thermophilic flat sour spores provides useful information in the first of the above situations in that a clue often is provided to the packer's operations; for example, a high percentage of containers showing the presence of spores may suggest that the operation may be permitting a spore buildup due to improper temperature control in some item of equipment. In the second situation, information is provided relative to the commercial sterility of a production lot of a specialized product.

Randomly selected warehouse samples of low acid canned foods, some of which are found to have undergone flat sour spoilage, can reveal information relative to the condition contributing to the development of spoilage. Spoilage confined to produce situated in the outer layers or rows of cases on a pallet suggests localized heating, e.g., due to close proximity to a space heater, or too close to the building roof during periods of hot weather. Finding spoilage confined to inner cases on a pallet is indicative of insufficient cooling, i.e., palletizing cases while the product was still in the thermophilic growth temperature range. Inner cases are insulated by exterior cases and thus may retain heat for several days.

21.7 REFERENCES

1. Breed, R. S., E. G. D. Murray, and N. R. Smith. 1957. Bergey's Manual of Determinative Bacteriology, 7th Ed: Williams and Wilkins Co., Baltimore, Md.
2. Cook, A. M. and R. J. Gilberg. 1968. Factors affecting the heat resistance of *Bacillus stearothermophilus* spores. J. Food Technol. **3**:285–293
3. Department of Defense. 1958. Military Standard. Bacterial standards for starches, flours, cereals, alimentary pastes, dry milks and sugars used in the preparation of canned foods for the armed forces. MIL-STD-900
4. Hersom, A. C. and E. D. Hulland. 1964. Canned Foods, An Introduction to Their Microbiology (Baugartner). 5th Ed. Chemical Publishing Company, Inc., New York, N.Y.
5. Horwitz, W., ed. 1975. Official Methods of Analysis of the Association of Official Analytical Chemists. 12th ed. p 920. AOAC, Washington, D.C.
6. National Canners Association Research Laboratories. 1968. Laboratory Manual for Food Canners and Processors. The Avi Publishing Company, Inc., Westport, Conn. **1**:102–108
7. Richmond, B. and M. L. Fields. 1966. Distribution of thermophilic aerobic sporeforming bacteria in food ingredients. Appl. Microbiol. **14**:623–626
8. Schmitt, H. P. 1966. Commercial sterility in canned foods, its meaning and determination. Quart. Bull. Assoc. Food and Drug Officials of the U.S. **30**:141–151
9. Smith, N. R., R. E. Gordon, and F. E. Clark. Aerobic Sporeforming Bacteria. Agriculture Monograph No. 16. U.S. Department of Agriculture, Washington, D. C.
10. Stumbo, C. R. 1973. Thermobacteriology in Food Processing. 2nd Ed. Academic Press, New York, N.Y.
11. Walker, P. D. and J. Wolf. 1971. The Taxonomy of *Bacillus stearothermophilus*, pp 247–262, in A. N. Barber, G. W. Gould, and J. Wolf, Eds., England Spore Research 1971. Academic Press, London, England

CHAPTER 22

THERMOPHILIC ANAEROBES

David H. Ashton

22.1 INTRODUCTION

The nonhydrogen sulfide producing, thermophilic anaerobic sporeformers are classified in the family *Bacillaceae*, genus *Clostridium*.[1] The type species of this group is *Clostridium thermosaccharolyticum*.[8] These organisms are obligately anaerobic, and are strongly saccharolytic, producing acid and abundant gas from glucose, lactose, sucrose, salicin, and starch. Proteins are not hydrolyzed and nitrites are not produced from nitrates.[4] Vegetative cells are long, slender gram negative rods. Spores are terminal and swollen. Neither toxins nor infections are produced and therefore, the organisms are of spoilage, but not of public health, significance.

The thermophilic anaerobes which do not produce hydrogen sulfide have been responsible for the spoilage of canned products such as spaghetti with tomato sauce, sweet potatoes, pumpkin, green beans, and asparagus.[7] Also they have caused spoilage of highly acid products such as fruit and farinaceous ingredient mixtures, and tomatoes. Spores of these organisms characteristically possess a z value (slope of thermal death time curve) of approximately 6.7 C (12 F) and thus have extreme resistance in the 105 to 113 C (220 to 235 F) range.[9] Their survival in canned foods is not unexpected therefore, but thermophilic anaerobes are rarely found in foods processed above 121 C (250 F). Only when the finished product is improperly cooled, or if the product is held for extensive periods at elevated temperatures, do the thermophilic anaerobes express themselves.

The growth temperature optimum of these organisms is 55 C. They seldom grow at temperatures below 32 C, but can produce spoilage in 14 days at 37 C if the spores are first germinated at a higher temperature. They have a pH optimum of 6.2 to 7.2, but grow readily in products at pH 4.7 or higher, and, on occasion, have been responsible for spoilage in tomato products at pH 4.1 to 4.5.[7]

Ingredients such as sugar, dehydrated milk, starch flour, cereals, and alimentary pastes have been found to be the predominant sources of thermophilic anaerobes. These organisms occur widely in soil, and therefore are found on raw materials, such as mushrooms and onion products. The thermophilic anaerobes do not multiply on equipment and handling sys-

CONTRIBUTOR: Keith Ito
IS/A Committee Liaison: Cleve B. Denny

tems unless an anaerobic environment containing nutrients and moisture and an elevated temperature is provided. Excessive populations of thermophilic anaerobes can develop in ingredients such as chicken stock, beef extract, or yeast hydrolysate if an incubation period in the thermophilic temperature range is provided during concentration or hydrolysis.

22.2 EQUIPMENT AND SUPPLIES

Provide equipment and supplies as needed in accordance with specifications in Chapters 2 and 21.2. The following additional supplies are recommended.

22.21 Culture Medium

The recommended substrate for growth of the nonhydrogen sulfide producing anaerobes is PE-2 medium[3] (Chapter 2). This medium is prepared by placing six dried, small Alaskan seed peas and approximately 12 ml of a 2% peptone solution containing 4 ml of 1% bromcresol purple per liter into 18 × 150 mm tubes.

Because of the abundant gas production by these organisms, metal closures are preferable to screw caps, if the medium is to be used within two weeks of preparation. The medium is allowed to stand for 1 hour to effect hydration, autoclaved 15 minutes at 121 C, and cooled to 55 C before using. Unless freshly prepared medium is used, tubes should be subjected before use to flowing steam for 20 minutes to exhaust before cooling to 55 C. After inoculating, tubes are stratified with 3 ml of sterile 2% agar tempered to 50 C. When the agar overlay has solidified, the tubes are preheated to 55 C and incubated. The medium should be supplemented to contain 0.3% yeast extract for the detection of severely heat stressed spores.

22.22 Incubator

An incubator which will maintain a uniform temperature of 55 ± 2 C is required. For evaluation of a finished product, a large capacity incubator with sturdy shelves is recommended.

22.3 SAMPLING

22.31 Ingredients

Samples of dry ingredients should consist of 200 g taken aseptically from five different bags or barrels per shipment or lot for lot sizes of 50 or less containers, from 10% of the containers for lot sizes 50 to 100, and from a number of containers equal to the square root of the lot size for shipments with greater than 100 containers, and placed in a sterile sealed container. Liquid sugar is sampled by drawing five 200 g portions per tank during transfer from tanks, or at the refinery during the tank filling operation. If preliminary analyses indicate considerable variability in a lot, the number of samples should be increased.

22.32 Equipment and Systems

Thermophilic anaerobes will not develop on equipment unless elevated temperatures are provided in a relatively anaerobic environment containing nutrients. Accumulated food materials in such locations should be sampled with a sterile spatula or similar device, placed in a sterile sealed container, cooled to room temperature, and cultured. Examination of food materials before and after exposure to processing equipment will help to reveal the contamination level of the equipment.

22.33 Product in Process

Periodically 200 g samples of product in process should be obtained to monitor the system. Sample timing should be arranged to coincide with the introduction of a new batch of ingredients, or a shutdown, which may have permitted an incubation period. The samples should be cooled immediately but slowly to room temperature, and the analysis conducted as soon as possible. Refrigeration is not recommended.

Representative containers of finished product should be obtained to reflect the condition of the entire population of containers in a production period. The need for sampling will be dictated by such considerations as the previous record of the product with respect to thermophilic spoilage, and the temperature stresses to which the product is expected to be subjected during transit and storage. The number of containers sampled should be of the order of one per thousand containers produced. If immediate postprocessing cooling to 43 C is not achievable, monitoring of surviving thermophiles becomes extremely important. Incubate at 55 C for 5 to 7 days.

22.4 PROCEDURE

The following procedures apply to the detection of spores only, rather than spores and vegetative cells.

22.41 Dry Sugar and Powdered Milk (Sugar: AOAC method[5])

Place 20 g of sample in a sterile flask and add sterile distilled water to a final volume of 100 ml. Stir to dissolve the sample and bring the contents of the flask rapidly to a boil. Boil for 5 minutes, cool by placing the flask in cold water, and bring the volume back to 100 ml with sterile distilled water. Divide 20 ml of boiled solution equally among six freshly exhausted tubes of PE-2 or liver broth medium. Stratify each tube with 3 ml of sterile 2% agar, allow the agar to solidify, preheat the tubes to 55 C, and incubate at 55 C for 72 hours.

22.42 Liquid Sugar

Place a sample containing 20 g dry sugar, determined on the basis of degrees Brix (29.41 g of 68 degrees Brix liquid sugar is equivalent to 20 g of dry sugar) in a sterile flask and proceed as in 22.41.

22.43 Fresh Mushrooms

Place 200 g of mushrooms in a sterile blender jar and dice with a sterile knife. Blend the diced sample until the pieces are finely chopped. Frequent shaking of the jar is essential to ensure proper blending. Place 20 g of blended sample in a sterile flask and proceed as in 22.41.

22.44 Starches and Flours[6]

Place 20 g of sample in a sterile flask containing a few glass beads, and add sterile distilled water to a final volume of 100 ml. Shake well to obtain a uniform suspension. Divide 20 ml of the suspension equally among six freshly exhausted tubes of PE-2 medium. Spin three tubes at a time in the hands immediately after adding the sample. Place the tubes in a boiling water bath and continue to spin the tubes for the first 5 minutes of heating. Continue the heating for an additional 10 minutes, remove the tubes, and place in cold water. Stratify the tubes with 3 ml of sterile 2% agar; allow to solidfy. Preheat the tubes to 55 C and incubate at 55 C for 72 hours.

22.45 Cereals and Alimentary Pastes[6]

Place 50 g of well mixed sample in a sterile blender jar and add 200 ml of sterile distilled water. Blend for 3 minutes to obtain a uniform suspension. Proceed as in 22.44. For calculations, assume that 10 ml of the blended materials contain 2 g of the original sample.

22.46 Product in Process

Place 20 g of product in a sterile blender jar and blend for 3 minutes. Distribute 20 ml of the blended sample equally among six freshly exhausted tubes of PE-2 medium and proceed as in 22.44. If the product in process is at a temperature of 65 C or greater when sampled, further heating will be unnecessary.

22.47 Finished Products

Representative samples of finished products should be incubated at 55 C for 5 to 7 days and observed daily for evidence of loss of vacuum or container distortion. Samples which show signs of spoilage should be removed from incubation, opened aseptically, and 3 g of the contents placed in each of two tubes of freshly exhausted PE-2 medium by means of a wide bore pipet. Product smears should be made for morphological confirmation. The conditions necessary for preventing laboratory contamination when subculturing cans of finished product are detailed in the literature.[2]

22.48 Spore Suspensions

A greater degree of quantitation than is described in 22.41 to 22.47 above is to be desired with spore suspensions prepared for thermal inactivation studies. Therefore, 10 ml of the desired dilution of the spore suspension is placed in a 18 × 150 mm screw cap tube and immersed in boiling water for

8 minutes followed by rapid cooling in cold water. A conventional five tube most probable number (MPN) dilution series of the boiled suspension is prepared in freshly exhausted PE-2 medium. The inoculated tubes are treated as in 22.41. The population of the original spore suspension is computed from MPN tables.

22.5 PRECAUTIONS

Every precaution should be taken to assure that the ingredients of the detection medium are free from growth inhibitors. For example, one should attempt to obtain peas with a pesticide free history. As an added precaution, each new lot of ingredients should be incorporated into the medium and tested for growth inhibitors with a known suspension of a thermophilic anaerobe. These precautions will eliminate or help to minimize the occurrence of false negatives.

The detection procedures described above (22.41 to 22.45) are, at best, only semiquantitative and should be regarded as such. The objective in surveying ingredients is the detection of the ingredient in a known quantity rather than absolute quantitation. It is important that the dried peas be soaked in the peptone solution before autoclaving to assure the proper sterilizing effect. Repeated steaming of unused tubes of medium does not reduce the effectiveness of PE-2 as a substance for the thermophilic anaerobes.

22.6 INTERPRETATION OF RESULTS

Tubes of PE-2 medium positive for growth of nonhydrogen sulfide producing thermophilic anaerobes show gas production with peas rising to the top of the liquid medium. Thermophilic flat sour bacteria may be present only if a color change from blue to yellow appears (Chapter 21).

22.61 Ingredients

22.611 For canners' use

Spores of nonhydrogen sulfide producing thermophilic anaerobes should not be present in more than 60% of the samples tested or in more than 66% of the tubes for any single sample.[4] Use of ingredients meeting this standard will minimize the possibility of spoilage in finished product. Canned foods with a pH below 4.0 are not involved in spoilage by the thermophilic anaerobes.

22.612 For other use

The presence of excessive numbers of spores of nonhydrogen producing thermophilic anaerobes in ingredients for use in other than canned products is of little significance unless a thermophilic incubation period is provided during processing. In such a case, the number of vegetative cells present after a processing step is important, and should be determined as outlined above, omitting the boiling step.

22.62 Equipment and Systems

The presence of detectable levels of spores of nonhydrogen sulfide producing thermophilic anaerobes on equipment and systems suggests that the equipment is in need of thorough sanitization or growth is occurring, or both. If proper sanitation is practiced, and if systems are properly designed, spore buildup should not occur. It is especially important to sample foams and their residues on equipment.

22.63 Product in Process

Excessive numbers of vegetative cells or spores in a product in process prepared from ingredients meeting the above requirements (22.611, 22.612) suggest that multiplication is occurring during one or more manufacturing steps. The manufacturing sequence should be sampled and the point of increase in population determined. Remedial steps should be taken immediately. The presence of vegetative cells suggests that sporulation can and will occur.

22.64 Finished Products

The presence of low numbers of spores of the nonhydrogen sulfide producing thermophilic anaerobes in processed canned foods is not unusual. These organisms possess extreme resistance to the center can temperatures achieved by many commercial processes.[9] An attempt to eliminate the spores by increased thermal treatments may endanger the nutritional and functional integrity of many products. If cooling of processed cans to a center can temperature of 43 C or less is effected immediately and cans are stored remote from heating ducts and the like at temperatures below 35 C, the presence of spores of thermophilic anaerobes is of no consequence. However, the potential for spoilage exists if temperature abuse of the cans occurs, and therefore, this situation should be avoided through the use of carefully selected ingredients carefully handled throughout the production sequence. The importance of efficient cooling followed by storage below 35 C cannot be overemphasized if thermophilic anaerobes are present.

The presence of detectable thermophilic anaerobes in canned foods destined for hot vend or tropical distribution constitutes an unacceptable spoilage hazard. The situation must be overcome by the use of thermophile free ingredients or by increasing the processing.

22.7 REFERENCES

1. Buchanan, R. E. and N. E. Gibbons. 1974. Bergey's Manual of Determinative Bacteriology. 8th Ed: The Williams and Wilkins Co., Baltimore, Md.
2. Evancho, G. M., D. H. Ashton, and E. J. Briskey. 1973. Conditions necessary for sterility testing of heat processed canned foods. J. Food Sci. **38**:185–188
3. Folinazzo, J. F. and V. S. Troy. 1954. A simple bacteriological medium for the growth and isolation of spoilage organisms from canned foods. Food Technol. **8**:280–281
4. National Canners Association Research Laboratories. 1968. Laboratory Manual for Food Canners and Processors. The Avi Publishing Co., Inc., Westport, Conn.
5. Official First Action 1972. Detecting and estimating numbers of thermophilic bacterial spores in sugars. J.A.O.A.C. **55**:445–446
6. Powers, E. M. 1973. Microbiological requirements and methodology for food in military and federal specifications. Technical Report 73-33-FL. U.S. Army Natick Labs., Natick, Mass.
7. Rhoads, A. T. and C. B. Denny. 1964. Spoilage Potentialities of Thermophilic Anaerobes. Research Report No. 3-64. National Canners Association, Washington, D. C.
8. Stumbo, C. R. 1973. Thermobacteriology in Food Processing. 2nd Ed. Academic Press, New York, N.Y.
9. Xezones, H., J. L. Segmiller, and I. J. Hutchings. 1965. Processing requirements for a heat-tolerant anaerobe. Food Technol. **19**:1001–1003

CHAPTER 23

SULFIDE SPOILAGE SPOREFORMERS

R. V. Speck

23.1 INTRODUCTION

Early studies on "sulfide stinker" spoilage in canned sweet corn and other vegetables were reported by Werkman and Weaver[13] and Werkman.[12] The cans involved showed no evidence of swelling. However upon opening, a decided odor of hydrogen sulfide was evident. The product had a blackened appearance due to the reaction between the sulfide and the iron of the container.

Sulfide spoilage, the presently preferred designation for this type of spoilage, is not common. It is nonexistent in acid foods because of the pH requirements of the causative organism. Spoiled products, although possessing a strong, disagreeable odor of hydrogen sulfide, exhibit no other putrefactive odor.

There has been no evidence of pathogenicity for man or laboratory animals associated with *Clostridium nigrificans,* the causative organism, or with product spoiled by this organism.

Cameron and Williams,[3, 4] and Cameron and Yesair[5, 6] found sugar and starch to be important sources of these organisms in canneries.

23.11 Classification

The type species of the sulfide spoilage group, *C. nigrificans,* was originally classified and named by Werkman and Weaver. In 1938, Starkey isolated cultures from mud, soil and sewage at both 30 and 55 C.[11] Those growing at 55 C were large, slightly curved, spore forming rods, while those isolated at 30 C were asporogenous short vibrios. Cultures isolated at 30 C failed to grow when transferred directly to 55 C. Those isolated at 55 C and transferred to 30 C underwent striking morphological changes, eventually resulting in small vibrios resembling those originally isolated at this temperature. As a result, Starkey proposed the new genus *Sporovibrio* for the anaerobic vibrio shaped cells which produced endospores. The organism *Sporovibrio desulfuricans* was later shown by Campbell, et al[7] to be identical to *C. nigrificans*. Since the latter had taxonomic priority, the thermophilic sporeforming, sulfate reducer was considered properly to

CONTRIBUTOR: Keith Ito
IS/A Committee Liaison: Cleve B. Denny

be named *C. nigrificans*. Campbell and Postgate later proposed the name *Desulfotomaculum nigrificans* for this organism.[8]

23.12 Occurrence

Although relatively rare, sulfide spoilage may occur in canned sweet corn, peas, mushroom products, and other nonacid foods. Spoiled sweet corn usually has a bluish gray colored liquor with many blackened grains throughout the can. Spoiled peas sometimes show no discoloration, but more frequently show blackening with a dark colored brine. In many instances, spangling of the enamel system of the can occurs and is the result of the interaction of the dissolved hydrogen sulfide with the iron of the container. This is evident through the enamel system because of the semitransparency of the coating.

The cause of sulfide spoilage is the combination of high spore numbers and holding of finished product at elevated temperatures. This latter factor may be the result of inadequate cooling of processed product.

23.13 Temperatures and pH Requirements

Most isolates from sulfide spoilage are obligate thermophiles in that optimum growth occurs at 55 C. Most of these strains will grow at 43 C but not at 37 C. Therefore, using the Cameron and Esty definition,[2] these would be considered as obligate thermophiles. Organisms resembling *C. nigrificans* have been isolated from soil, mud, sewage, etc., as well as from certain food ingredients. Such isolates may be classified as mesophiles, facultative thermophiles, or obligate thermophiles, using again the Cameron and Esty guidelines.

According to Breed, *et al.* the type species, isolated from canned corn showing "sulfur stinker" spoilage, will grow between 65 and 70 C, with optimum growth at 55 C.[1] Campbell and Postgate reported that the organism can be "trained" to grow slowly at 37 or 30 C.[8]

Optimum growth of *C. nigrificans* occurs between pH 6.8 and 7.3. Scanty growth occasionally occurs as low as 5.6; however, 6.2 is considered the lower limit. Maximum pH for growth has been recorded as 7.8. The pH values of most vegetables fall below 5.8, corn and peas being exceptions. This may be responsible for the limited and relatively uncommon occurrence of sulfide spoilage.

23.2 SPECIAL EQUIPMENT AND SUPPLIES (21.2)

A special water bath for the examination of sugar, starch, and flour, originally described by National Canners Association and later by Military Standard-Bacterial Standards for Starches, etc., MIL–STD–900, 8 September 1958, must be used. In addition, 250 ml Erlenmeyer flasks marked to indicate 100 ml must be prepared.

A quantity of common 6 d nails must be cleaned with hydrochloric acid and thoroughly washed in distilled water before placing head down in tubes of sulfite agar (Chapter 2).

23.3 SAMPLING

Since there is little evidence to indicate a serious in-plant buildup potential with *C. nigrificans* in modern food processing plants, recommended sampling will be limited to frequently used ingredients.[10]

23.31 Ingredients

One-half pound samples of sugar, starch, or flour, are taken from each of five 100 pound bags of a shipment or lot. In the case of bulk shipments of such ingredients, sampling will probably have to be carried out through a loading port or hatch at the top of the car or tank. A suitable trier should be used so that samples can be taken from various depths of the load. Samples of liquid sugar are obtained from tank trucks with the use of a sterile, long handle dipper.

It is recognized that the adequacy of sampling will vary with the size of a shipment; however, it is felt that when there is any significant variability in the shipment, individual tests on five samples will make this evident in the majority of cases.

23.4 PROCEDURE: METHODS OF ANALYSIS

23.41 Sugar, Starch, and Flour (Sugar: AOAC method[10])

23.411 Sample preparation

In the case of dry sugar, place 20 g of the sample into a dry sterile 250 ml Erlenmeyer flask with sterile rubber stopper. Add sterile water to the 100 ml mark and shake to dissolve. Replace the stopper with a sterile cotton plug and bring the solution rapidly to a boil, and continue boiling for 5 minutes. Replace the evaporated liquid with sterile water. Cool immediately in cold water.

Prepare samples of liquid sugar the same way, except the amount added to the sterile flask should be determined, depending upon the Brix, to be equivalent to 20 g of dry sugar (22.42).

In examining starch or flour, 20 g of the dry ingredients are placed in a dry sterile 250 ml Erlenmeyer flask, and sterile cold water added to the 100 ml mark, with intermittent swirling. Close the flask with the sterile rubber stopper and shake well to obtain a uniform, lump free suspension of the sample in water. Sterile glass beads added to the sample water mixture will facilitate thorough mixing during shaking.

23.412 Cultural methods

When examining sugar, divide 20 ml of the heated solution among six 20 × 150 mm screw cap tubes containing approximately 10 ml of sulfite agar and a nail. Make the inoculations into freshly exhausted medium and solidify rapidly by placing the tubes in cold water. Preheat the tubes to 50 to 55 C and incubate at that temperature for 24 and 48 hours.

In the case of starch or flour, divide 20 ml of the cold suspension among six 20 × 150 mm screw cap tubes containing approximately 10 ml of sulfite agar and a nail. The tubes should be swirled manually and gently inverted several times before heating and during the 15 minute heating period in a boiling water bath. The periodic swirling and inversion of the tubes will assure an even dispersion of the starch and flour in the tubes of medium. Following heating, cool the tubes immediately in cold water. Preheat the tubes to 50 to 55 C, and incubate at that temperature for 24 to 48 hours.

C. nigrificans will appear as jet black spherical areas, the color due to the formation of iron sulfide. No gas is produced. Certain thermophilic anaerobes not producing H_2S give rise to relatively large amounts of hydrogen which splits the agar and, in the case of sulfite agar, reduces the sulfite thereby causing general blackening of the medium. Total the colonies in the six tubes. Calculate, and report as number of spores per 10 g of ingredient.

An alternate method of analysis, using thioglycollate agar in lieu of sulfite agar, has been used by some laboratories.

23.42 Nonfat Dry Milk

23.421 Sample preparation

Weigh 10 g of the sample into a sterile 250 ml Erlenmeyer flask marked to 100 ml. Add N/50 sodium hydroxide to the 100 ml mark and shake to completely dissolve the sample. Heat 10 minutes at 5 lb steam pressure, then cool immediately.

23.422 Culturing methods

Transfer 2 ml of the heated solution to each of ten 20 × 150 mm screw cap tubes of freshly exhausted sulfite agar and nail. Gently invert several times and solidify rapidly by placing the tubes in cold water. Preheat the tubes to 50 to 55 C and incubate at that temperature for 24 and 48 hours ± 3 hours. Count colonies of *C. nigrificans* described earlier and report on the basis of 10 g of sample.

23.43 Cream

23.431 Sample preparation

Mix 2 g of gum tragacanth and 1 g of gum arabic in 100 ml of water in an Erlenmeyer flask. Sterilize in the autoclave for 20 minutes at 121 C. Transfer 20 ml of sample to a sterile 250 ml Erlenmeyer flask marked for 100 ml. Add the sterilized gum mixture to the 100 ml mark and carefully shake, using a sterile rubber stopper. Loosen stopper and autoclave for 5 minutes at 5 psi pressure.

23.432 Culturing methods

Transfer 2 ml of the sample and gum mixture to each of ten 20 × 150 mm screw cap tubes of freshly exhausted sulfite agar and nails. Gently invert several times and solidify rapidly by placing the tubes in cold water. Preheat the tubes to 50 to 55 C and incubate at that temperature for 24 and

48 hours ± 3 hours. Count colonies of *C. nigrificans* described earlier and report on the basis of 1 ml of sample.

23.5 PRECAUTIONS AND LIMITATIONS OF PROCEDURE

When analyzing ingredients by the above methods, thorough dispersion of the sample solution or slurry in each tube of medium is essential. More difficulty will be encountered in the analysis of starch or flour because of the thickening effect during heating. Frequent swirling or gentle inversion of the tubes during the first 10 minutes of heating will assure proper dispersion.

Since tubes containing numerous colonies of *C. nigrificans* may become completely blackened after 48 hours of incubation, a preliminary count should be made after 20 to 24 hours ± 3 hours.

23.6 INTERPRETATION OF RESULTS

A standard for sulfide spoilage spores need only apply to ingredients (sugar, starch, flour, etc.) to be used in low acid, heat processed canned foods.

Sulfide spoilage spores shall be present in not more than 2 (40%) of the 5 samples tested and in any 1 sample to the extent of not more than 5 spores per 10 g.[9] This would be equivalent to 2 colonies in the inoculated tubes.

23.7 REFERENCES

1. Breed, R. S., E. G. D. Murray, and N. R. Smith. 1957. Bergey's Manual of Determinative Bacteriology, 7th ed. p. 649–650
2. Cameron, E. J. and J. R. Esty. 1926. The examination of canned spoiled foods, 2. Classification of flat sour, spoilage organisms from nonacid foods, J. Infect. Diseases, **39**:89–105
3. Cameron, E. J. and C. C. Williams. 1928 a. The thermophilic flora of sugar in its relation to canning. Centbl. Bakt. (etc.) 2 abt. **76**:28–37
4. Cameron, E. J. and C. C. Williams. 1928 b. Thermophilic flora of sugar in its relation to canning. J. Bact. **15**:31–32
5. Cameron, E. J. and J. Yesair. 1931 a. About sugar contamination: its effect in canning corn. Canning Age, **12**:239–240
6. Cameron, E. J. and J. Yesair. 1931 b. Canning tests prove presence of thermophiles in sugar. Food Indus. **3**:265
7. Campbell, L. L., Jr., H. A. Frank, and E. R. Hall. 1957. Studies on thermophilic sulfate reducing bacteria. I. Identification of *Sporovibrio desulfurians* as *Chostridium nigrificans.* J. Bacteriol. **73**:516–521
8. Campbell, L. L. and J. R. Postgate. 1965. Classification of the spore-forming sulfate-reducing bacteria. Bact. Rev. **29**:359–363
9. National Canners Association Research Laboratories. 1968. Laboratory Manual for Food Canners and Processors. The AVI Publishing Co. Inc., Westport, Connecticut, Vol. **1**:104
10. Official First Action 1972. Detecting and estimating numbers of thermophilic bacterial spores in sugars. J.A.O.A.C. **55**:445–446
11. Starkey, R. L. 1938. A study of spore formation and other morphological characteristics of *Vibrio desulfuricans*. Arch. Mikrobiol. **9**:268–304
12. Werkman, C. H. 1929. Bacteriological studies on sulfide spoilage of canned vegetables. Iowa Agr. Exp. Sta. Pes. Bul. **117**:163–180
13. Werkman, C. H. and H. J. Weaver. 1927. Studies in the bacteriology of sulfur stinker spoilage of canned sweet corn. Iowa State College, J. Sci. **2**:57–67

INDICATOR MICROORGANISMS AND PATHOGENS; TOXIN DETECTION

CHAPTER 24

COLIFORMS, FECAL COLIFORMS, *E. COLI*, AND ENTEROPATHOGENIC *E. COLI*

Morris Fishbein,* I. J. Mehlman, Lee Chugg, and Joseph C. Olson, Jr.

24.1 INTRODUCTION

Bergey's Manual of Determinative Bacteriology defines the *Enterobacteriaceae* family as consisting of gram negative aerobic and facultatively anaerobic rods which produce acid from glucose and other carbohydrates, and are usually aerogenic.[2] The coliform group of indicator organisms comprise certain members of this family which are capable of fermenting lactose with the production of gas.

Escherich indicated the universal presence of *E. coli* in human stools.[12] Shardinger suggested that the organism be used as an index of fecal pollution since it could be more easily recovered than *Salmonella*.[41] The first edition of the "Standard Methods of Water Analysis" (1905) was directed towards the recovery of *E. coli*.[24] In 1914 the Public Health Service standard was changed from *E. coli* to the coliform group indicator.[24] This premise was based on the erroneous assumption that all members of the coliform group had equal sanitary significance.[5]

The present definition of the coliform group in Standard Methods for the Examination of Water and Wastewater states: "the coliform group includes all the aerobic and facultative anaerobic, gram negative, nonsporeforming, rod shaped bacteria which ferment lactose with gas formation within 48 hours at 35 C."[43] This definition is reiterated for other products such as seawater and shell fish[36] and dairy products.[42] In the latter procedure the temperature of incubation is established at 32 C instead of 35 C. The coliform group contains individual species whose habitat is intestinal and nonintestinal such as soil, water, and grain. Clemesha questioned the practice of applying the characteristics properly belonging to a species to a group of bacteria.[5] It has been stated also that the ability of certain organisms to ferment the same sugar does not cause them to possess other characteristics in common, such as fecal habitat.[3]

Eijkman employed a 46 C glucose fermentation test in an attempt to separate the coliforms of warm blooded animals from those of cold blood-

*Deceased.

IS/A Committee Liaison: Nino F. Insalata

ed animals.[11] MacConkey likewise proposed an incubation temperature of 42 C.[25] In America, the work of Perry and Hajna resulted in the development of EC medium which is commonly employed in the recovery of the fecal coliform group at the present time.[33] The literature contains examples showing that elevated temperature tests (which recover the fecal coliform group) more closely parallel the true sanitary conditions of soils, waters, shellfish and foods than do the lower temperature coliform group procedures.[17,26,44] Thus the fecal coliform group indicator has gained wide acceptance as a better indicator of fecal pollution than the coliform group indicator. Undoubtedly, this is due to the increased recovery of fecal *E. coli*.[16] It also results in a greater specificity.

24.2 THE *ENTEROBACTERIACEAE* GROUP AS AN INDICATOR

In Europe some use has been made of the total *Enterobacteriaceae* group as an indicator.[29,30] Glucose brilliant green bile broth is used as enrichment, and isolation of the organisms is on 1% glucose violet red bile agar. The reported advantage of this procedure is that it is designed to detect the presence of all of the *Enterobacteriaceae*. As such, it not only detects the presence of organisms that can or cannot utilize lactose, but may be used as a broad base in detecting organisms which may have recontaminated the product after processing. Various authors have reported results employing this procedure.[31,47]

24.21 Definitions

It is necessary to define the terms which will be employed. These include the coliform group, the fecal coliform group, *E. coli* (the fecal indicator) and *E. coli*, the human enteropathogenic form, **EEC**.

24.211 Coliform group

The coliform group includes aerobic and facultative anaerobic, gram negative, nonsporeforming rods which ferment lactose with acid and gas formation within 48 hours at 32 to 37 C. The test media include any lactose enrichment liquid or solid medium. However, it is essential that results of any test for the qualitative or quantitative presence of the coliform group be expressed in terms of the test procedure used. To do otherwise would result in confusion since the nature of the medium and the incubation conditions singly or collectively may affect the result materially. This group also may include any organism not a member of the *Enterobacteriaceae*, but which conforms to the above criteria. It is not necessary to identify the coliform group organisms by the IMViC pattern.

24.212 Fecal coliforms

Elevated temperature tests for the separation of organisms of the coliform group into those of fecal origin and those derived from nonfecal sources have been used in many parts of the world and with various modifications. None of these methods will differentiate rapidly and absolutely coliforms of fecal origin from those originating in nonfecal sources. Nevertheless, practical methods have been developed which favor the selection and growth of fecal coliform organisms while eliminating many but not all of the types generally considered to have little or no public health significance.

The following indicates the variations in test procedures for the enumeration of fecal coliforms. Standard Methods for the Examination of Water and Waste Water lists three coequal routes to fecal coliform determinations: from positive presumptive lauryl sulfate tryptose (LST) broth to EC broth with incubation at 44.5 ± 0.2 C for 24 ± 2 hours, or to boric acid lactose broth with incubation at 43 ± 0.2 for 48 ± 3 hours, or a membrane filter procedure utilizing M-FC broth with incubation at 44.5 ± 0.2 C for 24 ± 2 hours.[43] Canadian authorities specify incubation of EC broth at 44.7 ± 0.2 C for 24 hours for fecal coliform counts of fish, fishery products and shellfish, but for other foods 45 ± 0.2 C for 24 hours is used. In the United States the fecal coliform method used in the National Shellfish Sanitation Program also specifies EC broth incubation at 44.5 ± 0.2 C for 24 ± 2 hours.

It should be emphasized that the term "fecal coliform" like the term "coliform" has no taxonomic validity. Therefore, the meaning of a fecal coliform count becomes clear only when it is expressed in terms of the test procedure used.

24.213 *E. coli*

E. coli must conform to the coliform and fecal coliform group definitions. The organism may be further identified by the IMViC pattern, ++−− or −+−− (24.45). Its natural habitat is the lower part of the intestines of vertebrate animals.[12, 41]

24.214 Enteropathogenic *E. coli* (EEC)

Enteropathogenic *E. coli* is one of the causes of a form of gastroenteritis which may occur in man. In this role *E. coli* is referred to as enteropathogenic *E. coli*, abbreviated EEC. The majority of all EEC strains conform to the definition of the coliform group, the fecal coliform group, and *E. coli*. Some EEC organisms are unable to ferment lactose with gas production within 48 hours (delayed), and therefore do not conform to the definitions of the coliform, fecal coliform group and *E. coli* in the broad sense. Thus, the methods for the recovery of EEC are based on total growth recoveries as well as on gas in lactose broth. Although EEC conforms to the IMViC patterns ++−−or −+−−, this minimal biochemistry is inadequate for proper identification purposes. A more detailed biochemical identification is required in accordance with proper taxonomic procedures.[10] In addition, it is necessary to identify the isolated organism serologically according to its complete capsular, somatic, and flagellar components. Since certain *E.*

coli serotypes have been consistently associated with foodborne and waterborne disease,[20, 23] serology becomes a valuable tool in labeling these EEC types. Final identification of the isolate as an EEC type requires the confirmation of its pathogenicity by means of appropriate animal testing procedures.

24.3 SPECIAL EQUIPMENT FOR LIQUID AND AGAR ENRICHMENT METHODS FOR COLIFORMS, FECAL COLIFORMS AND *E. COLI*

Petri plates, pipets (1, 5, 10 ml), dilution bottles, balance, blender, air incubator (35 C).

Water bath, plastic gable covered, with mechanical circulation system, capable of maintaining a temperature of 45.5 C ± 0.05 C. A partial immersion thermometer approximately 45 to 55 cm in length, with smallest subdivision 0.1 C graduated in 1.0 C to 55 C and standardized against a National Bureau of Standards certified thermometer or equivalent.

An inoculating needle and a 3mm inside diameter inoculating loop.

24.31 Special Reagents and Media (Chapter 2)

Lauryl sulfate tryptose broth (LST)
Brilliant green bile broth (BGB)
EC broth
Levine's eosin methylene blue agar (EMB)
Tryptone (trypticase) broth
Buffered glucose broth (MR-VP Medium)
Koser's citrate medium
Violet red bile agar (VRBA)
Butterfield's phosphate buffer dilution water
Kovac's indole reagent
Methyl red indicator
Voges-Proskauer reagents
Gram stain reagents

24.32 Recommended Controls

Control cultures of *E. coli* (++−−) and *Enterobacter aerogenes* (−−++) should be maintained for the testing of the IMViC media and reagents and the Gram stain reagents. The productivity of all media employed under the conditions of use, including the elevated temperature bath, should be continually monitored. The control cultures should and should not produce gas from EC medium at 45.5 C, according to their classification.

24.33 Precautions

24.331 Dilutions

It is advisable to prepare no more sample dilutions than can be cultured within a 15 minute interval.

24.332 Water bath

To insure uniform incubation temperature throughout the culture tube the water bath fluid should be level with the inverted fermentation tubes in the various media.

24.333 Broth cultures

BGB broth absorbs air during cold storage and should be allowed to warm up to room temperature prior to use. This allows entrapped air to expand and form a visible bubble which can be noted before use.

24.34 General Limitations of the Liquid Enrichment Tests MPN

In applying the *Most Probable Number* technic to the multiple tube fermentation test, actually one is not enumerating the coliforms. Instead, one finds the index of the number of coliform bacteria, which, more probably than any other number, expresses the results which were obtained in the laboratory. Further details in the use of this procedure are found in Standard Methods for the Examination of Water and Wastewater.[43]

24.341 Minimal number of fecal coliforms

The only easily recognizable organism of fecal habitat in the fecal coliform group is *E. coli.* Other non *E. coli* fecal types may be present, but these are not identified easily by the IMViC tests, nor referrable to an original fecal habitat. Without true speciation of all isolates, the minimal number of probable fecal coliforms in a foodstuff corresponds to the number of *E. coli* present. The upper limit is unknown.

24.342 General limitations of EMB agar

Physical and subjective limitations are encountered in the use of EMB agar. The ability to spot one colony among many, and to recognize an *E. coli* colony among many coliform colonies, is critical to the success of the entire analytical procedure.

24.343 Specific limits of the coliform, fecal coliform and *E. coli* tests

In the coliform and fecal coliform group tests, only the lactose fermentations of total bacterial populations are observed and quantitated by means of the MPN technic. Individual organisms are not isolated in pure culture, transferred to lactose broth, or Gram stained. Thus, conformity with the definitions for the coliform and fecal coliform group is not obtained. Uncertain identity is exchanged for rapid results.

With the *E. coli* test all requirements of the original definitions are met. The organism is isolated in pure state, Gram stained, tested for lactose fermentation, and conformance to the IMViC tests.

24.4 PROCEDURE—LIQUID ENRICHMENT PROCEDURES

24.41 Preparation of Sample: Regular or Frozen or Prepared Foods

a. If the sample is frozen and must be thawed, hold in the refrigerator at 2 to 5 C for 18 hours prior to analysis.
b. Weigh 2.5 grams of regular food or thawed food sample aseptically into a sterile blender jar.
c. Add 225 ml of phosphate buffer diluent to the blender jar and blend for 2 minutes.
d. Prepare decimal dilutions in the range 1 : 10, 1 : 100, and 1 : 1000, and higher if necessary, by adding 10 ml of the previous dilution to 90 ml of the sterile diluent. Shake all dilutions 25 times in an arc of one foot for 7 seconds. Do not exceed 15 minutes from the time of blending the sample to the time of preparation of all of the dilutions.

24.42 Presumptive Test for Coliform Group MPN (Note: This Method is in conformance with the AOAC Method.[32])

a. Inoculate 3 replicate tubes of lauryl sulfate tryptose broth (LST) per dilution with one ml of the previously prepared 1 : 10, 1 : 100 and 1 : 1000 dilutions.*
b. Incubate tubes for 24 and 48 ± 2 hours at 35 ± 0.5 C.
c. Observe all tubes for gas production either in the inverted vial or by effervescence produced when the tube is gently shaken. Read tubes for gas production at the end of 24 hours. Reincubate negative tubes for an additional 24 hours.
d. Record all LST tubes showing gas within 48 ± 2 hours and refer to MPN tables for the 3 tube dilutions (Chapter 6) and report results as the presumptive MPN of coliform bacteria per g (or ml of liquid product).
e. In the examination of shellfish, except as indicated below, all presumptive test fermentation tubes showing any amount of gas at the end of 24 or 48 hours of incubation shall be subjected to the confirmed test (24.43).[36] When samples from sources are examined frequently on a routine basis, and when the frequency of false positive presumptive tests is known to be low, it may not be necessary to confirm all positive presumptive tubes, especially the 24 hour tubes. When 3 or more replicate portions of a series of 3 or more decimal dilutions of a sample are planted, an acceptable alternate procedure is as follows:
    1. Select the tubes of the highest dilution (smallest volume) in which all the tubes show gas production in 24 hours.

---

*Recommended Procedures for the Examination of Sea Water and Shellfish[36] specifies the use of 5 tubes of LST broth for the presumptive test for coliforms (24.42). It is further stated that in conducting bacteriological surveys of shellfish growing areas where stations are to be sampled on repeated occasions, the use of 3 tubes per dilution may be justified in order to save time and material. Decision concerning the choice of 3 or 5 tubes would depend to a great extent on the degree of accuracy desired. In no case should less than 3 tubes per dilution be used.
Serial dilutions of the sample should represent 1.0, 0.1, 0.01, and 0.001 g. In routine practice, either 90 or 99 ml dilution blanks are recommended.

2. Submit all these tubes to the confirmed test as well as every one of the gas positive tubes in all the higher dilutions.
3. If there are no dilutions in which less than all tubes show gas production, all gas positive tubes of the highest dilution and of the next to the highest shall be submitted to the confirmed test.
4. Submit to the confirmed test all tubes of all dilutions in which gas is produced only at the end of 48 hours incubation.
5. If fewer than 3 portions of any dilution or volume or if fewer than 3 decimal dilutions of the original sample are planted, all tubes producing gas during 24 or 48 hours incubation shall be submitted to the confirmed test.
6. All tubes producing gas that have not been submitted to the confirmed test shall be recorded as containing organisms of the coliform group, even though all the confirmed tests may yield negative results.

24.43 Confirmed Test for Coliform Group (24.42, See Note)

a. Subculture all positive LST tubes showing gas within 48 ± 2 hours (24.42) into BGB broth by means of the 3 mm loop.
b. Incubate all BGB tubes at 35 ± 0.5 C for 48 ± 2 hours.
c. Record all BGB tubes showing gas, and refer to the MPN tables for 3 or 5 tube dilutions, whichever is applicable (Chapter 6), and report results as confirmed MPN of coliform bacteria per g (or per ml, if liquid product).

24.44 Test for Fecal Coliforms (Note: This method is in conformance with Recommended Procedures for the Examination of Sea Water and Shellfish[36]).

a. Subculture all positive LST tubes showing gas within 48 ± 2 hours (24.42) into EC broth by means of the 3 mm loop.
b. Incubate the inoculated EC fermentation tubes at 44.5 C ± 0.2 C in a circulating covered water bath for 24 ± 2 hours. All EC tubes must be placed in the water bath within 30 minutes after inoculation. Submerge the broth tubes in the bath so that the water level is above the highest level of the medium. The thermometer used in the water bath shall be subdivided to at least 0.2 C, preferably 0.1 C, and shall be checked for accuracy at the test temperature with a Bureau of Standards thermometer or one of equivalent accuracy. It is desirable to equip the water bath with a recording thermometer to record automatically the temperature variations throughout the entire period of incubation. In the absence of a recording thermometer, positive and negative controls using a known 44.5 C gas positive *Escherichia coli* culture and a known 44.5 C gas negative *Enterobacter aerogenes* or other coliform biotype culture should be included in each water bath at time of use.
c. Record the presence or absence of gas formation in the EC tubes after 24 ± 2 hours incubation at 44.5 C ± 0.2 C. Gas production in

the fermentation tube within 24 ± 2 hours is considered a positive confirmed test for fecal coliform organisms. Tubes which fail to produce gas within 24 ± 2 hours are considered negative for fecal coliform organisms.

d. Refer to MPN tables for the 3 or 5 tube dilutions, whichever is applicable, (Chapter 6), and report results as MPN of fecal coliform bacteria per g (or per ml if liquid product).

24.45 Test for *E. coli* (Note: This method is in conformance with the AOAC Method[32])

a. Subculture all positive LST tubes showing gas within 48 ± 2 hours (24.42) into EC broth by means of the 3 mm loop.

b. Incubate all EC tubes in a circulating water bath for 48 ± 2 hours at 45.5 ± 0.05 C.

c. Subculture all EC tubes showing gas within 48 ± 2 hours by streaking on Levine's eosin methylene blue (EMB) plates and incubate 24 ± 2 hours at 35 C.

d. Examine plates for typical nucleated (dark centers) colonies with or without sheen.

e. If typical colonies are present, pick 2 from each EMB plate by touching needle to center of colony and transfer each to PCA slants (4.31). If typical colonies are not present, pick 2 or more colonies considered likely to be *E. coli* from every plate and transfer each to PCA slants.

f. Incubate slants at 35 C for 18 to 24 hours.

g. Transfer growth from plate count agar slants into following broths for identification by biochemical tests:

1. Tryptophane broth (Chapter 2). Incubate 24 ± 2 hours at 35 C and test for indole by adding 0.2 to 0.3 ml Kovac's reagent, to 24 hour culture. Test is positive if upper layer turns red.
2. MR-VP medium, (Chapter 2). Incubate 48 ± 2 hours at 35 C. Aseptically transfer 0.7 ml culture to porcelain spot plate to test for acetylmethylcarbinol. Add 0.1 ml 5% alcoholic α-naphthol solution, 0.1 ml of 40% KOH solution, and a few crystals of creatine. Let stand 2 hours. Test is positive if eosin pink develops.

   Incubate the remainder of MR-VP medium for additional 48 hours and test for the methyl red reaction by adding 5 drops of methyl red solution to culture. Test is positive if culture turns red, negative, if yellow. (Prepare methyl red solution by dissolving 0.1 g methyl red in 300 ml 90% alcohol and make up to 500 ml with H_2O.)
3. Koser citrate broth, (Chapter 2). Incubate 96 hours at 35 C and record growth as + or −.
4. Lauryl sulfate tryptose broth (Chapter 2) Incubate 48 ± 2 hours at 35 C. Examine tubes for gas formation.
5. Gram stain (Chapter 2).—Perform the Gram stain on a smear prepared from 18 hour agar slant. Coliform organisms will stain red (negative); gram positive organisms will stain blue black.

6. Classification. Classify biochemical types as follows:

Indole	*MR*	*VP*	*Citrate*	*Type*
+	+	−	−	Typical *E. coli*
−	+	−	−	Atypical *E. coli*
+	+	−	+	Typical Intermediate
−	+	−	+	Atypical Intermediate
−	−	+	+	Typical *E. aerogenes*
+	−	+	+	Atypical *E. aerogenes*

Other groupings may appear; in such cases cultures are usually mixed. Restreak to determine their purity.

Compute MPN of *E. coli* per g (or per ml if liquid product) considering gram negative, nonspore forming rods producing gas in lactose and producing ++−− or −+−− IMViC patterns as *E. coli*.

24.5 SIGNIFICANCE OF FINDINGS

24.51 The Coliform Group

The recovery of the coliform group from foods, with its unidentified individual members has less interpretive impact than the single indicator organism, *E. coli*, or the fecal coliform group. This is because the coliform group may contain such nonenteric members as *Serratia, Erwinia* and *Aeromonas.*

The specificity of the group as an indicator is diminished by the anonymity of its individual members. This may be considered a weakness in the coliform method. However, the presence of coliforms in processed foods is a useful indicator of postsanitization and postprocessing (pasteurization) contamination. Practices which permit their presence in such instances are not consistent with sanitation standards required for food processing operations.

24.52 The Fecal Coliform Group

Because the fecal specificity of the coliform group is low, bacteriologists have turned to the use of the narrower fecal coliform group. The fecal coliform group gives greater fecal specificity because of the high *E. coli* incidence within the group. The non *E. coli* members have doubtful fecal identity, and their presence within the fecal coliform group tends to dilute the group's specificity. Their presence may represent a failure in methods.

The test incubation period for fecal coliforms in foods may be limited to 24 hours as in water and shellfish bacteriology, or it may be extended to 48 hours as is called for in the classical identification of the coliform group. Usually it is used as a 48 hour test in order to recover a sizeable proportion of the *E. coli* isolates which develop after 24 hours.[16] It must be recognized that this extended 48 hour fecal coliform test for foods generally produces higher fecal coliform counts than do similar tests of only 24 hours.

24.53 *E. coli*

The recovery of *E. coli* from foods implies that:

a) Other organisms of fecal origin may be present;

b) Pathogenic forms could be present. It does not imply that they are present, nor in what form, nor in what numbers.

The failure to recover *E. coli* from foods does not assure wholly the absence of *E. coli*, fecal matter, or pathogens, because *E. coli* is not a perfect indicator organism. There is none. However, it is the best fecal indicator available at the present time.

24.6 COLIFORM GROUP SOLID MEDIUM METHOD—(Violet Red Bile Agar)

24.61 Introduction

Most of the media for the isolation and enumeration of coliforms were developed for testing water, sewage, and dairy products. Only a few of these methods have found practical application in the examination of foods. The foremost of these methods has been a pour plate method using violet red bile agar (VRBA).[42]

24.62 Procedure

Weigh a 25 gram sample of the food to be tested. Add the sample to 225 ml of sterile phosphate buffer and blend at high speed for 2 minutes.

The resulting 1 : 10 diluted sample may be serially diluted further in sterile phosphate buffer as necessary if the expected coliform count of the food is high.

Add 1 ml amounts of the sample dilutions to sterile Petri dishes. Pour the plate with 10 to 15 ml of VRBA, tempered to 45 C. To enhance sensitivity of the tests without unduly increasing the number of dishes used, plate portions up to 4 ml per dish and use 15 to 20 ml of medium per dish. Swirl plates to mix thoroughly and allow to cool. After solidification of the medium overlay plates with 3 to 4 ml of additional VRBA.

Incubate inverted plates at 35 C for 18 to 24 hours.*

Examine plates on a colony counter. Count all purplish red colonies which are surrounded by a reddish zone of precipitated bile, 0.5 mm in diameter or larger. Counts should be made of plates containing 30 to 150 colonies.

For coliform confirmation pick 10 colonies from VRBA and transfer to separate tubes of BGB containing fermentation tubes. The 10 colonies picked should be representative of all the colony types observed.

Incubate the BGB tubes at 35 C; examine for evidence of gas production after 24 hours and again after 48 hours.

Only those tubes showing gas production are confirmed as coliform organisms. Any tubes showing aerobic growth in a pellicle should be Gram

---

*Lawton (J. Milk & Food Technol. 18:288. 1955) found no significant difference in counts obtained with 32 or 35 C incubation for pasteurized milk.

stained, in order that gas forming, gram positive rods are not counted as coliforms. The number of coliforms per gram of sample is determined by multiplying the percentage of tubes confirmed as positive (gas production and gram negative rods) by the original VRBA count, multiplied by the dilution factor used.

24.63 Recommended Controls

Control food samples, with and without known numbers of coliforms, and also with known numbers of pure cultures of coliforms, shall be used in testing new lots of media, to ensure consistency of results.

24.64 Precautions and Limitations of Method

Coliform counts can differ significantly depending upon the food tested and upon the conditions of the medium used. Various conditions in food processing may cause cell injury. A resuscitation period in a nonselective broth is necessary for accurate enumeration of coliforms.[34] The temperature of tempered medium, 45 C, used in a pour plate technic imposes sublethal stress on many coliform cells. The repair and subsequent formation of colonies is inhibited in selective media such as VRBA.[35] (See Chapter 7).

24.65 Interpretation of Data

If there is a "processing step" involved in the preparation of the food which may destroy bacteria, the finding of coliforms may indicate recontamination of the food. Coliform counts are indicators of unsanitary production practices. The finding of coliforms in foods does not mean necessarily that the food was associated with fecal material. Coliforms occur naturally in soil, and most will grow in and on processing equipment in the presence of process material and moisture.

24.7 ENTEROPATHOGENIC *ESCHERICHIA COLI* (EEC)

24.71 Introduction

Enteric illness associated with the consumption of food or water containing *E. coli* has been suspected for about 70 years.[8] Several difficulties are encountered in the enrichment, recovery, and recognition of EEC. The standard procedures employed in sanitary microbiology may be unsuitable for some pathotypes either because of the elevated temperature or of their inability to ferment lactose with the production of gas within 48 hours. Recent findings have made possible the recognition of pathogenic biotypes not previously known. Following the development of serological methods,[21] many of the serotypes implicated in infantile diarrhea were found in food and waterborne outbreaks.[1, 6, 20, 23, 27, 38, 45, 46]

A recent development is the extended definition of EEC. Classical serogroups were isolated from children under three years of age. Subsequently, the group has been enlarged to include dysentery like strains[18] and cultures recovered from all age groups.[37] The procedure described in this section is designed to recover most serogroups described in the literature. These include:

Group A	Group B	Group C	Alkalescens-Dispar (A-D)
026 : B6	086 : B7	018 : B21	01
055 : B5	0119 : B14	020 : B7	02
0111 : B4	0124 : B17	020 : 84(B)	03
0127 : B8	0125 : B15	028 : B18	04
	0126 : B16	044 : K74	
	0128 : B12	0112 : B11	

24.72 Special Equipment and Supplies (Chapter 2)

a. Water baths
   1. Maintained at 44 ± 0.1 C.
   2. Maintained at 41.5 ± 0.1 C.
   3. Maintained at 49 ± 1 C.
b. Incubators
   1. Refrigeration at 4 ± 2 C.
   2. Maintained at 22 C. and at 35 C.
c. Blendor, Waring or equivalent. 2 speed standard model, with low-speed operation at 8000 rpm, equipped with 1 liter glass or metal jars. One sterile jar per sample is required.
d. Serological racks—Stainless steel, or equivalent, accommodating 12 × 75 mm tubes.
e. Tubes
   1. 12 × 75 mm
   2. 13 × 100 mm
   3. 16 × 150 mm, screw cap
f. Petri dishes
   1. 15 × 100 mm sterile plastic or glass dishes
   2. Clean, unscratched 15 × 150 or 20 × 150 mm glass dishes (for serological analysis)
g. Pipets
   1. Pasteur
   2. 0.1 or 0.2 ml serological
   3. Pipet filler
h. Micro concavity slide and cover slips (for hanging drop preparation)

24.73 Reagents (Chapter 2)

a. Kovacs' reagent
b. Methyl red solution
c. Voges-Proskauer reagents
d. Gram stain reagents
e. Nitrate reduction reagents
f. Sodium bicarbonate, 10% aqueous sterile
g. Saline solution, 0.5% aqueous sterile

h. Saline solution, 0.85% aqueous sterile
i. Formalinized saline solution
j. Cytochrome oxidase reagent
k. pH test paper, range 5.0–8.0
l. ONPG disks
m. *Escherichia coli* antisera
    1. polyvalent OB
    2. monovalent OB
    3. monovalent O
    4. polyvalent H
    5. monovalent H
    6. Alkalescens-Dispar polyvalent
    7. Alkalescens-Dispar monovalent
n. *Escherichia coli* O and OB antigens[40]
o. H_2SO_4, 1% aqueous solution
p. $BaCl_2$, 1% aqueous solution
q. Mineral oil, sterile
r. Paraffin
s. 36% formaldehyde
t. 5N acetic acid
u. HCl, concentrated

24.74 Media (Chapter 2)

a. Lauryl sulfate tryptose broth (LST)
b. MacConkey broth
c. Nutrient broth
d. Enteric enrichment (EE) broth
e. Levine's eosin methylene blue agar (EMB)
f. MacConkey agar
g. Triple sugar iron agar (TSI)
h. Urea broth
i. Brom cresol purple carbohydrate broth supplemented with the following carbohydrates:
    1. glucose 1%
    2. adonitol 0.5%
    3. cellobiose 0.5%
j. Tryptophane broth
k. MR-VP broth
l. Potassium cyanide broth (KCN)
m. Blood agar base (BAB)
n. Veal infusion agar
o. Veal infusion broth
p. Acetate agar
q. Lysine decarboxylase broth
r. Mucate broth
s. Mucate control broth
t. Indole nitrite broth

24.75 Standards

24.751 McFarland nephelometer[19]

Density of preparations must be controlled to increase reliability of serological data. Commercially available or prepared standards can be used.

24.76 Recommended Controls

To check the toxicity of EE broth select a culture of *E. coli* capable of visible growth within 18 hours in EE broth at 41.5 C from an initial inoculum of 100 cells/ml. Check each lot of EE broth.

Procure standard strains of human EEC or commercially available O and B antigens for evaluation of sera.

24.77 Precautions

The serotyping of EEC requires the complete identification of the O, B, and H antigens.

Rough strains are a critical problem. To avoid this, isolates should be subcultured on carbohydrate free media such as veal infusion agar. Analyses should proceed as soon as possible after isolation.

Slide agglutination reactions are more distinct on glass than on plastic surfaces.

False negative reactions may result from unsatisfactory quality of commercial sera. More rarely, certain pathotypes may not be included in the serogroups listed in Section 24.71.

To identify the somatic and capsular antigens of EEC, examination with both O and OB sera is required. Unencapsulated, partially capsulated, and fully capsulated strains will give positive reactions in homologous OB serum. Encapsulated culture may agglutinate only partially in O serum. Minimal criteria for positive presumptive serological identification include: agglutination of unheated culture in homologous OB sera, none or limited agglutination of unheated culture in homologous O serum, and agglutination of heated culture in homologous O serum. Slide agglutination reactions must be confirmed by quantitative tube methods.

24.78 Procedure (Figure 1)

The use of an enrichment procedure requires relatively nonspecific media, less inhibitory incubation temperatures and the selection of nonlactose fermenting cultures. Consequently, a wide variety of gram negative bacteria will be recovered from which *E. coli* must be separated. Enrichment media should be checked with appropriate positive and negative controls prior to use.

The method outlined permits the qualitative determination of the presence of EEC in a food sample. If quantitation is essential, the dilution end point or the MPN approach should be used.

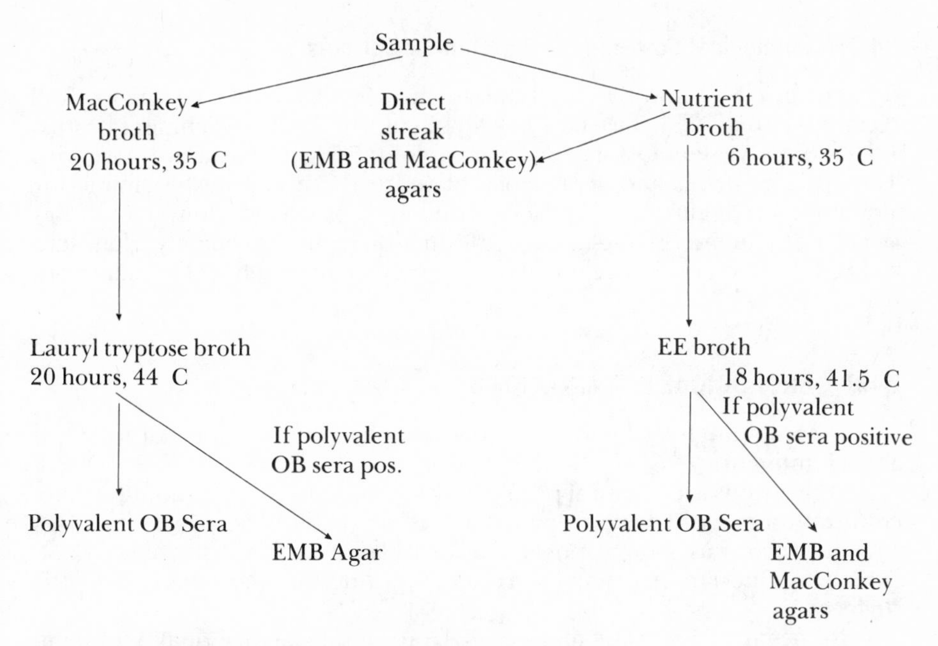

FIGURE 1. Enrichment for Enteropathogenic E. coli

24.781 Sample preparation

Samples should be analyzed as promptly as possible after arrival. Refrigeration of perishable material should be brief since most pathogenic biotypes lose viability at 6 C.

Using aseptic conditions, open sample container and weigh two 25 g portions into 225 ml MacConkey broth and 225 ml nutrient broth in beakers. Homogenize 30 seconds. Return contents to Erlenmeyer flasks. These constitute 1:10 dilutions.

24.782 Direct streak

Streak nutrient broth homogenate to EMB and MacConkey agars. Incubate 24 ± 2 hours at 35 C.

24.783 Enrichment

Incubate MacConkey broth 20 ± 2 hours at 35 C. Transfer with sterile 3 mm loop to 30 ml lauryl sulfate tryptose broth (LST). Incubate 20 ± 2 hours at 44 C.

Incubate nutrient broth 6 ± 1 hours at 35 C. Transfer with loop to 30 ml EE broth. Incubate 18 ± 2 hours at 41.5 C.

24.79 Serological Screening of Enrichment Cultures

Neutralize LST and EE enrichments with 10% $NaHCO_3$. To wax marked rectangles on glass surface add 1 drop of culture (from enrichment broths) 1 drop of polyvalent OB sera, and 1 drop of 0.5% saline in accordance with Table 1. Mix drops and gently rock plate for three minutes. Examine for agglutination against a dark background with overhead illumination. Reject all enrichments giving (a) no agglutination in any of the polyvalent sera within 3 minutes, or (b) agglutination in a polyvalent serum and in saline control.

24.8 SEROLOGICAL EXAMINATIONS

The serotyping of EEC requires the complete identification of the O, B and H antigens.

Rough strains are a critical problem. To avoid this, isolates should be subcultured on carbohydrate free media such as veal infusion agar. Analyses should proceed as soon as possible after isolation.

Slide agglutination reactions are more distinct on glass than on plastic surfaces.

False negative reactions may result from unsatisfactory quality of commercial sera. More rarely, certain pathotypes may not be included in the serotypes listed in Section 24.71.

To identify the somatic and capsular antigens of EEC, examination with both O and OB sera is required. Unencapsulated, partially capsulated, and fully capsulated strains will give positive reactions in homologous OB serum. Encapsulated culture may agglutinate only partially in O serum. Minimal criteria for positive presumptive serological identification include: agglutination of unheated culture in homologous OB sera, none or limited agglutination of unheated culture in homologous O serum, and agglutina-

Table 1: Serological Examination of Enrichment Cultures

Rectangle	Medium[1]	Poly A	Poly B	Poly C	Poly A-D	Saline
1	LST	X				
2	LST		X			
3	LST			X		
4	LST				X	
5	LST					X
6	EE	X				
7	EE		X			
8	EE			X		
9	EE				X	
10	EE					X

[1]Use 0.05 ml of culture and of serum or saline.

Analysis for Somatic (0) and Capsular (B) Antigens

Slide Test

Saline → Poly OB → Mono OB → Mono O →

Susp

↑

BAB

Tube Tests

4 Tube Rapid Test	Quantitative Test
0 ag[1] × 0 ser[2]	0 ag × 0 ser
B ag × 0 ser	B ag × 0 ser
B ag × OB ser	B ag × OB ser
0 ag × Saline	0 ag × OB ser

Analysis for Flagellar (H) Antigen

Indole Nitrite broth → Veal infusion broth → Poly H → Mono H

[1]Antigen
[2]Serum

FIGURE 2. Serological Characterization of Pure Culture Isolates

tion of heated culture in homologous O serum. Slide agglutination reactions must be confirmed by quantitative tube methods.

24.81 Procedure

Neutralize LST and EE enrichments with 10% $NaHCO_3$. To wax marked rectangles on glass surface add 1 drop of culture (from enrichment broths), 1 drop polyvalent OB sera, and of 0.5% saline in accordance with Table 1. Mix drops and gently rock plate for three minutes. Examine for agglutination against a dark background with overhead illumination. Reject all enrichments giving (a) no agglutination in any of the polyvalent sera within 3 min, or (b) agglutination in a polyvalent serum and in saline control.

24.82 Isolation on Selective Agars

Streak positive LST enrichment broths to EMB agar for recovery of lactose fermenting organisms. Streak positive EE enrichment broths to MacConkey and EMB agars for recovery of negative, slow, and rapid fermenting organisms. Incubate 24 ± 2 hours at 35 C. Typical colonies on EMB correspond to the description previously given. The colonies of nonlactose-fermenting organisms on MacConkey agar are colorless or slightly pink. Clinical laboratory procedures specify examination of 10 representative colonies from each plate. Each laboratory is advised to use the number of cultures according to its resources.

24.83 Biochemical Tests: Physiological Characterization

Because of the large number of bacterial species capable of growth in the recommended enrichment media, the standard method for biochemical recognition of *E. coli* is inadequate for EEC. Cultures possessing the following characteristics are *E. coli*:

Fermentative (TSI)	+ (acid)
H_2S (TSI)	−
Voges-Proskauer (22 C)	−
Indole (2 days)	+
ONPG ase (TSI)	+
Urease	−
KCN	−
Cellobiose	−
Adonitol	−
Cytochrome oxidase	−
Gram negative, short rods	+

A small percentage of *E. coli* isolates may give aberrant reactions in indole, H_2S, urease, cellobiose, and adonitol. Cultures deviating in two or more reactions should be examined by the standard criteria.[10]

Alkalescens-Dispar biotypes are characterized by anaerogenesis in glucose broth, and lack of motility in indole nitrite broth. These are differentiated from *Shigella sonnei* by a positive indole reaction, and from other *Shigella* species by a positive reaction in one of the following tests: acetate, mucate, and lysine decarboxylase.

24.84 Serological Examination of Pure Cultures

24.841 Slide agglutination (presumptive identification of O and B antigens).

Suspend growth from BAB slant in 5 ml of 0.5% saline to a density corresponding to McFarland standard 4. Discard rough cultures (those which fail to show homogeneous stable suspensions). Examine with polyvalent OB sera in accordance with Table 2. Reject cultures agglutinating in saline or in all four sera. (A culture may react in two sera because of the presence of a common factor). If negative in all sera, heat suspension 15 minutes at 100 C to destroy interfering mucoid material. Retest. If still negative, reject. If positive, examine in corresponding monovalent OB sera. If nega-

Table 2: Preliminary Serological Characterization of Isolates

Rectangle	Culture[1]	Poly A	Poly B	Poly C	Poly A-D[2]	Saline
1	X	X				
2	X		X			
3	X			X		
4	X				X	
5	X					X

[1]Use 0.05 ml of suspension and of serum or saline

[2]Poly A-D sera is used with Poly A, B, & C sera when the culture is nonmotile and anaerogenic.

tive, reject. If positive, examine in corresponding monovalent O sera. If positive, the culture lacks sufficient capsular factor. Either reject or restreak on BAB plate to select an encapsulated variant. If negative in O sera, heat one-half of the suspension for 1 hour at 100 C to destroy capsule. If negative, reject. If positive, perform tube agglutination tests.

24.842 Tube agglutination test (Confirmed identification of O and B antigens)

Ascertain from the manufacturer the potency of the sera used. Employ O sera with O titer of 1:180 to 1:360, and OB sera with B titer of 1:40 to 1:80. Dilute both O serum 1:40 and OB serum 1:20 in 0.5% saline. Dilute the O and B antigens previously prepared in 24.841 with 0.5% saline to a density corresponding to McFarland standard 3. Add 0.5 ml of the heated antigen (0), unheated antigen (B), 1:40 O serum, 1:20 OB serum, and 0.5% saline to clean 12 × 75 mm tubes as in Table 3. Gently agitate and cover with aluminum foil. Incubate tubes 1, 2 and 3 for 16 hours at 49 C. Chill for 1 hour at 4 C. Incubate tube 4, 2 hours at 35 C and then 16 hours at 4 C. Examine tubes carefully for agglutination with overhead illumination against a dark background. A positive culture will give agglutination in tubes 2 and 4, and possibly a weak reaction in tube 3. Reject all cultures showing complete agglutination in tube 1 (rough strain).

24.843 Quantitative tube agglutination (Completed identification)

The quantitative determination of O and B antigens require resources not found in the average analytical laboratory. See reference 28 for complete analysis.

24.844 Identification of flagellar (H) antigen

To determine motility of freshly isolated cultures, inoculate into semisolid medium such as indole nitrite broth. Inoculate medium within the central cylinder. Incubate at 35 C for 24 to 48 hours. Withdraw a loopful of culture from outside the central cylinder, and examine for motility in hanging drop preparation. If a majority of the cells are nonmotile, subculture repeatedly until 90% of the hanging drop cells are motile. Inoculate veal infusion broth. Incubate 18 ± 2 hours at 35 C. Adjust turbidity with

Table 3: Semiquantitative Determination of O and B Antigens

Tube #	O Serum	OB Serum	O Antigen	B Antigen	0.5% Saline
	0.5 ml (1:40)	0.5 ml (1:20)	0.5 ml	0.5 ml	0.5 ml
1			X		X
2	X		X		
3	X			X	
4		X		X	

Table 4: Flagellar Antigens Found in Human EEC

Serogroup	H Factors[1]
018:B21	NM[2], 6, 7, 10, 21
020:B7	NM
020:K84(B)	NM, 19, 26
026:B6	NM, 9, 11, 32, 33
028:B18	NM
044:K74(L)	12, 18, 34
055:B5	NM, 1, 2, 4, 6, 7, 10, 11, 12, 19, 27, 32, 33, 34
086:B7	NM, 7, 8, 9, 10, 11, 21, 27, 34, 47
0111:B4	NM, 2, 4, 6, 7, 12, 16, 21, 25
0112:B11	NM
0119:B14	NM, 1, 2, 4, 6, 8, 9, 18, 39
0124:B17	NM, 12, 19, 30, 32
0125:B15	NM, 6, 11, 12, 15, 21, 25, 30
0126:B16	NM, 2, 7, 10, 11, 12, 19, 20, 21, 27, 29, 30, 33
0127:B8	NM, 1, 5, 6, 9, 11, 19, 27, 33, 40
0128:B12	NM, 1, 2, 6, 7, 8, 10, 11, 12, 16, 35
Alkalescens-Dispar 01	NM
Alkalescens-Dispar 02	NM
Alkalescens-Dispar 03	NM
Alkalescens-Dispar 04	NM

[1]See reference 15
[2]Non-motile

formalinized saline to McFarland standard 2. This is the H antigen suspension. Hold 1 hour at ambient temperature prior to use. Place 11 clean unscratched 12 × 75 mm tubes in a serological rack. Add 0.5 ml of each of the 10 polyvalent H antisera to tubes 1 to 10. The final dilution of each monovalent serum after the addition of antigen, must be 1:1000, whether tested in pools or individually. To the eleventh tube add 0.5 ml of 0.85% saline. Add 0.5 ml of H antigen to each tube. Mix gently, cover with aluminum foil, and incubate at 49 C. Examine after 15, 30 and 60 minutes. A positive reaction in a polyvalent serum requires examination in the monovalent H sera. Align tubes as in the previous analysis, substituting monovalent H sera. A positive response in a single tube indicates the identity of the flagellar antigen. The flagellar antigens of recognized EEC serogroups are shown in Table 4. On rare occasions multiple responses in the monovalent sera may occur. These are due to mixed cultures or to serological interrelationships of flagellar antigens. See reference 10 for further clarification.

24.9 PATHOGENESIS: CURRENT STATUS

Cultures which have been established physiologically, biochemically, and serologically as potential EEC, must be evaluated for pathogenic capacity. The tests should be selected in accordance with the resources of the individual laboratory. There are two recognized mechanisms of pathogenesis, toxigenesis, and invasiveness. Toxigenic cultures attach to the epithelium of the small intestine, proliferate, and elaborate toxins which result in the loss of fluids and electrolytes from the body. Invasive cultures attack the epithelium of the large intestine and produce lesions, ulcerations, and loss of fluids and electrolytes. Heat labile and heat stable toxins have been differentiated among the toxigenic cultures. A culture may produce either the heat stable, heat labile, or a mixture of the heat stable and heat labile toxins. Classical serogroups associated with infantile diarrhea, such as 0111:B4 or 026:B6, are examples of the toxigenic type. Serogroups 0124:B17 or 028:B18 are examples of the invasive type. Some members of recognized EEC serogroups may be avirulent. New serogroups pathogenic for man have been reported.

Inasmuch as human volunteer feeding studies may be impractical, the following categories of model test systems, in decreasing order of human equivalence, have been described.

24.91 Primate Feeding Studies

The syndromes are analagous to those found in human cases. Toxigenic organisms colonize the small intestine of the primate, elaborate toxins and produce a condition resembling cholera. Invasive cultures penetrate the epithelial tissue of the large intestine, cause ulceration, and produce a condition resembling dysentery.[18]

24.92 Localized Tissue Organ Responses

24.921 Sereny test

A drop of the test culture is introduced into the keratoconjunctiva of the guinea pig eye. A positive response is an inflammation of the eye within 4 days. Only invasive cultures produce this response.[39]

24.922 Vascular permeability

A sterile filtrate is inoculated intradermally in rabbits. Eighteen hours later a solution of Evans' blue dye is inoculated intravenously. A positive response is a zone of blueing at the site of inoculation of the test filtrate. Only toxigenic strains producing a heat labile toxin are positive in this test.[13]

24.923 Ligated ileal loop

The test culture is introduced into ligated loops of the small intestine of rabbits, guinea pigs, rats, and chickens.[4] A positive response is dilation of the loop. Both toxigenic and invasive strains give positive results. The ileal loop response of toxigenic strains must be observed in duplicate animals, with observations at 6 hours in one animal, and 18 hours in the other.[14] Routine application is hindered by extensive false positive reactions.

24.924 Infant mouse test

The test culture is introduced intragastrally through the skin surface of the infant mouse. A positive response is distension of the abdomen within 6 hours. Toxigenic strains producing heat stable toxin give a positive result.[7]

24.93 Cell Culture Systems

24.931 Hamster adrenal cell culture.

Addition of a preparation containing heat labile toxin to cell culture produces both a morphological and a physiological change in the cells.[9]

24.932 HELA cell culture

Addition of invasive EEC cells results in cell culture destruction.[22]

24.933 Significance of pathogenicity data

Although the number of cultures to be examined for pathogenicity will be limited by the resources of the laboratory, perhaps only 1 out of 10 cultures from a patient ill with EEC diarrhea will possess characteristics capable of being measured in the above systems.

24.94 Interpretation of Data

The following criteria are required for the identification of EEC:

(a) Morphological, physiological, and biochemical identification of the culture as *E. coli*.

(b) The somatic, capsular and flagellar antigens must be identified and must correspond to *E. coli* serotypes involved in a significant number of cases of human enteric illness.

(c) The culture must possess one of the recognized characteristics of pathogenicity, toxigenesis, or invasiveness.

24.10 REFERENCES

1. Anonymous. 1973 (March 2). *Escherichia coli* Food Poisoning. Communicable Disease Report, Public Health Service (Great Britain)
2. BERGEY'S Manual of Determinative Bacteriology. 1974. Eighth Edition. Williams and Wilkins Co., Baltimore, Md.
3. BURKE-GAFFNEY, H. J. O'D. 1932. The classification of the colonaerogenes group of bacteria in relation to their habitat and its application to the sanitary examination of water supplies in the tropics and in temperate climates. A comparative study of 2500 cultures. J. Hyg. **32**:85–131
4. BURROWS, W. and G. M. MUSTEIKIS. 1966. Cholera infection and toxin in the rabbit ileal loop. J. Inf. Dis. **116**:183–190
5. CLEMESHA, W. W. 1912. J. Hyg. **12**:463. Quoted in Burke-Gaffney 1932
6. COSTIN, I. D., V. VOICULESCU, and V. GORCEA. 1964. An outbreak of food poisoning in adults associated with serotype 086:B7:H34. J. Path & Microbiol. **27**:68–78
7. DEAN, A. G., Y. CHING, R. G. WILLIAMS, and L. B. HARDEN. 1972. Test for *Escherichia coli* enterotoxin using infant mice. Application in a study of diarrhea in children in Honolulu. J. Inf. Dis. **125**:407–411
8. DELEPINE, S. 1903. Food poisoning and epidemic diarrhea. J. Amer. Med. Assoc. **40**:657
9. DONTA, S. T., H. W. MOON, and S. C. WHIPP. 1974. Detection of heat-labile *Escherichia coli* enterotoxin with the use of adrenal cells in tissue culture. Science. **183**:334–336
10. EDWARDS, P. R. and W. H. EWING. 1972. Indentification of *Enterobacteriacae*. 3rd Edition. Burgess Pub. Co., Minneapolis, Minn.
11. EIJKMAN, C. 1904. Die garungsprobe bei 46° als hilfsmittel bei der trink wasseruntersuch ung. Zentrl. Bakteriol. Parasitenk. Abt. **1 37**:347–348
12. ESCHERICH, TH. 1887. Central. Bacteriol. Parasitenk **1**:705
13. EVANS, D. J., JR., D. G. EVANS, and S. L. GORBACH. 1973. Production of vascular per-

meability factor by enterotoxigenic *Escherichia coli* isolated from man. Infect. and Immun. **8**:725–730

14. EVANS, D. G., D. J. EVANS, JR., and N. F. PIERCE. 1973. Differences in the response of rabbit small intestine to heat-labile and heat stable enterotoxins of *Escherichia coli*. Infect. 2nd Immun. **7**:873–880
15. EWING, W. H., B. R. DAVIS, and T. S. MONTAGUE. 1963. Studies on the Occurrence of *Escherichia coli* Serotypes Associated with Diarrheal Disease. USDHEW-PHS. Communicable Disease Center, Atlanta, Georgia
16. FISHBEIN, M. and B. F. SURKIEWICZ. 1964. Comparison of the recovery of *Escherichia coli* from frozen foods and nutmeats by confirmatory incubation in E. C. medium at 44.5° C. and 45.5° C. Appl. Microbiol. **12**:127–131
17. FISHBEIN, M., B. F. SURKIEWICZ, E. F. BROWN, H. M. OXLEY, A. P. PADRON, and R. J. GROOMES. 1967. Coliform behavior in frozen foods. 1. Rapid test for recovery of *Escherichia coli* from frozen foods. Appl. Microbiol. **15**:233–238
18. FORMAL, S. B., H. L. DUPONT, R. HORNICK, M. J. SNYDER, J. LIBONATI, and E. H. LABREC. 1971. Experimental models in the investigation of the virulence of dysentery bacilli and *Escherichia coli*. New York Acad. Sci. **176**:190–196
19. GRADWOHL'S Clinical Laboratory Methods and Diagnosis. 1970. 7th edition. C. V. Mosby Company, St. Louis, Mo.
20. HOBBS, B. C., M. E. M. THOMAS, and J. TAYLOR. 1949. School outbreak of gastroenteritis associated with a pathogenic paracolon bacillus. The Lancet. September **17**:530–532
21. KAUFFMANN, F. 1947. Serology of the coli group. J. Immunol. **57**:71–100
22. LABREC, E. H., H. SCHNEIDER, T. J. MAGNANI, and S. B. FORMAL. 1964. Epithelial cell penetration as an essential step in the pathogenesis of bacillary dysentery. J. Bacteriol. **88**:1503–1518
23. LANYI, B. J. SZITA, B. RINGELHANN, and K. KOVACH. 1959. A waterborne outbreak of enteritis associated with *Escherichia coli* serotype 124:72:32. Acta Microbiol. Acad. Sci. Hung. **6**:77–88
24. LEVINE, M. 1961. Facts and fancies of bacterial indices in standards for water and foods. Food Technol. **15**:29–38
25. MACCONKEY, A. T. 1901. Corrigendum et Addendum. Zentrl. Bakteriol. Parasitenk. (Abt 1) **29**:740
26. MACKENZIE, E. F. W. and F. C. HILTON-SERGEANT. 1938. The coliform bacilli and water supplies (Part II) J. R. Army Med. Cps. **70**:74
27. MARIER, R., J. G. WELLS, R. C. SWANSON, W. CALLAHAN, and I. J. MEHLMAN. 1973. An outbreak of enteropathogenic *Escherichia coli* foodborne disease traced to imported French cheese. The Lancet. **II**:1376–1378
28. MEHLMAN, I. J., A. C. SANDERS, N. T. SIMON, and J. C. OLSON, JR. 1974. Methodology for recovery and identification of enteropathogenic *Escherichia coli*. J.A.O.A.C. **57**:101–110
29. MOSSEL, D. A. A., W. H. J. MENGERINK, and H. H. SCHOLTS. 1962. Use of a modified MacConkey agar medium for the selective growth and enumeration of *Enterobacteriaceae*. J. Bacteriol. **84**:381
30. MOSSEL, D. A. A., M. VISSER, and A. M. R. CORNELISSEN. 1963. The Examination of foods for *Enterobacteriaceae* using a test of the type generally adopted for the detection of salmonellae. J. Appl. Bacteriol. **26**:444–452
31. MOUSSA, R. S., N. KELLER, G. CURIAT, and J. C. DEMAN. 1973. Comparison of five media for the isolation of coliform organisms from dehydrated and deep frozen foods. J. Appl. Bacteriol. **36**:619–629
32. Official Methods of Analysis of the Association of Analytical Chemists. 1975. Twelfth Edition. A.O.A.C. Washington, D.C.
33. PERRY, C. A. and A. A. HAJNA. 1944. Further evaluation of E. C. medium for the isolation of coliform bacteria and *Escherichia coli*. Am. J. Public Health **34**:735–738
34. RAY, B. and M. L. SPECK. 1973a. Enumeration of *Escherichia coli* in frozen samples after recovery from injury. Appl. Microbiol. **25**:499–503
35. RAY, B. and M. L. SPECK. 1973b. Discrepancies in the enumeration of *Escherichia coli*. Appl. Microbiol. **25**:494–498
36. Recommended Procedures for the Examination of Seawater and Shellfish. 1970. 4th Edition. American Public Health Association. New York, N.Y.

37. SACK, R. B., S. L. GORBACH, J. G. BANWELL, B. JACOBS, B. D. CHATTERJEE, and R. C. MITRA. 1971. Enterotoxigenic *Escherichia coli* isolated from patients with severe cholera-like disease. J. Inf. Dis. **123**:378–385
38. SCHROEDER, S. A., J. R. CALDWELL, T. M. VERNON, P. C. WHITE, S. I. GRANGER, and J. V. BENNETT. 1968. A waterborne outbreak of gastroenteritis associated with *Escherichia coli*. The Lancet **I**:737–740
39. SERENY, B. 1957. Experimental keratoconjunctivitis in shigellosa. Acta. Microbiol. Hung. **4**:367–376
40. Serological Identification of *Escherichia coli*. 1975. Bulletin No. 0154. Difco Laboratories, Detroit, Mich.
41. SHARDINGER, F. 1892. Ueber das vorkommen gahrung erregender spaltzpilze in trinkwasser und ihre bedeutung für die hygienische. Wien. Klin. Wochschr. **5**:403.
42. Standard Methods for the Examination of Dairy Products. 1972. 13th Ed. W. J. Hausler, Jr. Ed. American Public Health Association, Washington, D.C.
43. Standard Methods for the Examination of Water and Wastewater. 1971. 13th Ed. American Public Health Association, New York, N.Y.
44. TENNANT, A. D. and J. E. REID. 1961. Coliform bacteria in Seawater and Shellfish. 1. Lactose fermentation at 35.5° C. and 44° C. Canad. J. Microbiol. **7**:725–731
45. UEDA, S., S. SASAKI, and M. KABUTO. 1959. The detection of *Escherichia coli* 0-55 from an outbreak of food poisoning. Jap. J. Bacteriol. **14**:48–49
46. ULEWICZ, K. 1956. Food poisoning caused by *Escherichia coli* 026:B6 Przegl. Eped. **4**: 341–346
47. VAN SCHOTHORST, M., D. A. A. MOSSELL, E. H. KAMPELMACHER, and E. F. DRION. 1966, The estimation of the hygienic quality of feed components using an *Enterobacteriaceae* enrichment test. Zentr. Veterinarmed. **13**:273–285

Chapter 25

SALMONELLA

Paul L. Poelma and John H. Silliker

25.1 GENERAL BASIS OF METHODS

25.11 Introduction

The annual incidence of reported isolations of *Salmonella* from human sources has increased each year from 1967 to 1972, according to the *Salmonella* Surveillance Annual Summary 1972.[8] In 1972, 26,110 isolations of *Salmonella* from humans were reported. In the same year, there were 1,880 cases of foodborne illness due to *Salmonella* reported in 36 confirmed outbreaks.[7] The organism produces an infection after an incubation period of 12 to 74 hours.

Because of the health hazard involved when food products are contaminated with *Salmonella* and Arizona, many microbiological methods have been developed to detect these microorganisms and to isolate and identify them in foods. For an extensive review of these methods refer to "Reference Methods for the Microbiological Examination of Foods" by the Subcommittee on Food Microbiology, Food Protection Committee, National Research Council, National Academy of Sciences,[29] and "*Salmonella* and the Food Industry—Methods for Isolation, Identification and Enumeration," by Litchfield.[27] Several reviews of the entire *Salmonella* problem are available: "An Evaluation of the *Salmonella* Problem," prepared by the National Academy of Sciences,[28] and "World Problem of Salmonellosis" by Van Oye.[55]

25.12 General Description of Methods for the Isolation of *Salmonella* and Arizona in Foods.

25.121 Introduction

The examination of various types of food products for the isolation of *Salmonella* and Arizona often requires the use of different methods from those used in clinical and public health laboratories. Many methods used for examining foods are essentially similar in principle and employ the procedural steps of preenrichment, selective enrichment, differential and selective plating, isolation, and confirmation or identification of the selected isolates. The complexity of the analytical method is due chiefly to the fact that relatively small numbers of *Salmonella* or Arizona are usually present

IS/A Committee Liaison: John H. Silliker

in foods as compared to other competing microflora. Most laboratories conduct qualitative tests, but the methods can be adapted to enumerate these microorganisms by the most probable number (MPN) technic.[21, 30]

The reference procedures described later for the isolation and identification of *Salmonella* and Arizona, including the classification scheme, have been adapted from the Official Methods of Analysis of The Association of Official Analytical Chemists, Twelfth Edition.[2] The biochemical and serological tests used for the identification of *Salmonella* and Arizona are described in the reference procedure and listed in Table 2.

It is beyond the scope of this work to present the classical definition of *Salmonella* and Arizona. Anyone requiring further information is referred to "Identification of Enterobacteriaceae" by Edwards and Ewing,[13] "The Bacteriology of Enterobacteriaceae" by Kauffman,[26] and Bergey's Manual of Determinative Bacteriology.[5]

Cultures belonging to the Arizona group originally were placed in the genus *Salmonella* and designated as *Salmonella arizona*. The name "Arizona group" was applied to these particular bacteria by Edwards et al. [13, 15] In 1969, Ewing[16] proposed the name of *Arizona hinshawii* for these bacteria. F. Kauffmann[26] classified the Arizona as a biochemically defined subgenus III of the genus *Salmonella*. Arizona has caused illness in man similar to gastroenteritis produced by *Salmonella*. Approximately 61% (Ewing et al,[17] 1965) of the Arizona strains ferment lactose within 48 hours. If the investigator is unaware of this biochemical characteristic of Arizona, many of these cultures will be discarded at the isolation or preliminary identification stages of analysis, being mistaken for coliforms.

The authors recognize the fact that the eighth edition of Bergey's Manual[5] has included Arizona within the genus *Salmonella* as Subgenus III, *Salmonella arizonae*. *Salmonella* and Arizona are equally important from a public health aspect, but each is distinguished on an individual basis in this chapter. The reason for this is to distinguish them from each other on a practical basis, i.e., Arizona and *Salmonella* often resemble each other when grown on primary isolation media, but can be differentiated from each other by ascertaining their biochemical and/or serological characteristics. From a practical point of view, the analyst should recognize the divergent or similar biochemical and serological characteristics necessary for the differentiation of *Salmonella* and Arizona from other microorganisms encountered in food analysis.

25.122 Preenrichment.

To perform the preenrichment procedure the food sample is mixed with a nonselective broth prior to the use of selective broths and isolation procedure. The preenrichment procedure[30, 53] was devised for use with specified food products to promote the growth of injured cells of *Salmonella* and Arizona. This step favors the recovery of microorganisms from the injured state in which some are believed to exist in many foods. Many of these foods have been subjected to processes involving heat, desiccation, preservatives, high osmotic pressure, and changes in pH, factors known to cause sublethal damage to bacteria.

When attempting to initiate growth of these inactive bacteria, suitable conditions for their growth such as favorable temperature and pH values, and the necessary nutritional requirements should be maintained during analysis. North[30] recommended the use of lactose broth as a preenrichment broth for dried egg products. The subcommittee on Food Microbiology of the Food Protection Committee, National Academy of Sciences[29] recommended lactose broth in their reference methods. A lactose broth preenrichment method for the examination of egg products has been adopted as an Official AOAC method.[2] Most *Salmonella* do not ferment lactose but are able to multiply to a large population in lactose broth. Silliker et al[39] reported on the limited applicability of this procedure when there is an unfavorable coliform: *Salmonella* ratio in the sample. For improved recovery of *Salmonella* they recommended a selenite broth containing 10% sterile feces.[38] Later an extract of Milorganite was substituted for the extract of feces. The use of a preenrichment broth also permits dilution of water soluble inhibitory substances in the food sample prior to subculturing to selective enrichment broths. Silliker and Taylor[44] demonstrated that the watersoluble components in the sample could be removed by centrifugation of food suspensions in water, and reported improved recovery of *Salmonella* with subsequent enrichment of the sediment containing the contaminating microorganisms. Silliker and Gabis[42] reported that increased recovery of salmonellae from raw meats could be achieved with the use of lactose preenrichment followed by incubation of the selective broths at 43 C. Edel and Kampelmacher[10] likewise found that incubation at 43 C yielded significantly better results than incubation at 37 C.

Many different preenrichment media have been recommended, depending on the type of food provided for examination.[27, 29, 53] For dried milk, 0.002% brilliant green distilled water or 0.5% lactose broth have been recommended. For dried eggs, egg products and other dried food products, investigators have used lactose broth, lactose broth with 0.6% Tergitol 7, lauryl tryptose broth with polyvalent H antiserum, mannitol purple sugar broth, nutrient broth, and nutrient broth with polyvalent H antiserum. Also sterile distilled water, phosphate buffer, and Ringer's solution have been used to rehydrate dried foods before analysis.

The usual ratio of sample size to preenrichment broth volume has been 1:10.[30] Thus, as the sample size has increased, the preenrichment broth volume has been increased, i.e., 10g/90 ml, 25 g/225 ml. Other recommended ratios are: 30 g/100 ml, 20 g/100 ml, 30 g/200 ml, and 100 g/1000 ml.

The incubation temperatures usually recommended for the sample preenrichment mixture are 35 or 37 C. Subcultures are made to selective broths at 24 hours, and in some procedures again at 48 hours.

Many methods[21, 29, 53] specify the transfer of 1 ml of the preenrichment culture into 10 ml of selective broth. Price et al[33] examined the effect of 1 ml and 2 ml transfer volumes of the preenrichment cultures on *Salmonella* detection and found no significant differences in recoveries.

Price et al[33] and Silliker and Gabis[41] have demonstrated that pooling or wet compositing 1 ml subcultures from multiple individual incubated preenriched sample mixtures into one container of selective broth can be used in the analysis without loss of efficiency in the recovery of salmonellae.

25.123 Selective enrichment

Selective enrichment broths are employed in the analysis of foods for the purpose of increasing the *Salmonella* and Arizona populations, and inhibiting other organisms in the food sample. Unprocessed foods, raw or highly contaminated products usually are added directly to the selective broths without preenrichment.[21, 27, 29, 43, 50, 51, 52, 53] The type of food under examination affects the performance of the enrichment broth employed in the examination procedure.

Edel and Kampelmacher,[11] reporting on comparative studies conducted in nine European laboratories, found that preenrichment gave higher isolation rates than direct enrichment on both artificially contaminated and naturally contaminated minced meat. Gabis and Silliker[19] reported similar results in the analysis of various samples of frozen meats and eggs. These findings indicate that direct inoculation of unprocessed foods into selective enrichment broths is contraindicated, that preenrichment of the sample in a nonselective medium will yield higher rates of *Salmonella* recovery.

The food sample may impair the selectivity of the medium.[44] Some of these problems were discussed in 25.122. Selective broths have improved the recovery of *Salmonella* and Arizona when used as a secondary broth with the preenrichment procedure. When the selective broths are employed in this manner, the preenrichment cultures are subcultured into the selective broths.

The Leifson selenite broth was modified by North and Bartram[31] with the addition of cystine. Selenite cystine broth has given improved recovery of *Salmonella* when used as a direct enrichment broth, and when used as a secondary broth with the preenrichment procedure. Many other modifications of this broth have been suggested, and include the addition of brilliant green dye and the addition of sulfapyridine.[32] Silliker et al[38, 39] reported that the addition of 10% sterile feces enhances the selectivity of selenite broth. Dulcitol selenite broth was developed by Raj,[34] for the improved recovery of *Salmonella* from seafood.

Many modifications of tetrathionate broth have been recommended to improve the performance of this selective broth. The Mueller-Kauffman[25] tetrathionate broth with brilliant green dyes frequently is used in food analysis. This broth has been reported to inhibit *Salmonella cholerae-suis, Salmonella abortus-ovis*, and *Salmonella paratyphi-A*.[3, 21, 27, 45] Tetrathionate broth has been modified by the addition of one or a combination of the following: brilliant green dye, sulfathiazole, lauryl sulfate, bismuth sulfite, and Tergitol-7. The additives have been used in an attempt to inhibit the growth of the interfering and competing non-*Salmonella* organisms often found in foods. Some other media developed for use as selective enrichment broth media are: brilliant green MacConkey, GN (gram negative), crystal violet (0.004%) solution, brilliant green solution, brilliant green and bile salts, magnesium chloride malachite green, strontium chloride, and neutral red lysine iron.

Selective motility selenite, selective motility Rappaport,[35] selective motility brilliant green modified Rappaport[35] and shigella-salmonella are some of the different selective motility media[48] that have been used to separate *Salmonella* from other organisms by means of differential motility.

The effect of incubation temperature on the recovery of *Salmonella* when using various selective enrichment broths has received much attention.[19] Selective enrichment broths are usually incubated at 35 or 37 C.[2, 21, 29, 51, 52, 53] Georgala and Boothroyd[22] reported improved recovery of *Salmonella* from frozen meat samples when the selenite broths were incubated at 43 C. Silliker and Gabis[42] reported on the analysis of frozen raw meat for *Salmonella* using preenrichment in lactose broth followed by selective enrichment at 35 C and 43 C. Incubation of the selective enrichment broths at 43 C resulted in the detection of more *Salmonella* positive samples.

Selective enrichment broths usually are incubated for 24 hours; however, some investigators have recommended an incubation period of 8 hours, 48 hours, and 72 hours. Hobbs[23] obtained additional positive results after 72 hours of incubation where negative results were obtained when the 24 hour enrichment broths were tested.

25.124 Selective plating media

Numerous selective plating media are used in the conventional examination procedure for isolating *Salmonella* and Arizona. A subculture of the incubated selective enrichment broth is streaked or spread on one or several plates of selective agar media. Some of the agar media recommended for this purpose are: bismuth sulfite, brilliant green, SS (salmonella-shigella), desoxycholate, desoxycholate citrate, MacConkey's, eosin methylene blue, Endo, lactose sucrose urea, and XL (xylose lysine) with desoxycholate or brilliant green. Galton et al[20] added sulfadiazine to brilliant green agar to inhibit *Pseudomonas* organisms. Osborn and Stokes[32] found that sulfapyridine enhanced the selectivity of brilliant green agar by inhibiting coliform organisms. Many reference methods[2, 21, 29, 43, 50, 52, 53] recommend the use of two or more selective plating media to facilitate the recovery of strains of *Salmonella* that are inhibited by a particular plating medium. Some *Salmonella* stereotypes[3, 13, 27] grow poorly or fail to grow on certain selective agars, i.e., *Salmonella typhi* is inhibited by brilliant green agar.

Another function of the selective plating media is differentiation of *Salmonella* colonies from the non-*Salmonella* colonies that grow on the surface of the media. Colonies of *Salmonella* are usually distinguished from non-*Salmonella* colonies by their inability to ferment lactose. Several plating media contain lactose plus an indicator system so that the lactose and non-lactose fermenting colonies are distinguished from each other. Other plating media contain differential systems based on the production of hydrogen sulfide. XL agar[49] uses the fermentation of xylose and reversion to an alkaline pH (due to the decarboxylation of lysine) to distinguish *Salmonella* and other enteric organisms of health significance.

After the selective plating media are inoculated they are usually incubated at 35 C or 37 C for 24 hours and observed.[2, 21, 29, 51, 52, 53] Negative plates usually are reincubated for an additional 24 hours to permit the growth of additional colonies and further development of existing colonies. Wun et al[57] obtained best recovery of *Salmonella typhi* when the selective broths and selective agars were incubated at 41.5 C. Reed and Reyes[36] recorded the best recovery of salmonellae from raw milk using brilliant green sulfadiazine agar at 41.5 C for 30 hours.

25.125 Nonselective differential agar

In conventional examination methods, suspect colonies from the selective agar plates are transferred to tubes of nonselective agar for the preliminary biochemical characterization of the isolates. Two to three suspicious colonies from each selective agar plate are selected for the primary differentiation procedure. One of the media more commonly used for this purpose is triple sugar iron (TSI) agar, which demonstrates the ability or inability of the test culture to ferment dextrose, lactose, or sucrose, and to produce hydrogen sulfide. Another medium, lysine iron agar (LIA),[14] produces reactions which indicate the decarboxylation of lysine, or the deamination of lysine, and the production of hydrogen sulfide when inoculated with an appropriate culture. LIA is used often in conjunction with TSI agar.[2, 21, 52] The LIA medium was designed to detect Arizona, but it is also useful in the detection of other enteric organisms, such as *Salmonella* and *Proteus*.

Many other nonselective differential agar media have been developed and include dulcitol lactose iron, Gillies medium I and II, Kligler's iron, mannitol dulcitol lysine iron, lactose urea, and triple sugar iron urea.[27]

The differential screening media are useful because they differentiate the cultures that require confirmation from those that may be discarded. This may result in a significant reduction of the total number of cultures carried through the confirmation tests.

25.13 General Description for the Identification of *Salmonella* and Arizona

25.131 Confirmatory biochemical tests

An extensive number of biochemical tests are available for the characterization of cultures isolated from food products. These tests have been described in detail by Edwards and Ewing.[13] It is not necessary to employ all of the biochemical tests which are designed to identify cultures belonging to the family *Enterobacteriaceae*. When the food samples are examined for the presence of *Salmonella* and Arizona only, it is sufficient to use a series of differential tests including the following: urease, lysine decarboxylase, dulcitol broth, KCN broth, malonate broth, indole, and sometimes Voges-Proskauer, methyl red, citrate (Simmons), and lactose broth.[2] Identification of cultures can be accomplished with limited number of biochemical tests when serological tests with the appropriate antisera also are performed, and when atypical strains are not encountered. Some of the additional tests used for identifying *Salmonella* and Arizona are: O-nitrophenyl-beta-D-galactosidase (ONPG), arginine dihydrolase, ornithine decarboxylase, gelatinase, phenylalanine deaminase, and the fermentation of mannitol, salicin, inositol, sorbitol, rhamnose, adonitol, arabinose, and raffinose.[13, 26]

25.132 Confirmatory serological tests

The genus *Salmonella* and Arizona are characterized serologically by their antigenic composition.[13] The antigens are divided into somatic (O), flagellar (H), and capsule (K) antigens. The somatic (O) antigens are composed of phospholipid polysaccharide complexes that are heat stable, and resistant to alcohol and dilute acid. The flagellar (H) antigens are heat labile, protein in nature, and occur in the flagella. The K antigens are somatic (K) antigens that occur as capsules, and are polysaccharides. If the K antigens are present in sufficient amounts, they inhibit the agglutination of unheated bacterial suspensions when tested with O antisera. The mucoid (M) antigens belong to the general class of bacterial substances known as the K antigens and also can inhibit somatic (O) agglutination. For further infor mation concerning *Salmonella* and Arizona antigens and antisera, refer to Edwards and Ewing,[13] "Identification of *Enterobacteriaceae*," or "The Bacteriology of *Enterobacteriaceae*," by Kauffman.[26]

The somatic (O) antigens are determined by performing an agglutination test with somatic (O) antisera and growth from an agar culture of the test microorganism. A slide agglutination test usually is used. Refer to section 25.551. It is advisable to perform an initial polyvalent somatic (O) test and then perform additional agglutination tests using individual group somatic (O) antiserum representative of each group in the polyvalent somatic (O) antiserum.

The flagellar (H) antigens are determined by performing an agglutination test with flagellar (H) antisera and a formalized saline infusion broth culture. A tube of the antigen antiserum mixture is incubated for 1 hour in a 50 C water bath. Refer to sections 25.553 and 25.554. The culture usually is tested with a polyvalent flagellar (H) antiserum and then with more specific flagellar (H) antisera. A multiple polyvalent (H) antisera set, "Spicer-Edwards,"[9, 12, 43, 47] has been developed for determining 17 flagellar antigens. Some of these determinations denote complex antigens, and additional antisera are required for final serotyping. Organisms with flagella of sufficient motility must be used for antigens to yield positive agglutination tests.

Since some of the *Enterobacteriaceae* have related antigens, a few false positive serological tests may occur when testing isolates with *Salmonella* and Arizona antisera. This problem diminishes when more specific antisera are used. Final identification of a culture as to the specific serotype must be performed by a laboratory equipped for definitive serotyping.

25.14 Association and Agency Methods

25.141 AOAC methods[2]

The primary objective of the Association of Official Analytical Chemists in part is "to secure, devise, test, and adopt uniform, precise, and accurate methods for the analysis of foods, feeds, fertilizers, economic poisons, and other commodities." To be worthy of adoption as an official method of AOAC, the method must meet these criteria: it must be (1) reliable, (2) practical, (3) available to all analysts, and (4) substantiated. The recommended procedure in section 25.5 is taken from the AOAC methods.

25.142 Bacteriological Analytical Manual (BAM)[53] methods

The Bacteriological Analytical Manual for foods contains the bacteriological analytical methods commonly used in the Food and Drug Administration laboratories for the examination of foods. These methods currently are considered to be the most useful to the FDA in enforcing the provisions of the Food, Drug, and Cosmetics Act. The Third Edition is restricted to those methods primarily applicable to foods. Methods having official AOAC status are included. The BAM methods of preparing various foods for the isolation of Salmonella are listed in Table 1. The isolation and con firmation procedures follow the recommended procedure; refer to section 25.53.

25.143 Center for Disease Control methods

"Salmonellae in Foods and Feeds, Review of Isolation Methods and Recommended Procedures" by Galton et al[21] has been published by the Communicable Disease Center (Center for Disease Control), Atlanta, Georgia. Methods are given for the isolation of *Salmonella* from eggs, egg products, other dried foods, processed foods, meats, other raw nonprocessed foods, rendered animal byproducts, feeds, dried and liquid milk products, candy, candy coating, chocolate, animal glandular products (thyroid tablets, etc.), and water. Other procedures included are: screening from selective plating media, serological and biochemical confirmation, and enumeration of salmonellae. There are 2 schematic diagrams of the isolation procedures and 16 figures of colored photographs to illustrate characteristic reactions of enteric organisms when grown on different media.

25.144 United States Department of Agriculture methods

The Agriculture Research Service, United States Department of Agriculture, has published a "Recommended Procedure for the Isolation of *Salmonella* Organisms from Animal Feeds and Feed Ingredients."[51] A recommended test procedure using cultural techniques is presented, and tests for the serologic identification and biochemical confirmation of cultures are described. The appendix contains sampling procedures, fluorescent antibody (FA) screening procedures, and directions for preparing enrichment and plating media, and biochemical test reagents. Nine figures of colored photographs to illustrate the procedures are included.

A "Microbiology Laboratory Guidebook" (MLG) is published by the Scientific Services, Meat and Poultry Inspection Program, Animal and Plant Health Inspection Service, USDA.[52] This laboratory guidebook consists of laboratory methods of analysis for organoleptic examination, examination of fresh or prepared foods, isolation and identification of *Salmonella* from foods, examination of canned foods, determination of antibiotic residues in animal tissues, animal species determination, serological, and animal disease diagnosis. *Salmonella* isolation procedures are given for the following foods: precooked frozen foods, salted natural casings, raw meat, and food homogenate. Reference methods are listed for powdered eggs, breading mixes, dehydrated sauces, and dried milk. A comprehensive discussion which covers weighing samples, enrichment, plating, and rapid screening is included. Directions are given for the use of screening media, biochemical procedures, and serological tests for the identification of *Salmonella*.

25.145 International organizations methods

Many international organizations are concerned with microbiological criteria for foods.[29] Many are involved in an effort to develop, study, and standardize methods for the microbiological examination of foods. Several of these organizations are: International Commission on Microbiological Specifications for Foods, International Association of Microbiological Societies; Codex Alimentarius Commission, Expert Committee on Food Hygiene and Joint FAO/WHO Expert Committee on Food Hygiene (Food Microbiology), and Sub Committee 6 (Meat and Meat Products), International Organization for Standardization (ISO).

25.15 Screening Methods and Other Detection Methods

A procedure designed to permit the earliest possible serological testing was developed by Silliker et al.[40] The technique entails subculture of non-lactose fermenting colonies growing on differential agar plating media directly into infusion broth. These cultures are incubated for 4 to 6 hours at which time an H agglutination test is performed on this antigen using a polyvalent H antiserum prepared by mixing the individual sera in the Spicer-Edwards "kit".[9, 47] The authors reported over 1000 different *Salmonella* cultures were tested by this procedure, and only one false positive reaction occurred with an Arizona type paracolon. Less than 1% of the salmonellae encountered failed to agglutinate in the polyvalent H antisera, and the majority of these nonreacting cultures were nonmotile *Salmonella*. Both false positive and false negative reactions with the rapid polyvalent H agglutination procedure are detected when confirmatory somatic serological tests and biochemical tests are performed.

An accelerated procedure for *Salmonella* detection in dried feeds and foods involving only broth cultures and serological reactions has been developed by Sperber and Deibel.[46] The procedure includes preenrichment (18 hours), selective enrichment (24 hours), elective enrichment (6 to 8 hours), and serological testing (2 hours). The authors state that their method is as sensitive as the more time consuming traditional procedure. The new procedure does not necessitate the isolation of pure cultures, but the traditional procedure can be performed concurrently if desired.

For the rapid determination of *Salmonella* in samples of egg noodles, cake mixes, and candies, Banwart and Kreitzer[4] devised a test using a glass apparatus with three U tubes projecting into a central chamber. The sample is mixed with sterile lactose broth in the center chamber of the apparatus, then each of media, semisolid SIM agar, mannitol purple (MP) agar, and selenite brilliant green enrichment broth plus agar, is added to an indi-

Table 1: Bacteriological Analytical Manual for Foods[53]—Food and Drug Administration Method of Preparing Various Foods for the Isolation of *Salmonella*

Product	Sample Size	Preenrichment broth	Preparation	Enrichment[(a)] broth
Dried whole eggs, dried egg yolks, dried egg whites pasteurized liquid and frozen eggs, prepared powdered mixes (cake, cookie, doughnut, biscuit, and bread), infant formula. Coconut[(b)]	25 g	225 ml lactose broth	Mix	10 ml selenite cystine (SC) 10 ml tetrathionate with brilliant green dye (TT)[(e)]
Dried yeast	50 g	200 ml sterile distilled water	Mix	10 ml selenite cystine 10 ml tetrathionate with brilliant green dye (TT)[(e)]
Frosting and topping mixes	25 g	225 ml nutrient broth	Mix	10 ml S C 10 ml T T
Nonfat dry milk and dry whole milk	100 g	1,000 sterile distilled water containing 2 ml of 1% aqueous brilliant green dye	Mix	10 ml S C 10 ml T T
Egg containing foods (noodles, egg rolls, etc.)	25 g	225 ml lactose broth	Blend	10 ml S C 10 ml T T
Candy and candy coating	100 g	1,000 ml sterile reconstituted nonfat dry milk 2 ml of 1% aqueous brilliant green dye[(c)]	Blend	10 ml SC 10 ml TT

Table 1: Bacteriological Analytical Manual for Foods[53]—Food and Drug Administration Method of Preparing Various Foods for Isolation of *Salmonella* (Continued)

Product	Sample Size	Preenrichment broth	Preparation	Enrichment[(a)] broth
Dyes and coloring substances with pH of 6.0 or above (10% aqueous suspension)		225 ml lactose broth	Mix	10 ml SC 10 ml TT
Laked dyes and dyes with a pH below 6.0	25 g	None	Mix	tetrathionate 225 ml with brilliant green dye [(e)]
Non-pasteurized egg products	25 g 25 g	None	Mix Mix	225 ml selenite cystine 225 ml tetrathionate with brilliant green dye[(e)]
Raw and highly[(d)] contaminated meats, animal substances, glandular products & fish meal	25 g 25 g	None	Blend Blend	225 ml selenite cystine 225 tetrathionate with brilliant green dye[(e)]
Heated, processed & dried, meats, animal substances, glandular products & fish meal	25 g	225 ml lactose broth to which 2.2 ml Tergitol #7 is added after blending when analysing high fat products	Blend	10 ml SC 10 ml TT

[(a)]When preenrichment broth is used, subculture 1 ml of incubated sample, preenrichment broth mixture to each enrichment broth.
[(b)]Add 2.2 ml Tergitol #7 to lactose broth.
[(c)]Reconstituted nonfat dry milk (100g nonfat dry milk plus 1000 ml distilled H_2O). Four (4.0) ml of 1% aqueous crystal violet dye may be substituted for brilliant green dye.
[(d)]If the extent of contamination is unknown, then examine the third 25g sample as directed for heated processed products.
[(e)]Tetrathionate broth (TT) is prepared by adding 10 ml of 0.1% brilliant green dye to 1000 ml of tetrathionate broth base.

vidual U tube side arm. After each of the agars in the side tubes solidifies, sterile brain heart infusion (BHI) is added to the top of the agar plugs. The prepared flask plus sample is incubated at 37 C. After 24 and 48 hours, the fermentation of mannitol and production of hydrogen sulfide is noted. Growth from turbid BHI is tested with *Salmonella* H polyvalent antiserum. The detection of positive samples requires incubation for 48 hours, and a sample is considered to be negative if, after incubation for 48 hours, there is no fermentation of semisolid MP, no H_2S production, or no motility through the three different agars. The samples showing an agglutination with the *Salmonella* H antiserum are tested with selective agars for isolation and further identification.

A dulcitol selenite enrichment medium in a motility flask was used by Abrahamsson et al[1] for the detection of *Salmonella* in food. A drop in pH of the dulcitol selenite enrichment motility broth indicated the presence of *Salmonella*. This phenomenon was confirmed by fluorescent antibody staining. There was no significant difference in sensitivity as determined in this study between the new technique and a conventional *Salmonella* detection technique.

Wells et al[56] used a cotton gauze swab suspended in raw milk for ten minutes with subsequent incubation of the swab in brilliant green tetrathionate broth (TET) as a new method of isolating *Salmonella*. The swab method was found to be as sensitive as the North procedure for recovering *Salmonella* when incubated at 37 C, but more sensitive when incubated at 43 C. Incubation of the swab cultures in TET at the elevated temperature of 43 C gave good results when *Salmonella* was present at inoculated levels as low as one per liter.

25.2 TREATMENT OF SAMPLE

25.21 Collection (see Chapter 1)

Test samples should be drawn randomly and should be representative samples of the lot. In general the confidence or probability limits established for the food product determine the number of test samples required for analysis. Consideration should be given to the homogeneous or nonhomogeneous distribution of contaminating microorganisms in the food.

The *Salmonella* Committee of the National Research Council (National Academy of Science)[28] recommended the establishment of a salmonellae sampling plan based on several conclusions (1) that some processed foods have a greater salmonellae hazard than others, and (2) that it is impossible with certainty to assure the complete absence of salmonellae in a lot of food when performing a microbiological examination on part of the units in the lot. The Committee classified food products into categories according to risk based on the number of salmonellae hazards a food possesses and whether a food is consumed by infants, the aged, or the infirm. The three defined salmonellae hazards of foods are (1) the food or an ingredient of the food is a significant potential source of salmonellae, (2) the manufacturing process does not include a controlled step that destroys salmonellae, and (3) the food has significant potential for microbial growth if "abused" in distribution or by consumers. Considering the above hazards, the foods are placed in one of five categories as follows:

I. nonsterile foods (foods that are a significant potential source of salmonellae) intended for use in infants, the aged, and the infirm—the restricted population of high risk;
II. processed foods with all three salmonellae hazards;
III. processed foods with two of the three salmonellae hazards;
IV. processed foods with one of the three salmonellae hazards;
V. processed foods with none of the three salmonellae hazards.

The *Salmonella* committee proposed criteria for the acceptance of questioned food lots in the various categories as follows:

I. 60 units (25 g each) tested and found negative; this gives a 95% probability that there is one organism or fewer in any 500 g of the lot in question;
II. 29 units (25 g each) tested and found negative; 95% probability that there is one organism or fewer in 250 g;
III. 13 units (25 g each) tested and found negative; 95% probability that there is one organism or fewer in 125 g;
IV. and V. same as Category III, because tests of fewer than 13 units will not establish an estimate with 95% assurance.

Many 25 g units must be analyzed if the above criteria are to be fulfilled. It is advantageous to reduce the labor and cost of testing by pooling samples if the sensitivity of the test procedure is maintained and not diminished. If larger size samples can be analyzed without significant reduction in sensitivity, then greater degrees of confidence in control procedures can be realized. Studies by Huhtanen et al[24] compared the results of a single 300 g to ten 30 g samples, and Silliker and Gabis[41] compared the results of a 1500 g sample to (1) sixty 25 g subsamples, (2) fifteen 100 g subsamples, and (3) three 500 g subsamples. The overall results in both studies indicate that salmonellae may be detected with equal accuracy, and that there is no significant difference in sensitivity when testing a large sample versus multiple subsamples.

To reduce the analytical workload within the laboratories of the U.S. Food & Drug Administration, the analytical units of 25 g may be composited for analysis.[54] The maximum size of a composite unit is 375 g. Criteria for the minimum number of composite units tested for samples in each food category are:

Food category	Number of sample units (100 g) collected	Minimum number of composite units (375 g) tested
I	60	4 (1,500 g)
II	30	2 (750 g)
III	15	1 (375 g)

Another approach used to reduce the analytical work when analyzing multiple samples is "wet compositing" or pooling of the preenrichment broth cultures. Silliker[37] reported on a technique in which 1 ml from each of the 10 preenrichment cultures was introduced into 10 ml of double strength selective medium. The "wet composite," representing 10 randomly selected samples, was then analyzed as a single unit. In one study, the results of analysis from 106 "wet composites" agreed well with the results of analysis from 1,060 individual components. In eight instances, the results were in disagreement. In a later study, Silliker and Gabis[41] evaluated wet compositing of sixty 25 g and fifteen 100 g preenrichment samples in groups of five. The results gave the same assurance of detection of positive lots as the analysis of individual samples. Price et al[33] reported results of a study for *Salmonella* testing of pooled preenrichment broth cultures. As many as 25 preenrichment broth cultures were pooled without apparent loss in the sensitivity of *Salmonella* detection as compared to individual sample analysis.

25.22 Holding (see Chapter 1)

25.23 Mixing and Homogenization

If the food product is frozen, thaw a suitable portion with moderate temperature, as rapidly as possible, to prevent an increase in the number of organisms, or destruction of *Salmonella* or Arizona in the sample. If the sample is a powdered, ground, or comminuted food product, then a homogeneous suspension can usually be obtained by mixing the sample and the broth together while stirring the mixture with a sterile glass rod, or other suitable sterile implement. Homogeneous suspensions are obtained in some cases by shaking the sample broth mixture by hand or by using a mechanical shaker. Mechanical blending may be required if the sample consists of a large portion or large pieces. A blending time of 2 minutes at 8,000 rpm is usually satisfactory for most foods.

25.3 SPECIAL EQUIPMENT AND SUPPLIES

a. Balance, with weights, sensitivity of 5 mg
b. Culture dishes (100 × 15 mm), glass or plastic
c. Blender and sterile blender jars
d. Incubator 35 ± 1 C
e. Inoculating needle with nichrome or platinum wire, approximately 3 mm inside diameter loop, and inoculating needle without loop
f. Incubator 35 ± 1 C
g. Pipets, 1 ml capacity with 0.01 graduations, also 5 ml and 10 ml capacity with 0.1 ml graduations, and 0.2 ml capacity with 0.02 ml graduations
h. Sterile, 16 ounce (pint) wide mouth, screw capped jars or flasks
i. Sterile, 64 ounce (2 qt.) wide mouth, screw capped jars or flasks
j. Sterile spoons for weighing food specimens

25.4 PRECAUTIONS AND LIMITATIONS OF METHODS

a. A representative number of samples from each lot should be analyzed. The samples should be selected in accordance with a statistically valid sampling plan.

b. An effective method for analyzing the particular type of food product should be selected and used when analyzing food samples for *Salmonella* and Arizona.

c. A reference or standard method for the recovery of *Salmonella* or Arizona should not be modified unless a collaborative or comparative study shows that the modification improves the method.

d. An analysis should not be concluded after observation of suspicious colonies on selective plates. If the suspicious colonies are not tested, false positive or false negative results may be reported.

e. A sufficient number of colonies should be picked from the selective plates to maintain the sensitivity of the method.

f. A pure culture is required in order to establish valid biochemical test reactions. If positive serological test reactions and negative biochemical test reactions are obtained, the culture should be checked for purity. A contaminated triple sugar iron agar culture should not be discarded unless it has been purified and tested. A contaminating microorganism can mask *Salmonella* or Arizona, i.e., *Proteus* mixed with *Salmonella*.

g. Occasionally *Salmonella* and Arizona cultures showing atypical biochemical reactions may be isolated, e.g., salmonellae that are H_2S negative, lactose positive, or dulcitol negative. It should be appreciated that the classification of an isolate as an Arizona or *Salmonella* rests ultimately upon the antigenic structure of the organism, not unqualifiedly on its biochemical characteristics. It is a fact that most salmonellae (and Arizona isolates) show a relatively consistent pattern of biochemical characteristics, but variants are frequently encountered. There are no indications that biochemically atypical strains present reduced public health hazards.

h. Some foods contain multiple serotypes, and an extensive analysis is required for the detection of every serotype.

i. Some types of foods contain microbial inhibitors (natural or added) which may reduce the efficiency of the examination method.

j. Multicomponent foods, food products composed of several components such as, dry milk, dried egg and dried yeast mixed together, present a problem as to the selection of the best examination procedure.

k. Culture media and reagents (including antiserum) should be subjected to quality control procedures.[6]

25.5 PROCEDURE

25.51 Introduction

The procedure for the examination of dried whole egg, dried egg yolk, dried egg white, and dry milk is presented in this section.

For the examination of other types of food, refer to section 25.142 and Table 1.

25.52 Preenrichment

25.521 Dried whole egg, dried egg yolk, and dried egg white[2]

a. Aseptically open a sample container and aseptically weigh a 25 g sample into a sterile, empty wide mouth container with cap or suitable closure.
b. Add 225 ml of sterile 0.5% lactose broth. Add small portions (15 ml) of sterile lactose broth and stir with a sterile glass rod (or other suitable implement) to obtain a smooth suspension. After adding four portions of lactose broth, add the remainder of the broth for a total of 225 ml. Stir until the sample is suspended without lumps.
c. Cap the container securely and let stand at room temperature for 60 minutes.
d. Mix well by shaking and determine pH with the test paper.
e. Adjust the pH, if necessary, to 6.8 ± 0.2 with sterile 1 N sodium hydroxide or hydrochloric acid. Cap the jar securely and mix well before determining the final pH.
f. Loosen the container cap about 1/4 turn, and incubate 24 ± 2 hours at 35 C.
g. Continue as directed under section 25.53a

25.522 Nonfat dry milk and dry whole milk[2]

a. Aseptically open a sample container and aseptically weigh 100 g sample into a sterile, empty wide mouth container (2 liter) with cap or suitable closure.
b. Add 1 liter of sterile distilled water, and mix well.
c. Determine pH with test paper. If pH is below 6.6, adjust to 6.8 ± 0.2 with sterile 1 N sodium hydroxide.
d. Add 2 ml of 1% aqueous brilliant green and mix well.
e. Loosen the container cap about 1/4 turn, and incubate 24 ± 2 hours at 35 C.
f. Continue as directed under section 25.53a

25.53 Isolation of *Salmonella* and Arizona

a. Gently shake the incubated sample mixture, and transfer 1 ml portions each to 10 ml of selenite cystine broth and 10 ml of tetrathionate broth (with brilliant green and iodine).
b. Incubate 24 ± 2 hours at 35 C.
c. Streak a 3 mm loopful of incubated selenite cystine broth on selective media plates of brilliant green agar, salmonella-shigella agar,

and bismuth sulfite agar. Adjust inoculum to obtain isolated colonies.

d. Repeat with a 3 mm loopful of tetrathionate broth.
e. Incubate plates 24 ± 2 hours at 35 C.
f. Note appearance of *Salmonella* and Arizona colonies

1. Brilliant green agar
Colorless, pink to fuschia, translucent to opaque, with surrounding medium pink to red. Some salmonellae appear as transparent green colonies if surrounded by lactose or sucrose fermenting organisms which produce colonies that are yellow green or green.

2. Salmonella-shigella agar
Colorless to pale pink, opaque, transparent or translucent. Some strains produce black centered colonies.

3. Bismuth sulfite agar
Brown, black, sometimes with metallic sheen. Surrounding medium is usually brown at first, turning black with increasing incubation time. Some strains produce green colonies with little or no darkening of surrounding medium.

g. Pick with a needle or 2 more typical or suspicious colonies, if present, from brilliant green, salmonella-shigella or bismuth sulfite agar plates onto triple sugar iron (TSI) agar and lysine iron agar (LIA) slants. If plates do not have typical or suspicious colonies, or do not show growth, incubate additional 24 hours.
h. Inoculate TSI agar slants with a portion of each colony by streaking a slant and stabbing the butt of each agar. After inoculating TSI agar with a needle, do not obtain more inoculum from the colony, and do not heat the needle until the inoculation of the LIA agar is completed.
i. Retain the plates after picking selected colonies and store at 5 to 8 C or 25 C.
j. Incubate TSI agar slants 24 ± 2 hours at 35 C.* *Salmonella* positive cultures show alkaline (red) slants and acid (yellow) butts, with or without H_2S (blackening of the agar). Do not exclude H_2S negative TSI agar slants.
   Arizona may appear the same, due to late fermentation of lactose. With further incubation (holding) however, the tubes of TSI may show the appearance of coliform cultures, i.e., yellow butts and yellow slants.
k. Incubate LIA agar slants 24 ± 2 hours at 35 C.* *Salmonella* and Arizona positive cultures have an alkaline (purple) reaction throughout the medium. If H_2S is produced, the butt of the medium is blackened.

---

*Cap tubes loosely to maintain aerobic conditions while incubating slants to prevent excessive H_2S production.

l. If typical reactions are not observed, pick additional colonies from brilliant green, salmonella-shigella, and bismuth sulfite plates to TSI and LIA slants.

m. Retain for biochemical testing all suspect TSI and LIA slants in sections 25.53j and 25.53k (See footnote page 317). TSI cultures which otherwise appear not to be salmonellae (including yellow slants) cannot be excluded from further examination and should be treated as presumptive positive TSI agar cultures if corresponding LIA gives typical *Salmonella* reactions. LIA is useful in detection of lactose or sucrose fermenting *Salmonella* (atypical) or lactose fermenting Arizona organisms which give negative TSI reactions.

n. Apply biochemical and serological identification tests to 3 presumptive positive TSI agar cultures from selenite cystine, and 3 presumptive positive TSI agar cultures from tetrathionate broth if present. A minimum of 6 TSI cultures should be examined; if necessary, all 6 presumptive positive cultures may be selected from 1 set of selective agar plates.

25.54 Confirmation of Isolates

25.541 Purify mixed cultures

a. Streak any TSI agar slant culture, which appears to be mixed, on MacConkey's agar or brilliant green agar.

b. Incubate plates for 24 ± 2 hours at 35 C. Observe results as follows:

(1) MacConkey's agar—typical colonies appear transparent and colorless, sometimes with dark center.

(2) Brilliant green agar—refer to Section 25.53f.

The importance of obtaining pure cultures for biochemical testing cannot be overemphasized, since with mixed cultures the results of biochemical tests are meaningless, and indeed may lead to discarding a *Salmonella* or Arizona isolate. Proper pure culture isolation procedures demand that all colonies on the streaked plate be of the same type before subculture as directed in section 25.542. If the streaked plate shows a mixed culture, a typical, well isolated colony should again be streaked onto differential agar, and a colony from this plate should be selected for subculture as described in section 25.541b, only if all colonies on the plate are of the same type. Otherwise the procedure must be repeated until the differential plate shows only one colony type.

c. Pick with a needle at least 2 colonies onto separate TSI agar and LIA agar slants as directed in sections 25.53g and 25.53h.

d. Select for biochemical testing presumptive positive *Salmonella* and Arizona cultures as described in section 25.53m.

25.542 Pure cultures

a. **Urease test**

(1) Transfer 2 loopfuls (3 mm each) of growth from presumptive positive TSI agar culture to urea broth. Incubate 24 ± 2 hours at 35 C; or,

(2) inoculate rapid urea broth and incubate for 2 hours in water bath at 37 ± 0.5 C.
(3) Discard all cultures that give a positive test (purple red color).
(4) Retain for further testing all cultures that give a negative test (no change in orange color of medium).

25.543 Serological flagellar (H) screening test

a. Transfer one loopful (3 mm) of growth from a urease negative TSI agar culture to
  (1) brain heart infusion broth and (for test on the same day) incubate 4 to 6 hours at 35 C until visible growth occurs; or,
  (2) trypticase soy tryptose broth and (for test on following day) incubate 24 ± 2 hours at 35 C.
  (3) Select a minimum of 2 broth cultures and perform the polyvalent flagellar (H) test or "Spicer-Edwards" flagellar (H) test on each culture as directed in sections 25.553 or 25.554.

b. If flagellar broth cultures are positive, perform additional confirmation tests as directed.

c. If flagellar broth cultures are negative, perform flagellar (H) tests on 4 additional broth cultures. If possible, obtain a minimum of 2 positive cultures for additional testing with confirmation tests.

d. If all urease negative TSI cultures from sample are *Salmonella* flagellar (H) test negative then perform additional confirmation tests which follow.

25.544 Testing urease negative cultures

a. Lysine iron agar (LIA)
  (1) If LIA slant of this culture is positive, do not retest; perform additional tests which follow.
  (2) If LIA slant is negative, incubate culture an additional 24 ± 2 hours at 35 C, and observe final reaction.

b. Lysine decarboxylase broth. If the results of LIA test are unsatisfactory, then perform test using this broth.
  (1) Inoculate broth with a loopful of TSI agar culture.
  (2) Replace the cap tightly and incubate 48 ± 2 hours at 35 C.
  (3) Examine at least every 24 hours.
  (4) *Salmonella* and Arizona cause an alkaline reaction of purple color.
  (5) A negative test is permanent yellow color throughout the broth.

c. Phenol red dulcitol broth (or purple broth base with dulcitol.)
  (1) Inoculate broth with a loopful (3 mm) from a TSI agar culture.
  (2) Incubate 48 ± 2 hours at 35 C.
  (3) Examine at least every 24 hours.
    (a) Most salmonellae give a positive test: gas formation and acid reaction (yellow). Arizona is negative, alkaline (red).
    (b) A negative test is an alkaline reaction (red for phenol red dulcitol broth or purple for purple broth base with dulcitol broth), and no gas.

d. Tryptophane broth.
   (1) Inoculate the broth with a loopful (3 mm) of TSI agar culture.
   (2) Incubate 24 ± 2 hours at 35 C, and test as follows:
      (a) KCN broth
         (1) Transfer a loopful (3 mm) of tryptophane broth to KCN broth.
         (2) Heat the rim of tube to form a good seal when stoppered, and incubate 48 ± 2 hours at 35 C.
         (3) Neither *Salmonella* nor Arizona grow in this broth. A negative test is shown by lack of turbidity.
      (b) Malonate broth
         (1) Transfer a loopful (3 mm) of tryptophane broth to malonate broth.
         (2) Incubate 48 ± 2 hours at 35 C.
         A negative test, as shown by a green color (unchanged), results with most salmonellae.
         A positive test (alkaline reaction) is shown by a blue color. Arizona is malonate positive.
      (c) Indole test.
         (1) Transfer 5 ml of tryptophane broth to an empty test tube.
         (2) Add 0.2 to 0.3 ml Kovac's reagent. A positive test is shown by a deep red color. Both *Salmonella* and Arizona are indole negative.

e. *Salmonella* serological flagellar (H) test
   (1) If the flagellar broth culture of the TSI agar slant was tested as directed in Section 25.543, then perform the additional serological test in Section 25.544f.
   (2) Perform flagellar (H) test on untested cultures as directed in 25.553 or 25.554.

f. *Salmonella* serological somatic (O) plate test
   Perform test on TSI agar or brain heart infusion agar cultures as directed in sections 25.551 and 25.552.

25.545 Classification of *Salmonella* and Arizona. Table 2, one through 8.

a. Discard as not *Salmonella* or Arizona cultures that show either
   (1) positive indole test (red) and negative *Salmonella* serological flagella (H) test;
   (2) positive KCN broth test (growth) and negative lysine decarboxylase test (yellow).

b. Classify cultures that have characteristics shown in Table 2, one through 8, columns A, B, or C.

25.546. Additional biochemical tests Note. Perform additional tests on cultures that do not classify as *Salmonella* or Arizona (25.545), and were not discarded.

a. Phenol red lactose broth or purple lactose broth.

(1) Inoculate with loopful (3 mm) of growth from each unclassified TSI agar slant.
(2) Incubate 48 ± 2 hours at 35 C.
(3) Examine at least every 24 hours.
A positive test is shown by gas formation and/or acid reaction (yellow).
A negative test is shown by alkaline reaction and no gas.
(4) A negative test usually results with salmonellae. Arizona may show delayed formation of acid and gas, or acid, or may be negative.

b. Phenol red sucrose broth or purple sucrose broth. Same as directed in section 25.546a one to 4, above.

c. MR-VP medium.

Table 2: Characteristics of *Salmonella*, Arizona and Non-*Salmonella* Organisms

Test or substrate	*Salmonella*		Arizona	Non-*Salmonella*[(a)]
	A	B	C	
1. Urease	−	−	−	+ or −
2. Lysine decarboxylase[(d)]	+ (alk)	+ (alk)	+ (alk)	− (acid)
3. Phenol red dulcitol broth	AG	AG or A or −	−	−
4. KCN broth	NGR	NGR	NGR	+
5. Malonate broth	−	−	+	+ or −
6. Indole test	−	−	−	− or +
7. Polyvalent flagellar test	+[(b)]	+[(b)]	+[(b)]	−
8. Polyvalent somatic test	+[(b)]	+[(b)]	+[(b)]	−
9. Phenol red lactose broth	−[(c)]	−[(c)]	AG or A or −	AG
10. Phenol red sucrose broth	−[(c)]	−[(c)]	−	AG
11. Voges-Proskauer test	−	−	−	+ or −
12. Methyl red test	+	+	+	−
13. Simmons citrate	+	+ or −	+	+ or −

(a)Other non-Salmonella reactions occur.
(b)Negative reactions occur if corresponding agglutinins are not contained in antisera.
(c)Majority of strains give negative test, but atypical strains giving positive test have been reported.
(d)*Salmonella paratyphi A* is characteristically lysine decarboxylase negative.

Symbols: + = Positive test
− = negative test
A = acid
AG = acid and gas
Alk = alkaline reaction
NGR = no growth
GR = growth

(1) Inoculate with a loopful (3 mm) of growth from each unclassified TSI agar slant.
(2) Incubate 48 ± 2 hours at 35 C.
(3) Perform the VP test.
(a) Transfer 1 ml of 48 hour culture to a test tube.
(b) Add 0.6 ml alpha naphthol and shake.
(c) Add 0.2 ml of 40% NaOH solution and shake.
(d) Optionally add a few crystals of creatine.
(e) Read results after 4 hours.
A positive VP test is development of an eosin pink color.
A negative test results with *Salmonella* and Arizona.
(f) Reincubate remainder of MR-VP medium an additional 48 hours at 35 C.
(4) Perform the MR test.
(a) Transfer 5 ml of a 96 hour MR-VP culture to a test tube.
(b) Add 5 to 6 drops of methyl red solution.
(c) The red color results immediately.
Salmonella and Arizona give positive results (positive is red; yellow is negative).

d. Simmon's citrate agar.
(1) Inoculate with a needle from the growth of an unclassified TSI agar slant.
(2) Inoculate by streaking the slant and stabbing the butt.
(3) Incubate 96 ± 2 hours at 35 C.*
A positive test is indicated by growth, accompanied by a color change from green to blue.
A negative test is indicated by no growth and no color change.
(4) *Salmonella* and Arizona usually give positive results.

25.547 Recording of results

Record all results on appropriate record sheets.

25.548 Classification of cultures

a. Classify cultures according to results listed in Table 2, one through 13. If one TSI culture from a portion of the sample is classified as *Salmo nella* sp., or Arizona, then the sample is designated: *Salmonella* positive or Arizona positive.
(1) *Salmonella* sp. Cultures that have reaction patterns of Table 2, column A or B.
(2) Arizona. Cultures that have reaction pattern of Table 2, column C.
(3) Unclassified cultures.
(a) Discard as not *Salmonella* or Arizona, cultures that give results listed in any one subdivision of Table 3.
(b) Use tests described in Identification of *Enterobacteriaceae*, P. R. Edwards and W. H. Ewing,[13] to classify any unidentified or doubtful cultures (cultures not clearly identified as *Salmo*

---

*Cap tubes loosely to maintain aerobic conditions while incubating slants to prevent atypical reactions.

Table 3: Criteria for Discarding Cultures

A. Urease test	positive (purple-red)
B. Indole test Flagellar test (Polyvalent or Spicer-Edwards)	positive (red) negative (no agglutination)
C. Lysine decarboxylase test KCN broth	negative (yellow) positive (growth)
D. Phenol red lactose broth Lysine decarboxylase test	positive (gas and/or acid) negative (yellow)
E. Phenol red sucrose broth Lysine decarboxylase test	positive (gas and/or acid) negative (yellow)
F. KCN broth Voges-Proskauer test Methyl red test	positive (growth) positive (red) negative (yellow)

nella sp. or Arizona by classification schemes in Table 2, or not eliminated from these groups by test reactions listed in Table 3).

(4) If none of the TSI cultures carried through the confirmation tests confirm as *Salmonella* or Arizona, perform confirmation tests (Section 25.544) on all unconfirmed urease negative TSI cultures isolated from the same portion of the sample.

25.55 *Salmonella* Serological Tests (All *Salmonella* antisera should be pretested with known cultures.)

25.551. Polyvalent somatic (0) plate test

a. Using wax pencil, mark off 2 sections about 1 × 2 cm on the inside of a glass or plastic petri dish, or glass slide.
b. Place 1/2 of a 3 mm loopful of culture from a 24 to 48 hour TSI (or BHI) agar slant on the dish in upper part of each marked section.
c. Add 1 drop of saline solution to the lower part of each section.
d. Emulsify the culture in saline solution with a clean, sterile transfer loop or needle (or applicator sticks) in both sections.
e. Add 1 drop of *Salmonella* polyvalent somatic (0) antiserum to 1 section.
f. Mix antiserum and culture as in 25.551d above.
g. Tilt mixtures back and forth 1 minute and observe against a dark background in good illumination.**
Any degree of agglutination is a positive reaction.

**Handle viable cultures carefully to prevent contaminating the analyst and the environment.

h. Classify polyvalent somatic (0) test
   1) Positive; agglutination in test mixture; no agglutination in saline control;
   2) Negative; no agglutination in test mixture; no agglutination in saline control;
   3) Nonspecific; agglutination in both test and control mixtures.

25.552 Determination of somatic groupings

a. Test as shown in section 25.551a to h, using individual group somatic (0) antisera including Vi, instead of *Salmonella* polyvalent somatic (0) antiserum.
b. Heat suspensions of Vi positive cultures in boiling water for 20 to 30 minutes and cool. Retest heated suspension, using somatic group C_1, group D, and Vi antisera. Heated cultures which give somatic group C_1 or D positive reactions, and a Vi negative reaction, are classified as *Salmonella* if they have characteristics of *Salmonella* as listed in Table 2, column A or B.
c. Record cultures that give somatic (0) test with any individual somatic (0) antiserum as positive for that somatic (0) group.
d. Record cultures that do not react with any individual somatic (0) antiserum as negative for individual group somatic (0) group.

25.553 Polyvalent flagellar (H) test

a. Prepare a 4 to 6 hour brain heart infusion broth, or 24 ± 2 hour trypticase soy tryptose broth culture for the test by adding 2.5 ml formalized physiological saline solution to 5 ml of each broth culture.
b. Test each formalized broth culture with *Salmonella* flagellar antisera as follows:
   (1) Place 0.5 ml appropriately diluted *Salmonella* polyvalent flagellar (H) antiserum in a 10 × 75 or 13 × 100 mm serological test tube.
   (2) Add 0.5 ml of antigen to be tested (formalized broth culture) to a tube with antiserum.
   (3) Prepare saline control by mixing 0.5 ml of formalized saline with 0.5 ml antigen.
   (4) Incubate mixtures 1 hour in a water bath at 50 C.
   (5) Observe at 15 minute intervals and read results at 1 hour.
      (a) Positive; agglutination in test tube with antigen-antiserum mixture and no agglutination in control.
      (b) Negative; no agglutination in test tube with antigen-antiserum mixture and no agglutination in control.
      (c) Nonspecific; both test and control mixtures agglutinate. Cultures with such results require additional testing as in "Identification of *Enterobacteriaceae*", P. R. Edwards and W. H. Ewing.[13]

25.554 "Spicer-Edwards" flagellar (H) test

Perform this test on formalized broth cultures using seven "Spicer-Edwards" flagellar (H) antisera [6, 9, 47] per culture as directed in 25.553b, 1 to 5 above. Identity of antigens is obtained by comparing the pattern of agglutination reactions obtained in the test procedure with agglutinins known to be present in each of the seven "Spicer-Edwards" antisera. This information is supplied by the manufacturers of the antisera.

25.555 Treatment of cultures found negative by the flagellar (H) tests

a. Test cultures for motility to determine if sufficient flagellar (H) antigens were present when the flagellar (H) test was performed.
b. Test cultures as follows:
   (1) Inoculate motility test medium in a petri dish with a loopful (3 mm) of growth from a TSI slant;
   (2) Inoculate by stabbing the medium once, 10 mm from the edge of the plate to a depth of 2 to 3 mm;
   (3) Do not stab to the bottom of the plate or inoculate any other portion;
   (4) Incubate 24 ± 2 hours at 35 C; do not invert the plates.
   (5) When organisms have migrated 40 mm or more, retest as follows:
      (a) Transfer a loopful (3mm) of growth which migrated farthest from the inoculation point to trypticase soy tryptose broth or brain heart infusion broth;
      (b) Test as directed in section 25.553a to b.
   (6) If cultures are not motile after the first 24 hours, incubate an additional 24 hours at 35 C; if still nonmotile, incubate 5 days at 35 C;
   (7) Classify as nonmotile if unable to obtain a motile subculture after several passages through semisolid motility test medium.

25.556 Serotyping of *Salmonella* cultures

To obtain definitive serotyping of *Salmonella* cultures, send them to a qualified *Salmonella* serotyping laboratory for final identification.

25.56 Arizona Serological Tests

25.561 The somatic (O) and flagellar (H) tests for cultures of Arizona are performed in the same manner as for *Salmonella*.

a. The availability of commercially produced Arizona antisera is limited, thereby making the serological detection and characterization of these cultures difficult. The directions supplied by the manufacturers of the antisera should be followed. Additional information on Arizona can be found in Chapter 10. Identification of *Enterobacteriaceae*, P. R. Edwards and W. H. Ewing.[13] Also refer to Bergey's manual.[5]

(b) The somatic (0) and flagellar (H) antigens of Arizona are closely related to certain antigens of *Salmonella*. Some positive agglutination reactions are the result of cross reactions (the agglutination of Arizona antigens with *Salmonella* antisera).

25.6 INTERPRETATION OF DATA

25.61 See Table 2. Characteristics of *Salmonella*, Arizona, and Non-*Salmonella*

25.62 See Table 3. Criteria for Discarding Cultures

25.7 REFERENCES

1. Abrahamsson, K., G. Patterson, and H. Riemann. 1968. Detection of *Salmonella* by a single-culture technique. Appl. Microbiol. **16**:1695–1698
2. Association of Official Analytical Chemists. 1975. Official Methods of Analysis, 12th ed., Washington, D.C.
3. Banwart, G. J. and J. C. Ayres. 1953. Effect of various enrichment broths and selective agars upon the growth of several species of *Salmonella*. Appl. Microbiol. **1**:296–301
4. Banwart, G. J. and M. J. Kreitzer. 1969. Rapid determination of *Salmonella* in samples of egg noodles, cake mixes, and candles. Appl. Microbiol. **18**:838–842
5. Buchanan, R. E. and N. E. Gibbons. eds., 1974. Bergey's Manual of Determinative Bacteriology, 8th ed., Williams and Wilkins, Baltimore, Md.
6. Center for Disease Control. 1972. Specifications for *Salmonella* Antisera. Reagents Evaluation Unit, Biological Reagents Section, CDC. 23pp, Atlanta, Ga.
7. Center for Disease Control. 1973. Foodborne Outbreaks, Annual Summary 1972. DHEW Publication No. (CDC) 74-8185, Atlanta, Ga.
8. Center for Disease Control. 1973. *Salmonella* Surveillance Report, Annual Summary 1972. DHEW Publication No. (CDC) 74-8219, Atlanta, Ga.
9. Difco Laboratories. 1974. Serological Identification of *Salmonella*. Technical Information Publication No. 0168, Detroit, Mich.
10. Edel, W. and E. H. Kampelmacher. 1968. Comparative studies on *Salmonella*—isolation in eight European laboratories. Bull. World Health Org. **39**:487–91
11. Edel, W. and E. H. Kampelmacher. 1973. Comparative studies on the isolation of "sublethally injured" salmonellae in nine European laboratories. Bull. World Health Org. **48**:167–174
12. Edwards, P. R. 1962. Serologic Examination of *Salmonella* Cultures for Epidemiologic Purposes. Communicable Disease Center, Atlanta, Ga.
13. Edwards, P. R. and W. H. Ewing. 1972. Identification of *Enterobacteriaceae*, 3rd. ed., Burgess Publishing Co., Minneapolis, Minn.
14. Edwards, P. R. and M. A. Fife. 1961. Lysine-iron-agar in the detection of Arizona cultures. Appl. Microbiol. **9**:478–480
15. Edwards, P. R., M. G. West, and D. W. Bruner. 1947. Arizona Group of Paracolon Bacteria. Ky. Agric. Exp. Sta. Bull. No. 499, Lexington, Ky.
16. Ewing, W. H. 1969. *Arizona hinshawii* comb. nov. Int. J. System. Bacteriol. **19**:1
17. Ewing, W. H., M. A. Fife, and B. R. Davis. 1965. The Biochemical Reactions of *Arizona arizonae*. Communicable Disease Center, Atlanta, Ga.
18. Gabis, D. A. and J. H. Silliker. 1974a. ICMSF methods studies. II. Comparison of analytical schemes for detection of *Salmonella* in high-moisture foods. Can. J. Microbiol. **20**:663–668
19. Gabis, D. A. and J. H. Silliker. 1974b. ICMSF methods studies. VII. The influence of selective enrichment media and incubation temperatures on the detection of *Salmonella* in dried foods and feeds. Can. J. Microbiol. **20**:1509–1511
20. Galton, M. M., W. D. Lowery, and A. V. Hardy. 1954. *Salmonella* in fresh and smoked pork sausage. J. Infect. Dis. **95**:232–235

21. Galton, M. M., G. K. Morris, and W. T. Martin. 1968. Salmonellae in Foods and Feeds. Review of Isolation Methods and Recommended Procedures. Communicable Disease Center, Atlanta, Ga.
22. Georgala, D. L. and M. Boothroyd. 1965. A system for detecting salmonellae in meat and meat products. J. Appl. Bact. **28**:206–212
23. Hobbs, B. C., in Chemical and Biological Hazards in Foods, J. C. Ayres, A. A. Kraft, H. E. Snyder, and H. W. Walker, eds. 1962. Iowa State University Press, Ames, Ia.
24. Huhtanen, C. N., J. Naghski, and E. S. Dellamonica. 1972. Efficiency of *Salmonella* isolation from meat- and bone-meal of one 300-g sample versus ten 30-g samples. Appl. Microbiol. **23**:688–692
25. Kauffmann, F. 1935. Weitere erfahrungen mit dem kombinierten Anreicherungs verfahren fur Salmonellabacillen. J. Hyg. Infektionskr. **117**:26–32
26. Kauffmann, F. 1966. The Bacteriology of *Enterobacteriaceae*. The Williams and Wilkins Co. Baltimore, Md. 400 pp.
27. Litchfield, J. 1973. CRC Critical Reviews in Food Technology. *Salmonella* and the food industry—methods for isolation, identification and enumeration. **3**:415–456
28. National Academy of Sciences, National Research Council. 1969. An Evaluation of the *Salmonella* problem. Academy Pub. No. 1683, Washington, D.C.
29. National Academy of Sciences, National Research Council. 1971. Subcommittee on Food Microbiology, Food Protection Committee. Reference Methods for the Microbiological Examination of Foods. Washington, D.C.
30. North, W. R., Jr. 1961. Lactose pre-enrichment method for isolation of *Salmonella* from dried egg albumin. Appl. Microbiol. **9**:188–195
31. North, W. R., Jr. and M. T. Bartram. 1953. The efficiency of selenite broth of different compositions in the isolation of *Salmonella*. Appl. Microbiol. **1**:130–134
32. Osburn, W. W. and J. L. Stokes. 1955. A modified selenite brilliant green medium for the isolation of *Salmonella* from egg products. Appl. Microbiol. **3**:295–299
33. Price, W. R., R. A. Olsen, and J. E. Hunter. 1972. *Salmonella* testing of pooled pre-enrichment broth cultures for screening multiple food samples. Appl. Microbiol. **23**:679–682
34. Raj, J. 1966. Enrichment medium for selection of *Salmonella* from fish homogenate. Appl. Microbiol. **14**:12–20
35. Rappaport, F. and N. Konforti. 1959. Selective enrichment medium for paratyphoid bacteria. Appl. Microbiol. **7**:63–66
36. Reed, R. B. and A. L. Reyes. 1968. Variation in plating efficiency of salmonellae on eight lots of brilliant green agar. Appl. Microbiol. **16**:746–748
37. Silliker, J. H. 1969. "Wet compositing" as an approach to control procedures for the detection of salmonellae. Appendix D, pp. 206–207, in An Evaluation of the *Salmonella* Problem. Committee on *Salmonella*, National Research Council—National Academy of Sciences. Publication No. 1683, Washington, D.C.
38. Silliker, J. H., R. H. Deibel, and P. T. Fagan. 1964a. Enhancing effect of feces on isolation of salmonellae from selenite broth. Appl. Microbiol. **12**:100–105
39. Silliker, J. H., R. H. Deibel, and P. T. Fagan. 1964b. Isolation of salmonellae from food samples. VI. Comparison of methods for the isolation of *Salmonella* from egg products. Appl. Microbiol. **12**:224–228
40. Silliker, J. H., P. T. Fagan, J. Y. Chiu, and A. Williams. 1965. Polyvalent H agglutination as a rapid means of screening non-lactose-fermenting colonies for *Salmonella* organisms. Am. J. Clin. Pathol. **43**:548–554
41. Silliker, J. H. and D. A. Gabis. 1973. ICMS methods studies. I. Comparison of analytical schemes for detection of *Salmonella* in dried foods. Canadian Jour. of Microbiol. **19**:475–479
42. Silliker, J. H. and D. A. Gabis. 1974. ICMSF methods studies. V. The influence of selective enrichment media and incubation temperatures on the detection of salmonellae in raw frozen meats. Canadian Jour. of Microbiol. **20**:813–816
43. Silliker, J. H. and R. A. Greenberg. 1969. Laboratory Methods, Analysis of Foods for Specific Food Poisoning Organisms—*Salmonella*. pp. 467–475 in Food-Borne Infections and Intoxications, H. Riemann, ed., Academic Press, New York, N.Y.
44. Silliker, J. H. and W. I. Taylor. 1958. Isolation of salmonellae from food samples. II.

The effect of added food samples upon the performance of enrichment broths. Appl. Microbiol. **6**:228–232

45. Smith, H. W. 1959. The isolation of salmonellae from the mesenteric lymph nodes and faeces of pigs, cattle, sheep, dogs and cats, and from other organs of poultry. J. Hyg. **57**:266–273
46. Sperber, W. H. and R. H. Deibel. 1969. Accelerated procedure for *Salmonella* detection in dried foods and feeds involving only broth cultures and serological reactions. Appl. Microbiol. **17**:533–539
47. Spicer, C. C. 1956. A quick method of identifying *Salmonella* "H" antigens. J. Clin. Path. **9**:378–379
48. Stuart, P. F. and H. Pivnick. 1965. Isolation of salmonellae by selective motility systems. Appl. Microbiol. **13**:365–372
49. Taylor, W. I. 1965. Isolation of Shigellae. I. Xylose lysine agars, new media for isolation of enteric pathogens. Amer. J. Clin. Path. **44**:471–475
50. Thatcher, F. S. and D. S. Clark. 1968. Microorganisms in Foods: Their Significance and Methods of Enumeration. pp. 90–106. University of Toronto Press, Toronto, Can.
51. U.S. Department of Agriculture. 1968. Recommended Procedure for the Isolation of *Salmonella* Organisms from Animal Feeds and Feed Ingredients. ARS 91-68. Animal Health Division, Agriculture Research Service, Hyattsville, Md.
52. U.S. Department of Agriculture. 1974. Scientific Services, Meat and Poultry Inspection Program, Animal and Plant Health Inspection Service. Microbiology Laboratory Guidebook, Washington, D.C.
53. U.S. Food and Drug Administration. 1972. Bacteriological Analytical Manual, 3rd ed. Chapter 8, pp. 1 to 25, Washington, D.C.
54. U.S. Food and Drug Administration. 1972. Compliance Program Guidance Manual. Salmonellae Sampling Plans. Washington, D.C.
55. Van Oye, E. 1964. World Problem of Salmonellosis. W. Junk, The Hague, Holland.
56. Wells, J. G., G. K. Morris, and P. S. Brachman. 1971. New method of isolating salmonellae from milk. Appl. Microbiol. **21**:235–239
57. Wun, C. K., J. R. Cohen, and W. Litsky. 1972. Evaluation of plating media and temperature parameters in the isolation of selected enteric pathogens. Health Lab. Sci. **9**:225–232

CHAPTER 26

FLUORESCENT ANTIBODY DETECTION OF SALMONELLAE

Berenice M. Thomason

26.1 INTRODUCTION

Salmonellae may be detected by fluorescent antibody (FA) because proteins, including serum antibodies, may be labeled by chemical combination with fluorescent dyes, such as fluorescein isothiocyanate (FITC). Labeled antibodies, when applied to smears of *Salmonella*, will attach to the cell walls of the organisms. These labeled antibodies become visible when illuminated by appropriate wavelengths of near ultraviolet or blue light. *Salmonella* stained by FA are seen under a fluorescence microscope as yellow green fluorescent rods against a dark background. FA staining can be considered the first stage of an agglutination test. One major advantage of the FA test over conventional serological procedures is that the FA test does not require pure cultures. Small numbers of pathogenic organisms can be detected by the FA test within smears containing large numbers of contaminants.[1]

The first report of the application of the FA technique to the detection of salmonellae in food products appeared in a Russian journal in 1962.[2] Since then, over 30 publications describing the application of the technique to a variety of food and feed products have appeared. Most of these papers and their data are tabulated by Cherry et al.[3] In the past, the variety of procedures and reagents used by the various investigators made it difficult to assess the value of the FA test for salmonellae. This may be one of the reasons why the technique has not been used widely in food laboratories. In their excellent reviews of the scientific literature, Ayres[4] and Goepfert and Insalata[5] evaluated the FA technique for detecting salmonellae, and cited as the critical problem the lack of a reliable commercial conjugate for performing this test. In the last few years, polyvalent conjugates suitable for the detection of salmonellae have become available from commercial sources. These reagents have been evaluated and found satisfactory.[6]

Finally, in October 1974, the Association of Official Analytical Chemists (AOAC) adopted as an "Official First Action" a fluorescent antibody method for the detection of salmonellae.[7] Now diagnostic laboratories have available a procedure recommended by the AOAC for screening specimens for these pathogens. Sufficient data should have been available by October

IS/A Committee Liaison: John H. Silliker

1975 for this procedure to be adopted as a "Final Action" or modified to improve the efficiency of the method.

Research is needed to improve the postenrichment broth which is the critical part of the AOAC procedure. The broth recommended, selenite cystine, works well for specimens contaminated with large numbers of salmonellae or for those specimens requiring preenrichment in a nonselective broth. Salmonellae in meat, poultry, water, and in environmental specimens which are not preenriched, may not be present in sufficient numbers for detection after 4 hour postenrichment in selenite cystine broth. This is pointed out in the precautions accompanying the AOAC method, but needs reemphasizing. Several modifications may overcome this problem without changing the recommended postenrichment broth. For instance, elevated temperatures have been found to be beneficial in isolating salmonellae.[8, 9, 10] For water samples, 48 hour incubation in tetrathionate broth at 41.5 to 43 C increases the number of salmonellae.[11, 12] Morris and Dunn[13] found that incubation at 43 C for 48 hours with plates streaked at both 24 and 48 hours yielded more isolates of salmonellae from fresh pork sausage than did lower temperatures. These and other modifications need to be investigated.

26.2 TREATMENT OF SAMPLE

Follow methods recommended in the AOAC procedure 26.8.

26.3 SPECIAL EQUIPMENT AND SUPPLIES

See Chapter 25 plus the following:

26.31 Microscope Slides (see 26.8)

Slides may be prepared as shown in Figure 1 by spotting cleaned slides with glycerol to cover a circle 5 to 6 mm in diameter and then spraying with Fluoroglide.[14] Caution: do not spray too heavily (opaque white) or the Teflon will flake off during staining and rinsing.

26.32 Coverslips (see 26.7)

Use coverslips of the thickness recommended by the manufacturer of the fluorescence microscope used. Some require a No. 1 coverslip whereas others recommend No. 1 ½.

26.33 Inoculating Loops

The AOAC recommends delivering about .0075 ml of the post enrichment broth with a 2 mm loop into a well of a multiwell slide (26.7). The .0075 ml amount specified is too much inoculum for a 6 mm well. The amount delivered should be standardized by delivering about .001 ml with one or the other loops described below. The loop recommended in the APHA Standard Methods for Examination of Diary Products is made of B & S #26 gauge wire with an internal diameter (I.D.) of 1.45 mm, and is designed to deliver .001 ml. A 2 mm (I.D.) loop made of B & S #26 gauge wire delivers slightly more than .001 ml, but is suitable for use with multiwell slides containing 5 or 6 mm wells. Either loop is satisfactory, but one or the other should be selected and used consistently for preparing smears for FA staining.

26.34 Fluorescence Equipment

A detailed account of the nature of fluorescence and the theory of fluorescence microscopy are not within the scope of this article. Furthermore, Chadwick and Fothergill,[15] Goldman,[16] Koch,[17] Richards,[18] Ploem,[19] and Kraft[20] have written excellent reviews on these subjects. Therefore, only the basic requirements in equipment for observing the FITC fluorescence of salmonellae will be discussed.

26.341 Illuminating methods

a. **Transmitted light.**

For years the only method of observing the fluorescence of microorganisms was by transmitted light. In this method the specimen was illuminated from below through a darkfield condenser which, in most cases, had to be oiled to the bottom of the slide. The proper centering and focusing of the condenser was very critical for observing maximal fluorescence.

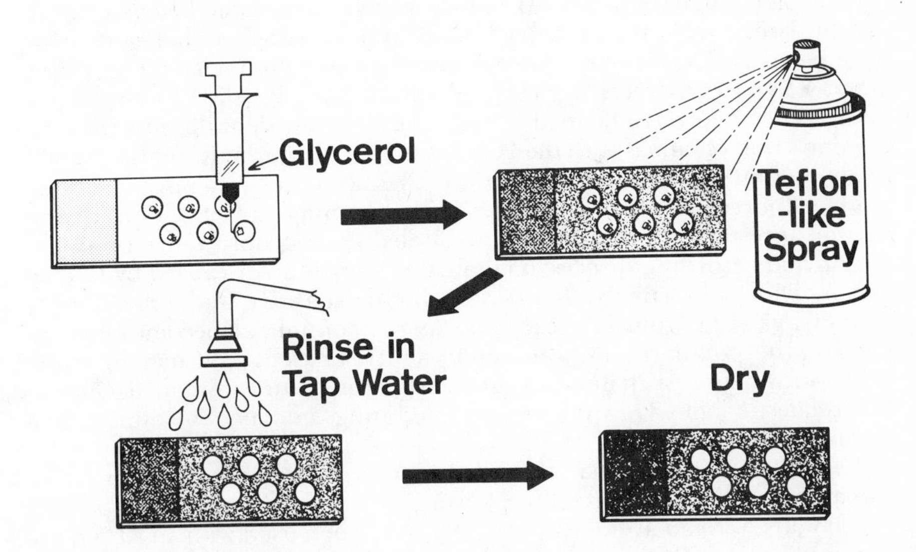

FIGURE 1. Preparation of multiwell, water repellant slides for fluorescent antibody staining**

**Reproduced from Thomason[14] with permission of publisher.

b. **Incident light.**

The most recent development in fluorescence excitation is the use of incident light in which the specimen is illuminated from above. This requires a special illuminator containing a dichroic beamsplitting mirror located between the microscope nosepiece and the binocular or monocular tube. The light passes through the illuminator where the beamsplitting mirror deflects the excitation wave lengths at right angles down through the objective and onto the specimen. The emitted fluorescent light is reflected back through the objective, the beamsplitting mirror, and the suppression filters to the microscope ocular. Since a darkfield condenser is not required, this method gives far greater fluorescence intensity than the transmitted light system, and the illuminator and microscope are much easier to align and focus. When the image is in sharp focus, both the excitation and emission lights are at maximum intensities. Another advantage of the incident light system is that it permits one to use the full numerical aperture of an objective without having to use funnel stops or objective diaphragms. This system also allows one to observe fluorescent organisms on opaque surfaces. For detecting unstained organisms, light transmitted from a low voltage tungsten lamp is admitted to an added darkfield condenser.

26.342 Light sources

Several light sources excite FITC. These are (a) ultrahigh pressure mercury burners, (b) high pressure zenon burners, (c) tungsten halogen lamps, and (d) lasers. Only the mercury burner and the tungsten halogen lamps have been used extensively for FA detection of salmonellae. The xenon lamp, XBO-150, is receiving more attention since Ploem[19] reported that fluorescence intensity obtained with this lamp by incident illumination was 2½ times that obtained with the HBO 200 mercury lamp. Recently, the use of another intense light source, the laser, has been recommended for quantitative fluorescence studies.[21, 22, 23] The advantage of this type of excitation appears to be that when it is coupled to electronic fluorescence detection and recording devices, it eliminates human error caused by fatigue and individual variability. It provides highly sensitive and reproducible quantitative data from FA reactions. As the equipment becomes less expensive, the popularity of both xenon and laser excitation may increase. However, whether such intensity is necessary for routine salmonellae detection is questionable. Only the mercury and tungsten halogen lamps will be discussed here.

The HBO 200 lamp manufactured by Osram in Germany is the most commonly used light source for FA work. It is a high intensity lamp rated at 200 watts, with an average life of 200 hours when used with an AC power supply. The average life can be prolonged to 400 hours if a DC power supply is used. Frequent starts tend to shorten the life of the bulb; once ignited it should burn at least one hour. Keep a record of the number of starts and burning times to anticipate the need for a new lamp, because emission decreases during the life of the lamp. Recently, the HBO 50 W mercury lamp has been used for FITC excitation. This lamp is smaller and easier to change than the HBO 200 W; however, the price is higher and its life is rated at only 100 hours. For incident light illumination, the HBO 100 W is the most intense light source, but it requires the use of a DC power supply.

The 12 V, 50 and 100 W tungsten halogen lamps are adequate for FITC excitation when used with either a 490 nm or a 500 nm interference filter and appropriate barrier filters. These lamps are inexpensive and have a rated life of 50 hours.

The various light sources, power supplies, and approximate costs are listed in Table 1.

26.343 Filters

a. **Exciting filters**

These also are called primary filters. They should transmit light of approximately 490 nm, which is the excitation maximum for FITC. Over the past few years, new filters have been manufactured which transmit 80% to 90% of the light in this region of the spectrum. These are the selective band interference filters, the KP490 and the KP500. Such filters, made of vapor film coatings sandwiched between glass, are highly susceptible to heat, and never should be placed close to the lamp. On microscopes utilizing transmitted light, usually they are mounted under the darkfield condenser.

Before interference filters were available, the Schott BG-12 or UG-1 filters were the most commonly used exciting filters. These filters transmit only 15% or less of the light near 490 nm, but have high transmission at wavelengths 365 and 400 nm respectively. Thus, they should be used with a light source, such as the mercury arc HBO-200, which has intense emission spectra at these wavelengths. These filters also may be used with xenon lamps which have high intensity emission at all wavelengths in the visible range.

b. **Suppression filters.**

The suppression or barrier filter is mounted between the microscope objective and the eye lens of the ocular. It minimizes the residual exciting light reaching the eye. For FITC fluorescence, either the K510 or the K530 suppression filter should be used. The K510 filter gives an apple green fluorescence; the K530, a yellow orange fluorescence.

c. **Heat filters.**

Protective heat filters must be used to prevent the exciting filter from overheating and cracking as a result of absorbing a large portion of the lamp emission. There are usually two types of heat filters. One is an almost colorless, heat resistant glass that has a high absorbance beyond 700 nm. Usually this is installed permanently in the light path in a position nearest to the lamp collecting lens. The other is a red ab-

Table 1: Light Sources and Power Supplies for Excitation of FITC Fluorescence

Light source	Rated life in hours	Approx. cost	Power supply	Approx. cost	Suitable for: transmitted light (A) incident light (B)
Mercury Vapor					
HBO-50*	100	$65	AC	$200	A & B
HBO-100**	200	$80	DC	$1000	B
HBO-200	200 AC	$65	AC	$250	A & B
	400 DC		DC	$650	A & B
Tungsten Halogen					
12V 50W*	50	$5	AC	$225	A & B
12V 100W*	50	$6	AC	$225	A & B
Xenon					
XBO-75**	400	$180	DC	$1000	B
XBO-150	1200	$100	DC	$800	A & B

*These light sources are adequate for most FITC excitation if used on appropriate microscopes. The larger research models require more intense light sources.
**The light sources are recommended for incident light illumination when quantitative measurements of fluorescence are required.

sorbing filter which is a blue green glass with high broad band transmission at shorter wavelengths and reduced transmission above 500 nm. This red absorbing filter is not very heat resistant and must be protected by the heat absorbing filter if it is used with an intense excitation source. The red absorbing filter must be used with the interference filter to block the transmission of red light.

The filter combinations listed in Table 2 are recommended for FA detection of the salmonellae.

26.344 Microscope

The microscope should be of high optical and mechanical quality. An expensive research model is not needed because most laboratory microscopes are satisfactory for routine FA work when equipped with the proper light source and filter combinations. It is much wiser to purchase a simple instrument for use in FA work only, and to become thoroughly proficient in its use.

a. Observation tubes

The choice of a monocular or binocular tube is determined by the light source and filter combinations. With the high pressure lamps and the new interference filters, a binocular tube should be satisfactory for observing fluorescence from weakly fluorescent organisms. When the tungsten halogen lamp is used, the loss of light through a binocular tube may result in false negative readings of fluorescence in the 1+, 2+, or 3+ range.

Table 2: Light Source and Filter Combinations for Obtaining Optimal Fluorescence of FITC

Light source	Heat filters	Excitation filters transmitted light (A) incident light (B)	Suppression filter
Mercury Vapor			
HBO50	KG1 + BG38	BG12 (B)	K510 or K530
		KP490 (A)	"
		2 × KP490* or KP500 (A & B)	"
HBO100	KG1 or Bl/K2** + BG38	BG-12 (B)	K510 or K530
		2 × KP490 or	"
		KP500 (B) + K450 or K480 edge filter***	"
HBO200	KG1 or Bl/K2 + BG38	BG12 (A & B)	K510 or K530
		KP490 (A)	"
		2 × KP490 or	"
		KP500 + K450 or K480 edge filter	"
Tungsten Halogen			
12V 50W or 12V 100W	KG1 + BG38	KP490 (A & B)	K510 or K530
		KP500 (A & B)	"
Xenon			
XBO75	B1/K2 + BG38	BG12(B)	K510 or K530
		2 KP490 or KP500 (B) + K450 or K480 edge filter	"
XBO150	Bl/K2 + BG38	BG12 (A & B)	K510 or K530
		KP490 (A)	"
		2 KP490 or	"
		KP 500 (A & B) + K450 or K480 edge filter	"

*2 × KP-490 equals two separate KP-490 filters.
**The B1/K2 filter is a more effective heat filter than the KG-1.
***When using incident light with the Xenon and the Mercury lamps, the edge filter will help eliminate background fluorescence and give better contrast.

b. **Condenser**

An oil darkfield condenser of the bispheric or cardioid type is recommended for use with transmitted light.

c. **Objectives**

Generally, most microscope objectives are suitable for observing fluorescence by transmitted darkfield illumination. With this system, the field of view is illuminated only partially under the 10x objective; therefore, this objective is used solely for focusing and centering the oil immersion darkfield condenser, focusing the specimens, and rapid scanning. For medium power, a 40x to 50x objective is very useful. For higher magnifications, a 90x to 100x oil immersion objective is required, and this should be fitted with a funnel stop or adjustable iris diaphragm for maximal contrast. For incident illumination, complete sets of oil, glycerol, and even distilled water, immersion objectives are available, and these perform much better than the dry objectives. Dry objectives, in general, have not proven as satisfactory as the immersion objectives because of lower light gathering properties. Achromat objectives are suitable for most FA work, since they are corrected both chromatically and spherically for green light.

d. **Oculars**

Compensating oculars of 8 to 10x are the most useful for FA work. Oculars of higher magnifications should be used only with low power objectives. Many workers prefer widefield oculars, although they reduce fluorescence slightly. Oculars fitted with eyeshields reduce the need for a darkened room for fluorescence observation.

26.345 Immersion oil

A nonfluorescent mineral oil, or one with very, very low fluorescence such as Cargille's Type A, should be used for oiling the condenser to the slide and the slide to the oil objectives.

26.4 SPECIAL REAGENTS AND MEDIA

See 26.7

26.41 Trypticase Soy Tryptose Broth (TST) (Chapter 2)

26.5 RECOMMENDED CONTROLS

26.51 FA Reagent

a. To adequately determine the sensitivity of a polyvalent reagent, it should be checked with all of its homologous antigens. Thus, each laboratory should have stock strains of *Salmonella* O groups A to S for testing the polyvalent reagents. These strains should be kept in nutrient agar stabs sealed with waxed cork stoppers. The stock stabs should be stored in the dark at room temperature.

The stocks are transferred to TST broth, incubated 12 to 18 hours at 35 C, and killed by adding an equal volume of 0.85% NaCl containing 0.6% formalin. The somatic antigens of each broth culture should be stained to maximum fluorescence (4+) with a conjugate of known potency. At this point, a set of smears can be made, fixed, and frozen at –20 or –70 C. These will remain stable for a year or more and can be used to check various batches of conjugates.[24] Each user must titer any conjugates received as undiluted reagents. Twofold dilutions of the rehydrated conjugate should be made in PBS and used to stain the homologous *Salmonella* serotypes. Titers may vary from one laboratory to another, depending upon the type of fluorescence assembly used. The working dilution should be the one just below the highest dilution that stain all strains at a 4+.

b. A sample known to contain salmonellae should be included in every run. This sample should be carried through all enrichment steps and stained along with the unknown specimens. It serves as a check on cultural methods, FA staining procedures, and equipment. If a known positive sample cannot be included when smears are being stained, a smear of a pure culture of *Salmonella* should be included.

26.6 PRECAUTIONS AND LIMITATIONS OF METHODS

See section 26.7

26.61 Specificity of Reagents

Testing of commercial *Salmonella* polyvalent reagents with heterologous enteric bacteria[6] showed that the major cross reacting genera among the *Enterobacteriaceae* were *Arizona, Citrobacter,* and *E. coli*. The percentages of cross reactions obtained are summarized in Table 3. Only fluorescence reactions of 2+ or better were considered positive.

Table 3. Staining Reactions of *Salmonella* Conjugates with Arizona, *Citrobacter*, and *E. coli* O Groups*

Conjugate	Arizona (36)	*Citrobacter* (33)	*E. coli* (145)
Difco Panvalent (A – 064)	89%	27%	14%
Difco polyvalent (A – S)	42%	15%	13%
Clinical Sciences Fluoro-kit (A – S)	39%	12%	14%
Sylvana polyvalent (A – S)	39%	18%	11%

**Premarket evaluation of these reagents showed that none of the conjugates stained any organism of other Enterobacteriaceae tested with more than a 2+ intensity.*

These data emphasize the fact that FA positive results can be considered presumptive evidence only of the presence of salmonellae. Their presence must be confirmed by cultural isolation and serologic identification.

26.62 FA Staining of Selective Enrichments

Salmonella organisms may be stained in smears prepared from either selenite or tetrathionate broths. The ease of detection depends upon the nature of the sample. Smears made from broths inoculated with fat containing foods such as pork sausage, raw meat, or poultry will contain fat droplets, and other sample debris that may make interpretation difficult, if not impossible. In such cases, the specimen must be enriched further to dilute the sample material and allow multiplication of the salmonellae (26.7).

26.63 Influence of Large Numbers of Organisms in Smears

Large numbers of salmonellae present in a smear will depress the intensity of fluorescence because of the excess of antigen over available antibody. Investigators should be careful to make thin smears of the broths to avoid this possibility. If a smear contains large numbers of organisms, areas with fewer cells should be looked for. It may be necessary to dilute the broth 1:10 to 1:20 and restain the smear.

26.64 Influence of Small Numbers of Organisms in Smears

A minimum of approximately 10^5 salmonellae per milliliter is required to see one *Salmonella* per oil immersion field. The recommended procedures for the enrichment of samples usually result in smears with approximately 20 *Salmonella* organisms per field when the original specimen contains approximately one viable *Salmonella* per 25 grams as determined by the most probable number (MPN) test.

26.7 PROCEDURE (Taken from AOAC Fluorescent Antibody (FA) Method Official First Action)

26.71 Precautions

This method is a screening test only for the presence of *Salmonella*; it is not a confirmatory test, since the conjugate will react with some other members of the *Enterobacteriaceae*.

Enrichment broths from samples positive by the FA method must be streaked on selective media as in AOAC 46.016, and typical or suspicious colonies identified as in AOAC 46.017–46.026.

The method must be followed rigorously since errors in preparation of the sample, smears, conjugate, and other reagents can lead to invalid results. Microscopic observation of stained smears must be performed with critically aligned and properly functioning equipment.

Visual estimation of degree of fluorescence of stained cells is somewhat subjective and should be conducted by an analyst with prior training or experience in both FA methodology and in cultural technic for detection of *Salmonella*.

If the sample preparation does not normally include a preenrichment step (as with meat, poultry, and certain environmental samples), the 4 hour postenrichment incubation period may not be sufficient for development of the minimum number of *Salmonella* cells required for detection by the FA method. Therefore, include the preenrichment step or extend the post-enrichment incubation time. In some cases, when the preenrichment step is not used, the sample is not adequately diluted, and carryover of debris into postenrichment broth may interfere with observation of FA stained cells.

26.72 Apparatus

a. *Multiwell coated slides.*—Clean thin (1.0–1.2 mm) slides thoroughly with detergent and rinse with distilled water and alcohol. Apply a double row of 4 separate drops of glycerol (8 drops total) to each of the series of slides and spray with fluorocarbon coating material (Fluoroglide, Fisher Scientific Co.). After a few minutes, rinse off each slide individually under the tap and then with distilled water, and stand on end in a rack to dry. (Prepared slides are available from Cell-Line Associates, Minotola, NJ 08341 and Clinical Sciences, Inc., 30 Troy Rd, Whippany, NJ 07981.)
b. *Fluorescent microscope.*—With exciter filter with wavelength transmission of 330–500 nm and barrier filter with wavelength reception > 400 nm.

26.73 Reagents

a. *Phosphate-buffered saline (PBS) solution.*—pH 7.5; 0.01*M*; 0.85% NaCl. Dissolve 12.0 g anhydrous Na_2HPO_4, 2.2 g $NaH_2PO_4 \bullet H_2O$, and 85.0 g NaCl in water and dilute to 1 liter. Dilute 100 ml of this solution to 1 liter with water. Adjust pH to 7.5 with 0.1*N* HCl or 0.1*N* NaOH, if necessary.
b. *Carbonate buffer.*—pH 9.0. Mix 4.4 ml 0.5*M* Na_2CO_3 (5.3 g in 100 ml water) with 100 ml 0.5*M* $NaHCO_3$ (4.2 g in 100 ml water). pH should be 9.0; if not, adjust by addition of 0.5*M* Na_2CO_3.
c. *Glycerol saline solution.*—pH 9.0. Mix 9 ml glycerol with 1 ml carbonate buffer, (b). pH decreases on storage; prepare weekly.
d. *Salmonella polyvalent fluorescent antibody conjugate.*—Fluorescein isothiocyanate-labelled Salmonella OH globulin, polyvalent, containing antibodies for all antigens within *Salmonella* O groups A-S, and meeting specifications of the Center for Disease Control, Atlanta, GA 30333 (1975). (Available from Difco Laboratories; Clinical Sciences, Inc., 30 Troy Rd, Whippany, NJ 07981; Sylvana Co., 22 E. Willow St., Millburn, NJ 07041.) Before use, titer each lot to determine appropriate routine test dilution (RTD). Use pure cultures of *Salmonella* representative of several somatic groups. Prepare 5 dilutions (1 : 2, 1 : 4, 1 : 8; 1 : 16, and 1 : 32) of conjugate in PBS solution, (a). Stain duplicate smears from cultures with each dilution and determine intensity of fluorescence. RTD is that dilution one less than the highest dilution

giving 4+ fluorescence with representative *Salmonella* cultures. Freeze and store stock (undiluted) conjugate of known titer, and dilute when needed. Diluted conjugate can be stored at 4° for a few weeks as long as control cultures remain positive.

26.74 Determination

a. *Pre-enrichment.*—Pre-enrich the product in noninhibitory broth to initiate growth of salmonellae. Methods used vary with the product as in 1–8. In all cases loosen jar caps ¼ turn and incubate 24 ± 2 hr at 35 C. Except where selenite cystine and tetrathionate broths, AOAC 46.013(b)(*1*) or (*2*) and (c), respectively, have already been used (2(*b*) and 5), transfer 1 ml of incubated mixtures to selenite cystine broth and tetrathionate broth for selective enrichment as in AOAC 46.016(a). Where these broths have already been used (2(*b*) and 5), proceed directly to postenrichment (b).

1. *Dried yeast (inactive).*—Weigh 50 g into sterile, wide-mouth, screw-cap, 500 ml (pt) jar, add 200 ml sterile distilled water, and mix well. If pH is <6.6, adjust to 6.8±0.2 with 1*N* NaOH.
2. *Meats, animal substances, glandular products, and fish meal.*
   (*a*) *Heated, processed, and dried products.*—Weigh 25 g into sterile blending jar, add 225 ml sterile lactose broth, AOAC 46.005(f), and blend 2 min at 8000 rpm. If product is powdered, ground, or comminuted, blending may be omitted. Transfer aseptically to a sterile, wide-mouth, screw-cap, 500 ml (pt) jar and adjust pH to 6.8 ± 0.2 with 1*N* NaOH. If product contains a large amount of fat, add 2.2 ml of steamed (15 min) Tergitol Anionic 7, AOAC 44.003(jj)(*1*).
   (*b*) *Raw and highly contaminated products.*—Weigh duplicate 25 g samples into separate sterile blending jars. Add 225 ml of selenite cystine broth to one jar and 225 ml of tetrathionate broth to the other, and blend 2 min. Transfer aseptically to sterile, wide-mouth, screw cap, 500 ml (pt) jars.
   (*c*) *Raw frog legs.*—Aseptically place 2 legs into single wide-mouth, screw cap, 500 ml (pt) jar containing 225 ml sterile lactose broth, AOAC 46.005(f).
3. *Dry nonfat and dry whole milk.*—Weigh 100 g into sterile 2 liter flask, add 1 liter sterile distilled water, and mix well. Adjust pH to 6.8±0.2 with 1*N* NaOH, if necessary. Add 2 ml 1% aqueous brilliant green solution and mix well.
4. *Dried whole eggs, yolks, and whites; pasteurized liquid and frozen eggs; prepared powdered mixes (cake, cookie, donut, biscuit, and bread); and infant formula.*—If product is frozen, thaw rapidly at ≤45 C for ≤15 min or overnight at 5–10 C. Weigh 25 g into a wide-mouth, screw cap jar. Add 225 ml lactose broth, a little at time with mixing, cap jar, and let stand at room temperature 60 min. Mix well and adjust to pH 6.8±0.2 with 1*N* NaOH or HCl.
5. *Nonpasteurized frozen egg products.*—Thaw as in 4. Weigh duplicate 25 g samples into seperate sterile, wide-mouth, screw cap, 500 ml

(pt) jars. Add 225 ml selenite cystine broth to one jar and 225 ml tetrathionate broth to the other, and mix well. Adjust pH to 6.8±0.2 with 1*N* NaOH.

6. *Egg-containing foods (noodles, egg rolls, etc.)*—Proceed as in 2(*a*).
7. *Coconut.*—Proceed as in 2(*a*), using Tergitol Anionic 7, but omitting blending.
8. *Candy and candy coatings.*—Weigh 100 g into sterile blending jar. Add about 300 ml from 1 liter sterile reconstituted skim milk (100 g nonfat dry milk dispersed in 1 liter water), blend 2 min, and add remainder of skim milk. Adjust pH to 6.8±0.2 with 1*N* NaOH, if necessary. Add 4 ml 1% aqueous brilliant green solution and mix well.

b. *Postenrichment.*—Transfer 1 ml of incubated selenite cystine enrichment broth to 10 ml of sterile selenite cystine broth as post-enrichment. (Other volumes may be used if 1 : 10 dilution ratio is maintained.) Take an aliquot from upper third of selective enrichment cultures to minimize product carryover. Similarly transfer 1 ml of incubated tetrathionate enrichment broth to 10 ml of sterile selenite cystine broth. Incubate 4 hr in 35 C water bath.

c. *Staining.*—Transfer about 0.0075 ml of each post-enrichment medium with a sterile 2 mm loop into separate wells of a multiwell coated slide, and dry thoroughly in air at room temperature. Fix by immersion in a bath of alcohol-$CHCl_3$-formalin (60 + 30 + 10) for 3 min. Rinse 2 or 3 times in alcohol, and air dry at room temperature. Change alcohol periodically to prevent cell carryover (250 ml alcohol will rinse 5 to 10 slides). Slides may also be fixed and rinsed by flooding. Apply solutions to one end of slide and allow to flow into wells.

Cover dried smears with titered *Salmonella* polyvalent FA conjugate and let stain in moist chamber 15 to 30 min. *FA conjugate must not dry on smear.* (A covered plastic petri dish containing a piece of filter paper moistened with water is an excellent staining chamber.) Drain excess conjugate by standing slide on its edge a few seconds. (Avoid mixing conjugate from one well to another on each slide.) Immediately rinse slides in PBS solution, AOAC 46.A05(a). Then soak slides 10 min in fresh PBS solution and rinse briefly with water. Air-dry smears again at room temperature and then mount by placing a drop of glycerol saline solution (b), directly onto each smear and cover with No. 1 glass cover slip. Add enough glycerol saline solution to each smear to ensure adequate, but not excessive, coverage of all wells after cover slips have been emplaced. Do not trap air bubbles under cover slip.

d. *Examination.*—Examine smears with the fluorescent microscope. Scan the entire smear using a 40 to 50 magnification oil immersion objective to locate fluorescent cells. When found, change the objective to the 100 magnification oil immersion lens for definitive determination of cell morphology and fluorescence. Objectives with the iris diaphragm for adjusting the numerical aperture are helpful for control of contrast between cells and background. Estimate the degree of fluorescence of cells on a scale of negative to 4 + as follows:

4 + = Maximum fluorescence; brilliant yellow-green; clear-cut cell outline; sharply defined cell center.
3 + = Less brilliant yellow-green fluorescence; clear-cut cell outline; sharply defined cell center.
2 + = Definite but dim fluorescence; cell outline less well defined.
1 + = Very subdued fluorescence; cell outline indistinguishable from cell center in most instances.
+ = Negligible or complete lack of fluorescence (neg).

Typical positive smears for salmonellae exhibit ≥ 2 short to medium rod-shaped cells per field, using 100 magnification objective. Cells should be distributed throughout the entire smear. The intensity of fluorescence should be in a range of 3 + to 4 +. Occasionally cells are observed with proper morphology and cell distribution, but fluorescence is rated 2 +. Sometimes 3 + to 4 + fluorescence is observed, but distribution is poor and not all fields contain cells, due to improper processing of slides. Score both cases positive and subject to confirmatory tests.

Each time samples are tested, carry culture of known *Salmonella* strain through all cultural, staining, and observation steps as a control.

Report: (1) morphological characteristics of fluorescent cells; (2) number of typical cells per field under 100 magnification oil immersion objective; and (3) degree of fluorescence of cells (1 + to 4 +).

26.8 ACKNOWLEDGMENTS

Wayne Vashaw, southeastern district representative, E. Leitz, Inc., technical information on fluorescence equipment.

26.9 REFERENCES

1. THOMASON, B.M., M.D. MOODY, and M. GOLDMAN. 1956. Staining bacterial smears with fluorescent antibody. II. Rapid detection of varying numbers of *Malleomyces pseudomallei* in contaminated materials and infected animals. J. Bacteriol. **72**:362–367
2. ARKHANGELSKII, I.I. and V.M. KARTASHOVA. 1962. Accelerated methods of detecting *Salmonella* in milk. Veterinariya **9**:74–78
3. CHERRY, W.B., B.M. THOMASON, J.B. GLADDEN, N. HALSING, and A.M. MURLIN. 1975. Detection of Salmonellae in foodstuffs, water, and feces by immunofluorescence. Ann. N.Y. Acad. Sci. **254**:350
4. AYRES, JOHN, C. 1967. Use of fluorescent antibody for the rapid detection of enteric organisms in egg, poultry, and meat products. Food Technol **21**:145–154
5. GOEPFERT, J.M. and N.F. INSALATA. 1969. Salmonellae and the fluorescent antibody technique: A current evaluation. J. Milk & Food Technol. **32**:465–473
6. THOMASON, BERENICE M. and G. ANN HEBERT. 1974. Evaluation of commercial conjugates for fluorescent antibody detection of salmonellae. Appl. Microbiol. **27**:862–869
7. Fluorescent antibody (FA) method. 1975. Official First Action. Section 46.A03-46.A06-J.A.O.A.C. **58**:417–419
8. HARVEY, R.W.S. and S. THOMSON. 1953. Optimum temperatures of incubation for isolation of Salmonellae. Man. Bull. Min. Health (London) **12**:149–150
9. DIXON, J.M.S. 1961. Rapid isolation of Salmonellae from feces. J. Clin. Path. **14**:397–399
10. CARLSON, V.L., G.H. SNOEYENBOS, B.A. MCKIE, and C.F. SNYSER. 1967. A comparison of incubation time and temperature for the isolation of *Salmonella*. Avian Dis. **11**:217–225

11. Spino, D.F. 1966. Elevated temperature technique for the isolation of salmonellae from streams. Appl. Microbiol. **14**:591–596
12. Cheng, Chu Ming, W.C. Boyle, and J.M. Goepfert. 1971. Rapid quantitative method for *Salmonella* detection in polluted waters. Appl. Microbiol. **21**:662–667
13. Morris, G.K. and C.G. Dunn. 1970. Influence of incubation temperature and sodium heptadecyl sulfate (Tergitol No. 7) on the isolation of Salmonellae from pork sausage. Appl. Microbiol. **20**:192–195
14. Thomason, Berenice M. 1971. Rapid detection of *Salmonella* microcolonies by fluorescent antibody. Appl. Microbiol. **22**:1064–1069
15. Chadwick, C. S. and J. E. Fothergill. 1962. Fluorescent Protein Tracing. Ed. R. C. Nairn. E. & S. Livingston, Ltd., Edinburgh & London, England, pp 4–30
16. Goldman, Morris. 1968. Fluorescent Antibody Methods. Academic Press, New York, N.Y. pp 21–86
17. Koch, K. F. 1971. Lichtquellen fur die Fluoreszenmikroskopie. 1. FITC Immunofluoreszenz. Leitz-Mitt Wiss u Techn V., Wetzlar, Germany
18. Richards, O. W. 1965. Fluorescence Microscopy Analytical Cytology. R. C. Mellors, ed. McGraw-Hill, New York, N.Y. pp 1–37
19. Ploem, J.S. 1971. A study of filters and light sources in immunofluorescence microscopy. Ann. of the N.Y. Acad. Sciences **177**:414–429
20. Kraft, W. 1973. Fluorescence Microscopy and Instrument Requirements. Scientific and Technical Information. E Leitz, Wetzlar. **11**:97–109
21. Kaufman, G.I., J.F. Nester, and D.E. Wasserman. 1971. An experimental study of lasers as excitation sources for automated fluorescent antibody instrumentation. J. Histochem. Cytochem. **19**:469–476
22. Bergquist, N.R. 1973. The pulsed dye laser as a light source for the fluorescent antibody technique. Scand. J. Immunol. **2**:37–44
23. Kasatiya, S.S., N.G. Lambert, and R.A. Laurence. 1974. Use of tunable, pulsed dye laser for quantitative fluorescence in syphilis serology (FTA-ABS test). Appl. Microbiol. **27**:838–843
24. Thomason, Berenice M. 1974. Evaluation of frozen fixed smears for use in fluorescent antibody studies of salmonellae. Appl. Microbiol. **27**:418–419

CHAPTER 27

SHIGELLA

George K. Morris, M. J. Nakamura, and Joy G. Wells

27.1 INTRODUCTION

27.11 Description of the Genus

The genus *Shigella* consists of gram negative, aerobic, nonsporulating, nonmotile bacteria in the family *Enterobacteriaceae*. They do not decarboxylate lysine, and only one serotype, *Shigella flexneri* 6, produces gas. Lactose usually is not fermented, but when it is, the fermentation usually is delayed for several days. The *Shigella* genus consists of 4 species: *S. dysenteriae* (serogroup A), *S. flexneri* (serogroup B), *S. boydii* (serogroup C), and *S. sonnei* (serogroup D). The organisms formerly described as *S. alkalescens* and *S. dispar* now constitute the Alkalescens-Dispar group of *Escherichia coli* and are nonmotile, anaerogenic biotypes.[6]

27.12 Distribution and Epidemiology

Shigellae are host adapted to man and higher primates. Shigellosis is an infectious disease, most commonly spread by person to person transmission, which affects primarily younger children because of poor personal hygiene and those of the lower socioeconomic strata because of inadequate sanitation and crowded living conditions. Water is sometimes a mode of common source spread, and several reports of foodborne shigellosis have been published.[5, 8, 9, 10, 20] Twenty-one foodborne outbreaks of shigellosis were reported to the Center for Disease Control between 1964 and 1968.[5] In all outbreaks in which the vehicle was identified, 7 were traced to salads, 5 of the 7 to potato salad.

27.13 Contamination of Foods

Foods become contaminated from contact with fecal material containing *Shigella*, most likely from human hands. The handling of the multiple ingredients in salads probably accounts for salads as prominent vehicles of foodborne shigellosis. *Shigella* organisms usually do not compete well when mixed with large numbers of other bacteria, but in foods like salads, in

IS/A Committee Liaison: George J. Herman

which many ingredients such as eggs, potatoes, macaroni, and shrimp are cooked prior to mixing, *Shigella* may be the predominant flora.

27.2 TREATMENT OF SAMPLE

Foods should be analyzed as soon as possible after collection. The sample should be held at refrigerator temperature if it is to be analyzed within 24 hours, and frozen if it is to be held longer than 24 hours. *S. sonnei* survived in stewed apples (pH 3.2) for a week at 20 C, but died off in 24 hours at 37 C.[2] The same organism reportedly survived in cheese stored at 4 C for 19 to 72 days.[2] *S. sonnei* and *S. flexneri* 2a persisted in flour and milk for over 100 days at 25 C; in eggs, clams, oysters, and shrimp for 30 days at –20 C; and in orange juice, tomato juice, cooking oil, root beer, and ginger for shorter periods.[18] Nakamura and Taylor[14] reported that *S. flexneri* and *S. sonnei* survived in human feces, human urine, and human blood for over 100 days when these specimens were stored at –20 C.

27.3 RECOMMENDED CONTROLS

27.31 Media Controls

A known *Shigella* isolate should be included as a control for all isolation and biochemical tests. Lot to lot variation in commercial media has been reported by several workers.[3, 11, 15]

27.32 Antiserum Controls

Because of the frequency of serologic cross reactions between *Shigella* and other enteric bacteria, and the questionable quality of some commercial *Shigella* antiserum, all antiserum should be tested against all 4 *Shigella* species frequently.

27.4 PRECAUTIONS AND LIMITATIONS OF THE METHODS

27.41 Variations in Growth on Bacteriologic Media

S. sonnei has been shown to be suppressed more than other species by salmonella-shigella (SS) agar[12, 21] and deoxycholate citrate agar.[21] *S. dysenteriae* 1 is inhibited by SS agar and slightly suppressed by xylose-lysine-deoxycholate (XLD) agar. It is isolated best with relatively nonselective media such as Tergitol 7 agar or MacConkey's agar.[1, 4]

27.42 Sensitivity of Procedure

Methods used for isolating *Shigella* are not sensitive procedures. This is indicated by the high frequency with which an isolation is made from only 1 plate when multiple plating media are used. Many other bacteria are indistinguishable from *Shigella* on commonly used enteric plating media. Also, *Shigella* species undergo colonial variation on usual culture media. Occasionally, therefore, a species of *Shigella* may not exhibit the expected colonial morphology. Thus, it is essential that multiple plating media be used and that multiple colonies be picked for testing.

27.43 Disease caused by Few Organisms

The number of *Shigella* organisms that constitutes an infective dose is very low, possibly as low as 10 organisms. This increases the challenge to the microbiologist to detect low numbers of *Shigella* that may be infective doses in foods.

27.44 Survival of *Shigella* in Foods

Large numbers of shigellae experimentally added to food preparations are known to survive for long periods of time, but small numbers of shigellae do not survive readily in foods, particularly in those in the acid pH range. It is suspected that food preservatives inhibit *Shigella* species. For example, sodium benzoate, even at concentrations of 0.1%, may be inhibitory. This may reflect increased nutritional requirements of debilitated or injured cells. For example, shigellae injured by freezing require a more complex medium for recovery.[13]

27.5 METHODS FOR ISOLATING SHIGELLAE FROM FOODS

An enrichment broth is recommended for isolating *Shigella* from foods. Although direct plating is preferred for fresh fecal specimens,[12] enriching specimens in Gram negative (GN) broth has been effective in isolating *Shigella* from stool specimens received in clinical laboratories[19] and from food samples.[7]

Three types of plating media should be used to isolate *Shigella* from foods: one each of high selectivity, intermediate selectivity, and low selectivity. XLD agar (intermediate selectivity) should be used because it has been shown in many reports to be superior to other plating media for isolating *Shigella*.[12, 16, 17] It contains xylose as a differentiating agent, and since most *Shigella* do not ferment xylose, they appear as alkaline (red) colonies on the plate. *Shigella* isolates which ferment xylose rapidly, however, can be missed on this plating media. Therefore, additional plating media containing lactose as the differentiating sugar are recommended. Some shigellae ferment lactose slowly, but freshly isolated *Shigella* do not ferment lactose.[6] A medium of high selectivity should be used to prevent overgrowth by competitive organisms in foods, and one of low selectivity to isolate more fragile shigellae (e.g., *S. dysenteriae* 1).

27.51 Enrichment Procedure

a. If food is frozen, thaw a portion of the specimen for analysis.
b. Weigh a 25 gram sample into a jar (capacity approximately 500 ml). Products such as meat or vegetables may be cut into small pieces with scissors.

c. Add 225 ml of GN broth and thoroughly mix sample with broth.
d. Incubate at 35 to 37 C for 18 hours.

27.52 Plating on Selective Agar

a. Prepare dried plates of 3 selective agar media; one plate should be selected from each of the following groups:

(1) Low selectivity	Tergitol 7 agar or MacConkey's agar (Chapter 2)
(2) Intermediate selectivity	XLD agar (Chapter 2)
(3) High selectivity	SS agar, Hektoen enteric agar, or deoxycholate citrate agar (Chapter 2)

b. Transfer a 5 mm loopful of each enrichment broth culture to the surface of the 3 media and streak to obtain isolated colonies.

c. Invert and incubate plates at 35 to 37 C for 24 hours.

(1) Typical *Shigella* on XLD agar appear as red or pink colonies, usually about 1 mm in diameter. Colonies with black centers are not *Shigella*, but most likely *Proteus*, *Salmonella*, or Arizona. False positive colonies are frequently observed with *Providencia*, *Proteus*, and *Pseudomonas*.
(2) Typical *Shigella* appear as opaque or transparent colonies on deoxycholate citrate agar, MacConkey's agar, and SS agar.
(3) On Hektoen enteric agar *Shigella* appear as blue colonies without black centers.
(4) On Tergitol 7 agar *Shigella* appear as blue colonies.

27.53 Identifying Shigella

27.531 Biochemical screening

a. Inoculate each suspect colony to a tube of triple sugar iron (TSI) agar by streaking the slant and stabbing the butt.
b. Incubate the cultures overnight at 35 to 37 C.
c. Purify the cultures that appear contaminated on TSI agar by streaking each culture on separate plates of MacConkey's agar to obtain isolated colonies, incubating plates at 37 C for 24 hours, and transferring cells from an isolated colony to a tube of TSI agar. Incubate these tubes at 37 C for 24 hours.
d. Discard cultures that do not give reactions typical of *Shigella* in TSI agar. Typical reactions are indicated by a red slant (alkaline reaction) and a yellow butt (acid; glucose fermentation), with no H_2S (indicated by blackening of the medium) or gas*. Typical cultures are confirmed by further testing biochemically and serologically.
e. The minimum number of characters needed to identify *Shigella* is shown in Table 1. Further characteristics are described elsewhere.[6]

---

*Certain biotypes of *S. flexneri 6* produce gas.

Cultures may be screened in urea agar and motility medium. Urease positive or motile cultures may be discarded without further examination.

27.532 Serologic screening

a. Using a wax pencil, mark off 2 sections about 1 × 2 cm on the inside of a glass Petri dish or on a 2 × 3 inch glass slide.
b. Place a small amount (1.5 mm loopful) of culture from a nutrient agar or TSI agar slant (after 24 or 48 hours' incubation) directly on the dish or slide in the upper part of each marked section.
c. Add 1 drop of 0.85% sodium chloride or mercuric iodide solution to the lower part of each marked section. With a sterile transfer loop or needle, emulsify the culture in the saline solution for 1 section and repeat for the other.
d. Add a drop of *Shigella* antiserum to 1 section of emulsified culture and mix with a sterile loop or needle. The other section is the autoagglutination control.
e. Tilt the mixture in both sections back and forth for 1 minute and observe against a dark background. A positive reaction is indicated by a rapid, strong agglutination.

Table 1: Biochemical Characteristics of *Shigella*

Media or Test	Reaction	Percent of Shigellae Positive
Acetate	–[1]	0
Gas from glucose	–[2]	2.1
Voges-Proskauer	–	0
Indol	+ or –[3]	37.8
Methyl red	+	100
Lysine	–	0
Arginine	+ or –	7.6 (5.6)
Ornithine	+ or –[4]	20
Christensen's citrate	–	0
Lactose	–[4]	0.3 (11.4)
Mannitol	+ or –[5]	80.5
Urease	–	0
Motility	–	0

[1]Some strains of *S. flexneri* 4a utilize sodium acetate.
[2]Certain types of *S. flexneri* 6 may produce a small amount of gas.
[3]Group D are always negative but Groups A, B, and C may be positive.
[4]*S. sonnei* cultures ferment lactose slowly and decarboxylate ornithine.
[5]Group A are negative, but Groups B, C, and D may ferment this substrate.

f. Test each culture with antiserum to groups A, B, C, D, and A-D (*E. coli*, Alkalescens-Dispar group). Cultures that appear to be shigellae, but which agglutinate poorly or not at all, should be heat treated by heating a suspension of the organisms in a water bath at 100 C for 15 to 30 minutes. After such treatment, the suspension is cooled and retested for agglutination on a slide.

27.533 Identify cultures with the aid of Table 1 based on the above biochemical and serologic tests.

27.6 INTERPRETATION OF DATA

Contamination of foods with *Shigella* indicates contamination with human feces, probably from the hands of a human carrier. Because of the labile nature of this organism, a positive culture probably indicates recent contamination or contamination with a large inoculum. Foods contaminated with *Shigella* are considered unfit for human consumption.

27.7 REFERENCES

1. Block, N. B. and W. Ferguson. 1940. An outbreak of Shiga dysentery in Michigan, 1938. Amer. J. Public Health **30**:43–52
2. Bryan, F. L. 1969. Infections due to miscellaneous microorganisms. In H. Riemann, Food-borne Infections and Intoxications. Academic Press, New York, N.Y.
3. Center for Disease Control. 1968. *Salmonella* Surveillance Report No. 74, p. 3, Atlanta, Ga.
4. Center for Disease Control. 1970. Morbidity and Mortality Weekly Report **19**:269–270, Atlanta, Ga.
5. Donadio, J. A. and E. J. Gangarosa. 1969. Foodborne shigellosis. J. Infect. Dis. **119**:666–668
6. Edwards, P. R. and W. H. Ewing. 1972. Identification of the *Enterobacteriaceae*. Third edition. Burgess Publishing Co., Minneapolis, Minn.
7. Fishbein, M., I. J. Mehlman, and B. Wentz. 1971. Isolation of *Shigella* from foods. J.A.O.A.C. **54**:109–111
8. Kaiser, R. L. and L. D. Williams. 1967. Tracing two bacillary dysentery outbreaks to a single food source. Penn. Med. J. **65**:351–354
9. Keller, M. D. and M. L. Robbins. 1956. An outbreak of *Shigella* gastroenteritis. Public Health Rep. **71**:856–862
10. Lewis, J. N., M. S. Loewenstein, L. C. Guthrie, and M. Surgi. 1972. *Shigella sonnei* outbreak on the island of Maui. Amer. J. Epidemiol. **96**:50–58
11. McCormack, W. M., W. E. DeWitt, P. E. Bailey, G. K. Morris, P. Soeharjono, and E. J. Gangarosa. 1974. Evaluation of thiosulfate-citrate-bile salts-sucrose agar, a selective medium for the isolation of *Vibrio cholerae* and other pathogenic vibrios. J. Infect. Dis. **129**:497–500
12. Morris, G. K., J. A. Koehler, E. J. Gangarosa, and R. G. Sharrar. 1970. Comparison of media for direct isolation and transport of shigellae from fecal specimens. Appl. Microbiol. **19**:434–437
13. Nakamura, M. and D. A. Dawson. 1962. Role of suspending and recovery media in the survival of frozen *Shigella sonnei*. Appl. Microbiol. **10**:40–43
14. Nakamura, M. and B. C. Taylor. 1965. Survival of *Shigella* in biological materials. Health Lab. Sci. **2**:220–226
15. Read, R. B. and A. L. Reyes. 1968. Variation in plating efficiency of salmonellae on eight lots of brilliant green agar. Appl. Microbiol. **16**:746–748
16. Taylor, W. I. 1965. Isolation of Shigellae. I. Xylose lysine agars: New media for isolation of enteric pathogens. J. Clin. Path. **44**:471–475

17. TAYLOR, W. I. and B. HARRIS. 1965. Isolation of shigellae. II. Comparison of plating media and enrichment broths. J. Clin. Path. **44**:476–479
18. TAYLOR, B. C. and M. NAKAMURA. 1964. Survival of shigellae in food. J. Hyg. **62**:303–311
19. TAYLOR, W. I. and D. SCHELHART. 1968. Isolation of shigellae. V. Comparison of enrichment broths with stools. Appl. Microbiol. **16**:1383–1386
20. WEISSMAN, J. B., S. V. WILLIAMS, A. R. HINMAN, G. R. HAUGHIE, and E. J. GANGAROSA. 1974. Foodborne shigellosis at a country fair. Amer. J. Epidemiol. **100**:178–185
21. WHEELER, K. M. and F. L. MICKLE. 1945. Antigens of *Shigella sonnei*. J. Immunol. **51**:257–267

CHAPTER 28

YERSINIA ENTEROCOLITICA

James C. Feeley, W. H. Lee, and George K. Morris

28.1 INTRODUCTION

28.11 Description of the Organism

The organism now known as *Yersinia enterocolitica* has been called a variety of names, *Pasteurella* "x", *Pasteurella pseudotuberculosis b.*, and *Pasteurella pseudotuberculosis-like* bacterium. It has been classified recently as a member of the family *Enterobacteriaceae*, a change that is reflected in the eighth edition of Bergey's Manual of Determinative Bacteriology.[1] *Y. enterocolitica* closely resembles *Yersinia (Pasteurella) pseudotuberculosis* and, to a lesser extent, *Yersinia (Pasteurella) pestis.* Thus, it must be destinguished from these zoonotic pathogens as well as other enteric bacteria and bacteria of the genera *Vibrio* and *Aeromonas*.

Y. enterocolitica is a gram negative, facultative anaerobic bacillus, 1.0 to 3.5μm by 0.5 to 1.3μm in size. When incubated at 25 C, young cultures may be coccoid. The organism is motile at 25 C via peritrichous flagella, but is not motile at 36 C.[2] Several biotypes have been established.[2, 3] Most strains are urease positive. All are nonpigmented.

Serologically, many factors have been recognized,[4, 5, 6] some of which are also common to the genera *Brucella, Vibrio,* and *Salmonella.*

28.12 Isolation Methods

28.121 Incubation at 25 C

Y. enterocolitica is similar to other enteric bacteria except that it grows better at 25 C than at 36 C. This offers a selective advantage for isolating it. Primary cultures should be incubated at 25 C for 48 hours. These colonies of *Y. enterocolitica* are approximately 1 mm in diameter, while those grown at 36 C are pinpoint size.

28.122 Cold enrichment at 4 C

Another property of the organism is its ability to grow at 4 C.[2] Many other enteric organisms either do not grow, or die at this temperature. The frequency of isolating *Y. enterocolitica* has been increased by holding specimens in saline or 1/15 M disodium phosphate buffer for 21 days at 4 C.[7, 8]

IS/A Committee Liaison: George J. Herman

28.123 Use of common enteric media

Most of the common enteric plating media have been used successfully to isolate *Y. enterocolitica*. Examples are MacConkey's agar, salmonella shigella (SS) agar, eosin methylene blue (EMB) agar, Tergitol 7 agar, lactose sucrose urea agar, and desoxycholate citrate agar. Enrichments such as magnesium chloride-malachite green-carbenicillin medium, selenite F broth,[10] and tetrathionate broth,[11] have been used with success. Note that the growth of a strain of *Y. enterocolitica* may be supported by one medium and not by another. Knowing that a specific serotype can grow on a more selective medium has been used to advantage.[9]

28.2 RECOMMENDED CONTROLS

It is recommended that MacConkey's agar be used for duplicate plating media. Strains may vary in their growth on various media, but most will grow on MacConkey's agar. A known *Y. enterocolitica* strain should be included as a control.

28.3 TREATMENT OF SAMPLE

For holding and shipping, samples should be placed in sealed containers to prevent dehydration, and refrigerated until analyzed. Environmental swab samples may be placed in a tube containing sterile 1/15 M sodium phosphate buffer at pH 7.6 and held at 4 C.

28.4 PRECAUTIONS AND LIMITATIONS OF THE METHODS

Y. enterocolitica is inhibited by commonly used enteric media containing brilliant green dye, such as brilliant green agar or tetrathionate broth (with dye added). The use of potassium tellurite as a selective agent should be done with care because it is reported to be toxic for *Y. enterocolitica*.[8] Strains of *Y. enterocolitica* that produce a metallic sheen, similar to that produced by *Escherichia coli* grown on EMB, may be mistaken for *E. coli*. Other peculiarities may exist that have not yet been reported.

Lactose usually is not fermented by *Y. enterocolitica* on commonly used enteric media, but occasional organisms may be encountered that do ferment lactose. For this reason, several lactose positive colonies should be selected for identification from primary plating media when they closely resemble *Y. enterocolitica* in size and morphology.

28.5 PROCEDURE

28.51 Sample Preparation

Weigh a 25 g sample into a tared blender jar and add 225 ml of 1/15 M sodium phosphate buffer to the jar. Blend the sample for 2 minutes.

28.52 Isolation

28.521 Liquid enrichment

A sample in phosphate buffer should be inoculated immediately to both selective enrichment broth and plating media as described in sections 28.522 and 28.523. It is then held at 4 C for 14 or 21 days, and again inoculated to primary plating media, and to selective enrichment broth. A nonselective enrichment such as cooked meat medium may also be inoculated.[10]

28.522 Selective broth enrichment

A 0.1 ml amount of the sample in phosphate buffer (28.51) is inoculated into 10 ml of magnesium chloride-malachite green-Carbinicillin medium 9. Incubate at 25 C for 48 hours, and plate onto selective agar. Other selective broths may be used such as selenite F.[10]

28.523 Plating on selective agar media

a. Prepare dried plates of 2 selective agar media, SS agar and MacConkey's agar.

b. Transfer a 5 mm loopful of each liquid culture, after appropriate enrichment, to the surfaces of the 2 selective agar media, and streak to obtain isolated colonies.

c. Incubate plates inverted at 25 C for 48 hours. Typical *Y. enterocolitica* on both media will appear as round, opaque, or colorless colonies. Colonies on SS agar are usually about 1 mm in diameter, whereas those on MacConkey's agar may be as large as 3 mm in diameter after 48 hours' incubation at 25 C. A dissecting microscope with oblique lighting may be used to advantage.[9, 10]

d. Select 5 lactose negative colonies from each selective agar medium and inoculate triple sugar iron (TSI) agar. Lysine iron agar (LIA) may also be inoculated simultaneously. The combined use of TSI and LIA media is of benefit in the presumptive differentiation of *Y. enterocolitica*.[12, 13]

28.524 Biochemical screening

a. TSI reaction

Inoculate each suspect colony to a tube of TSI agar by streaking the slant and stabbing the butt. Incubate the cultures overnight at 25 C. *Y enterocolitica* cultures will have an acid (yellow) slant and an acid butt, but no gas, and no blackening (H_2S production) in the butt. Growth on the slants will be sparse, but becomes much denser after further incubation. Also, strains fermenting sucrose rapidly may cause the slants to revert to alkaline (red) after a day of incubation. Save cultures showing reactions typical of *Y. enterocolitica* for further testing. Purify the cultures that appear contaminated on TSI agar by streaking each culture on separate plates of MacConkey's agar to obtain isolated colonies, incubating plates at 25 C for 24 hours, and transferring organisms from a separate colony in each case to a tube of TSI agar. Incubate these tubes at 25 C for 24 hours.

b. Screening of TSI cultures

TSI cultures showing typical reactions for *Y. enterocolitica* should be further screened by inoculating urea and motility agar with the growth.

1. Streak a slant of Christensen's Urea agar and incubate at 25 C. An alkaline (red) slant should result after 24 hours' incubation.
2. Stab the tops of 2 tubes of semisolid motility agar to a depth of 5 mm with a bacteriologic needle. Incubate both tubes for 24 hours, one at 25 C—and the other at 36 C. Spreading growth from the stab line (indicating motility) in the tube held at 25 C and no motility in the tube held at 36 C is typical of *Y. enterocolitica*. In rare instances, some strains of *Y. enterocolitica* may be weakly motile at 25 C, and require additional incubation, or passage through semisolid agar in U tubes.

28.525 Definitive biochemical identification

Cultures showing appropriate TSI, urea, and motility reactions should be tested as shown in Table 1. Incubate at 36 C because the biochemical reactions for other enteric bacteria have been recorded at this temperature. All media, reagents, and tests are the same as those used for identifying enteric bacteria[14] and are listed in Chapter 2.

28.526 Typing systems

a. Niléhn was the first to separate strains of *Y. enterocolitica* into 5 biotypes by the tests listed in Table 2.[2] Later Wauters attempted to simplify this system by reducing the number of tests;[3] however, a separate biotyping system resulted as shown in Table 3. The biotyping systems shown in Tables 2 and 3 may be useful in epidemic investigations when it is important to trace the source of the epidemic strains.

b. Thirty-four "O" factors and 19 "H" factors have been recognized serologically.[3, 4, 5, 6] Some of these antigens are shared with other bacteria; for example, *Y. enterocolitica* serotype 9 has crossreactions with certain *Brucella* species.[15] Weaver and Jordan, who studied 29 isolates recovered from a variety of clinical sources in the United States, found that the majority of strains were serotype 8.[16] They noted that this serotype has not been reported in Europe, and suggested that there may be differences in the epidemiology of *Y. enterocolitica* infections in the United States and in Europe.

c. A phage typing system has been developed in Europe.[17] However, most of the isolates from the United States are insensitive to the European typing bacteriophages.[16]

28.6 INTERPRETATION OF DATA

Y. enterocolitica has been isolated from humans with a variety of clinical symptoms ranging from mild gastroenteritis to appendicitis and terminal ileitis. Consequently, the presence of this bacterium in foodstuffs is potentially hazardous to human health. Isolates already have been recovered from several types of food, including pork,[18] ice cream,[3] mussels,[19] and oysters.[10] The organism also has been found in drinking water.[20] Undoubt-

Table 1: Reactions of *Yersinia enterocolitica* and Other Closely Related Bacteria (Tests Performed at 36 C Unless Otherwise Noted)

Tests	*Yersinia enterocolitica*	*Yersinia pestis*	*Yersinia pseudo-tuberculosis*	*Vibrio cholerae*	*Aeromonas hydrophila*	*Serratia*	*Enterobacter*	*Citrobacter diversus*	*Klebsiella*	*Proteus morganii*	*Proteus rettgeri*	*Providencia*	*Chromobacter violaceum*
Oxidase	−	−	−	+	+	−	−	−	−	−	−	−	−(W+)
Christensen's urea	+	−	+	−	−(+)	−(+)	−(+)	+,−	+,−	+	+	−	−(L+)
Lactose	*	−	−	+(L)	−(+)	−(+)	+(−)	+,−	+(−)	−	−	−	−
Maltose	+(L)	+	+	+	+	+	+(L)	+	+	−	−	−	−
Sucrose	+	−	−	+	+(−)	+	+(−)	+,−	+(−)	−	−(+)	+L(−)	+,−
Simmon's citrate	−	−	−	+(−)	+(−)	+	+	+	+,−	−	+	−	+(L)
Motility–25 C	+	−	+	+	+	+	+	+	−	+	+	+	+
Motility–36 C	−	−	−	+	+	+	+	+	−	+	+	+	+
Arginine dihydrolase	−	−	−	−	+	−	−(+)	+	−	−	−	−	+
Lysine decarboxylase	−	−	−	+	−(+)	+	+(−)	−	+,−	−	−	−	−
Ornithine decarboxylase	+	−	−	+	−	+	+(−)	+	−	+	−	−	−
Phenylalanine deaminase	−	−	−	−	−	−	+,−	−	−	+	+	+	−

( ) minority of reactions
W weak
L late
* negative with enteric base medium, positive with Hugh-Leifson O-F medium
\+ positive
− negative

Table 2: Niléhn's Biotype Schema

	Biotypes				
	1	2	3	4	5
Tests at 36 C					
Salicin	+	–	–	–	–
Esculin	+	–	–	–	–
Indole	+	+	–	–	–
Nitrate	+	+	+	+	–
Trehalose	+	+	+	+	–
Sorbitol	+	+	+	+	–*
Sorbose	+	+	+	+	–*
Tests at 25 C					
Lactose (O/F)	+	+	+	–	–
Xylose	+	+	+	–	–
Ornithine decarboxylase	+	+	+	+	–
Voges-Proskauer	+	+	+*	+	–
B. galactosidase	+	+	+	+	–
Sucrose	+	+	+	+	–*

* Reactions may vary for specific strains, particularly in those biotypes that are asterisked.
+ positive
– negative

Table 3: Wauter's Biotype Schema

	Biotypes				
Tests*	1	2	3	4	5
Lecithinase	+	–	–	–	–
Indol	+	+	–	–	–
Lactose (O/F medium)	+	+	+	–	–
Xylose (48 hours)	+	+	+	–	–
Nitrate	+	+	+	+	–
Trehalose	+	+	+	+	–
Ornithine decarboxylase	+	+	+	+	–
B-galactosidase	+	+	+	+	–

* All test cultures are incubated for 48 hours at 25 C except Indol which is incubated at 29 C.
+ positive
– negative

edly, isolates will be cultured from a wide variety of foods as they are examined more extensively. Any foodstuff (not to be subsequently cooked) containing *Y. enterocolitica* should be considered unfit for human consumption.

28.7 REFERENCES

1. Buchanan, R. E. and N. E. Gibbons. 1974. Bergey's Manual of Determinative Bacteriology, 8th Edition, American Society for Microbiology. The Williams and Wilkins Co., Baltimore, Md.
2. Niléhn, B. 1969. Studies on *Yersinia enterocolitica* with special reference to bacterial diagnosis and occurrence in human acute enteric disease. Acta Path. Microbiol. (Scand.) Supplement **206**:5–48
3. Wauters, G. 1970. Contribution à l'étude de *Yersinia enterocolitica*. These d'agrégation. Vander, Louvain, Belgium
4. Winblad, S. 1968. International Symposium on Pseudotuberculosis, Paris, Symposium Series Immunobiol. Stand. Karger Basel/New York **9**:337–342
5. Wauters, G., L. Le Minor, and A. M. Chalon. 1971. Antigènes somatiques et flagellaires des *Yersinia enterocolitica*. Ann. Inst. Pasteur **120**:631–642
6. Wauters, G., L. Le Minor, A. M. Chalon, and J. Lassen, 1972. Supplément au schéma antigénique de *Yersinia enterocolitica*. Ann. Inst. Pasteur **122**:951–956
7. Patterson, J. S. and R. Cook. 1963. A method for recovery of *Pasteurella pseudotuberculosis* from feces. J. Path. and Bacter. **85**:241–242
8. Tsubokura, M., K. Otsuki, and K. Itagaki. 1973. Studies on *Yersinia enterocolitica* 1. Isolation of *Y. enterocolitica* from swine. Jap. J. Vet. Sci. **35**:419–424
9. Wauters, G. 1973. Improved Methods for the Isolation and the Recognition of *Yersinia enterocolitica*. In Contributions to Microbiology and Immunology, Vol. 2. *Yersinia, Pasteurella* and *Francisella*. pp 68–70 Karger, Basel; Switz.
10. Toma, S. 1973. Survey on the incidence of *Yersinia enterocolitica* in the province of Ontario. Canad. J. Pub. Hlth. **64**:477–487
11. Wetzler, T. F. 1970. Pseudotoberculosis. In Bodily, H. L., E. L. Updyke, and J. O. Mason, eds. Diagnostic Procedures for Bacterial, Mycotic and Parasitic Infections. American Public Health Association, New York, N.Y.
12. Hall, C. T. 1973. Proficiency Testing. Bacteriology I. 1973. Summary Analysis. Center for Disease Control, Atlanta, Ga.
13. Sonnenwirth, A. C. 1974. Yersinia. In Lennette, E. H., E. H. Spaulding, and J. P. Truant, eds. Manual of Clinical Microbiology, 2nd Edition. American Society for Microbiology, Washington, D. C.
14. Edwards, P. R. and W. H. Ewing. 1972. Identification of Enterobacteriaceae, 3rd Edition. Burgess Publishing Co., Minneapolis, Minn.
15. Hurvell, B., P. Ahvonen, and E. Thal. 1971. Serological cross-reactions between different *Brucella* species and *Yersinia enterocolitica*. Agglutination and Complement fixation. Acta. Vet. Scand. **12**:86–94
16. Weaver, R. E. and J. G. Jordan. 1973. Switz. Recent human isolates of *Yersinia enterocolitica* in the United States. In Contributions to Microbiology and Immunology, Vol. 2. *Yersinia, Pasteurella* and *Francisella*. pp. 120–125 Karger, Basel, Switz.
17. Nicolle, P., H. H. Mollaret, Y. Hamon, and J. F. Vieu. 1967. Etude lysogenique, bacteriocinogénique et lysotypique de l'espece *Yersinia enterocolitica*. Ann. Inst. Pasteur **112**:86–92
18. Zen-Yoji, H. and S. Sakai. 1974. Isolation of *Yersinia enterocolitica* and *Yersinia pseudotuberculosis* from swine, cattle and rats. Japan J. Microbiol. **18**:103–105
19. Spadaro, M. and M. Infortuna. 1968. Isolamenta di *Yersinia enterocolitica* in *Mitilus galaprovincialis Lamk.* Boll. Soc. Ital. Brol. Sper. **44**: 1896–1897
20. Lassen, J. 1972. *Yersinia enterocolitica* in drinking water. Scand. J. Infect. Dis. **4**:125–127

CHAPTER 29

VIBRIO

George K. Morris, Morris Fishbein,* John A. Baross, and Wallis E. DeWitt

29.1 INTRODUCTION

29.11 Description of Genus

Members of the genus *Vibrio* are defined as gram negative, asporogenous rods which are straight or have a single, rigid curve. They are motile with a single, polar flagellum. All vibrios produce oxidase and catalase, and ferment glucose without producing gas.[14] Two species, *Vibrio cholerae* and *Vibrio parahaemolyticus*, are well documented human pathogens.[3, 12]

V. cholerae, the type species of the genus *Vibrio*, is characterized by various biochemical properties and antigenic types. There are *Vibrio* organisms closely resembling *V. cholerae* which fail to agglutinate in cholera antiserum. These organisms may be identical to, or slightly different from, *V. cholerae* in biochemical characteristics, and commonly are referred to as nonagglutinable (NAG) vibrios or noncholera vibrios (NCV). The current taxonomic trend is to include these in the species *V. cholerae*[14]. At this time, only those strains that are agglutinable in Inaba or Ogawa cholera antiserum are well documented human pathogens. However, increasing evidence indicates that NAG strains sometimes are involved in cholera like diarrheal disease.

V. parahaemolyticus is a halophilic, marine microorganism. All strains share a common H antigen, but possess 12 O type and 52 K (capsular) antigens. Pathogenic strains of *V. parahaemolyticus* are differentiated usually from nonpathogenic strains by the ability to produce a special hemolysin (Kanagawa phenomenon).

29.12 Distribution and Sources of Contamination

29.121 *Vibrio cholerae*

This organism is excreted in great numbers in the feces of both cholera patients and carriers. Disease is transmitted by the fecal oral route, indirectly through contaminated water supplies. Direct person to person spread is not common. Food supplies may be contaminated by the use of human excreta as fertilizer, or by freshening vegetables for market with

*Deceased

IS/A Committee Liaison: George J. Herman

contaminated water. Recent cholera epidemics in Italy and on Guam are thought to have resulted from the consumption of raw, contaminated seafood.[5, 6]

29.122 *Vibrio parahaemolyticus*

This organism is isolated frequently from coastal waters and seafoods from all parts of the world and is the most frequent cause of foodborne disease in Japan, where many residents eat fish raw. A number of common source gastroenteritis outbreaks attributed to *V. parahaemolyticus* occurred in the United States in 1971 and 1972. Eight of these outbreaks were laboratory confirmed by the isolation of *V. parahaemolyticus* from stool specimens, foods, or both.[4] Foods implicated in this country are crab, shrimp, and lobster, which, unlike fish in Japan, were cooked prior to eating in all instances.[4] The outbreaks in this country due to cooked seafoods probably were caused by gross mishandling practices: improper refrigeration, insufficient cooking, cross contamination, and recontamination.

29.13 Methods of Isolation

Vibrio species, like many other gram negative bacteria, grow in the presence of relatively high levels of bile salts. They are facultatively aerobic and grow best on an alkaline medium. The strict halophilic nature of *V. parahaemolyticus* probably accounts for the fact that diseases caused by this organism were not documented in the U.S. until workers began examining foods and feces on appropriate media containing added salt. Thiosulfate citrate bile salts sucrose (TCBS) agar has proved to be an excellent medium for the isolation of both *V. cholerae* and *V. parahaemolyticus*. Both species grow well on this medium, and most nonvibrios are inhibited.

29.2 STORAGE OF SAMPLE

The sample should be held refrigerated (4 to 10 C). There is less die off of vibrios when refrigerated than when frozen.[18, 24]

29.3 SPECIAL MEDIA

Isolation from foods is facilitated by the use of special media. Glucose salt teepol broth[1] and bismuth sulfite salt broth[23] both have been recommended as enrichment broths for isolating *V. parahaemolyticus* from foods; alkaline peptone water is used commonly for isolating *V. cholerae* from foods.[9, 20] TCBS agar is designed especially for the isolation of *Vibrios*. Media used for testing the biochemical reactions of *V. parahaemolyticus* should contain 3% added NaCl.

29.4 RECOMMENDED CONTROLS

Duplicate plating media should be used for *V. cholerae* because strains may vary in their growth. It is advisable in conducting the biochemical tests for *V. parahaemolyticus* and *V. cholerae* to inoculate known positive control organisms to insure appropriate reactions.

29.5 PRECAUTIONS AND LIMITATIONS

In *V. parahaemolyticus* outbreaks, the importance of stool, rectal swab, vomitus specimens, and samples of incriminated foodstuffs cannot be overemphasized. Usually, clinical specimens will contain Kanagawa positive organisms. Clinical specimens must be obtained as early in the disease as possible because the duration of excretion is short.

A good selective enrichment broth has not been developed for *V. cholerae*. Alkaline peptone water provides suitable enrichment for incubation periods of 6 to 8 hours, but other competing microflora may overgrow *V. cholerae* during longer enrichment periods.

29.6 LABORATORY PROCEDURES

29.61 *Vibrio cholerae*

29.611 Enrichment and plating

a. Weigh 25 g of sample into a tared jar (capacity approximately 500 ml). Products such as seafood or vegetables may be blended or cut into small pieces with scissors.
b. Add 225 ml alkaline peptone water. Thoroughly mix sample with broth.
c. Prepare dried plates of two media; one should be TCBS agar, the other a nonselective medium, such as gelatin agar.
d. Transfer a 5 mm loopful of broth suspension to the surface of each of the 2 plating media, and streak in a manner that will yield isolated colonies.
e. Incubate broth at 35 to 37 C for 6 to 8 hours.
f. Transfer a 5 mm loopful of incubated broth suspension to the surface of each of the two plating media, and streak in a manner that will yield isolated colonies.
g. Incubate both sets of plating media for 18 to 24 hours at 35 to 37 C.
h. Subculture three or more typical colonies from each plating medium to nutrient agar slants.

    1. Typical colonies of *V. cholerae* on TCBS agar are large, smooth, yellow, and slightly flattened with opaque centers and translucent peripheries.
    2. Typical colonies of *V. cholerae* on gelatin agar are transparent, and usually have a characteristic cloudy zone around them which becomes more definite after a few minutes of refrigeration.

29.612 Confirmation

a. Triple sugar iron (TSI) reaction

Inoculate each suspect culture to a tube of TSI agar by streaking the slant and stabbing the butt. Incubate the cultures overnight at 35 to 37 C. *V. cholerae* cultures will have an acid (yellow) slant and an acid butt, no gas, and no blackening (H_2S production) in the butt.

b. Serologic agglutination test

Serotyping of *V. cholerae* is generally applied to the somatic or O antigens. Two major serotypes, Ogawa and Inaba, and one rarely encountered serotype, Hikojima, are recognized as human pathogens. All three serologic types are seen in both the classical *V. cholerae* and the El Tor biotype.

1. Mark off two sections about 1 × 2 cm on the inside of a glass Petri dish or on a 2 × 3 inch glass slide.
2. Place a small amount of culture from a 6 to 18 hour agar slant directly onto the dish or slide in the upper part of each marked section.
3. Add one drop of 0.85% saline solution to the lower part of each marked section. With a sterile transfer loop or needle, emulsify the culture in the saline solution for one section, and repeat for the other section.
4. Add a drop of polyvalent *V. cholerae* O antiserum to one section of emulsified culture and mix with a sterile loop or needle. (The other section containing only antigen is the autoagglutination control.)
5. Tilt the mixtures back and forth for one minute and observe against a dark background. A positive reaction is indicated by a rapid, strong agglutination.
6. Test each culture with polyvalent antiserum and, if positive, test with Ogawa and Inaba antisera. The Hikojima serotype reacts with both antisera.

c. String test

The string test as described by Smith is a useful presumptive test for suspected strains of *V. cholerae*.[22] All *V. cholerae* including strains commonly referred to as NAG or NCV are positive. The string test consists of emulsifying a large colony or other inoculum from an agar culture in a large drop of 0.5% aqueous sodium desoxycholate. Within 60 seconds a mucoid mass forms, and this material strings when a loop is lifted from the slide.

d. Biochemical tests

The minimal number of characters needed to identify *V. cholerae* strains is shown in Table 1.[15]

29.613 Differentiation of El Tor and classical biotypes

a. Bacteriophage susceptibility

A modification of the method described by Mukerjee is used.[19] Inoculate 4 hour broth cultures to Mueller-Hinton agar plates (Chapter 2) using a wire loop or cotton swab, to yield confluent growth. With a 3 mm platinum loop, superimpose one loopful of an appropriate test dilution of Phage IV on the inoculated agar surface. Incubate overnight at 37 C. Classical strains are sensitive, El Tor strains are resistant to the action of Mukerjee's Phage IV.

Table 1: Minimal Number of Characters Needed to Identify *Vibrio cholerae* Strains

	Reaction	Percent positive
Gram negative, asporogenous rod	+	100
Oxidase	+	100
Glucose, acid under a petrolatum seal	+	100
Glucose, gas	–	0
D-mannitol, acid	+	99.8
L-inositol, acid	–	0
Hydrogen sulfide, black butt on TSI	–	0
L-lysine decarboxylase	+	100
L-arginine dihydrolase	–	0
L-ornithine decarboxylase	+	98.8
Growth in 1% tryptone broth*	+	99.1

*No sodium chloride added

b. Polymyxin-B sensitivity

A modification of the technique of Han and Khie is used.[13] Inoculate Mueller-Hinton agar plates with 4 hour broth cultures to yield confluent growth. Dry the plates and place a 50 unit disc of Polymyxin-B on the surface. Record the results after 18 to 24 hours. Classical strains are sensitive; El Tor strains are resistant.

c. Other tests

Some additional tests useful for biotype differentiation are: hemolysis of 1% sheep red blood cells[8], chicken red blood cell agglutination[10], and the Voges-Proskauer (VP) test.[7] The El Tor biotypes are usually hemolytic to 1% sheep red blood cells, agglutinate chicken red blood cells, and give a positive VP reaction at 22 C.

29.62 *Vibrio parahaemolyticus*

The following analytical scheme for culturing *V. parahaemolyticus* is divided into two parts, one for seafood and another for epidemic samples. The procedures differ only in the way the specimens are handled on the first day in the enrichment procedure. Thereafter they are identical.

29.621 Seafood samples

a. Enrichment, isolation, enumeration

1. Weigh 50 g of seafood sample into a blender. Obtain surface tissues, gills, and gut of fish. Sample the entire interior of shellfish. For crustaceans, such as shrimp, use the entire animal if possible; if it is too large, select the central portion, including gill and gut.

2. Add 450 ml of 3% NaCl dilution water and blend for one minute at 8,000 rpm. This constitutes the 1:10 dilution.
3. Prepare 1:100, 1:1,000, 1:10,000 dilutions or higher if necessary.
4. Inoculate 3 × 10 ml portions of the 1:10 dilution into 10 ml of glucose salt teepol broth (GSTB)—2 × concentration. This represents the 1 gm portion. Similarly, inoculate 3 × 1 ml portions of the 1:10, 1:100, 1:1,000, and 1:10,000 dilutions into 10 ml of single strength GSTB.
5. Incubate the tubes overnight (18 hours or less) at 37 C.
6. Streak a 3 mm loopful from the 3 highest dilutions of GSTB showing growth onto TCBS agar.
7. Incubate TCBS agar plates 18 hours at 37 C.
8. *V. parahaemolyticus* appears as round, green, or bluish colonies, 2 to 3 mm in diameter. Interfering, competitive, *V. alginolyticus* colonies are larger and yellow.
9. When these blue green colonies are finally identified biochemically as *V. parahaemolyticus* (see below), refer to the original positive dilutions on GSTB and apply the 3 tube MPN tables (Chapter 6) for final enumeration of the organism.

b. Biochemical identification of isolates

Pick two or more typical or suspicious colonies with a needle to:

1. TSI agar slants
   a) Streak the slant, stab the butt, and incubate overnight at 37 C.
   b) *V. parahaemolyticus* produces an alkaline slant and an acid butt but no gas or H_2S in TSI.

2. Trypticase soy broth—3% NaCl, and Trypticase soy agar—3% NaCl (Chapter 2).
   a) Inoculate both media and incubate overnight at 37 C. These cultures provide inocula for other tests as well as material for the Gram stain and for microscopic examination.
   b) *V. parahaemolyticus* is a gram negative, pleomorphic organism exhibiting curved or straight rods with polar flagella.

3. Motility test medium
   a) Inoculate a tube of motility test medium by stabbing the column of the medium to a depth of approximately 5 mm. Incubate for 18 hours at 37 C. A circular outgrowth from the line of stab constitutes a positive test. *V. parahaemolyticus* is motile.
   b) Preliminary screening data.

      Only motile, gram negative rods which produce an acid butt and an alkaline slant on TSI and do not form H_2S or gas are examined further. The identifying characteristics of *V. parahaemolyticus* are shown in Table 2.

c) Differentiating *V. parahaemolyticus* from other marine vibrios.

In addition to *V. parahaemolyticus*, *V. alginolyticus*, and *V. anguillarum*, there are many poorly identified vibrios which are abundant in all marine environmental samples, particularly shellfish. Some of these are mesophilic and therefore can be a problem when culturing *V. parahaemolyticus* from seafood. In general, however, *V. parahaemolyticus* and *V. alginolyticus* can be distinguished from *V. anguillarum* and most undefined marine vibrios by growth at 42 C, decarboxylation of lysine, and the inability to grow in liquid media without salt. *V. parahaemolyticus* can be distinguished from *V. alginolyticus* by the latter's ability to form swarming colonies on TSA (Chapter 2) at 35 C, grow in 10% NaCl, produce acetoin, and ferment sucrose on TSI slants. A scheme for differentiating *V. parahaemolyticus* from related marine vibrios is presented in Table 3.

Table 2: Identifying Characteristics of *Vibrio parahaemolyticus*

Tests	Reactions
Gram stain	gram negative
Morphology	curved/straight rods
Motility	+
TSI	K/A*, H_2S(−), gas (−)
Hugh-Liefson glucose (O/F medium)	fermentation (+), gas (−)
Oxidase	+
Arginine dihydrolase	−
Lysine decarboxylase	+
Ornithine decarboxylase	+
Gelatin	+
Halophilism (NaCl)	6%, 8% (+) 0%, 10% (–)
Growth at 42 C	+
Voges-Proskauer	−
Indole	+
Cellobiose	−
Sucrose	−
Maltose	+
Mannitol	+
Trehalose	+

*Alkaline slant

Table 3: Scheme for Differentiating *Vibrio parahaemolyticus* from Related Marine Vibrios

Test	*V. parahaemolyticus*	*V. alginolyticus*	*V. anguillarum*	Undefined vibrios
TCBS (green colony)	+	−	− +	+ −
Growth at 42 C	+	+	−	− +
Growth in media with				
0% NaCl	−	−	− +	− +
8% NaCl	+	+	− +	−
10% NaCl	−	+	−	−
Lysine decarboxylase	+	+	−	+ −
Sucrose fermentation	−	+	+ −	− +
Voges-Proskauer	−	+	−	−
Swarming colony on TSA with 3% NaCl at 35 C	−	+	−	−

29.622 Epidemic specimens

a. If the transit time from field to laboratory is more than 8 hours, place rectal swabs or swabs of fresh stool specimens in Cary-Blair transport medium.
   1. In the laboratory, inoculate the swab onto TCBS agar (Chapter 2) and streak for isolation. Incubate the plate at 37 C for 24 hours.
   2. Place the swab in a 10 ml tube of alkaline peptone water containing 3% NaCl. After 8 hours' incubation at 37 C, streak a loopful onto TCBS agar. Incubate the plate at 37 C for 24 hours.
   3. Proceed as indicated under seafood samples (29.621).

b. If transit time from field to laboratory is less than 8 hours, place stool specimen or rectal swab in alkaline peptone water containing 3% NaCl.
   1. After 8 hours' incubation, streak a loopful onto TCBS agar. Incubate the plate at 37 C for 24 hours.
   2. Proceed as indicated under seafood samples (29.621).

29.623 Serologic typing

V. parahaemolyticus possesses three antigenic components, H, O, and K. The H antigen is common to all strains of *V. parahaemolyticus*, and is of little value in serotyping. The K, or capsular antigen, may be removed from the bacterial body by heating the isolate for one or two hours at 100 C. This process exposes the O or somatic antigen which is thermostable. Since the K antigen masks the O antigen, it is necessary to remove this antigen by heating before performing the O agglutination tests.

There are 12 O group and 52 K antigens. Two of the K antigens have been found to occur with either of two antigens; therefore, there are 54

Table 4: *Vibrio parahaemolyticus* Antigen Combinations

O group	K Type
1	1, 25, 26, 32, 38, 41, 56
2	3, 28
3	4*, 5, 6, 7, 29, 30*, 31, 33, 37, 43, 45, 48, 54, 57
4	4*, 8, 9, 10, 11, 12, 13, 34, 42, 49, 53, 55
5	15, 17, 30*, 47
6	18, 46
7	19
8	20, 21, 22, 39
9	23, 44
10	24
11	36, 40, 50, 51
12	52
Total 12	52

*Occurs with more than one O group.

known serotypes, as seen in Table 4.[21, 26] Serologic tests by themselves are not used to identify *V. parahaemolyticus* because of cross reactions with many other marine organisms. On the other hand, during investigations of food-borne outbreaks, serologic tests become an indispensible epidemiologic tool.

Commercial Japanese *V. parahaemolyticus* diagnostic antiserum is available* as follows: 8 polyvalent K antisera, 52 monovalent K antisera, and 12 monovalent O antisera.

Because the antiserum is expensive, it is not recommended for small laboratories; biochemical identification alone should be sufficient. For those who wish to serotype, use the following procedure.

a. Determination of the O group

1. Inoculate two trypticase soy agar (TSA) (Chapter 2) slants and incubate at 37 C for 24 hours. Wash one slant with 1 ml of a 3% NaCl 5% glycerol solution. Autoclave this suspension at 121 C at 15 lb for one hour. This yields the prepared O antigen.
2. On a plastic or glass plate, mark off 13 one cm squares with a wax pencil and place a small drop of heated O antigen in each box with a Pasteur pipette.

*Nichimen Co., 1185 Avenue of the Americas, 31st floor, New York, N. Y. 10036.

3. Place a drop of O group antiserum in all 12 boxes representing all 12 groups. The 13th box containing only O antigen is the autoagglutination control.
4. Mix all drops with a wire needle and rock back and forth for at least two minutes. A positive agglutination should be produced rapidly and completely. Weak and delayed reactions are considered negative. Frequently, all 12 O groups will be negative, and the culture must be reported as O group-untypable.

b. Determination of the K type

1. The second TSA slant from a.1. is washed down with 1 ml of 3% NaCl solution. This is the K antigenic suspension.
2. The foregoing O group determination establishes the number and identity of the individual K types that could be associated with the specific O group. For example, if group O8 were found, one would test against monovalent K20, 21, 22, and 39 antisera. Thus 4 boxes plus one autoagglutination control box would be drawn with wax pencil on a plastic plate.
3. Repeat as in a.2., and deliver a drop of K antigen to each with a Pasteur pipette.
4. Add one drop of specific antiserum to each box except the control box.
5. Repeat as in a.4. The K type reactions, if positive, are fairly rapid (1 to 2 minutes).
6. Our experience has been that only 40% of all American seafood isolates are serotypable.

29.624 Determination of pathogenicity

Kato[17] showed that *V. parahaemolyticus* isolates from the stools of patients with enteric infections are hemolytic on a special high salt blood agar, while *V. parahaemolyticus* isolates from seafood and marine water are not. This special agar later was modified by Wagatsuma.[25] To avoid confusion with the regular normal hemolytic activity of *V. parahaemolyticus* on ordinary blood agar, the special agar was named Wagatsuma agar and the special hemolytic response the Kanagawa phenomenon (Chapter 2).

The correlation has been well established that *V. parahaemolyticus* strains which cause illness in humans are almost always Kanagawa positive and isolates recovered from seafood are almost always Kanagawa negative. The test should be performed with fresh isolates since this characteristic may be lost by an organism. During a *V. parahaemolyticus* epidemic, it is essential to perform a Kanagawa test on isolates from all patients and from the incriminated food to establish the causative serotype.

a. Kanagawa test

1. Inoculate suspect organism in trypticase soy broth (TSB) 3% NaCl (Chapter 2) and incubate overnight at 37 C.

2. Drop a loopful of culture on a previously well dried Wagatsuma blood agar plate. Several drops in a circular pattern may be made on a single plate.
3. Incubate at 37 C and read in less than 24 hours. A positive test consists of beta hemolysis, a zone of transparent clearing of the blood cells around the area of growth.
4. No observation beyond 24 hours is valid.

29.7 INTERPRETATION OF DATA

The isolation of *V. parahaemolyticus* from seafood is not unusual.[2, 11, 16] *V. parahaemolyticus* is a normal saprophytic inhabitant of the coastal marine environment and multiplies during the warm summer months. In this period the organism is readily recovered from most of the seafood species harvested in these areas. The Japanese have separated the virulent from the avirulent strains of *V. parahaemolyticus* by means of the Kanagawa test.[25] No isolates from American seafoods involved in outbreaks have proved to be Kanagawa positive thus far. However, these types have occurred in Japan and England, and probably occur here too in extremely low numbers. In most instances *V. parahaemolyticus* Kanagawa negative seafood strains do not cause human gastroenteritis.

In contrast, contamination of food with *Vibrio cholerae* (NAG vibrios not included) constitutes an important finding from the standpoint of public health. The entire lot of contaminated food should be withheld from distribution until appropriate health authorities are notified, and an epidemiologic investigation can be undertaken.

29.8 REFERENCES

1. Bacteriological Analytical Manual. 1972. Division of Microbiology, Food and Drug Administration, pp. X1–X16, Washington, D. C.
2. Bartley, C. H. and L. W. Slanetz. 1971. Occurrence of *Vibrio parahemolyticus* in estuarine waters and oysters of New Hampshire. Appl. Microbiol. **21**:965–966
3. Barua, D. and W. Burrows. 1974. Cholera. W. B. Saunders Company, Philadelphia, Pa.
4. Center for Disease Control. 1973. Morbidity and Mortality Weekly Report **22**:231
5. Center for Disease Control. 1973. Morbidity and Mortality Weekly Report **22**:300
6. Center for Disease Control. 1974. Morbidity and Mortality Weekly Report **23**:277–278
7. Feeley, J. C. 1965. Classification of *Vibrio cholerae* (*Vibrio comma*), including El Tor vibrios, by infrasubspecific characteristics. J. Bacteriol. **89**:665–670
8. Feeley, J. C. and M. Pittman. 1963. Studies on the haemolytic activity of El Tor vibrios. Bull. WHO **28**:347–356
9. Felsenfeld, O. 1965. Notes on food, beverages and fomites contaminated with *Vibrio cholerae*. Bull. WHO **33**:725–734
10. Finkelstein, R. A. and S. Mukerjee. 1963. Hemagglutination: A rapid method for differentiating *Vibrio cholerae* and El Tor vibrios. Proc. Soc. Exp. Biol. Med. **112**:355–359
11. Fishbein, M., I. J. Mehlman, and J. Pitcher. 1970. Isolation of *Vibrio parahaemolyticus* from the processed meat of Chesapeake Bay blue crabs. Appl. Microbiol. **20**:176–178
12. Fujino, T., G. Sakaguchi, R. Sakazaki, and Y. Takeda. 1974. International Symposium on *Vibrio parahaemolyticus*. Saikon Publishing Co. Ltd., Tokyo, Japan
13. Han, G. K. and T. D. Khie. 1963. A new method for the differentiation of *Vibrio comma* and *Vibrio* El Tor. Amer. J. Hyg. **77**:184–186

14. HUGH, R. and J. C. FEELEY. 1972. Report (1966–1970) of the subcommittee on taxonomy of vibrios to the International Committee on Nomenclature of Bacteria. Int. J. Systematic Bacteriol. **22**:123
15. HUGH R. and R. SAKAZAKI. 1972. Minimal number of characters for the identification of *Vibrio* species, *Vibrio cholerae* and *Vibrio parahaemolyticus*. Publ. Health Lab. **30**:133–137
16. KAMPELMACHER, E. H., L. M. VAN NOORLE JANSEN, D. A. A. MOSSEL, and F. J. GROEN. 1972. A survey of the occurrence of *Vibrio parahaemolyticus* and *V. alginolyticus* on mussels and oysters and in estuarine waters in the Netherlands. J. Appl. Bacteriol. **35**:431–438
17. KATO, T., Y. OBARA, H. ICHINOE, K. NAGASHIMA, S. AKIYAMA, K. TAKIZAWA, A. MATSUCHIMA, S. YAMAI, and Y. MIYAMOTO. 1965. Grouping of *V. parahaemolyticus* with a hemolysis reaction. Shokuhin Eisei Kenkyu **15**:83–86 (In Japanese)
18. MATCHES, J. R., J. LISTON, and L. P. DANEAULT. 1971. Survival of *Vibrio parahaemolyticus* in fish homogenate during storage at low temperatures. Appl. Microbiol. **21**:951–952
19. MUKERJEE, S. 1961. Diagnostic uses of cholera bacteriophages. J. Hyg. **59**:109–115
20. PRESCOTT, L. M. and N. K. BHATTACHARJEE. 1969. Viability of El Tor vibrios in common foodstuffs found in an endemic cholera area. Bull. WHO **40**:980–982
21. SAKAZAKI, R., S. IWANAMI, and K. TAMURA. 1968. Studies on the enteropathogenic, facultatively halophilic bacteria, *Vibrio parahaemolyticus*. II. Serological characteristics. Jap. J. Med. Sci. Biol. **21**:313–324
22. SMITH, H. L. 1970. A presumptive test for vibrios: the "String" test. Bull. WHO **42**:817–818
23. THATCHER, F. S. and D. S. CLARK. 1968. Microorganisms in Food: Significance and Methods of Enumeration. pp. 107–114, University of Toronto Press, Toronto, Canada
24. VANDERZANT, C. and R. NICKELSON. 1972. Survival of *Vibrio parahemolyticus* in shrimp tissue under various environmental conditions. Appl. Microbiol. **23**:34–37
25. WAGATSUMA, S. 1968. A medium for the test of the hemolytic activity of *Vibrio parahaemolyticus*. Media Circle **13**:159–161
26. ZEN-YOJI, H., S. SAKAI, Y. KUDOH, T. ITOH, and T. TERAYAMA. 1970. Antigenic schema and epidemiology of *Vibrio parahaemolyticus*. Health Lab. Sci. **7**:100–108

CHAPTER 30

THE ENTEROCOCCI

R. H. Deibel and Paul A. Hartman

30.1 INTRODUCTION

The classification of the enterococci is based primarily on their serology and secondarily on their physiology. Heretofore, *S. bovis* and *S. equinus* were excluded from this division; however, the demonstration of group D antigen in these species necessitates their inclusion. *Streptococcus avium,* a species that commonly inhabits the chicken gut, also contains the group D antigen. Thus, the five recognized species of group D streptococci now include these three species in addition to *S. faecalis* and *S. faecium*.[1] The traditional Sherman criteria employed for primary delineation (growth at 10 and 45 C) must be amended to conform to the serological grouping.

Two varieties of *S. faecalis* (*S. faecalis* var. *liquefaciens* and *S. faecalis* var. *zymogenes*), and one of *S. faecium* (*S. faecium* var. *durans*) have been proposed, and in reality these are biotypes based on proteolytic and hemolytic characteristics. The proteolytic characteristic appears to be stable; however, the hemolytic ability is quite unstable, and often is lost upon subculture in laboratory media. The wisdom of continued use of the varietal epithets is questionable.

Some difficulties have arisen with the use of the terms enterococci, group D streptococci, and "true enterococci"—the latter referring specifically to *S. faecalis* and *S. faecium*. Some investigators use the terms enterococci and group D streptococci interchangeably; others restrict enterococci to the *S. faecalis* and *S. faecium* species only. In reality all group D streptococci are enterococci, and this aspect of the problem rapidly degenerates to one of semantics. In food microbiology, *S. faecalis* and *S. faecium* are definitively the most common group D species encountered. This undoubtedly reflects the rationale of employing KF streptococcal agar for the quantitative estimation of group D streptococci in foods, since this medium appears to be selective for these two species. For purposes of this review, the terms group D streptococci and enterococci will be considered synonymous.

Generally, the enterococcus habitat is in the intestinal contents of animals, including warm and cold blooded varieties, as well as of insects. Documented reports indicate that on occasion these bacteria can establish them-

IS/A Committee Liaison: Edward A. Powers

selves in an epiphytic relationship on growing vegetation, and hence grow outside of their native habitat.[5] Usually only *S. faecalis, S. faecium* and their varieties are found under these conditions. None of the species can be considered as absolutely host specific, although some species evidence a degree of host specificity.

Many foods contain from small to large numbers of enterococci, especially *S. faecalis* and *S. faecium*. Certain varieties of cheese, and occasionally, fermented sausage, may contain in excess of 10^6 organisms per gram. Relatively low levels, 10^1 to 10^3 organisms per gram, commonly are encountered in a wide variety of other foods. The shelf life of sliced, prepackaged ham (and sometimes other similarly prepared cured meats) frequently is dictated by controlling the initial numbers of contaminating enterococci. No acceptable levels of these bacteria can be stated because enterococcus counts vary with the product, the holding conditions, the time of storage, and other factors. In general, enterococci serve as a good index of sanitation and proper holding conditions. However, the entire history of each product must be established prior to setting specific standards.

Many investigators have reported a lack of correlation between enterococcus and *E. coli* counts, and the unreliability of enterococcus counts as a reflection of fecal contamination is established rather well. The ability of enterococci to grow in food processing plants, and possibly other environments, long after their introduction, as well as the observation that enterococci can establish epiphytic relationships, reinforces these observations.

Various species of enterococci are noted for their ability to initiate growth under a variety of adverse conditions. Most of the species are salt tolerant; all are facultatively anaerobic, and all grow well at 45 C. *Streptococcus faecium* and *S. faecalis* are relatively heat resistant, and characteristically may survive traditional milk pasteurization procedures. *Streptococcus faecium* is markedly heat tolerant, and it is a spoilage agent in marginally processed canned hams. Of the enterococci, only *S. bovis* and *S. equinus* are unable to grow at relatively low temperatures (i.e., about 7 to 10 C). Most of the enterococci are relatively resistant to freezing and, unlike *E. coli*, they readily survive this treatment.

The enterococci may be identified by physiological as well as serological methods. When the former are employed, a spectrum of characteristics must be examined because there are no one, two, or three traits that will establish a definitive identification. As with other streptococci, the demonstration of the group antigen, if possible, provides adequate identity.

On occasion, in debilitated or compromised patients, the enterococci may be associated with disease. Although this is not uncommon, the potential pathogenicity of this group generally is considered marginal. In the past, some investigators have associated food poisoning outbreaks with these bacteria, but definitive experiments wherein unequivocal results were obtained are lacking. Thus, the elusive food poisoning potential continues to be a stigma even after many failures to incriminate these bacteria.

To say that a plethora of media have been advocated for the selective isolation and/or quantitation of the enterococci would be an understatement.[3] Many selective agents, incubation conditions, and combinations of these have been advanced, but all have one or more shortcomings. The media and methods that are available presently lack selectivity, differential ability, quantitative recovery, relative ease of manipulation, or a combination of these deficiencies to various degrees. Essentially, a compromise must be reached in the selection of a general purpose medium for the recovery of enterococci from foods. Although selectivity is not absolute, quantitative recovery is less than ideal, and the preparation of the medium necessitates an aseptic addition of an indicator. KF streptococcal agar has been accepted by most industry and regulatory agencies for the quantitative estimation of enterococci in nondairy foods. For dairy products, a more selective medium or higher incubation temperature (45 C) may be necessary to reduce background growth of other streptococci or lactobacilli.

KF streptococcal agar is a selective differential medium that employs sodium azide as the chief selective agent and triphenyltetrazolium chloride (TTC) for differential purposes.[4] The medium contains a relatively high concentration of maltose (2.0%) and a small amount of lactose (0.1%). Most, but not all, streptococci ferment these sugars, as do most group D strains. Variation in the intensity of TTC reduction is noted among the streptococci; *S. faecalis* and its varieties reduce the compound to its formazan derivative, imparting a deep red color to the colony. The other group D species, as well as *S. mitis* and *S. salivarius*, are feebly reductive and the colonies appear light pink, in contrast to those of *S. faecalis*. Most other lactic acid bacteria are partially or completely inhibited; however, some strains of *Pediococcus, Lactobacillus* and *Aerococcus* may grow, producing light pink colonies.

KF streptococcal medium is available commercially with or without agar. The broth is available for the MPN procedure, but inasmuch as the detection of low numbers by the MPN procedure is rather insignificant in foods, a discussion of the methodology involved is precluded.

30.2 TREATMENT OF SAMPLE

Prepare the sample for culturing by the pour plate method as directed in Chapter 1.

30.3 SPECIAL REAGENTS AND MEDIA (Chapter 2)

a. KF streptococcal agar
b. Filter sterilized, triphenyltetrazolium chloride (TCC 1%)
c. Hydrogen peroxide (3%)
d. 6.5% salt medium (BHI + 6.0% NaCl)
e. Bile-esculin agar[2]

30.4 PROCEDURE

a. Dispense 1 ml of decimal dilutions into duplicate Petri plates. If the count is expected to be low, the accuracy and sensitivity may be increased by plating 1 ml of a 1:10 dilution into each of 10 Petri plates, in which case the total number of colonies on the 10 plates will represent the count per gram of food.

b. Incubate the plates for 48 ± 2 hours at 35 ± 1 C.

c. With the aid of a dissecting microscope with a magnification of 15 diameters, or a colony counter, count all red and pink colonies. This is reported as the "KF streptococcal" count.

30.41 Confirmation of Enterococci

a. If confirmation is desired, pick 5 to 10 red or pink colonies and transfer into brain heart infusion (BHI) broth.

b. Incubate at 35 C for 18 to 24 hours.

c. Prepare gram stained smears of the culture and observe for typical enterococcal morphology (gram positive cocci, elongated, in pairs and occasionally short chains).

d. Catalase test. Test for catalase activity by adding 1 ml of 3% hydrogen peroxide to the culture and observe for the generation of oxygen bubbles. Streptococci are catalase negative and no reaction should occur. *Caution: do not test for catalase activity on azide-containing media such as KF streptococcal agar*.

e. Growth on bile-esculin agar.[2] Incubate at 35 C for 24 hours.

f. Test for growth in BHI broth containing 6.5% NaCl. Incubate at 35 C for 72 hours.

g. Test for growth at 45 C in BHI which has been tempered in an air incubator prior to incubation. Note: If growth in the salt containing medium and growth at 45 C are to be determined, subcultures must be inoculated prior to testing for catalase to avoid contamination introduced with the peroxide.

30.5 INTERPRETATION OF DATA

S. equinis and *S. bovis* are the only enterococci which will not grow in media containing 6.5% NaCl, but all 5 species should grow at 45 C. There may be a few exceptions. An excellent confirmatory test is the ability of the pure culture isolate to grow on bile-esculin agar. All group D streptococci tolerate bile and hydrolyze esculin.[2] Although typical morphology, lack of catalase activity, the ability to grow in medium containing 6.5% NaCl, growth at 45 C, and growth on bile-esculin agar do not definitively characterize the enterococci, these tests are sufficient to establish a presumptive characterization. Demonstration of a serological group D reaction will identify definitively the isolate as an enterococcus.

12.6 REFERENCES

1. Deibel, R. H. 1964. The group D streptococci. Bacteriol. Rev. **28**: 330–366
2. Facklam, R. R. and M. D. Moody. 1970. Presumptive identification of group D streptococci: the bile-esculin test. Appl. Microbiol. **20**:245–250
3. Hartman, P. A., G. W. Reinbold, and D. S. Saraswat. 1966. Media and methods for the isolation and enumeration of the enterococci. Adv. Appl. Microbiol. **8**:253–289
4. Kenner, B. A., H. F. Clark, and P. W. Kabler. 1961. Fecal streptococci. I. Cultivation and enumeration of streptococci in surface waters. Appl. Microbiol. **9**:15–20
5. Mundt, J. O. 1970. Lactic acid bacteria associated with raw plant food materials. J. Milk and Food Tech. **33**:550–553

CHAPTER 31

METHODS FOR THE ISOLATION AND ENUMERATION OF *STAPHYLOCOCCUS AUREUS*

Edward F. Baer,* Rodney J. H. Gray, and Donald S. Orth

31.1 INTRODUCTION

The growth of *Staphylococcus aureus* in foods presents a potential public health hazard since many strains of *S. aureus* produce enterotoxins which cause food poisoning if ingested. Among the reasons for examining foods for *S. aureus* are a) to confirm that this organism may be the causative agent of foodborne illness; b) to determine whether a food or food ingredient is a potential source of staphylococcal food poisoning; and c) to demonstrate postprocessing contamination, which usually is due to human contact with processed food, or exposure of the food to inadequately sanitized foodprocessing surfaces. Foods subjected to postprocess contamination with enterotoxigenic types of *S. aureus* represent a significant hazard because of the absence of competitive organisms which normally restrict the growth of *S. aureus*.

In processed foods, the presence of *S. aureus* usually indicates contamination from the skin, mouth, or nose of food handlers. This contamination may be introduced directly into foods by process line workers, with lesions caused by *S. aureus* on hands and arms, coming into contact with the food, or by the coughing and sneezing which is common during respiratory infections. Contamination of processed foods also may occur when deposits of contaminated food collect on or adjacent to processing surfaces to which food products are exposed. When large numbers of *S. aureus* are encountered in processed food, it may be inferred that sanitation, temperature control, or both were inadequate.

In raw foods, especially animal products, the presence of *S. aureus* is common and may not be related to human contamination. Staphylococcal contamination of animal hides, feathers, and skin is common, and may or may not result from lesions or bruised tissue. Contamination of dressed animal carcasses by *S. aureus* is common and often unavoidable. Raw milk and unpasteurized dairy products frequently contain large numbers of *S. aureus*, usually as a result of staphylococcal mastitis.

---

*Deceased

IS/A Committee Liaison: Edmund A. Powers

The significance of the presence of *S. aureus* in foods should be made with caution. Significance normally relates to the capacity of certain types of *S. aureus* to produce enterotoxins when conditions permitting growth prevail. As a result of such conditions, large numbers of *S. aureus* usually will be present in the food. The presence of large numbers of the organism in foods is not, however, sufficient cause to incriminate a food as the vector of food poisoning. The potential for causing staphylococcal intoxication cannot be assured without testing the enterotoxigenicity of the *S. aureus* isolate, or demonstrating the presence of staphylococcal enterotoxin in food. Neither is the absence or presence of small numbers of *S. aureus* complete assurance that a suspect food is safe. Conditions inimical to the survival of *S. aureus* may result in diminished population or death of viable cells, whereas sufficient toxin remains to elicit symptoms of staphylococcal food poisoning. For example, outbreaks of illness with convincing epidemiological evidence to suggest staphylococcal food poisoning have been attributed to dried skim milk from which viable cells of *S. aureus* could not be isolated.[3]

The choice of method to be used for the detection and enumeration of *S. aureus* depends to some extent upon the reason for conducting the testing. Foods suspected to be vectors of staphylococcal food poisoning frequently contain a large population of *S. aureus*, in which case a highly sensitive method will not be required. A more sensitive method may be required to demonstrate an insanitary process, or postprocess contamination, since small populations of *S. aureus* may be expected. In most cases which require testing for *S. aureus*, this organism may not be the only or the predominant species present in the food; therefore, selectively inhibitory media generally are employed for isolation and enumeration.

Selective media employ various toxic chemicals to achieve selectivity, and these media are to a varying extent inhibitory for *S. aureus* as well as for the competitive species. The adverse effect of selective agents may be more acute in processed foods containing injured cells of *S. aureus*. On the other hand, a relatively toxic medium may be advantageous in preventing overgrowth of *S. aureus* by competing species.

Commonly used characteristics for discriminating colonies of *S. aureus* from other species are: a) the ability of *S. aureus* to grow in the presence of specific concentrations of selectively toxic chemicals; b) the form and appearance of colonies of *S. aureus*; c) microscopic morphology; d) the capacity of metabolites produced by *S. aureus* to hydrolyze substrates such as egg yolk or DNA; and e) the production of a substance which coagulates the plasma of humans and other animal species. Media used in the detection and enumeration of *S. aureus* may employ one or more of these diagnostic features.

Ancillary diagnostic features frequently used in confirming the identity of *S. aureus* are: a) the ability of *S. aureus* to utilize glucose and mannitol anaerobically[6]; b) the production of catalase; c) the production of a number of hemolytic, dermonecrotic and lethal toxins[19]; d) the elaboration of micrococcal nuclease[17]; e) sensitivity to lysostaphin[34]; f) patterns of resistance to certain antibiotics; g) DNA composition[4]; and h) the presence of certain components in cell walls.[5, 6] A wide variety of other physiological characteristics also may be useful in studying the taxonomy and genetics of *S. aureus*, but such extensive physiological testing usually is not required in routine detection and enumeration procedures.

Patterns of sensitivity of *S. aureus* to certain bacteriophages may be useful in epidemiological investigations of food poisoning outbreaks,[9, 10] but such patterns generally are not helpful in determining enterotoxigenicity of the species.[33]

31.2 TECHNIQUES FOR ISOLATION AND ENUMERATION

The two most commonly used approaches to detection and enumeration of *S. Aureus* in foods are enrichment isolation and direct plating.

Enrichment procedures may be selective[2] or nonselective.[23] Nonselective enrichment procedures may be especially useful for demonstrating the presence of injured cells, the growth of which is inhibited by toxic components of selective enrichment broth. Enumeration by enrichment isolation may be achieved by determining either an indicated number or the most probable number (MPN) of *S. aureus* present. Commonly used MPN procedures are 3 tube (3 tubes of each dilution) and 5 tube (5 tubes of each dilution) MPNs.[2, 23] The 5 tube MPN has somewhat greater sensitivity and precision than the 3 tube MPN.

In enumeration procedures samples may be applied to a variety of selective media in various ways: surface streaking, drop plates and pour plates have all been used in direct plating procedures. Surface streaking and drop plate procedures are advantageous in that the form and appearance of surface colonies are somewhat more characteristic than the subsurface colonies encountered with pour plates. A greater amount of sample can be accommodated with surface streaking techniques than by drop plates; thus the surface streaking technique is more sensitive than drop plates. The principal advantage of pour plates is that greater sample volumes can be used. The relative precision of the various direct plating techniques for enumeration of *S. aureus* has not been established.

Since the same types of selective media frequently are employed in both enrichment and in direct plating, the relative sensitivity of the two procedures is largely dependent upon sample volumes employed. Larger volumes of sample normally are used in enrichment tubes, but equivalent volumes can be utilized in direct plating procedures whenever it is feasible to use the required number of replicate plates. The relative precision of the two procedures for enumeration of *S. aureus* has not been established, but generally it is conceded that plate counting procedures are somewhat more precise than MPN procedures.

Procedures for demonstrating the presence of *S. aureus* in foods based upon detecting the presence of micrococcal nuclease[17] have been described.[13, 20, 27] The enzyme is produced by most strains of *S. aureus*, but at different rates by different strains of the species. Large numbers of cells are required to produce detectable amounts of the enzyme; thus, the procedure is not suitable for demonstrating the presence of small numbers of *S. aureus*. The procedure is not satisfactory for enumeration, but may be useful for screening to determine the necessity for more elaborate analytical procedures such as enterotoxin testing.

31.3 TYPES OF MEDIA COMMONLY USED FOR DETECTION AND ENUMERATION

The two selectively toxic chemicals most frequently used in staphylococcal isolation media are sodium chloride (NaCl) and potassium tellurite (K_2TeO_3). Various concentrations of these agents have been used, ranging from 5.5% to 10% NaCl and from .0025% to .05% K_2TeO_3. Other chemicals such as ammonium sulfate, sorbic acid, glycine, lithium chloride, and polymyxin frequently are combined with NaCl and K_2TeO_3. Sodium azide alone, or in combination with NaCl, and Neomycin also have been used in selective isolation media. Media containing identical selective agents at the same concentrations vary further in respect to pH. Additional sources of differences in media are contributed by combinations of selective agents and by the inclusion of different, or different combinations of diagnostic features.

The principal diagnostic features used on contemporary media include: a) ability of *S. aureus* to grow in the presence of 7.5% or 10% NaCl;[2, 12, 28] b) ability of *S. aureus* to grow in the presence of .01% to .05% K_2TeO_3 in combination with 0.2% or 0.5% lithium chloride, and from 0.12 to 1.26% glycine[7, 21, 29, 40–43] or 40 μg/ml polymyxin;[15] c) ability of *S. aureus* to reduce K_2TeO_3, producing black colonies; d) colonial form and appearance; e) pigmentation of colonies; f) coagulase activity and acid production in solid medium;[30] g) ability of *S. aureus* to hydrolyze egg yolk;[40] h) production of phosphatase.[41] Ancillary tests usually are required to establish the identity of *S. aureus*.

The confirmatory procedure most frequently used to establish the identity of *S. aureus* is the coagulase test. Coagulase is a substance which clots the plasma of human and other animal species. Differences in suitability among plasmas from the various animal species have been demonstrated.[19, 31] Human or rabbit plasma is most frequently used for coagulase testing, and both are available commercially. The use of pig plasma has been found advantageous in certain cases, but this product is not as widely available. Coagulase production by *S. aureus* may be affected adversely by physical factors such as culture storage conditions, the pH of the medium, and desiccation. The extent to which production of coagulase may be impaired by toxic components of selective isolation media has not been demonstrated clearly.

Genetic changes also may affect the capability of *S. aureus* to produce coagulase. Cultures producing typical diagnostic reactions in selective isolation media, and failing to give a positive coagulase test, should be subjected to further testing. Some simple ancillary tests by which the identity of *S. aureus* may be established are: a) production of catalase;[6] b) anaerobic utilization of glucose, and usually of mannitol;[6] c) sensitivity to lysostaphin;[34] and d) production of micrococcal nuclease.[17] Coagulase testing with plasma from additional animal species also may be useful. If identity of *S. aureus* cannot be ascertained by the above testing procedures, more elaborate determinations such as bacteriophage sensitivity,[9, 10] enterotoxin production, cell wall composition,[5] DNA composition,[4] and DNA homology studies may be required.

31.4 TREATMENT OF SAMPLES

Procedures for sample collection, shipment and preparation described in Chapter 1 should be observed.

Conclusions regarding the potential hazard of foods in noncommercial, or opened commercial containers, in which the presence of *S. aureus* has been detected, should be made with considerable caution. Correlation of biotypes isolated from food containers and from food poisoning victims should be established.

31.5 EQUIPMENT AND SUPPLIES

In addition to the equipment listed in Chapter 2, the following equipment, reagents and media are required.

31.51 Equipment and Supplies

31.511 Pipets-20 μl

31.512 Glass streaking rods: sterile, fire polished, hockey or hoe shaped; approximate measurements: 3 to 4 mm diameter, 15 to 20 cm long, with an angled spreading surface 45 to 55 mm long.

31.513 Drying cabinet or incubator for drying surface of agar plates.

31.52 Reagents

31.521 Coagulase plasma containing EDTA. (Plasma derived from blood for which EDTA was used as the anticoagulant, or to which is added 0.1% EDTA (w/v).

a) Rabbit plasma—fresh or desiccated.
b) Porcine plasma—fresh or desiccated.

31.522 Gram stain reagents. (Chapter 2)

31.53 Media (Chapter 2)

Baird-Parker agar
Brain heart infusion agar
Brain heart infusion broth
KRANEP agar
Staphylococcus medium number 110
Tellurite polymyxin egg yolk agar
Trypticase soy or tryptic soy agar

Trypticase soy or tryptic soy broth
Trypticase soy or tryptic soy broth—double strength
Trypticase soy or tryptic soy broth containing 10% NaCl
Trypticase soy or tryptic soy broth containing 20% NaCl
Vogel and Johnson agar

31.6 PREPARATION OF STANDARDS

31.61 Stock Cultures

Stock cultures of the following properly identified organisms should be maintained for testing the quality of media and reagents.

S. aureus. A coagulase positive biotype with the combined features of egg yolk hydrolysis and pigment production is preferable.

S. epidermidis.

***Micrococcus spp.-M. caseolyticus* is acceptable.**

Prolonged storage resulting in desiccation and frequent serial transplantation of stock cultures should be avoided to lessen risk of loss of certain diagnostic traits.

31.7 RECOMMENDED CONTROLS

Each separately prepared batch of medium used for isolation and enumeration of *S. aureus* should be tested for sterility, productivity, and suitability of diagnostic criteria.

To test the sterility of the media, the melted solid media should be poured into plates of assured sterility and incubated 45 to 48 hours at 35 to 37 C. Liquid media also should be incubated 45 to 48 hours at 35 to 37 C.

Media productivity testing may be accomplished by determining counts of *S. aureus* obtained in 18 to 24 hour broth cultures grown in a noninhibitory medium, such as brain heart infusion broth (Chapter 2). Enumeration should be accomplished on noninhibitory solid plating media such as brain heart infustion agar (Chapter 2). The isolation medium being tested for productivity should give counts not significantly less (20%) than the noninhibitory medium.

Each separately prepared batch of medium should be streaked with known cultures of *S. aureus* to test for appropriate diagnostic criteria, such as colony size and appearance, pigmentation, egg yolk reaction, etc.

Each lot of coagulase plasma should be tested with known cultures of *S. aureus* and *S. epidermidis* to determine its suitability for distinguishing positive and negative coagulase reactions.

31.8 PRECAUTIONS AND LIMITATION OF METHODS

There are many factors affecting the usefulness and reliability of *S. aureus* detection and enumeration procedures. Among the more important factors are a) physiologic state of the organism; b) competitive position of *S. aureus* in the sample menstruum; c) limitations of isolation media.

It has been demonstrated that growth of injured cells of *S. aureus* is restricted by many of the selective isolation media used. Media satisfactory for detecting the presence of *S. aureus* in animal lesions, excretory products, and nonprocessed foods may not be adequate for analyzing processed foods. Factors such as heating, freezing, desiccation, and storage, which are common elements of food processing have been shown to affect adversely growth of *S. aureus*.[8] The extent of cellular injury inflicted during process manipulation depends upon the type and severity of treatment.

The importance of the physiologic state of *S. aureus* to the selection of media for use in isolation and enumeration procedures is receiving increased attention. Among frequently used staphylococcal isolation media which have been found to restrict growth of sublethally heated cells of *S. aureus* are: mannitol salt agar, Vogel and Johnson agar, egg yolk azide agar, phenolphthalein phosphate agar containing Polymyxin, milk-salt agar, tellurite glycine medium, staphylococcus medium number 110, and tellurite polymyxin egg yolk agar.[14, 22] Metabolically impaired cells surviving the toxic chemicals of selective media also may fail to show typical form and appearance.

Limitations in detection and enumeration methods are generally those associated with limitations of the isolation media in supporting growth of *S. aureus* and in suppressing growth of competing species. In addition to variations contributed by the competition for growth media nutrients, procedural efficiency may be affected by other factors such as acid base changes and production of growth limiting products (antibiotics, bacteriocins, and bacteriophage) by the extant flora of food products. It is generally conceded that none of the staphylococcal isolation media will prevent growth of all competing species while failing to restrict growth of some *S. aureus*. Among the sources of variation shown to affect media efficiency significantly are a) type of food examined; b) the relative competitive position of *S. aureus*; c) the strain of *S. aureus* involved.[16]

The inadequacy of the diagnostic criteria utilized in most staphylococcal isolation media in definitely discriminating colonies of *S. aureus* from other species contributes further to variation in procedural efficiency. The physiology of *S. aureus* is diverse, and not all strains of the species demonstrate similar activity. For example, not all biotypes have the capacity to hydrolyze egg yolk, a common diagnostic feature utilized in many detection and enumeration procedures.[18, 26] Considerable divergence also has been demonstrated in the response of various strains to the chemical agents used in selective isolation media. This diversity may lead to considerable confusion regarding the suitability of various isolation media.

Instability has been shown in certain of the physiological traits demonstrated by this species. Variability has been attributed to both physiologic and genetic factors. The frequency with which physiological traits may change, and the elements stimulating such change, have not been demonstrated clearly. In applying the customary procedures for detection and enumeration of *S. aureus*, consideration should be given to the possibility of variation in certain physiological traits. Certain of the usual diagnostic features characteristic of the species may, or may not, be displayed in a given instance.

31.9 PROCEDURES FOR ISOLATION AND ENUMERATION

31.91 Nonselective Enrichment Procedure[23]

This procedure is recommended for testing processed foods likely to contain a small population of injured cells.

Prepare food samples by the procedure described in the section on Preparation and Dilution of Food Homogenate (Chapter 1).

Transfer 50 ml of a 1:10 dilution of the sample into 50 ml of double-strength trypticase soy bean broth. (Chapter 2).

Incubate 2 hours at 35 C.

Add 100 ml of single strength trypticase soy broth containing 20% NaCl.

Incubate for 24 ± 2 hours at 35 C.

Transfer 0.1 ml aliquots of culture to each of duplicate plates of either Baird-Parker agar, Vogel and Johnson agar, or staphylococcus medium number 110, and spread inoculum so as to obtain isolated colonies.

Incubate the plates for 46 ± 2 hours at 35 C.

Select two or more colonies suspected to be *S. aureus* (see media specifications section for description of typical colonies) from each plate and subject to coagulase test (31.95).

Report results as *S. aureus* present or absent in 5g of food, as indicated by results of coagulase testing.

31.92 Selective Enrichment Procedure[1, 2]

This procedure is recommended for detecting small numbers of *S. aureus* in raw food ingredients and nonprocessed foods expected to contain a large population of competing species.

Prepare food samples by procedure described in the section on Preparation and Dilution of the Food Homogenate (Chapter 1).

Inoculate 3 tubes of trypticase soy broth containing 10% NaCl (Chapter 2) at each test dilution with 1 ml aliquots of decimal sample dilutions. Maximum dilution of sample tested must be high enough to yield a negative end-point.

Incubate 45 to 48 hours at 35 to 37 C.

Using a 3 mm inoculating loop, transfer one loopful from each turbid tube to previously prepared plates of either Baird-Parker agar (Chapter 2), Vogel and Johnson agar (Chapter 2), or staphylococcus medium 110 (Chapter 2), and streak plates to obtain isolated colonies.

Incubate for 45 to 48 hours at 35 to 37 C.

Pick one or more colonies from each plate containing colonies suspected to be *S. aureus* (see media specifications section for description of typical colonies) and subject to coagulase test (31.95).

Report most probable number (MPN) of *S. aureus* per gram of sample from tables of MPN values (Chapter 6).

31.93 Surface Plating Procedure[1]

This procedure is recommended for the detection of *S. aureus* in raw or processed food. The sensitivity of this procedure may be increased by using larger inoculum volumes (> 1 ml) distributed over > 3 replicate plates.

Prepare food samples by the procedure given in the chapter on Preparation and Dilution of the Food Homogenate (Chapter 1).

Plating of two or more decimal dilutions may be required to obtain plates with the desired number of colonies per plate.

For each dilution to be plated, aseptically transfer 1 ml of sample suspension to triplicate plates of Baird-Parker agar (Chapter 2) and distribute the 1 ml inoculum equitably over the triplicates plates (e.g., 0.4 ml—0.3 ml—0.3 ml).

Spread the inoculum over the surface of the agar using sterile, bentglass streaking rods. Avoid the extreme edges of the plate. Retain the plates in an upright position until the inoculum is absorbed by the medium (about 10 minutes on properly dried plates). If the inoculum is not readily absorbed, plates may be placed in an incubator in an upright position for about 1 hour before inverting.

Invert plates and incubate 45 to 48 hours at 35 to 37 C.

Select plates containing 20 to 300 colonies unless plates at only lower dilutions (> 200 colonies) have colonies with the typical appearance of *S. aureus* (see media specifications for description of typical colonies). If several types of colonies are observed which appear to be *S. aureus*, count the number of colonies of each type and record counts separately. When plates at the lowest dilution plated contain < 20 colonies, they may be used. If plates containing > 200 colonies have colonies with the typical appearance of *S. aureus* and typical colonies do not appear on plates at higher dilutions, use these plates for enumeration of *S. aureus*, but do not count nontypical colonies.

Select one or more colonies of each type counted and test for coagulase production (31.95). Coagulase positive cultures may be considered to be *S. aureus*.

Add the number of colonies on triplicate plates represented by colonies giving a positive coagulase test, and multiply the total by the sample dilution factor. Report this number as "*S. aureus* per gram of product tested."

Note: Many laboratories have reported the successful use of tellurite polymyxin egg yolk agar,[15] (Chapter 2) and KRANEP agar,[35] (Chapter 2) in place of Baird-Parker agar for this procedure. However, comparative testing of these three media has not been extensive.

31.94 Spot Plate Procedure

This procedure is recommended only for foods in which the number of *S. aureus* is likely to be > 5000/g.

Prepare food samples by procedure given in the chapter on Preparation and Dilution of the Food Homogenate (Chapter 1).

Prepare dried plates of Baird-Parker agar (Chapter 2) and divide plates by marking the bottom into two or more equal segments. Indicate for each segment the dilution to be plated, using two segments for each dilution.

Using a microliter pipet, deliver duplicate 20 μl aliquots of each dilution tested onto the surface of indicated segments of the plate.

Allow the plates to dry for about 10 minutes at room temperature before inverting plates.

Incubate the plates 45 to 48 hours at 35 to 37 C.

Select dilutions producing ≤ 10 individual colonies appearing to be *S. aureus* (see media specification section for description of typical colonies), and count the number of suspect colonies.

Select one or more of the suspect colonies and perform the coagulase test (31.95).

Calculate the content of *S. aureus* by adding the number of coagulase positive colony types observed on duplicate plate segments. Determine the average of the duplicate counts, multiply by 50 and then by the sample dilution factor. Report number as "*S. aureus* per gram of sample."

31.95 Coagulase Test

a. Transfer colonies suspected to be *S. aureus* to small tubes containing about 0.3 ml brain heart infusion broth (Chapter 2), and emulsify thoroughly.

b. Withdraw 1 loopful of resulting culture suspension, and transfer to trypticase or tryptic soy agar (Chapter 2) slants.

c. Incubate culture suspensions and slants 18 to 24 hours at 35 to 37 C.

d. Retain slant cultures at room temperature for ancillary or repeat tests in case the coagulase test results are questionable.

e. Add 0.5 ml coagulase plasma with EDTA to broth cultures and mix thoroughly. Incubate at 35 to 37 C and examine periodically over a 6 hour interval for clot formation. A 3+ or 4+ clot formation (Fig. 1) is considered a positive reaction for *S. aureus*.[32, 36] Small or poorly organized clots (1+ and 2+) should be confirmed[32, 36] by performing the ancillary tests listed below. Recheck doubtful coagulase test results on broth cultures which have been incubated at 35 to 37 C for more than 18 hours but less than 48 hours. Assure culture purity before rechecking coagulase test results. Do not store rehydrated plasma longer than 5 days.

31.96 Ancillary Tests

If anomalies are encountered during testing (31.91–31.94), additional testing may be required to establish speciation of *S. aureus,* or its role as an agent of foodborne illness. The following tests are usually adequate for speciation of *S. aureus*:

a. Microscopic examination

A Gram stain (Chapter 2) of *S. aureus* cultures will produce gram positive cocci, 0.8 to 1.0 μm in diameter occurring singly, in pairs, or most typically in irregular clusters resembling clusters of grapes.

b. Catalase reaction (Chapter 2.)

Cultures of *S. aureus* are catalase positive.

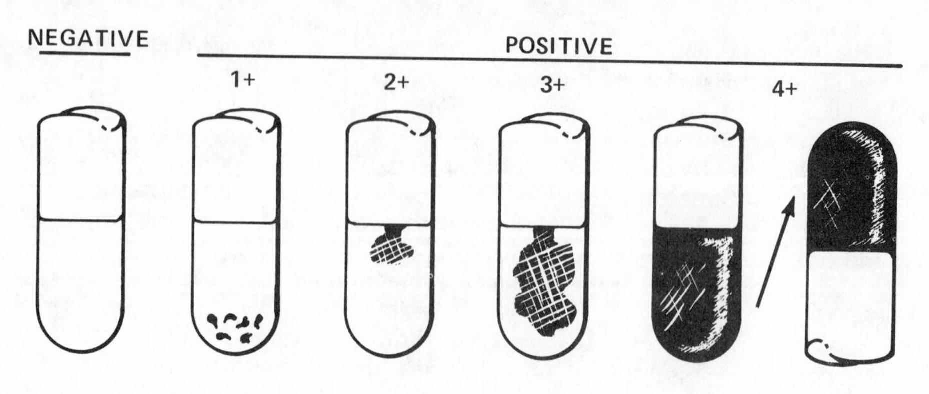

NEGATIVE	NO EVIDENCE OF FIBRIN FORMATION
1+ POSITIVE	SMALL UNORGANIZED CLOTS
2+ POSITIVE	SMALL ORGANIZED CLOT
3+ POSITIVE	LARGE ORGANIZED CLOT
4+ POSITIVE	ENTIRE CONTENT OF TUBE COAGULATES AND IS NOT DISPLACED WHEN TUBE IS INVERTED

FIGURE 1. Types of Coagulase Test Reactions

c. Anaerobic utilization of glucose and mannitol

When tested by procedures recommended by the Subcommittee on Taxonomy of Staphylococci and Micrococci,[39] cultures of *S. aureus* will utilize glucose and usually mannitol anaerobically.[6]

d. Production of micrococcal nuclease

Cultures of *S. aureus* generally produce a heat stable nuclease which hydrolyzes partially denatured DNA, RNA, and certain oligonucleotides. Micrococcal nuclease may be detected by 1) the slide procedure of Lachica, Hoeprich, and Genigeorgis;[27] 2) the spectrophotometric procedure of Cunningham, Catlin, and Privat De Garilhe;[17] 3) the turbidimetric assay of Erickson and Deibel.[26]

e. Susceptibility to lysostaphin

Cells of *S. aureus* are lysed by lysostaphin when tested by the procedure of Schindler and Schuhardt.[34]

f. Bacteriophage sensitivity

Bacteriophage typing of *S. aureus* by the methods of Blair and Carr[9] or Blair and Williams[10] may be useful in elucidating the epidemiology of staphylococcal food poisoning outbreaks. A Subcommittee on Phage-Typing of Staphylococci[38] has recommended a basic set of 22 phages for typing cultures of *S. aureus* of human origin.

31.10 REFERENCES

1. AOAC. 1975. Changes in official methods of analysis made at the eighty-eighth annual meeting, October 14–17, 1974. J.A.O.A.C. **58**:416–417
2. AOAC. 1970. Official Methods of Analysis, 11th. Ed., Association of Official Analytical Chemists, Washington, D.C. pp 843–844
3. ARMIJO, R., D. A. HENDERSON, R. TIMOTHEE, and H. B. ROBINSON. 1957. Food poisoning outbreaks associated with spray-dried milk—an epidemiologic study. Amer. J. Pub. Health **47**:1093–1100
4. AULETTA, A. E. and E. R. KENNEDY. 1966. Deoxyribonucleic acid base composition of some members of the *Micrococcaceae*. J. Bacteriol. **92**:28–34
5. BAIRD-PARKER, A. C. 1970. The relationship of cell wall composition to the current classification of staphylococci and micrococci. Int. J. Sys. Bacteriol. **20**:483–490
6. BAIRD-PARKER, A. C. 1965. The classification of staphylococci and micrococci from worldwide sources. J. Gen. Microbiol. **38**:363–387
7. BAIRD-PARKER, A. C. 1962. An improved diagnostic and selective medium for isolating coagulase positive staphylococci. J. Appl. Bacteriol. **25**:12–19
8. BAIRD-PARKER, A. C. and E. DAVENPORT. 1965. The effect of recovery medium on the isolation of *Staphylococcus aureus* after heat treatment and after storage of frozen or dried cells. J. Appl. Bacteriol. **28**:390–402
9. BLAIR, J. E. and M. CARR. 1960. The techniques and interpretation of phage typing of staphylococci. J. Lab. Clin. Med. **55**:650–662
10. BLAIR, J. E. and R. E. O. WILLIAMS. 1961. Phage typing of staphylococci. Bull. World Health Org. **24**:771–784
11. BUSTA, F. F. and J. J. JEZESKI. 1963. Effect of sodium chloride concentration in an agar medium on growth of heat-shocked *Staphylococcus aureus.* Appl. Microbiol. **11**:404–407
12. CHAPMAN, G. H. 1945. The significance of sodium chloride in studies of staphylococci. J. Bacteriol. **50**:201–203
13. CHESBRO, W. R. and K. AUBORN. 1967. Enzymatic detection of the growth of *Staphylococcus aureus* in foods. Appl. Microbiol. **15**:1150–1159
14. COLLINS-THOMPSON, D. L., A. HURST, and B. ARIS. 1974. Comparison of selective media for the enumeration of sublethally heated food-poisoning strains of *Staphylococcus aureus.* Canad. J. Microbiol. **20**:1072–1075
15. CRISLEY, F. D., R. ANGELOTTI, and M. J. FOTER. 1964. Multiplication of *Staphylococcus aureus* in synthetic cream fillings and pies. Pub. Health Rep. **79**:369–376
16. CRISLEY, F. D., J. T. PEELER, and R. ANGELOTTI. 1965. Comparative evaluation of five selective and differential media for the detection and enumeration of coagulase-positive staphylococci in foods. Appl. Microbiol. **13**:140–156
17. CUNNINGHAM, L., B. W. CATLIN, and M. PRIVAT DE GARILHE. 1956. A deoxyribonuclease of *Micrococcus pyogenes.* J. Amer. Chem. Soc. **78**:4642–4645
18. DEWAART, J., D. A. A. MOSSEL, R. TENBROEKE, and A. VAN DE MOOSDIJK. 1968. Enumeration of *Staphylococcus aureus* in foods with special reference to egg-yolk reaction and mannitol negative mutants. J. Appl. Bacteriol. **31**:276–285
19. ELEK, S. D. 1959. *Staphylococcus pyogenes* and Its Relation to Disease. E. and S. Livingstone Ltd., London, England, pp 178–312
20. ERICKSON, A. and R. H. DIEBEL. 1973. Turbidimetric assay of staphylococcal nuclease. Appl. Microbiol. **25**:337–341
21. GIOLITTI, G. and C. CANTONI. 1966. A medium for the isolation of staphylococci from foodstuffs. J. Appl. Bacteriol. **29**:395–398
22. GRAY, R. J. H., M. A. GASKE, and Z. J. ORDÅL. 1974. Enumeration of thermally stressed *Staphylococcus aureus* MF 31. J. Food Sci., **39**:844–846
23. HEIDELBAUGH, N. D., D. B. ROWLEY, E. M. POWERS, C. T. BOURLAND, and J. L. MCQUEEN. 1973. Microbiological testing of skylab foods. Appl. Microbiol. **25**:55–61
24. IANDOLO, J. J. and Z. J. ORDAL. 1966. Repair of thermal injury of *Staphylococcus aureus*. J. Bacteriol. **91**:134–142
25. JACKSON, H. and M. WOODBINE. 1963. The effect of sublethal heat treatment on the growth of *Staphylococcus aureus*. J. Appl. Bacteriol. **26**:152–158

26. Koskitalo, L. D. and M. E. Milling. 1969. Lack of correlation between egg yolk reaction in staphylococcus medium 110 supplemented with egg yolk and coagulase activity of staphylococci isolated from cheddar cheese. Canad. J. Microbiol. **15**:132–133
27. Lachica, R. V. F., P. D. Hoeprich, and C. Genigeorgis. 1972. Metachromatic agar-diffusion microslide technique for detecting staphylococcal nuclease in foods. Appl. Microbiol. **23**:168–169
28. Lewis, K. H. and R. Angelotti. 1964. Examination of Foods for Enteropathogenic and Indicator Bacteria. PHS Pub. No. 1142, p 98–99. Supt. Documents, U.S. Gov. Printing Off., Washington, D.C. pp 98–99
29. Moore, T. D. and F. E. Nelson. 1962. The enumeration of *Staphylococcus aureus* on several tellurite-glycine media. J. Milk Food Technol. **25**:124–127
30. Orth, D. S. and A. W. Anderson. 1970. Polymyxin-coagulase-mannitol-agar. I. A selective isolation medium for coagulase-positive staphylococci. Appl. Microbiol. **19**:73–75
31. Orth, D. S., L. R. Chugg, and A. W. Anderson. 1971. Comparison of animal sera for suitability in coagulase testing. Appl. Microbiol. **21**:420–425
32. Ryman, M. K., C. E. Park, J. Philpott, and E. C. D. Dodd. 1975. Reassessment of the coagulase and thermostable nuclease tests as means of identifying *Staphylococcus aureus*. Appl. Microbiol. **29**:451–454
33. Saint-Martin, M., G. Charest, and J. M. Desranleau. 1951. Bacteriophage typing in investigations of staphylococcal food-poisoning outbreaks. Canad. J. Pub. Health. **42**:351–358
34. Schindler, C. A. and V. T. Schuhardt. 1964. Lysostaphin: A new bacteriolytic agent for the staphylococcus. Proc. Nat. Acad. Sci., **51**:414–421
35. Sinell, H. J. and J. Baumgart. 1966. Selektionährboden zur isolierung von staphylokokken aus lebensmitteln. Zentralbl. Bakteriol. Parasitenk. Infektionskr. Hyg: Abt. Orig. **197**:447–461
36. Sperber, W. H. and S. R. Tatini. 1975. Interpretation of the tube coagulase test for identification of *Staphylococcus aureus*. Appl. Microbiol. **29**:502–505
37. Stiles, M. E. and L. D. Witter. 1965. Thermal inactivation, heat injury, and recovery of *Staphylococcus aureus*. J. Dairy Sci. **48**:677–681
38. Subcommittee on Phage-Typing of Staphylococci. 1970. Report to the international committee on nomenclature of bacteria. Int. J. Sys. Bacteriol. **21**:167–170
39. Subcommittee on Taxonomy of Staphylococci and Micrococci. 1965. Recommendations. Int. Bull. Bacteriol. Nomenclature Taxonomy. **15**:109–110
40. Tirunarayanan, M. O. and H. Lundbeck. 1968. Investigations on the enzymes and toxins of staphylococci. Acta Pathol. Microbiol. Scandinav. **73**:429–436
41. Turner, F. S. and B. S. Schwartz. 1958. The use of lyophilized human plasma standardized for blood coagulation factors in the coagulase and fibrinolytic tests. J. Lab. Clin. Med. **52**:888–894
42. Williams, M. L. B. 1972. A note on the development of a polymyxin-mannitol-phenolphthalein diphosphate agar for the selective enumeration of coagulase positive staphylococci in foods. J. Appl. Bacteriol. **35**:139–141
43. Zebovitz, E. J., J. B. Evans, and C. F. Niven, Jr. 1955. Tellurite-glycine agar: A selective plating medium for the quantitative detection of coagulase-positive staphylococci. J. Bacteriol. **70**:686–690

CHAPTER 32

STAPHYLOCOCCAL ENTEROTOXINS

Merlin S. Bergdoll and Reginald W. Bennett

32.1 INTRODUCTION

32.11 General Information

The staphylococcus enterotoxins are proteins produced by some strains of staphylococci.[1] If these strains are allowed to grow in foods, sufficient enterotoxin may be produced to cause illness when the food is consumed (staphylococcus food poisoning). The true incidence of this disease is not known but it is one of the leading causes of foodborne illness. The most common symptoms are vomiting and diarrhea which develop 2 to 6 hours after ingestion of the toxin. The illness is relatively mild, normally lasting only a few hours to one day; however, in some instances the illness is sufficiently severe to require hospitalization.

The need to detect enterotoxins in foods encompasses two areas (1) foods that have been implicated in food poisoning outbreaks; (2) foods that are suspected of containing enterotoxin. In the former case, the detection of enterotoxin in the foods is confirmatory of staphylococcal food poisoning. In the latter case, the presence or absence of enterotoxin determines the marketability of the product. The latter use cannot be overemphasized because it is difficult to prevent the presence of staphylococci in some types of food materials.

The methods for detection of the enterotoxins involve the use of specific antibodies to each of them. The fact that there are several antigenically different enterotoxins (A to E) complicates their detection because each one must be assayed separately. One other problem is that unidentified enterotoxins exist, and since specific antibodies to these are not available, it is not possible to detect them. These, however, appear to be responsible for only a few percent of food poisoning outbreaks.

32.12 Sensitivity Requirements

Limits have not been set on the minimum amount of enterotoxin to insure the safety of a food, but are dependent on the sensitivity of the detection methods employed. Whether this sensitivity is adequate is dependent on the amount of enterotoxin required to cause illness in man. This is not known, although information from food poisoning outbreaks indicates

IS/A Committee Liaison: Edmund A. Powers

that those made ill probably consumed less enterotoxin A than 1 μg total, the enterotoxin most frequently involved in food poisoning outbreaks. The minimum possible to detect with the microslide technique is 0.1 to 0.25 μg of enterotoxin per 100 g of food. This is the guide that is used for the testing of new methods, and should be the guide for those using the methods outlined in this chapter.

32.13 Serological Detection Methods

A number of methods employing specific antibodies to the enterotoxins have been used for detection and measurement of the enterotoxins. To be useful for detection in food extracts at the minimum level needed, the method should be equal in sensitivity to the microslide method. This is around 0.1 μg enterotoxin per ml, for the competent operator. This sensitivity requires that extracts from 100 g of food be concentrated to about 0.2 ml; thus any method less sensitive than the microslide will be inadequate. Methods that are more sensitive than the microslide, such as radioimmunoassay and hemagglutination techniques, require less concentration of the food extracts; thus, are less time consuming. However, we are not including these methods in this chapter because to do so would be premature. After several years work on the reversed passive hemagglutination technique,[14] we have been unable to resolve all of the problems encountered. The two main problems are that it is impossible to adsorb enterotoxin antibodies from all antisera preparations onto the red blood cells, and, secondly, some food materials give nonspecific agglutination of the cells.

The radioimmunoassay method[4, 6, 9] is one that probably will be used widely in the future; however, at the present time methods have not been adequately standardized, nor has there been enough experience with it to make recommendations. The major drawback to its use is that enterotoxins of at least 90% purity are required. As of this date not all of the enterotoxins have been purified, and those that have been are not readily available.

32.14 Enterotoxigenicity of Staphylococcal Strains

Examination of staphylococci for enterotoxin production is a helpful step in the detection of enterotoxin in foods, as well as desirable in the examination of strains isolated from various sources. Sometimes this involves dozens of strains; hence, a simple method for doing this is very useful. The methods outlined here have been used for hundreds of cultures and are designed to detect the minimum amount of enterotoxin production that might be important if such strains were food contaminants.

32.15 Enterotoxin in Foods

The major problem in the detection of enterotoxins in foods is the small amount that may be present in food poisoning outbreaks, plus the fact that marketable foods should contain essentially none. To detect these small amounts, either a very sensitive detection procedure or a satisfactory means of concentrating the food extract must be available. The sensitivity of the radioimmunoassay technique[9] is such that the concentration of food extracts are seldom needed; however, with the microslide, methods must be available to concentrate the extract from 100 g of food to 0.2 ml. At the same time, interfering substances must be removed from the extract, or the precipitate lines on the microslides may be obscured.

32.2 TREATMENT OF SAMPLES

Samples of food for analysis should be kept refrigerated or frozen, and should not be allowed to stand at room temperature except when handling. This is particularly true of foods that contain live organisms. In the case of food poisoning outbreaks, it is desirable to collect the samples and refrigerate them as soon as possible to avoid their being mishandled.

32.3 SPECIAL EQUIPMENT, SUPPLIES, REAGENTS, AND MEDIA

32.31 Microslide Method[3]

a. Electricians tape

Although either Scotch Brand vinyl plastic electrical tape no. 33, 3M Company, or Homart plastic tape ¾ inch wide (Sears Roebuck and Company) are recommended, any good quality electrician's plastic tape should be satisfactory. The major criterion is the ability to stick to the glass slides and not be readily removable with repeated washings.

b. Plexiglas template

The templates are made from plexiglas according to the diagram below.

c. Silicone grease

Silicone grease (Dow Corning), silicone lubricant spray (available at hardware stores), Lubriseal, or similar lubricant is used for coating the template so it can be removed from the microslide after development without disrupting the layer of agar on which it is resting.

d. Petri dishes

20 × 150 mm size is convenient for incubation of slides.

e. Platform for filling templates

A Cordis Laboratories viewer (P.O. Box 684, Miami, Fla. 33137) with a piece of frosted glass over the lighted area is ideal for observing the presence of, and removing, any bubbles in the wells.

f. Capillary pipets

Pasteur capillary pipets (9 inch) are essential for applying the enterotoxin reagents and samples to microslides.

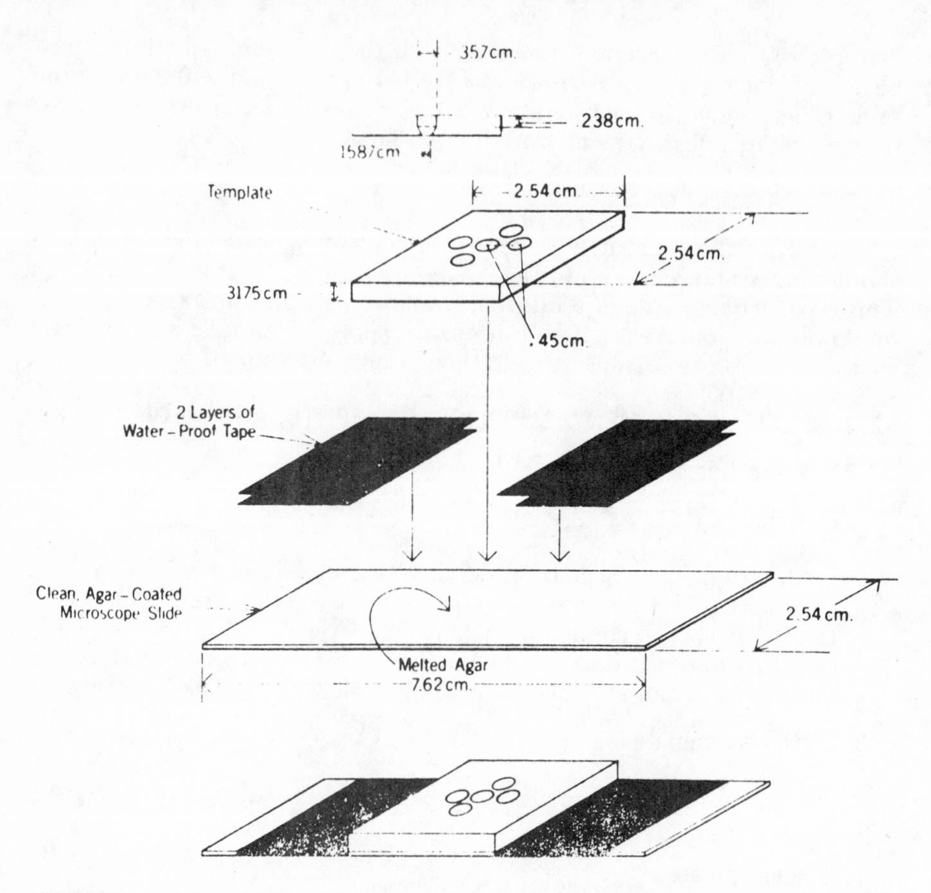

FIGURE 1. Microslide Gel Diffusion Assembly with Specifications for Plexiglas Template.

g. Incubator

Any incubator or storage device that can be held constant at temperatures from 25 to 37 C will suffice.

h. Fluorescent lamp

Any fluorescent desk lamp to which the microslide can be held at an oblique angle is adequate.

i. Staining equipment

A Wheaton horizontal staining dish with removable slide rack, or Coplin jars can be used for staining the microslides.

j. Enterotoxins

Crude enterotoxin preparations are adequate for the microslide test as long as only one enterotoxin precipitate line is obtained with the specific antiserum to that enterotoxin. Purified enterotoxins eliminate the possibility of extra precipitate lines, but they are not readily available.

k. Enterotoxin antisera

Specific antisera to each of the enterotoxins that give only one line in the microslide with the respective crude enterotoxin is necessary. These materials are not readily available; however, they can be obtained from the Food Research Institute, University of Wisconsin, and the Food and Drug Administration Laboratories, Washington, D.C.

l. Agar

Difco purified agar, Noble agar, or Agarose agar (Oxoid) are recommended, although any highly purified agar can be used. The main concern is that the solidified agar in the microslide be clear.

m. Thimersol
(Merthiolate; ethylmercurithiosalicylate)

Thimersol (sodium salt) can be obtained from Sigma Chemical Co. (P.O. Box 14508, St. Louis, Mo. 63178).

n. Sodium barbital

Any reagent grade sodium barbital can be used.

o. Thiazine Red-R

This stain is used to enhance the precipitate lines in the microslide and can be obtained from a number of sources. An alternate stain is Woolfast Pink RL (American Hoechst Corp., Route 202-206, Bridgewater, N.J. 08807), which some investigators prefer to the Thiazine Red-R.

32.32 Optimum Sensitivity Plate[13]

a. Petri dishes

Plastic Petri dishes, 50 × 12 mm with tight lids, are used (Falcon Plastics No. 1006 Petri dish).

b. Plexiglas

Template: A template is made from plexiglas with the hole pattern indicated below. It is made slightly larger than the Petri dish and a groove is made around the edge so the template can be placed in a fixed position while the holes are made in the agar with cork borers. The area of the larger holes is approximately double that of the smaller holes. This in effect gives the same results with half the concentration that is used in the small holes, thus increasing the sensitivity two-fold.

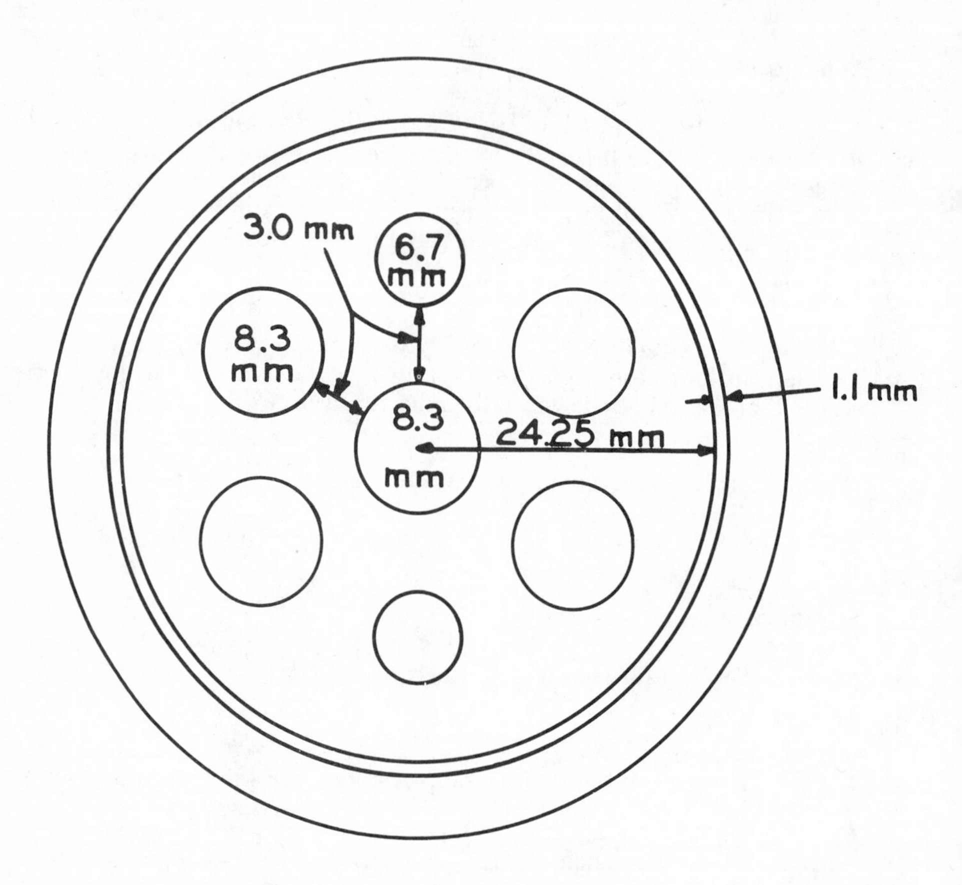

FIGURE 2. Optimum Sensitivity Plate Template for Making the Wells in the Agar Layer

c. Hole cutters

Ordinary cork borers with the outside diameters equal to the diameters of the holes in the template are used. If this size is not available, cutters to fit should be made, or the holes in the template made to fit the cutters available, keeping in mind that the area of the larger holes should be approximately double that of the smaller holes.

d. Reading device

The reading device suggested for use is the Cordis Laboratories viewer used for filling microslide templates (without the frosted glass) (32.31e). Any similar light device can be used, the main essential being that the light comes through the bottom of the plate at an angle.

e. Storage boxes

A nearly airtight plastic rectangular box, such as is commonly sold for refrigerator storage, is used, with 5 ml beakers containing water placed in each corner.

f. Enterotoxins

See 32.31j

g. Enterotoxin antisera

See 32.31k

h. Agar

See 32.31l

32.33 Enterotoxin Production by Staphylococcus Isolates

32.331 Soft agar method[2]

a. Test tubes

Test tubes (25 × 200 mm) are used for sterilization of the medium in 25 ml lots. The tubes containing the medium can be stored until the medium is needed.

b. Centrifuge

A high speed refrigerated centrifuge such as a Sorvall RC-2B is ideal; however, centrifuges of lower speed and unrefrigerated can be used, but require longer centrifuging times.

c. Centrifuge tubes

50 ml centrifuge tubes that will withstand the speed at which the centrifugation is done are adequate (e.g., polycarbonate tubes).

d. Agar

Any ordinary agar such as Bacto-agar.

e. Media

The medium normally used is brain heart infusion (BHI) broth although other media such as 3% protein hydrolysate powder (Mead

Johnson International, Evansville, Indiana), plus 3% N-Z amine NAK (Sheffield Chemical Co., Lyndhurst, N.J.), plus 10 mg niacin and 0.5 mg thiamin per liter of medium at pH 6.8 (3 + 3 medium), and 4% N-Z Amine NAK supplemented with niacin and thiamin (see above), 0.2% yeast extract, and 0.2% glucose are satisfactory.

f. Barium chloride and sulfuric acid

$BaCl_2$ and H_2SO_4 are used to make the No. 1 McFarland standard for inoculum comparison.

32.332 Cellophane-over-agar-method[8, 13]

a. Cellophane

Of the membranes tested, 3¼ inch dialysis tubing from A.H. Thomas Co. (P.O. Box 779, Philadelphia, Pa. 19105) was found to be the most suitable. It is essential that this be obtained fresh and kept refrigerated to minimize aging.

b. Filter paper

Filter paper (9 cm disks) is used in the sterilization of the membranes.

c. Centrifuge

See 32.331b

d. Centrifuge tubes

Any 15 ml centrifuge tubes that will withstand the speed at which the centrifuging is done are adequate (e.g., polycarbonate tubes)

e. Agar

See 32.331d

f. Media

See 32.331e

32.333 Sac culture method (Donnelly)[7]

a. Dialysis tubing

Dialysis tubing of approximately 3 cm flat width has been found to give good results.

b. Media

The recommended media is double strength BHI broth, although other media can be used, but in higher concentrations than is recommended for normal production.

c. Shaker

A rotary shaker such as a gyrotory shaker from New Brunswick Scientific Co., Inc. (1130 Somerset St., New Brunswick, N.J. 08903), is desirable for shaking the flasks containing the sacs. Growth and enterotoxin production can be obtained without shaking, but a longer period of time is necessary, and the amount of enterotoxin produced is lower.

d. Centrifuge

See 32.331b

e. Centrifuge tubes

See 32.331c

32.334 Enterotoxin detection in foods

a. Extraction for microslides (Casman and Bennett Method)[15]

1. Refrigerated cabinet or cold room

Part of the extraction procedures, such as the carboxymethyl cellulose (CMC) column extraction, is done at about 5 C, primarily because the column is allowed to run overnight. Also, this type of space is useful for storing the food materials, extracts, etc., thus eliminating the need for a refrigerator.

2. Waring blender

The food samples must be ground into a slurry for adequate extraction of the enterotoxin.

3. pH meter

The pH at which the extraction is done is important, as well as the pH's of the buffers used in the CMC column. The pH adjustments usually are made within ± 0.1 pH unit.

4. Refrigerated centrifuge

The food extracts are centrifuged at relatively high speeds at 5 C using a refrigerated centrifuge, such as a Sorval RC-2B which can reach speeds of 20,000 rpm. The lower the speed at which the centrifuging is done, the more difficult is the clarification of the extracts.

5. Centrifuge tubes

285 ml stainless steel centrifuge bottles (Sorvall No. 530) generally are used.

6. Magnetic stirrer

A magnetic stirrer is a convenient device to keep samples agitated during pH adjustments, dialysis, etc.

7. Filter cloth

Filtering the food extracts at various stages in the procedures is done with a coarse material such as Miracloth (Chicopee Mills, Inc., 1450 Broadway, New York, N. Y. 10018), or several layers of cheesecloth placed in a funnel. Wetting the Miracloth before placing in the funnel reduces the loss of sample due to adhering to the cloth. Use of the coarse material allows for rapid flow with efficient removal of food particles, the chloroform layer, etc.

8. Carboxymethyl cellulose

The extract is purified partially by adsorption onto carboxymethyl cellulose (CMC) Whatman CM 22, 0.6 mequiv/g, H. Reeve Angel, Inc., 9 Bridewell Place, Clifton, New Jersey). A quantity of soluble extractants are removed by this step.

9. Polyethylene glycol

The most convenient method for concentrating the food extracts is with polyethylene glycol (Carbowax 20,000; Union Carbide Corp., Chemical Division, 230 North Michigan Ave., Chicago, Illinois 60638).

10. Lyophilizer

The final concentration of the extract is done by freeze drying because this is the only convenient method for reducing the volume to 0.2 ml, and at the same time recovering all of the extract.

11. Chromatographic tube

Partial purification of the enterotoxin in the food extracts is done most efficiently using CMC, with elution in a chromatographic tube. For this purpose a 19 mm inside diameter column (e.g., chromaflex plain with stopcock, size 234, Kontes Glass Company, Vineland, New Jersey) is recommended.

12. Dialysis tubing

Cellulose casing of 1⅛ inches flat width and an average pore diameter of 48 angstrom units is used.

13. Separatory funnels

Separatory funnels of varying sizes are needed in the $CHCl_3$ extractions and with the chromatographic column.

14. Glass wool

Glass wool makes ideal plugs for use in the chromatographic column.

15. Chloroform

The food extract is treated with $CHCl_3$ (several times in some instances) to remove lipids and other substances that interfere with the concentration of the extract to small volumes.

b. Extraction for microslides (Reiser and Conaway method)[12]

1. Refrigerated cabinet or cold room

See 32.334a1

2. pH meter

See 32.334a3

3. Omnimixer

An Omnimixer (Sorvall) is convenient for grinding food samples. This can be done directly in the stainless steel centrifuge tubes.

4. Refrigerated centrifuge

See 32.334a4

5. Centrifuge tubes

See 32.334a5

6. Magnetic stirrer

See 32.334a6

7. Filter cloth

See 32.334a7

8. Amberlite Cg-50 ion exchange resin

The extract is purified partially by adsorption onto Amberlite CG-50 ion exchange resin (100 to 200 mesh) (Mallinckrodt Chemical Works, St. Louis, Mo. 63160).

9. Polyethylene glycol

See 32.334a9

10. Lyophilizer

See 32.334a10

11. Dialysis tubing

Cellulose casing of ¾ inch flat width is used.

12. Chloroform

See 32.334a15

13. Trypsin

The extracted sample is treated with trypsin to eliminate extraneous proteins carried through the concentration procedures (crude trypsin Type II, Sigma Chemical Co., P.O. Box 14508, St. Louis, Mo. 63178).

32.4 PREPARATION OF STANDARDS

32.41 Microslide Method

32.411 Enterotoxins

Dilute the enterotoxin preparations according to the specific instructions supplied with them. It is important that the reagents be balanced, that is, that the concentration of the enterotoxin be adjusted to the concentration of the antiserum, so that a line of precipitate will appear approximately halfway between the antigen and antiserum wells in the microslide. If they are out of balance, lines may not appear or may be difficult to observe. The buffer used in dissolving and diluting the enterotoxin is 0.37% dehydrated BHI in normal physiological saline, pH 7.0, or 0.3% proteose peptone in physiological saline.

32.412 Enterotoxin antisera

Dilute the antisera according to the specific instructions supplied with the antisera. The buffer recommended for dilution is the same as for the enterotoxins. The antisera can be maintained by keeping them in the refrigerator, although for long range storage, freeze drying or freezing is recommended.

32.42 Optimum Sensitivity Plate Method (OSP)

32.421 Enterotoxins

See 32.411. Use the enterotoxin at a concentration of 4 μg/ml in the smaller wells. The balance of the reagents for the OSP is not as critical as it is for the microslide, primarily due to the thickness of the agar layer.

32.422 Enterotoxin antisera

See 32.412

32.5 RECOMMENDED CONTROLS

32.51 Microslide Method

It is recommended that the enterotoxin reagents be used at concentrations which will give a visible line halfway between the antigen and antibody wells. The test can be made more sensitive by reducing the concentration of the reagents so that a line is visible only after staining or enhancement. The amount of enterotoxin necessary to give visible lines varies with the antisera but, normally, a visible line can be obtained with 0.25 to 0.50 μg enterotoxin per ml of enterotoxin solution. It is essential that, under the conditions of the test, only one line be observed with the control reagents. A set of slides should be prepared which contains the antiserum and reference enterotoxin only.

32.52 Optimum Sensitivity Plate Pethod

The control enterotoxin is placed in the smaller wells at a concentration of 4 μg/ml. The concentration of the antisera is adjusted to give a line of precipitate approximately halfway between the enterotoxin and antisera wells.

32.53 Extraction for Microslides (Casman and Bennett method)[3]

Each time food extractions are to be made, it is recommended that a small amount of enterotoxin (0.25 to 0.50 μg) be added to 100 g of uncontaminated food of the same type that is being examined for enterotoxin, and determine whether this can be detected. This will indicate whether the test as it is being performed is sufficiently sensitive to detect the small amounts of enterotoxin that may be present in the food under examination. If it is known what enterotoxin may be present in the food, then this enterotoxin type can be used as the control. If no information is available as to what enterotoxin might be present, then enterotoxin A is the recommended control. Crude enterotoxins can be used as the controls.

32.54 Extraction for Microslides (Reiser and Conaway Method)[12]

See Sec. 32.53.

32.6 PROCEDURES

32.61 Microslide Analysis[3]

32.611 Preparation of materials

a. Agar for coating slides

Add 2 g of agar to 1 liter of boiling, distilled water and heat with stirring until a clear solution is obtained. Pour 30 ml into 6 oz prescription bottles or other suitable containers. Store at room temperature. The agar can be remelted until all is used.

b. Gel diffusion agar for slides

Add 12 g agar to 988 ml of boiling, sodium barbital saline buffer (0.9% NaCl, 0.8% sodium barbital, 1:10,000 merthiolate, adjusted to pH 7.4), continue boiling until the agar dissolves, filter quickly with suction through two layers of filter paper, and store in 15 to 25 ml quantities in 4 oz prescription bottles. Remelting more than twice may break down the purified agar.

c. Slides

Wash microscope slides, 3 × 1 inch, in detergent solution, rinse thoroughly in tap water followed by distilled water, airdry and store in dust free containers. It is essential that the slides be scrupulously clean, or else the agar will not adhere to the glass and will tear when the template is removed.

d. Taping the slides

Wrap a 9.5 to 10.5 cm length (length needed will depend on how much the operator stretches the tape in applying it) of black plastic electrician's insulating tape twice around each end of the slide, about 0.5 cm from each end with a 2 cm space between the two strips. Stretch the tape slightly and press down firmly while wrapping the tape around the slide to avoid air bubbles. The finishing edge should end where the starting edge begins.

e. Rewashing slides

Wash the slides after taping as described in 32.31a

f. Precoating

Wipe the area between the tapes with 95% ethanol using a small stick with a piece of absorbent cotton twisted on the end. Cover the surface between the tapes with two drops of 0.2% agar (approximately 100 C) with a 1 ml pipet and rotate the slide to cover the surface evenly. If the agar forms beads instead of spreading evenly, rewash the slide. Airdry the slides on a flat surface in a dust free atmosphere.

g. Petri dishes for slides

Place two strips of synthetic sponge (approximately 0.5 by 0.5 by 2.5 inches), or two 3 inch strips of absorbent cotton opposite each other around the periphery of 15 cm Petri dishes. Saturate the strips with distilled water, pouring off any excess. The dishes will hold up to four slides and can be used repeatedly as long as the sponge or cotton is kept saturated with water.

h. Template

Spread a thin uniform film of silicone grease or other suitable lubricant on the bottom side of the template. If too little lubricant is applied, the template cannot be removed easily from the agar layer, and if too much is applied, the template will not stay firmly in place and development of the precipitate lines will be distorted. If silicone spray is used, spray the silicone on a piece of cotton on the end of a stick and wipe the bottom surface of the template, keeping the coat as thin as possible. Before reusing the templates, wash them in a warm detergent solution with a piece of cheesecloth to remove the silicone, being careful not to nick or scratch the template. Rinse them thoroughly with tap and distilled water to remove any traces of detergent.

i. Preparation of microslide

Place 0.35 to 0.40 ml of 1.2% melted agar (80 to 90 C) on the precoated slide area between the tape and immediately lay the silicone coated template on the agar by placing one edge on the edge of the tape on one side and bringing it down onto the edge of the tape on the other side. After the agar solidifies, place the slides in the Petri dishes. The temperature of the agar should be such that the agar does not begin to harden before the template can be put in place. Care should be taken that agar is not forced into the bottom of the wells of the template.

j. Stain solution

If Thiazine Red-R stain is used, dissolve 100 mg Thiazine Red-R in 100 ml of 1% acetic acid. If Woolfast Pink RL is used, dissolve 1 g in 100 ml of the following solution: 5% trichloracetic acid, 1% acetic acid, and 25% ethanol in distilled water.

k. Preparation of record sheet

Draw the hole pattern of the template in a notebook or use a rubber stamp of the hole pattern. Indicate the materials that are placed in each well.

32.612 Addition of reagents and test materials to slides

Partially fill a capillary pipette by capillary action with the solution to be added. Remove excess liquid by touching the pipette to the edge of the sample tube. Slowly lower the pipette into the well to be filled until it touches the agar surface. (This leaves a small drop of liquid in the bottom of the well.) Refill the capillary pipette with more reagent, lower into the well, and fill the well to convexity. (Do not overfill because of the danger of mixing reagents from different wells.) The antiserum is placed in the center well of the template, the standard enterotoxin reagent in the upper left and lower right hand wells, and the unknown samples in the other two wells.

Careful application of the reagents is necessary to avoid the formation of air bubbles in the bottom of the wells. If such are formed, it is essential that they be removed, so that the reagents will make proper contact with the agar layer. If air bubbles are formed, remove them by inserting the end of a heat sealed capillary pipet to the bottom of the well. This operation and the filling of the wells is best done against a lighted background such as described under equipment (32.31). Place the slides in the Petri dishes as soon as possible to avoid undue evaporation of liquid from the wells. An alternate method is to add the reagents to the slides without removing them from the Petri dishes.

32.613 Slide incubation

Incubate the slides for 48 to 72 hours at room temperature or for 24 hours at 37 C.

32.614 Reading the slide

a. Preliminary examination

Remove the template by sliding it to one side; clean the slide with a momentary dip into water. Examine the slide by holding it at an oblique angle against a fluorescent desk lamp.

b. Enhancement of precipitate lines

If the precipitate lines are faint, or none are visible, immerse the slides in 1% cadmium acetate for 5 to 10 minutes, or immerse the rinsed slide directly into Thiazine Red-R solution for 5 to 10 minutes. Alternate procedure: Place slides in a Wheaton jar filled with distilled water and extract with stirring for 30 minutes. Place slides in Woolfast Pink RL stain for 20 minutes at room temperature. Rinse off excess dye and destain in distilled water for 30 minutes or until stain is adequately washed out.[10]

32.615 Interpretation of data

a. Typical results (Figure 3)

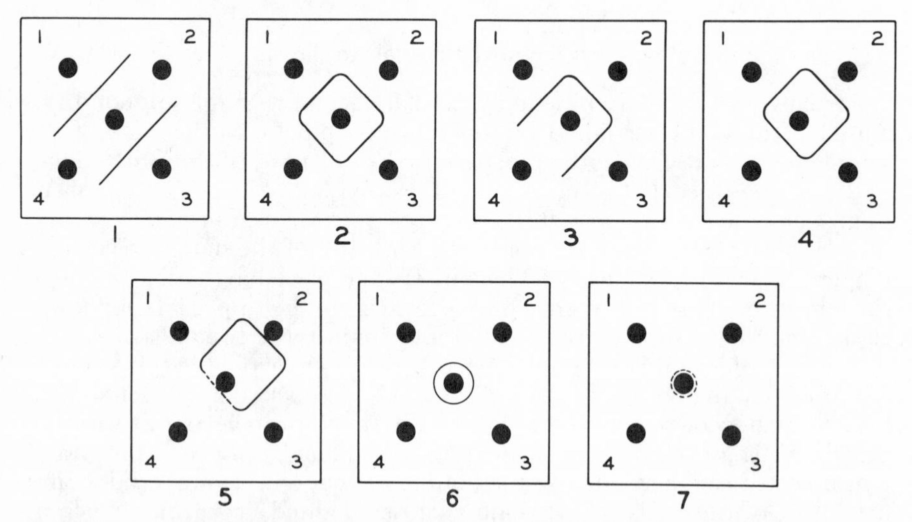

FIGURE 3. Typical Results Obtained with the Microslide test. ***See Interpretation of Data***

The control toxin always should give one precipitate line between the antigen and antibody wells as shown in diagram one with the toxin in wells 1 and 3, and water or unknowns containing no toxin in wells 2 and 4. The formation of a line by the unknown sample(s) that joins with the control line as illustrated in diagrams two (wells 2 and 4) and three (well 2) shows the unknown to contain toxin at the concentration in the controls. The unknown sample in well 4 of diagram three contains none of the type of toxin present in the controls. The unknown sample in well 2 of diagram four contains a lesser amount of enterotoxin than is present in the controls, and the unknown in well 4 contains a larger amount than is present in the controls. The unknown sample in well 2 of diagram five contains much less toxin than is in the controls, while the unknown in well 4 contains a much larger amount than is in the controls. Both the unknowns in diagram six contain more enterotoxin than is present in the controls, and both the unknowns in diagram seven contain much more. In the latter case the unknown samples should be diluted and the slide rerun.

b. Atypical results

Occasionally results are obtained which are difficult to interpret, especially for the less experienced operator. Some of these are illustrated in Figure 4. In diagram one the reference lines do not meet or cross the lines produced by the unknown in wells 2 and 4. It is impossible to interpret these results positively; hence, the slide should be redone. Frequently, extraneous lines are produced by the unknowns such as is shown in diagrams two and three. The lines from the un-

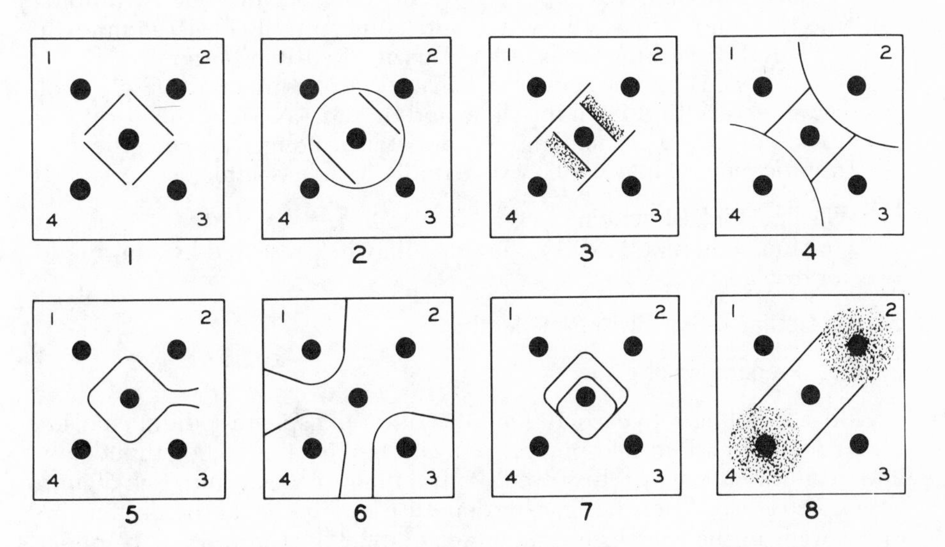

FIGURE 4. Atypical Results Obtained with the Microslide Test. ***See Interpretation of Data***

knowns (wells 2 and 4) that join with the control lines (wells 1 and 3) shows the unknowns to be positive. The unknowns in diagram three (wells 2 and 4) are negative because there are no joining lines. The extraneous lines are artifacts which are sometimes produced by unknown samples. The results illustrated in diagram four cannot be interpreted as positive because the sample lines extend beyond the lines produced by the controls, even though the latter lines do not cross the sample lines. This type of slide should be redone. The results shown in diagram five and six indicate channeling of the serum under the template due to improper contact of the template with the agar layer. The results shown in diagram five indicate that the unknowns in wells 2 and 4 are positive because the lines produced by them do join with the control lines. The results shown in diagram six cannot be interpreted as positive for the unknown in well 4, and the slide should be redone. The partially double ring shown in diagram seven is a result of movement of the template sometime during filling or development. The results indicate the presence of enterotoxin in the unknowns (wells 2 and 4), but to be certain, the slide should be redone. The haze around the unknowns in diagram eight (wells 2 and 4) is often encountered with food extracts. Such results are very difficult, if not impossible, to interpret. The haze is the result of the presence of contaminating substances (usually proteins) remaining in the concentrated extract. This may be remedied by additional $CHCl_3$ extractions, or possibly longer treatment with trypsin.

32.616 Preserving the slide[5,10]

a. Thiazine Red-R staining

Rinse away any reactant liquid remaining on the slide by dipping the slide momentarily in water and immersing it for 10 minutes in each of the following baths: 0.1% Thiazine Red-R in 1% acetic acid, 1% acetic acid, 1% acetic acid, and 1% acetic acid containing 1% glycerol. Drain excess fluid from the slide, and dry it in a 35 C incubator if it is to be stored as a permanent record. After prolonged storage, the lines of precipitation may not be visible until the slide is immersed in water.

b. Woolfast Pink RL staining

After staining (32.614b), the slides may be preserved by allowing to air dry.

32.62 Optimum Sensitivity Plate Method[13]

32.621 Preparation of plate

Add 1.2 g of agar to each 100 ml of sodium phosphate-sodium chloride buffer (0.02M sodium phosphate, 0.9% NaCl, pH 7.4; 1:10,000 thimersol), and heat until the agar is dissolved. Pour 3 ml of the agar into each 50 mm plastic Petri dish. After the agar hardens, use cork borers of the proper size to cut wells in the agar using the template (32.32b, Figure 2.) as a guide.

Remove the agar plugs with suction. If the plates are not to be used soon after they are poured, store them uncut in the plastic boxes used for incubation.

32.622 Addition of reagents and test materials

Place the enterotoxin controls in the two smaller wells at a concentration of 4 μg/ml. Place the antiserum in the center well, and the test samples in the four larger outer wells.

32.623 Incubation

Place the plates in the plastic storage boxes and incubate overnight at 37 C or at 25 C for 24 hours. Incubation can be continued for longer periods of time, e.g., over the weekend without appreciable loss of liquid from the wells. Longer incubation enhances the lines formed by low concentrations of enterotoxin.

32.624 Reading the plate

a. Preliminary examination

Examine the plates by use of a Cordis Laboratories viewer or a similarly lighted device.

b. Enhancement of precipitate reaction

After initial reading of the plates, rinse them with water and flood with 0.1 M H_3PO_4. Observe the plate to see if any lines are enhanced; this should take place in 1 to 3 minutes. If the plate is to be held for an additional period of time, e.g., for photographing, remove the H_3PO_4 and rinse the plate with water to minimize the development of a haze due to extraneous proteins from both the antisera and the antigen solutions. Lines and broader precipitate areas that may not be visible otherwise will become visible by the enhancement procedure.

32.625 Interpretation of data

a. Typical results

The different types of results are illustrated in Figures 5 and 6. Unknown U1 (Figure 5) contains approximately 0.5 μg enterotoxin A (indicated by the hook at the 1X concentration and confirmed by concentration to 5X) and 6 μg B per ml; unknown U2 contains no enterotoxin A, B, or C; unknown U3 contains approximately 2 μg A per ml; unknown U4 contains approximately 16 μg C per ml.

Frequently positive reactions can be confirmed by enhancement of the precipitate lines by treatment with 0.1 M H_3PO_4 as shown in Figure 6. The hook given by U1 (No. 1) is confirmed as a positive by treatment with H_3PO_4. This is particularly helpful when the sample is a concentrated one as is the case here. Unknown U2 (No. 2) has cut off the standard line indicating that this sample contains large amounts of

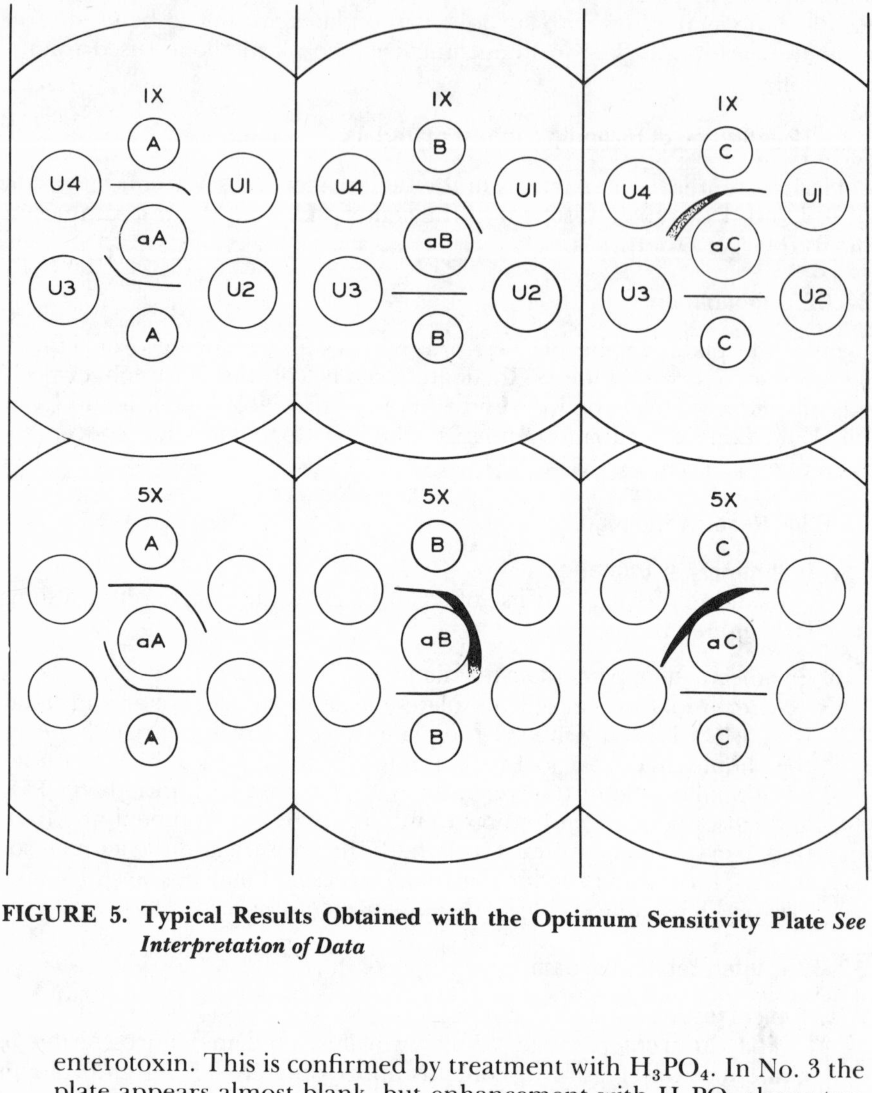

FIGURE 5. Typical Results Obtained with the Optimum Sensitivity Plate ***See Interpretation of Data***

enterotoxin. This is confirmed by treatment with H_3PO_4. In No. 3 the plate appears almost blank, but enhancement with H_3PO_4 shows two very short curved standard lines and a diffused haze, which may or may not be visible around the serum well. These results indicate that all four unknowns contain large amounts of enterotoxin B. This can be confirmed by rerunning the unknowns at dilutions of 1:10 and 1:40.

b. Nonspecific reactions

Sometimes precipitate lines are observed that cross the control line as illustrated in Figure 7 (No. 1). Unknown 4 is negative for enterotoxin A. Occasionally results such as those illustrated in No. 2 are

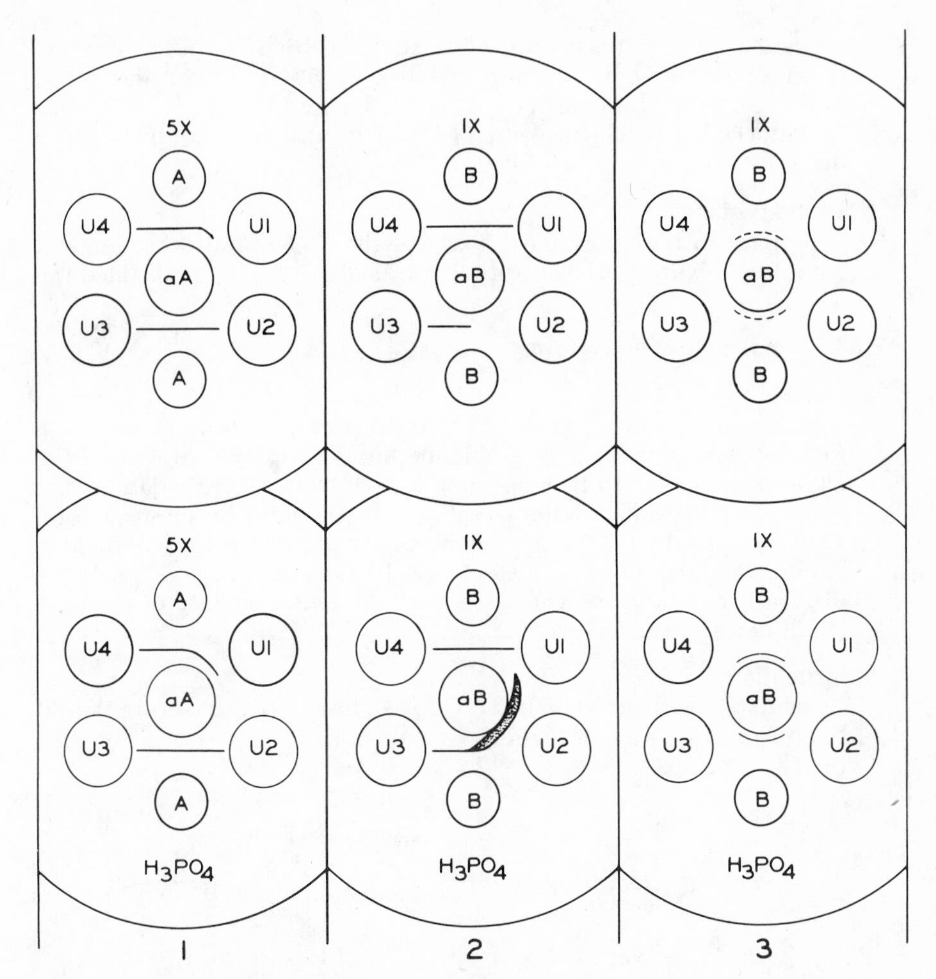

FIGURE 6. Enhancement of Precipitate Lines in Optimum Sensitivity Plate with H_3PO_4 *See Interpretation of Data*

observed. All the unknowns are negative for enterotoxin A because the control lines extend beyond the lines given by the unknowns. In No. 3, the general haze observed around the serum well indicates negative results, because it has had no effect on the control lines.

32.63 Enterotoxin Production of Staphylococcus Isolates by Semisolid Agar Method[2]

32.631 Preparation of materials

a. Agar medium

Add 0.7% agar to BHI broth at pH 5.3 (0.7g/100 ml). Dissolve agar by minimal boiling. Distribute the medium in 25 ml quantities in test tubes (25 × 200 mm), and autoclave at 121 C for 10 minutes. The medium is poured aseptically into Petri dishes (15 × 100 mm), and stored in the test tubes until needed. (Alternate procedure: dissolve the agar by autoclaving batchwise, and pour 25 ml quantities into the Petri dishes.)

b. Turbidity standard

Prepare turbidity standard No. 1 of the McFarland nephelometer scale[11] by mixing 1% $BaCl_2$ with 99 parts of 1% H_2SO_4 in distilled water.

32.632 Production of enterotoxin

a. Inoculum

Pick representative colonies (5 to 10 for each culture), transfer each to a nutrient agar (or comparable medium) slant, and grow for 18 to 24 hours at 35 C. Add a loopful of growth from the agar slant to 3 to 5 ml of sterile distilled water or saline. The turbidity of the suspension should be approximately equivalent (by visual examination) to the turbidity standard (approximately 3×10^8 organisms/ml). Spread 4 drops of the aqueous suspension over the entire surface of the agar medium with a sterile spreader.

b. Incubation

Incubate the plates at 35 to 37 C for 48 hours (pH of culture should be approximately 8.0 or higher).

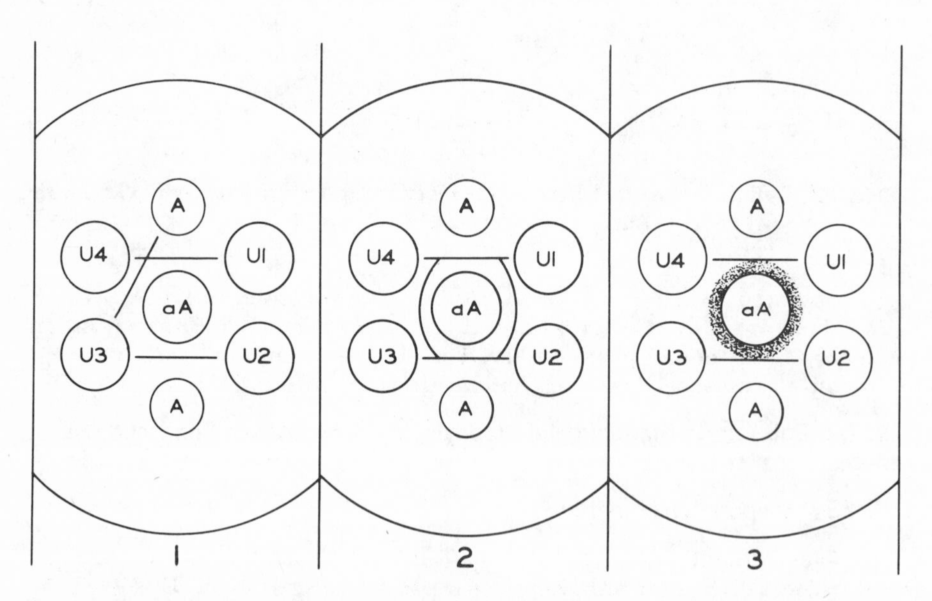

FIGURE 7. Atypical Results Obtained with the Optimum Sensitivity Plate ***See Interpretation of Data***

c. Enterotoxin recovery

Transfer the contents of the Petri dish to a 50 ml centrifuge tube with the aid of an applicator stick or equivalent. Centrifuge for 10 minutes at 32,800 × g.

32.633 Testing for enterotoxin

Test the supernatant fluid for enterotoxin by the microslide or OSP plate methods.

32.634 Interpretation of data

Negative results with the concentrated supernatants show the cultures to be nonproducers of those enterotoxins for which tests were run.

32.635 Limitations and precautions

The volume of extract from this method is greater than from the cellophane-over-agar method; hence, it is not as sensitive a technique without concentration of the supernatant fluid. Higher concentrations of some of the enterotoxins are produced with some media, i.e., more enterotoxin B is produced in the 3 + 3 medium than in BHI. The best all around medium appears to be the 3 + 3 medium; however, the BHI medium is easier to obtain than are the ingredients for the 3 + 3 medium.

32.64 Enterotoxin Production of Staphylococcus Isolates by Cellophane-Over-Agar Method

32.641 Preparation of materials

a. Agar medium

Add 1.5% agar to the selected medium (1.5 g/100 ml) and autoclave. Pour 25 ml quantities into 100 mm. Petri dishes.

b. Cellophane

Cut disks from the 3¼ inch dialysis tubing using 9 cm filter paper as a template. Place the disks alternately with filter paper disks in a glass Petri dish (moistened with distilled water to eliminate wrinkling). Autoclave for 20 minutes at 121 C. Transfer the sterile membrane disks aseptically to the Petri dishes containing the agar medium.

32.642 Production of enterotoxin

a. Inoculum

Spread 0.1 ml of inoculum (32.632a) over the surface of the membrane with an applicator.

b. Incubation

Incubate the plates at 37 C for 24 hours.

c. Enterotoxin recovery

Wash the culture from the membrane with 2.5 ml of 0.01 M Na_2HPO_4. Centrifuge at 39,100 × g for 10 minutes, or at a speed and time to clarify the solution.

32.643 Testing for enterotoxin

Test the supernatant fluid for enterotoxin by the OSP plate or microslide methods.

32.644 Interpretation of data

See 32.634.

32.645 Limitations and precautions

It is advisable to check the procedure with a known low enterotoxin producer occasionally to determine whether adequate growth and enterotoxin production are obtained. If not, the problem may be the membrane, in which case new membrane should be obtained and checked.

32.65 Enterotoxin Production of Staphylococcus Isolates by the Sac Culture Method of Donnelly[7]

32.651 Preparation of sacs

Wash a 40 to 50 cm piece of dialyzing tubing (or whatever length is required to hold 100 ml of medium) in distilled water, knot one end, and inflate to make a sac. Insert the knotted end into a 250 ml Erlenmeyer flask so that it rests on the bottom of the flask. Place 100 ml of medium into the sac and knot the open end. Tie the two knotted ends together with a rubber-band and position the sac in the flask in a U shape with the two knotted ends located in the neck of the flask. Autoclave the flask and contents at 121 C for 15 minutes. Pour off any excess liquid in the flask.

32.652 Enterotoxin production

a. Inoculum

Inoculate 18 ml of sterile phosphate saline buffer (0.02M sodium phosphate, pH 7.4, in 0.9% NaCl) with a suitable inoculum (see 32.632a) and transfer the mixture to the Erlenmeyer flasks containing the sacs.

b. Incubation

Incubate the flasks on a shaker (200 rpm) at 37 C for 24 hours.

c. Enterotoxin recovery

Remove the culture fluid from the flasks and centrifuge (see 32.642c).

32.653 Testing for enterotoxin

Test the supernatant fluid for enterotoxin by the OSP plate or microslide methods.

32.654 Interpretation of data

See 32.634.

32.655 Precautions

Shaking the flasks at too rapid a rate may result in breakage of the sacs.

32.66 Extraction of Enterotoxin from Food for Microslides by the Casman and Bennett Method[15]

32.661 Preparation of materials

a. Polyethylene glycol solution

Prepare a 30% polyethylene glycol 20,000 (PEG) solution by adding 3 g of PEG for each 7 ml of distilled water.

b. Dialysis tubing

Cut a piece of dialysis tubing (1⅛ inch flat width) of sufficient length to accommodate the volume of food extract to be concentrated. Soak the tubing in two changes of distilled water to remove the coating of glycerol. Tie one end of the tubing with two knots close together. Fill the tube with distilled water, and test for leaks by squeezing the filled sac while the untied end is held tightly closed with the fingers. Empty the sac and place in distilled water until it is used.

c. Separatory funnel set-up

Attach a piece of latex tubing (approximately 24 inches) to the lower end of a 2 liter separatory funnel. Attach a piece of glass tubing in a No. 3 rubber stopper to the other end of the latex tubing. Place the separatory funnel in a ring stand above the chromatographic column.

d. Carboxymethyl cellulose column

Suspend 1 g of carboxymethyl cellulose (CMC) in 100 ml of 0.005 M sodium phosphate buffer, pH 5.7, in a 250 ml beaker. Adjust the pH of the CMC suspension to 5.7 with 0.005 M H_3PO_4. Stir the suspension intermittently for 15 minutes, recheck the pH, and readjust to 5.7 if necessary. Pour the suspension into the 1.9 cm chromatographic tube, and allow the CMC particles to settle. Withdraw the liquid from the column through the stopcock to within about 1 inch of the surface of the settled CMC. Place a loosely packed plug of glass wool on top of the CMC. Pass 0.005 M sodium phosphate buffer, pH 5.7, through the column until the washing is clear (150 to 200 ml). Check the pH of the last wash. If the pH is not 5.7, continue the washing until the pH of the wash is 5.7. Leave sufficient buffer in the column to cover the CMC and the glass wool to prevent the column from drying out.

32.662 Extraction of enterotoxin from food

a. Grind a 100 g sample of food in a Waring Blender at a high speed for three minutes with 500 ml of 0.2M NaCl.
b. Adjust the pH to 7.5 with 1N NaOH or HCl if the food material is highly buffered and 0.1N NaOH or HCl if the food material is weakly buffered (e.g., custards). Let the slurry stand for 10 to 15 minutes, recheck the pH, and readjust if necessary.

c. Transfer the slurry to two 285 ml stainless steel centrifuge bottles. Centrifuge at 27,300 × g for 20 minutes at 5 C. Lower speeds with longer centrifuging time can be used, but clearing of some food materials is not as effective. Separation of fatty materials is ineffective unless centrifugation is done at refrigeration temperature.
d. Decant the supernatant fluid into an 800 ml beaker through miracloth or other suitable filtering material placed in a funnel.
e. Reextract the residue with 125 ml of 0.2M NaCl by blending for three minutes. Adjust the pH to 7.5 if necessary.
f. Centrifuge at 27,300 × g for 20 minutes at 5 C. Filter the supernatant through miracloth, and pool the filtrate with the original extract.
g. Place the pooled extracts into a dialysis sac. Immerse the sac into 30% PEG at 5 C until the volume is reduced to 15 to 20 ml or less (normally an overnight procedure).
h. Remove the sac from the PEG and wash the outside thoroughly with cold tap water to remove any PEG adhering to the sac. Soak in distilled water for 1 to 2 minutes and 0.2M NaCl for a few minutes.
i. Pour the contents into a small beaker. Rinse the inside of the sac with 2 to 3 ml of 0.2 ml NaCl by running the fingers up and down the outside of the sac to remove material adhering to the sides of the tubing. Repeat the rinsing until the rinse is clear. Keep the volume as small as possible.
j. Adjust the pH of the extract to 7.5. Centrifuge at 32,800 × g for 10 minutes. Decant the supernatant fluid into a graduated cylinder to determine the volume.
k. Add the extract with ¼ to ½ volume of $CHCl_3$ to a separatory funnel. Shake vigorously 10X through an arc of 90 degrees.
l. Centrifuge the $CHCl_3$ extract mixture at 32,800 × g for 10 minutes at 5 C. Return the fluid layers to the separatory funnel. Draw off the $CHCl_3$ layer from the bottom of the separatory funnel, and discard.
m. Measure the volume of the water layer and dilute with 40 volumes of 0.005M sodium phosphate buffer, pH 5.7. Adjust the pH to 5.7 with 0.005 M H_3PO_4 or 0.005M Na_2HPO_4. Place the diluted solution in the 2 liter separatory funnel.
n. Place the stopper (attached to the bottom of the separatory funnel) loosely into the top of the chromatographic tube, and slowly fill the tube nearly to the top with the liquid from the separatory funnel. Tighten the stopper in the top of the tube, and open the stopcock of the separatory funnel. Allow the fluid to percolate through the CMC column at 5 C at 1 to 2 ml per minute by adjusting the flow rate with the stopcock at the bottom of the column. The flow rate is adjusted so that the percolation can be completed overnight. If all the liquid has not passed through the column during the night, stop the flow when the liquid level reaches the glass wool layer. If all the liquid has passed through overnight, rehydrate the column with 25 ml of distilled water.
o. After percolation is complete, wash the CMC column with 100 ml of the 0.005 M sodium phosphate buffer (1 to 2 ml/minute) stopping the flow when the liquid level reaches the glass wool layer. Discard the wash.

p. Elute the enterotoxin from the CMC column with 150 ml of 0.2M sodium phosphate buffer, pH 7.4 (0.05M phosphate—0.5M NaCl buffer, pH 6.5, if the food contains large amounts of protein), at a flow rate of 1 to 2 ml per minute at room temperature. Force the last of the liquid from the CMC by applying air pressure to the top of the chromatographic tube.
q. Place the eluate in a dialysis sac. Place the sac in 30% PEG at 5 C and concentrate almost to dryness.
r. Remove the sac from the PEG and wash as in 32.662h. Soak the sac in the 0.2M phosphate buffer, pH 7.4. Remove the concentrated material from the sac by rinsing 5X with 2 to 3 ml of 0.01 M sodium phosphate buffer, pH 7.4 to 7.5.
s. Extract the concentrated solution with $CHCl_3$ as described in 32.662k. Repeat the $CHCl_3$ extractions until the precipitate is so lacy that it falls apart in the $CHCl_3$ layer in the miracloth.
t. Place the extract in a short dialysis sac (approximately 6 inches). Place the sac in 30% PEG, and allow it to remain until all the liquid has been removed from inside the sac (usually overnight).
u. Remove the sac from the PEG and wash the outside with tap water. Place the sac in distilled water for 1 to 2 minutes.
v. Remove the contents of the sac by rinsing the inside of the sac with 1 ml portions of distilled water. Keep the volume below 5 ml.
w. Place the rinsings in a test tube (18 × 100 mm) or other suitable container and freeze dry.
y. Dissolve the freeze dried sample in as small an amount of 0.2M NaCl as possible (0.1 to 0.15 ml).
z. Check the sample for enterotoxin by the microslide method.

32.663 Precautions and limitations

It is essential that the food to be tested be ground to a very fine consistency to effect adequate extractions of the toxin. This is particularly important for foods, such as cheese, where the enterotoxin may be present throughout the food and not just on the surface. It is advantageous to remove as much of the nonsoluble food materials as possible to eliminate interference of these substances later in the extraction procedures, such as in the CMC adsorption process. The number of chloroform extractions needed to remove extraneous materials in the food extracts is dependent upon whether the extract can be concentrated to 0.2 ml and be of a consistency that can be applied readily to the microslides.

The method cannot be used to recover the enterotoxin quantitatively, particularly at the levels usually found in foods involved in food poisoning outbreaks. At the level of 1 μg or less per 100g of food, the recovery is very low, probably 10 to 20%. One need not be too concerned about the amount recovered as long as a sufficient quantity is recovered to be detectable. There is a variation from food to food, but one should be able to detect the presence of 0.125 to 0.250 μg enterotoxin in 100g food.

32.67 Extraction of Enterotoxin from Food for Microslides by the Reiser and Conaway Method[12]

32.671 Preparation of materials

a. Polyethylene glycol solution

See 32.661a.

b. Dialysis tubing

See 32.661b.

c. CG-50 Ion exchange resin

Suspend 100 g resin in 1.5 liters of distilled water, adjust to pH 12 with 5N NaOH, and stir for one hour at room temperature. Allow the resin to settle, and decant the supernatant fluid. Resuspend the resin in distilled water, allow the resin to settle, and decant. Repeat this washing until the pH is that of the distilled water. Resuspend the resin in distilled water, adjust the pH to 2.0 with 6N HCl, and stir for one hour. Allow the resin to settle, and decant the supernatant fluid. Resuspend the resin in distilled water, allow the resin to settle, and decant. Repeat this washing procedure until the pH is that of the distilled water. Suspend the resin in 0.005M sodium phosphate buffer at pH 5.6. If the pH is below 5.4, add 5N NaOH with stirring until the pH remains unchanged.

32.672 Extraction of enterotoxin from food

a. Place a 100 g sample of food in a 285 ml stainless steel centrifuge bottle with 140 ml distilled water, and grind to even consistency (1 to 2 minutes) with a Sorvall Omnimixer.
b. Blend 10 ml 1.0 M HCl into the mixture (pH 4.5 to 5.0). If the sample is a meat product, add only 5 ml of HCl as the pH should not drop below 4.5.
c. Centrifuge the sample at 27,300 $\times$ g at 5 C for 20 minutes. Pour the supernatant fluid (some of which may be quite viscous) into a beaker.
d. Adjust the pH to 7.5 with 5 N NaOH. Extract the sample with $CHCl_3$ (1 ml/10 ml sample) by mixing with a magnetic stirrer for 3 to 5 minutes.
e. Transfer the sample to the stainless steel centrifuge bottle, and centrifuge at 27,300 $\times$ g at 5 C for 20 minutes. Filter slowly through miracloth placed in a glass funnel to trap the $CHCl_3$.
f. Adjust the pH to 4.5 with 6 N HCl. Centrifuge the sample as above to remove any precipitate. Remove the supernatant fluid.
g. Adjust the pH to 7.5 with 5N NaOH and recentrifuge if any precipitate forms.
h. Add about 20 ml settled Amberlite CG-50 ion exchange resin, pH 5.4 to 5.9 (32.671c) per 100 g of original sample. Measurement of the resin can be done by decanting the buffer, and transferring the settled resin to a 20 ml beaker with a spatula. After enough resin has been removed from the stock supply, the decanted buffer is added back and the resin stored at 5 C.
i. Stir the mixture with a magnetic stirrer at 5 C for one hour. Remove the resin from the liquid by filtration through Miracloth in a Buchner funnel (Coors No. 1) with suction.

j. Wash the resin in the funnel with 200 ml of a 1 to 10 dilution of 0.15 M sodium phosphate buffered 0.9% NaCl, pH 5.9. Discard the wash.
k. Resuspend the resin in 30 ml of 0.15M Na_2HPO_4 in 0.9% NaCl. Adjust the pH to 6.8 with 5N NaOH.
l. Stir the mixture with a magnetic stirrer at 5 C for 45 minutes to elute the enterotoxin from the resin. Filter with suction through Miracloth in a Buchner funnel. Wash the resin on the miracloth with a small amount of the phosphate NaCl buffer. Discard the resin.
m. Stir into the eluate 1 g of purified agar and continue stirring at 5 C for one hour. Filter the mixture through a glass fiber filter (Type E, Gelman Instrument Co., P.O. Box 1448, Ann Arbor, Mich.) in a Buchner funnel. Discard the agar.
n. Place the eluate into a dialysis tube (3 cm flat width). Place the dialysis tube into PEG overnight.
o. Remove the sacs from the PEG and wash the outside of the sacs with warm tap water. Place the sacs in warm tap water for 10 to 20 minutes to loosen the toxin adhering to the walls of the sac.
p. Empty the contents of the sacs into 15 ml centrifuge tubes (Corex high speed, Corning No. 8441). Rinse the sacs three times with distilled water, keeping the final volume to 5 ml or less.
q. Add 0.5 ml of $CHCl_3$ to the centrifuge tube and mix contents vigorously (a Vortex mixer serves this purpose very well).
r. Centrifuge the sample at 34,800 × g at 5 C for 10 minutes. Decant the water layer to a test tube (18 × 100 mm), and freeze dry.
s. Dissolve the dried sample in 0.2 ml of a 1% trypsin suspension in distilled water. Digest for at least 30 minutes at room temperature.
t. Test the sample for enterotoxin by the microslide method.

32.673 Precautions and limitations

One of the critical points is the ionic strength of the enterotoxin solution to which the CG-50 resin is added (32.672h). Instructions should be followed closely, because if the ionic strength is too high, the toxin will not be absorbed completely, or possibly not at all. See 32.663.

32.7 REFERENCES

1. Bergdoll, M. S. 1972. The Enterotoxins. In J. O. Cohen (ed.), The Staphylococci, pp. 301–331. John Wiley and Sons, Inc. New York, N.Y.
2. Casman, E. P. and R. W. Bennett. 1963. Culture medium for the production of staphylococcal enterotoxin A. J. Bacteriol. **86**:18–23
3. Casman, E. P., R. W. Bennett, A. E. Dorsey, and J. E. Stone. 1969. The microslide gel double diffusion test for the detection and assay of staphylococcal enterotoxins. Health Lab Sci. **6**:185–198
4. Collins, W. S., A. D. Johnson, J. F. Metzger, and R. W. Bennett. 1973. Rapid solid-phase radioimmunoassay for staphylococcal enterotoxin A. Appl. Microbiol. **25**:774–777
5. Crowle, A. J. 1958. A simplified micro double-diffusion agar precipitin technique. J. Lab. Clin. Med. **52**:784–787
6. Dickie, N., Y. Yano, C. Park, H. Rohern, and S. Stavric. 1973. Solid-phase Radio Immunoassay of Staphylococcal Enterotoxins in Food. Proceedings of Staphylococci in Foods Conference. Pennsylvania State University, University Park, Pa.
7. Donnelly, C. B., J. E. Leslie, L. A. Black, and K. H. Lewis. 1967. Serological identification of enterotoxigenic staphylococci from cheese. Appl. Microbiol. **15**:1382–1387

8. JARVIS, A. W. and R. C. LAWRENCE. 1970. Production of high titers of enterotoxin for the routine testing of staphylococci. Appl. Microbiol. **19**:698–699
9. JOHNSON, H. M., J. A. BUKOVIC, and P. E. KAUFMANN. 1973. Staphylococcal enterotoxins A and B: Solid-phase radioimmunoassay in food. Appl. Microbiol. **26**:309–313
10. LUMPKINS, E. D., SR. 1972. Methods for staining and preserving agar gel diffusion plates. Appl. Microbiol. **24**:499–500
11. MCFARLAND, J. 1907. The nephelometer: an instrument for estimating the number of bacteria in suspensions used for calculating the opsonic index and for vaccines. J. Am. Med. Assoc. **49**:1176–1178
12. REISER, R., D. CONAWAY, and M. S. BERGDOLL. 1974. Detection of staphylococcal enterotoxin in foods. Appl. Microbiol. **27**:83–85
13. ROBBINS, R., S. GOULD, and M. S. BERGDOLL. 1974. Detecting the enterotoxigenicity of *Staphyloccocus aureus* strains. Appl. Microbiol. **28**:946–950
14. SILVERMAN, S. J., A. R. KNOTT, and M. HOWARD. 1968. Rapid, sensitive assay for staphylococcal enterotoxin and a comparison of serological methods. Appl. Microbiol. **16**:1019–1023
15. ZEHREN, V. L. and V. F. ZEHREN. 1968. Examination of large quantities of cheese for staphylococcal enterotoxin A. J. Dairy Sci. **51**:635–644

CHAPTER 33

BACILLUS CEREUS

J. M. Goepfert

33.1 INTRODUCTION

Bacillus cereus is widely distributed in nature and can be isolated readily from a wide variety of foods in which it may be present normally. Its presence is insignificant, however, unless it is able to grow. Consumption of food containing millions of viable *B. cereus* cells per gram has resulted in outbreaks of food poisoning,[1] and in the establishment of standards for one type of food by a European country. Foods incriminated in past outbreaks of *B. cereus* poisoning include vanilla pudding, cooked meat and vegetable dishes, boiled and fried rice.

33.11 Taxonomic Position of *Bacillus cereus*

Within the genus *Bacillus*, a great diversity of strains and species exists. The works of Smith, Gordon and Clark[7, 8] and Gordon[2, 3] have brought a measure of order to the classification of bacilli and greatly facilitated the identification of the more common species. Their suggestion, which has general acceptance, is that the bacilli be classified into groups based on the shape of the spore, and whether or not the sporangium (vegetative cell) is swollen by the spore. Thus Group 1 (which includes *Bacillus cereus*) is defined as those bacilli forming ellipsoidal or cylindrical spores which do *not* appreciably bulge the sporangium. Most of the commonly encountered bacilli are classified as Group 1 organisms. Within Group 1, a dichotomy can be made on the basis of the width of the vegetative cells. Organisms whose cell width equals or exceeds 0.9 μm are called "large-celled species" and include *B. cereus, B. mycoides, B. thuringiensis, B. anthracis,* and *B. megaterium.* Organisms less than 0.9 μm in width are "small-celled species" and include *B. subtilis, B. licheniformis, B. coagulans, B. firmus,* and *B. pumilus.*

It is the "large-celled species" that this chapter is concerned with, and a brief word on the taxonomy of these organisms is warranted. There has been considerable argument (which will not be perpetuated here) about whether *B. anthracis, B. thuringiensis* and *B. mycoides* should be accorded species status, or considered as varieties of the "parent" species, *B. cereus*. Each of these organisms, for all practical purposes, differs from *B. cereus* only by a single characteristic. That is to say that a) loss of pathogenicity for ani-

IS/A Committee Liaison: E. M. Foster

mals by *B. anthracis,* b) lack of production of a crystalline parasporal inclusion by *B. thuringiensis,* and c) absence of rhizoidal growth by *B. mycoides* would render each organism indistinguishable from *B. cereus*. For the purpose of this chapter, each will be considered as a species, and since the examination of foods for *B. anthracis* is not done in this country often, a discussion of methods employed to demonstrate its pathogenicity will be omitted.

33.12 Characteristics of the "Large-Celled" Group 1 Bacilli

The most salient features of the Group 1 bacilli are summarized in Table 1. Some caution is necessary in applying and interpreting the tests for identification purposes, since even within the species, variability and strain heterogeneity are relatively common. It should be noted also that some tests are more valuable or more easily performed than others, and for that reason are advocated below in section 33.82. The term "egg yolk reaction" is used to describe the turbidity developed in egg yolk or in agar containing egg yolk. The responsible agent is an extracellular substance(s) referred to in some cases as egg yolk turbidity factor (EYTF), lecithinase, or phospholipase.

Although it has been established that *B. cereus* does produce a phospholipase C, there is some evidence that the turbidity developed in egg yolk may be due to a more complex series of events than the action of a single enzyme.[5] Quantitation of *B. cereus* in a given sample is performed by a simple surface plating technique. In the United States and many parts of Europe use is made of a) the ability of *B. cereus* organisms to produce turbidity surrounding colonies growing on agar containing egg yolk, and b) resistance of *B. cereus* to the antibiotic Polymyxin B to create a selective and differential plating medium. Two such media have been described and, in one, mannitol has been incorporated to enhance differentiation.[4, 6] In Great Britain it is not uncommon for blood agar to be used as the primary plating medium, particularly for the examination of stool specimens. Colonies of appropriate morphology which show the characteristic alteration of the surrounding medium are considered as "presumptive *B. cereus*." Confirmatory tests, when desired, include microscopic examination of sporulating cultures, gelatin hydrolysis, nitrate reduction, the Voges-Proskauer reaction, and anaerobic utilization of glucose.

33.13 Treatment of Sample

a. Collection

The object is to obtain a representative sample of the material to be examined (Chapter 1).

b. Holding

There are no data available to suggest that refrigeration of samples will cause a reduction in the number of viable *B. cereus* (Chapter 1).

Table 1: Some Characteristics of Large Celled Group 1 Bacilli

Feature	*B. megaterium*	*B. cereus*	*B. thuringiensis*	*B. mycoides*	*B. anthracis*
Gram reaction	+	+ [a]	+	+	+
Catalase	+	+	+	+	+
Motility	±	± [b]	±	− [d]	−
Reduction of nitrate	∓ [c]	+	+	+	+
Hydrolysis of starch	+	±	+	+	+
Hydrolysis of gelatin	+	+	+	+	+
Egg yolk reaction	−	+	+	+	+
Anaerobic utilization of glucose	−	+	+	+	+
V-P reaction	−	+	+	+	+
Acid produced from mannitol	+	−	−	−	−
Hemolysis (rabbit RBC)	−	+	+	+	±

[a] + 90–100% of strains are positive
[b] ± 50–90% of strains are positive
[c] ∓ 10–50% of strains are positive
[d] − < 10% of strains are positive

c. Homogenization and dilution

The surface plating procedure is employed. Refer to Chapter 1 for details on dilution of fluid materials, homogenization, and dilution of solid or semisolid samples. Also refer to Chapter 4 for making surface plates.

d. Heat shocking

It appears that the spores of many, if not most, strains of *B. cereus* germinate readily on the plating media employed for enumeration. In most cases there does not seem to be a need for heat shocking treatment to enhance germination. Sometimes the investigator may desire a spore count only, or for other reasons may wish to employ a heat shock procedure. In such cases, the time temperature treatment of 70 C for 15 minutes is recommended.

33.14 Special Equipment and Supplies

a. Microscope

It is sometimes necessary or desirable to ascertain to which group a particular *Bacillus* sp. isolate belongs. This is done by examining a smear made from a sporulating culture to determine whether the spore has distended the sporangium. This can be determined most easily by darkfield examination of a smear under oil immersion at 600 to 1200×. Alternatively, a routine laboratory brightfield microscope equipped with a 90 to 100 × oil immersion objective can be used to examine methylene blue stained preparations.

33.15 Special Reagents and Media (Chapter 2)

a. KG agar[4]
b. MYP agar[6]
c. Egg yolk emulsion
d. Colbecks egg yolk broth
e. Concentrated egg yolk emulsion

33.16 Additional Media and Reagents (Chapter 2)

a. Nutrient gelatin
b. Nitrate broth
c. Modified VP broth
d. OF medium
e. α Naphthol
f. Sulfanilic acid
g. 40% Potassium hydroxide
h. Creatine
i. Sterile glucose solution (20%)
j. Vaspar

33.17 Standards

None

33.18 Controls

It is advisable to have a typical strain of *B. cereus* available for several purposes. First, this is useful to acquaint those unfamiliar with the reactions of *B. cereus* on, or in, the various plating and biochemical test media. Second, the strain can serve as a test organism to measure the performance of either laboratory prepared or commercially available media. Typical *B. cereus* strains are available from various sources including the American Type Culture Collection, Rockville, Md.

33.19 Precautions and Limitations

a. General

There are several limitations common to the plating media described above that are mentioned prior to discussing the specific advantages and disadvantages of each individual medium. *B. cereus* colonies on both KG and MYP agar may show various types of colonial morphology. Most commonly seen are the round, flat, dry, ground glass appearing colonies that may be translucent to creamy white in color. Less commonly, but by no means rarely, colonies may be rather amorphous in shape with highly irregular edges. In these cases the central portion of the colony is usually white, while the perimeter is translucent.

Most commonly, *B. cereus* produces a very strong reaction in egg yolk agar that is characterized by a wide zone of turbidity surrounding the individual colonies after 20 to 24 hours incubation. Quite frequently zones from individual colonies will coalesce if many *B. cereus* are present, and estimation of the true number of "zone forming colonies" is difficult. This situation often develops when more than 30 *B. cereus* colonies appear on a plate. For this reason, the countable zone forming colony range on these media is reduced to 3 to 30 rather than the 30 to 300 normally employed in most quantitative plating analyses. It should be noted also that these media are not one-hundred percent selective and that other organisms (mostly *Bacillus* spp) are often encountered. Frequently, these colonies will be moist or almost mucoid in appearance which, when viewed from the underside of the plate, will appear to have a zone of precipitate immediately beneath (though not extending beyond) the border of the colony. These are not *B. cereus*, and the investigator should be concerned only with colonies having the typical morphology with (and infrequently without) a zone of turbidity surrounding them. Neither of these media is especially proficient in enumerating *B. cereus* in feces, particularly if these are fewer than 100,000 cells per gram of feces. It has been suggested (Dr. R. Gilbert, personal communication) that blood agar is a more appropriate plating medium for the isolation of *B. cereus* from fecal samples, taking advantage of the very frequent characteristic of hemolysin production by these organisms. Although human blood has been employed successfully (Dr. R. Gilbert, personal communication) the experience of this author with both rabbit and sheep blood indicates that these are suitable alternatives.

b. Specific

1. KG agar (Chapter 2)

This medium was formulated in a manner to promote sporulation and free spore formation within the incubation period of the plate, i.e., 20 to 24 hours. This feature allows a) the direct confirmation of the zone forming organism as a Group 1 *Bacillus* by means of microscopic examination, and b) the immediate differentiation of *B. cereus* from *B. thuringiensis* by visualization of the parasporal inclusion body in sporulated cells of the latter organism. It is also helpful in determining whether a nonzone forming colony of typical morphology is a Group 1 *Bacillus*, and should be taken through the confirmatory biochemical tests. An additional advantage to this medium is that certain Group 2 bacilli, which produce lecithinase and may appear as presumptive positive colonies on MYP agar, are unable to form the lecithinase under the rather nutritionally poor conditions imposed by KG agar.

2. MYP agar (Chapter 2)

This medium has equal sensitivity and selectivity to KG agar. It has, in addition, a differential system, i.e., mannitol fermentation that may aid in differentiating nonzone forming *B. cereus* (mannitol negative) from other Group 1 bacilli. In practice however, it is seen often that the acid produced by non *B. cereus* colonies diffuses throughout the agar making it, at best, very difficult to distinguish mannitol fermenting from nonfermenting organisms.

Certain members of Group 2 bacilli, e.g., *B. polymyxa*, produce lecithinase and do form a zone of precipitate on the MYP medium. Although these organisms are mannitol fermenters, this may not be immediately apparent due to the problems of acid diffusion from other colonies as described above. The problem is further complicated by the fact that *B. cereus* sporulates poorly, if at all, on this medium, thereby necessitating a subsequent transfer to a medium promoting sporulation so that identification of the group can be made.

33.2 PROCEDURE

33.21 Sample Collection, Preparation and Dilution (Chapter 1)

Plate counts should be made on unheated as well as heat shocked preparations. One-tenth ml aliquots of suitable dilutions should be placed on the surface of predried plates of KG or MYP agar. The inoculum should be spread evenly over the entire surface of the plate with a bent glass rod or "hockey stick." A separate sterile hockey stick should be employed for each series of plates representing a single sample. This method is advocated, rather than dipping the rod in alcohol and flaming, since the spores of the bacilli will survive both the alcohol and flaming procedures, and may be transferred from the plates representing one sample to those of another. After spreading, the plates are inverted and incubated at 32 C for 20 to 24 hours.

Following incubation, the plates are examined for typical colonies which are surrounded by a zone of turbidity or precipitate. The number of such colonies is multiplied by the reciprocal of the dilution that the countable plate represents. This is the "presumptive" *B. cereus* count.

33.22 Confirmatory Tests

As discussed above, there are several reasons why confirmatory tests are sometimes necessary. These include a) encountering (rarely) certain strains of *B. cereus* that have proper colonial morphology, but are egg yolk turbidity negative and b) occasional egg yolk turbidity producing strains of Group 2 bacilli. The following tests are suggested as being easiest to perform while yielding most pertinent information. It is advised that five separate colonies be chosen for the confirmatory tests.

a. Examination for sporangium morphology

Use a sterile inoculating needle to transfer a small amount of growth from the center of a colony to a small drop of distilled water on a slide. Emulsify the growth well, and either cover with a cover slip and examine directly by dark field microscopy, or allow to air dry, fix, and perform a spore stain for visualization by light microscopy. Group 1 bacilli should show spores within the sporangium without appreciable swelling of the sporangium. Note that this test may be performed directly on colonies from KG agar, but not from MYP agar. If the latter medium is used, a transfer to a sporulation medium (KG or nutrient agar slant) is necessary prior to performing this test.

b. Gelatinase (Chapter 2)

c. Nitrate reduction (Chapter 2)

d. Voges-Proskauer test (Chapter 2)

e. Anaerobic utilization of glucose (Chapter 2)

33.3 INTERPRETATION OF DATA

The number of zone forming colonies times the reciprocal of the dilution in the counted plate yields the "presumptive *B. cereus*" count. This number, times the fraction (i.e., x/5) of colonies that conform to the microscopic and biochemical criteria described, gives the 'confirmed *B. cereus*' count.

33.4 REFERENCES

1. GOEPFERT, J. M., W. M. SPIRA, and H. U. KIM. 1972. *Bacillus cereus*: Food poisoning organism. A review. J. Milk Food Technol. **35**:213–227
2. GORDON, R. E. 1973. The Genus *Bacillus*. Handbook of Microbiology. Vol. 1, p. 71–88 The Chemical Rubber Co., Cleveland, Ohio
3. GORDON, R. E., W. C. HAYNES, and C. HOR-NAY PANG. 1973. The Genus *Bacillus*. USDA Agriculture Handbook, No. 427, Washington, D.C.
4. KIM, H. U. and J. M. GOEPFERT. 1971. Enumeration and identification of *Bacillus cereus* in foods. I. 24-hour presumptive test medium. Appl. Microbiol. **22**:581–587
5. KUSHNER, D. J. 1957. An evaluation of the egg yolk reaction as a test for lecithinase activity. J. Bacteriol. **73**:297–302
6. MOSSEL, D. A. A., M. J. KOOPMAN, and E. JONGERIUS. 1967. Enumeration of *Bacillus cereus* in foods. Appl. Microbiol. **15**:650–653
7. SMITH, N. R., R. E. GORDON, and F. E. CLARK. 1946. Aerobic Mesophilic Sporeforming Bacteria. USDA Misc. Publ. No. 559, Washington, D.C.
8. SMITH, N. R., R. E. GORDON, and F. E. CLARK. 1952. Aerobic Spore-forming Bacteria. USDA Monograph No. 16, Washington, D.C.

CHAPTER 34

CLOSTRIDIUM BOTULINUM

Donald A. Kautter and Richard K. Lynt

34.1 INTRODUCTION

Clostridium botulinum is a species of anaerobic, sporeforming, rod shaped bacteria, producing a protein with a characteristic neurotoxicity. The severe food poisoning, botulism, results from consumption of botulinum toxin produced in a food in which growth of this organism occurred.

Antigenic types of *C. botulinum* are identified on the basis that their toxins are neutralized completely by the homologous type of antitoxin only, and that cross neutralization by heterologous antitoxin types is absent or minimal. The 7 recognized types are designated A, B, C, D, E, F, and G. Cultures of 5 of these apparently produce only one type of toxin, but all are given type designations corresponding to their toxin production. Types C and D cross react with antitoxins to each other because they each produce more than one toxin and have at least one common toxic component. Type C produces predominantly C_1 toxin with lesser amounts of D and C_2 ($C\alpha$) or only C_2 ($C\beta$), and type D produces predominantly type D toxin along with smaller amounts of C_1 and C_2 toxins. The production of more than one type of toxin may be a more common phenomenon than previously realized. There is a slight reciprocal cross-neutralization of types E and F, and recently one strain of *C. botulinum* has been shown to produce a mixture consisting mostly of type A toxin but also a small amount of type F toxin.

Botulism is a rare type of human food poisoning, but case fatality is high. In the U.S.A., 672 outbreaks have been recorded from 1899 through 1972. These have involved 1,731 cases and caused 963 deaths. Of outbreaks in which the toxin type was determined 148 were due to type A, 39 to type B, 19 to E, and 1 to type F. The implicated foods of two outbreaks contained both A and B toxins. The limited number of reports which consider C or D toxin to be the causative agent of human botulism have not received general acceptance. On the other hand, all except types F and G, about which little is known, are important causes of animal botulism.

The organism is distributed widely in soils and in sediments of oceans and lakes, so that foods may be contaminated with *C. botulinum* from many possible sources. The finding of type E in aquatic environments by many investigators correlates with the majority of the cases of type E botulism

IS/A Committee Liaison: John T. Graikoski

being traceable to contaminated fish or other seafoods. Types A and B are most commonly encountered terrestrially, and the important vehicles of botulism due to these two types are foods liable to contamination with soils. In the U.S.A. these foods have been primarily home canned vegetables, but in Europe meat products also have been important vehicles.

By properties other than toxin types, *C. botulinum* cultures fall into three distinct groups with each group composed only of strains having similar cultural and physiological characteristics. Organisms of types C and D are nonproteolytic, being unable to digest coagulated egg white or meat, and have a common metabolic pattern which sets them apart from all the others. Cultures of the other types fall into two additional groups distinct from one another on the basis of proteolysis. All type A strains and some B and F strains are proteolytic, whereas, all type E strains and the remaining type B and F strains are nonproteolytic. Type G has been studied insufficiently for adequate characterization.

Optimum temperature for growth and toxin production of the proteolytics is close to 35 C while that of the nonproteolytics is approximately 26 C. Nonproteolytic types B, E, and F are able to produce toxin at refrigeration temperature (3 C to 4 C). Toxins of the nonproteolytics do not manifest maximum potential toxicity until they are activated with trypsin; toxins of the proteolytics generally occur in fully, or close to fully, activated form. These and other differences are important in epidemiological and laboratory considerations of botulism outbreaks.

Measures to prevent botulism include reduction of the microbial contamination level, acidification, reduction of moisture level, and destruction of all botulinum spores in the food. The most common method of destruction is by heat; properly processed canned foods will not contain viable *C. botulinum*. The greater incidence of botulism from home canned foods as compared to commercially canned foods undoubtedly reflects the commercial canners' greater awareness and better control of the heating required.

A food can contain viable *C. botulinum* and still not be capable of causing botulism. As long as the organisms do not grow, toxin is not produced. Nutritional requirements of *C. botulinum* are satisfied by many foods, but not all provide the necessary anaerobic conditions. Both nutritional and anaerobic requirements are met by many canned foods and many meat and fish products. However, growth in otherwise suitable foods is prevented if the product, naturally or by design, is acid (of low pH), has low water activity, a high sodium chloride concentration, an inhibitory sodium nitrite concentration, or two or more of these in combination. Unless the temperature is very precisely controlled and kept below 3 C, refrigeration will not prevent growth and toxin formation by nonproteolytic strains. Moreover, the usual vehicles of botulism are those foods which are processed to prevent spoilage, and which usually are not refrigerated.

34.2 TREATMENT OF SAMPLES

Clinical diagnosis of botulism is most effectively confirmed in the laboratory by demonstrating botulinum toxin in the blood, feces, or vomitus of the patient. *Specimens must be collected before the patient is treated with botulinum antitoxin.* Identifying the causative food is most important to prevent further cases.

34.21 Food Samples

Samples should be refrigerated until tested except for unopened canned foods, which need not be, unless badly swollen and in danger of bursting.

Before testing record identifying data such as:

a. product
b. manufacturer or home canner
c. source
d. type of container and size
e. labeling
f. manufacturer's batch, lot, or production code
g. condition of container

Clean and mark container with laboratory identification, disinfect, and open aseptically for sampling. Be careful to avoid aerosols.

Check for ingredients which, by their presence or concentration in the product, could be lethal for mice by the intraperitoneal route of administration; for example, a high salt concentration (as in anchovies).

34.22 Clinical Specimens

All clinical specimens should be collected as soon as botulism is suspected and before botulinum antitoxin is administered.

34.221 Serum

Collect enough blood to provide at least 10 ml of serum (preferably 15 to 20 ml) to allow for toxin neutralization tests. Allow blood to clot, centrifuge, and remove serum to a sterile vial or test tube with a leakproof cap. Examine immediately or refrigerate at 4 C until tested. Examination of posttreatment samples is also helpful to evaluate antitoxin therapy.

34.222 Feces

Collect at least 50 grams of the patient's feces in a sterile, unbreakable, leakproof container. Preferably use a screwcap, wide mouth, plastic bottles. Seal caps with waterproof tape. Cardboard containers are not satisfactory. Refrigerate specimens at 4 C until examined. A "soap suds" enema should not be employed before the feces are collected, since inactivation of toxin by the soap may occur.

34.223 Miscellaneous clinical materials

Specimens of materials such as vomitus, gastric washings, cerebrospinal fluid, or tissues obtained at autopsy, should be collected in sterile, leakproof containers and refrigerated at 4 C.

34.3 SPECIAL EQUIPMENT AND SUPPLIES

34.31 Culture and Isolation of *C. botulinum*

Refrigerator
Clean dry towels
Bunsen burner
Sterile can opener (bacti-disc or puncture type)
Sterile mortar and pestle
Sterile forceps
Sterile pipets
Sterile sample jars
Sterile culture tubes (at least a few should have screw caps)
Anaerobic jars, i.e., Gaspak or Case (nitrogen replacement)
Transfer loops
35 C incubator
26 C incubator
Culture tube racks
Microscope slides
Microscope (phase contrast or bright field)
Sterile 100 mm Petri dishes

34.32 Toxin Detection and Assay

(Sterility of equipment is desirable, but not absolutely necessary except as noted)
Sterile vials for storage of serum
Mortar and pestle
Centrifuge tubes (some sterile for separation of patient's serum from clot)
Refrigerated, high speed centrifuge
Trypsin (Difco 1:250)
37 C waterbath
Sterile 25 to 50 ml syringes with 18 to 20 gauge 1½" needles for obtaining blood from patients
1 ml or 2.5 ml inoculating syringes with 25 gauge ⅝" needles for inoculation of mice.
Mice (ca 15 to 20 gm)
Mouse cages, feed, etc.

34.4 MEDIA AND REAGENTS

34.41 Culture and Isolation Procedures

Alcoholic solution of iodine (or other suitable disinfectant)
Sterile culture media (Chapter 2)
Cooked meat (liver or beef heart)
Trypticase peptone glucose yeast extract broth with trypsin (TPGYT)
Liver veal egg yolk agar or anaerobic egg yolk agar
Sterile gel phosphate buffer, pH 6.2
Absolute alcohol
Gram stain, crystal violet, or methylene blue solutions

34.42 Toxin Detection and Assay

Physiological saline
Gel phosphate buffer pH 6.2
Trypsin solution (prepared from Difco 1:250)
1 N sodium hydroxide
1 N hydrochloric acid
Monovalent antitoxins, types A through F. (may be obtained from the Center for Disease Control, Atlanta, Georgia 30333)

34.5 PRECAUTIONS AND LIMITATIONS OF METHODS

There is a slight degree of cross neutralization between types E and F. For example, type E antitoxin at a concentration of 1000 anti-MLD will commonly neutralize 2 or 3 MLD of type F toxin. If low levels of type F toxin are present in the sample, mice protected with monovalent type E and those protected with monovalent type F antitoxin may all survive. It may also happen, if low levels of toxin are present in a sample, that the sample may be toxic on initial testing, but may be nontoxic on subsequent testing. Botulinum toxin is heat labile. Samples and cultures should be refrigerated or frozen for storage. In addition, the pH of the toxic material must be controlled to keep it slightly on the acid side as botulinum toxin is less stable at alkaline pH.

When performing toxicity tests, care must be taken to distinguish between the symptoms of botulism and other causes of death, such as high concentrations of salt, acid, protein degradation products, or other toxic substances which may be present in the food being tested.

Be careful to avoid creation of aerosols.

Avoid pipetting by mouth. Use one of several mechanical pipetting devices available for this purpose.

All glassware and utensils coming in contact with contaminated or potentially contaminated samples should be autoclaved before they are handled by personnel who are not immunized.

Botulinum toxins are among the most lethal proteins known; the specific toxicity of type A neurotoxin is 1×10^8 mouse LD_{50}/mg protein. The syndrome of botulism (paralysis of throat, eyes, respiratory musculature, etc.) is the result of toxin inhibiting the release of the acetylcholine neurotransmitter at peripheral synapses. Laboratory personnel who expect to be exposed to the toxins should be immunized with toxoid (available from the Center for Disease Control, Atlanta, Georgia). Botulinum toxin is heat labile, therefore, botulism will not occur if all foods are heated thoroughly, such as boiling for 10 minutes, before the time of serving.

34.6 PROCEDURE FOR DETECTION OF VIABLE *C. BOTULINUM*

34.61 Opening of Canned Foods

Sanitize the uncoded end of the can with an effective disinfectant. Allow a contact time of a few minutes, then remove the disinfectant and wipe the sanitized area with a sterile, dry towel. If the can is swelled, position the can so that the side seam is away from the analyst. A container with buckled ends should be chilled before opening and flamed with extreme caution to avoid bursting the can. Flame sterilize the sanitized can end with a Bunsen burner by directing the flame down onto the can until the visible moisture film evaporates. Avoid excessive flaming, indicated by scorching and blackening of the inside enamel coating. Remove a disc of metal from the center area of the flamed end with a flame sterilized or autoclaved bacti-disc cutter. A disc about 5 cm in diameter is removed, except from 202 diameter cans, where a 3 cm disc is satisfactory.

34.62 Solid Foods

Foods with little or no free liquid are transferred aseptically to a sterile mortar. Add an equal amount of gel phosphate buffer solution and grind with a sterile pestle in preparation for inoculation. Alternatively, small pieces of the product may be inoculated directly into the enrichment broth with sterile forceps.

34.63 Liquid Foods

Inoculate liquid foods directly into the culture media, using sterile pipets.

34.64 Reserve Sample

After culturing, aseptically remove a reserve portion of the sample to a sterile sample jar for later tests.

34.65 Examination of Product for Appearance and Odor

Note any evidence of decomposition, but **do not** taste the product under any circumstance. Record findings.

34.66 Preparation of Enrichment Cultures

34.661 Preparation of broth media

Before inoculation, unless the enrichment media are freshly prepared, heat broth media in flowing steam or boiling water for 15 minutes. After heating, cool rapidly in cold water without agitation to room temperature before inoculation.

34.662 Inoculation of enrichment media

Inoculate 1 to 2 g of solid or 1 to 2 ml of liquid food per 15 ml of enrichment broth. Inoculate two tubes of cooked meat medium, either beef heart or chopped liver broth, and two tubes of TPGYT broth. Introduce the inoculum slowly below the surface of the broth.

34.663 Incubation

Incubate the inoculated cooked meat medium at 35 C and the TPGYT enrichment broth at 26 C.

34.664 Examination of cultures

After 5 days of incubation, examine each culture for turbidity, the production of gas, and the digestion of the meat particles. Note the odor. Examine the culture microscopically, by a wet amount preparation under high power, phase contrast microscopy, or a stained smear (Gram stain, crystal violet, or methylene blue) with bright field illumination. Observe the morphology of the organisms in the preparation and note the existence of clostridial cells, the occurrence and relative extent of sporulation, and location of spores within the cells. At this time, test each culture for toxin, and, if demonstrated, determine the toxin type according to the procedure described in 34.72 and 34.73. The highest concentration of botulinum toxin is usually present after the period of active growth. In general five days of incubation are necessary to reach maximal toxin levels. An enrichment culture showing no growth at five days should be incubated an additional ten days to detect possible delayed germination of spores of *C. botulinum*, before the culture is discarded as sterile.

For pure culture isolation, gently mix, transfer, at peak sporulation, 1 to 2 ml of the culture to a sterile screwcap tube and refrigerate.

34.67 Isolation of Pure Cultures

34.671 Alcohol treatment

The possibility of isolating *C. botulinum* in pure culture from a mixed flora in the enrichment culture is greatly improved if its cells sporulate well. To 1 to 2 ml of a culture showing some sporulated cells (or the retained sample), add an equal volume of filter sterilized absolute ethanol in a sterile screwcap tube. Mix the alcohol with the culture, and incubate the mixture at room temperature for one hour. Streak plates of the recovery medium as described in 34.673.

34.672 Heat treatment

An alternative procedure to the alcohol method is to heat 1 to 2 ml of the enrichment culture enough to destroy the vegetative cells present but not the spores of *C. botulinum*. For a nonproteolytic type, however, do not use heat; for a proteolytic type, heat at 80 C for 10 to 15 minutes.

34.673 Plating

Streak the alcohol or heat treated culture on Petri dishes containing either liver veal egg yolk agar or anaerobic egg yolk agar (Chapter 2) in order to obtain well separated colonies. Dilution of the culture may be necessary before plating in order to select well isolated colonies. To prevent spreading of the colonies, the plates must be well dried. Use Petri plates with unglazed porcelain covers, aluminum covers with absorbent discs, or dry the plates at 35 C for about 24 hours before streaking.

34.674 Incubation

Incubate the inoculated streak plates anaerobically at 35 C for about 48 hours. A Case Anaerobic Jar or the Gaspak system is adequate to obtain anaerobiosis. Other anaerobic systems also can be used.

34.675 Selection of typical colonies

After anaerobic incubation, select about ten well separated and typical colonies per plate. Colonies of *C. botulinum* are raised or flat, smooth or rough; they commonly show spreading, and have an irregular edge. On egg yolk medium the colonies usually exhibit a surface irridescence when examined by oblique light. This luster zone often is referred to as a pearly layer, and usually extends beyond, and follows, the irregular contour of the colony. Besides the pearly layer, colonies of *C. botulinum* types C, D, and E are ordinarily surrounded by a wide zone (2 to 4 mm) of yellow precipitate. Colonies of types A and B generally show a smaller zone of precipitation. However, considerable difficulty in picking toxin producing colonies may be experienced since certain other members of the genus *Clostridium*, which do not produce toxin, produce colonies which are characteristic of *C. botulinum*.

Inoculate each colony into a tube of sterile broth by means of a sterile transfer loop. For *C. botulinum* type E, inoculate TPGYT broth; for the other toxin types, inoculate cooked meat medium. Incubate the inoculated tubes for five days as previously described, then test for toxin as described in 34.72. If toxin is demonstrated, determine the toxin type (34.73). Restreak the toxin producing culture in duplicate on egg yolk agar medium. Incubate one of the plates anaerobically and the other aerobically at 35 C. If colonies typical of *C. botulinum* only are found after 48 hours on the plate incubated anaerobically, and no growth is found on the plate incubated aerobically, the culture may be pure. Failure to isolate *C. botulinum* from at least one of the colonies selected means that its prevalence among the mixed flora in the enrichment culture is probably low. Sometimes the numbers can be increased enough to permit isolation by repeated serial transfer through additional enrichment steps.

34.676 Storage

The pure culture should be stored in the sporulated state under refrigeration, on glass beads, frozen, or lyophilized.

34.7 DETECTION OF BOTULINUM TOXIN

34.71 Preparation of Food Sample

After a sample has been removed for detection of viable *C. botulinum*, remove a portion of the food for toxicity testing. The remainder should be stored in a refrigerator. Samples containing suspended solids should be centrifuged in the cold, and the supernatant fluid used for the toxin assay.

Extract solid foods with an equal volume of gel phosphate buffer. The food and buffer should be macerated with a prechilled mortar and pestle. Centrifuge to remove the solids. Wash out emptied containers suspected of having contained toxic foods with a few ml of gel phosphate buffer. Do not use too much liquid as the toxin may be diluted below the detection level.

34.72 Toxicity Determinations in Food Samples or Culture

Toxins of nonproteolytic types, if present in the sample, may need trypsin activation to be detected. Therefore, treat a portion of the food supernatant fluid, liquid food, or cooked meat culture with trypsin before testing for toxin. Do not treat TPGYT cultures with trypsin. At the same time, test another portion of supernatant fluid extract, or cooked meat culture, for toxin without trypsin treatment, since the fully active toxin of a proteolytic strain, if present, may be degraded by trypsin. The same is true if TPGYT containing the fully activated toxin of a nonproteolytic strain is further trypsinized.

To trypsinize, adjust a portion of the supernatant fluid, if necessary, to pH 6.2 with 1N sodium hydroxide or HCl. To 1.8 ml of each supernatant fluid, add 0.2 ml of an aqueous trypsin solution and incubate at 37 C with occasional gentle agitation for one hour. Prepare trypsin solution by placing 1 gm of Difco 1:250 trypsin in a clean culture tube and adding 10 ml of distilled water. Agitate from time to time and keep at room temperature until as much of the trypsin as possible has been dissolved.

Dilute a portion of the untreated sample fluid or culture to 1:2, 1:10, and 1:100 in gel phosphate buffer. Make the same dilutions of each trypsinized sample fluid or culture. Inject separate pairs of mice intraperitoneally (I.P.) with 0.5 ml of the undiluted fluid and with 0.5 ml of each dilution using a 1.0 ml or 2.5 ml syringe with a 25 gauge ⅝" needle. Repeat this for the trypsinized samples. Heat 1.5 ml of the untreated supernatant fluid or culture at 100 C for 10 minutes. Cool the heated sample and inject each of a pair of mice with 0.5 ml of the undiluted fluid.

Observe all of the mice periodically for 72 hours for symptoms of botulism, and record deaths. If all of the mice die at the dilutions used, repeat using higher dilutions to determine the end point, or the minimum lethal dose (MLD) as an estimate of the amount of toxin present. The MLD is contained in the highest dilution killing both mice (or all mice inoculated). From this, the number of MLD/ml may be calculated.

It is important to observe the mice closely for signs of botulinum intoxication during the first 24 hours after inoculation. Death of mice without clinical signs of botulism is not sufficient evidence that the material injected contained botulinum toxin.

Typical botulism symptoms in mice in sequence are: ruffling of the fur, labored breathing, not rapid, weakness of the limbs, gasping for breath (opening of lower jaw), and death due to respiratory failure. Mice that die immediately after injection usually are injured or react to ammonia in the sample. Mice that die after 12 hours with closed, matted eyes are generally killed by infection not toxin, and do not have the typical symptoms of botulism.

34.73 Typing of Toxin

In determining the type of toxin, either the untreated or trypsin treated fluid may be used, provided that it was lethal to mice. Use the preparation which gave the greater number of MLD. If the trypsinized fluid is to be used, prepare a freshly trypsinized fluid. The continued action of trypsin may destroy the toxin.

Dilute monovalent antitoxins to types A, B, E, and F in physiological saline to contain one international unit per 0.5 ml. Prepare enough of this dilution to inject each of two mice with 0.5 ml of the antitoxin for each dilution of the toxic preparation to be tested.

Protect separate groups of mice by injecting each mouse with 0.5 ml of one of the above antitoxins 30 minutes to 1 hour before injecting them with the suspected toxic preparation. Inject both the unprotected and protected mice with a sufficient number of dilutions to cover a range of at least 10,100, and 1,000 MLD below the previously determined end point of toxicity. Observe the mice for 72 hours for symptoms of botulism, and record deaths.

An alternative procedure is to perform the test as described in 34.81, substituting the supernatant fluid for the patient's serum and employing monovalent type F antiserum for the polyvalent.

Whichever method is used, if the toxin is not neutralized, repeat the test using monovalent antitoxins to types C and D, and a pool of types A through F.

34.8 TESTING CLINICAL SAMPLES FOR TOXIN

34.81 Toxin Neutralization Tests

Botulinum toxins in clinical specimens are identified by injecting mice with the suspect specimen alone, and the same specimen mixed with one or more botulinum antitoxins. If botulinum toxin is present in sufficient quantity to be detected, mice receiving the unneutralized toxin will die, and mice receiving the toxin neutralized by the specific antitoxin will survive. Although botulinum intoxication usually kills mice within 6 to 24 hours after inoculation, deaths may be delayed if the quantity of toxin is near the minimum detectable concentration. Therefore, the animals should be observed for 72 hours before recording as negative, as in the case of food samples.

34.82 Identification of Botulinum Toxin in Serum

Prepare serum-antitoxin mixtures as follows:

Patient's Serum	Normal serum or antitoxin
1.2 ml	0.3 ml normal serum
1.2 ml	0.3 ml anti-A
1.2 ml	0.3 ml anti-B
1.2 ml	0.3 ml anti-E
1.2 ml	0.3 ml polyvalent (anti-A, B, C, D, E, F)

Incubate mixtures in a 37 C waterbath for 30 minutes.

Inoculate 0.5 ml of the serum-antitoxin mixtures intraperitoneally into two mice (15 to 20 grams) for each test mixture. Each mouse, therefore, receives 0.4 ml of the patient's serum and 0.1 ml of normal serum or antitoxin.

Observe mice at intervals for 72 hours, note clinical signs, and record deaths and survivors.

If no toxin is detected using 0.4 ml of patient's serum per mouse, repeat the test using larger mice (25 to 30 grams) inoculated with 0.8 ml of patient's serum mixed with 0.1 ml of antitoxin or normal serum per mouse. To do this, use 2.4 ml of patient's serum to the same amount of normal serum or antitoxin used before and inject each mouse with 0.9 ml of the mixture using the same scheme as outlined above. Do not inject mice with a total volume larger than 1.0 ml since excessive amounts of normal serum can cause death. Trypsinization of the patient's serum is not necessary for demonstration of toxin.

If all of the mice inoculated with the test mixtures described above develop signs suggestive of botulism and die, it is possible that type G, or an unidentified toxin, is involved. In this case, the patient's serum should be submitted to a reference laboratory for additional tests. At present, type G antitoxin is not available commercially.

If mice were protected by neutralization of the toxin by the polyvalent antitoxin only, repeat the test using monovalent type C, type D, and Type F, the polyvalent antitoxin, and the normal test mixtures. If the mice receiving the polyvalent antitoxin mixture are the only survivors again, then other combinations of monovalent antitoxins should be tested to determine if multiple toxin types are present. To estimate the quantity of botulinum toxin in the patient's serum, prepare dilutions (1:2, 1:4 etc.) with gel phosphate buffer and determine the MLD. Inoculate groups of mice with each dilution of the patient's serum, 0.5 ml intraperitoneally per mouse.

34.83 Identification of Botulinum Toxin in Feces

Recent experience from the investigation of several botulism outbreaks, has revealed that examination of feces for botulinum toxin is a valuable diagnostic procedure.

Homogenize 15 to 50 grams of feces in an equal quantity (W/V) of gel phosphate buffer with a chilled mortar and pestle, cover and refrigerate at 4 C for 12 to 18 hours to extract toxin.

After refrigeration, centrifuge (12,000 × g) at 4 C for 20 minutes and remove the liquid for performance of a toxin neutralization test. If the extract is not clear, clarify by repeating the centrifugation procedure.

Prepare the following extract-antitoxin mixtures:

Extract of Feces	Normal serum or antitoxin
1.2 ml	0.3 ml normal serum
1.2 ml (heated)*	0.3 ml normal serum
1.2 ml	0.3 ml anti-A
1.2 ml	0.3 ml anti-B
1.2 ml	0.3 ml anti-E
1.2 ml	0.3 ml anti-F
1.2 ml	0.3 ml polyvalent (anti-A, B, C, D, E, F)

Incubate the mixture in a 37 C waterbath for 30 minutes.

Inoculate 0.5 ml of the extract-antitoxin mixtures intraperitoneally into two mice (15 to 20 grams) for each test mixture.

*a portion of the extract is heated in a boiling water bath for 10 minutes; botulinum toxin is heat labile.

Observe mice at intervals for 72 hours; note clinical signs and record deaths and survivors. Mice injected with the heated extract should survive. Interpret the results of the tests and perform repeat tests, if required, as described above, for identification of botulinum toxin in serum samples.

34.84 Miscellaneous Clinical Specimens

Tests for botulinum toxin in cerebrospinal fluid, urine, and other body fluids, are performed as described for serum samples (34.82) after centrifugation, if necessary, to clarify the liquids.

Dilute vomitus and gastric washings 1:1 with gelatin diluent (Chapter 2) before clarification by centrifugation and performance of the toxin neutralization tests.

Grind tissues with an equal volume of gelatin diluent using a sterile chilled mortar and pestle. Cover and refrigerate at 4 C for 12 to 18 hours. After refrigeration, centrifuge (12,000 × g) at 4 C, remove the liquid and perform a toxin neutralization test as described for feces (34.83). Omit the heated extract-serum mixture.

34.9 INTERPRETATION OF DATA

Laboratory tests required in an actual or potential botulism outbreak are designed to identify botulinum toxin and/or the organism in foods and clinical specimens. Toxin in a food means that the product, if consumed without thorough heating, could cause botulism. In clinical specimens, it means that a clinical diagnosis of botulism is consistent with the laboratory findings. Viable *C. botulinum,* but no toxin in specimens, has a different meaning. In foods, the observation by itself is not proof that the food in question caused botulism, and in clinical specimens does not necessarily mean that the patient has botulism. The presence of toxin in the food is required for an outbreak of botulism to occur. Ingested organisms may be found in the alimentary tract, but they are considered to be unable to multiply and produce toxin *in vivo.* Assuming proper handling of specimens to prevent toxin inactivation, failure to find toxin means a detectable level is not present.

Presence of botulinum toxin and/or organisms in low acid (i.e., above pH 4.6) canned foods means that the items were underprocessed or were contaminated through postprocessing leakage. These failures occur more frequently in home processed than in commercially canned food. Swollen containers are more likely than normal containers to have botulinum toxin since the organism produces gas during growth. The rare occurrence of toxin in a flat can may imply that the seams were loose enough to allow gas to escape. Botulinum toxin in canned foods is usually of a type A or of a proteolytic type B strain, since spores of the proteolytics can be among the more heat resistant of bacterial spores. Spores of the nonproteolytics, types B, E, and F, generally have low heat resistance and do not normally survive even mild heat treatment.

The protection of mice from botulism and death with one of the monovalent botulinum antitoxins, confirms the presence of botulinum toxin and determines the serological type of toxin in a sample.

If the mice are not protected by one of the monovalent botulinum antitoxins, there may be (1) too much toxin in the sample, (2) more than one type of toxin, or (3) deaths may be due to some other cause. This requires retesting at a higher dilution of toxin and the use of mixtures of A, B, C, D, E, and F antisera in place of the monovalent antiserum.

If both the heated and unheated supernatant fluids of the sample are lethal to mice, the deaths are probably not due to botulinum toxin. However, it is possible that a toxic substance not destroyed by heat may mask the presence of botulinum toxin in such a sample. If the botulinum toxin is present in large quantities, further dilution of the sample may eliminate the heat stable toxin and still allow the botulinum toxin to act.

34.10 GENERAL REFERENCES

1. Dowell, V. R. and T. M. Hawkins. 1968. Laboratory Methods in Anaerobic Bacteriology, CDC Laboratory Manual. PHS Pub No. 1803, U.S. Government Printing Office, Washington, D.C.
2. Smith, L. DS. and L. V. Holderman. 1968. The Pathogenic Anaerobic Bacteria. Charles C Thomas, Springfield, Ill.
3. Stumbo, C. R. 1973. Thermobacteriology in Food Processing, 2nd ed. Academic Press, Inc., New York, N.Y.
4. Thatcher, F. S. and D. S. Clark. 1968. Microorganisms in Foods, University of Toronto Press, Toronto, Canada
5. Anonymous. 1974. Botulism in the United States, 1899–1973. U.S. Department of Health, Education, and Welfare, Public Health Service, DHEW Publication No. (CDC) 74–8279
6. Lewis, K. H. and K. Cassel, Jr. (Eds.) 1964. Botulism. U.S. Government Printing Office, Washington, D. C.
7. Ingram, M. and T. A. Roberts (Eds.) 1967. Botulism 1966, Chapman and Hall, London, England
8. Herzberg, M. (Ed.) 1970. Toxic Microorganisms. U.S. Government Printing Office, Washington, D.C.

CHAPTER 35

CLOSTRIDIUM PERFRINGENS

Charles L. Duncan and Stanley M. Harmon

35.1 INTRODUCTION

The methods included in this chapter are useful for direct quantitation of colony forming units of *Clostridium perfringens* or for obtaining an indication of the presence of viable or nonviable *C. perfringens* cells. These methods are in conformance with those adopted as AOAC Official First Action.

35.11 Food Poisoning. Enterotoxin Formation

C. perfringens food poisoning is one of the most common types of human foodborne illnesses. The vehicles usually involved are cooked meat or poultry products containing large numbers of viable cells. A heat labile enterotoxin produced only by sporulating cells induces the major symptom of diarrhea in perfringens poisoning. The enterotoxin appears to be released *in vivo* in the intestine by the sporulating organism. Although the enterotoxin is generally not thought to be preformed in the food vehicle, foods in which conditions are favorable for sporulation conceivably may contain enterotoxin.

35.12 Importance of Cell Numbers

C. perfringens is not uncommon in raw meats, poultry, dehydrated soups and sauces, raw vegetables, and certain other foods or food ingredients. Thus, its mere presence in foods may be unavoidable. Large numbers in a food may, however, be indicative of mishandling. In food poisoning outbreaks, demonstration of hundreds of thousands or more per gram in a suspect food supports a diagnosis of perfringens poisoning when substantiated by clinical and epidemiological evidence.

A loss in the viability of *C. perfringens* cells may occur if foods are frozen or held under prolonged refrigeration. This may lead to difficulty in incriminating *C. perfringens* in food poisoning outbreaks if the suspect food vehicle has been frozen or refrigerated for extended periods of time prior

IS/A Committee Liaison: John T. Graikoski

to analysis for the organism. In such cases, the alpha toxin indicator method[4] may prove useful as an indicator of the presence of *C. perfringens* cells.

Spores of different strains of *C. perfringens* may vary widely in their heat resistance. Some may withstand 100 C for several hours, whereas others are inactivated by a few minutes or less at the same temperature. In most environments the heat sensitive strains outnumber heat resistant strains. Both heat resistant and heat sensitive types may cause food poisoning.

35.13 Selective Differential Media

Several solid media have been devised for quantitation of *C. perfringens*. These have included neomycin blood agar,[12] sulfite polymyxin sulfadiazine (SPS) agar,[1] tryptone sulfite neomycin (TSN) agar,[9] Shahidi Ferguson perfringens (SFP) agar,[11] D-cycloserine blood agar,[2] oleandomycin polymyxin sulfadiazine perfringens (OPSP) agar,[3] tryptose sulfite cycloserine (TSC) agar,[5] and egg yolk free tryptose sulfite cycloserine (EY-free TSC) agar.[7] The selectivity of these media is derived from the incorporation of one or more antibiotics inhibitory to certain anaerobes or facultative anaerobes. With the exception of the blood agars, the media contain added iron and sulfite. The sulfite is reduced by clostridia to sulfide which reacts with the iron to form a black iron sulfide precipitate. Black colonies are presumptive *C. perfringens* and must be confirmed by additional tests.

The SFP, OPSP, and TSC also contain egg yolk for differential purposes. The alpha toxin (phospholipase C or lecithinase) of *C. perfringens* hydrolyzes egg yolk lecithin and produces an opaque halo around the black colonies. However, other sulfite reducing clostridia or other facultative anaerobes may produce a similar reaction. In some instances the egg yolk reaction of *C. perfringens* alpha toxin may be masked by that of other organisms. In addition, false negative *C. perfringens* without detectable halos may occur on the plates.[7] EY-free TSC agar which is not dependent on alpha toxin production for its differential utility has been proposed as an improvement over TSC agar.[7]

The selectivity of TSN and SPS media is sufficient to inhibit many strains of *C. perfringens*. SPS also suffers from the failure of many strains to form colonies that are distinguishably black. Although the selectivity of SFP and neomycin blood agars is limited, they may be adequate when *C. perfringens* is the predominant organism. The selectivity of OPSP agar also may be limited with some facultative anaerobes. D-cycloserine blood agar may be useful for the selective isolation of *C. perfringens*, although it has not been tested for routine isolation of the organism from foods. TSC agar or its modified form EY-free TSC agar have been documented as the most useful of the media for quantitative recovery of *C. perfringens*, with accompanying adequate suppression of the growth of practically all facultative anaerobes. Other anaerobes can grow on these media. TSC agar contains egg yolk and must be used for surface plating. EY-free TSC agar is used in pour plates. The methods described here for quantitation of *C. perfringens* employ TSC agar or EY-free TSC agar. The EY-free TSC may be employed with equally good, if not better, results than the TSC agar.

35.14 Indication of the Presence of *C. perfringens* on the Basis of Alpha Toxin Formation

A decrease of three to five log cycles in the plate count of this organism is not uncommon when foods are stored frozen or refrigerated for several days or shipped frozen to the laboratory. In such cases, population levels in foods associated with an outbreak can be estimated by extracting and quantitating the alpha toxin produced in the food during growth of the organism. A detectable quantity of alpha toxin is produced by about 10^6 *C. perfringens* cells per g of substrate, and the amount increases in proportion to the cell population. This method has been shown to be effective for estimating population levels of *C. perfringens* with a variety of foods associated with food poisoning outbreaks.[4, 6]

35.2 TREATMENT OF SAMPLE

Generally it is recommended that food samples to be tested for *C. perfringens* be analyzed immediately, or refrigerated and tested as soon as possible, but not frozen. Loss of viability of some strains that occurs on refrigeration may be even greater when the cells are frozen. It has been reported that the smallest loss in viable count occurs when foods are mixed 1 : 1 (w : v) with 20% glycerol and kept in dry ice at −55 to −60 C until analysis can be done.[7]

35.3 SPECIAL EQUIPMENT AND SUPPLIES

35.31 Isolation and Quantitation

a. Anaerobic containers or anaerobic incubator with equipment and materials for obtaining anaerobic conditions. These may be anaerobic devices in which the air is replaced 3 to 4 times with 90% N_2 + 10% CO_2, or those in which oxygen is catalytically removed.
b. Sterile blender, container and motor, or sterile mortar and pestle and sterile sand. The blender is preferable.
c. Air incubator 35 to 37 C
d. Water bath 46 C
e. Colony counter with a piece of white tissue paper over the counting background area to facilitate counting black colonies

35.32 Alpha Toxin Method as an Indicator of the Presence of *C. perfringens* Cells

a. Seitz filters; 100 to 250 ml capacity with sterilizing filter pads
b. Blender
c. High speed centrifuge and centrifuge tubes
d. Dialysis tubing
e. Filter paper
f. Refrigerator 4 C
g. Air incubator 35 to 37 C
h. Stainless steel tubing or well punch, 3mm O.D.

i. Vacuum source
j. Pasteur pipets

35.4 Special Reagents and Media

See Chapter 2 for formulations of the media and reagents used below.

35.41 Isolation and Quantitation

a. Fluid thioglycollate medium
b. 0.1% Peptone water diluent
c. Motility nitrate medium
d. Reagents for nitrate reduction
e. Lactose gelatin medium
f. Tryptose sulfite cycloserine (TSC) agar or egg yolk free tryptose sulfite cycloserine (EY-free TSC) agar

35.42 Alpha Toxin Indicator Method

a. N-2-hydroxyethylpiperazine N′-2-ethane sulfonic acid (HEPES)
b. Polyethylene glycol 20,000 m.w.
c. Purified agar
d. Fresh eggs
e. Human red blood cells
f. *C. perfringens* type A diagnostic serum (Welcome Reagents Division, P.O. Box 1887, Greenville, N.C. 27834)

35.5 RECOMMENDED CONTROLS

35.51 Direct Quantitation

In using the selective differential media for quantitation of *C. perfringens*, it may be useful to employ a control strain of the organism to validate the performance of the media. This also may be useful when one has not observed typical *C. perfringens* colonies on the medium before.

35.52 Alpha Toxin Indicator Method

A filter sterilized culture supernatant of an 18 to 24 hour cooked meat culture of *C. perfringens* can be used as a positive control for the hemolytic and lecithinase activities of alpha toxin in the hemolysin plate and the lecithovitellin test. The culture filtrate containing alpha toxin should be diluted and tested in the same manner as that described for the concentrated food extracts. The hemolytic and lecithinase activities of alpha toxin are neutralized by the addition of a suitable amount of *C. perfringens* type A diagnostic serum. Saline used to dilute the food extracts should be tested in the hemolysin plates as a negative control.

35.6 PRECAUTIONS AND LIMITATIONS OF METHODS

35.61 Isolation and Quantitation

a. Both TSC and EY-free TSC appear to be suitable for enumeration of

C. perfringens. However, some strains of *C. perfringens* may not produce distinguishable halos via the egg yolk reaction on TSC agar. Therefore, the absence of a halo around a black colony does not eliminate the possibility of the strain being *C. perfringens*. The halo of one colony also may be masked by the halo of another colony.

b. Contamination with sulfite reducing clostridia other than *C. perfringens*.

Several other sulfite reducing clostridia that may produce black colonies such as *Clostridium bifermentans, Clostridium sporogenes,* or *Clostridium botulinum* can grow on the TSC and EY-free TSC agars.

c. Contamination with *Streptococcus*.

Enterococci may be present in high numbers in some foods. In media other than those containing D-cycloserine, the overgrowth of these organisms may interfere or prevent the isolation of pure cultures from typical colonies of *C. perfringens*. The incorporation of D-cycloserine in TSC and EY-free TSC agars effectively inhibits growth of enterococci.

d. Confirmation of the presumptive plate count.

Presumptive sulfite reducing colonies of *C. perfringens* from the selective differential media often are confirmed on the basis of their nonmotility and ability to reduce nitrate to nitrite. However, other clostridia with these characteristics have been reported, such as *Clostridium filiforme*, and unidentified *Clostridium* species isolated from fecal specimens.[7] These other clostridial species may be distinguished from *C. perfringens* by their inability to liquefy gelatin within 44 hours.[7] It is therefore important to include gelatin liquefaction as a confirmatory test for *C. perfringens*.

35.62 Alpha Toxin Indicator Method

a. Variation in alpha toxin production in outbreak foods

The alpha toxin method has been used successfully for estimating population levels of *C. perfringens* in a variety of foods associated with food poisoning outbreaks. However, it should be kept in mind that results obtained by this method are not strictly quantitative, but are estimates only. A number of factors will undoubtedly affect, to some extent, the amount of alpha toxin that will be formed. Among these are the variation in toxin producing ability of different strains, the type of food, and the time and temperature at which the food has been held. Nevertheless, the presence of any detectable alpha toxin is an indication that the *C. perfringens* population was probably sufficient to cause food poisoning, and higher titers indicates that a very large viable population was present in the food.

Data obtained with incriminated foods suggests that whenever conditions are suitable for growth of the organism, alpha toxin usually will be formed also. However, the absence of alpha toxin does not preclude the possibility that *C. perfringens* was responsible for a food poisoning outbreak. In a few instances, strains of this organism which produce very little alpha toxin have been isolated from incriminated foods. A false negative result, or at least an underestimation of the population level, can be expected with foods contaminated by these strains alone.

b. Interference with the detection of alpha toxin by other bacteria

Interference with the detection and quantitation by hemolysins and lecithinases of other bacteria rarely occurs, but it must be recognized that if they are present in the food extract, they can interfere with performance of the alpha toxin test. With the possible exception of the hemolytic activity of *C. bifermentans* lecithinase, none of these should be neutralized by *C. perfringens* type A diagnostic serum.

The problem most frequently encountered is due to the production of hemolysins or lecithinases by other bacteria growing in the test substrates during the performance of the alpha toxin assay. Hemolytic or lecithinase activities which do not diminish proportionally with dilution of the extract are most likely due to bacterial growth in the substrate. This can be eliminated by filter sterilization of the extract (with either a Seitz filter or a membrane with a prefilter) or the addition of a broad spectrum antibiotic such as penicillin or D-cycloserine to the extract, and repeating the alpha toxin test.

35.7 PROCEDURE

35.71 Isolation and Quantitation

35.711 Enrichment procedure to determine the presence of low numbers

In some instances it may be desirable to isolate *C. perfringens* from samples contaminated with very low numbers of cells. An enrichment procedure employing fluid thioglycollate medium may be used. Inoculate about 2 grams of food sample into 25 ml of fluid thioglycollate medium contained in screw capped tubes. Incubate in a water bath at 46 C for 4 to 6 hours. Positive tubes show turbidity and gas production. Prepare dilutions from each positive tube, and plate as described in 35.72 to obtain presumptive *C. perfringens* colonies. Proceed to confirm as indicated in 35.73.

35.72 Anaerobic Plate Count

a. Preparation of food homogenate

Blend for 2 minutes at slow speed (8,000 to 10,000 r.p.m.) (or macerate with sterile sand) a 10 to 20 g food sample combined with 0.1% peptone to obtain a 1 : 10 dilution of the sample.

b. Preparation of dilutions

Prepare serial decimal dilutions (through at least 10^{-7}) using 0.1% peptone dilution blanks.

c. Plating procedures

Make duplicate surface platings of each dilution using 0.1 ml amounts on TSC agar. After the agar has dried slightly, overlay the surface with 5 or more ml of EY-free TSC agar. If EY-free TSC agar is used as the plating medium, duplicate pour plates of each dilution are made employing 1.0 ml of diluted culture per plate. After the pour plates solidify, cover with an additional 5 or more ml of EY-free TSC agar.

d. Incubation

Incubate the plates upright and anaerobically for 18 to 20 hours at 35 to 37 C.

e. Presumptive *C. perfringens* plate count

Select plates containing black colonies (preferably 30 to 300) which may be surrounded by a zone of precipitate on the TSC agar but not on the EY-free TSC agar. Count all black colonies and calculate the average number of colonies in the duplicate plates.

35.73 Confirmation of *C. perfringens*

Select at least 10 representative black colonies obtained by the methods described in X.712 and determine the ratio of *C. perfringens* among these colonies.

a. Obtaining pure cultures

Inoculate a portion of each selected black colony into a tube of fluid thioglycollate medium. Incubate for 4 hours in a water bath at 46 C, or overnight at 35 to 37 C. After incubation, examine for the presence of gram positive rods with blunt ends; endospores usually are not produced in this medium. The culture may also be examined by phase contrast microscopy. Streak the culture onto TSC agar and incubate anaerobically for 24 hours at 35 to 37 C to obtain isolated colonies. Typical colonies are yellowish gray, 1 to 2 mm in diameter, usually surrounded by an opaque zone due to lecithinase production. These then may be picked into fluid thioglycollate medium.

b. Motility nitrate reduction test

Stab inoculate each fluid thioglycollate medium culture into two tubes of motility nitrate medium. The medium recommended contains 0.5% each of glycerol and galactose which improves the consistency of the nitrate reduction reaction with different strains of the organism.[7] Incubate the inoculated medium at 35 to 37 C for 24 hours. Read for motility. Since *C. perfringens* is nonmotile, growth should occur only along the line of inoculum and not diffuse throughout the entire medium. Test one culture of each pair for reduction of nitrate to nitrite. A red or orange color indicates reduction of nitrate to nitrite. If nitrate has not been reduced, incubate the second tube for an additional 24 hours and repeat the test.

c. Lactose gelatin medium

Stab inoculate an isolated colony into lactose gelatin medium. Incubate at 35 to 37 C for 24 to 44 hours. Lactose fermentation is indicated by gas bubbles and a change in color of the medium from red to yellow. Gelatin usually is liquefied by *C. perfringens* within 24 to 44 hours.[7]

d. Subculture isolates which do not liquefy gelatin within 44 hours, or are atypical in other respects, into fluid thioglycollate medium. Incubate the cultures for 18 to 24 hours at 35 to 37 C, make a Gram stained smear, and check for purity. If pure, inoculate a tube of peptone yeast extract (PY) medium containing 1% salicin and a tube of PY medium containing 1% raffinose with 0.1 ml of the fluid thioglycollate culture of each isolate. Incubate inoculated PY medium at 35 to 37 C for 24 hours and check for production of acid and gas. To test for acid, transfer 1 ml of culture to a test tube or test plate and add 2 drops of 0.04% phenol red. A yellow color indicates that acid has been produced. Reincubate PY cultures for an additional 48 hours and retest for production of acid. Salicin is rapidly fermented with the production of acid and gas by closely related species such as *C. paraperfringens* (*C. barati*) and *C. absonum*, but usually not by *C. perfringens*.[10] Acid is usually produced from raffinose within 3 days by *C. perfringens* but not produced by closely related species.

35.74 Alpha Toxin Indicator Method

35.741 Preparation of hemolysin indicator plates

a. Red blood cells

Wash packed human red blood cells three times by mixing with 4 volumes of sterile physiological saline. Centrifuge at 2500 rpm to sediment cells and remove supernatant using a vacuum flask. Resuspend the cells in additional saline and repeat these steps twice. After the final wash, resuspend the cells in an equal volume of saline. Use sterile precautions throughout the washing procedure.

b. Pouring plates and cutting test wells

Melt 100 ml of saline agar base, cool to 50 C, and add 11 ml of washed red cells. Mix thoroughly and dispense 7 ml in 15 × 100 mm sterile plastic or glass Petri dishes. Dry plates overnight at room temperature and store at 4 C. Just before use, cut test wells by applying vacuum to 3 mm O.D. stainless steel tube and plunging the tube into the agar. Using a template, space 9 test wells 3 cm apart and 2 cm from the edge of the plate. Make 2 additional wells 3 cm apart near the center of the plate.

35.742 Extraction of toxin from foods

a. Extraction

Homogenize 25 g of food (do not include fat) in 100 ml of HEPES buffered salt solution for one minute in a high speed blender. Centrifuge the homogenate for 20 minutes at 20,000 × g at 5 C. Filter supernatant extract through filter paper to remove fat (chill extracts for 1 or 2 hours in the refrigerator before filtering). Discard the solids. Rinse the pad of a sterile Seitz filter with 15 ml of saline. Discard the saline and filter sterilize the extract. Rinse the filter pad with 10 ml of additional saline to remove residual alpha toxin.

b. Concentration

Soak 3 feet of dialysis tubing in distilled water. Tie one end and fill with sterile saline solution. Check for leaks and rinse out twice with additional sterile saline. Transfer sterile extract to a dialysis sac and concentrate to less than 10 ml by dialysing against 15% to 30% polyethylene glycol in the refrigerator. Rinse the outside of the sac with tap water and collect the concentrated extract in a sterile tube.

35.75 Toxin Assay

a. Dilution

Adjust the volume of the concentrated extract to 10 ml with saline. Make two-fold dilutions of the extract from undiluted to 1:256 (9 dilutions containing 0.5 ml each) in 13 × 100 mm tubes. Change pipets after 3 dilutions to prevent excessive carryover. Mix 0.25 ml of ext, 0.25 ml of saline, and 0.1 ml of *C. perfringens* type A diagnostic serum diluted 1:10 in a separate tube.

b. Hemolysin plates

Fill one well of duplicate hemolysin plates with each dilution of extract, using a fine tipped Pasteur pipet. Fill one center well of the plate with the extract antiserum mixture and the other with saline solution. Incubate plates for 24 hours at 35 C in a plastic bag.

c. Lecithovitellin test

Add 0.5 ml of lecithovitellin solution to the remainder of the diluted extract in each tube including the extract antiserum mixture. Shake tubes and incubate for 24 hours at 35 C.

35.8 INTERPRETATION OF DATA

35.81 Quantitation of *C. perfringens* Populations Based on Confirmed Anaerobic Plate Counts

Cultures obtained from presumptive *C. perfringens* black colonies on selective differential TSC or EY-free TSC medium are confirmed as *C. perfringens* if they are nonmotile, reduce nitrate, ferment lactose, and liquefy gelatin within 44 hours, and produces acid from raffinose. Calculate the number of viable *C. perfringens* per gram of food sample by multiplying the presumptive plate count by the reciprocal of the dilution plated and then by the ratio of the colonies confirmed as *C. perfringens* to total colonies tested. In the case of surface plated dilutions, 0.1 should be multiplied times the dilution plated to obtain the proper dilution factor.

35.82 Indication of the Presence of *C. perfringens* Based on the Presence of Alpha Toxin in Food

35.821 Alpha toxin titer

a. Hemolysin plates

After incubation, refrigerate the plates for 2 hours. Measure hemolytic zones (width from edge of the well in mm). The last three dilutions before the endpoint should exhibit about a 1 mm reduction in width for each 2 fold dilution. A hemolytic zone 1 mm in width is considered the endpoint of the titration. Record this dilution and refer to Table 1 for estimating the population. Hemolysis due to alpha toxin will be neutralized by *C. perfringens* type A diagnostic serum.

Table 1. Correlation between the Population of *Clostridium perfringens* and the Amount of Alpha Toxin Produced in Food

Alpha toxin titer[a]		Estimated *C. perfringens* population per gram ($\times 10^6$)[b]
Hemolysin plate	Lecithovitellin test	
undiluted		1.2
1:2	undiluted	2.5
1:4	1:2	6.5
1:8	1:4	9.5
1:16	1:8	25
1 32	1:16	55
1:64	1:32	80
1:128	1:128	150
1:256	1:256	210

[a]Dilution which produces a 1 mm zone of hemolysis in the hemolysin plate, or minimum activity in the lecithovitellin test.

[b]Based on viable counts obtained with 6 strains in chicken broth.

b. Lecithovitellin test

Examine extract-lecithovitellin mixture for lecithinase activity. Maximum reaction is a white pellicle 4 to 5 mm thick over a clear liquid. Activity decreases with dilution to minimum activity which is evidenced by an opaque solution with no pellicle. This dilution is the endpoint for the lecithovitellin test. Lecithinase activity due to alpha toxin will be neutralized in the tube containing *C. perfringens* type A diagnostic serum.

35.822 Population estimate

To estimate the population of *C. perfringens* in the food on a per gram basis, compare the titer of alpha toxin present in the food with the data presented in Table 1 for correlation of population levels and alpha toxin titer.

The hemolysin plate titer is preferred for this because the test is more sensitive and less subject to interference by other components of the extract.

35.9 REFERENCES

1. Angelotti, R., H. E. Hall, M. J. Foter, and K. H. Lewis. 1962. Quantitation of *Clostridium perfringens* in foods. Appl. Microbiol. **10**:193–199
2. Fuzi, M. and Z. Csukas. 1969. New selective medium for the isolation of *Clostridium perfringens*. Acta Microbiol. Acad. Sci. Hung. **16**:273–278
3. Handford, P. M. and J. J. Cavett. 1973. A medium for the detection and enumeration of *Clostridium perfringens (welchii)* in foods. J. Sci. Food Agric. **24**:487
4. Harmon, S. M. and D. A. Kautter. 1970. Method for estimating the presence of *Clostridium perfringens* in food. Appl. Microbiol. **20**:913–918
5. Harmon, S. M., D. A. Kautter, and J. T. Peeler. 1971. Improved medium for enumeration of *Clostridium perfringens*. Appl. Microbiol. **22**:688–692
6. Harmon, S. M. and D. A. Kautter, 1974. Collaborative study of the alpha toxin method for estimating population levels of *Clostridium perfringens* in food. J.A.O.A.C. **57**:91–94
7. Hauschild, A. H. W. and R. Hilsheimer. 1974. Evaluation and modifications of media for enumeration of *Clostridium perfringens*. Appl. Microbiol. **27**:78–82
8. Hauschild, A. H. W. and R. Hilsheimer. 1974. Enumeration of foodborne *Clostridium perfringens* in egg yolk-free tryptose-sulfite cycloserine (TSC) agar. Appl. Microbiol. **27**:521–526
9. Marshall, R. S., J. F. Steenbergen, and L. S. McClung. 1965. Rapid technique for the enumeration of *Clostridium perfringens*. Appl. Microbiol. **13**:559–563
10. Nakamura, S., T. Shimamura, M. Hayase, and S. Nishida. 1973. Numerical taxonomy of saccharolytic clostridia, particularly *Clostridium perfringens*-like strains: Description of *Clostridium absonum* sp.n. and *Clostridium paraperfringens*. Int. J. Syst. Bacteriol. **23**: 419–429
11. Shahidi, S. A. and A. R. Ferguson. 1971. New quantitative, qualitative, and confirmatory media for rapid analysis of food for *Clostridium perfringens*. Appl. Microbiol. **21**:500–506
12. Thatcher, F. S. and D. S. Clark. 1968. Microorganisms in Foods. Their Significance and Methods of Enumeration. p. 128. University of Toronto Press, Toronto, Canada

MICROORGANISMS AND FOOD SAFETY: FOOD-BORNE ILLNESS

CHAPTER 36

FOODBORNE ILLNESS—SUGGESTED APPROACHES FOR THE ANALYSIS OF FOODS AND SPECIMENS OBTAINED IN OUTBREAKS

Arvey C. Sanders, Frank L. Bryan, and Joseph C. Olson, Jr.

Toward the end of the last century food poisoning was thought to be caused by ptomaines that had been formed in foods as a consequence of protein decomposition. This hypothesis was invalidated later when it was determined that many foodborne diseases were caused by microorganisms.

About 300 outbreaks and about 20,000 cases of foodborne diseases are reported each year in the United States.[3] The real incidence, however, may be 10 to 1000 times higher than these numbers. This is because many of the afflicted persons do not seek treatment or complain to health agencies. Also, state and local health agencies always do not report the occurrence of outbreaks. A person seeking treatment by a physician or reporting to an investigator can appear to have sporadic diarrheal illness. Therefore, a detailed investigation often is not made. Although mortality is usually low (except for cases of botulism and a few other rare diseases), the enteric illnesses do account for considerable morbidity.

The greatest foodborne disease hazard now appears to be pathogenic bacteria. Although the available statistics are deficient in showing the real incidence of foodborne illness, they are useful in determining trends, and do reflect certain points of interest regarding etiologic agents and their epidemiology. For example, the 1973 statistics[1] indicate that the great majority of foodborne disease incidents are attributable to errors in food service in the home, in public and private institutions, or in commercial eating establishments. From these data, it becomes apparent that many persons are either careless or uninformed about the microbiological hazards of food preparation and handling. Increased application of good sanitary and personal hygiene practices could reduce markedly the incidence of foodborne illness.

Foodborne disease is an inclusive term for many syndromes. Acute gastroenteritis with sudden onset of vomiting or diarrhea, or both, with accompanying abdominal pain is typical. Some episodes include fever, prostration, shock, or neurological effects. The incubation period (time between eating and onset of first symptom), the type of symptoms, and the duration of symptoms is variable, and depends upon the etiologic agent. Instability

IS/A Committee Liaison: Joseph C. Olson, Jr.

and irritability of the gastrointestinal tract during convalescence are frequent sequelae of such illnesses.

The principal causes of foodborne diseases fall into three categories:

1. metallic or poisonous chemicals;
2. toxicants occurring naturally in plants or animals; and
3. microbiological agents (bacteria, viruses, protozoa, helminths, molds) or their toxins

Food poisoning outbreaks caused by chemicals are relatively uncommon. Metallic poisonings can occur when high acid foods (such as fruit juices and carbonated beverages) are stored in, or allowed to flow through, metal-containing (copper or cadmium), or metal coated (zinc, antimony, or lead) vessels or pipe lines. Other chemical poisonings occur because workers inadvertently mistake poisons (such as sodium flouride or other pesticides) for food stuffs or condiments, add excessive amounts of flavor intensifiers (monosodium glutamate), curing or color intensifying agents (nitrites), or preservatives (benzoic acid) to foods, or inadvertently contaminate foods when applying pesticides. Such contamination may occur at any point in the food chain.

Illnesses from inherently poisonous foods occur because uninformed persons mistake toxic fish, shellfish, crustacea, mushrooms, or uncultivated plants for edible varieties. Fortunately, such foods are excluded from the commercial food chain.

Bacterial contamination of food is the most frequent cause of foodborne disease. Many bacterial pathogens that are conveyed by foods invade the intestinal mucosa (salmonellae, some shigellae, some enteropathogenic strains of *Escherichia coli*) causing true infections. Others release enterotoxins during growth or lysis (*Vibrio cholerae*, some enteropathogenic *E. coli*), or during sporulation, *Clostridium perfringens* in the gut. Other bacteria such as *Clostridium botulinum* and *Staphylococcus aureus*, produce toxins as they proliferate within a food, and, when the food is eaten, cause an intoxication.

Staphylococcal intoxication, salmonellosis, and *Clostridium perfringens* gastroenteritis are the foodborne diseases reported most frequently in the United States. However, other agents have been found recently to cause foodborne disease (such as *B. cereus*, enteropathogenic *E. coli, Vibrio parahaemolyticus*, and *Yersinia enterocolitica*), and must be considered during outbreak investigations. Other agents that have caused foodborne outbreaks are listed and discussed by Bryan.[1,2] Hence, the testing of foods incriminated in outbreaks should not be limited to those methods that are selective for the commonly recognized etiologic agents.

Occasionally outbreaks of viral hepatitis have been traced to foods, but there is little epidemiologic or laboratory evidence of foodborne transmission of other viral agents. The lack of this evidence can be attributed to inept investigations, inadequate methods for detecting pathogens in foods, the absence of viruses in foods, or the relative unimportance of foods as vehicles of viral diseases. The proof that viruses are important agents of foodborne disease awaits more frequent and more exhaustive epidemiological investigations of outbreaks, and better methods of recovering viruses from foods, and routine use of these improved methods in laboratories.

The incidence of foodborne parasitic (protozoan and helminthic) diseases is declining in technically developed countries. Anisakiasis (a disease caused by round worms known as anisakine nematodes) and angiostrongyliasis have recently been recognized as foodborne problems in these areas, however. Changes in food preferences, health fadism, and increased international food trade may heighten the risk of exposure to parasites. Using sewage to fertilize crops, and using animal waste as animal feed, may increase the potential for some foods to be contaminated by parasitic agents as well as by other pathogens.

Some molds previously considered to be of economic significance have been found to produce mycotoxins in foods. This has stimulated an awareness of the hazard of molds to health because some of these toxins have caused illness in individuals who ate large quantities of moldy foods.

Many foodborne disease outbreaks are of unknown etiology. The causative agents may not be identified because no samples of foods as specimens, or inadequate samples of foods or specimens are obtained by investigators, or because inadequate or inappropriate laboratory methods are used. Sometimes the incriminated food is heavily contaminated by a broad spectrum of bacteria other than the usual pathogens. Frequently it is difficult to determine whether, for example, *Proteus* sp or enterococci found in epidemiologically incriminated foods caused the illness or overgrew the pathogens after the foods were served, but before they were sampled. Good microbiological technique and knowledgeable interpretation of results are needed to reduce the number of outbreaks attributed to unknown etiologies.

Verification of a foodborne illness requires the coordinated effort of the attending physician, an epidemiologist, and a microbiologist or chemist. Depending upon the circumstances, materials collected for examination should include such items as (a) leftover food from the incriminated or suspected meal; (b) vomitus and stool or rectal swab from patients (blood serum and feces of patients suspected of having botulism); (c) blood, spleen, liver, and intestinal contents of fatal cases*; (d) specimens such as stool, blood, or nasal, throat, and rectal swabs, or swabs of infected lesions from persons who handled the food; (e) swabs of equipment used to process epidemiologically incriminated foods; (f) portions, scraps, or swabs of raw foods which may have introduced pathogens into the kitchen or plant environments; and (g), when applicable, rectal or cloacal swabs of animals, swabs of animal droppings, or samples or swabs of environmental contacts of animals that may have introduced pathogens into processing plants. All materials should be delivered expeditiously to the laboratory under proper conditions.

The type of tests and the order in which they are to be conducted should be determined by the clinical signs and symptoms and the incubation period if known). (See Table 1. for a guide for making such determinations.) Six categories of foodborne diseases are listed in Table 1:

---

*Whenever symptomatology suggests a specific illness collect the specimens listed in Table 1.

Table 1: Guide for Laboratory Tests Indicated by Certain Symptoms and Incubation Periods

Incubation Periods	Predominant Symptoms	Specimens to Analyze	Organism or Toxin
Upper gastrointestinal tract symptoms (nausea, vomiting) occur first or predominate			
Less than 1 hour	Nausea, vomiting, unusual taste, burning of mouth	Vomitus, urine, blood, stool	Metallic chemicals[a]
	Nausea, vomiting, retching, diarrhea, abdominal pain	Vomitus	Mushrooms[b]
1 to 2 hours	Nausea, vomiting, cyanosis, headache, dizziness, dyspnea, trembling, weakness, loss of consciousness	Blood	Nitrites[c]
1 to 6 hours, mean 2 to 4 hours	Nausea, vomiting, retching, diarrhea, abdominal pain, prostration	Vomitus, stool	*Staphylococcus aureus* and its enterotoxins
8 to 16 hours (2-4 hours rarely)	Vomiting, abdominal cramps, diarrhea, nausea	Vomitus, stool	*Bacillus cereus*
6 to 24 hours	Nausea, vomiting, diarrhea, thirst, dilation of pupils, collapse, coma	Urine, blood	Amanita mushrooms[d]
Sore throat and respiratory symptoms occur			
12 to 72 hours	Sore throat, fever, nausea, vomiting, rhinorrhea, sometimes a rash	Throat swab	*Streptococcus pyogenes*
2 to 5 days	Inflamed throat and nose, spreading grayish exudate, fever, chills, sore throat, malaise, difficulty in swallowing, edema of cervical lymph node	Throat swabs, blood	*Corynebacterium diphtheriae*
Lower gastrointestinal tract symptoms (abdominal cramps, diarrhea) occur first or predominate			
8 to 22 hours, mean 10 to 12 hours	Abdominal cramps, diarrhea, putrefactive diarrhea associated with *C. perfringens*	Stool	*Clostridium perfringens, Bacillus cereus, Streptococcus faecalis, S. faecium*
12 to 74 hours, mean 18 to 36 hours	Abdominal cramps, diarrhea, vomiting, fever, chills, malaise	Stool	*Salmonella,* Arizona, *Shigella,* Enteropathogenic *Escherichia coli,* Oth-

Table 1: Guide for Laboratory Tests Indicated by Certain Symptoms and Incubation Periods *(Continued)*

Incubation Periods	Predominant Symptoms	Specimens to Analyzer	Organism or Toxin
			er *Enterobacteriacae, Vibrio parahaemolyticus, Yersinia enterocolitica, Aeromonas (?), Pseudomonas aeruginosa (?)*
3 to 5 days	Diarrhea, fever, vomiting, abdominal pain, respiratory symptoms	Stool	Enteric viruses
1 to 6 weeks	Mucoid diarrhea (fatty stools), abdominal pain, weight loss	Stool	*Giardia lamblia*
1 to several weeks, mean 3 to 4 weeks	Abdominal pain, diarrhea, constipation, headache, drowsiness, ulcers, variable—often asymptomatic	Stool	*Entamoeba histolytica*
3 to 6 months	Nervousness, insomnia, hunger pains, anorexia, weight loss, abdominal pain, sometimes gastroenteritis	Stool	*Taenia saginata, T. solium*
Neurological symptoms (visual disturbances, vertigo, tingling, paralysis) occur			
Less than 1 hour	Tingling and numbness, giddiness, staggering, drowsiness, tightness of throat, incoherent speech, respiratory paralysis		Shellfish toxin[e]
	Gastroenteritis, nervousness, blurred vision, chest pain, cyanosis, twitching, convulsions	Blood, urine, fat biopsy	Organic phosphate[f]
	Excessive salivation, perspiration, gastroenteritis, irregular pulse, pupils constricted, asthmatic breathing	Urine	Muscaria-type mushrooms[g]
	Tingling and numbness, dizziness, pallor, gastroenteritis, hemorrhage, and desquamation of skin, eyes fixed, loss of reflexes, twitching, paralysis		Tetraodon toxin[h]

Table 1: Guide for Laboratory Tests Indicated by Certain Symptoms and Incubation Periods *(Continued)*

Incubation Periods	Predominant Symptoms	Specimens to Analyze	Organisms or Toxin
1 to 6 hours	Tingling and numbness, gastroenteritis, dizziness, dry mouth, muscular aches, dilated eyes, blurred vision, paralysis		Ciguatera toxin[i]
	Nausea, vomiting, tingling, dizziness, weakness, anorexia, weight loss, confusion	Blood, urine, stool, gastric washings	Chlorinated hydrocarbons[j]
12 to 72 hours	Vertigo; double or blurred vision; loss of reflex to light; difficulty in swallowing, speaking, breathing; dry mouth; weakness, respiratory paralysis	Blood, stool	*Clostridium botulinum* and its neurotoxins
More than 72 hours	Numbness, weakness of legs, spastic paralysis, impairment of vision, blindness, coma	Urine, blood, stool, hair	Organic mercury[k]
	Gastroenteritis, leg pain, ungainly high stepping gait, foot and wrist drop		Triorthocresyl phosphate[l]
	Allergic symptoms (facial flushing, itching) occur		
Less than 1 hour	Headache, dizziness, nausea, vomiting, peppery taste, burning of throat, facial swelling and flushing, stomach pain, itching of skin	Vomitus	Histamine[m]
	Numbness around mouth, tingling sensation, flushing, dizziness, headache, nausea		Monosodium glutamate[n]
	Flushing, sensation of warmth, itching, abdominal pain, puffing of face and knees	Blood	Nicotinic acid[o]
	Generalized infection symptoms (fever, chills, malaise, prostration, aches, swollen lymph nodes) occur		
4 to 28 days, mean 9 days	Gastroenteritis, fever, edema about eyes, perspiration, muscular pain, chills, prostration, labored breathing	Muscle biopsy	*Trichinella spiralis*
7 to 28 days, mean 14 days	Malaise, headache, fever, cough, nausea, vomiting, constipation, abdominal pain, chills, rose spots, bloody stools	Stool, blood	*Salmonella typhi*

Table 1: Guide for Laboratory Tests Indicated by Certain Symptoms and Incubation Periods *(Continued)*

Incubation Periods	Predominant Symptoms	Specimens to Analyze	Organism or Toxin
10 to 13 days	Fever, headache, myalgia, rash	Lymph node biopsy, blood	*Toxoplasma gondii*
10 to 50 days, mean 25 to 30 days	Fever, malaise, lassitude, anorexia, nausea, abdominal pain, jaundice	Urine, blood	Etiological agent not yet isolated—probably viral
Varying periods (depends on specific illness)	Fever, chills, head- or joint-ache, prostration, malaise, swollen lymph nodes, and other specific symptoms of disease in question[p]	Blood, stool, urine, sputum, lymph node, gastric washings (one or more depending on organism)	*Bacillus anthracis, Brucella melitensis, Br. abortus, Br. suis, Coxiella burnetii, Francisella tularensis, Listeria monocytogenes, Mycobacterium tuberculosis, Mycobacterium spp., Pasteurella multocida, Streptobacillus moniliformis, Vibrio fetus*

[a]Consider chemical tests for such substances as zinc, copper, lead, cadmium, arsenic, antimony.[7]

[b]Many mushrooms irritate the gastrointestinal tract and cause gastroenteritis.[8]

[c]Consider nitrites, test for discoloration of blood.[7]

[d]*Amanita* mushroom poisoning should be considered. Identify mushroom species eaten; test urine and blood for evidence of renal damage (SGOT, SGPT enzyme tests).[8]

[e]Shellfish poisoning should be considered.[5, 7]

[f]Organic phosphate insecticide poisoning should be considered.[4, 7]

[g]*Muscaria* species of mushrooms should be considered.[8]

[h]Tetraodon (puffer) fish poisoning should be considered.[6]

[i]Ciguatera fish poisoning should be considered.[6]

[j]Chlorinated hydrocarbons insecticides should be considered.[7]

[k]Organic mercury poisoning should be considered.[7]

[l]Triorthocresyl phosphate should be considered.[7]

[m]Scombroid poisoning should be considered. Examine foods for proteus species or other organisms capable of decarboxylating histidine into histamine and for histamine.[6, 7]

[n]Chinese restaurant syndrome caused by monosodium glutamate, a flavor intensifier, should be considered.[7]

[o]Nicotinic acid should be considered.[7]

[p]For more information on the rarer diseases.[2]

1. upper gastrointestinal tract symptoms (nausea, vomiting) occur first or predominate;
2. sore throat and respiratory tract symptoms occur;
3. lower gastrointestinal tract symptoms (abdominal cramps, diarrhea) occur first or predominate;
4. neurological symptoms (visual disturbances, vertigo, tingling, paralysis) occur;
5. allergic symptoms (facial flushing, itching) occur; and
6. general infection symptoms (fever, chills, malaise, prostration, aches, swollen lymph nodes) occur.

A Gram or other appropriate stain of liquid foods, or of the initial 1 : 10 homogenate of solid foods, should be made as soon as feasible after samples are received in the laboratory. The Gram reaction and cellular morphology of predominant organisms provide information indicative of the kind of microorganisms to be sought by definitive bacteriological methods. (See Table II for a guide for the interpretation of staining results.)

Bacteriological studies and epidemiological investigations have established common sources, reservoirs, and vehicles of many foodborne pathogens. When a food that has a history of being a vehicle of a specific illness has been ingested, and when the symptoms of the ill are compatible with that illness (Table I), or when contaminants have the characteristics of specific agents (Table II), examine the suspect food for the agent causing that illness, or for the agents' toxic byproducts. See Table III for tests to run when examining foods that have been alleged, suspected, or epidemiologically incriminated as vehicles of foodborne illness. This Table includes the classes of foods that are covered in this manual.

Food products and food processing operations are diverse; this diversity affects the number and types of microorganisms that may be introduced into or survive and grow in a food product. These factors also influence the methods that can be used in isolating pathogens that may be present in the foods and in the interpretation of results. Pathogens are often easier to isolate from clinical materials than from foods. For example, pathogenic members of the *Enterobacteriaceae*, when present in the human gut, enjoy an enrichment often greater than that provided by foods. In foods, these same pathogens sometimes occur in small numbers because of process injury, death caused by storage, or overgrowth by spoilage organisms. Clinically oriented laboratories should become especially sensitive to the idea that clinical methods may be inadequate for the detection of pathogens in foods.

After laboratory determinations have been made, it is imperative that the findings be interpreted professionally. The difference between a food that is an immediate hazard to a consumer and food that is a potential hazard only is often a matter of the number of pathogens present. For example, 100 enterotoxigenic staphylococci per milliliter of cheese milk in a vat are of little consequence. If starter culture activity, however, is impaired for several hours during cheese making, the staphylococci may grow and produce enough enterotoxin to cause an outbreak of staphylococcal food poi-

Table 2: Microscopic Observations for Determining Morphology of Predominant Organism[a] in Suspect Foods

Gram Reaction	Cell Shape	Cell Groupings	Flagella Arrangement[b]	Spore[b]	Organism[c]
Positive	Rods	Singly, in pairs or short chains	Peritrichous	Ovoid; central, subterminal, terminal spores swelling the cell	*Clostridium botulinum*
Positive	Rods	Singly, in pairs, or short chains	Atrichous	Ovoid; subterminal spores not swelling cell	*Clostridium perfringens*
Positive	Rods	Singly, in pairs, but frequently form tangled chain	Peritrichous	Ellipsoidal; central or paracentral spores	*Bacillus cereus*
Positive	Cocci	Clusters	Atrichous	None	*Staphylococcus aureus*
Positive	Cocci	Long chains	Atrichous	None	*Streptococcus pyogenes*
Positive	Cocci	Short chains	Atrichous	None	*Streptococcus faecalis*, *S. faecium*
Negative	Rods	Singly and in pairs	Peritrichous (most types)	None	*Salmonella*, Arizona, *Escherichia coli*
Negative	Rods	Singly and in pairs	Atrichous	None	*Salmonella pullorum*, *S. gallinarum*, *Shigella*, *Escherichia coli* (few types atrichous)
Negative	Rods, cocoid forms	Singly, in pairs, or short chains	Peritrichous	None	*Yersinia enterocolitica*
Negative	Slightly curved rods	Singly and in pairs	Monotrichous (polar)	None	*Vibrio parahaemolyticus*
Negative	Short curved rods, pleomorphic	Singly and in spiral chains	Monotrichous (polar)	None	*Vibrio cholerae*

[a] As disclosed by Gram, or otherwise stained food sample, or 1 to 10 dilution food sample homogenate
[b] Features are not always seen in Gram stain.
[c] If organisms are not listed, review Bryan, 1969; 1975.[1, 2]

Table 3: Guide for Tests to Perform the Examination of Foods Alleged, Suspected or Epidemiologically Incriminated as Vehicles of Foodborne Illness

Food	Organism or Toxin
Soft drinks, fruit juices, and concentrates (in metallic containers or vending machines)	Test for chemicals such as copper, zinc, cadmium, lead, antimony, tin
Canned foods	*Clostridium botulinum* and its neurotoxins
Cereals, rice, foods containing corn starch	*Bacillus cereus,* mycotoxins
Cream-filled baked goods	*Staphylococcus aureus* and its enterotoxins *Salmonella*
Confectionery products	*Salmonella*
Egg and egg products	*Salmonella*
Molluscan shellfish	*Vibrio parahaemolyticus*, shellfish toxin (Saxotoxin) *Vibrio cholerae*, hepatitis-A virus (epidemiological implication only)
Raw fruits and vegetables	Parasites, *Shigella*
Mixed vegetable, meat, poultry, fish salads	*Staphylococcus aureus* and its enterotoxins *Salmonella,* beta-hemolytic streptococci, *Shigella,* enteropathogenic *Escherichia coli*
Meat and poultry, and mixed foods containing meat and poultry	*Salmonella, Clostridium perfringens, Staphylococcus aureus* and its enterotoxins
Ham	*Staphylococcus aureus* and its enterotoxins
Fermented meats	*Staphylococcus aureus* and its enterotoxins
Fish	*Vibrio parahaemolyticus,* histamine *(Proteus spp.),* fish poisons
Crustaceans	*Vibrio parahaemolyticus,* fish poisons
Cheese	*Staphylococcus aureus* and its enterotoxins, *Brucella* sp. enteropathogenic *Escherichia coli*
Dry milk	*Salmonella, Staphylococcus aureus* and its enterotoxins

soning when the cheese is eaten. The final cheese product may have a relatively low staphylococcal count due, for example, to die-off during the ripening period, and it thus could escape incrimination as a suspect food, but at the same time possess undetected enterotoxin sufficient to cause illness.

Awareness and concern of the foodborne disease problem by laboratory personnel, adequate specimens and samples, and the choice of appropriate methods are necessary to identify the specific agents responsible for foodborne illness.

REFERENCES

1. Bryan, F. L. 1969. Infections due to miscellaneous organisms. In H. Reiman, ed. Foodborne Infections and Intoxications. Academic Press, New York, N.Y. pp 224–276
2. Bryan, F. L. 1975. Diseases Transmitted by Foods. A Classification and Summary. Center for Disease Control, Public Health Service, HEW, Atlanta, Ga.
3. Foodborne and Waterborne Disease Outbreaks. 1973. Annual Summary. Center for Disease Control, Public Health Service, HEW, Atlanta, Ga.
4. Gleason, M. W., R. E. Gosselin, H. C. Hodge, and R. P. Smith. 1969. Clinical Toxicology of Commercial Products: Acute Poisoning. 3rd edition. Williams & Wilkins, Baltimore, Md.
5. Halstead, B. W. 1965. Poisonous and Venomous Marine Animals of the World. Vol. 1, Invertebrates. U.S. Govt. Printing Office, Washington, D.C.
6. Halstead, B. W. 1967. Poisonous and Venomous Marine Animals of the World. Vol. II, Vertebrates, U.S. Govt. Printing Office, Washington, D.C.
7. Official Methods of Analysis of the Association of Official Analytical Chemists. Horwitz, W. (ed.) 1975. Assoc. Official Anal. Chemists, Washington, D.C.
8. Lampe, K. F. and R. Fagerstrom. 1968. Plant Toxicity and Dermatitis. Williams & Wilkins, Baltimore, Md.

CHAPTER 37

FOODBORNE VIRUSES

D. O. Cliver

37.1 INTRODUCTION

Detection of a foodborne virus ordinarily requires that a discernible infection be produced in tissue culture. The need to use tissue cultures seems to distress many microbiologists who are unfamiliar with this technique, but this procedure is not difficult.

37.11 Viruses

As they occur in foods, viruses are particles ranging upward in size from 25 nm to perhaps 250 nm diameter. They are potentially infectious (only those infectious for man will be considered here), but they replicate only in suitable living cells, and so cannot multiply in food. Once the virus has entered the food, it can either persist or be inactivated on the way to the consumer.

The virus particle comprises either deoxyribonucleic acid or ribonucleic acid (depending upon the kind of virus) coated with protein. Some viruses have an outer envelope containing lipid, and particles of certain viruses include enzymes. Inactivation of a virus, defined as loss of its ability to produce infection, results from denaturation of one or more of the constituents of the particle.

37.12 Foods

The epidemiologic record shows that virtually any food may become contaminated with virus.[1] Not all foods are likely to transmit virus to the consumer. Canned goods, if thermally processed in the final, hermetically sealed container so as to be shelf stable at room temperature, are essentially above suspicion.

37.121 Modes of contamination

If a food is an animal product (meat, milk, milk product, egg, etc.) it may contain virus because the donor animal was infected. Only a few of the viruses which enter food in this way are of concern to human health.

IS/A Committee Liaison: George J. Herman

Shellfish present a special case. Bivalve mollusca, such as clams and oysters, can concentrate viruses of human origin from even mildly polluted waters. The mollusca are not infected, but they accumulate the virus by filtration as they feed.

Water polluted with human feces, if used for washing or irrigation, can carry viruses to other foods. Mechanical vectors, such as flies, roaches, and perhaps even rodents, also may transport viruses shed in human feces to foods. In the record to date, polluted water appears more often than vectors as the probable source of viruses in foods.

Man is the most common source of foodborne virus infections for man. Infected food handlers, some of whom worked with food while visibly ill, have introduced viruses into foods in many ways. Most of the viruses in question are those shed principally in feces.

37.122 Vehicle effects

The composition of a contaminated food, as well as the thermal events involved in its processing and distribution, determine the fate of the virus. Foodborne viruses are likely to be destroyed by high temperature processes and preserved by low temperature storage. Pasteurization of milk and milk products probably will destroy viruses (except perhaps the agent of infectious hepatitis) in any quantity which is likely to occur in food.[2] The effects of specific food constituents upon thermal inactivation of virus need further study. Shellfish appear to be extremely protective,[3] and some other foods mildly protective, of virus.

37.123 Risk assessment

A food is a high risk vehicle for virus if it is likely to become contaminated, and if the virus contaminant is unlikely to be inactivated before the food is consumed. Shellfish and ground beef are adjudged high risk vehicles, with the result that several methods have been described for detecting viruses in each.

Shellfish accumulate virus from their environment and appear to protect the virus during cooking; shellfish are also eaten raw. Ground beef is subject to contamination by human mishandling. Virus is likely to be inactivated by thorough cooking, but ground beef, too, may be eaten raw or nearly so. There are certainly other high risk vehicles. However, shellfish and ground beef are the only foods for which several virus detection methods are available.

37.13 Significance

Viruses which infect people as a result of transmission through foods do not always cause frank, clinical disease. When disease does occur, the drugs which are available usually treat the symptoms rather than the infection itself. Inapparent virus infections benefit the host only insofar as they induce immunity. Thus, one would rather prevent virus infections; forestalling virus transmission through foods is one means to this end.

37.131 Enteric viruses

Viruses produced in the intestines usually are transmitted by a fecal oral cycle. Though they are found first in the intestines, these viruses may be transmitted to, and affect, other sites in the body. The virus of infectious hepatitis is the best known of the enteric viruses transmitted through foods. As knowledge of the viruses causing acute nonbacterial gastroenteritis develops, their transmission through foods may be demonstrated. The enteroviruses (polio, coxsackie, and echo) are being detected in foods, and reoviruses and adenoviruses seem likely to be found in the future. All of these viruses are quite stable outside of the host, notably in many kinds of foods. They are a threat to the health of the consumer, and their presence in foods suggests direct or indirect fecal contamination.

37.132 Other cytopathogenic viruses

Not all foodborne viruses are of intestinal origin. For example, dairy animals bitten by infected ticks may shed the virus of tickborne encephalitis in their milk. The virus is fairly stable in unpasteurized milk and milk products,[4] and has caused encephalitis (sometimes fatal) in consumers. This agent has not been encountered in the United States. Most other nonenteric viruses are likely to be less stable outside the host's body.

37.133 Oncogenic viruses

Some kinds of tumors and other cancers are caused by viruses. Oncogenic or cancer viruses just are beginning to be identified in man, but are well known in domestic, food-source animals. In addition to those causing tumors in potentially edible muscle tissue, oncogenic viruses have been shown to occur in milk and in eggs. Some of these viruses may infect man, but they are not yet known to cause human cancer. The means of detecting such viruses in foods certainly are worth developing.

37.14 Detection

Detection by current methods requires that the virus produce a discernible infection. This is true because no method of virus detection based on properties of the particle other than its infectivity is as sensitive. However a suitable laboratory host must be available.

37.141 Infectivity tests

The host in which virus is to be detected must show some abnormality as a result of the infection. Sickness or death may be seen in an animal host, but indirect tests are sometimes required. Effects on tissue cultures may be directly visible, microscopically visible, or discernible only by indirect means.

37.142 Available test hosts

A susceptible tissue culture is to be preferred to alternative hosts for detecting a known virus. If the type of virus which may be present in a food sample is not known, a properly chosen tissue culture will be susceptible to more viruses than any other single alternative. No single type of tissue culture, no laboratory animal, and no other living substrate is susceptible to all of the viruses which may be transmitted to man through foods. No tissue culture or other practical laboratory host is presently available for the detection of foodborne infectious hepatitis virus.

The choice of a host system is dictated by availability and cost. There is no "best" tissue culture, cost factors aside. Two complementary test hosts may do the job better and more economically than a single, broad-spectrum one.

37.143 Sensitivity of test host

A test host is not simply susceptible or insusceptible to a virus. Susceptible test hosts may vary significantly in sensitivity to a given virus. Strains of a single virus may differ genetically in infectivity for a given tissue culture, just as tissue cultures or strains of a single cell line may differ genetically in susceptibility to the same virus.

The ideal case is one in which a single physical particle of virus can initiate a discernible infection. Another susceptible host culture may require a thousand times larger inoculum, either because of genetic factors in the cells or in the virus, or because the manner in which the test is performed inhibits expression of the virus effect. The choice among tissue culture methods for infectivity testing is still cause for disagreement among those who are studying this problem.

37.2 GENERAL APPROACH TO DETECTION

The premises of testing are simple: virus is either present in a food sample or not, and a test either detects the virus or it does not. Negative test results mean little, but a positive finding in a properly controlled test indicates that virus was present in the suspect food. The ideal method is one that minimizes false negative test results as economically as possible.

37.21 Two Basic Steps

In spite of innumerable possible variations, there are two basic steps common to every detection method. First, the food sample must be made fluid. Second, the fluid derivative of the sample must be inoculated into a test host. The way in which these steps are performed, and whether any other steps are added to the procedure, depend upon how important it is to avoid false negative test results.

37.22 Sensitivity Goals

A test method should be sensitive enough to detect any virus in a food which may infect a consumer. This is an obviously impossible goal, but not all of the reasons why it is impossible are equally obvious.

We do not know the quantity of virus required to cause a consumer infection, nor whether the same quantity of virus is enough to cause disease.

Some evidence suggests that a dose of poliovirus vaccine sufficient to infect a tissue culture can cause a human infection.[5] This may be more than a single physical particle of virus, as was discussed above. The experiments by which this evidence was obtained probably are not pertinent to most instances of virus transmission through foods. Additional research needs to be done. At present, there is no basis for stating a *tolerable* level of foodborne virus.

The sensitivity of a test is limited absolutely by the size of the sample. Virus shows a quantal distribution and is either present in the tested sample or not. The likelihood of a false negative test result is greater if a small sample only is tested. On the other hand, the size of the sample is subject to some constraints. The quantity of food available for testing is never infinite, and testing facilities are certain to be limited, as well.

37.23 Deciding Factors

The actual design of any method is based upon practical considerations. It is not practical to design a procedure to detect all of the viruses which may infect man. In most instances, one wants the capability to detect enteric viruses, but not all of these, either. If a type of tissue culture is selected which is highly sensitive to these viruses, then the first constraint on the sensitivity of the test is the size of the food sample. A kilogram of food must be tested to achieve a test sensitivity of one virus per kilogram.

One can homogenize, dilute, or suspend the sample in fluid and inoculate as many tissue cultures as required. If the cost or availability of tissue cultures is limited, one may prefer to concentrate the fluid suspension before inoculation. This can be done by ultrafiltration, ultracentrifugation, or some other method, depending upon available apparatus. Any significant degree of concentration must be preceded by removal of some food solids from the suspension. The volume of the suspension may then be reduced to less than that of the original food sample.

37.3 SPECIFIC STEPS

The general principles involved in detecting foodborne viruses have been discussed. Now we shall consider the procedure with a real food sample to determine if the food contains virus.

37.31 Sampling

Standard statistical principles apply to sampling foods for virus contamination. One assumes a quantal distribution which is summarized by the Poisson formula. Sample sizes in published methods have ranged roughly from 1 to 100 grams.

Samples which are expected to be more than two to four hours in transit to the laboratory should be transported at, or below, the temperature at which the food is stored. The same principle applies to the storage of samples at the laboratory: foods requiring refrigeration should be refrigerated or frozen, whereas food stable at room temperature can be stored that way for reasonable periods of time. Viruses which do not persist in a food at its normal storage temperature probably are a minimal hazard to the consumer. On the other hand, viruses are generally less likely than bacteria to be damaged by freezing and thawing; freezing temperatures (the lower, the better) are appropriate for short term storage and mandatory for long term storage of samples. Vapors of CO_2 should not contact the samples if "dry ice" is used for refrigeration. Polymers for sample containers should be chosen with this in mind.

37.32 Liquefaction

Conversion of the sample to a liquid suspension (for other than liquid foods) must be done in a way that is compatible with the whole of the detection process. The fluids used may be autogenous (as the shell liquor in an oyster sample[6]), or added. Added fluids may include water, buffers,[7] and tissue culture media,[8] enhanced at times with antibiotics,[8] blood serum, bentonite, or trichlorotrifluoroethane.[7]

Mechanical means of liquefaction include shaking,[8] stirring,[9] trituration with sand, high speed homogenization,[7] etc. The less rigorous treatments are appropriate if the substance of the food is easily dispersed, or if the food solids are to be removed in a later step. A meat sample thought to contain intracellular virus (because the donor animal was infected) may require trituration or homogenization; the viruses which have been detected in market samples of ground beef[8] were probably of human rather than animal origin, and ought not to have been intracellular.

37.33 Separation of Food from Virus

Although not one of the basic steps, the separation of food from virus must be included if the fluid suspension is to be concentrated before inoculation into tissue cultures. The bacteria present in the sample may also be removed or reduced at this stage, either purposely or incidentally. This is often necessary, for food contaminated by virus will usually contain bacteria, and the antibiotics in tissue culture may not be sufficient to prevent growth of the bacteria.

The centrifuge has long been a standard instrument for removing solids from fluid suspensions. A centrifuge which will clarify a homogenized food suspension must be capable of exerting several thousand times the force of gravity. If only a low speed centrifuge is available, one may either avoid homogenizing the food suspension, so that the solids are less finely dispersed, or modify the suspension so that the solids are selectively precipitated. The second approach has been to use acid[10] or trichlorotrifluoroethane with serum or bentonite.[7]

Filtration is another means of removing solids from suspensions. Many food suspensions are extremely difficult to filter, so that satisfactory filtration methods for clarifying food suspensions have been reported only recently. Filter materials have included cotton gauze (cheesecloth),[8] glass wool,[11] woven glass fiber,[11] and a blend of cellulose and glass fibers with carbon.[9] A polycationic sewage flocculant may be used in conjunction with the last of these. Conventional paper filters are not successful; whereas the filtrates from the materials described above are sometimes sufficiently clear to permit further filtration through a membrane filter of 0.45 μm[11] or even 0.20 μm[9] porosity to remove bacterial contaminants.

37.34 Concentration

The purpose of concentration is to reduce the volume of the suspension, with little or no loss of virus, before inoculation into tissue cultures. Addition of a concentration step does not make a detection method more sensitive. Such a step should be included in an otherwise sensitive detection procedure only if the saving in cost of tissue cultures and associated labor will outweigh the cost of the concentration.

The preparative ultracentrifuge has served reasonably well for concentration. It may cause very little loss of virus.[7] The liabilities of the machine are its high initial cost and limited volume capacity. It should be capable of forces from 100,000 to 200,000 x gravity, or more, to insure efficient sedimentation of the smaller viruses.

Ultrafiltration is a relatively new technique derived from older methods which used dialysis tubing. The mean pore size of ultrafilter membranes ranges downward from 10 nm. Pore size varies a great deal in some of these membranes, so a mean pore size should be selected which is much smaller than the smallest virus to be retained.[10] Positive pressures at the membrane may be 2 atmospheres or greater, and the suspension being concentrated must be kept in constant agitation to prevent the pores from plugging. Ultrafiltration has been used successfully with extracts of ground beef and other foods. The cost of the equipment is somewhat lower than that of an ultracentrifuge, but the cost of labor and materials may be slightly higher per sample tested.

37.35 Testing

When the fluid food suspension or extract is ready to be tested in tissue culture, the selection of cell types is based upon the principles discussed above. To detect viruses infectious to man, cells of primate (monkey or human) origin usually are used.

The cells form a monolayer on the inside surface of a tissue culture vessel. The vessel may be a culture tube, Petri dish, prescription bottle, or any of several flasks designed especially for tissue culture. These are produced of glass or polystyrene; the choice of vessel is governed substantially by the kind of virus effect one hopes to see. The virus may cause cytopathic effects (CPE), visible by brightfield or phase contrast microscopy, or gross areas of cell killing (plaques) under semisolid media. Cell suspensions or prepared cultures can be purchased commercially.

The absolute limit on the volume of sample suspension which can be tested in a single tissue culture is the capacity of the vessel itself, with rare exceptions.[10] Thus, a prescription bottle in which 25 to 30 cm^2 of cell monolayer will grow has a volume capacity of ~ 90 ml. To observe for plaque formation, one may inoculate 0.1 to 0.5 ml in such a vessel. A CPE test in the same vessel may permit inoculation of 5 ml or more of food extract.[9] Too large a volume of inoculum may decrease, delay, or prevent the expression of virus effects in a culture. Where the limit lies (between 5 and 90 ml in the example given) is not yet known.

The medium chosen to maintain the cells after the virus has been inoculated may exert some important selective effects. Media for the plaque test may impose limits on the susceptibility or sensitivity of a tissue culture to a given virus.[12] A positive test usually is obtained within 2 to 7 days, but a test is judged negative only after incubation for 10 to 14 days. Both positive and negative results may require verification by further passage in tissue culture. Elapsed times are a good deal longer in viral testing than in bacterial testing. The tissue cultures should be observed on the day after inoculation to determine whether the food extract has had a toxic effect upon the cells. Later observations can be made at two day intervals to save personnel time.

37.36 Identification

Given a positive test result, preliminary characterization of a virus can be based upon particle size, ether and acid stability, type of CPE, etc. Final identification is done serologically; determination of the exact antigenic type may not be necessary.

Antisera for virus identification can be obtained from commercial and, in some instances, government sources. There are laboratories at various levels of government and in the private sector which will identify viruses serologically. Until simpler serologic methods have been devised, a newcomer to food virology may better delegate the final identification task to specialists.

37.4 CURRENT METHODS

The principles and techniques of detecting foodborne viruses have been presented without describing any complete method for detecting any single virus in any single food. Complete methods for detecting enteroviruses in ground beef and in shellfish abound.

Five procedures for ground beef seem especially significant. These include both centrifuge[7, 10, 11] and filter[8, 9, 11] clarification, ultracentrifuge[8], and ultrafilter[9, 10, 11] concentration procedures. None of these has been evaluated with decomposed ground beef. Alkaline buffers are often used to separate viruses from ground beef. Jay's studies[13] suggest that these fluids may be very difficult to recover if the meat is spoiled.

Four of the many published methods for detecting viruses in shellfish are cited.[6, 7, 9, 14] Extracts of shellfish are often toxic to tissue cultures; these papers describe different ways of dealing with the problem. The methods have been evaluated for the detection of enteroviruses, and one of them also detects reoviruses in naturally contaminated samples.[6]

There are no standard methods for the detection of viruses in foods. A number of investigators, however, have modified procedures used in clinical and veterinary virology to recover viruses from foods. The reported yield of viruses varies but appears acceptable, especially if human enteroviruses are sought. Recently a collaborative study was completed on the glass wool filtration method for the recovery of viruses from ground beef. This method was accepted as official first action at the 88th annual meeting of the AOAC. It is hoped that this first attempt at standardization will be followed by the submission of other methods, or that changes in the proposed method will be suggested, so that a number of standard methods will be made available. Only then will sufficient market samples be analyzed so that an evaluation may be made of the role that viruses play in foodborne disease.

Each of the cited methods has its advocates, including investigators other than those who devised them. Comparative evaluations of some of them are under way in several laboratories. A combination of selected steps from different methods may result in a procedure best suited to the reader's needs.

37.5 SUMMARY

Most foodborne viruses, with the exception of the agent of infectious hepatitis, probably can be detected using methods which already have been published. The reader is advised to use the principles presented and adapt existing methods to his individual needs.

37.6 REFERENCES

1. Cliver, D. O. 1971. Transmission of viruses through foods. Critical Rev. in Environmental Control **1**:551–579
2. Sullivan, R., J. T. Tierney, E. P. Larkin, R. B. Read, Jr., and J. T. Peeler. 1971. Thermal resistance of certain oncogenic viruses suspended in milk and milk products. Appl. Microbiol. **22**:315–320
3. DiGirolamo, R., J. Liston, and J. R. Matches. 1970. Survival of virus in chilled, frozen, and processed oysters. Appl. Microbiol. **20**:58–63
4. Grešíková-Kohútová, M. 1959. The persistence of the virus of the tick-borne encephalitis in milk and milk products. (in Slovak, English abstract) Čs. Epidemiol., Mikrobiologie, Immunologie **8**:26–30
5. Plotkin, S. A. and M. Katz. 1967. Minimal infective doses of virus for man by the oral route. In Transmission of Viruses by the Water Route, pp. 151–166. G. Berg, ed. Interscience, New York, N.Y.
6. Metcalf, T. G. and W. C. Stiles. 1968. Enteroviruses within an estuarine environment. Am. J. Epidemiol. **88**:379–391
7. Herrmann, J. E. and D. O. Cliver. 1968. Methods for detecting food-borne enteroviruses. Appl. Microbiol. **16**:1564–1569
8. Sullivan, R., A. C. Fassolitis, and R. B. Read, Jr. 1970. Method for isolating viruses from ground beef. J. Food Sci. **35**:624–626
9. Kostenbader, K. D., Jr. and D. O. Cliver. 1973. Filtration methods for recovering enteroviruses from foods. Appl. Microbiol. **26**:149–154
10. Konowalchuk, J. and J. I. Speirs. 1973. An efficient ultrafiltration method for enterovirus recovery from ground beef. Canad. J. Microbiol. **19**:1054–1056
11. Tierney, J. T., R. Sullivan, E. P. Larkin, and J. T. Peeler. 1973. Comparison of methods for the recovery of virus inoculated into ground beef. Appl. Microbiol. **26**:497–501
12. Cliver, D. O. and R. M. Herrmann. 1969. Economical tissue culture technics. Health Lab. Sci. **6**:5–17
13. Jay, J. M. 1964. Release of aqueous extracts by beef homogenates, and factors affecting release volume. Food Technol. **18**:1633–1636
14. Konowalchuk, J. and Speirs, J. I. 1972. Enterovirus recovery from laboratory-contaminated samples of shellfish. Can. J. Microbiol. **18**:1023–1029

CHAPTER 38

FOODBORNE PARASITES

George R. Healy, George J. Jackson, J. Ralph Lichtenfels, Glen L. Hoffman, and Thomas C. Cheng

38.1 INTRODUCTION

38.11 Present Status and Importance of Foodborne Parasites

Man may acquire all manner of parasitic animals in the process of ingesting various types of foods.[1] These infectious organisms include the protozoa, the trematodes or flukes, the cestodes or tapeworms, and the nematodes or roundworms. On very rare occasions, man has acquired the acaonthacephala or thorny headed[2] worms from food. In addition, there are numerous examples of myiasis, or infestation of fly larvae, as well as other arthropods infecting the consumer by way of the food chain.[3] A detailed list of the parasites is included below.

Although meat from mammals is thought of as a prime food for acquiring parasites, some recent reports have catalogued the enormous variety of helminths that infect man after ingestion of fish and the aquatic invertebrates from both marine and fresh water environments.[4,5,6,7] In a comprehensive review of human helminth zoonoses, Sprent[8] listed a vast source of human infections, with a considerable number being of food origin.

38.12 Types of Foods from Which Parasites Can be Acquired

Two contradictory trends are, perhaps, responsible for parasitism from foods becoming a more important matter. Food shortages and the increasing prices of many food items, expecially animal products containing protein, have forced people to eat unusual foods of both plant and animal origin. These may contain parasites, and they are a health hazard when the methods for their proper preparation are unfamiliar. On the other hand, affluence, acquired tastes for exotic dishes, and rapid transportation have placed foodstuffs from one area of the globe on the menus of distant restaurants.

One may categorize parasites on the basis of how they become associated with foods and how they are transmitted. Many parasites are encysted, encapsulated, or free in or upon the food, not as adventitious contaminants, but as a part of their life cycle. Examples of this type include the cyst stages of various helminths and protozoa in domesticated beef, pork, veal, chicken,

IS/A Committee Liaison: George J. Herman

as well as in game meats such as bear, deer, and pheasant. Encysted, encapsulated, or free parasites are also found in fish (marine, brackish, and fresh water), prawns, crabs, crayfish, oysters, and clams, as well as in various other molluscs. Terrestrial invertebrates that may transmit parasites to humans by the food chain include slugs and adult beetles, as well as various larval stages of arthropods. These invertebrates, usually the contaminants of foods such as cereal or lettuce, are not ingested intentionally.

The second type of parasite involvement in the food chain is by the contamination of vegetables and such fluids as milk and water. The latter is ingested universally, and is easily contaminated with parasites from feces and other wastes. In endemic areas, vegetables such as parsley and water caltrops serve as strata upon which infective stages of the parasites, especially of the flukes, encyst as a part of their life cycle.

36.13 Parasitic Organisms Involved in Foodborne Diseases

Most of the parasites acquired from foods, the methods of infection, and the status of pathogenesis are listed below:

Endoparasites Transmissible through Foods

Parasite	Potential Pathogenicity	Method of Infection
Protozoa		
Entamoeba histolytica	+	Food and drink contaminated with cysts
Entamoeba coli	−	Food and drink contaminated with cysts
Endolimax nana	−	Food and drink contaminated with cysts
Iodamoeba butschlii	−	Food and drink contaminated with cysts
Dientamoeba fragilis	+	Food and drink contaminated with trophozoites
Retortamonas intestinalis	−	Food and drink contaminated with cysts
Retortamonas sinensis	+	Food and drink contaminated with cysts
Chilomastix mesnili	−	Food and drink contaminated with cysts
Enteromonas hominis	−	Food and drink contaminated with cysts
Enteromonas hervei	−	Probably via food and drink contaminated with cysts
Trichomonas hominis	−	Food and drink contaminated with trophozoites
Trichomonas tenax	−	Food and drink contaminated with trophozoites
Giardia lamblia	+	Food and drink contaminated with cysts
Isospora hominis	+	Food and drink contaminated with oocysts
Isospora belli	+	Food and drink contaminated with oocysts
Isospora natalensis	+	Food and drink contaminated with oocysts
Balantidium coli	+	Food and drink contaminated with cysts
Sarcocystis lindemanni	+	Red meat enclosing cysts
Toxoplasma gondii	+	Meats and milk contaminated with cysts

Endoparasites Transmissible through Foods (*Continued*)

Parasite	Potential Pathogenicity	Method of Infection
Nematoda		
Trichinella spiralis	+	Meats enclosing larvae
Trichuris trichiura	+	Food and drink contaminated with eggs
Capillaria hepatica	+	Food and drink contaminated with eggs
Capillaria philippinensis	+	Ingestion of fish harboring larvae
Dioctophyma renale	+	Ingestion of fish harboring larvae
Rhabditis sp.	−	Contaminated drink; Molluscs
Syngamus laryngeus	−	Food and drink contaminated with larvae
Trichostrongylus sp.	+	Vegetables contaminated with larvae
Haemonchus contortus	+	Vegetables contaminated with larvae
Enterobius vermicularis	+	Food and drink contaminated with eggs
Syphacia obvelata	−	Food and drink contaminated with eggs
Ascaris lumbricoides	+	Food and drink contaminated with eggs
Toxocara cati	+	Food and drink contaminated with eggs
Toxocara canis	+	Food and drink contaminated with eggs
Gnathostoma spinigerum	+	Fish and frog legs containing larvae
Echinocephalus sp.	+	Oysters containing larvae
Phocanema sp.	+	Fish containing larvae
Anisakis sp.	+	Fish containing larvae
Porrocaecum sp.	+	Fish containing larvae
Contracaecum sp.	+	Fish containing larvae
Angiostrongylus cantonensis	+	Vegetables contaminated with larvae or small molluscs harboring larvae; prawns and other paratenic hosts
Angiostrongylus costaricensis	+	Vegetables contaminated with slugs harboring larvae
Trematoda		
Watsonius watsoni	+	Vegetables contaminated with metacercariae
Gastrodiscoides hominis	+	Vegetables contaminated with metacercariae
Fasciola hepatica	+	Vegetables contaminated with metacercariae
Fasciola gigantica	+	Vegetables contaminated with metacercariae
Fasciolopsis buski	+	Vegetables contaminated with metacercariae
Echinostoma ilocanum	+	Molluscs harboring metacercariae
Himasthla muehlensi	+	Clams harboring metacercariae
Echinochasmus perfoliatus	−	Fish harboring metacercariae
Opisthorchis felineus	+	Fish harboring metacercariae
Clonorchis sinensis	+	Fish harboring metacercariae
Heterophyes heterophyes	+	Fish harboring metacercariae
Metagonimus yokagawai	+	Fish harboring metacercariae

Endoparasites Transmissible through Foods (*Continued*)

Parasite	Potential Pathogenicity	Method of Infection
Metagonimus minutus	+	Fish harboring metacercariae
Centrocestus armatus	+	Fish harboring metacercariae
Centrocestus formosanus	+	Fish and frog legs harboring metacercariae
Haplorchis pumilio	+	Fish and frog legs harboring metacercariae
Haplorchis microrchia	+	Fish harboring metacercariae
Haplorchis yokogawai	+	Fish and shrimp harboring metacercariae
Haplorchis taichui	+	Fish harboring metacercariae
Diorchitrema formosanum	+	Fish harboring metacercariae
Diorchitrema amplicaecale	+	Fish harboring metacercariae
Diochitrema pseudocirratum	+	Fish harboring metacercariae
Nanophyetus salmincola	+	Fish harboring metacercariae
Paragonimus westermani	+	Crabs and other crustacea harboring metacercariae
Isoparorchis hypselobagri	−	Fish harboring metacercariae
Cestoda		
Spirometra sp.	+	Water containing procercoid infected *Cyclops*
Diphyllobothrium latum	+	Fish harboring plerocercoids
Diplogonoporus grandis	+	Fish harboring plerocercoids
Digramma brauni	−	Fish harboring plerocercoids
Ligula intestinalis	−	Fish harboring plerocercoids
Braunia jasseyensis	−	Probably fish harboring plerocercoids
Taenia solium	+	Pork containing cysticerci
Taeniarhynchus saginatus (= *Taenia saginata*)	+	Beef containing cysticerci
Multiceps multiceps (larva)	+	Food and drink contaminated with eggs
Multiceps serialis (larva)	+	Food and drink contaminated with eggs
Echinococcus granulosus (larva)	+	Food and drink contaminated with eggs
Echinococcus multilocularis (larva)	+	Food and drink contaminated with eggs
Hymenolepis nana	+	Food contaminated with eggs
Hymenolepis diminuta	+	Food contaminated with cysticercoid infected beetles

38.2 METHODS OF EXAMINATION AND IDENTIFICATION

38.21 Gross Examination

The examination of foods that are suspected of containing, or of being contaminated with, parasites depends on the type of food one is examining, and the type of parasite one is looking for. In many instances, a human case is suspected after the food has been consumed, and only food that has not been eaten is available to be examined for parasites. Unfortunately, there are no readily available cultivation procedures for "multiplying" the number of organisms suspected in foodborne parasitoses.

Direct examination of the food by the unaided eye or with a hand lens or binocular dissecting microscope of low magnification is useful. The large *Taenia* tapeworm larvae (cysticerci) found in pork and beef are several millimeters in diameter. The white slender larvae of the broadfish tapeworm *Diphyllobothrium* in fish, and also in fish the larvae of such nematodes as the anisakines are visible, being several millimeters long.

38.22 Dissection

The suspected food should be carefully dissected with forceps and dissecting needles, if possible. Small samples should be selected, and the plant or animal food material should be teased apart with the aid of a magnifier or a dissection microscope. Small organisms (trematode cysts or small nematode larvae such as *Trichinella*) can be viewed by pressing the food between two pieces of glass and holding it against a lamp to permit light to penetrate the material. If no organisms are seen on gross examination or dissection and magnification, the techniques described below can be used.

38.23 Recovery, Concentration, and Digestion

Even when parasites are apparent in foods by the direct macroscopic or microscopic observation of natural and dissected surfaces of press preparations, or of fixed and stained smears, more must be done to determine quantity. Investigators want to know, for various reasons, how many parasites there are per unit of a food item, or how many of the parasites are alive. Often, of course, they suspect the presence of parasites when they are not directly apparent, or are so few in number that they may be overlooked. Consequently, parasites are concentrated by various means. In general, however, these methods are not as standardized as the methods in food bacteriology. Many are ad hoc procedures, adapted for foods from those methods used for obtaining parasites from clinical specimens or invertebrates from soil samples. There are, probably, as many methods in use as there are species of parasites multiplied by the number of foods in which they occur, and again by the number of laboratories doing the work.

38.231 Washing procedures.

Washing and scrubbing procedures for parasites that adhere to the surfaces of foods vary considerably in the efficiency of recovery. Some parasites are stickier than others; some foods retain parasites and other small bodies more avidly than others. An effective method must be sufficiently forceful to free a significant number of parasites from the food, but no so forceful that it damages the parasites to such an extent that their viability or even their identity becomes problematic. Neither should the washing method recover an obscuring number of other bodies. If it does, steps may be needed to isolate the parasites sufficiently for identification or counting.

For instance, to recover *Ascaris* eggs and hookworm larvae from vegetables, the produce must be scrubbed.[9] This is followed by 10 hours of sedimentation for the wash and rinse water. The sediment is strained, concentrated further by centrifugation, and examined with a microscope. Although this procedure works, fewer parasites are obtained than expected from seeded samples. Attempts to improve the method have not been successful.[10]

38.232 Tissue separation

To obtain parasites that are embedded in foods, it is often necessary to dissect, grind, or homogenize the samples before further processing. Homogenization is appropriate for small pathogens with tough outer membranes, such as bacteria, but damages the larger organisms. The larvae of *Trichinella spiralis* in muscle are killed when the meat in which they are encysted is homogenized, but not when the meat is ground. Anisakine larvae, which are larger than those of *T. spiralis*, may be destroyed by grinding of the fish host flesh. However, fish are dissected before their tissues are digested artifically, or placed in a Baermann apparatus (see below) for the recovery of anisakines.

38.233 Baermann elution

Modifications of Baermann's[11] techniques are probably the most common of elution procedures. Baermann originally devised the method for freeing *Ancylostoma* larvae from soil samples, but the procedure is effective for other nematodes and for a great variety of microinvertebrates found in plant or animal tissues or in the soil.

Samples are placed in a mesh or gauze container within a funnel. Lukewarm water is added to the level at which it wets the bottom of the sample. Under these conditions many invertebrates tend to migrate into the warmer water. Once in the liquid, some species settle to the bottom of the funnel, whereas others, active swimmers, concentrate at the surface. In practice, dead organisms from the sample also may be found in the water. Obviously, these have been passively washed out of the sample.

38.234 Flotation, sedimentation, and centrifugation

In clinical diagnosis the exclusive use of sedimentation or centrifugation methods for obtaining protozoan cysts and helminth eggs or larvae from feces has been inefficient. As a consequence, flotation methods have been developed. These may be adapted for use with foods or soil. Faust et al[12, 13] diluted feces with 10 parts of lukewarm water, strained the suspension (ca 10 ml) through a funnel with one layer of wet cheesecloth, centrifuged the strained materials (25,000 rpm for 45 to 60 seconds), poured off the supernatant, and resuspended the sediment. Centrifugation and resuspension were repeated 3 to 4 times until the supernatant became clear. After the last supernatant was discarded, 3 to 4 ml of a 33% zinc sulphate solution (specific gravity 1.180) was added to resuspend the packed sediment. After centrifugation, the surface film was removed with a bacteriological loop. Lugol's iodine stain was added with mixing to aid in the identification of helminth eggs, larvae, and protozoan cysts. This preparation was then examined microscopically for parasites.

Simplified zinc sulfate flotation techniques have been devised by various laboratories. Flotation solutions other than zinc sulfate include sodium chloride, (Willis[14]) calcium chloride and sugars. Ether sometimes is used for centrifugation or sedimentation (Ritchie[15]) to remove obscuring lipids. However, all of these chemicals may damage microorganisms and thus may be impractical for determining the viable number of parasites in food samples.

A dilution egg count technique was developed in order to obtain quantitative data for hookworm surveys.[15, 16] It was intended for use with fecal samples, but may be adapted for foods with similar consistencies. It is useful primarily for heavily contaminated samples. The (fecal) sample is weighed precisely to 4 g, mixed with decinormal sodium hydroxide to 60 ml, shaken with glass beads, allowed to stand overnight so that the sample disintegrates adequately, and remixed; 0.15 ml is then removed to a counting slide, covered with a 22 × 40 mm glass slip, and examined for helminth eggs. The total number of helminth eggs is multiplied by 100 to obtain the number per gram of sample.

38.235 Digestion

After dissection, grinding, or homogenation, artificial gastric juice (a 1% pepsin solution, adjusted to pH 3 with 1 N hydrochloric acid after 200 g of the food material has been added per liter of solution and stirred at 35 to 37 C for a minimum of 4 hours, is often used to recover parasites from foods. This method serves not only to free parasites from cyst walls and host tissues but also, hopefully, to select the potential parasites of warm-blooded animals. Presumably, microinvertebrates that are not infective by way of the mammalian or avian gastrointestinal tract would be killed under these conditions.

The larvae of *Trichinella spiralis* in muscles classically are recovered from infected meat by such digestion, followed by sedimentation. When edible snails are ground and digested, nematodes like *Angiostrongylus cantonensis* or *A. costaricensis* survive, but *Rhabditis* sp. are killed. The former infect mammals by the gastrointestinal tract, but the latter are soil dwellers or the external parasites and saprophytes of snails.

Correlation between survival in artificial gastric juice and gastrointestinal infectivity for warm blooded animals is far from perfect, but the method does serve to eliminate a background of varied microinvertebrates when potential pathogens are being sought in foods.

38.24 Morphological Identification

When a specimen suspected of being a parasite is obtained, the first question, logically, is, what is it? Several additional questions occur, especially to workers unaccustomed to dealing with parasites. In what should the specimen be placed for fixing and preserving? What keys, monographs, or other references are available to aid in identifying the parasite? Where may expert assistance be provided? This section answers these questions. No attempt is made to provide detailed instructions for handling specimens or keys for identification. Instead, brief directions and useful references to more complete instructions are appended.

38.241 Fixing, labeling, preserving, staining, or clearing

Fresh or living specimens should be washed free of mucus or other debris in physiological solution, or if that is unavailable, in water. Before fixation, an attempt should be made to determine the parasites' viability by observing them for motility in water or saline. Protozoan cysts and helminth eggs are fixed by the techniques described above for flotation and sedimentation to concentrate the parasites.

Most nematodes should be dropped into steaming 70% alcohol or acetic alcohol (1 part glacial acetic acid and 3 parts 95% alcohol). The heat usually causes the nematodes to straighten somewhat and die in that position. The nematodes can be stored in 70% alcohol. Adding a few drops of glycerine will prevent drying should the alcohol evaporate during storage. Small adult nematodes and larvae (2 or 3 mm or less in length) will be fixed more advantageously by killing them in a solution of hot 0.5% acetic acid and fixing in a mixture of 4% formalin and 1% or 0.5% acetic acid for 24 hours, after which the nematodes can be stored in the fixative. Nematodes are studied usually in temporary mounts and may be cleared for study in glycerine, in phenol-alcohol (80 parts melted phenol crystals and 20 parts absolute alcohol), or in lacto-phenol (20 parts distilled water, 40 parts glycerine, 20 parts lactic acid, 20 parts melted phenol crystals). After clearing in phenol-alcohol or lacto-phenol, nematodes should be washed in 70% alcohol before being returned to the preservative.

Trematodes and cestodes should be placed in distilled water for 10 to 15 minutes before fixing to relax them. Trematodes can be fixed satisfactorily by dropping them into hot 10% formalin which has been heated until it starts to bubble. The 10% formalin should be replaced by 5% formalin for storage after 24 hours. In addition, trematodes to be used for identification should be fixed under the pressure of a slightly weighted No. 1 coverglass. If this technique is used, the fixative is introduced from the edge of the coverglass.

Tapeworms should be killed by plunging them into water heated to 70 C and redipping several times rapidly in the hot water before fixing. The cestodes should be transferred promptly to AFA (85 parts ethyl alcohol, 10 parts formalin, commercial strength, 5 parts glacial acetic acid) for 24 hours after which the AFA should be replaced with 70% alcohol for storage. Both cestodes and trematodes should be stained and permanently mounted on slides for further study and identification. The references given at the end of this section contain procedures for staining and mounting the parasites.

Living acanthocephalans should be placed in a small dish of water until the proboscis remains fully everted. Acanthocephalans can be fixed in steaming 70% alcohol to which a few drops of glacial acetic acid have been added. Acanthocephala should be preserved in 70% alcohol until they can be stained and permanently mounted as described in the references given below.

Arthropods (fleas, lice, parasitic copepods, larval flies, etc.) may be fixed in hot water or 70% alcohol and stored in 70% alcohol.

Any parasitic specimen is almost worthless without a record of its host's name (scientific and common) and other data, such as location in or on the host, geographic locality of the host, date the host was collected, date the parasite was collected, collector's name, presumptive identification, and, when the parasite is authoritatively identified, the name of the identifier. Notes on the condition of the host and its source and history also should be included when available. Be as detailed as possible. As much of this data as practicable (preferably all) should be written on a small slip of paper, with pencil or an alcohol-proof ink, placed in the vial with the specimen as a specimen label.

More detailed instructions for processing specimens of parasites can be found in many laboratory manuals and texts, including Meyer and Olsen[18] and Cable.[19]

38.242 Keys and other aids to morphological identification

The following keys, texts, manuals, monographs, and reviews are some of the most widely available and useful printed aids for identifying parasites. They are grouped by kind of parasite or host group covered.

General

Cheng, T. C. 1973. General Parasitology. Academic Press, New York, N.Y.

Chitwood, M. B. and J. R. Lichtenfels. 1972. Identification of parasitic metazoa in tissue sections. Experimental Parasitology **32**:407–519

Spencer, G. M. and L. S. Monroe. 1961. The Color Atlas of Intestinal Parasites. Charles C Thomas, Springfield, Illinois

Shellfish Hosts

Cheng, T. C. 1967. Marine Molluscs as Hosts for Symbioses: With a Review of Known Parasites of Commercially Important Species. Advances in Marine Biology, Vol. 5. Academic Press, London & New York, N.Y.

Fish Hosts

Hoffman, G. L. 1967. Parasites of North American Freshwater Fishes. Univ. California Press, Berkeley, Cal.

Hoffman, G. L. and F. P. Meyer. 1974. Parasites of Freshwater Fishes: A Review of Their Control and Treatment. T. F. H. Publications, Neptune, N.J.

Protozoa

Levine, N. D. 1973. Protozoan Parasites of Domestic Animals and of Man. 2nd Ed. Burgess Publishing Co., Minneapolis, Minn.

Nematoda

Chitwood, B. G. and M. B. Chitwood. 1975. Introduction to Nematology. University Park Press, Baltimore, Md. (Reprint of parts first printed in 1937, 1938, 1941, and 1950)

Levine, N. D. 1968. Nematode Parasites of Domestic Animals and of Man. Burgess Publishing Co., Minneapolis, Minn.

McDonald, M. E. 1974. Key to Nematodes of Waterfowl. Resource Publication No. 122. Fish and Wildlife Service, U.S. Department of Interior, Washington, D. C.

Myers, B. J. 1970. Nematodes transmitted to man by fish and aquatic mammals. J. Wildlife Dis. **6**:266–271

Yorke, W. and P. A. Maplestone. 1926. The Nematode Parasites of Vertebrates. J. and A. Churchill, London. Reprinted 1962, Hafner Publishing Co., New York, N.Y.

Cestoda

Meyer, M. C. 1970. Cestode zoonoses of aquatic animals. J. Wildlife Dis. **6**:249–254

Schmidt, G. D. 1970. How to Know the Tapeworms. Wm. C. Brown Co., Dubuque, Iowa

Wardle, R. A. and J. A. McLeod. 1952. The Zoology of Tapeworms. The University of Minnesota Press, Minneapolis, Minn.

Trematoda

Healy, G. R. 1970. Trematodes transmitted to man by fish, frogs, and crustacea. J. Wildlife Dis. **6**:255–261

Skrjabin, K. I. et al. (English translation, 1964). Keys to the Trematodes of Animals and Man. University of Illinois Press, Urbana, Ill.

Schell, S. C. 1970. How to Know the Trematodes. Wm. C. Brown Co., Dubuque, Iowa

Arthropoda

CDC Pictorial Keys. Arthropods, Reptiles, Birds, and Mammals of Public Health Importance. Center for Disease Control, Atlanta, Georgia

38.25 Expert Assistance

There are more than 2000 parasitologists in the United States and many other veterinary and medical workers who can identify commonly encountered parasites. In seeking help, therefore, the inexperienced worker should first look to public health laboratories, local colleges, universities, hospitals, agricultural extension agents, U.S. Department of Agriculture meat inspectors, etc. If one cannot identify the parasite, institutions in the following list may be willing to help. Their help should be solicited only after the previously mentioned local sources have been consulted.

a. Any foodborne parasite

General Parasitology Branch
Parasitology Division
Bureau of Laboratories
Center for Disease Control
Atlanta, Georgia 30333
(Specimens should be submitted through the local State Health Department.)

Laboratory of Parasitology (HFF-124)
Division of Microbiology
Food and Drug Administration
200 C St., S.W.
Washington, D. C. 20204

b. Parasites of meat foods

Meat and Poultry Inspection Program
Animal and Plant Health Inspection Service
Beltsville Agricultural Research Center East #318
U.S. Department of Agriculture
Beltsville, Maryland 20705

c. Parasites of fish (U.S. Fish Hatchery Biologists)

Alchesay-Williams Creek National Fish Hatchery
Whiteriver, Arizona 85941

Fish Hatchery Biologist
Route 3, Box 38
Heber Springs, Arkansas 72543

Fish Disease Control Center
P.O. Box 917
Fort Morgan, Colorado 80701

Dworshak National Fish Hatchery
Ahsahka, Idaho 83520

Craig Brook National Fish Hatchery
East Orland, Maine 04431

Fish Cultural Development Center
Bozeman, Montana 59715

National Fish Hatchery
Pisgah Forest, North Carolina 28768

National Fish Hatchery
Lamar, Pennsylvania 16848

Little White Salmon National Fish Hatchery
P.O. Box 17
Cook, Washington, 98605

Abernathy Salmon Cultural Development Center
1440 Abernathy Road
Longview, Washington, 98632

National Fish Hatchery
Kearneysville, West Virginia 25430

Fish Hatchery Biologists Laboratory
Genoa, Wisconsin 54632

d. Parasites of marine fish

Laboratory of Parasitology (HFF-124)
Division of Microbiology
Food and Drug Administration
200 C St., S.W.
Washington, D.C. 20204

e. Parasites of livestock

Veterinary Services Laboratories
Animal and Plant Health Inspection Service
Beltsville Agricultural Research Center East #320
U.S. Department of Agriculture
Beltsville, Maryland 20705

Systematic Entomology Laboratory
Insect Identification and Beneficial
Insect Introduction Institute
Agricultural Research Service
Beltsville Agricultural Research Center West #003
U.S. Department of Agriculture
Beltsville, Maryland 20705

f. Helminths

Parasite Classification and Distribution Unit
Animal Parasitology Institute
Agricultural Research Service
Beltsville Agricultural Research Center East #120
U.S. Department of Agriculture
Beltsville, Maryland 20705

38.26 Animal Inoculation

Animal inoculation of parasitic material found in food is not carried out routinely. Among the protozoa only *Toxoplasma gondii* is searched for in suspected meat by animal inoculation. The procedure for detecting *Toxoplasma* in meat samples has been described by Jacobs et al.[17] Among the nematodes none are routinely detected by animal inoculation, although selected species have been maintained in experimental animals. In like manner, the many trematode parasites whose larval metacercariae are found in and on plant and animal material are not routinely isolated or identified by animal inoculation. Cestode parasites also do not lend themselves to routine animal inoculation. If a laboratory does diagnose one of the various parasitic species in food, a reference source should be consulted, particularly if the foodstuff is fresh or has been refrigerated. Food that has been frozen, heated, or placed in preservative or fixative so as to destroy the parasites obviously cannot be used for animal inoculation.

38.3 RESEARCH NEEDS

Considering the worldwide increase in travel, exchange of recipes, and transport of foods, the imminent use of incompletely treated sewage for fertilizing and watering crops and for feeding animals, and the consumers' tendency to prefer "natural," i.e., fresh, raw foods, there is a need to standardize and improve the methods of food parasitology.

Recovery methods must be made more efficient and uniform, if comparisons of results obtained in different laboratories are to be valid.

Identification methods must be simplified from the standpoints of time and ease, so that laboratory technicians doing routine work can be employed in the diagnosis of food contaminated with parasites. Essentially, this means that morphological criteria must be supplemented with quick physiological or serological tests.

No selective culture media for foodborne parasites exist. Food samples can be inoculated into culture media which tell the observer, without microscopy, that the samples do or do not contain *Salmonella* or other bacteria. Eventually, we must be able to do this for parasitic animals.

38.4 REFERENCES

1. HEALY, G. R. and N. N. GLEASON. 1969. Parasite Infections. *In* Food-Borne Infections and Intoxications. Ed. HANS RIEMANN. Academic Rev. New York, N.Y. and London, England, pp 175–222
2. MOAYEDI, B., M. IZADI, M. MALEKI, and E. GHODIVIAN. 1971. Human infection with *Moniliformis moniliformis* (Bremser, 1811). Travassos, 1915 (Syn. *Moniliformis dubius*). Am. J. Trop. Med. Hyg. **20**:445–448
3. ZUMPT, F. 1965. Myiasis in Man and Animals in the Old World. Butterworth & Co., Ltd. London
4. MEYER, M. C. 1970. Cestode zoonoses of aquatic animals. J. of Wildlife Dis. **6**:249–254
5. HEALY, G. R. 1970. Trematodes transmitted to man by fish, frogs, and crustacea. J. of Wildlife Dis. **6**:255–261
6. MYERS, B. J. 1970. Nematodes transmitted to man by fish, and aquatic mammals. J. of Wildlife Dis. **6**:266–271
7. CHENG, T. C. 1973. Human parasites transmissible by seafood and related problems. In Microbial Safety of Fishery Products. C. O. CHICHESTER and H. D. GRAHAM, eds. Academic Press, New York, N.Y. pp 163–189
8. SPRENT, J. F. A. 1969. Helminth "Zoonoses": An analysis. Helminthol. Abst. **38**:333–351
9. CHOI, DONG WIK and SUP LEE. 1972. Incidence of parasites found on vegetables collected from markets and vegetable gardens in Taegu area. The Korean J. of Parasit. **10**:44–51
10. KIM, JAE WOUN and DONG WIK CHOI. 1973. A comparison of methods for the demonstration of ascarid eggs and hookworm larvae from vegetables. The Kyungpook Univ. Med. J. **14**:379–385
11. BAERMANN, G. 1917. Eine Einfache Methode zur Auffindung von Ankylostomum (Nematoden) Larven in Erdproben. Meded. geneesk. Lab. Weltvreden, Feestbundel, Batavia. pp 41–47
12. FAUST, E. C., J. S. D'ANTONI, V. ODOM, M. J. MILLER, C. PERES, W. SAWITZ, L. F. THOMEN, J. TOBIE, and J. H. WALKER. 1938. A critical study of clinical laboratory technics for the diagnosis of protozoan cysts and helminth eggs in feces. Am. J. Trop. Med. **18**:169–183
13. FAUST, E. C., W. SAWITZ, J. TOBIE, V. ODOM, C. PERES, and D. R. LINCICOME. 1939. Comparative efficiency of various technics for the diagnosis of protozoa and helminths in feces. J. Parasitol. **25**:241–262
14. WILLIS, H. H. 1921. A sample levitation method for the detection of hookworm ova. Med. J. Austral. **ii**:375–376
15. RITCHIE, L. S. 1948. An ether sedimentation technique for routine stool examinations. Bull. U.S. Army Med. Dept. **8**:326
16. STOLL, N. R. and W. C. HAUSHEER. 1926. Accuracy in the dilution egg-counting method. Am. J. Hygiene **6**, March Suppl.:80–133
17. JACOBS, L., J. REMINGTON, M. L. MELTON. 1960. A survey of meat samples from swine, cattle, and sheep for the presence of *Toxoplasma*. J. Parasit. **46**:23–28
18. MEYER, M. C. and O. W. OLSEN. 1971. Essentials of Parasitology. Wm. C. Brown Co., Dubuque, Iowa
19. CABLE, R. M. 1961. An Illustrated Laboratory Manual of Parasitology. Burgess Publishing Co., Minneapolis, Minn.

CHAPTER 39

TOXIGENIC FUNGI

Joseph V. Rodricks and Joseph Lovett

39.1 INTRODUCTION

39.11 Definition of Terms

The fungi have adapted to a wide variety of substrates and are noted for their biochemical activity. Because some substrate changes and metabolites produced from fungal growth are desirable, fungi have been used for centuries in food processing and more recently in commercial production of chemicals and other products. Conversely, man has recognized for centuries that fungi frequently invaded his stored foods and materials, making them unpalatable or useless. Often, he observed an association between consumption of fungi, or foods contaminated with fungi, and altered health states. Until recently, observations of the toxicity of moldy foods and feeds did not include the identification of a specific fungal metabolite as the etiologic agent. Even today, the technology for isolation and identification of fungal toxins from foods has not kept pace with the growing awareness of the need for such technology.

Fungal metabolites demonstrating toxicity, regardless of other properties they may possess, are termed mycotoxins. These chemical compounds are diverse in structure and biological effect. Hundreds of toxins are produced by more than a dozen fungal genera.[1,2] Some mycotoxins were discovered originally as antibiotics, but have been found unsuitable for therapeutic use because of toxicity. The disease state produced by ingestion, inhalation, or other contact with a mycotoxin is called a mycotoxicosis (plural, mycotoxicoses).

39.12 History of Mycotoxicoses

39.121 Ergotism

Possibly the earliest recorded mycotoxicosis was ergotism. While mentioned in the 9th century A.D., it was not investigated until the 17th century when the toxic principle was found to be associated with the growth of *Claviceps purpurea* on cereal grains.[3,4] Ergot is not a single compound but a group of pharmacologically active alkaloidal compounds[5]. These mycotox-

IS/A Committee Liaison: George J. Herman

ins can produce restriction of blood flow to the extremities, resulting in gangrene, or convulsions and hallucinations. Either form can be fatal. Although an episode involving contaminated bread occurred in 1951,[6] ergotism is not known to exist today.

39.122 Toxic mushrooms

A more common mycotoxicosis results from ingestion of toxic mushrooms such as the deadly *Amanita* or the hallucinogenic *Psilocybe*. Mushroom consumption is increasing, but mycotoxicosis from this source is rare. Most mushrooms eaten today are the nontoxic *Agaricus campestris* grown commercially under carefully controlled conditions.

39.123 Alimentary toxic aleukia

Alimentary toxic aleukia (ATA), a mycotoxicosis involving humans, has been under investigation since the 1930s. This disease was prevalent in the Russian grain belt from 1942 to 1947, and was the result of the consumption of moldy grain. Mayer[7, 8] thoroughly reviewed the literature available. The pathology, involving the hematopoietic system, resembled aplastic anemia, hemorrhagic aleukia, or panmyelopathy.[9]

Based on a review of the epidemiological evidence and experimental data, Rubinshteyn[10] refers to ATA as "Fusariotoxicosis." A model for ATA was established with cats and monkeys. Fusarial strains grown on grains could produce ATA symptoms in these animals. Usually, toxigenic strains were isolated from barley and oats, sometimes from wheat, and least often from rye.

An alimentary mycotoxicosis termed "Urov disease," because it was endemic in that region, also was noted in China and Korea for at least 100 years before it was described in 1949.[10] Pathology involved changes in enchrondial bones and their joints. The disease was produced in growing puppies and rats by toxic fusarial species endemic to the affected areas. Epidemiological, clinical, and experimental observations were the basis for calling Urov disease a mycotoxicosis.

39.124 Stachybotryotoxicosis

A noncommunicable disease of horses in the Ukraine (USSR) was found to be caused by moldy hay. Irritation of the mouth, nose, throat, and lips, and swelling and soreness of the submaxillary glands was accompanied by leukopenia, necrotic lesions of the respiratory and digestive tracts, and hemorrhages in all organs.[11] A similar illness was found among human residents of the area. Symptoms were produced in human subjects with hay contaminated with *Stachybotrys atra*, or with spores of the fungus.

39.13 Importance of Toxigenic Fungi

39.131 Agricultural significance

Before the discovery of the important group of mycotoxins known as the aflatoxins, little significance was attached to the occasional and isolated episodes of animal mycotoxicoses. They were noteworthy enough, however, to produce three reviews.[12, 13, 14] Some reviewers saw a potential for human health damage from fungal metabolites in agricultural products, basing their hypotheses on the variety and ubiquity of toxigenic fungal strains in animal and human foodstuffs.

In 1960, thousands of turkeys died in England of a disease of unknown etiology called "turkey X disease." The causative agent was found to be a toxic factor in imported peanut meal. The initial report of disease in turkeys was followed by reports of toxicity to chickens, swine, and cattle. Reports that animals fed toxic peanut meal developed carcinomas were verified experimentally.[15] A toxigenic *Aspergillus flavus* was discovered as a contaminant of the toxic diets. The toxic factor produced by the fungus and isolated from the toxic meal was found to be a group of metabolites characterized by their blue or green fluorescence under ultraviolet light. Originally they were called "aflatoxin B" or "G," based on their fluorescence.

Using a crystalline aflatoxin mixture, Dickens and Jones[16] established that purified toxin was an active carcinogen for subcutaneous rat tissue. Butler and Barnes[17] showed that the sequence of histological changes in male rats fed aflatoxin in their diet closely resembled those induced by other oral carcinogens. Continuous feeding produced malignant tumors in 25 to 88 weeks.

Fungi capable of aflatoxin production, and the aflatoxins themselves, have been isolated from legumes and grains from many of the world's producing areas. In addition, other compounds with carcinogenic, hemorrhagic, hepatotoxic, neurotoxic, and uterotrophic properties have been isolated from foodstuffs, and identified as metabolites of fungi common to a variety of agricultural commodities. Subacute levels of many of these toxins may contribute to loss of animal condition, low weight gains, loss of fecundity, and subsequent economic loss. Some compounds, or their toxic metabolites, may be retained in edible tissues or excreted in milk. The potential for economic loss and human health effects make mold contamination of agricultural products doubly significant.

39.132 Human health significance

While human mycotoxicoses have been observed and demonstrated for centuries, the technology for isolation and identification of specific metabolites is a recent development. Therefore, much of the information accumulated on human illness cannot be linked to a specific mycotoxin. ATA, ergotism, stachybotryotoxicosis, and mushroom poisoning already have been discussed as human illnesses and are well documented. ATA is the least well characterized as to etiology, but the evidence for mycotoxicosis is overwhelming.

It was the finding that the aflatoxins were potential carcinogens that produced the initial surge of research interest. Evidence followed that other mycotoxins were capable of producing carcinoma when included in animal diets.[18] As assay methods for individual toxins were developed, it was discovered also that the ubiquitous fungal strains involved could utilize a wide variety of foodstuffs for toxin production. Several mycotoxins have been verified as naturally occurring in foods and feeds.

It is not yet known to what degree the various mycotoxins pose a health hazard to man. The information accumulating on the pathology and toxicology of the purified compounds leads to speculation that mycotoxins play a role in diseases of unknown etiology, particularly those where the symptoms tend to rule out infective agents. Such speculation has centered on the aflatoxins and their correlation with liver disease in the tropical and subtropical areas of the world.[9, 19] Conditions are frequently favorable for mycotoxin production before harvest and during postharvest storage. In addition, dietary preferences and methods of food preparation and storage may provide unusual exposure or opportunities for fungal growth, and for toxin production in foods.

The distribution of toxigenic fungi and aflatoxins in some of the foods of Thailand was studied extensively.[20, 21, 22, 23] The objective of the studies was to determine the relationship between foodborne aflatoxin and the incidence of primary liver cancer in man. Foods from the Ratburi area were contaminated extensively with toxigenic *Asperigillus flavus* and aflatoxins. The Sangkhla area showed only low level contamination. A comparison of the incidence of primary liver cancer in the two areas showed that the incidence in the Ratburi area was three times greater.

In another study[24] autopsy specimens from Thai children with an acute disease of unknown etiology were examined. The disease involved acute encephalopathy and fatty degeneration of the viscera, and is referred to as EFDV. Ninety-nine autopsy specimens collected from 23 children who died of EFDV, plus urine from an additional 51 EFDV cases, were analysed for aflatoxins. Autopsy specimens from 15 children and adolescents who died of causes unrelated to EFDV, plus urine specimens from 39 healthy children, served as controls. All controls resided in the same area as the EFDV cases. Low levels of aflatoxins were found in all tissues (1 to 4 μg/kg). This may represent an artifact, or reflect the low level occurrence of aflatoxins in the diet of the area. None of the control urine samples contained aflatoxins. Aflatoxin was found in 22 of 23 EFDV cases. The levels found in liver were comparable to those found in monkeys receiving approximately 2 LD_{50} aflatoxin B_1.

An aflatoxin etiology for infantile liver cirrhosis in India has been postulated, and some supporting studies have been done. Robinson[25] found aflatoxins in the diets, milk, and urine of mothers of children with liver cirrhosis. Parpia and Sreenivasamurthy[26] observed children suffering from kwashiorkor who accidentally consumed for periods of one week to one month peanut meal containing aflatoxin. Liver pathology showed development of the infant cirrhosis commonly observed in the area.

The relationship between malnutrition and the toxic effects of aflatoxins is not established, but there is evidence that the two factors may make mycotoxin contamination of foods a particular problem in those areas where protein deficiency is endemic. Experimental animals sustain more liver damage when the diet is low in protein. There is, in addition, some evidence that the transition from an undernourished to an adequately nourished state may induce carcinoma from a dormant precancerous liver. Both these possibilities, and the supporting literature, are reviewed by Kraybill and Shapiro.[19] The problem is compounded by the fact that mycotoxins are common in foods in many of the same geographical areas where dietary deficiencies exist.

39.2 MYCOTOXINS AND TOXIGENIC FUNGI

39.21 Toxigenic Fungi Associated with Foods and Feeds

The fungi contaminating foods may be divided into two groups, parasitic and saprophytic, sometimes called endophytic and epiphytic.[27] It is more convenient for our purposes to relate the two fungal groups to agricultural practices, and call them "field" and "storage" fungi.[28] Field fungi are those invading forage or seeds in the field during or after development. They are characterized by their high moisture requirement. They die as the moisture content of the product is reduced during storage. Storage fungi have adapted to low moisture substrates. They grow on almost any organic substrate and become the dominant genera in stored, low moisture, agricultural commodities. The toxigenic storage fungi are primarily *Aspergillus* and *Penicillium*, while some, like *Fusarium*, may be either field or storage organisms. The Fusaria may appear in stored grains as advanced decay fungi.[28] The storage fungi are most important among the toxigenic molds.

The toxigenic fungi discussed below do not exhaust all known mycotoxin producing strains. Hundreds of species of more than a dozen fungal genera are known to produce metabolites with some degree of toxicity. The genera discussed will be those elaborating the mycotoxins which are now considered more important in foods and feeds.

39.211 Aspergillus toxins

a. Aflatoxins

There are more than a dozen closely related chemical compounds designated by the term aflatoxin. Some of these are fungal metabolites, and others are metabolites produced by animals which have ingested aflatoxin contaminated feed, or by bacteria metabolizing a substrate containing aflatoxins. The most important member of the aflatoxins group is known as aflatoxin B_1 (Figure 1); each of the other aflatoxins, variously known as B_2, G_1, G_2, M_1, etc., exhibits some minor structural variation on the difuranocoumarin system of B_1. The fungal produced aflatoxins, including B_1, are derived primarily from the *Aspergillus flavus* group. Members of this group are widely distributed in nature and commonly are isolated from soils, forage, decaying vegetation, and stored foods and seeds. They are used extensively in the alcoholic and fermented food industries of the Far East. Because of the lipolytic enzymes of this group, they thrive on oil seeds such as cotton seed and peanuts, coconut, olives, castor beans, and sunflower seed.[29] Although 15 *Aspergillus* and *Penicillium* sp have been reported to produce aflatoxins, only *A. flavus* and *A. parasiticus* have been confirmed consistently as sources of aflatoxin.

The aflatoxins have molecular weights in the range 312 to 346. They are isolated from natural sources by extraction with organic solvents (most are poorly soluble in water), and can be purified by a

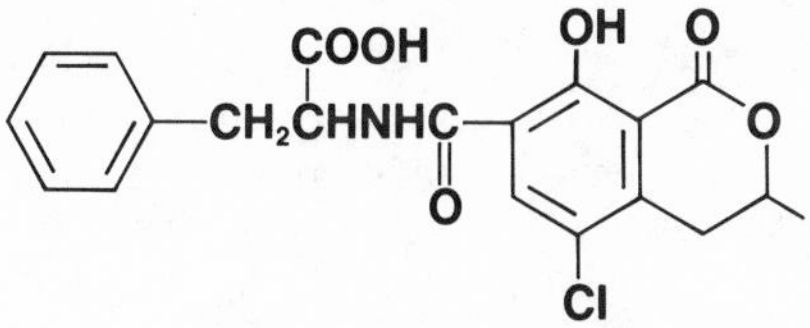

OCHRATOXIN A

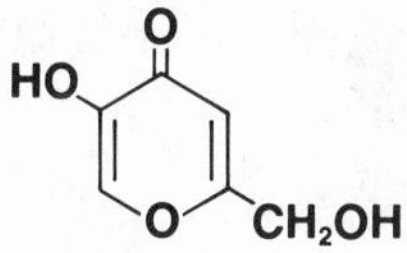

KOJIC ACID

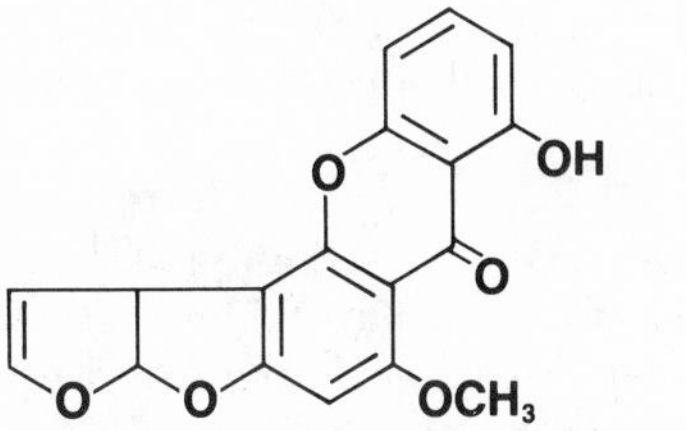

STERIGMATOCYSTIN

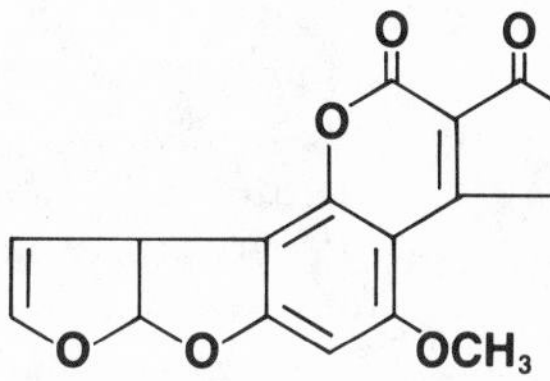

AFLATOXIN B_1

FIGURE 1. Some Important *Aspergillus* Toxins

variety of chromatographic techniques.[30] The aflatoxins are colorless compounds with relatively high melting points (>250 C). The heat stability of purified, crystalline aflatoxins is high and most are stable in the pH range of 3 to 10.[30] All are highly fluorescent substances, and analytical techniques have been developed to take advantage of this characteristic.

Most of the information in regard to health hazards from aflatoxin derives from experimental studies with animals. Aflatoxin B_1 has proved to be toxic to every animal species tested, and in some, it is the most potent hepatocarcinogen known. The LD_{50} values of B_1 in several animal species are: rat, 5.5 to 7.4 mg/kg; guinea pigs, 1.4 mg/kg; dogs, 1 mg/kg; hamsters, 10.2 mg/kg; mice, 9 mg/kg; and rainbow trout, 0.5 to 1.0 mg/kg.[5] In rats, a high liver tumor incidence is found when a diet containing 15 ppb aflatoxin is fed for a prolonged period.[5] The review by Detroy, Lillehoj, and Ciegler gives more detailed information on the toxic and biochemical effects of these compounds in animals.[30] The importance of these compounds in human health has been discussed in 39.132. The presence of the aflatoxins in the feeds of food producing animals is an important part of the aflatoxin problem, not only because they may harm the animal, but because they can appear as residues in meat and eggs, and can be excreted in milk.[31]

b. Ochratoxins

These substances are a group of closely related compounds produced by *Aspergillus ochraceus*, *A. sulphureus*, and *A. melleus*. The *A. ochraceus* group is common in soils and decaying vegetation. These organisms also appear in grains and are adapted to low moisture substrates, as evidenced by their growth in the dried and salted fermented foods of the Orient. The *A. ochraceus* group may invade wheat, corn, cotton seeds, legumes, peppers, surface injured apples, onions, and pears.[29]

All of the ochratoxins contain the isocoumarin moiety. The predominant member of the group is called ochratoxin A and its structure is shown in Figure 1.[32] Although the ochratoxins have chemical characteristics radically different from the aflatoxins, they are, like the aflatoxins, highly fluorescent substances. Again, assay procedures depend on this property.

Ochratoxin A has an oral LD_{50} in rats of 20 to 22 mg/kg, and this compound causes fatty infiltration of parenchymal liver cells. It is also nephrotoxic. Other ochratoxins produce similar effects. One study indicates that ochratoxin A is not a carcinogen,[33] but the effects of chronic ingestion have yet to be fully evaluated.

c. Sterigmatocystin

Another common food contaminant is *Aspergillus versicolor*, a mold which produces many highly colored metabolites. A major product of this species is sterigmatocystin, a compound bearing some structural resemblance to the aflatoxins in that it also contains the unusual difuran structural moiety (Figure 1).[34, 35] Sterigmatocystin is obtained from natural sources by extraction with organic solvents and can be crystallized as pale yellow needles. Its stability to heat, acid, and alkali is similar to that of the aflatoxins. This mycotoxin also has been isolated from *A. nidulans* and a *Bipolaris* sp. Upon ingestion, sterigmatocystin produces liver and kidney damage and is, like the aflatoxins, a hepatocarcinogen, although it is far less potent, and its histopathology differs.[36, 37]

d. Other Aspergillus toxins

There are a number of other metabolites of this genus which have been shown to be toxic to animals, at least in an acute sense, which are potential food contaminants.

Common in soil, particularly soil enriched with organic matter, *A. clavatus* attracted attention in 1942 when it was found to produce a broad spectrum antibiotic named clavacin.[38] Others have isolated the antibiotic from this fungus and called it clavatin.[39] This mycotoxin, produced primarily by the penicillia, is now known as patulin and is discussed more fully in the section on penicillia, below.

The *Aspergillus terreus* group, common to soil, is composed of a single species, subdivided into varieties. It is the most important patulin producer of the Aspergilli because it has been isolated from a variety of foodstuffs. It has been found in stored grains, straw and forage, cotton, pineapples, and has been implicated in the storage rot of apples.[29] Other important mycotoxins from the Aspergilli are kojic acid (Figure 1) from *A. flavus*, fumagillin, gliotoxin, and helvolic acid from

A. fumigatus, xanthocillin from *A. chevalieri*, and oxalic acid from a number of species. The chemistry and toxicology of these miscellaneous *Aspergillus* toxins have been reviewed by Wilson.[5]

39.212 Penicillium toxins

a. Patulin

Among the more interesting and important of the large number of mycotoxins produced by the penicillia is the potent antibiotic, patulin. The following species have been described as patulin producers: *P. claviforme, P. divergens, P. expansum. P. griseofulvum, P. patulum, P. novozealandiae, P. lapidosum.* Some of these were shown to produce patulin during the search for antibiotics that began with the discovery of the penicillins, and this accounts for the synonyms of patulin such as expansin, claviformin, and leucopin. Of these nine species, *P. expansum* and *P. patulum (urticae)* are best known. Both are excellent patulin producers on a variety of substrates. *P. expansum* is the cause of storage rot in apples, and the finding of toxigenic strains and preformed patulin in fungal rotted apples is common.[40] *P. patulum (urticae)* is found in stubble mulched grain fields[41] and patulin producing strains have been isolated from two mixed feeds, one commercial, and one corn soybean meal mix, suspected of causing the death of farm animals.[42]

Patulin is a highly polar, water soluble organic compound containing an α,β-unsaturated lactone function (Figure 2). It has an oral LD_{50} in mice of 30 to 35 mg/kg, and has been implicated as a carcinogen.[43] Patulin is a phytotoxic agent and also a potent inhibitor of the growth of many microorganisms.[44]

b. Penicillic acid

Eleven species of penicillia have been reported to produce penicillic acid. They are *P. cyclopium, P. janthinellum, P. viridicatum, P. baarnense, P. puberulum, P. thomii, P. martensii, P. madriti, P. stoloniferum, P. suavolens,* and *P. palitans.* Of these, *P. cyclopium* is the most frequently isolated toxigenic species. *P. martensii,* which produces blue eye mold of corn, has been isolated from toxic high moisture yellow dent corn, and found capable of toxin production on corn of 12.7 mg penicillic acid per gram of substrate.[45] Toxigenic strains of *P. janthinellum, P. viridicatum,* and *P. cyclopium* have been isolated from mold fermented sausages. No penicillic acid could be detected in experimentally inoculated sausages.[46]

Penicillic acid bears some structural relationship to patulin in that it too contains an α,β-unsaturated lactone moiety (Figure 2), and for this reason is grouped together with patulin and a series of similar compounds as potential carcinogenic agents. Some experiments with penicillic acid indicate that it is a carcinogen, although there are no data regarding its activity as an oral carcinogen.[43]

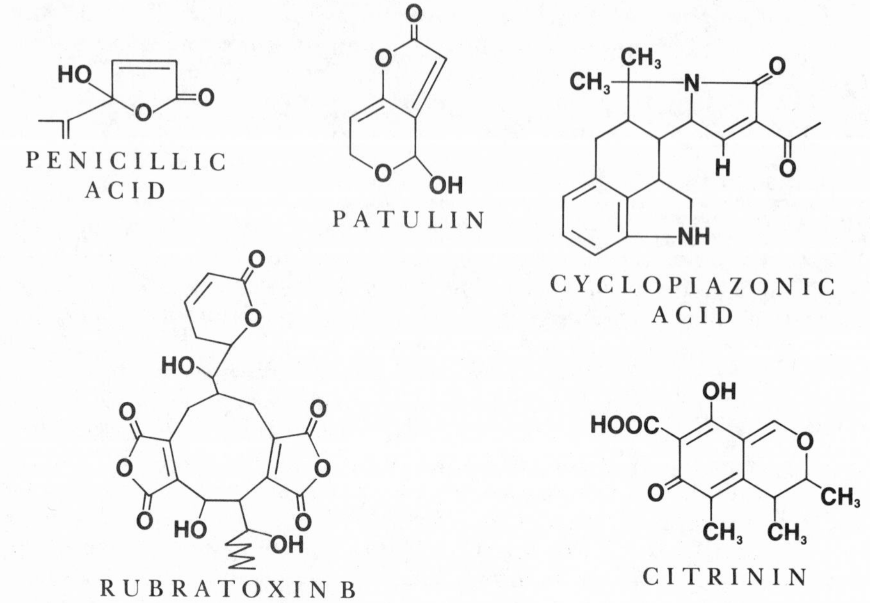

PENICILLIC ACID

PATULIN

CYCLOPIAZONIC ACID

RUBRATOXIN B

CITRININ

FIGURE 2. Some Important *Penicillium* Toxins

Like patulin, penicillic acid is an antibiotic, but is less potent as a toxin to mammals (LD_{50} in mice = 250 mg/kg i.v.). It is a water soluble compound, and in solution exists as an equilibrium mixture of the form shown in Figure 2, and the open chain keto-acid derived from lactone ring opening.[43]

c. Rubratoxin

One of the few instances where the only species reported to produce a particular toxin are all of the same series, the *P. purpurogenum* series of the Biverticillata symmetrica, *P. rubrum* and *P. purpurogenum* are the only known rubratoxin producers. Both are widespread in nature, and have been isolated from cereal and legume sources.

Although the closely related rubratoxins (A and B) were not characterized until 1962, toxigenic *P. rubrum* and *P. purpurogenum* were implicated in animal mycotoxicosis toxicosis as early as 1952.[47] Moldy corn implicated in farm animal toxicosis repeatedly has yielded toxigenic strains of these two species. They are also reported to be capable of producing toxins in poultry feeds, and are believed to play a role in poultry hemorrhagic disease. In addition to the hemorrhagic effects, the rubratoxins cause liver necrosis. Rubratoxin B has an oral LD_{50} in mice of 400 mg/kg. The compounds are not carcinogenic.[47]

Both rubratoxins are macrocylic compounds containing α,β-unsaturated lactone and acid anhydride groups. Rubratoxin B is depicted in Figure 2.

d. Tremorgens and cyclopiazonic acid

Penicillium cyclopium, P. crustosum, and *P. granulatum* have been reported to elaborate a toxin, as yet uncharacterized, which produces acute neurotoxic effects in experimental animals. Animals fed this substance show sustained trembling and convulsions. These fungi have been isolated from feeds involved in illness of farm animals.[48,49]

After the discovery of a tremorgen produced by *P. palitans,* and isolated from moldy commercial cow feed believed to have caused the death of dairy cows, Ciegler and Pitt[48] surveyed *Penicillium* cultures from their collection for tremorgen production on laboratory media. Tremorgens detected by thin layer chromatography were confirmed in animals. Confirmed producers were *P. olivino-viride, P. palitans, P. cyclopium, P. puberulum.* Unconfirmed traces were found in *P. gladioli* and *P. expansum.* Cole et al[50] have isolated another tremorgenic compound from *P. verruculosum,* and have proposed the name "verruculogen." The toxin was produced in mycological broth.

Another compound derived from *P. cyclopium* is an indole alkaloid named cyclopiazonic acid (see Figure 2). The metabolite has an LD_{50} in rats of 2.3 mg/kg i.p., and produces acute neurotoxic effects. The mold which produces this mycotoxin is a frequent contaminant of foods, but, as yet, cyclopiazonic acid has not been found in foods.[51]

e. Rice toxins

Storage fungi proliferate in improperly stored rice. Most are of the genera *Aspergillus* and *Penicillium,* and about 10% of the isolates tested are toxigenic. Of these toxigenic strains, *P. islandicum* has been investigated most fully, but more than a dozen species are involved in elaborating eight or more toxins (Table 1). Among the important metabolites are citrinin (Figure 2), an acidic compound found as a product of at least seven penicillia species, a chlorine containing cyclic peptide known as islanditoxin, a number of dimeric anthraquinones (luteoskyrin, rugulosin, skyrin), a polyenic compound called citreoviridin, and an acidic compound known as citreomycetin. Some of these toxins affect the liver and kidney, some are neurotoxic, and two are tumorogenic. This has led to the suggestion in the Orient that they may play a role in cancer. The toxic effects of these substances, interactions among them, and their natural occurrence have been reviewed exhaustively.[52]

39.213 Fusarium toxins

a. Zearalenone

While the fusaria may be both field and storage fungi, their potential for human health effects is probably realized by growth of *Fusarium* sp on grains after harvest in a high moisture condition. One of the toxic metabolites is F-2 toxin or zearalenone. This compound is produced primarily by *Fusarium graminearum (Gibberella zeae),* but also has been reported from *F. tricinctum* and *F. roseum culmorum.*[53]

Table 1: The Rice Toxins[52]

Mycotoxin	Mycotoxin producers
Citreoviridin	*P. citreo-viride, P. ochrosalmoneum, P. toxicarium*
Citrinin	*P. citreo-viride, P. citrinum, P. expansum, P. fellutanum, P. implicatum, P. lividum, P. jenseni, P. notatum, Aspergillus terreus, A. niveus, A. candidus*
Cyclochlorotin	*P. islandicum*
Erthroskyrine	*P. islandicum*
Islanditoxin	*P. islandicum*
Luteoskyrin	*P. islandicum*
Rugulosin	*P. brunneum, P. rugulosum, P. tardum, P. variabile, P. wortamanni*

Zearalenone is a rescorcylic, macrocylic lactone (Figure 3) which is not soluble in water. The compound exhibits an intense blue fluorescence on TLC plates, and this property is used for identification.

This mycotoxin has been associated with outbreaks of vulvovaginitis in swine. It is probably one of the more common mycotoxin contaminants of food and feed, but there is as yet no full assessment of its effects in humans.[53]

b. 12,13-epoxy-Δ^9-trichothecenes

There is a series of close to thirty naturally occurring sesquiterpene mold metabolites which have structural features similar to those of the compound known as diacetoxyscirpenol (Figure 3.) Some of these compounds, all of which can be categorized as 12,13-epoxy-Δ^9-trichothecenes, are metabolites of a variety of fusaria species including *F. scirpi, F. tricinctum, F. nivale, F. concolor,* and *F. equisete.* Some of the compounds of this series are produced by species in the genera *Myrothecium, Stachybotrys, Trichoderma,* and *Trichothecium.*

Compounds of this series which have undergone toxicological study have been found to cause severe dermal irritation when applied to the skin of experimental animals. Some cause hemorrhage on the lips and in the mouth, throat, and entire gastrointestinal tract. The acute toxicity of some of the trichothecenes is high: the LD_{50} in rats of diacetoxyscirpenol is 0.75 mg/kg i.p., and, when administered orally, is 7.3 mg/kg.

The trichothecenes have been suspected in alimentary toxic aleukia (ATA) and in stachybotryotoxicosis. A valuable review of this important group of fungal metabolites has been published.[54]

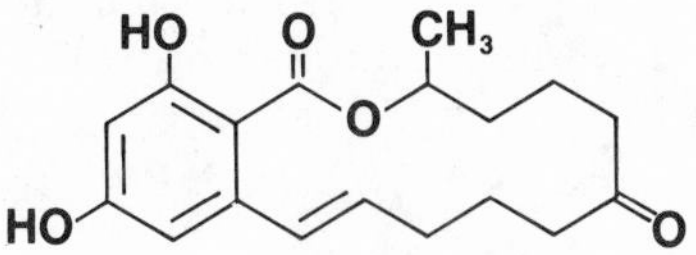

ZEARALENONE

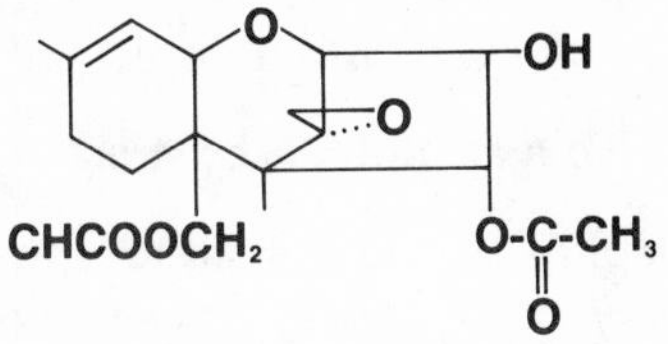

DIACETOXYSCIRPENOL

FIGURE 3. Some Important *Fusarium* Toxins

39.214 Miscellaneous mycotoxins

a. *Pithomyces chartarum*

There is a group of structurally related natural products generally categorized as epipolythiadioxopiperazines. The best known mold metabolites in this group are the sporidesmins, produced by *Pithomyces chartarum.* Most of the research done on these toxins has come from New Zealand, where facial eczema in ruminants has been related to these substances. The fungus is a saprophyte growing on dead grass in grazing areas.[55]

b. *Rhizoctonia leguminicola*

For many years, farmers observed that animals eating red clover from pasture or hay silage developed excessive salivation, diarrhea, frequent urination, and loss of appetite. The salivation factor has been found to be a mycotoxin, slaframine, produced by *Rhizoctonia leguminicola* growing on red clover foliage.

c. Other toxigenic fungi

There are hundreds of fungal species which have been shown to be toxigenic. Christensen et al[56] screened 943 isolates, primarily from animal feeds, peanuts, and seeds, including some isolates from flour, spaghetti, and black and red peppers, for lethality to poultry following growth on a laboratory medium. The following genera showed toxic isolates: *Alternaria, Aspergillus, Chaetomium, Cladosporium, Colletotrichum, Curvularia, Fusarium, Gliocladium, Myrothecium, Neocosmospora, Papulospora, Penicillium, Phoma, Rhizoctonia, Sclerotium, Scopulariopsis, Thielavia, Trichoderma,* and *Trichothecium.* No toxic metabolites were identified.

39.22 Control of Fungal Growth and Toxin Production

39.221 pH control

The fungi are highly active biochemically and able to utilize a wide variety of substrates. Although each species has an optimum pH for growth, fungi are noted for their wide pH tolerance. Many are able to change the pH of the substrate rapidly by the action of their metabolites. Control of pH is of little value in the control of fungal growth in food.

39.222 Oxygen and CO_2 control

Control of the gaseous atmosphere in which fresh fruits and vegetables are stored is a common way to retain quality foods, free from decay. Apples, for instance, are stored in controlled atmospheres where oxygen is kept at approximately 3% and carbon dioxide at 1 to 5%. This atmosphere, coupled with refrigeration temperature, maintains the quality of stored apples for several months. It does not protect entirely from fungal growth and toxin production, since some species, such as the patulin producing *Penicillium expansum,* grow and produce toxin under these conditions.

39.223 Temperature control

This long has been a standard means of preserving food. It does not protect from all toxigenic molds, however, for many will grow at refrigeration temperatures. Joffee[57] found several toxigenic species capable of growth and toxin production at temperatures down to –10 C. There is evidence that some strains may be more toxigenic at low temperatures than at optimum growth temperatures.

39.224 Water activity control

Adjustment of water activity, or equilibrium relative humidity, is the best means of controlling growth of microorganisms in foods which can take such a storage regimen. This includes most of the nuts, grains, and legumes that have become major concerns since recognition of the mycotoxin problem. The effectiveness of low water activity in controlling microbial growth is the basis for the recent intermediate moisture foods industry. Control of water activity for control of fungal growth has been discussed at length elsewhere.[58, 59, 60]

39.225 Fungicides and radiation

Fungicides and radiation have been explored as means of inactivating fungal spores and preventing germination and growth. Radiation levels permissible under FDA regulation are inadequate for the inactivation of storage fungi. Fumigation as a means of controlling fungi is discussed by Majumder et al.[27]

39.23 Assay Procedures

Since the arousal of interest in mycotoxins brought about by the discovery of the aflatoxins in the early 1960s, much activity has been devoted to the development of assay procedures aimed at detecting mycotoxins in foods and feeds. The presence of known toxigenic mold on food is not proof that its toxin is also present; conversely, the absence of obvious mold growth on a commodity is no demonstration that mycotoxin is not present. Thus, assay procedures must be aimed at the mycotoxin(s) and not at the toxigenic fungus. The problems of assaying foods and feeds for mycotoxins are similar to those encountered in the assay of foods and feeds for other organic pollutants (pesticides, PCBs, other chemicals of industrial origin). The procedures require four basic steps: 1) extraction of the compound of interest from the food or feed; 2) separation of the compound from other substances ("cleanup"); 3) quantitative or qualitative determination of the compound; and 4) confirmation of its identity. The extraction step involves treatment of the sample with a solvent which can penetrate plant tissue, that is, a hydrophilic solvent, and also can dissolve the mycotoxin contaminant. Two-phase systems are often employed for this purpose. Part of the extraction separation procedure involves removal of lipid components by partition of the polar phase against a lipid solvent. Most mycotoxins are highly polar organic compounds, and will not move into a lipid solvent phase.

Column chromatography procedures are used for the "clean up" step. Liquid-liquid partition has been used also as a separation tool for some mycotoxins. The determinative step almost always involves gas-liquid or thin-layer chromatography. Recently, high speed liquid chromatography has found some acceptance, and probably will come into greater use in the future. For certain mycotoxins, biological tests are useful as qualitative determination procedures. Confirmation of compound identity is necessary before the finding of mycotoxin can be considered definite. This usually involves preparation of certain chemical derivatives, or the application of biological tests of high specificity.

A number of tests for mycotoxins have been developed for rapid field use, usually under rather primitive laboratory conditions. These tests are ordinarily not quantitative, but will allow rapid analyses of a large number of samples to screen out all "negatives."

Caution must be taken with the sample used for assay. Mycotoxin contamination is frequently highly localized in a lot, and no system of sampling has been devised which assures that any particular sample is representative of a lot (except in the case of liquids). The best sampling method is the retrieval of a large number of randomly scattered subsamples. These should be finely ground and well mixed before the analytical sample is taken.

A number of methods for mycotoxins have been selected, reviewed, collaboratively studied, and adopted by the Association of Official Analytical Chemists (AOAC), the American Oil Chemists' Society (AOCS), the American Association of Cereal Chemists (AACC), and the International Union of Pure and Applied Chemistry (IUPAC). The AOAC, AOCS, and AACC each publishes its own book of Official Methods.[61, 62, 63] The IUPAC publishes information bulletins describing certain aspects of the problem of mycotoxin methodology.[64] The methods for aflatoxins in a number of different commodities are thoroughly covered in these publications and should be consulted for details. Methods for some of the other mycotoxins are also available. Following is a list of important references for these analytical procedures: sterigmatocystin,[65] ochratoxin A,[66] patulin,[67] penicillic acid,[68] citrinin,[42] zearalenone,[69] 12,13-epoxy-Δ^9-trichothecenes.[70] An excellent review of all methods has appeared.[71]

Not as much work has been done on bioassay procedures as on chemical approaches to mycotoxin identification and quantitation, although the earliest work on ATA depended heavily on biological tests. Good agreement between toxicity to man of fungi isolated from grains involved in ATA, and the rabbit dermal toxicity test have been reported.[72, 73] This test more recently has been used as an indicator of the presence of 12,13-epoxy-Δ^9-trichothecenes.

Embryonated eggs have been used in screening of fungal isolates for toxigenicity.[73, 74] The method also has been used for specific toxin assay. The LD_{50} for aflatoxin B_1 in the air cell of an unincubated egg is 0.025μg.[75]

The duckling is particularly susceptible to aflatoxin. As little as 2 μg B_1 given over a period of 5 days causes reproducible bile duct lesions[9] and the LD_{50} for the day old duckling is 0.5 mg/kg. The dog, rabbit, guinea pig, pig, and rainbow trout have similar LD_{50} values, but are less convenient for assay.

Cell cultures, both line and primary, are sensitive assay systems for mycotoxins.[76] A linear dose response relationship exists for inhibition of rat fibroblast growth over a range of 0.025 to 0.25 μg aflatoxin B_1 per ml of culture medium. The concentration which gives 50% growth inhibition is 0.06 μg/ml. The mitotic rate of human embryonic lung cells is suppressed 50% by 0.03 μg aflatoxin B_1 per ml of culture medium. B_1, in concentrations of 1 to 5 μg/ml culture medium, causes destruction of Chang and HeLa cells, as well as primary duck and chick embryo cells.[76]

Bacterial inhibition, by procedures common for antibiotics, also has application to some mycotoxins. The simplest methods[77] employ zone inhibition techniques. This test has the advantage that, with the proper organism and medium combination, assays may be read in less than 8 hours and always within 18 hours.

39.231 Standards

One problem besetting the analyst about to embark on mycotoxin assays is the availability of analytical standards. Although this has been a great problem in the past, the recent interest in mycotoxins has moved a number of commercial producers of chemicals to offer some of the more important mycotoxins for sale. Mycotoxins not commercially available can sometimes be obtained from research investigators active in the subject area. Criteria for mycotoxin standards have been published.[78]

39.24 Detoxification Procedures

39.241 Foods and feeds

The problem of the control of mycotoxins in foods and feeds is, obviously, best approached by the exercise of care in the harvest and storage of agricultural commodities. However, agronomic and technological breakdowns do occur, and the alternatives remaining after contamination has taken place are (1) removal of contaminated portions of a lot and (2) destruction of the mycotoxin. The removal of contaminated portions can be accomplished by the sorting out of mold damaged portions, as is done in the processing of peanuts to remove aflatoxin-containing nuts, or by solvent extraction of the mycotoxin. Removal of mycotoxins can occur during the processing of some foods; for example, the wet milling of aflatoxin-contaminated corn results in the concentration of aflatoxin in some of the streams destined for nonfood uses. Destruction of mycotoxins can be accomplished by heating, as in the roasting of peanuts, or by chemical treatment, as in the ammoniation of cottonseed meal. Each of the destruction techniques has attendant difficulties, among them the question of the chemical fate of mycotoxins and the formation of toxic alteration products of the mycotoxins, and of the normal food components. The question of detoxification has been reviewed in depth.[15]

39.242 Laboratory safety and equipment detoxification

Laboratory workers should exercise extreme care in working with mycotoxins. Crystalline materials should be handled in a glove box. Vinyl examination gloves provide effective protection when workers are handling solutions of mycotoxins. Laboratory glassware should be well rinsed with organic solvents before assay, since residues on the glass can invalidate analyses, especially those aimed at the ppb level. Hypochlorite solution (5%) is an effective decontaminating agent for aflatoxin residues on glassware and laboratory benches; it is probably effective for other mycotoxins as well.

39.3 MYCOTOXINS AS FOOD AND FEED CONTAMINANTS

The availability of analytical methods has permitted a limited amount of the food and feed surveillance activity necessary to determine the importance of individual mycotoxins as contaminants. This type of surveillance is necessary, since the experimental production of mycotoxins on certain commodities is no assurance that natural field conditions will be conducive to mycotoxin production.[15, 18] In Table 2 are summarized those commodities in which mycotoxins have been found under experimental and field conditions. The finding of mycotoxin under field conditions is meaningful as an evaluation of the potential health hazard. The data derived from experimental studies are helpful in providing the background against which field surveys are made. Some of the mycotoxin findings reported in Table II represent extensive surveys; others are very limited. Regardless of survey size, these data point out which commodities are potential problems.

39.4 SUMMARY

Most mycotoxicoses of man or animals have been recognized by observation of the toxicity of moldy foods and feeds, without identification of a specific fungal metabolite as the etiologic agent. Even today the technology for isolation and identification of mycotoxins from foods has not kept pace with the growing awareness of the need for such technology. The control and regulation of mycotoxins, known or suspected to be present in foods, awaits the development of specific and sensitive assay methods.

The mycotoxins presently considered to present the most potential for human health hazard are the toxins of the storage fungi in the genera *As-*

Table 2: Selected Mycotoxins Found in Various Foods*

Mycotoxin	Foods found as naturally contaminated	Foods supporting experimental production
Aflatoxins	Peanuts, corn, tree nuts, copra, cottonseed, rice. Milk (transfer from animal feed).	Grains, fruits, meats, spices, cheeses.
Sterigmatocystin	Wheat, green coffee	Corn meal, shredded wheat
Ochratoxin	Barley, corn	Grains, peanuts, legumes
Patulin	Rice, apple juice	Rice
Penicillic acid	Corn, dried beans	Corn
Rubratoxins	Corn	Grains
Citrinin	Rice	Rice, grains
Tremorgenic toxin	Corn	Grains
Yellow rice toxins	Rice	Rice
Zearalenone	Corn, other grains	Grains, rice
12,13-epoxy-Δ^9-trichothecenes	Grains, rice	Grains, rice
Stachybotrys toxins	Oats, hay	Grains

*See references mentioned in text as sources for this information.

pergillus, Penicillium and *Fusarium*. They are known to affect the kidney and liver in domestic animals or to be neurotoxic, uterotrophic, tremorogenic, or tumorogenic. Although the human health effects have not been evaluated completely, experimental and epidemiological data indicate the potential hazards.

The worldwide discovery of the carcinogenic aflatoxins in a variety of agricultural products is responsible for the present research effort and regulatory agency interest in the potential for health hazard presented by mold contaminated foods. The problem is considered acute in the tropical and subtropical areas of the world where malnutrition and widespread contamination of foods with mycotoxins coexist. Research has shown that protein deficient animals sustain more tissue damage than well fed animals when mycotoxins are a part of their dietary exposure. Additionally, aflatoxins have been shown to induce cancer in the livers of experimental animals when the toxins are a part of the diet used to correct malnutrition.

The most effective means of eliminating human exposure to mycotoxins in foods is by the prevention of toxin formation. This requires agricultural and industry practices designed to reduce the opportunity for fungal growth from harvest to ultimate commodity use. Prevention of mycotoxicoses must become a cooperative effort on the part of all involved in food production.

39.5 REFERENCES

1. Kenosita, R. and T. Shikata. 1965. On toxic moldy rice. Mycotoxins in Foodstuffs. G. N. Wogan, editor, The M.I.T. Press, Cambridge, Mass., pp 111–132

2. MILLER, MAX W. 1961. The Pfizer Handbook of Microbial Metabolites. McGraw-Hill Book Company, Inc., New York, N.Y.
3. PIDOPLICHKE, N. M. and V. I. BILAY. 1960. Toxic fungi which develop in food products and fodder. Mycotoxicoses of Man and Agricultural Animals. V. I. Bilay, editor, U.S. Joint Publication Research Service, Washington, D.C., pp 3–36
4. BARGER, G. 1931. Ergot and Ergotism. Gurney and Jackson, London, England
5. WILSON, B. J. and A. W. HAYES. 1973. Microbial toxins. Toxicants Occurring Naturally in Foods, Second Edition, National Academy of Science—National Research Council, Washington, D.C., pp 372–423
6. ANONYMOUS. 1951. Bread of madness infects a town. Life **31**:25
7. MAYER, C. F. 1953. Endemic panmyelotoxicosis in the Russian grain belt. Part I. The Military Surgeon **113**:173–189
8. MAYER, C. F. 1953. Endemic panmyelotoxicosis in the Russian grain belt. Part II. The Military Surgeon **113**:295–315
9. WOGAN, G. N. 1969. Alimentary mycotoxicosis. Food Borne Infections and Intoxications. Hans Rieman, editor, Academic Press, New York, N.Y. pp 395–451
10. RUBINSHTEYN, YU. I. 1960. Food fusariotoxicosis. Mycotoxicoses of Man and Agricultural Animals. V. I. Bilay, editor, U.S. Joint Publication Research Service, Washington, D.C., pp 97–110
11. DROBOTKO, V. G. 1945. Stachybotryotoxicosis. A new disease of horses and humans. Am Rev Soviet Med **2**:238–242
12. FORGACS, J. 1962. Mycotoxicosis—the neglected disease. Feedstuffs (May 5) pp. 124–134
13. FORGACS, J. and W. T. CARLL. 1962. Mycotoxicosis. Adv in Vet Science **7**:273–280
14. AINSWORTH, G. C. and P. K. C. AUSTWICK. 1959. Fungal Disease of Animals. Review Series #6 of the Commonwealth Agricultural Bureaux, Farnham Royal, Bucks, England, pp 97–110
15. GOLDBLATT, L. A. 1969. Aflatoxin. Academic Press, New York, N.Y. 472 pp
16. DICKENS, F. and H. E. H. JONES. 1964. The carcinogenic action of aflatoxin after its subcutaneous injection in the rat. Brit J Cancer **17**:691–698
17. BUTLER, W. H. and J. M. BARNES. 1964. Toxic effects of ground nut meals containing aflatoxins to rats and guinea pigs. Brit J Cancer **17**:697–710
18. CIEGLER, A., S. KADIS, and S. J. AJL. 1971. Microbial toxins. Vol. VI, Fungal Toxins, Academic Press, New York, N.Y. 563 pp
19. KRAYBILL, H. F. and R. E. SHAPIRO. 1969. Implications of Fungal Toxicity to Human Health. Aflatoxin. L. A. Goldblatt, editor, Academic Press, New York, N.Y. pp 401–441
20. SHANK, R. C., G. N. WOGAN, and J. B. GIBSON. 1972. Dietary aflatoxins and human liver cancer. I. Toxigenic molds in foods and foodstuffs in tropical Southeast Asia. Food and Cosmetic Toxicology **10**:51–60
21. SHANK, R. C., G. N. WOGAN, J. B. GIBSON, and A. NONDASUTA. 1972. Dietary aflatoxins and human liver cancer. II. Aflatoxins in market foods and foodstuffs of Thailand and Hong Kong. Food and Cosmetic Toxicology **10**:61–70
22. SHANK, R. C., J. E. GORDON, G. N. WOGAN, A. NONDASUTA, and B. SUBHAMANI. 1972. III. Field survey of rural Thai families for ingestive aflatoxins. Food and Cosmetic Toxicology **10**:71–78
23. SHANK, R. C., N. BHAMARAPRA VATI, J. E. GORDON, and G. N. WOGAN. 1972. Dietary aflatoxins and human liver cancer. IV. Incidence of primary liver cancer in two municipal populations of Thailand. Food and Cosmetic Toxicology **10**:171–179
24. SHANK, R. C., C. H. BOURGEOIS, N. KESCHAMRAS, and P. CHANDA VIMOL. 1971. Aflatoxins in autopsy specimens from Thai children with an acute disease of unknown aetiology. Food and Cosmetic Toxicology **9**:501–507
25. ROBINSON, P. Infantile cirrhosis of the liver in India. Clin Ped **6**:57–62
26. PARPIA, H. A. B. and V. SREENI VASMURTHY. 1971. Importance of Aflatoxin in Foods with reference to India. In Proceedings SOS/70, Third International Congress, Food Science and Technology, Institute of Food Science, pp. 701–704
27. MAJUMDER, S. K., K. S. NARCISIMHAN, and H. A. PARPIA. 1965. Microecological Factors of Microbial Spoilage and the Occurrence of Mycotoxins on Stored Grains. Mycotoxins in Foodstuffs. G. N. Wogan, editor. The M.I.T. Press, Cambridge, Mass., pp 27–47
28. CHRISTENSEN, C. M. 1965. Fungi in cereal grains and their products. Mycotoxins in Foodstuffs. G. N. Wogan, editor, The M.I.T. Press, Cambridge, Mass., pp 9–14

29. RAPER, K. B. and D. I. FENNELL. 1965. The Genus Aspergillus, Williams and Wilkins Company, Baltimore, Md.
30. DETROY, R. W., E. B. LILLEHOJ, and A. CIEGLER. 1971. Aflatoxin and related compounds. Microbial Toxins, A Comprehensive Treatise. Vol. VI, Fungal Toxins, A. Ciegler, S. Kadis, and S. J. Ajl, editors, Academic Press, New York, N.Y. pp 3–178
31. ARMBRECHT, B. 1971. Aflatoxin residues in food and feed derived from plant and animal sources. Residue Reviews, Vol. 41, F. A. Gunther, editor, Springer-Verlag, New York, N.Y. pp. 13–41
32. STEYN, P. S. 1971. Ochratoxin and other dihydroisocoumarins. Microbial Toxins, A Comprehensive Treatise, Vol. VI, Fungal Toxins, A. Ciegler, S. Kadis, and S. J. Ajl, editors, Academic Press, New York, N.Y.
33. PURCHASE, I. F. H. and J. J. VAN DER WATT. 1971. The long-term toxicity of Ochratoxin A to rats. Food and Cosmetic Toxicology **9**:681–682
34. BULLOCK, E., J. C. ROBERTS, and J. G. UNDERWOOD. 1962. Structure of isosterigmatocystin and an amended structure for sterigmatocystin. J Chem Soc. 4179–4182
35. RODRICKS, J. V. 1968. Fungal metabolites which contain substituted 7,8-dihydrofurofurans and 2,3,7,8-tetrahydrofurofurans. J of Agri Food Chem **17**:457–461
36. PURCHASE, I. F. H. and J. J. VAN DER WATT. 1968. Carcinogenicity of sterigmatocystin. Food and Cosmetic Toxicology **6**:55–556
37. PURCHASE, I. F. H. and J. J. VAN DER WATT. 1969. Acute toxicity of sterigmatocystin to rats. Food and Cosmetic Toxicology **7**:135–137
38. WAKSMAN, S. A., E. S. HORNING, and E. L. SPENCER. 1942. The production of two antibacterial substances, fumigacin and clavasin. Science **96**:202–203
39. J. L. WARD. 1943. An antibacterial substance from *Aspergillus clavatus* and *Penicillium claviforme*, and its probable identity with patulin. Nature **152**:750
40. HARWIG, J., Y-K CHEN, B. P. C. KENNEDY, and P. M. SCOTT. 1973. Occurrence of patulin and patulin producing strains of *Penicillium expansum* in natural rots of apple in Canada. Can Inst Food Sci and Technol J. **6**:22–25
41. NORSTADT, F. A. and T. M. MCCALLA. 1969. Patulin production by *Penicillium urticae* Bainer in batch culture. Appl Microbiol **17**:193–196
42. SCOTT, P. M., W. WALBEEK, B. KENNEDY, and D. ANYETI. 1972. Mycotoxins (Ochratoxin A, Citrinin and Sterigmatocystin) and toxigenic fungi in grains and other agricultural products. J of Agri Food Chem. **20**:1103–1109
43. CIEGLER, A., R. W. DETROY, and E. B. LILLEHOJ. 1971. Patulin, penicillic acid and other carcinogenic lactones. Microbial Toxins, Vol. VI, Fungal Toxins, A. Ciegler, S. Kadis, and S. J. Ajl, editors, Academic Press, New York, N.Y. pp 409–434
44. SINGH, J. 1967. Patulin. Antibiotics I. Mechanisms of Action. D. Goulieg and P. D. Shaw, editors, Springer, New York, N.Y. pp 621–630
45. KURTZMAN, D. P. and A. CIEGLER. 1970. Mycotoxin from blue-eye mold of corn. Appl Microbiol **20**:204–207
46. CIEGLER, A., H. J. MINTZLAFF, D. WEISLEDER, and L. LEISTNER. 1972. Potential production and detoxification of penicillin in mold-fermented sausage (Salami). Appl Microbiol **24**:114–119
47. MOSS, M. O. 1971. The rubratoxins, toxic metabolites of *Penicillium rubrum* stoll. Microbial Toxins, Vol. VI, Fungal Toxins, A. Ciegler, S. Kadis, and S. J. Ajl, editors, Academic Press, New York, N.Y. pp 381–407
48. CIEGLER, A. and J. I. PITT. 1970. Survey of the genus *Penicillium* for tremorgenic toxin production. Mycopathologia et Mycologia Applicata **42**:119–124
49. WILSON, B. J., C. H. WILSON, and A. W. HAYES. 1968. Tremorgenic toxic from *Penicillium cyclopium* grown on food materials. Nature **220**:77
50. COLE, R. J., J. W. KIRKSEY, J. H. MOORE, B. R. BLANKENSHIP, U. L. DIENER, and N. D. DA VIS. 1972. Tremorgenic toxin from *Penicillium verruculosum*. Appl Microbiol **24**:248–256
51. PURCHASE, I. F. H. 1971. The acute toxicity of the mycotoxin cyclopiazonic acid to rats. Toxicology Applied Pharmacology **18**:114
52. SAITO, M., M. ENOMOTO, and T. TATSUMO. 1971. Yellowed rice toxins. Luteoskyrin and related compounds, chlorine-containing compounds and citrinin. Microbial Toxins, Vol. VI, Fungal Toxins. A. Ciegler, S. Kadis, S. J. Ajl, editors, Academic Press, New York, N.Y. pp 299–380

53. MIROCHA, C. J., C. M. CHRISTENSEN, and G. H. NELSON. 1971. F-2 (Zearalenone) estrogenic mycotoxin from *Fusarium*. Microbial toxins, Vol. VII, Algal and Fungal Toxins, S. Kadis, A. Ciegler, S. J. Ajl, editors, Academic Press, New York, N.Y. pp 107–138
54. BAMBURG, J. R. and F. M. STRONG. 12,13-Epoxytrichothecenes. Microbial Toxins, Vol. VII, Algal and Fungal Toxins, S. Kadis, A. Ciegler S. J. Ajl, editors, Academic Press, New York, N.Y. pp 207–292
55. TAYLOR, A. 1971. The toxicology of sporidesmin and other epipolythiadioxopiperazines. Microbial Toxins, Vol. VII, Algal and Fungal Toxins, Academic Press, New York, N.Y. pp 337–376
56. CHRISTENSEN, C. M., G. H. NELSON, C. J. MIROCHA, and F. BATES. 1968. Toxicity to experimental animals of 943 isolates of fungi. Cancer Res **28**:2293–2295
57. JOFFEE, A. Z. 1965. Toxin production by cereal fungi causing toxic alimentary aleukia in man. Mycotoxins in Foodstuffs. G. N. Wogan, editor, The M.I.T. Press, Cambridge, Mass., pp 77–85
58. DIENER, U. L. and N. D. DA VIS. Aflatoxin formation by *Aspergillus flavus*, Aflatoxin. L. A. Goldblatt, editor, Academic Press, New York, N.Y. pp 13–75
59. JAR VIS, B. 1972. Mold spoilage of foods. Process Biochemistry **7**:11–14
60. GOLUMBIC, C. and M. M. KULIK. 1969. Fungal spoilage in stored crops and its control. Aflatoxin, L. A. Goldblatt, editor, Academic Press, New York, N.Y. pp 307–332
61. Aflatoxins in corn and soybeans. 1972. Approved Methods of the American Association of Cereal Chemists. Am. Assoc. Cer. Chem., St. Paul, Minn.
62. Aflatoxins in peanuts. 1975. Official and Tentative Methods of the Am. Oil Chem. Soc., The Am. Oil Chem. Soc., Champaign, Ill. Abstract 6–68
63. Association of Official Analytical Chemists, Official Methods of Analysis, 12th edition, Association of Official Analytical Chemists, Washington, D.C., Chapter 26
64. RODRICKS, J. V. and L. STOLOFF. 1971. Tech. Rep. #1. Inter. Union of Pure and Appl. Chem. Info. Bull. IUPAC Secretariat Oxford OX437K, England
65. STACK, M. E. and J. V. RODRICKS. 1973. Collaborative study of the quantitative determination and chemical confirmation of sterigmatocystin in grains. J Assoc of Official Analyt Chem. **56**:1123–1126
66. NESHEIM, S. 1973. Analysis of ochratoxins A and B and their esters in barley, using partition and thin-layer chromatography. II. Collaborative study. J Assoc Official Analyt Chem **56**:822–826
67. POHLAND, A. E., K. SANDERS, and C. W. THORPE. 1970. Analysis and chemical confirmation of patulin in grains. J Assoc of Official Analyt Chem **53**:686–687
68. THORPE, C. W. and R. JOHNSON. 1974. Method for the determination and confirmation of penicillic acid. J of Assoc Official Analyt Chem **57**:861–865
69. EPPLEY, R. M. 1968. Screening method for zearalenone, aflatoxin and ochratoxins. J Assoc Official Analyt Chem **51**:74–78
70. IKEDIOBI, C. O., J. BAMBURG, and F. M. STRONG. 1971. Determination of 12, 13-epoxytrichothecane mycotoxins by gas-liquid chromatography. Analyt Biochem **43**:327–340
71. STOLOFF, L. 1972 Analytical methods for mycotoxins. Clin Toxicology. **5**:465–494
72. JOFFEE, A. Z. 1963. Toxicity of overwintered cereals. Plant and Soil **18**:31–44
73. BURNSIDE, J. E., W. L. SIPPEL, J. FORGACS, W. T. CARLL, M. B. ATWOOD, and E. R. DOLL. 1957. A disease of swine and cattle caused by eating moldy corn. II. Experimental production with pure culture of molds. Am J Vet Res. **18**:817–824
74. LO VETT, J. 1972. Toxigenic fungi from poultry feed and litter. Poultry Science **51**:309–313
75. VERRETT, J. J., J. P. MARLIAC, and J. MCLAUGHLIN, JR. 1974. Use of the chicken embryo in the assay of aflatoxin toxicity. J Assoc Official Analyt Chem. **47**:1003–1006
76. GABLICKS, J., W. SCHAEFFER, L. FRIEDMAN, and G. N. WOGAN. 1965. Effect of aflatoxin B_1 on cell cultures. J of Bacteriol **90**:720–703
77. BROWN, R. F. 1970. Some bioassay methods for mycotoxins. Proc First U.S.-Japan Conf Toxic Micro-Organisms. Mendel Herzberg, editor. UJNR Joint Panels on Toxic Microorganisms and the U.S. Department of the Interior, pp 12–18
78. RODRICKS, J. 1973. Criteria for mycotoxin standards. J Assoc Official Analyt Chem **56**:1290

FOODS AND THE MICROORGANISMS INVOLVED IN THEIR SAFETY AND QUALITY

CHAPTER 40

FISH, CRUSTACEANS, AND PRECOOKED SEAFOODS

J. Liston and J. R. Matches

40.1 INTRODUCTION

Fish and crustaceans appear on the market in many forms, and microbiological methods are needed to determine quality and to detect spoilage in the various forms. Raw and semiprocessed fish and crustaceans are highly perishable protein foods with rather limited refrigerated shelf lives. These products contain the normal bacterial flora from their environments in addition to the contaminants picked up during harvesting and handling of the products. Precooked crustaceans such as crab and lobster are practically sterile after cooking, but subsequently can be recontaminated during handling. Other precooked and prepared seafood products such as breaded shrimp, fish sticks, crab cakes, etc., are breaded and cooked during processing. This reduces the normal flora of the product and the organisms present in the breading ingredients. These products also can be contaminated subsequently by handling. Smoked and salted fish are sometimes, although not desirably, prepared from fresh fish with high bacterial counts and a short refrigerated shelf life. In addition to the normal flora, which may survive the process, contaminants are picked up from the salt ingredients, and during handling of the finished product. The salts may select for specific types of microorganisms which may grow on the product.

Since fish and processed seafoods are held under refrigeration (or frozen), a low incubation temperature is more realistic for the detection of cold tolerant organisms. An incubation temperature of 20 C is used in this chapter for aerobic plate counts as an index of spoilage. An incubation of 35 C is used with some samples as a measure of contamination, and especially for organisms associated with warm blooded animals.

The methodology suggested for this chapter is that used by the Food and Drug Administration.[2] These methods will be followed except where modified as noted in this manual. In addition, the methodology and sampling for microbiological analysis published by the International Commission on Microbiological Specifications for Foods (ICMSF) of the International Association of Microbiological Societies may be considered for reference.[17, 20]

Microbiological standards have been established for some seafoods by several eastern states and cities, and a partial list is appended at the end of this chapter.

IS/A Committee Liaison: John T. Graikoski

40.2 FRESH AND FROZEN FISH AND CRUSTACEANS

40.21 Normal Flora

Living fish and crustaceans carry populations of predominantly gram negative psychrotrophic* bacteria on their external surfaces. The internal tissues and blood system of healthy fish and crustaceans are usually sterile. However, it has been reported recently that crabs caught in shallow inshore waters may have a bacteremic condition. Since fish and crustaceans are harvested normally in waters more or less remote from the processing sites, the microflora at the point of processing will be influenced by the conditions of handling and by the storage of the raw product prior to processing. Moreover, fish processing involves extensive human handling, and the microflora of the product immediately after processing reflects this contact. Since fish and crustaceans typically are held under chilling or refrigerated conditions during storage and transport, the dominant microflora usually is composed of gram negative psychrotrophic bacteria related to the types originally present on the living fish. Aquatic animals also may accumulate bacteria from their environment. Fish and crustaceans taken from polluted waters may carry bacteria derived from human and animal sources. Naturally occurring, potentially pathogenic bacteria are limited to *Clostridium botulinum* Types A, B, E, and F and *Vibrio parahaemolyticus*.

Fresh caught fish and crustaceans typically carry populations of 10^2 to 10^3 bacteria per square centimeter of skin surface or per gram of gill tissue. The numbers in the alimentary canal depend on the amount of food present, ranging from a very few to in excess of 10^7 bacteria per gram of gut contents. Bacteria of the genera *Pseudomonas, Acinetobacter-Moraxella* (*Achromobacter*), *Flavobacterium,* and *Vibrio* (particularly in gut samples) are normally most abundantly present. Coryneform bacteria (mostly *Arthrobacter*) frequently are present in low numbers. However, they have been reported to be a significant component of the microflora of some crustaceans. Other gram positive bacteria such as *Micrococcus, Bacillus*, and *Clostridium* occur variably and in small numbers. The microflora is typically psychrotrophic and most of the organisms do not grow well at 35 to 37 C. The most convenient temperature of incubation is 20 to 25 C, although higher counts may be obtained at 0 C. The fish bacteria do show halotrophism, but are not obligately halophilic, except for *Vibrio parahaemolyticus,* so that growth of most of the organisms occurs readily on normal laboratory media. Addition of 0.5% NaCl to such media may produce better results and usually does not alter their productivity vis-á-vis terrestrial bacteria. Coliforms should be absent, or present in very low numbers only, and *Salmonella, Shigella,* and other enteric pathogens should not occur, since these organisms are not part of the normal flora of the animals or of their environment. However, because of handling or localized contamination on the fishing vessel, small numbers of coliforms are not uncommon.

---

*Psychrotrophic refers to bacteria which grow well at temperatures near 0 C but have an optimum temperature above 20 C, with a maximum temperature in the range 30 to 35 C.

Fish and crustaceans normally are transported from the harvesting areas to processing plants in ice or refrigerated sea water. While the growth of bacteria is slowed by this procedure, the population, because of its psychrotrophic nature, does increase, and the count on the raw material may be expected to be between 10^4 and 10^6 bacteria per square centimeter of surface or of gram on arrival at the processing plant. Good quality material held under good conditions has counts in the range of 10^4 to 10^5 bacteria per gram or per square centimeter. Proper in-plant processing should reduce the count, or at least hold it stable. However, handling of the product frequently will bring about an increase in the proportion of gram positive bacteria, particularly staphylococci, on the seafood and may introduce coliform bacteria. Subsequent storage of the product at refrigerating temperatures may lead to a progressive disappearance of these bacteria, due to overgrowth by psychrotrophic spoilage bacteria. Insanitary processing conditions may lead to contamination of the product with potentially harmful bacteria of human or warm blooded animal origin.

40.22 Floral Change During Spoilage

The nature of the bacterial flora developing during spoilage is related to the bacteria initially present on the product and to the temperature of storage. *Pseudomonas* species rapidly become dominant in the microflora of chilled fish, reaching from 70 to 90% in a few days. *Acinetobacter-Moraxella* (*Achromobacter*) and *Flavobacterium* make up the bulk of the remainder. Small numbers of other bacterial species may be present, particularly in the early stages of spoilage, but they rapidly become diluted out by the predominant flora. Raw chilled crustaceans generally have a similar spoilage flora, but there have been reports of dominance of *Acinetobacter-Moraxella* (*Achromobacter*) and even, at somewhat higher holding temperatures, of coryneform bacteria on these products. Spoilage occurs principally as the result of the surface growth and the biochemical activity of bacteria. However, in the case of whole fish, there may be some penetration of bacteria into the blood system through the gills during death, and into tissue adjacent to the gut cavity in eviscerated fish. Significant penetration of bacteria into deep muscle tissues does not occur until spoilage is advanced. Penetration occurs more rapidly in fillets than in whole fish and, of course, varies with size. Thus the spoilage rate under equivalent conditions of holding is highly variable for fish of different sizes and for different raw product forms.

40.23 Health Related Bacteria

Fish and crustaceans harvested from offshore marine waters should be free of the potentially pathogenic bacteria normally associated with warm blooded land animals.[23, 25] However, when taken from estuarine waters or rivers and lakes, their microflora may reflect the degree of contamination in the environment. The only naturally occurring, potentially pathogenic bacteria are *C. botulinum* Types A, B, E, and F, present usually in very low numbers, mainly in gut contents,[4, 5, 6, 9, 11, 22, 27] and *Vibrio para-*

haemolyticus, present also in very low numbers on most fish and crustaceans, but occasionally more abundantly present on crustaceans taken from warm inshore waters.[3] Neither of these organisms represents a major hazard at their normal level of occurrence and usually they become significant only where improper processing or handling conditions are applied to the dead animal. Insanitary processing procedures can lead to contamination of fish or crustaceans with potentially pathogenic bacteria (e.g., staphylococci and salmonellae), and the dead tissues of these animals will support survival and growth of such microorganisms at suitable temperatures.

40.24 Recommended Methods

40.241 Preparation of samples

Care must be exercised in taking samples so that they are representative of the items to be tested, and to make sure that no contamination or change occurs prior to examination. Samples must be taken under aseptic conditions with sterilized instruments.

Cleaned sampling instruments should be sterilized prior to use by autoclaving for 15 minutes at 121 C. When instruments must be reused several times, they may be cleaned and then sterilized by alcohol flaming three times before each use, or by heating in a flame and cooling before use. Attention must be given to the cleaning of tissue from sampling instruments prior to flaming, especially from the rough gripping areas of forceps.

Collected samples should be placed in sterile containers and transported to the laboratory with sufficient speed and refrigeration to prevent microbial multiplication during transit. Frozen samples must be kept solidly frozen and unfrozen samples transported under wet ice or refrigeration. Unfrozen samples should not be frozen, especially if *Vibrio parahaemolyticus* and other organisms are to be estimated, because of possible freezing damage with reduction in numbers. When samples must be transported over long distances, freezing may be necessary. The results obtained then must be interpreted in the light of possible freezing damage to the microorganisms.

For whole fish, surface samples should be obtained of a sufficient total area to be representative of the fish. The areas to be sampled can be outlined by templates. Templates may be prepared from light weight aluminum sheet. The round hole which defines the area to be sampled may be cut with snips or a gasket punch. A rim of ½ to 1 centimeter wide should be left around the hole. A tab left on one side of the template and bent up at a slight angle is handy to hold and handle the sterile template. The size of the hole is dependent upon the area to be sampled; however, the hole size should be known, so that the area sampled can be calculated. A convenient template size for large flat surfaces is 10 centimeters.[2] Sterilize templates in a Petri dish.

a. Skin samples

Incisions are made around a measured area as defined by a template. A corner of skin is lifted with sterile forceps, and removed by cutting adhering muscle, using a sterile scalpel. The excised sample is placed quickly in a sterile dilution bottle.

b. Swab samples

With sterile forceps, select a sterile template and position firmly against the surface to be sampled. The template can be held in place with the forceps, or the tab on the template can be held with the fingers. When several samplings are to be made from a number of fish, select the same approximate area on each fish.

Remove a sterile swab (absorbant cotton or alginate) from a test tube, moisten in sterile water, and swab the entire skin surface exposed in the opening of the template. The swab should be applied firmly, and slowly twirled, so that it travels in all directions thoroughly covering the surface exposed, removing the slime and adhering organisms. Remove the closure from a dilution bottle and insert the swab about 2 inches within the neck of the bottle. Break the tip off below where the operator has touched the applicator stick, thereby dropping the cotton tip into the dilution water. Shake the dilution blank adequately with a mechanical shaker or by hand. Skin or swab samples should be shaken or blended with 0.1% peptone water (generally 100 ml). If a shaking procedure is followed for skin samples, sterile sand or glass beads should be included with the peptone water to ensure abrasive removal of the microorganisms. Shaking can be accomplished best by using a mechanical shaker, but shaking by hand can be used. In each case, a 3 minute period of agitation is recommended.

c. Tissue samples

For fillets or other semiprocessed raw fish, shrimp, picked crabmeat, and lobster, a representative 50 gram sample of the meats should be blended with 450 ml. of cold 0.1% peptone water in a sterile cold blender jar for 1 to 2 minutes. The samples should not be blended for more than 3 minutes to avoid overheating. Dilutions should be made in cold 0.1% peptone water.

Frozen fish and crustaceans such as cooked or uncooked, peeled or unpeeled shrimp, uncooked lobster tails, cooked and picked or unpicked crabmeat may be handled in the same manner as fillets or other semiprocessed raw fish. Meat from cooked unpicked crab and crab legs will have to be removed aseptically from the shell for analysis.

A brace or electric drill and a sterile bit may be employed for taking samples from large frozen fish or frozen blocks. Small pieces of tissue may be cut from IQF fillets, shrimp, etc., using a sterile knife.

40.242 Counts

a. Aerobic plate count: Use Chapter 4 modified by the incorporation of 0.5% sodium chloride in the medium, and an incubation temperature in the range 20 to 25 C for 4 days, and 35 C for 48 hours ± 2 hours.

b. Coliform group: Use the method described in Chapter 24.

c. ***E. coli:*** As outlined in Chapter 24.

d. ***S. aureus:*** Use MPN technique, Chapters 6 and 31. For samples suspected of containing large numbers, spread 0.1 ml volumes of each dilution directly on Baird-Parker agar plates.

e. ***V. parahaemolyticus:*** Use procedure described by Fishbein and Wentz[8] or Chapter 29. Do not freeze the sample. Make dilutions in 3% sodium chloride solution.

f. ***C. botulinum*** Types A, B, E, and F: Chapter 34.

40.243 Interpretation of data

An aerobic plate count at 35 C may be useful to provide an index of insanitary handling or high temperature storage. In most fresh raw seafoods, the count at 35 C should be 1/10th of the count at 20 C. Equivalent counts at 20 and 35 C suggest mishandling and contamination from human or warm blooded animal sources.

Fish and shellfish of good quality will have counts of less than $10^5/cm^2$ (or per gram of fish or crustacean tissue) at 20 C. However, counts up to 10^6 may be found on products of acceptable quality, although such seafoods will have a limited shelf life in the raw chilled state. Counts in excess of 10^6 should be taken as evidence of incipient spoilage requiring further analysis of the raw material.[17]

Since neither coliform nor fecal coliform bacteria nor staphylococci are naturally present on raw seafoods in most cases, the presence of bacteria of these types in significant numbers (fecal coliform in excess of 10/gram and staphylococci in excess of 100/gram) should be taken as indication of contamination from human sources. However, the presence of these bacteria on fish and crustaceans harvested from fresh water or from estuarine areas may be representative of the condition of these environments, indicating pollution from human or land animal sources. Fish or crustaceans harvested from water suspected to be polluted by human sewage should be tested for *Salmonella,* since these organisms will survive on fish and grow at temperatures in excess of 8 C.[14, 15] Routine analysis for *C. botulinum* serves no useful purpose, since these organisms present no real hazard in the case of chilled raw fish or crustaceans. *V. parahaemolyticus* can be expected to occur, particularly on crustaceans harvested from inshore waters during the warm period of the year. Essentially all food poisoning outbreaks due to this organism which have occurred in the USA have resulted from consumption of crustaceans. Some type of surveillance of the occurrence of these organisms on fresh products is thus indicated. It is suggested that the discovery of numbers in excess of 10^2/gram should be considered cause for alarm and for further investigation.

40.3 PRECOOKED CRUSTACEAN PRODUCTS

40.31 Normal Flora

Normally in processing, crab and lobster are cooked by steam or immersion in boiling water. This inactivates the enzymes and facilitates the separation of the meat from the shell. The products are marketed as picked meat or as the whole cooked animal. Shrimp also may receive a cooking procedure prior to peeling or before freezing, but much shrimp is retailed in the uncooked state. The precooking process inactivates the majority of the naturally occurring psychrotrophic bacteria, and the microflora on the finished product is derived principally from recontamination after the cooking step. Consequently it is quite common to find significant occurrence of gram positive bacteria and other organisms of terrestrial origin on precooked crustacean products. If the material is held for a period of more than a few days at chilling temperatures, the gram negative psychrophilic population once more will become dominant. Crab meats may be separated from shell fragments by a brine flotation process which provides the opportunity for a buildup of salt tolerant bacteria, particularly staphylococci. Recontamination of crustacean meats after cooking with *V. parahaemolyticus* is a particular hazard because of the very rapid growth rate of these organisms on such products.

40.32 Spoilage

Spoilage of the cooked products held under chilling conditions follows the same pattern as for the raw product, with *Pseudomonas* dominating the flora in most cases. Heat pasteurized picked crabmeat may show a very long shelf life under refrigeration, and little is known of the spoilage flora of this product.

40.33 Health Related Bacteria

These are of particular importance in precooked crustacean products, since the heat treatment destroys the rapidly growing spoilage bacteria which normally overgrow mesophilic bacteria and reduce the hazard of food poisoning. Staphylococci and coliform (preferably fecal coliform) bacteria are useful indicators of hazard in such products and should be tested for routinely. *V. parahaemolyticus* is extremely heat sensitive and should be destroyed readily by the precooking process. Therefore, its presence on the cooked products is indicative of postprocess contamination. The possibility of growth of *C. botulinum* Type E and F in chilled pasteurized crabmeat has been a cause of concern because of its presence on the raw product. Although these products have not been incriminated, it is advisable to check for these organisms.

40.34 Recommended Methods

As for fresh fish and crustaceans, use 35 C count routinely, together with 20 C.

40.35 Interpretation of Data

Total counts in excess of 10^6/gram at 20 C should be considered undesirable, though much frozen raw shrimp in commerce does show counts in excess of this. Cooked picked crabmeat normally should be expected to have total counts less than 10^5/gram. Staphylococci should not exceed 10^3/

gram in good quality products, and generally will be less than 10^2/gram. Fecal coliforms should be low, and in most cases will be undetectable in a one gram sample, but in any case should not exceed 40/gram in products of good quality. *V. parahaemolyticus* should be absent or at most present to less than 10^2/gram.

40.4 PRECOOKED AND PREPARED SEAFOOD PRODUCTS

40.41 Normal Flora

This category includes breaded shrimp and fish products, crab cakes, fish cakes, etc. The characteristic difference between these and other seafood products is the addition of material from nonmarine sources, including breading, spices, milk products, and egg products. This raises the possibility of contamination by organisms which are exotic to fish but more common on terrestrial plants and animals, such as *Salmonella* and *Staphylococcus.* Indeed, if a precooking step is involved, as in the case of fish sticks, breaded shrimp, fish cakes, etc., then the final microflora may be more representative of the heat tolerant bacteria in the added material than the generally heat sensitive bacteria of fish and crustaceans.

The microflora of prepared products might, therefore, be expected to include fish spoilage bacteria, gram positive cocci, including enterococci, *Bacillus*, *Lactobacillus*, and other gram positive rods, as well as variable numbers of coliforms or other members of the *Enterobactereaceae* and even molds and yeasts. In uncooked chilled products, the fish spoilage types (mainly *Pseudomonas*) might be expected to be dominant, but in cooked products the gram positive bacteria will be most abundant unless the product is stored for more than a few days at chilling temperatures.

40.42 Spoilage

For chilled products, floral changes are similar to those for other fish products, though in certain cases low temperature lactobacilli and possibly yeasts may be present in significant numbers.

40.43 Health Related Bacteria

These may include any of the species known to occur on the added nonfish components, as well as the fish related types of *C. botulinum* types A, B, E, F, and *V. parahaemolyticus*. Of greatest concern because of the likelihood of their presence in added ingredients are *Staphylococcus aureus* and *Salmonella* species. *C. botulinum* and *V. parahaemolyticus* are of significance, particularly in products unlikely to be heated strongly before consumption.

40.44 Recommended Methods

As for fresh fish and crustaceans;
Total aerobic plate count, 35 C and 20 C, for spoilage;
Coagulase positive staphylococci;
Fecal coliforms (*E. coli*);

V. parahaemolyticus in the case of shrimp and crab products produced during the summer;
C. botulinum.

40.45 Interpretation of Data

Prepared precooked products usually are prepared for eating by mild heating, or may not be heated at all. Consequently the potential hazard to the consumer from contaminated products of this type is much higher than for raw products. It is important, therefore, to evaluate carefully the results of indicator organism (fecal coliform and *Staphylococcus*) counts, and to undertake tests for enteric pathogens where count values are in excess of those considered normal for the product. In the absence of indicator organism data, high counts at 35 C (> 500,000/gram) should be considered evidence of a potentially hazardous situation indicative of faulty processing, contamination of the processed product, or poor temperature control during storage. Discovery of high counts in the final product should be followed by an evaluation of the processing procedure, and of the microbiological quality of the breading and batter mix or other components added to the fish or crustaceans.

40.5 SMOKED AND SALTED FISH PRODUCTS

40.51 Normal Flora

Smoked and salted fish are prepared in a variety of ways, each of which affects the final microflora composition. Smoking is preceded by a variable amount of hand processing which can add bacteria of human or terrigenous origin to the fish. The brining process may be wet or dry, and may result in an intermediate product for smoking, with salt content from 1% to 5%, or a fully salted product containing in excess of 20% salt. Fully salted fish commonly are dried and shipped in a stable, dry condition. Smoked products will be dried to a greater or lesser extent, depending on the smoking process, and may be essentially uncooked, as in cold smoking, or may be more or less fully cooked, as in hot smoked or barbecued products. The bacteriological stability of the final product will depend on the amount of unbound water remaining in it (a_w).

It is common to find a great variety of bacterial types on smoked fish, with gram positive microorganisms, including *Micrococcus*, *Bacillus*, coryneform bacteria, and yeast predominating, though gram negative bacteria also will be present. Indeed, in a lightly brined cold smoked fish, gram negative bacteria, including *Pseudomonas*, *Flavobacterium*, and *Acinetobacter-Moraxella* (*Achromobacter*) may approximate in number the gram positive types. Bacteria may be derived from the brine or the wooden containers in which salting frequently is carried out. Mold spores commonly occur on smoked fish, and molds may grow during storage of such fish under refrigeration or ambient temperature. Halophilic molds and bacteria may be present on fully salted fish and may grow during storage.

40.52 Spoilage

Spoilage of smoked fish may be due to a variety of microorganisms, depending on the residual a_w of the fish, the smoking temperature, and the storage conditions. Spoilage of lightly brined cold or hot smoked fish of high a_w held under refrigeration is due mainly to gram negative bacteria; however, gram positive types also may be involved. Frequently, however, molds also will grow during storage, though their role in actual spoilage is questionable. Hard smoked products of low a_w normally will spoil due to growth of molds, although halophilic bacteria also may be involved. Spoilage of salt dried fish may occur due to the growth of halobacteria (pink) or molds (particularly *Sporadonema* (Dun). The halobacteria may grow in the fish during aerobic stages of the curing process (e.g., wet stack stage) and are derived frequently from solar salts.[12]

40.53 Health Related Bacteria

Because of the extensive contact handling of the raw material in the preparation of smoked and salted fish, it may be expected that considerable contamination of the product with bacteria from human and terrigenous sources will occur. The low a_w and relatively high salt content of many of these products are not conducive to growth of many health related bacteria, a notable exception being *Staphylococcus aureus*. *Salmonella* will survive on smoked fish and can represent a hazard. *C. botulinum*, particularly Type E, represents a potential hazard in hot smoked fish products. The Eh of the flesh of these products is sufficiently low to permit outgrowth of *C. botulinum*, which naturally can be present on fish, and elimination of most of the heat sensitive spoilage flora provides conditions for growth to occur when held at abuse temperature. U.S. Federal and several state regulations require that the process be sufficient to destroy these organisms or to have adequate concentrations of salts (sodium chloride and sodium nitrite) and holding temperatures to prevent outgrowth if present.

40.54 Recommended Methods

Total aerobic count, 35 C and 20 C;
Fecal coliforms (*E. coli*);
Coagulase positive *Staphylococcus*;
C. botulinum Types A, B, E, and F (Smoked Fish);
Halophilic microorganisms. Chapter 12.

40.55 Interpretation of Data

Cold smoked fish should have total count values less than 10^5/gram, staphylococci less than 10^2/gram, and fecal coliforms less than 40/gram. Total counts of 10^6/gram and greater would suggest spoilage, and *Staphylococcus* counts in excess of 10^3/gram and fecal coliforms much over 40/gram would indicate faulty handling procedures. Similar values might be expected for hot smoked products, although counts should be lower in their case because of the lethal effect of the heating process.

The presence of *C. botulinum* Types A, B, E, and/or F in smoked fish should be taken to indicate a potentially hazardous product resulting from contamination.

40.6 MISCELLANEOUS FISH PRODUCTS

40.61 Normal Flora

In this section are handled fish and fish products which are processed in different ways. As a result, the normal flora will be dependent upon the type of process and methods of handling and storage.

40.611 Dried fish

Dried fish produced from uncooked fish will contain a population of bacteria present at the time of catch, as well as contaminants picked up during processing. Dried fish produced from precooked fish flesh will contain a normal flora composed almost entirely of contaminants picked up after cooking. Although the flora will be composed of gram negative bacteria from the aquatic environment and mainly gram positive contaminants, this population will change gradually as drying results in the death of vegetative forms of bacteria.[12] Bacterial and mold spores will remain viable for long periods of time but will not multiply as long as the a_w is low (moisture content of 20% or lower). Naturally or sun dried fish would be expected to contain higher numbers of bacteria than fish dried in sanitary driers; however, little is known about the microbiology of naturally dried fish.

40.612 Fermented fish

The variety of fermented fish products available has been reviewed by Amano.[1] Fermented fish products such as fish sauces and pastes are produced by the action of natural fish enzymes, and probably, by enzymes hydrolyzing the fish proteins. These fermentations are carried out at high salt concentration, which retards microbial growth. During this process, which takes several months or longer, the normal bacterial flora is retarded or destroyed by the high salt concentration, and a halophilic or halotolerant population composed of gram positive bacilli remains. After fermentation for one year, aerobic plate counts on media containing 0.5 to 20.0% sodium chloride range from 10^1 to 10^3 bacteria/ml.[21]

40.613 Pickled fish

Pickled or cured fish is prepared from salted fish which are washed to remove the salt. The freshened fish are then repacked into smaller containers, and usually a hot pickle solution consisting of vinegar, sugar, and spices is added.[12] A fermentative population can develop during storage; however, the concentration of vinegar will determine the numbers and types of organisms. Erichsen[7] studied the microbiology of herring tidbits and found the population to be composed of gram positive cocci and rods. Pickled fish unless spoiled normally carry very low levels of bacteria in the range of 10^1 to 10^3/gram.

40.614 Fish protein concentrate

In the production of FPC, fish enzymes are heat denatured, and most of the lipids are removed by hot isopropanol extraction. This process also destroys most of the bacteria present in the raw material.[10] However, the FPC may become cross contaminated from raw fish or from the environment, and some bacteria may survive the processing operation. Federal regulations (1967) require that FPC shall be free of *Escherichia coli* and pathogenic organisms, including *Salmonella*, and shall have a total bacterial plate count of not more than 10,000/gram. This flora may be made up of both gram negative and gram positive organisms. The organisms found in FPC usually have been *Bacillus* and in a few cases *Staphylococcus* and *Micrococcus.*

40.615 Fish sausage

Fish paste or jelly (Kamaboko) and fish sausage are major fish products in Japan. The heat treatment given these products destroys many of the "normal fish bacteria" and selects for the sporeforming bacteria. In studies reported by Amano,[1] Okitsu et al,[18] the deterioration is characterized by the formation of spots or speckles on the surface, or by formation of gas in fish sausage. The causative organisms were identified as members of the genus *Bacillus*. Therefore imported fish products of this type may be expected to contain a microflora of *Bacillus*.

40.62 Spoilage

Of the miscellaneous fish products mentioned in this chapter, none is produced in the United States in large volumes with the exception of pickled herring. The spoilage of these products will be different from fresh fish because the processes used are designed to produce products with an extended shelf life as well as other properties. With normal handling, dried, fermented, and pickled fish products and FPC are shelf stable for long periods. The spoilage of these products will be caused by a breakdown of proper processing and storage, such as an increase in the moisture content of dried fish products and FPC, low salt levels in fermented or pickled fish, or too high a pH in pickled fish. Fish sausage, on the other hand, is a highly perishable product without the inhibitors used in Japan. The spoilage, therefore, may be expected to be due to the growth of both gram negative and gram positive organisms.

40.63 Health Related Bacteria

The potential for seafoods and other fishery products to act as causes of foodborne diseases is very large; however, the data available in the literature show that such products, except molluscan shellfish, infrequently are involved. Catching from polluted waters, improper handling, and contamination are potential sources of disease producing bacteria. These may include any of the species known to occur on fisheries products and include *Clostridium botulinum* Types A, B, E, and F, *Vibrio parahaemolyticus*, *Staphylococcus aureus*, and *E. coli.*

40.64 Recommended Methods

As for precooked fish and crustaceans;
Total aerobic plate count, 35 C and 20 C;
C. botulinum Types A, B, E, and F, pickled fish, and fish sausage;
Coagulase positive *Staphylococcus*;
Vibrio parahaemolyticus, pickled fish, and fish sausage;
Molds. Chapter 16;
E. coli;
Halophilic microorganisms. Chapter 12.

40.65 Interpretation of Data

High total counts of 10^6/gram and greater would suggest spoilage, unsanitary processing conditions, or the use of unsound fish. *Staphylococcus* counts in excess of 10^3/gram and fecal coliforms much over 40/gram would indicate poor sanitation and handling procedures. The presence of *C. botulinum* or *Vibrio parahaemolyticus* can be taken to indicate a potentially hazardous situation.

Table 1: States and Municipalities Having Laws, Regulations, Codes, or Ordinances Specifying Microbiological Limits for Seafoods (1975)

State or Municipality	Product	Microbiological Limits	
Connecticut	Shellfish growing water	Coliform organism MPN	≤ 70/100 ml
	Shellfish	Standard plate count	≤ 500,000/g
		Escherichia coli	≤ 250/100g
Kentucky	Oysters	Coliform organism MPN	160,000/100g
		Standard plate count	1,000,000/g
New York, N.Y.	Crabmeat	*Staphylococcus aureus*	< 100/g
		Coliform organisms	< 100/g
		Enterococci	< 1000/g
		Standard plate count	< 100,000/g
	Shellfish	Fecal coliforms	< 230/100g
		Standard plate count	≤ 500,000/g
	Perishable or potentially hazardous foods*	Standard plate count	≤ 100,000/g
		Fecal coliforms	none
		Salmonella	none
Memphis-Shelby County, TN	Cooked foods	Standard plate count	100,000/g
	Seafood, uncooked	Standard plate count	1,000,000/g
Eau Claire City/ County, WI	Ready-to-eat foods	Standard plate count	100,000/g
		Coliform organisms	100/g

*Includes ready-to-eat fish and crabmeat other than pasteurized or sterilized crabmeat.

Table 2: States and Municipalities Having Administrative Regulations or Guidelines Specifying Microbiological Limits for Seafoods (1975)

State or Municipality	Product	Microbiological Limits	
Georgia	Shellfish, etc.	Specified as "For use of State only."	
Indiana	Oysters	*Escherichia coli*	≤ 250/100g
		Standard plate count	≤ 500,000g
Rhode Island	Seafood, fresh	Standard plate count	1,000,000/g
	Shellfish, fresh	Standard plate count	500,000/g
		Fecal coliform	2.3 for raw shellfish
	Smoked fish	Standard plate count	100,000/g
	Prepared foods, including seafood	Standard plate count	100,000/g
		Coliform organisms	100/g
	All prepared foods	Fecal streptococci	1000/g
		Staphylococci	100/g
		Staphylococcus aureus	none
		Salmonella and *Shigella*	none
		Beta hemolytic streptococci	none
		Vibrio parahaemolyticus	none
		Toxigenic mold	none
NOTE: Action taken when arithmetic average of 10 samples are substandard or when 1 of 4 samples is substandard.			
Virginia	Crabmeat	Standard plate count	100/g
		Fecal coliforms	50/100g
		Coliform organisms	4900/g
	Oysters and clams	Standard plate count	500,000/g
		Fecal coliforms	230/100g
Nebraska	Ready-to-eat foods	Standard plate count	100,000/g
		Coliform organisms	100/g
Chicago, IL	Shellfish	Standard plate count	500,000/g
		Fecal coliforms	230/100g
St. Louis, MO	All foods	Standard plate count	≤ 100,000/g
		Coliform organisms	≤ 100/g
		Staphylococcus aureus	none
New York	All foods	Standard plate count	< 1,000,000/g
		Escherichia coli	< 3/g
		Staphylococcus aureus	< 100/g
		Salmonella	none

40.7 REFERENCES

1. Amano, K. 1961. The influence of fermentations on the nutritive value of fish with special reference to fermented fish products of southeast Asia. Fish in nutrition, *eds*. R. Kreuzer and E. Heen. In Fishing News, Ltd., London, England, pp 180–200
2. Bacteriological Analytical Methods for Foods (BAM). 1972. Superintendent of Docu-

ments, U.S. Government Printing Office. Washington, D.C.

3. BAROSS, J. and J. LISTON. 1970. Occurrence of *Vibrio parahaemolyticus* and related hemolytic vibrios in marine environments of Washington State. Appl. Microbiol. **29**:179–186
4. CRAIG, J. M. and K. S. PILCHER. 1966. *Clostridium botulinum* type F; Isolation from salmon from the Columbia river. Science, **153**:311–312
5. CRAIG, J. M., S. HAYES, and K. S. PILCHER. 1968. Incidence of *Clostridium botulinum* type E in salmon and other marine fish in the Pacific Northwest. Appl. Microbiol. **16**:553–557
6. EKLUND, M.W. and POYSKY, F. 1967. Incidence of *C. botulinum* Type E from Pacific coast of the United States, *In* Botulism. Ed. M. Ingram and T. A. Roberts, Chapman and Hall, London, England. pp 49–55
7. ERICHSEN, I. 1967. The microflora of semi-preserved fish products. III. Principal groups of bacteria occurring in tidbits. Antonie von Leeuwenhoek, J. Microbiol. Ser. **33**:107–112
8. FISHBEIN, MORRIS and BARRY WENTZ. 1973. *Vibrio parahaemolyticus* methodology for isolation from seafoods and epidemic specimens. J. Milk & Food Technol. **36**:118–123
9. GANGAROSA, E. J., J. A. DONADIO, R. W. ARMSTRONG, K. F. MEYER, P. S. BRACHMAN, and V. R. DOWELL. 1971. Botulism in the United States, 1899–1969. Am. J. Epidemiol. **93**:93–101
10. GOLDMINTZ, DANIEL and J. C. HULL. 1970. Bacteriological aspects of fish protein concentrate production. Develop. Ind. Microbiol. **11**:335–340
11. GRAIKOSKI, J. T. 1971. Seafoods and botulism. In Fish Inspection and Quality Control. Ed. Rudolf Kreuzer, Fishing News (books) LTD, Ludgate House, London, England. pp 249–255
12. GRAIKOSKI, JOHN T. 1973. Microbiology of cured and fermented fish. Microbial Safety of Fishery Products. Academic Press, Inc., New York, N.Y. pp 98–110
13. LISTON, J. 1957. The occurrence and distribution of bacterial types on flatfish. J. Gen. Microbiol. **16**:205–216.
14. LISTON, J., J. R. MATCHES, and J. BAROSS. 1971. Survival and growth of pathogenic bacteria in seafoods. In Fish Inspection and Quality Control. Ed. Rudolf Kreuzer, Fishing News (books) LTD. Ludgate House, London, England, pp 246–249
15. MATCHES, JACK R. and J. LISTON. 1968. Growth of salmonellae on irradiated and non-irradiated seafoods. J. Food Sci. **33**:406–410
16. NEUFELD, N. 1971. Influence of bacteriological standards on the quality of inspected fish products. In Fish Inspection and Quality Control. Ed. Rudolf Kreuzer, Fishing News (Books) LTD, Lundgate House, London, England, pp 234–240
17. Microorganisms in foods. 1974. Sampling for Microbiological Analysis: Principles and Specific Applications. International Commission on Microbiological Specifications for Foods (ICMSF) of the International Association of Microbiological Societies, Univ. of Toronto Press, Toronto, Canada
18. OKITSU, TOMOAKI, TOSHIHARU KAWABATA, and TSUNEO KOZIMA. 1964. Features of the causative organisms of deterioration of fish sausage. I. Effect of some food preservatives on the growth of causative bacteria. Bull. Jap. Soc. of Sci. Fish. **30**:63–68
19. PASKELL, STEFAN L. and DANIEL GOLDMINTZ. 1973. Bacteriological aspects of fish protein concentrate production from a large-scale experiment and demonstration plant. Develop. Ind. Microbiol. **14**:302–309
20. SAISITHI, PRASERT, BUNG-ORN KASEMSARN, J. LISTON, and ALEXANDER M. DOLLAR. 1966. Microbiology and chemistry of fermented fish. J. Food Sci. **31**:105–110
21. SAKUGUCHI, G. 1969, Botulism type E. *In* Food-born infections and Intoxications. Edited by H. Riemann, Academic Press, New York, N.Y. p 329
22. SHEWAN, J. M. 1961. The microbiology of sea water fish. In Fish as Food, Vol. 1, Edited by G. Borgstrom, Academic Press, New York, N.Y. p 487
23. SHEWAN, J. M. 1970. Bacteriological standards for fish and fishery products. Chemistry and Industry, **6**:193–197
24. THATCHER, F. S. (Ed.) 1969. Hygienic and safety aspects of quality control. In fish Inspection and Quality Control. Ed. Rudolf Kreuzer, Fishing News (Books) LTD, Lundgate House, London, England pp 222–225
25. THATCHER, F. S. and D. S. CLARK (Eds). 1968. Microorganisms in Foods. Their Significance and Methods of Enumeration. Univ. of Toronto Press, Toronto, Canada
26. WALLS, N. W. 1968. *Clostridium botulinum* type F. Isolation from blue crabs. Science, **155**:375–376

CHAPTER 41

MOLLUSCAN SHELLFISH, FRESH OR FRESH FROZEN OYSTERS, MUSSELS, OR CLAMS

Daniel A. Hunt, John Miescier, James Redman, and Arnold Salinger

41.1 INTRODUCTION

Bivalved molluscs such as oysters, mussels, and clams are economically important marine food species found in abundance in estuarine and marine waters. These organisms are filter feeders and commonly are eaten whole in the living state. While feeding on plankton and other microflora of the estuary, shellfish may concentrate pathogenic bacteria and viruses from waters affected by domestic pollution. Historically, *Salmonella typhi* has been the most significant microbial contaminant from the point of view of shellfish-borne epidemics. Based upon numbers of cases during the last decade, infectious hepatitis has been the major public health problem associated with the consumption of raw shellfish. *Vibrio cholera* and *V. parahaemolyticus* have been incriminated to a lesser degree.

Colwell and Liston[9] have reported that gram negative, asporogenous rods of *Pseudomonas*, *Vibrio*, and *Flavobacterium* predominate the normal flora of the pacific oyster, *Crassostrea gigas*, in Washington waters, and that gram positive organisms constitute less than 20 percent of isolates. They further reported that the coliform count as calculated from the total viable count and MPN values never constituted more than 0.5 percent of the total viable count in their samples of Pacific oysters. A similar pattern was demonstrated by Vanderzant et al[25] in studies of the Eastern oyster *Crassostrea virginica* from Galveston Bay. Lovelace et al[15] reported *Vibrio, Pseudomonas, Achromobacter* and *Cytophaga/Flavobacterium* as the predominant organisms found in the Eastern oyster from two natural bars in Chesapeake Bay. Seven percent of the isolates from one bar with either *Enterobacter* or *Proteus* species, but neither of these two species, were reported in oysters from the second bar. This supports the Colwell and Liston report of low coliform populations in oysters from the West Coast, and indicates that the presence of elevated levels of coliforms, including "fecal" types, in shellfish, immediately after harvesting, may be indicative of the presence of a potential public health problem, although the presence of nonfecal coliforms can confuse the issue.

In the United States, sanitary control of the shellfish industry is based primarily upon the classification and control of the harvest areas through comprehensive sanitary surveys of the shore line, bacteriological monitor-

IS/A Committee Liaison: John T. Graikoski

ing of growing area waters, and prohibition of harvesting from areas not meeting "approved" growing area criteria. Routine control procedures are based upon guidelines in the National Shellfish Sanitation Program Manual of Operations.[16]

The safety of shellfish has been predicated upon the levels of indicator organisms, primarily the coliform group, present in the growing waters. In 1964, the National Workshop adopted the fecal coliform criterion for the wholesale market standard for shucked oysters,[17] and in 1968, for all species of fresh and fresh frozen shellfish.[18] There were no official NSSP procedures for the isolation of pathogenic bacteria from shellfish prior to the 1971 National Workshop.

Bacteriological quality of shellfish meats may be determined at several marketing levels, as harvested, in-plant sampling, in storage, upon receipt at the wholesale market, and at the retail market level. The only microbiological standard for shellfish meats developed by the NSSP is for product quality as received at the wholesale market. Fresh or fresh frozen shellfish generally are considered to be satisfactory at the wholesale market if the fecal coliform MPN does not exceed 230/100 grams, and the 35 C plate count is no more than 500,000/gram of sample.

41.2 EXAMINATION OF SHELLFISH

41.21 Collection and Transportation of Samples

Samples of shell stock and shucked unfrozen shellfish should be examined within 6 hours after collection, and in no case shall they be examined if they have been held more than 24 hours after collection. The report of the examination shall include a record of the time elapsed between collection and examination.

Individual containers of shellfish samples shall be marked for identification and the same mark shall be put in its proper place on the descriptive form which accompanies the sample.

A history and description of the shellfish shall accompany the sample to the laboratory. This shall include:

a. date, time, and place of collection;

b. the area from which the shellfish were harvested;

c. the date and time of harvesting;

d. the conditions of storage between harvesting and collection.

All of this information may not be obtainable for shellfish samples collected in market areas. In such case, the identification of the shipper, the date of shipment, and the harvesting area should be determined as well as the date, time, and place of collection.

41.22 Shell Stock (Shellfish in the Shell)

Samples of shellfish shall be collected in clean containers. The container shall be waterproof, and shall be durable enough to withstand the cutting action of the shellfish, and abrasion during transportation. Waterproof paper bags, paraffined cardboard cups or plastic bags are suitable types of containers. A tin can with a tight lid is also suitable.

Shell stock samples shall be kept in dry storage at a temperature above freezing but lower than 10 C (50 F) until examined. Shell stock shall not be allowed to come in contact with ice.

An adequate number of shellfish shall be collected in order to obtain a representative sample to allow for the selection of sound animals suitable for shucking. With most species, 10 to 12 shellfish should be examined, regardless of their individual size or weight.

Because of their large size, 10 to 12 shellfish of certain species such as the Pacific oyster, *Crassostrea gigas*, the surf clam, *Spisula solidissima*, and certain larger sizes of the hard clam, *Mercenaria mercenaria*, may require use of blender jars of a larger size than those ordinarily used. Certain blenders will accept ordinary mason jars and thus, a two or four quart container may be used for the blending of these species.*

On the other hand, 10 or 12 shellfish of certain other species, such as the Olympia oyster, *Ostrea lurida*, and small sizes of the Pacific little neck clams, *Protothaca staminea* and *Tapes japonica*, may produce much less than 100 g of shell liquor and meats. Blender containers of smaller size are indicated, but even when one pint or half-pint jars are used, as many as 20 to 30 of these species will be required to produce an adequate volume for proper blending.

41.23 Shucked Shellfish

A sterile widemouth jar of a suitable capacity with a watertight closure is an acceptable container for samples of shucked shellfish taken in shucking houses, repacking establishments, or bulk shipments in the market. The shellfish may be transferred to the sample jar with sterile forceps or spoon. Samples of the final product of shucking houses or repacking establishments may be taken in the final packing cans or containers. The comments pertaining to species of various sizes in the section on shell stock applies to shucked shellfish. Consumer size packages are acceptable for examination provided that they contain an adequate number of animals.

Samples of shucked shellfish shall be refrigerated immediately after collection by packing in crushed ice, and they shall be so kept until examined.

41.24 Frozen Shucked Shellfish

If the package contains an adequate number of animals (10 to 12), one or two packages may be taken as a sample. Samples from larger blocks may be taken by coring with a suitable instrument, or by quartering, using sterile technic. Cores or quartered samples shall be transferred to sterile widemouth jars for transportation to the laboratory.

It is desirable to keep samples of frozen shucked shellfish in the frozen state at temperatures close to those at which the commercial stock was main-

---

*Where two quart containers are not available, the 10 to 12 shellfish should be ground for 30 seconds. Then 200 g of this meat homogenate should be blended with 200 g sterile buffered phosphate water or 0.5% sterile peptone water for 60 seconds.

tained. When this is not possible, samples of frozen shucked shellfish shall be packed in crushed ice and kept so until examined.

41.3 PREPARATION OF SAMPLE FOR EXAMINATION. SHELLFISH IN THE SHELL

41.31 Cleaning the Shells

The hands of the examiner must be scrubbed thoroughly with soap and water.

Scrape off all growth and loose material from the shell, and scrub the shell stock with a sterile stiff brush under running water of drinking water quality, paying particular attention to the crevices at the junctions of the shells. Place the cleaned shell stock in clean containers or on clean towels and allow to drain in the air.

41.32 Removal of Shell Contents

Before starting the removal of shell contents, the hands of the examiner must be scrubbed thoroughly with soap and water and rinsed with 70% alcohol.

Open the shellfish as directed below, collecting the appropriate quantities of shell liquor and meats in a sterile blender or other suitable sterile container.

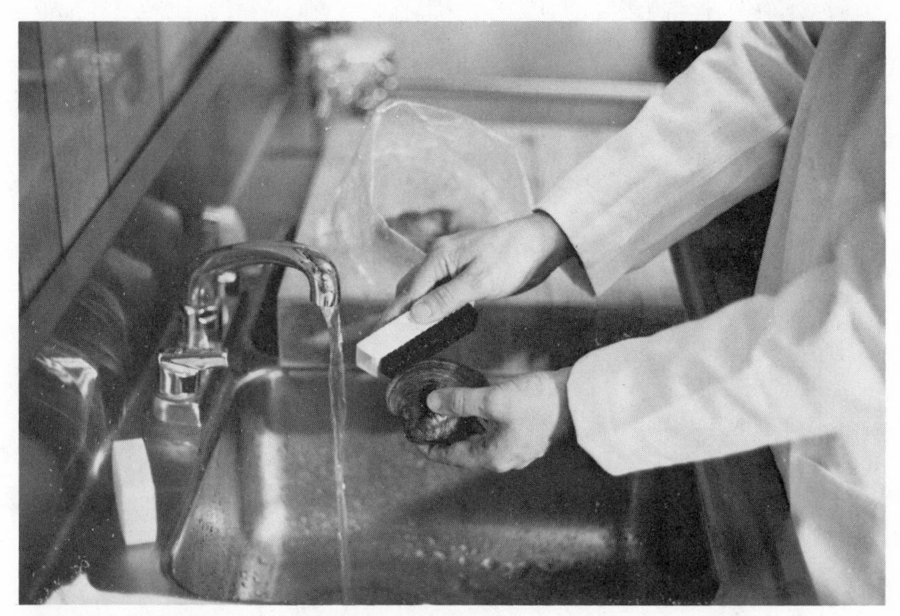

FIGURE 1. Scrubbing Shellfish with Sterile Brush after Washing Hands and Rinsing with 70% Alcohol

FIGURE 2. Shucking with Sterile Knife into Sterile Beaker

a. Oyster

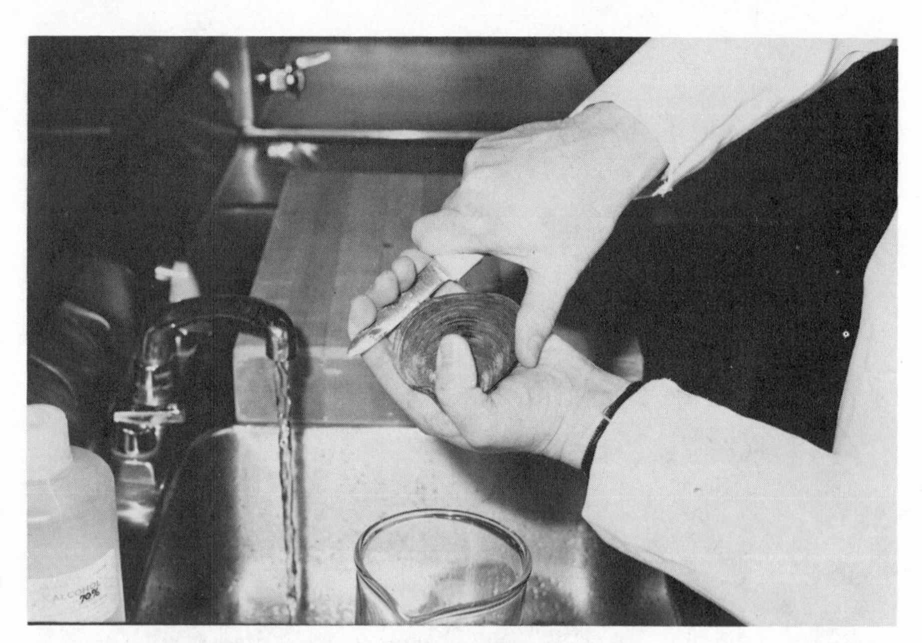

b. Hard Shell Clam

FIGURE 2. Shucking with Sterile Knife into Sterile Beaker

c. Sea Clam

d. Soft Shell Clam

41.321 Oysters

Hold the oyster in the hand or on a fresh clean paper towel on the bench with the deep shell on the bottom. Using a sterile oyster knife, insert the point between the shells on the ventral side (at the right when the hinge is pointed away from the examiner), about one-fourth the distance from the hinge to the bill. Entry also may be made at the bill after making a small opening with a sterile instrument similar to bonecutting forceps.

Cut the adductor muscle from the upper flat shell and pry the shell wide enough to drain the shell liquor into a sterile tared beaker, widemouth jar, or blender jar. The upper shell then may be pried loose at the hinge, discarded, and the meats transferred to the beaker or jar after severing the muscle attachment to the lower shell.

41.322 Hard clams

Entry into the hard clam, *Mercenaria mercenaria*, or the Pacific little neck clam, is best done with a sterile, thin bladed knife similar to a paring knife. To open the clam, hold it in the hand, place the edge of the knife at the junction of the bills, and force it between the shells with a squeezing motion. An alternative method is to nibble a small hole in the bill with sterile bonecutting forceps, and sever the 2 adductor muscles with the knife.

Drain the shell liquor into the sample container. Cut the adductor muscles from the shells and transfer the body of the animal to the sample container.

41.323 Other clams

The soft clam, *Mya arenaria*, the Pacific butter clam, *Saxidomus giganteus*, the surf clam, *Spisula solidissima*, and similar species may be shucked with a sterile paring knife, entering at the siphon end and cutting the adductor muscles first from the top valve and then from the bottom valve.

Mussels, *Modiolus*, and *Mytilus* species may be entered at the byssal opening. The byssal threads should be removed during the cleansing of the shell. The knife may be inserted and the shells spread apart with a twisting motion, allowing the draining of the shell liquor. Cut away the many attachments from the shell.

41.324 Shucked shellfish

Transfer a suitable quantity from a sample jar to a sterile tared blender jar or other container, using a sterile spoon.

41.33 Dilution and Grinding

Weigh the sample to the nearest gram. Transfer the weighed sample to a sterile blender jar, and add an equal amount, by weight, of sterile phosphate buffered dilution water or 0.5% sterile peptone water. Grind for 60 to 120 seconds in a laboratory blender operating at approximately 14,000 RPM. Two ml of this mixture contains 1 gram of shellfish meat. The optimum grinding time will vary with make of machine, condition of machine, spe-

FIGURE 3. Using Bone Cutters to Break Shell of Hard Shell Clam

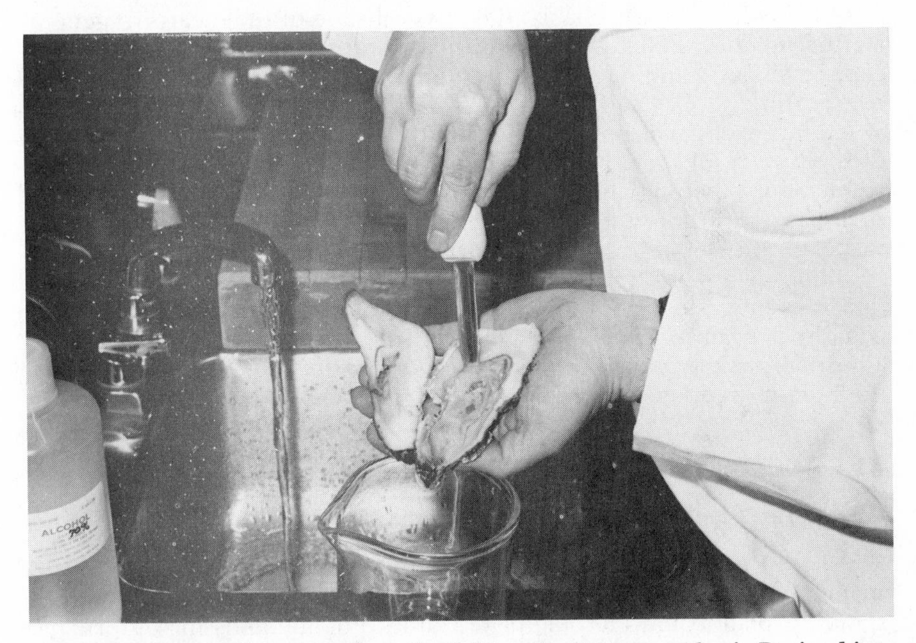

FIGURE 4. Removing Meat from Shell (Note Shell Liquor Also is Drained into Beaker)

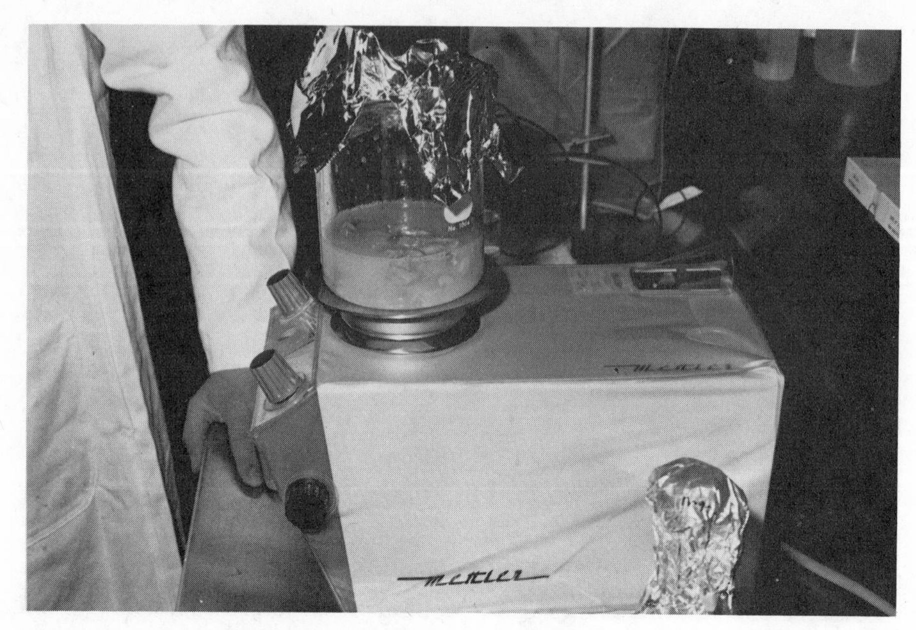

FIGURE 5. Weighing Shucked Meat and Shell Liquor

cies of shellfish, and, probably, the physical state of the meats. In general, a grinding time of 60 to 90 seconds will be found to be optimum for all species. Excessive grinding in small containers should be avoided to prevent overheating.

A dilution of equal amounts by weight of shucked packs of certain species of shellfish results, after grinding, in a mixture which is too heavy a consistency for pipetting and transferring to culture tubes. Meats of the hard shell clam, surf clam, and the butter clam are often in this category. In these cases, the use of a greater proportion of dilution water is permissible. Addition of 3 parts by weight of dilution water to 1 part of the weighed sample is suggested. With such dilutions, 4 ml of the ground sample will be equal to 1 gram portions of shellfish. If the 1:4 dilution method is used, adjustment in the concentration of presumptive broth in the tubes receiving the 1 gram portions should be made accordingly.

41.4 MICROBIOLOGICAL PROCEDURES

41.41 Tests for Members of Coliform Group

Procedures for the Examination of Sea Water and Shellfish[1] recommends the examination of shellfish for coliforms, fecal coliforms, and the standard plate count (aerobic plate count). For methods in conformance with these recommendations, refer to Chapter 4 (Aerobic Plate Count) and Chapter 24 (Coliforms and Fecal Coliforms).

FIGURE 6. Pouring Shucked Meats and Shell Liquor into Blender Jar

41.42 *Salmonella*

In 1950, Bidwell and Kelly[6] reported the isolation of *Salmonella typhimurium* from New York oysters. Duck farm wastes were judged to be the source of the organism. Brezenski and Russomanno[7] successfully isolated 12 *Salmonella* serotypes from hard clams harvested from Raritan Bay. Six serotypes were isolated from oysters taken from New Hampshire waters by Slanetz, Bartley, and Stanley.[24] During the Eighth National Shellfish Sanitation Workshop, Andrews et al reported the isolation of five *Salmonella* serotypes in hard clams and 11 serotypes in oysters from Northeast and Gulf Coast waters.[2]

It is doubtful that any single procedure can be recommended that will detect all salmonellae especially when occurring in low numbers, a situation likely to be encountered with shellfish.

However, procedures developed for foods in general may prove adequate for shellfish. Refer to Chapter 25.

41.43 *Vibrio parahaemolyticus*

Vibrio parahaemolyticus is a marine occurring organism which is reported to be the cause of "summer diarrhea" in Japan, and of gastrointestinal outbreaks from the consumption of blue crabs in Maryland, and shrimp in Texas. The organism was listed as the probable cause of an oyster-borne, gas-

FIGURE 7. Adding Equal Weight of Sterile Buffered Distilled Water

trointestinal outbreak in the State of Washington in 1969. An unidentified *Vibrio* has been associated with a gastrointestinal outbreak in England from the consumption of depurated oyxters. In consideration of the volume of oysters, mussels, and clams eaten raw in the United States and the low incidence of disease, it would appear that the potential hazard to consumers of raw molluscan shellfish from *V. parahaemolyticus* is minor. The organism has been recovered from numerous marine food species from the Atlantic, Gulf, and Pacific Coasts, and is reported to be distributed in the marine environment.

In consideration of the potential hazard from this organism, the Seventh National Shellfish Sanitation Workshop[19] recommended that the Food and Drug Administration's Bacteriological Analytical Manual (BAM[5]) procedure be accepted as the official reference of the NSSP for the examination of shellfish for *V. parahaemolyticus*. It was recommended further that state laboratories conduct the test for this organism when routine tests of marine foods suspected in foodborne outbreaks fail to demonstrate other enteric pathogens or bacterial toxins. For the detection of *V. parahaemolyticus* refer to the current edition of BAM and to Chapter 29 of this compendium.

41.44 *Vibrio cholera*

There are no official procedures for the analysis of shellfish for this bacterium. Refer to the methodology listed in Chapter 29.

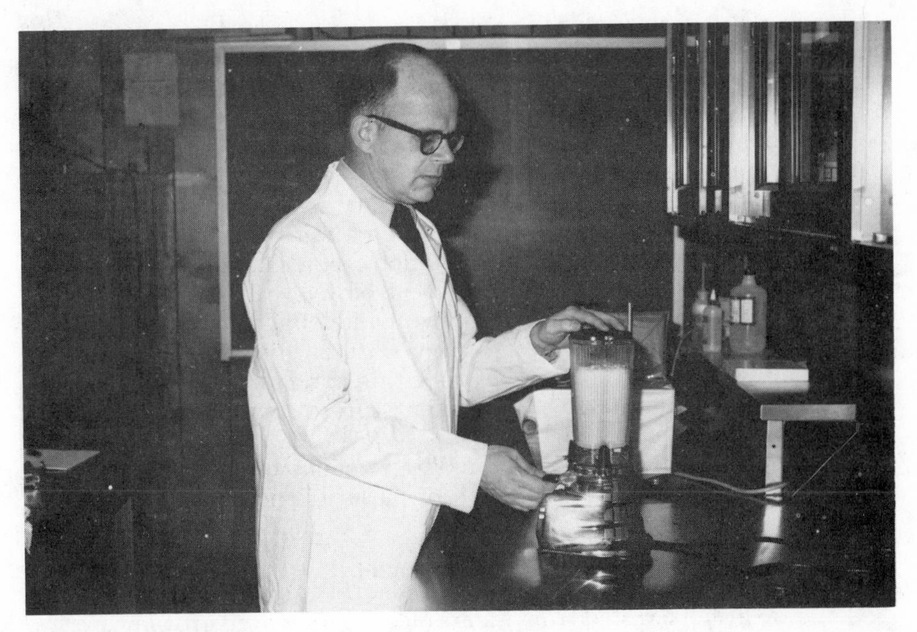

FIGURE 8. Blending

41.45 Interpretation of Data

The NSSP microbiological quality standard for fresh and fresh frozen oysters, mussels, and clams was developed as a guideline of acceptable quality for shellfish harvested, processed, and shipped according to recommended practices, and sampled upon receipt at the wholesale market, or retail markets or restaurants, if shipped directly to the receiver by the producer. The standard, a maximum fecal coliform MPN of 230/100 grams of sample, and a standard plate count not exceeding 500,000/gram can be used by receiver agency programs to determine product quality as delivered by shipper, in case of excessive counts, or forwarded to the producer state control agency for investigative and corrective purposes. The standard does not include a sampling plan, but in practice a single sample may represent a complete shipment.

As the bacteriological quality may reflect growing area water quality, the bacteriological water quality standards should be mentioned briefly. The coliform standard for approved growing areas, a median MPN value of 70, with not more than 10 percent of samples exceeding an MPN of 230 for a 5 tube, three dilution method, or an MPN of 330 for a 3 tube, three dilution method, has been unchanged essentially since 1946. In an effort to develop a growing area standard more indicative of fecal pollution, the Food and Drug Administration proposed a fecal coliform standard at the 1974 National Workshop.[13] The proposal stated: "The median fecal coliform MPN value for a sampling station shall not exceed 14 per 100 ml of sample and not more than 10 percent of the samples shall exceed 43 for a 5 tube, 3 dilution test, or 49 for a 3 tube, 3 dilution test." The methodology is the same as for meats except that 10 ml, 1 ml and 0.1 ml volumes of sample usually are used for routine samples. The fecal coliform criterion and proposed standard are being used by a limited number of state, federal, and Canadian agencies to classify shellfish waters.

Shelf life and bacteriological quality will vary according to species, salinity and bacteriological quality of growing area waters, climate, processing controls, refrigeration, and other conditions. For example, the northern quahog (*Mercenaria mercenaria*) harvested from New England waters normally will have a lower bacteriological count and longer shelf life than a soft shell clam (*Mya arenaria*) harvested from warmer less saline waters of the Mid-Atlantic States. Therefore, the proper interpretation of wholesale market data requires knowledge of the sanitary control conditions of the product from growing area to market.

The National Shellfish Sanitation Program has offered the following guide as an aid in the interpretation of bacteriological results of market shellfish:

a. Low fecal coliform/low standard plate count

1. Shellfish harvested from waters meeting approved growing area criteria;
2. Harvesting, processing, storage, and shipping practices conform to high sanitary standards;
3. Bacteriologically, a high quality product.

b. Low fecal coliform/high standard plate count

1. Shellfish harvested from waters meeting approved growing area criteria;
2. Excessive storage time a some point in processing or transit to market;
3. Product probably pathogen free but of inferior quality;
5. Producer state should be notified for investigative and corrective action.

c. High fecal coliform/low standard plate count

1. Shellfish probably harvested from improperly classified area;
2. Direct fecal contamination indicated;
3. Product potentially hazardous, and producer state agency should be notified for investigative and corrective action.

d. High fecal coliform/high standard plate count

1. Shellfish may or may not have been harvested from water meeting approved growing area criteria;
2. Direct contamination from handlers/or dirty equipment at any point in processing;
3. Inadequate refrigeration, mixing of old and fresh shellfish, poor repacking practices, and excessive storage times;

4. May contain pathogenic bacteria; overall poor quality;
5. Product potentially hazardous, and producer state agency should be notified for investigative and corrective action.

Experience with the shellfish product is valuable when interpreting results, and since this is not always easily attainable, the analyst is advised to consult with the appropriate agency of the state or country in which the product originated.

41.46 Rapid Methods

There has been considerable interest in recent years in rapid methods, particularly for use in monitoring depuration or controlled purification systems for shellfish harvested from marginally polluted waters.

Heffernan and Cabelli[12] developed a 24 hour plate method for the determination of elevated temperature coliforms using a modified MacConkey medium incubated at 45.5 C in an air incubator for 24 hours. The method was recommended as an official procedure for monitoring depuration processing of the soft shell clam (*Myo arenaria*) at the 8th National Shellfish Sanitation Workshop in 1974. This procedure was tested against the standard method using a 12 tube, 3 dilution system. It works well with the soft shell clam but has not proven to be effective with oysters.

Using a variety of frozen foods including seafoods, Fishbein et al[11] effectively recovered *E. coli* using LST broth incubated at 44 C in a waterbath for 24 hours.

More recently, Andrews et al[3, 4] have evaluated a multiple tube test for the recovery of *E. coli* from shellfish and shellfish waters using the A-1 medium incubated at 44.5 C for 24 hours. This test is receiving further evaluation in state and Canadian provincial laboratories for the analysis of shellfish growing area waters and shellfish meats.

Dussault et al in a paper: "An Activated Method for the Bacteriological Control of Clam Depuration," presented at the 17th Annual Atlantic Fisheries Technological Conference in 1972, described Vincent's method[23] which is based upon indole production in a culture medium incubated at 41.5 C for 48 hours. Working on depurated clam studies in Quebec, Dussault did a comparison of Vincent's method with A.P.H.A. Standard Methods[1] for the detection of fecal coliforms. The report states that the results obtained were as accurate and reliable as the standard method.

Qadri et al[22] have developed two rapid methods for the determination of fecal contamination in oysters which are reported to be as sensitive and accurate as the standard A.P.H.A. method.[1] One involves incubation of MacConkey broth for 2 hours at 37 C, then 22 to 24 hours at 44 C. The second method uses the same incubation system but includes the inoculation of peptone broth as well as MacConkey broth.

The roll tube method used in the United Kingdom is referred to later in this chapter.

With increases in cost for analytical services and greater demand upon laboratory personnel, the search for acceptable rapid test procedures should be continued.

41.5 INTERNATIONAL MICROBIOLOGICAL CRITERIA, STANDARDS, AND METHODS FOR THE EXAMINATION OF MOLLUSCAN SHELLFISH

Foreign countries utilize various microbiological criteria, standards, and methods to assess the sanitary quality of molluscan shellfish. The microbiological methods are used to enumerate criteria such as the total coliform group, or the fecal coliform group, or *Escherichia coli*, and/or a total aerobic plate count (standard plate count). In addition, some countries require that shellfish be tested for the presence of bacterial enteric pathogens.

The Federal Republic of Germany assesses the sanitary quality of shellfish by determining the total number of bacteria, the coliform score, and the presence of pathogenic enteric bacteria by serological methods.[10]

In Canada, Japan, and the Republic of Korea, both the fecal coliform MPN test (EC Medium, 24 hours at 44.5 ± 0.2 C) and the 2 day, 35 C Standard Plate Count are used to assess the sanitary quality of shellfish to be exported to the United States.[26] New South Wales, Australia, requires that shellfish be submitted to both the fecal coliform MPN test and a plate count.[21]

The use of *E. coli* as an indicator of fecal contamination in shellfish has been used in Europe for many years. In the United Kingdom, estimates of *E. coli* levels in shellfish are made by using a roll tube colony count method which employs modified MacConkey agar incubated at 44 ± 0.2 C for 18 hours.[26] Preincubation for 2 hours at 37 C has been shown to improve the yield and specificity of the method.[27] The methods for examining shellfish in Holland and Belgium are identical to the method used in the United Kingdom.[26]

The enumeration of *E. coli* in shellfish is accomplished by the use of an MPN method in France.[20] The method involves a presumptive test which requires the inoculation of portions of macerated shellfish into tubes of brilliant green lactose bile broth. The tubes are then incubated at 30 C for 24 to 48 hours. The identification of *E. coli* is made according to the method of Mackenzie, Taylor, and Gilbert for each presumptive culture fermenting lactose with the production of gas.

In Denmark, *E. coli* Type I densities in shellfish are determined by plating on violet red bile agar followed by incubation at 45 C for 48 hours. In addition, shellfish samples are submitted to a 5 day, 20 C plate count and a *Salmonella* test.[20]

E. coli densities in shellfish are determined by the use of an MPN method in Italy. The method utilizes a 3 tube, 3 dilution test in lactose broth incubated at 37 C for 48 hours. This is followed by confirmation in brilliant green lactose bile broth and tryptone broth incubated at 44 C for 48 hours.[8]

Shellfish for export from Sweden must meet the requirements of the importing countries.[10]

A summary of the microbiological criteria and standards used by these countries for the sanitary control of molluscan shellfish is presented in Table 1.

Table 1: Summary of International Microbiological Criteria and Standards for Molluscan Shellfish

Country	Microbiological Criteria	Microbiological Standard
Federal Republic of Germany	1. Total number of bacteria 2. Coliform Score 3. Presence of Pathogenic Bacteria	
Canada Japan Korea New South Wales	1. Fecal Coliform MPN 2. Standard Plate Count per gram	1. Satisfactory—230 per 100 grams Conditional—>230 per 100 grams 2. Satisfactory—500,000 per gram Conditional—>500,000 per gram
United Kingdom Holland Belgium	1. Fecal *E. coli* Roll tube count	1. Acceptable—90% up to 200 per 100 ml 10%—200 to 500 per 100 ml Unacceptable—>500 per 100 ml
France	1. *E. coli* MPN	1. Oysters and molluscans ordinarily eaten raw—<1 per ml Mussels and molluscs ordinarily eaten cooked—not >2 per ml.
Denmark	1. *E. coli* Type I plate count 2. Plate Count per gram 3. *Salmonella*	1. Each of 10 shellfish should be negative. 2. Each of 10 shellfish should not exceed 100,000 per gram. 3. Each of 10 shellfish should be negative.
Italy	1. *E. coli* MPN	1. Approved Area Shellfish: —160 per 100 ml of sample shall not be exceeded in 90% of samples during one year. —500 per 100 ml of sample shall not be exceeded in 10% of samples during one year. Market Shellfish: —600 per 100 grams of sample shall not be exceeded
Sweden	1. Same as for countries importing Swedish shellfish.	1. Same as for countries importing Swedish shellfish.

41.6 REFERENCES

1. Recommended Procedures for the Examination of Sea Water and Shellfish (Fourth Edition) American Public Health Association. 1970. New York, N.Y.
2. ANDREWS, W. H., C. D. DIGGS, M. W. PRESNELL, J. J. MIESCIER, C. R. WILSON, C. P. GOODWIN, W. N. ADAMS, S. A. FURFARI, and J. F. MUSSELMAN. 1975. Comparative validity of members of the total coliform and fecal coliform groups for indicating presence of *Salmonella* in the Eastern oyster, *Crassostrea virginica*. J. Milk Food Techn. **38**:453–56
3. ANDREWS, W. H., C. D. DIGGS, and C. R. WILSON. 1975. Evaluation of a medium for the rapid recovery of *Escherichia coli* from shellfish. Appl. Microbiol. **29**:130–31
4. ANDREWS, W. H. and M. W. PRESNELL. 1972. Rapid recovery of *Escherichia coli* from estuarine water. Appl. Microbiol. **23**:521–23
5. Bacteriological Analytical Manual, 1971. U. S. Department of Health, Education, and Welfare, Food and Drug Administration, Division of Microbiology, Washington, D.C.
6. BIDWELL, M. H. and C. B. KELLY, JR. 1950. Ducks and shellfish sanitation. J.A.P.H.A. **40**:923
7. BREZENSKI, F. T. and R. RUSSOMANNO. 1969. The detection and use of *Salmonellae* in studying polluted tidal estuaries. W.P.C.F. **41**:725
8. Codex Alimentarious Commission. 1973. Codex Committee on Food Hygiene Italian Delegation Proposal to the Molluscan Shellfish Codex. Washington, D.C.
9. COLWELL, R. R. and J. LISTON. 1960. Microbiology of shellfish. Bacteriological study of the natural flora of Pacific oysters (*Crassostrea gigas*). Appl. Microbiol. **8**:104–109
10. COPPINI, R. 1965. Sanitary Regulations for Molluscs. General Fisheries Council for the Mediterranean Studies and Reviews. **29**: 1–16
11. FISHBEIN, M., B. F. SURKIEWICZ, E. F. BROWN, H. M. OXLEY, A. P. PADRON, and R. J. GROOMES. 1967. Coliform behavior in frozen foods. I. Rapid test for the recovery of *Escherichia coli* from frozen foods. Appl. Microbiol. **15**:233–238
12. HEFFERNAN, W. P. and V. J. CABELLI. 1970. Elimination of bacteria by the northern quahog, (*Mercenaria mercenaria*): Environmental parameters significant to the process. J. Fisheries Research Board of Canada. **27**:1569–1577
13. HUNT, D. A. and J. SPRINGER. Preliminary report on a comparison of total coliform and fecal coliform values in shellfish growing area waters and a proposal for a fecal coliform growing area standard. 1974. Proceedings of the 8th National Shellfish Sanitation Workshop. DHEW Publication No. (FDA) 75-2025. Washington, D.C.
14. KELLY, C. B. Results of a cooperative study to evaluate present bacteriological criteria of market oysters. 1958. Proceedings Shellfish Sanitation Workshop. pp 16–21 (FDA) Washington, D.C.
15. LOVELACE, T. F., H. TUBIASH, and R. R. COLWELL. 1968. Quantitative and qualitative commensal bacterial flora of *Crassostrea virginica* in Chesapeake Bay. Proc. Natl. Shellfish Assoc. **58**:82–87
16. Manual of Operations, National Shellfish Sanitation Program, Part 1, Sanitation of Shellfish Growing Areas, Public Health Service Publication No. 33, 1965 Revision. (FDA) Washington, D.C.
17. Proceedings, Fifth National Shellfish Sanitation Workshop. 1964. U.S. Department of Health, Education, and Welfare
18. Proceedings, Sixth National Shellfish Sanitation Workshop. 1968. U.S. Department of Health, Education, and Welfare
19. Proceedings, Seventh National Shellfish Sanitation Workshop. 1971. U.S. Department of Health, Education, and Welfare
20. Provisional Codex of Hygienic Practice for Molluscan Shellfish. Draft. 1972
21. Pure Food Act—Standards for Shellfish. 1908. Health Commission of New South Wales, Sydney, N.S.W., Australia
22. QADRI, R. B., K. A. BUCKLE, and R. A. EDWARDS. 1974. Rapid methods for the determination of faecal contamination in oysters. J. Appl. Bact. **37**:7–14
23. Revue des Travaux de l'Institut des Peches Maritimes, Tome XXVII, fasc. 1, p. 24 (Mars 1963)
24. SLANETZ, L. W., C. H. BARTLEY, and K. W. STANLEY. 1968. Coliforms, fecal streptococci and *Salmonella* in seawater and shellfish. Health Lab. Sci. **5**:66

25. Vanderzant, C., C. A. Thompson, Jr., and S. M. Ray. 1973. Microbial flora and level of *Vibrio parahaemolyticus* of oysters (*Crassostrea virginica*), water, and sediment from Galveston Bay. J. Milk Food Technol. **36**
26. Wood, P. C. 1970. The principles and methods employed for the sanitary control of molluscan shellfish. FAO Technical Conference on Marine Pollution and Its Effects on Living Resources and Fishing. FIR:MP/70/R-12, Rome, Italy
27. Wood, P. C. The Sanitary Control of Molluscan Shellfish. Some Observations on Existing Methods and Their Possible Improvement. International Council for the Exploration of the Sea. Shellfish Committee. No. 24. C.M. 1963.

CHAPTER 42

MEAT AND POULTRY PRODUCTS

Ralph W. Johnston and R. Paul Elliott

42.1 RAW RED MEAT AND POULTRY MEAT

42.11 Normal Flora

Slaughter operations remove the vast majority of bacteria present on the surfaces and in the intestinal tract of animals. Those remaining on the eviscerated and skinned or defeathered carcass derive from the animal itself, and to a lesser extent from soil and water.

42.111 Intestinal tract

The gut is the most important source of bacteria, contributing *Clostridium perfringens,* coliforms, *Salmonella,* and *Staphylococcus* to the meat surfaces. In cattle, the rumen may contribute *Salmonella.*[11]

42.112 Outer surfaces

Hides, feathers, and skin harbor massive numbers of intestinal and soil and water bacteria. The killing, dressing, and washing operations vastly reduce this microbial load. Yet a few of the bacteria from the outside of the animal unavoidably become part of the surface flora of the carcass. Most are harmless mesophiles originally from the intestinal tract, but some of the food poisoning or animal disease pathogens may be present in low numbers. Harmless psychrotrophic bacteria from soil and water are also present in low numbers.

42.113 Lymph system, red meat animals

Lymph nodes frequently contain animal pathogens filtered from the lymph fluid.[21] Sometimes there is no evidence of general illness or even visible lesions, but laboratory tests show bacteria to be present frequently in these nodes.

Lymph nodes may contain any of the animal disease organisms, such as members of the genera *Staphylococcus, Clostridium, Streptococcus, Bordetella, Corynebacterium, Mycobacterium, Salmonella* and *Pseudomonas*.[29]

IS/A Committee Liaison: R. Paul Elliott

Some lymph nodes are too closely associated with edible parts to be removed by slaughter operations to rendered, nonedible products. Thus a few pathogenic bacteria pass inspection and remain in the raw meat.

42.114 Lesions

Staphylococci, and to a lesser extent other bacteria, cause localized infections in animal tissues, much as in human tissues. Bruising of the live animal encourages such infections.[16] Inspectors remove visible lesions from carcasses, but as in humans, staphylococci are present in normal mucous membranes and skin of animals;[4, 23] they are unavoidably present in small numbers in raw meats.[22, 27]

42.115 Air sacs

Poultry breathe not only through their lungs but also through air sacs, some of which are within the edible muscle of wings and thighs. Bacteria in the scald tank sometimes enter these air sacs, particularly if a dying bird flaps its wings; thus birds should be dead before they enter the scald tank. Cutting the trachea of the bird at time of kill reduces the number of bacteria. Scald water even at the usual 50 to 52 C is heavily contaminated with fecal and other bacteria.[28]

42.116 Blood system

Scalding introduces a wide variety of both aerobes and anaerobes[15] into the blood stream and the giblets of poultry. Continued heart action sometimes pulls a few bacteria into the blood stream of red meat animals from the slaughter knife. Characteristically, these bacteria lodge in the deep tissues.[13]

42.12 Floral Change in Spoilage

Mesophiles, including pathogens, cannot grow on chilled carcasses, but psychrotrophs of the *Pseudomonas-Achromobacter* group grow readily, and eventually spoil the meat.

When meat and poultry products are held at refrigeration temperatures, mesophilic bacteria grow very slowly or not at all. The psychrotrophic bacteria grow to predominance as spoilage progresses. Spoilage of whole cuts of meat or poultry under refrigeration is primarily a surface phenomenon resulting in slime formation and odor.[10]

Since spoilage is a surface problem, techniques that maintain high volume to surface ratios reduce spoilage rates. For example, the rate of spoilage of beef is slower in an entire carcass than it is in the cuts, or in ground meat. Spoilage by psychrotrophic bacteria is not hazardous to human health.[17] It is rather a problem of esthetics, and, for the meat industry, a problem of shelf life and thus of economics. The shelf life of chilled meats is enhanced by various factors affecting the rate of growth of the psychrotrophs: dry surface, low initial level of psychrotrophs, oxygen limitation, and temperature. Wrapping meats in oxygen impermeable films reduces surface growth. The conditions in a well wrapped piece of meat encourage the growth of the *Lactobacillaceae* at the expense of the *Pseudomonas-Achromobacter* group.[5]

The bacterial level in chilled meats after transportation and storage at the retail level has little or no relationship to that at the processor's level because bacterial growth has continued.

42.13 Disease Bacteria

Staphylococcus aureus, Clostridium perfringens, and *Salmonella* frequently are present in low numbers on raw meat surfaces. *Clostridium botulinum* occurs infrequently. These species are most hazardous when they grow without competition as in cooked foods.

Other animal disease bacteria can cause human disease with close animal-man contact. Some of these are *Brucella*, enteropathogenic *E. coli, Corynebacterium, Mycobacterium, Leptospira, Coxiella burneti,* and *Clostridium tetani.* Theoretically some of these could be transmitted to man by contact with raw meat, but epidemiological evidence is lacking.

42.14 Recommended Methods

For most raw meats, weigh a representative 50 gram sample into a sterile blender jar, add 450 ml. of diluent and proceed as directed in Chapters 1, 2 and 3. For nondestructive surface counts applicable to small poultry and meat cuts, a rinse technique[26] may be used. For large carcasses, when a rinse technique is impractical, use swabs.[2]

The usual methods employed in the analysis of raw meat and poultry are discussed in detail in other chapters of this compendium. A list of tests frequently performed and their appropriate compendium chapters is as follows:

Aerobic Plate Count (mesophilic). Chapter 4
Aerobic Plate Count (psychrotrophic). Chapter 4
  Incubate plates at 20 C for 4 days. Where lactic acid bacteria are expected, substitute APT or similar vitamin enriched agar. Chapter 8.
Coliforms, *E. coli* and Fecal Coliforms. Chapter 24
Staphylococcus aureus. Chapter 31
Salmonella. Chapter 25
Yeasts and Molds. Chapter 16
Enterococci. Chapter 30

42.15 Interpretation of Data

Most meat bacteria occur on the surface, but a few occur in deep tissue, having been deposited there from the slaughter cut by continuing heart action. In the past, anaerobic spores such as those of *Clostridium perfringens*, which had lodged in deep tissues, caused "bone taint" in carcasses not properly chilled.[13]

In the United States bone taint is no longer a problem, except for country cured hams;[18] efficient carcass chilling is standard procedure in slaughter operations.

The surface flora on freshly slaughtered carcasses, usually about 10^3 bacteria per square inch, are primarily mesophilic, having originated from the internal and external surfaces of the live animal. Psychrotrophic organisms originating from soil and water are also present at a level of about 10^1 per square inch.[10] The mesophiles are important because they indicate the degree of sanitation at slaughter. Coliforms, *E. coli*, enterococci, *S. aureus, C. perfringens*, and *Salmonella* are often present on fresh tissues since the slaughter process does not include a bactericidal step. The frequency and levels of these bacteria will vary, depending upon farm, climatic, and processing conditions. In general, all of them except *Salmonella* may be present at levels of about 10^1 to 10^2. *Salmonella* if present at all, is usually at a level of one cell in about 25 grams of tissue.[27]

42.2 READY-TO-EAT RED MEAT AND POULTRY MEAT

42.21 Cooked Uncured Meats

The microbiology of cooked meats begins with the cooking process. The majority of cooked products is given a thorough cook, and only spores survive, usually only a few hundred per gram. Some meat products such as flame seared beef patties and cooked beef are processed at lower temperatures. These temperatures are sufficient to destroy pathogens, but the final bacterial counts include some of the more heat resistant vegetative bacteria such as the enterococci. For these products, the final bacterial numbers normally may be at levels of about 10^3 to 10^4. Unless cooked products are packaged hot and immediately frozen, recontamination invariably occurs from equipment, food handlers, raw product, or dust. The amount of recontamination does not exceed a few thousand bacteria per gram in good practice.

In recognition of the seriousness of recontamination from raw meats, the USDA Meat and Poultry Inspection Program (MPIP) requires the separation of cooked meat products from raw products, and requires sanitization of hands and equipment that are in contact with the cooked products.[30]

Most cooked meat products freshly prepared and entirely cooked have 35 C plate counts of about 10^4 or less per gram. Unavoidable contamination usually will add coliforms at levels of 10^1 to 10^2 per gram. The presence of *E. coli* indicates possible insanitary conditions, and therefore warrants investigation of the conditions of manufacture to determine whether the source of *E. coli* is objectionable. Without such investigation, *E. coli* at low levels has less significance. Human contact with cooked food, as in slicing or deboning, invariably adds *S. aureus* at levels of 10^1 or 10^2 to many of the sample units.[25] Such levels are harmless, but offer sufficient inoculum for growth to hazardous levels under conditions of time-temperature abuse.

Cooked meat products offer ideal menstrua for microbial growth, for they are highly nutritious, have a neutral pH, high water, and low salt content. Given time and favorable temperatures, contaminants, including pathogens, will grow rapidly. In recognition of this, MPIP requires that precooked meat products in inspected establishments be held no longer than two hours between 4 C and 60 C.[30] Most such foods are frozen for shipment and distribution. Assuming the package does not thaw in commerce, the microbiological level at retail will relate fairly well to that shortly after the product was first frozen. Such foods, held too long above the freezing point, will spoil from a wide variety of organisms, including enterococci, pseudomonads, lactic acid bacteria, and yeasts.

Foods in which the level of *S. aureus* or *C. perfringens* has reached 10^6 per gram may be toxic; high levels of salmonellae are more hazardous than are low levels. Botulism rarely has been a problem in commercially precooked foods.

Adding uncooked ingredients such as celery, spices, or cheese to a cooked meat product changes the bacteriological picture so that the above estimates do not apply. A full description of formulation and processing procedure is essential for proper interpretation of laboratory data.

The hazard potential from foods precooked in commercial establishments is high,[19] but the incidence of outbreaks has been low.[31] The Center for Disease Control reports that although more than half of all food-borne disease outbreaks can be traced to meat and poultry products, there was a serious departure from good practices at the serving level (homes, restaurants, institutions) in nearly all instances.

42.22 Cooked Cured Meats

The heating step in the production of cooked cured meats destroys the typical raw meat flora except for the spores. Salt and nitrite in the cure inhibit the growth of survivors and contaminants somewhat selectively. The American Meat Institute has reviewed the use of nitrates and nitrites in meat products.[1]

Upon prolonged refrigeration, lactic acid bacteria, micrococci, enterococci, *Bacillus,* and yeasts may grow and form slime.[10] If the product is in a tight, gas impermeable package, the package may swell. Products of bacterial action sometimes combine with meat pigments to form a green color.

Cooked, nonfermented, cured meats have counts of 10^4 or less per gram; higher levels in products from retail cases indicate the time temperature history of storage. Under good refrigeration, such meats do not support the growth of pathogens, so that high aerobic plate counts are unrelated to health hazard. Coliforms, present as unavoidable contaminants at low levels ($\pm$ 10^1), fail to grow in refrigerated product. Human contact may sometimes introduce a few *E. coli* or *Staphylococcus aureus.* If the level of *S. aureus* remains low, it is harmless; but given a favorable temperature (7 to 46 C), the salt content of the cured meat permits selective growth of *S. aureus*.

The Center for Disease Control of the U.S. Public Health Service reported that in 1971, of a total of 40 *Staphylococcus* outbreaks for which the vehicle could be found, 18 (45%) incriminated cured meats—mostly ham, served warm. Slicing, followed by warm temperatures, favors contamination and growth of *S. aureus* on ham.

On the other hand, cured cooked luncheon meats seldom cause *Staphylococcus* food poisoning. For one thing, *S. aureus* fails to grow anaerobically in the presence of nitrite[6] (luncheon meats are usually vacuum packaged);

and for another, the *Staphylococcus* does not grow below 44 F. (Luncheon meats are usually well refrigerated.)

Cured meats seldom, if ever, cause other types of food poisoning. The heat process destroys salmonellae. The low incidence, and low numbers of *Clostridium botulinum,* coupled with the presence of nitrite and the prevalence of refrigeration preclude growth and toxin formation. Similarly, surviving spores of *C. perfringens* fail to grow under refrigeration.

42.23 Canned Cured Meats

Canned cured meats are prepared in two forms, "shelf stable" and "perishable, keep refrigerated." They owe their stability to the presence of curing salts and the heat process. No contamination occurs after the cooking of the closed can, and the only survivors of the heating are aerobic and anaerobic spores, usually at 10^1 per gram or lower. The incidence of *C. botulinum* spores in raw meat is low,[12] but when present, they will survive the low canning cook required of canned cured meats. Curing salts consisting of nitrite and salt are effective in inhibiting the outgrowth of *C. botulinum* in shelf stable products[8, 9, 24] to the extent that these products are as safe from a botulism standpoint as uncured canned meats given a 12D cook.[14] The pasteurization step destroys other foodborne disease organisms. Curing agents are also similarly[7] effective for perishable cured meats; however, these are labeled "keep refrigerated" for added safety. Nonshelf stable, canned cured meats are packed in containers up to 22 pounds while shelf stable cured meats usually are packed in containers of one pound or less.

Shelf stable, canned cured meats heated during processing to about 98 C do not spoil unless the curing process, retorting process, or can seal is faulty. A faulty cure will permit sporeforming bacteria that survive a normal retorting to grow in canned meat with low salt or low nitrite content. A gross underprocessing will permit survival of enterococci, micrococci, and other nonsporulating spoilage bacteria. A faulty can seal will permit a wide variety of nonsporulating organisms to enter and spoil the meat.

Nonshelf stable ("Keep Refrigerated") canned, cured meats processed to 65 C and above[30] may contain aerobic and anaerobic spore formers capable of growth in unrefrigerated cans. Enterococci which are widespread in slaughtering and processing operations, survive mild heating (pasteurizing) and grow under refrigeration. Therefore, they frequently survive and grow in nonshelf stable, canned, cured meats. A slight increase in temperature and/or time of cook usually will correct this problem.

42.24 Fermented Sausages

Such products as thuringer, summer sausage, pepperoni, cervelat, and Genoa salami depend upon lactic fermentation and a relatively low water activity for preservation. At the end of the fermentation period, the lactic organisms are normally at about 10^{10} per gram, members of the *Enterobacteriaceae* are absent, and *S. aureus* counts are less than 10^3 per gram. During subsequent storage, the organic acids, low pH, and drying may destroy many of the bacteria present, yet it is not uncommon to find 10^6 per gram of surviving lactic organisms in the product at the retail level. Some of these products are smoked and/or heated after the fermentation period, thus greatly reducing the bacterial levels of the final product.

In a poorly controlled process, staphylococci may grow and produce toxin during the fermentation period. The staphylococci gradually die out during drying and storage, but the toxin remains. Thus the level of staphylococci in the final product may not measure the health hazard. The only answer is control of the process to inhibit *Staphylococcus* growth during fermentation.[3, 20] Fermentation acids and curing salts control other food-poisoning bacteria.

Yeasts and molds may grow on the surface of such sausages during production and subsequent storage. Some of the wild molds found on sausages are mycotoxigenic, but no one has yet found a mycotoxin in the meat itself. Some processors cultivate a nontoxigenic surface mold growth to impart a characteristic flavor.

Information about formulation and processing procedure is essential for the proper interpreta.ion of laboratory data.

42.25 Vinegar Pickled Meats

Pickled pigs feet, pickled sausages, and similar products immersed in vinegar brine owe their stability to low pH, acetic acid, the absence of a fermentable sugar, and an airtight package. Lactic acid bacteria in moderate numbers and spores up to 10^4 per gram may be present. Aerobic plate counts are variable and unpredictable; coliforms and *S. aureus* are rarely present.

Lactic acid bacteria in numbers exceeding 10^7 per gram may make the brine cloudy; in hot weather the container may build up internal pressure.[2]

Foodborne disease organisms cannot survive, but mold can grow if the container is not properly sealed. This offers a potential but unproven hazard from mycotoxins.

42.26 Dried Meats

Most procedures for the preparation of dried meat include a cooking step that destroys the normal vegetative flora of the raw meat, and an efficient, rapid, drying procedure which will then reduce the water activity (a_w) to an inhibitory level before microorganisms of interest can grow.

Molds and sometimes yeasts may grow on the surface of dried meats including dried hams. Drying to a lower moisture level or using sorbic acid will control such growth. Mycotoxin formation is a possibility.

Bacterial levels on dried meats are highly variable and depend upon the nature of the product, its ingredients, and the process.

Elevated plate counts are normal on the surfaces of dry cured hams. Micrococci normally predominate; coliforms and *S. aureus* are occasionally present at 10^1 to 10^2 per gram; *E. coli* is rarely present.

Commercially dried meat products are generally free of hazard from foodborne disease bacteria, although salmonellae, cross contaminated from raw meat to the dried surface, can remain viable. Also, improper home preparation of the dried product may permit growth of any of the food poisoning organisms.

42.27 Recommended Methods

Most cooked meat and poultry products are prepared for analysis by the procedures and techniques listed in Chapters 1, 2, and 3 of this compendium. The usual methods employed in the analysis of cooked meat and poultry products are discussed in detail in other chapters. A list of tests frequently performed and their appropriate compendium chapters is as follows:

Aerobic Plate Count (mesophilic). Chapter 4

Aerobic Plate Count (psychrotrophic). Chapter 4

Incubate plates at 20 C for 4 days. Where lactic acid bacteria are expected, substitute APT or similar vitamin enriched agar. Chapter 8

Coliforms, *E. coli*, and Fecal Coliforms. Chapter 8

Staphylococcus aureus. Chapter 31

Salmonella. Chapter 25

Yeasts and Molds. Chapter 16

Enterococci. Chapter 30

Clostridium perfringens. Chapter 35

Bacillus cereus. Chapter 33

Shelf Stable Canned Products. Chapters 52 and 53 (Caution—viable mesophilic spores may be cultured from normal, shelf stable, cured meats.)

In certain specific investigations, other methods not cited in this chapter may be required. Many of these are discussed elsewhere while others such as protein, salt, water, fat, and water activity are not included in this compendium. To perform these tests the analyst should review other publications.

42.3 REFERENCES

1. American Meat Institute. 1971. The use of nitrate and nitrite in the meat industry. Meat Science Rev **5**:1–14
2. Recommended Methods for the Microbiological Examination of Foods. 1966. Second Edition. J. M. Sharf, ed. American Public Health Association, Inc., 1790 Broadway, New York, NY 10019
3. Anonymous. 1973. Controlling staph in sausage manufacturing. Western Meat Industry **19**:13–15, 24
4. Armstrong, C. H. and J. B. Payne. 1969. Bacteria recovered from swine affected with cervical lymphadenitis (jowl abscess). Am Jour of Veter Res **30**:1607–1612
5. Brown, W. L. and A. Hoffman. 1972. Microbiology of fresh beef in vacuum. Proceedings of the Meat Industry Research Conference. American Meat Institute Foundation, Chicago, Illinois 60605, pp 45–52
6. Christiansen, Lee N. and E. M. Foster. 1965. Effect of vacuum packaging on growth of *Clostridium botulinum* and *Staphylococcus aureus* in cured meats. Applied Microbiol **13**:1023–1025
7. Christiansen, L. N., R. W. Johnston, D. A. Kautter, J. W. Howard, and W. J. Aunan. 1973. Effect of nitrite and nitrate on toxin production by *Clostridium botulinum* and on nitrosamine formation in perishable canned comminuted cured meat. Applied Microbiol **25**:357–362
8. Duncan, Charles L. and E. M. Foster. 1968. Role of curing agents in the preservation of shelf-stable canned meat products. Applied Microbiol **16**:401–405
9. Duncan, C. L. 1970. Arrest of growth from spores in semipreserved foods. Jour of Applied Bacteriol **33**:60–73

10. Elliott, R. Paul and H. David Michener. 1965. Factors Affecting the Growth of Psychrophilic Microorganisms in Foods—A Review. Technical Bulletin No. 1320, Agricultural Research Service, U.S. Department of Agriculture, Superintendent of Documents, U.S. Government Printing Office, Washington, DC 20402
11. Grau, F. H., L. E. Brownlie, and E. A. Roberts. 1968. Effect of some preslaughter treatments on the *Salmonella* population in the bovine rumen and faeces. Jour of Applied Bacteriol **31**:157–163
12. Greenberg, R. A., R. B. Tompkin, B. O. Bladel, R. S. Kittaka, and A. Anellis. 1966. Incidence of mesophilic *Clostridium* spores in raw pork, beef and chicken in processing plants in the U.S. and Canada. Applied Microbiol **14**:789–793
13. Haines, R. B. 1941. The isolation of anaerobes from tainted meat. Chem and Indust **60**:413–416
14. Ingram, M. and T. A. Roberts. 1971. Application of the D-concept to heat treatments involving curing salts. Jour Food Technol **6**:21–28
15. Lilliard, H. S. 1973. Contamination of blood system and edible parts of poultry with *Clostridium perfringens* during water scalding. Jour Food Sci **38**:151–154
16. May, K. N. and M. K. Handy. 1966. Bruising of poultry—A review. World's Poultry Science Jour **22**:316–322
17. Michener, H. David and R. Paul Elliott. 1964. Minimum growth temperatures for food poisoning, fecal-indicator and psychrophilic microorganisms. Advan Food Res **13**:349–396
18. Mundt, J. Orvin and H. Mell Kitchen. 1951. Taint in southern country-style hams. Food Res **16**:233–238
19. National Research Council. 1964. An Evaluation of Public Health Hazards from Microbiological Contamination of Foods. Printing and Publication Office, National Academy of Sciences, 2101 Constitution Avenue, Washington, DC 20418.
20. Rose, M. J., B. F. Surkiewicz, R. W. Johnston, and R. Paul Elliott. 1973. Some microbiological considerations in the production of Genoa salami. Abstracts of the Annual Meeting, American Society for Microbiology, Miami Beach, Florida, Abstract E 65, page 11
21. Rubin, Harvey L., M. Sherago, and R. H. Weaver. 1942. The occurrence of *Salmonella* in the lymph glands of normal hogs. Am Jour of Hyg **36**:43–47
22. Sinell, H. J. and D. Kusch. 1969. Selektiv zuchtung von koagulase-positiven staphylokokken aus hackfleisch, Archiv fur Hyg and Bakteriol **153**:56–66
23. Smith, H. W. and W. E. Crabb. 1960. The effect of diets containing tetracyclines and penicillin on the *Staphylococcus aureus* flora of the nose and skin of pigs and chickens and their human attendants. Jour of Path and Bacteriol **79**:243–249
24. Spencer, F. 1966. Processing factors affecting stability and safety of non-sterile canned cured meats. Food Manufacture **41**:33–43
25. Surkiewicz, Bernard F., M. E. Harris, and R. W. Johnston. 1973. Bacteriological survey of frozen meat and gravy produced at establishments under Federal inspection. Applied Microbiol **26**:574–576
26. Surkiewicz, B. F., R. W. Johnston, A. B. Moran, and G. W. Krumm. 1969. A bacteriolog ical survey of chicken eviscerating plants. Food Technol **23**:80–85
27. Surkiewicz, B. F., R. W. Johnston, and R. Paul Elliott. 1972. Bacteriological survey of fresh pork sausage produced at establishments under Federal inspection. Applied Microbiol **23**:515–520
28. Thomson, J. E. and A. W. Kotula. 1959. Contamination of the air sac areas of chicken carcasses and its relationship to scalding and method of killing. Poultry Sci **38**:1433–1437
29. Laboratory Reports of the Medical Microbiology Group, Microbiology Staff, Meat and Poultry Inspection Program, Animal and Plant Health Inspection Service. 1973. U.S. Department of Agriculture, Washington, DC 20250. Unpublished.
30. Manual of Meat Inspection Procedures of the United States Department of Agriculture. 1973. Superintendent of Documents, U.S. Government Printing Office, Washington, DC 20402
31. Foodborne Outbreaks. 1972. Center for Disease Control, U.S. Public Health Service, Atlanta, GA 30333

CHAPTER 43

EGGS AND EGG PRODUCTS

W. M. Hill, Richard T. Carey, James M. Gorman, and Margaret Huston

The billions of hen eggs produced annually in the United States are utilized in a variety of forms. The major user is the food industry, but there is also a demand for eggs by manufacturers of cosmetics, shampoos, vaccines, and other nonfood products.

A substantial number of shell eggs are converted to liquid whole egg, albumen, and yolk at egg breaking plants. The liquid egg may be marketed fresh or may be frozen or dried. These egg products are used primarily as ingredients in other foods such as cake mixes, confectionery, noodles, bakery goods, and salad dressing. Certainly the microbiology of eggs and egg products has a definite impact on the microbiology of many food products.

The U.S.D.A. inspects the operations of egg breaking plants and requires most egg products to be pasteurized. Only salt yolk for use in salad dressings, containing at least 1.4% acetic acid (pH 4.1 or lower), is exempt from pasteurization.

43.1 NORMAL FLORA

43.11 Shell Eggs

The shell and contents of eggs at the time of oviposition are either sterile or harbor very low numbers of microorganisms.[9, 17] Contamination of the egg shell occurs after laying from nesting material, dirt, and fecal matter. Board[4] reported that the flora of egg shells is dominated by gram positive cocci while the gram negative rods are present in low numbers. The contents of shell eggs may become contaminated with organisms from the shell surface by improper washing and storage methods. The physical and/or chemical barrier provided by the egg shell membrane[10, 11] and the antimicrobial nature of the albumen[3] favor the penetration and multiplication of the gram negative bacteria over the gram positive cocci.

43.12 Liquid Eggs

The most common genera of bacteria found in liquid eggs are gram negative types, including *Pseudomonas*, *Alcaligenes*, *Proteus* and *Escherichia*. In

IS/A Committee Liaison: John H. Silliker

commercial egg breaking operations the egg shell is a source of contamination and may contribute large numbers of gram positive cocci (micrococci) to the liquid egg.

The numbers of bacteria in the liquid product depend upon the bacteriological condition of shell eggs used for breaking, plant sanitation,[12] and conditions under which the liquid may be stored.

Pasteurization of liquid egg, which is mandatory for all egg products except salt yolks to be used for salad dressings at pH 4.1 or lower, results in a reduction of 90 to 99% in standard plate counts.[7, 16] Surviving bacteria are usually species of heat resistant *Bacillus*, enterococci, and micrococci.[15]

43.13 Frozen and Dried Eggs

Pasteurized frozen or dried egg products have total counts ranging from less than one-hundred to a few thousand per gram. The dominant flora consists of micrococci and species of *Bacillus.*[15]

43.2 MICROFLORA CHANGES IN SPOILAGE

Shell eggs are not highly susceptible to penetration and growth of spoilage organisms. This is borne out by the observation that less than 1% of clean nest eggs spoils during storage.[6] Spoilage of shell eggs is promoted by cracking the egg shell, rubbing the shell with abrasives, and improper washing and storage techniques.

Colonization of the shell contents is characterized by a mixed flora of gram negative bacteria. The most common contaminants are the coliforms, *Achromobacter*, *Pseudomonas*, *Serratia*, *Proteus*, *Alcaligenes*, and *Citrobacter.* Spoilage caused by these organisms usually results in characteristic off odors and colors.[8] Spoiled eggs are separated from sound eggs by candling, and by visual and "sniff" testing at the time of breaking.

43.3 HUMAN DISEASE BACTERIA

The major pathogen associated with eggs and egg products is *Salmonella.* The numbers of salmonellae isolated from nonpasteurized liquid egg are less than 100 per gram, and most often less than 1 per gram. Commercial pasteurization techniques are capable of causing a 6 to 8 log reduction of salmonellae in liquid egg products. Thus, the presence of salmonellae in pasteurized egg products is due almost exclusively to recontamination.

Toxigenic staphylococci commonly are not found in either fresh or pasteurized egg products, though on occasion they may present a problem in salt yolks, due to the fact that salt selectively favors their presence.

43.4 RECOMMENDED METHODS

43.41 Sampling

43.411 Shell eggs

Select shell eggs at random from cases or cartons representative of the lot. Transfer the eggs to clean cartons or cases for transport to the laboratory, and maintain the eggs at temperatures below 50 F during transport and until analysis is made. Avoid sweating of the eggs. This results from transferring shell eggs from cold storage temperatures of around 30 F to room temperature without a tempering period. Returning sweating eggs to cold storage facilitates the penetration of bacteria through the shell. Also, the moisture on the shell surface greatly increases the chances of contamination during removal of the egg contents. Therefore, eggs should not be handled until warmed to room temperature.

43.412 Liquid eggs

Obtain samples from vats or tanks at the plant or from containers as delivered to the user. Make sure that the product has been mixed thoroughly by mechanical agitation or by hand mixing in the container. Use a sterilized dipper or sampling tube to withdraw the sample. Sterile pint mason jars or friction top cans are satisfactory for holding the sample. Obtain about ¾ pint of sample and hold it below 40 F for no more than four hours if possible. Avoid freezing as this will destroy many of the bacteria present. If there is doubt as to the homogeneity of the liquid in a vat, take several ¾ pint samples and composite them in a two quart jar to give a more representative sample.

Record the temperature of the containers from which the samples were taken; this is often the key to abnormal bacterial populations. Temperatures above 45 F indicate improper handling of the liquid and are likely to lead to rapid deterioration.

43.413 Frozen eggs

Select a number of cans representative of the lot. Open the cans, and with a sterile spoon remove any ice or frost on the frozen egg. The area selected for drilling should not be humped or peaked. With a high speed electric drill make a drilling starting about one inch from the edge of the can. Slant the bit so it will go through the center of the frozen egg to within an inch or two of the opposite lower edge of the can. Transfer the drillings to a sample container with a sterile spoon. Keep the drillings frozen at all times. Pack in an insulated box with dry ice for transport to the laboratory.

43.414 Dried eggs

Select a number of drums representative of the lot. Sample whole egg and yolk with a sterilized trier, taking at least three cores to the bottom of the container. Take samples of powdered or spray dried albumen with a trier while tilting the container. Sample flake or granular albumen using a sterilized spoon or scoop. Discard the bottom 1 inch of the egg in the drier and transfer the rest of the core to a sterile jar or friction top can. Do not fill the container more than ⅔ full in order to leave room for mixing.

If the product is in small packages, select several unopened packages. Open them at the laboratory under aseptic conditions and transfer a liberal quantity to a sterile can or beaker. Thoroughly mix with a sterilized spoon to obtain a homogeneous mixture.

43.42 Preparation of Sample

43.421 Shell eggs

Wash each egg with a brush using soap and water 20 F warmer than the egg but at least 90 F. Drain and immerse in 70 percent alcohol for 10 minutes. Remove from alcohol, drain, and flame.

Handle the alcohol flamed eggs with sterile gloves and aseptically remove the egg meat by cracking the egg with a sterile breaking knife.

Shake the contents of the sample container with glass beads, or beat with a sterile spoon, or with a sterile electric mixer until the sample is homogeneous.

For separate examination of egg white and egg yolk use a sterile commercial separator or a sterile spoon. Free the yolk of excess egg white by rolling it on a sterile towel. Weigh 11 grams of the homogeneous sample into a 99 ml saline, phosphate, or peptone water dilution blank with glass beads. Shake 25 times by hand through a 1 foot arc for 7 seconds. Prepare appropriate serial dilutions using similar dilution blanks.

43.422 Liquid eggs

Thoroughly mix the contents of the sample jar with a sterile spoon or electric mixer, taking care to avoid excessive foaming of the sample. Prepare initial dilutions as described in 43.421.

43.423 Frozen eggs

Hold the sample container ⅔ submerged in cool tap water, rotating frequently, until the sample thaws. Open the container aseptically, and stir thoroughly with a sterile spoon. Do not test samples which have arrived at the laboratory in a thawed condition.

Prepare initial dilutions as described in 43.421.

43.424 Dried eggs

Open sample containers under aseptic conditions and thoroughly mix the contents with a sterile spoon. If the pieces of flake albumen are thick and of large size, crush in a sterilized mortar until pieces are smaller than ¼ inch.

Weigh 11 gm of product into a 99 ml dilution blank (43.421) with glass beads, and shake until there is no evidence of lumps. If lumps persist, crush them against the side of the bottle with a sterile spatula and continue shaking.

Flake or lump albumen dissolves slowly. Shake dilution blank thoroughly and place in a cool place. Shake once or twice every few minutes until the flakes have dissolved, but do not hold more than 10 minutes.

Prepare serial dilutions by transferring 11 ml of the 1 : 10 dilution to a 99 ml dilution blank.

43.43 Aerobic Plate Count

43.431 Aerobic plate count—35 C (Chapter 4)

43.432 Aerobic plate count—32 C (Chapter 4)

43.44 Yeast and Mold Count (Chapters 4 and 16)

43.45 Coliform Group (Chapters 4 and 24)

43.46 *E. coli* (Chapters 4 and 24)

43.47 *Salmonella* (Chapters 4 and 24)

43.48 Direct Microscopic Count

43.481 Liquid and frozen eggs (Chapters 5 and 43)

43.482 Dried eggs (Chapters 5 and 43)

43.5 INTERPRETATION OF DATA

Fresh shell eggs usually contain less than 10 microorganisms per gram, and seldom contain more than 100 microorganisms per gram. Improper washing and sanitizing of the shell surface, followed by long periods of storage (>2 weeks), and inadequate temperatures, can increase the numbers of microorganisms. Storing eggs in environments of high humidity may encourage mold growth on the surface of shell eggs.

The microbiology of fresh liquid egg depends upon the bacteriological condition of the eggs used for breaking and the sanitation program of the plant. Aerobic plate counts of liquid egg from commercial breaking plants generally are in the range of 10^4 to 10^6 per gram. Aerobic plate counts exceeding 10^7 per gram indicate the use of low quality eggs for breaking,poor plant sanitation, and/or inadequate storage of the liquid. Coliform counts generally range from 10^3 to 10^5 per gram; counts routinely exceeding 10^6 coliforms per gram in untreated liquid egg should not be considered satisfactory. Salmonellae are found in untreated liquid eggs in low numbers (1 to 100/gm). The incidence of salmonellae in untreated liquid egg increases with the use of poor quality eggs for breaking, and with inadequate plant sanitation.[12]

Pasteurization is mandatory by law for all egg products except those intended for use in some high acid foods. The production of dried egg albumen presents a unique problem, since sugar must be removed before drying. The removal of sugar may be accomplished through bacterial fermentation using pure cultures or natural fermentation, or alternatively through the use of enzymes. During the removal of sugar process there may be substantial increase in microorganisms, including salmonellae. It is common industry practice to treat dried egg albumen with dry heat at temperatures between 130 to 140 F after the product has been dried. This treatment may require many days to many weeks in order to eliminate salmonellae from the dried product. The length of time necessary is a function of the numbers of these organisms in the dried product. Aside from the particular problem with dried egg albumen, pasteurization usually results in liquid egg, frozen egg, and dried egg products having aerobic plate counts of less than 25,000 per gram. The coliform counts on pasteurized egg range from less than 1 to 10 per gram, and salmonellae are routinely absent in 25 grams of sample. The presence of coliforms indicates inadequate pasteurization or postpasteurization contamination. The presence of salmonellae in pasteurized egg products is due almost exclusively to postpasteurization contamination.

Direct microscopic counts give an indication of the type of raw material used and the type of sanitary handling, as the microscope reveals the dead as well as the viable organisms in the sample. Unusually high microscopic counts are indicative of poor quality, regardless of the viable count, unless the product underwent bacterial fermentation to remove sugar prior to drying. Egg products that have been treated or held in such a manner as to show very low viable counts may be revealed by the direct microscopic technique to have had very high counts originally. Lepper et al[13] have pointed out that with liquid and frozen eggs, a microscopic count of over five million per gram with formic, acetic, or lactic acid in excess of seven mg per 100 grams demonstrates the presence of decomposed eggs. They also concluded that a microscopic count of over one hundred million per gram of dried egg with formic or acetic acid over 65 mg/100 gm or lactic acid over 50 mg/100 gm (on dry basis) was indicative of the presence of decomposed eggs.

43.6 REFERENCES

1. Recommended methods for the Microbiological Examination of Foods. 1966. Second Edition. SHARF, J. M., ed. American Public Health Association. Inc., 1790 Broadway, New York, N.Y.
2. Official Methods of Analysis of the AOAC, 1970. Eleventh Edition. Published by the AOAC, Box 540, Benjamin Franklin Station, Washington, D.C.
3. BOARD, R. G. 1964. The growth of gram-negative bacteria in the hen's egg. J. Appl. Bacteriol, **27**:350–364
4. BOARD, R. G. 1973. The microbiology of eggs. In Egg Science and Technology. W. J. STADELMAN and O. J. COTTERILL, ed. The AVI Publishing Company, Inc., Westport, Conn.
5. BREED, R. S. 1911. The determination of the number of bacteria in milk by directo microscopic examination. Zentralbl. F. Bakt., II, Abt. **30**:337–340
6. BROOKS, J. and D. I. TAYLOR, 1955. Eggs and Egg Products. Rep. Fd. Invest, Bd. 60, H.M.S.O. London, England
7. COTTERILL, O. J. 1968. Equivalent pasteurization temperatures to kill salmonellae in liquid egg white at various pH levels. Poultry Sci. **47**:354–365
8. FLORIAN, M. L. E. and P. C. TRUSSELL. 1957. Bacterial spoilage of shell eggs. IV. Identification of spoilage organisms. Food Technol. **11**:56–60
9. FORSYTHE, R. H., J. C. AYRES, and J. L. RADLO. 1953. Factors affecting the microbiological populations of shell eggs. Food Techno. **37**:49–56
10. HARTUNG, T. E. and W. T. STADELMAN. 1962. Influence of metallic ions on the penetration of the egg shell membranes by *Pseudomonas fluorescens.* Poultry Sci. **41**:1590–1596.
11. KRAFT, A. A., L. E. ELLIOTT, and A. W. Brant. 1958. The shell membrane as a barrier to bacterial penetration of eggs. Poultry Sci **37**:238–240
12. KRAFT, A. A., G. S. TORREY, J. C. AYRES, and R. H. FORSYTHE. 1967. Factors influencing bacterial contamination of commercially produced liquid egg. Poultry Sci. **46**:1204–1210
13. LEPPER, H. A., M. T. BARTRAM, and F. HILLIG. 1944. Detection of decomposition in liquid, frozen, and dried eggs. J. Off. Agr. Chem. **27**:204–223
14. ROSSER, F. T. 1942. Preservation of eggs. II. Surface contamination on egg shell in relation to spoilage. Canad. J. Res. **20D**:291

15. Shafi, R., O. J. Cotterill, and M. L. Nichols. 1970. Microbial flora of commercially pasteurized egg products. Poultry Sci. **49**:578–585
16. Speck, M. L. and F. R. Tarver. 1967. Microbiological populations in blended eggs before and after commercial pasteurization. Poultry Sci. **46**:1321
17. Stuart, L. S. and E. H. McNally. 1943. Bacteriological studies on the egg shell. U.S. Egg Poultry Mag. **49**:28–31
18. USDA Laboratory Methods for Egg Products. 1973. Agri. Marketing Serv., Poult. Div., Washington, D.C.

CHAPTER 44

FRUITS AND VEGETABLES

D. F. Splittstoesser and J. Orvin Mundt

44.1 INTRODUCTION

Under consideration in this chapter are fresh produce and nonsterile, processed fruits, and vegetables. Examples of the latter are the products preserved by freezing or dehydration. Some of these foods will be cooked prior to consumption; others will be eaten raw.

Products that might be classed with both fresh and processed vegetables are the chopped salad ingredients (lettuce, cabbage, carrots, etc.) sold in the grocery store and to the institutional trade. Although essentially fresh produce, contamination during processing, and changes in microbial growth patterns during storage, may alter the microflora of these foods quantitatively and qualitatively. At present little published information is available regarding the microbiology of packaged salad ingredients.

44.2 FRESH PRODUCE

44.21 Numbers of Microorganisms

44.211 Sources

On fruits and vegetables as received from orchards, vineyards, and growing fields there is a great variation in the numbers of organisms. All green plants possess an epiphytic microflora which normally subsists on the slight traces of carbohydrates, protein, and inorganic salts which dissolve in the water exuding from, or condensing on, the epidermis of the host.[14] Other important factors are the opportunity for contamination from soil, water, dust, and other natural sources, the extent of contact during harvest with the soiled surfaces of harvesters and containers, intrinsic properties of the fruit or vegetables, such as a thick skin that protects against surface damage, and subsequent growth of saprophytic organisms.

44.212 Populations

Some of the highest aerobic counts have been reported for tubers and other vegetables that are in contact with the soil, for example: potatoes,

IS/A Committee Liaison: Carl Vanderzant

28×10^6; cucumbers, 16×10^6; and beets, 3.2×10^6 per gram.[1, 16] By contrast, cabbage, which is not in direct contact with soil and has a large weight to surface ratio, gave an average count of only 42×10^3 organisms per gram.[16]

44.22 Types of Microorganisms

44.221 Predominant flora

Whether the population resulted primarily from growth on the produce or on product contaminated surfaces, or whether it originated from an unrelated source is an important factor. A soil contaminated fruit or vegetable will possess a microflora composed primarily of sporeformers, coryneforms and other soil types, while one that had supported microbial growth will yield the kinds of organisms which compete best on that particular substrate. The high acid and sugar content of fruits often permits yeasts and molds to predominate, while the high carbohydrate content of many vegetables favors the lactic acid bacteria. Some low saccharine vegetables have predominantly an aerobic, gram negative flora, both hyaline and yellow pigmented.

44.222 Market diseases

Decay caused by molds and certain bacteria accounts for much of the spoilage of fresh fruits and vegetables. Many of these organisms are true plant pathogens in that they can invade healthy plant tissue. While bacterial rot is caused mainly by one genus, *Erwinia*, numerous molds such as *Alternaria, Botrytis,* and *Phytophthora* are responsible for a variety of market diseases.

44.223 Indicator organisms

Nonfecal coliform bacteria and enterococci are a part of the naturally occurring microflora of many plants.[3, 6] The presence of *Escherichia coli* can be related to the use of polluted irrigation waters during growth,[2] contamination through human handling, the use of contaminated containers, or washing after harvest with polluted water.

44.224 Human pathogens

The paucity of published information related to pathogenic organisms on raw fruits and vegetables suggests that presently this is not a serious problem in the United States. An expanding population with concomitant shortages of resources, however, could effect changes in production and marketing that would reduce the safety of these foods. Thus the use of poor quality water for irrigation could increase the incidence of enteric pathogens. The reuse of pathogen contaminated containers is another potentially hazardous practice; for example, the shipping of lettuce in crates that previously had held chicken carcasses.

44.3 PROCESSED FRUITS AND VEGETABLES

44.31 Reducing Contamination

In the preparation of fruits and vegetables for freezing, fermentation, or drying, various procedures may be used that destroy or remove contaminating microorganisms.

44.311 Washing

This treatment removes many of the truly contaminating microorganisms, i.e., those that are not protected by the native mucilaginous material of the plant surface. With peas, for example, 72 to 94% of the microorganisms picked up at the viner are removed by the first washing.

44.312 Heat

Most vegetables and a few fruits are blanched as one of the early processing steps. The temperatures employed, 86 to 98 C, destroy all but the most heat resistant microorganisms; only bacterial spores usually survive. Examples of other processes that employ a lethal temperature are the heating of Concord grapes to 60 C, prior to pressing to extract pigment, and the immersion of various fruits and vegetables in a hot lye bath to facilitate skin removal.

44.313 Germicides

Chlorine in flume and spray water reduces the number of viable contaminants. The relatively high concentrations of sulfur dioxide, often 2500 ppm or more, used to treat apple splices before freezing, and various fruits and vegetables prior to dehydration, destroys most vegetative microorganisms. In winemaking, the addition of 100 ppm sulfur dioxide reduces the number of wild yeasts by 99.9%.

44.314 Freezing and dehydration

Although some organisms are destroyed by these processes, large numbers including many vegetative forms manage to survive.[15, 16] A further decrease in viable numbers occurs during the storage of frozen or dried fruits and vegetables. The rate of this decrease is influenced by many factors such as storage conditions, the food type, and the predominant microflora.

44.4 FROZEN VEGETABLES

44.41 Sources of Contamination

As discussed previously, at the time of harvest vegetables may be contaminated with large numbers of microorganisms. These populations may increase even higher due to growth if the product is held a number of hours before processing.

Common early processing steps are various cleaning and washing procedures. Although these treatments remove some of the microorganisms, the counts often remain in the millions per gram. Blanching is the operation that destroys most of the contaminating organisms, thus the microflora of frozen vegetables usually reflects postblanching contamination.

The major source of the organisms on frozen vegetables is contaminated equipment.[8, 18] Operations that have been especially troublesome are choppers, slicers, conveyor and inspection belts, and filling machines. Some of these units possess surfaces that are difficult to reach for proper cleaning. Belts, which generally are more accessible, may present problems because of the tenacity with which organisms adhere to certain surfaces, and because some fabrics absorb moisture and thus permit a microbial buildup within the belt interior. The latter occurs and becomes a source of contamination when fractures develop on the belt surface.

The degree of difficulty in controlling postblancher contamination is related to vegetable type. With corn, large amounts of starch and other organic material are released onto equipment surfaces and into flume water, while in the processing of green beens and peas, minimal soluble solids are lost from the vegetable.

44.42 Types of Microorganisms

44.421 Predominant flora

A wide variety of saprophytic bacteria can be isolated from frozen vegetables. The predominant flora is influenced by vegetable type, and perhaps even by geographic location. Gram negative rods predominate on certain vegetables such as greens. On others, lactic acid cocci belonging to the genera *Streptococcus* and *Leuconostoc* are most numerous. The proportion of lactic acid bacteria was found to increase as the processing season progressed, with corn, for example, from 30% of the aerobic count on the first day to above 90% after 15 days of processing.[11] Many of the cultures of streptococci from vegetables differ in some characteristics from described species.[7, 12]

44.422 Indicator organisms

Coliforms and enterococci are common contaminants of frozen vegetables and may be present in relatively large numbers, sometimes thousands per gram.[9] Their presence usually does not indicate fecal contamination. It appears that they are introduced onto equipment surfaces, perhaps via the air, and become a part of the microflora of the processing line, along with the more numerous contaminants such as the lactic acid cocci. *Escherichia coli* and fecal coliforms, on the other hand, are relatively rare contaminants of many frozen vegetables and thus may have special sanitary significance.

44.423 Pathogens

Frozen vegetables appear to present few problems with respect to organisms that cause foodborne illnesses. Although coagulase positive staphylococci can be recovered, their numbers are usually very low, under 10 per gram.[10] Attempts to isolate salmonellae from certain frozen vegetables have been unsuccessful[13] while data regarding the incidence of anaerobic pathogens such as *Clostridium perfringens* seem to be lacking.

44.5 FROZEN FRUITS

44.51 Sources of Contamination

Although fruits generally are not blanched, much of the orchard and harvest acquired microflora is removed by various washing and peeling steps. Thus, as with vegetables, many of the organisms contaminating frozen fruits and fruit products originate from the processing line.

44.52 Types of Microorganisms

44.521 Predominant flora

Because of the low pH of many fruits, aciduric organisms are the predominant contaminants. Yeasts and molds are often the most numerous microorganisms. Of acid-tolerant bacteria, the lactic acid group is most common although species of *Acetobacter* also may develop in the acid environment of fruit processing lines.

44.522 Indicator organisms

Coliforms can be recovered from various fruits even though the pH may be too low to support growth of these organisms. As with vegetables, the presence of indicator organisms on frozen fruits and fruit products usually does not indicate a public health problem.[17]

44.523 Pathogens

Fruits are generally too acid for growth of the more common foodborne pathogens. Many organisms do not survive in a low pH environment; for example, *Salmonella* and *Shigella* die off rapidly in citrus juices.[4]

A potential problem is toxigenic fungi since they can grow on fruits and other high acid foods. Patulin, a mycotoxin produced by *Penicillium expansum* and a number of other molds, has been found in apple juice. It appears that the most effective control is to exclude moldy fruit from the product.[5]

44.6 RECOMMENDED METHODS

44.61 Standard Plate Count Chapter 4

44.62 Yeasts and Molds Chapter 16

44.63 Fecal Coliforms Chapter 24

44.7 INTERPRETATION OF DATA

The microbiology of fresh fruits and vegetables often has little relationship to quality or safety of these foods. Sound vegetables, for example, may yield extremely high aerobic plate counts because of contamination from soil or other natural sources. The routine microbiological examination of most fresh fruits and vegetables, therefore, is not recommended.

Because most of the organisms on frozen and other processed fruits and vegetables often originate from equipment surfaces, aerobic plate counts provide a means of assessing sanitation of the processing line. Problems in controlling contamination may differ with the type of fruit or vegetable; a given population on one vegetable may signify poor sanitation, while the same count on another may indicate excellent conditions. Many frozen fruits and vegetables have aerobic plate counts under 50×10^3 per gram.

Coliforms and enterococci are part of the normal vegetable processing line flora and populations of 10^2 or 10^3 per gram are not uncommon. Fecal coliforms are more rare and counts of under 10 per gram usually can be achieved.

Coagulase positive *Staphylococcus aureus* may be present on vegetables, but usually in low numbers, under 10 per gram. The routine culturing of fruits and vegetables for staphylococci probably is not justified.

44.8 REFERENCES

1. Etchells, J. L., R. N. Costilow, T. A. Bell, and H. A. Rutherford. 1961. Influence of gamma radiation on the microflora of cucumber fruit and blossoms. Appl. Microbiol. **9**:145–149
2. Geldreich, E. E. and R. H. Bordner. 1971. Fecal contamination of fruits and vegetables during cultivation and processing for market. A review. J. Milk and Food Technol. **34**:184–195
3. Griffin, A. M. and C. A. Stuart. 1940. An ecological study of the coliform bacteria. J. Bacteriol. **40**:83–100
4. Hahn, S. S. and M. D. Appleman. 1952. Microbiology of frozen orange concentrate. I. Survival of enteric organisms in frozen orange concentrate. Food Technol. **6**:156–158
5. Harwig, J., Y-K Chen, B. P. C. Kennedy, and P. M. Scott. 1973. Occurrence of patulin and patulin-producing strains of *Penicillium expansum* in natural rots of apple in Canada. Can. Inst. Food Sci. J. **6**:22–25
6. Mundt, J. O. 1961. Occurrence of enterococci: bud, blossom, and soil studies. Appl. Microbiol. **9**:541–544
7. Mundt, J. O., W. F. Graham, and I. E. McCarty. 1967. Spherical lactic acid-producing bacteria of southern-grown raw and processed vegetables. Appl. Microbiol. **15**:1303–1308
8. Splittstoesser, D. F., W. P. Wettergreen, and C. S. Pederson. 1961. Control of microorganisms during preparation of vegetables for freezing. II. Peas and corn. Food Technol. **15**:332–334
9. Splittstoesser, D. F. and W. P. Wettergreen. 1964. The significance of coliforms in frozen vegetables. Food Technol. **18**:134–136
10. Splittstoesser, D. F., G. E. R. Hervey II, and W. P. Wettergreen. 1965. Contamination of frozen vegetables by coagulase-positive staphylococci. J. Milk and Food Technol. **28**:149–151
11. Splittstoesser, D. F. and I. Gadjo. 1966. The groups of microorganisms composing the "total" count population in frozen vegetables. J. Food Sci. **31**:234–239
12. Splittstoesser, D. F., M. Mautz, and R. R. Colwell. 1968. Numerical taxonomy of catalase-negative cocci isolated from frozen vegetables. Appl. Microbiol. **16**:1024–1028
13. Splittstoesser, D. F. and B. Segen. 1970. Examination of frozen vegetables for salmonellae. J. Milk and Food Technol. **33**:111–113
14. Thaysen, A. C. and L. D. Galloway. 1930. The Microbiology of Starch and Sugars. p 191, Oxford University Press, New York, N.Y.

15. Van Eseltine, W. P., L. F. Nellis, F. A. Lee, and G. J. Hucker. 1948. Effect of rate of freezing on bacterial content of frozen vegetables. Food Res. **13**:271–280
16. Vaughn, R. H. 1951. The microbiology of dehydrated vegetables. Food Res. **16**:429–438
17. Vaughn, R. H. and D. I. Murdock. 1956. Sanitary significance of microorganisms in frozen citrus products. Amer. J. Public Health **46**:886–894
18. Wolford, E. R., A. D. King, Jr., and H. D. Michener. 1965. Variations in bacterial count in commercial corn freezing. J. Milk and Food Technol. **28**:183–187

CHAPTER 45

FRUIT DRINKS, JUICES, AND CONCENTRATES

D.I. Murdock

45.1 INTRODUCTION

Of the large variety of fruit grown in the U. S. only a few are processed on a large commercial scale to produce fresh and frozen juices and concentrates. The citrus industry is the largest processor. Sanitation plays an important role in processing operations. Spoilage problems virtually can be eliminated by frequent cleaning and sanitizing of equipment, and by monitoring microbial activity during processing and in the final product. Citrus products are marketed as reconstituted single strength juices and as frozen citrus concentrates, both of which may not be sterile. In the concentrate form they are not sterile, and the concentrates must be stored at 0 F or lower to maintain product quality.

Federal Standards of Identity have been established for 11 orange juice products, including orange juice, orange juice from concentrate, and frozen concentrated orange juice. They state in part:

"Orange juice (27.105) is the unfermented juice obtained from mature oranges of the species *Citrus sinensis*. Seeds (except embryonic seeds and small fragments of seeds that cannot be separated by good manufacturing practice) and excess pulp are removed. The juice may be chilled, but it is not frozen. . . ."

"Frozen concentrated orange juice (27.109) is the food prepared by removing water from the juice of mature oranges as provided in 27.105. . . ." "The concentrate so obtained is frozen. . . ."

For a complete identification of orange juice products, refer to Code of Federal Regulations Title 21 Part 27.105 through 27.115.[4]

45.2 NORMAL FLORA

Microorganisms enter the processing plant on the surface of fruit, having originated from soil, untreated surface water, dusty air, and decomposed fruit. The degree of contamination varies, depending upon how the fruit was handled from the field to the processing plant. Proper grading, washing, and sanitizing of the fruit contribute materially to good product quality. In the citrus industry, sound fruit entering the extractors is rela-

IS/A Committee Liaison: Nino F. Insalata

tively free of microorganisms, while unsound fruit, afflicted with soft deteriorated spots, splits, etc., is heavily contaminated.[14]

45.3 MICROORGANISMS FOUND IN JUICE PRODUCTS

The low pH of juices and concentrates limits the organisms that can survive and grow. For example, the pH range of lemon or lime juice is 2.2 to 2.6, and none of the spoilage bacteria survive; in orange juice pH 3.4 to 4.0, *Lactobacillus* and *Leuconostoc* can survive and grow.[3, 17, 18] These bacteria frequently cause abnormal flavors and odors, sometimes similar to buttermilk.[10, 11, 13] However, they fail to grow in concentrates at 45 Brix* or higher, or below 40 F (5 C). *Leuconostoc mesenteroides* and *Lactobacillus brevis* are two organisms known to produce off flavors.

Acetic acid bacteria, yeasts and molds also may be present, but generally will not grow rapidly enough to build up large populations under conditions normally prevailing during concentration of the juice. Yeasts, on the other hand, will grow in citrus concentrates when held in the pH and temperature range for the growth of these organisms. Yeasts are primarily responsible for spoilage of chilled juice that is not sterile, which has spoiled as a result of being held in storage beyond its shelf life, or at temperatures above those recommended.

Coliform bacteria and related types (e.g., *Erwinia*) are frequently present on or in oranges before they are harvested.[20,21] Coliform organisms sometimes are present in orange juice in spite of washing and sanitizing the fruit before extraction.[2, 6] Routine presumptive tests have shown a very low incidence of coliform bacteria in orange concentrate;[20, 21] however, very high percentages of false positives occur.[20, 21, 22, 23, 24, 25] Martinez and Appleman[12] have encountered false positive coliform reactions caused by yeasts. It has been shown that false positive presumptive coliform tests are due primarily to the carryover of sugars (other than lactose) with the inoculum, and its subsequent fermentation by noncoliforms. There are no data to indicate that coliform bacteria actually grow in citrus juices. On the other hand there is considerable evidence to show that they can retain their viability for extended periods in frozen orange concentrate, but die off rapidly in fresh or reconstituted juices.[6, 9, 19] Coliform organisms are of little or no public health significance in frozen citrus products.[6, 20, 21]

45.4 HUMAN DISEASE BACTERIA

Coliform tests of frozen concentrate indicate neither fecal contamination, nor the presence of *Salmonella*, especially when the product has been stored before shipment. *Salmonella* and *Shigella* cannot survive for sufficiently long periods in the acid environment of citrus juices or concentrates, nor can spores of *C. botulinum* germinate and grow.[20, 21] Inability of food poisoning organisms to grow in citrus products does not rule out the importance of maintaining high sanitary standards. The rapidity at which lactic acid organisms grow during processing requires a constant vigil in regard to cleaning and sanitizing to prevent spoilage.

*Percent by weight of sucrose in a solution. In citrus it means "% total soluble solids".

Bacteria of the genus *Lactobacillus* and yeasts are also of concern in processing canned, nonrefrigerated shelf stable fruit juices, drinks, and concentrates such as apple, pineapple, apricot, etc. Products of this type have sanitary problems similar to those encountered in processing frozen concentrated orange juice. Many acid food products such as tomato juice, puree and paste, and fruit nectars and concentrates, according to Collier and Townsend,[5] are "presterilized" before canning, and filled at sufficiently hot temperatures to kill spoilage organisms on the container and cover during the short holding period between closing and cooling. For further information consult NCA's publication "Container Sterilization for Acid Products by Hot Fill-Hold-Cool Procedures" by Collier and Townsend.[5] Also refer to Chapters 52 and 53.

45.5 RECOMMENDED METHODS

The procedures for making total plate counts, yeast and mold counts, and direct microscopic examination of fruit juices and concentrates are described in Chapter 16 of Recommended Methods for the Microbiological Examination of Foods, 2nd Edition, American Public Health Association, 1966, New York, N.Y.[1] These methods are outlined in detail in Chapters IV and V of this compendium.

The "diacetyl test"[16] is a colorimetric procedure requiring about 30 minutes; it detects diacetyl and acetylmethylcarbinol in orange juice, as end products of microbial growth, principally those organisms belonging to the genera *Lactobacillus* and *Leuconostoc*.[15] The test has also been used for apple juice and concentrates.[7, 8] It can be used as an index of juice quality and of poor sanitation in processing frozen concentrated orange juice. When unsanitary conditions exist, diacetyl values usually increase.

45.51 Performance of Diacetyl Test[16]

45.511 Preparation of reagents

a. Potassium hydroxide (40%) containing creatine.

Dissolve 40 g of potassium hydroxide pellets in sufficient quantity of distilled water to make 100 ml of solution. Add 0.3 g of creatine (reagent or technical grade). Stir mixture until dissolved. Caution: Reagent breaks down very rapidly at room temperature. Store at 40 F (5 C) not over 3 days, or 21 days at 0 F (–18 C).

b. Alpha-naphthol reagent.

Dissolve 5 g of alpha-naphthol (Eastman 170 or C.P.) in 100 ml of 99% isopropyl alcohol. Temperature of storage not critical.

c. Antifoam solution.

Mix 1 g of Dow-Corning Antifoam AF emulsion in 10 ml distilled water.

45.512 Procedure

a. Add 300 ml of reconstituted juice (12 Brix) to a large boiling flask connected to a condenser for rapid distillation.

b. Add 1 or 2 drops of antifoam solution to each 300 ml sample.

c. Distill off 25 ml distillate into a graduated cylinder, filter through No. 1 Watman filter paper (used to eliminate cloud or oil interference).

d. Pour distillate (25 ml) into a 125 ml Erlenmeyer flask. Add 10 ml of alpha-naphthol solution and 4 ml of KOH solution containing creatine. Order of adding reagents is not critical.

e. Mix by swirling gently for 20 seconds, and immediately transfer to a colorimetric tube. Record reading 60 seconds after addition of reagents as % light transmission (L. T.), or as optical density (O.D.). Use a Lumetron colorimeter with a 530 μ filter.

Null the colorimeter by holding a blank containing distilled water and the two reagents for the same period as the test sample.

f. Prepare a calibration curve by applying this method to samples of distilled water to which known amounts of diacetyl have been added. Use a sealed, unopened bottle of diacetyl, recently purchased, if possible.

45.6 INTERPRETATION OF DATA

High plate counts on products in various stages of preparation prior to concentration may indicate improperly cleaned equipment, operation for an extended period without cleaning and/or sanitizing, or the processing of poor fruit.

Most juice concentrators now use a high temperature evaporator, commonly referred to as TASTE (Thermal Accelerated Short Time Evaporator). Products leaving this unit are almost free of microorganisms.

High plate counts in the finished product generally indicate improperly cleaned equipment or product abuse in the processing operation between the evaporators and packaging.

A direct microscopic count may be made at any stage of the processing operation. It should be remembered that the test does not differentiate between living and dead organisms. Repetitive differential counts of yeasts and bacteria are valuable; an increase in the number and proportion of one group over the other indicates growth of the predominating organism during the processing operation.

The diacetyl test is now being used by most citrus juice concentrators as a rapid method for detecting microbial activity during processing prior to juice concentration. Diacetyl producing microorganisms exhibit little or no growth after the product reaches 45 Brix. A slow steady increase in diacetyl indicates a buildup of microbial growth, usually in the juice extraction equipment. A sudden surge in diacetyl concentration may indicate either (1) a portion of spoiled juice being diverted into the processing stream, (i.e., from an unused pipeline improperly cleaned prior to use), or from a product remaining too long in a holding vessel, or (2) poor fruit (containing numerous splits (partially decomposed fruit), being processed. A sudden increase in diacetyl is usually followed by a corresponding decrease, usually in a half hour or less, as the portion of contaminated product passes through the system. If diacetyl does not return to normal, a flushdown or cleanup is indicated.

45.7 REFERENCES

1. Recommended Methods for the Microbiological Examination of Foods. 1966. 2nd Edition. American Public Health Association. New York, N.Y.
2. Beisel, C. C. and V. S. Troy. 1949. The Vaughn-Levine boric acid medium as a screening presumptive test in the examination of frozen concentrated orange juice. Food Products J. **28**:356–357
3. Cameron, E. J. and J. R. Esty. 1940. Comments on the microbiology of spoilage in canned foods. Food Res. **5**:549–557
4. Code of Federal Regulations Title 21 Food and Drugs. (Revised) 1973. Parts 10 to 129. pp 166–167, 168–169, 178
5. Collier, C. P. and C. T. Townsend. 1954. Container Sterilization for Acid Products by Hot Fill-Hold-Cool Procedures. Proc. Tech. Sessions 47th Annual Con. National Canners Assn., Atlantic City, N.J.
6. Dack, G. M. 1955. Significance of enteric bacilli in foods. Am. J. Pub. Health. **45**:1151–1156
7. Fields, M. L. 1962. Voges-Proskauer test as a chemical index to the microbial quality of apple juice. Food Technol. **16**:98–100
8. Fields, M. L. 1964. Acetylmethylcarbinol and diacetyl as chemical indexes of microbial quality of apple juice. Food Technol. **18**:114–118
9. Hahn, S. S. and M. D. Appleman. 1952. Microbiology of frozen orange concentrate. I. survival of enteric organisms in frozen orange concentrate. Food Technol. **6**:156–157
10. Hays, G. L. 1951. The isolation, cultivation and identification of organisms which have caused spoilage in frozen orange juice. Proc. Fla. State Hort. Soc. **64**:135–137
11. Hays, G. L. and D. W. Riester. 1952. The control of "off-odor" spoilage in frozen concentrated orange juice. Food Technol. **6**:386–389
12. Martinez, N. B. and M. D. Appleman. 1949. Certain inaccuracies in the determination of coliforms in frozen orange juice. Food Technol. **3**:392–394
13. Murdock, D. I., V. S. Troy, and J. F. Folinazzo. 1952. Development of off flavor in 20 Brix orange concentrate inoculated with certain strains of Lactobacilli and *Leuconostoc*. Food Technol. **6**:127–129
14. Murdock, D. I., J. F. Folinazzo, and C. H. Brokaw. 1953. Some observations of gum forming organisms found on fruit surfaces. Proc. Fla. State Hort. Soc. pp. 278–281
15. Murdock, D. I. 1967. Diacetyl test as a quality control tool in processing frozen concentrated orange juice. Food Technol. **22**:90–94
16. Murdock, D. I. 1967. Methods employed by the citrus concentrate industry for detecting diacetyl and acetylmethylcarbinol. Food Technol. **21**:157–161
17. Murdock, D. I. 1968. Sanitation problems in the production of frozen citrus concentrate. J. Milk & Food Tech. **31**:245–250
18. Pilcher, R. W. 1947. The Canned Food Reference Manual, 3rd. Ed. American Can Company, New York, N.Y.
19. Troy, V. S. 1954. Coliforms in Citrus Products Report to Florida Canners Association, Winter Haven, Fla. (Unpublished)
20. Vaughn, R. H. and D. I. Murdock. 1956. Sanitary significance of microorganisms in frozen citrus products. Amer. J. Pub. Health. **46**:886–894
21. Vaughn, R. H., D. I. Murdock, and C. H. Brokaw. 1957. Microorganisms of significance in frozen citrus products. Food Technol. **11**:92–95
22. Wolford, E. R. 1950. Bacteriological studies on frozen orange juice stored at –10 F. Food Technol. **4**:241–245
23. Wolford, E. R. 1954. Comparison of boric acid and lactose broth for isolation of *Escherichia coli* from citrus products. Applied Microbiol. **2**:223–227
24. Wolford, E. R. 1955. Significance of the presumptive coliform test as applied to orange juice. Applied Microbiol. **3**:353–354
25. Wolford, E. R. 1956. Certain aspects of the microbiology of frozen concentrated orange juice., Amer. J. Public Health, **46**:709

CHAPTER 46

SPICES AND CONDIMENTS

Philip A. Guarino and Henry J. Peppler

46.1 INTRODUCTION

Spices and herbs, like most raw agricultural products, are subject to microbial contamination after harvesting.[18, 20, 35, 37] The microbial flora of whole spices, herbs, and spice blends is dominated by members of the genus *Bacillus*, for example, *Bacillus subtilis, Bacillus licheniformis* and *Bacillus megaterium.*[14] Anaerobic sporeforming bacteria within the genus *Clostridium* are also found in small numbers. Thermophilic anaerobes and aerobes[25] are found occasionally, sometimes in moderate numbers. Enterococci and gram negative bacteria, including members of the family *Enterobacteriaceae,* occur occasionally.[5, 15] A variety of molds[4, 18, 20] may be found; spore densities range from insignificant levels to many millions. Yeasts are found infrequently. Pathogens, such as *Salmonella, Shigella* and coagulase-positive staphylococci are found rarely in spices.[10, 15, 18, 20] *Escherichia coli,* an indicator of fecal contamination, only occasionally is found.[15] *Clostridium perfringens* spores are present usually at low levels, i.e., from ten to several hundred per gram.[23]

The most frequently used spices, black and white pepper, usually contain very high microbial populations.[20, 29] Plate counts from ten million to 100 million per gram have been obtained. Members of the genus *Bacillus* usually predominate, generally as spores rather than as vegetative cells.[18, 20] Other spices containing relatively high microbial populations include celery seed, paprika, and ginger.[35]

Cassia, mace, nutmeg and most of the herbs contain moderately low numbers of bacteria.[15] Aerobic plate counts usually are not greater than a few hundred thousand organisms per gram, and aerobic sporeforming bacteria predominate.

Some spices, because of the marked bactericidal effect of their essential oil content, harbor few bacteria.[3, 35] The antibacterial action of cloves, for example, is due to the presence of eugenol. In mustard seed, the essential oil, allyl isothiocyanate, is inhibitory to bacteria. Onion and garlic inhibit bacterial growth because of their sulfur containing compounds.[11, 28, 32] Oregano and many other herbs are bactericidal also.[15, 35]

IS/A Committee Liaison: Edmund A. Powers

In suspensions with relatively high concentrations of these spices it is not uncommon to find low colony counts on the low dilution plates[28, 35] because of the carryover of inhibitory substances with the inoculum, while moderately high counts may be obtained on the higher dilution plates.

Plate counts of spices containing antibacterial compounds or oils seldom exceed several thousand organisms per gram and only a few microbial types are present. Pathogens or organisms of sanitary significance rarely are found in such spices.[18]

While aerobic plate count values have no public health significance,[19, 30] they do have a bearing on the degree to which a spice may contribute toward the spoilage of a product in which that spice is used.[7, 18, 20, 36] Plate count densities may range from several thousand organisms in some spices, such as cloves, to several million per gram in a variety of other spices.[15, 18, 20]

Cleaned spice materials show variable but a moderately low incidence of sanitary indicator microorganisms. Root items, berries, and herbs, generally carry a slightly greater microbiological load than do bark and seeds,[15] but organisms of sanitary significance occur only occasionally while pathogens rarely occur.

Spices are grown and harvested in areas of the world where often sanitary practices are poor by comparison with those usually considered acceptable in the U.S. Moreover, many of the spices are grown in warm, humid areas where a wide variety of fungi and bacteria are supported readily, and are easily distributed during the handling of the commodities. Consequently, uncleaned spices may be grossly contaminated and occasionally carry organisms of sanitary or public health significance.[18, 20] For these reasons, spices are generally not used in the form in which they are imported. They usually undergo extensive cleaning[15, 17, 19] by a variety of means including sifting, aspiration and other methods that take advantage of the shape or density of the spice particles.

Fumigation of spices with ethylene oxide[2] or with propylene oxide[8] can effectively reduce the microbial population as much as 99.9%. This includes organisms which are public health hazards.[15, 27, 34, 35] The majority of residual bacteria are sporeforming bacilli. Current FDA regulations limit the permissible residue of ethylene oxide to 50 ppm[9] and of propylene oxide to 300 ppm in spices.[8] Irradiation with gamma rays[13, 14, 34] has been attempted also in an effort to rid spices of undesirable bacteria, or to reduce total microbial levels. However, at this time FDA does not permit irradiation of any spices.

In general, processed spices should be examined according to recommended methods for aerobic plate count, yeasts, molds, coliforms and, if coliforms are present, for *E. coli.* In instances where there is an indication that the spice was held under unsanitary conditions, examination for *Salmonella* and *Shigella* should be carried out. Additional tests may be required for special use. The user may request enumeration of coagulase positive staphylococci, anaerobic sporeformers (mesophiles and thermophiles), *Clostridium perfringens,* enterococci, and lactobacilli.

46.2 TREATMENT OF SAMPLE

Measure 200 g samples from a minimum of five containers representing the shipment or lot to be examined, or use another suitable sampling plan. (Chapter 1)[19, 31] Use aseptic technique; sterilize the sampling equipment after use. A "trier" or "thief" may be used to sample many spice materials. Devices such as clean spoons, scoops, or similar implements may also be used to sample whole nutmeg, whole ginger, and other root items, as well as cassia bark, which cannot be sampled conveniently by means of a trier or thief. Sampling of all ground spices is done most readily by means of a clean tablespoon. Place the sample in sterile plastic bags, label clearly, and forward to the microbiological laboratory without delay. Examine all five sample units separately, if possible. Composite only as a last resort when lack of facilities or time prevent the recommended amount of testing. Retain the individual samples for additional testing when indicated.

Sample preparations and their initial dilution, as outlined below, are applicable to most of the routine microbiological examinations. Exceptions are certain procedures not done routinely, and methods used to detect *Salmonella* and *Shigella*.

46.3 WHOLE SPICES

46.31 Berries, Seeds and Herbs

a. Aseptically weigh the 10 g sample into a 4 or 8 oz wide mouth polypropylene or glass bottle containing 90 ml sterile phosphate buffer.

b. Shake 5 minutes at 200 strokes per minute with a mechanical shaker, such as Burrell wrist-action shaker.

c. Inoculate appropriate media within fifteen minutes.

46.32 Roots and Bark

a. Weigh 100 g of representative sample into a sterile, clean, and dry blender jar; blend the sample at the lowest setting for 30 seconds, or until a moderately fine particle size is attained.

b. Aseptically weigh 10 g of sample into a 4 or 8 oz wide mouth polypropylene or glass bottle containing 90 ml sterile phosphate buffer.

c. Shake at least 25 times in a 1 foot arc.

d. Inoculate appropriate media within 15 minutes.

46.4 GROUND SPICES, STANDARD AND SPECIAL BLENDS

46.41 Spices, Spice Blends, and Seasonings

a. Aseptically weigh 10 g of sample into a 4 or 8 oz wide mouth polypropylene or glass bottle containing 90 ml sterile phosphate buffer.

b. Shake at least 25 times in a 1 foot arc.

c. Inoculate appropriate media within 15 minutes.

46.5 MICROBIOLOGICAL ANALYSIS

46.51 Aerobic Plate Count (Chapter 4)

Since many spices contain essential oils, and some are inhibitory to bacteria, it is necessary to prepare decimal dilutions that are sufficiently high to overcome the effects of such compounds. Low dilutions may yield erroneous results when examining naturally inhibitory spices (ground mustard seed, cinnamon, cloves, onion powder and garlic powder, for example).

46.52 Yeast and Molds (Chapter 16)

46.53 Coliform Bacteria (Chapter 24)

a. MPN Method
b. Violet Red Bile Agar Count

46.54 *Escherichia coli* (Chapter 24 and References 1, 6, 19, 26, 30)

46.55 *Salmonella* (Chapter 25 and 26 and References 6, 10, 12, 24, 33)

46.56 Aerobic Sporeformers

a. Mesophiles (Chapter 18)
b. Thermophiles, including Flat sours (Chapters 20 and 21)

46.57 Anaerobic Sporeformers

a. Mesophiles (Chapter 19)
b. Thermophiles, including H_2S Producers (Chapter 22)

46.58 *Clostridium perfringens* (Chapter 35)

46.59 Supplementary Tests

46.591 Coagulase positive staphylococci (Chapter 31)

46.592 Enterococci (Chapter 30)

46.593 Lactobacilli (Chapter 15)

LBS agar is suitable for enumeration of lactobacilli in spices.

46.6 REFERENCES

1. Microbiological Examination of Precooked Frozen Foods. 1966. Quarterly Bulletin, Supplemental Issue, Association of Food and Drug Officials of the United States, pp. 1–77
2. Blake, D. F. and C. R. Stumbo. 1970. Ethylene oxide resistance of microorganisms important in spoilage of acid and high-acid foods. J. Food Science **35**:26–29
3. Blum, H. B. and F. W. Fabian. 1943. Spice oils and their components for controlling microbial surface growth. Food Prod. J. **22**:326–329
4. Christensen, C. M., H. A. Fanse, G. H. Nelson, F. Bates, and C. J. Mirochia. 1967. Microflora of black and red pepper. Appl. Microbiol. **15**:622–626
5. Clark, W. S., G. W. Reinbold, and R. S. Rambo. 1966. Enterococci and coliforms in dehydrated vegetables. Food Technol. **20**:113–116
6. Edwards, P. R. and W. H. Ewing. 1972. Identification of *Enterobacteriaceae*. Third Edition. Burgess Publishing Co., Minneapolis, Minn.

7. Fabian, F. W., C. F. Krehl, and N. W. Little. 1939. The role of spices in pickled food spoilage. Food Res. **4**:269–286
8. Federal Register. 1972. Sec. 121.1076, Propylene oxide, **37**:9463
9. Fine, S.D., 1970. Ethylene oxide, Part 121, Food Additives, Sec. 121.1231, Fed. Reg. **35** (145):12062; Sec. 121.1232, Fed. Reg. **38**:5342 (1973).
10. Foster, E. M. 1971. The control of salmonellae in processed foods: A classification system and sampling plan. JAOAC **54**:259–266
11. Frazer, W. C. 1967. Food Microbiology. Second Edition. McGraw-Hill, New York, N.Y.
12. Galton, M. M., G. K. Morris, and W. T. Martin. 1968. Salmonellae in Foods and Feeds. Review of Isolation Methods and Recommended Procedures. U.S. Department of Health, Education and Welfare, Public Health Service, National Communicable Disease Center, Atlanta, GA.
13. Gerhardt, U. 1969. Sterilization of spices (Ger.) Gordian **69** (1631):427–432. (From Chem. Abstr. **72**:65475)
14. Goto, A., K. Yamazaki, and M. Oka. 1971. Bacteriology of radiation sterilization of spices. Food Irrad. (Shokuhin-Shosha) **6**:35–42
15. Guarino, P. A. 1973. Microbiology of spices, herbs and related materials. Proceedings of 7th Annual Symposium. Fungi and Foods. Special Report No. 13, pp. 16–18. New York State Agricultural Experiment Station, Geneva, N.Y.
16. Hauschild, A. H. W. 1973. Food poisoning by *Clostridium perfringens*. Can. Inst. Food Sci. Technol. J. **6**:106–110
17. Huth, H. 1972. Method for manufacturing practically sterile, concentrated aromas of spices, vegetables and mushrooms. U.S. Pat. 3,681,090
18. Julseth, R. M. and R. H. Deibel. 1974. Microbial profile of selected spices and herbs at import. J. Milk Food Technol. **37**:414–419
19. Kramer, A. and B. Twigg. 1973. Quality Control for the Food Industry. Third Edition. Volumes 1 and 2. AVI Publishing Company, Westport, Conn.
20. Krishnaswamy, M. A., J. D. Patel, and N. Parthasarathy. 1971. Enumeration of microorganisms in spices and spice mixtures. J. Food Sci. Technol. **8**:191–194 (Food Sci. Technol. Abstr. **4**:T305)
21. Lee, W. H., P. B. Mislivec, and C. T. Dieter. 1973. Elimination of molds in peppercorns with methyl bromide. Abstracts of the Annual Meeting of the American Society for Microbiology, E64, Miami Beach, Fla.
22. Mislivec, P. B., C. T. Dieter, and A. C. Sanders. 1973. Mycotoxin potential of mold flora from dried beans. Abstracts of the Annual Meeting of the American Society for Microbiology, E119, Miami Beach, Fla.
23. National Academy of Sciences. 1969. An Evaluation of the *Salmonella* Problem. Publication 1683. National Research Council, Washington, D.C.
24. National Academy of Sciences. 1971. Reference Methods for the Microbiological Examination of Foods: Report of the Food Protection Committee, National Research Council, Washington, D.C.
25. Richmond, B. and M. L. Fields. 1966. Distribution of thermophilic aerobic sporeforming bacteria in food ingredients. Appl. Microbiol. **14**:623–626
26. Roche Diagnostics. 1973. Enterotube Numerical Coding and Identification System for *Enterobacteriaceae*. Hoffman-LaRoche, Inc., Nutley, N.J.
27. Sair, L. 1972. Ground spice product. U.S. Patent 3,647,487
28. Sheneman, J. M. 1973. Survey of aerobic mesophilic bacteria in dehydrated onion products. J. Food Science **38**:206–209
29. Surkiewicz, B. F., R. W. Johnston, R. P. Elliott, and E. R. Simmons. 1972. Bacteriological survey of fresh pork sausage produced at establishments under Federal inspection. Appl. Microbiol. **23**:515–520
30. Thatcher, F. S. and D. S. Clark (Editors) 1968. Microorganisms in Foods. I. Their significance in methods of enumeration. Univ. of Toronto Press, Toronto, Ontario, Canada
31. Thatcher, F. S. and D. S. Clark (Editors) 1974. Microorganisms in Foods II. Sampling for microbiological analysis: Principles and specific applications. Univ Toronto Press, Toronto, Ontario, Canada
32. Tynecka, Z. and Z. Gos. 1973. Inhibitory action of garlic *(Allium sativum)* on growth and

respiration of some microorganisms. Acta Microbiol. Pol. Ser. B. **5**:51–62. (From Chem. Abstr. **79**:63491)

33. U.S. Food and Drug Administration. 1972. Compliance Program Guidance Manual. Chapter 20—General Programs—Foods. No. 7320-15. Bureau of Foods, Division of Compliance Program, Washington, D.C.
34. Vajdi, M. and R. R. Pereira. 1973. Comparative effects of ethylene oxide, gamma irradiation and microwave treatments on selected spices. J. Food Science **38**:893–895
35. Weiser, H. H., G. J. Mountney, and W. A. Gould. 1971. Practical Food Microbiology and Technology. Second Edition. AVI Publishing Co., Westport, Conn.
36. Wright, W. J., C. W. Bice, and J. M. Fogelberg. 1954. The effect of spices on yeast fermentation. Cereal Chem. **31**:100–112
37. Yesair, J. and M. H. Williams. 1942. Spice contamination and its control. Food Res. **7**: 118–126

CHAPTER 47

PICKLE PRODUCTS*

J. L. Etchells and T. A. Bell

47.1 INTRODUCTION

Brined, salted, and pickled vegetables are classified into the following groups:

47.11 Cucumber Pickles and Similar Pickle Products

47.111 Salt stock for cured pickle products

a. Cucumbers (and onions, peppers, tomatoes, cauliflower, carrots, cabbage, melon rinds, etc.)
b. Genuine dill pickles (from cucumbers or tomatoes)
c. Green olives (Spanish type), whole or stuffed
d. Naturally ripe (black) olives

47.112 Finished pickle products from brine cured stock

a. Sweets
b. Sours
c. Mixed
d. Relishes
e. Processed dills
f. Hamburger slices

47.113 Types of pasteurized pickles (not brine cured)

a. Dills (sliced or whole)
b. Sweets (sliced or whole)
c. Relishes (mixed vegetables)
d. Vegetables other than cucumbers (onions, peppers, green tomatoes, okra, carrots, etc.)
e. Green olives, not fully fermented

47.114 Overnight dill pickles (refrigerated)

a. Cucumbers (primarily)
b. Green tomatoes

---

*The methods described herein are based in part on procedures referred to in reference 26.

IS/A Committee Liaison: John H. Silliker

47.12 Brined and Salted Vegetables for Nonpickle Use

47.121 Brined

a. Okra (whole)
b. Celery (cut)
c. Sweet pepper hulls
d. Citron (peel)

47.122 Dry salted

a. Corn
b. Lima beans
c. Peas
d. Green snap beans
e. Okra (cut)
f. Celery (cut)

47.2 NORMAL FLORA

The normal microflora that usually predominates during the natural fermentation of brined cucumbers, other vegetables, and green olives is similar, particularly, if the fermentation takes place under suitable and comparable conditions of temperature, salt content, percentage of solids to brine (by weight). The lactic acid bacteria usually reach the ascendancy, but they may be preceded by growth of the coliform bacteria, and followed by fermentative yeast development, depending on the available brine nutrients. From the briner's standpoint, the desired brine microflora is established early and continued predominance by the homofermentative lactic acid bacterial species—initiated by *Pediococcus cerevisiae* followed by *Lactobacillus plantarum*. Gas forming (heterofermentative) lactics such as *L. brevis* and *Leuconostoc mesenteroides* may also be present in certain vegetable brines. The extent of their growth is highly dependent on brining conditions, essentially those described above. For example, Pederson et al.[37] reported good acid formation by *L. mesenteroides* in ten days in kraut at about 50 F and 2 to 2.5% salt by weight. In contrast, we have been unable to initiate growth of *L. brevis* in brined olives of the Manzanillo variety. However, neither species is expected to predominate in brine strengths employed for commercially brined cucumbers destined for brine stock purposes. Pure culture fermentation of cucumbers, olives and other vegetables has recently been proposed by Etchells et al.[14, 17, 18]

47.3 FLORA CHANGES IN SPOILAGE

47.31 Salt Stock Vegetables and Genuine Dills

Vigorous activity in the cover brine by coliform bacteria, obligate halophiles, heterofermentative lactic acid bacteria, and fermentative species of yeast is associated with gaseous fermentation. This may bring about a physical "spoilage" condition in salt stock cucumbers and dill pickles known as "bloaters" or hollow cucumbers. Although most of these groups of organisms are extremely salt tolerant, the coliform bacteria and halophiles are not found usually in brines having appreciable acidity. An exception may be found in cases of highly buffered material, such as dry salted peas or beans.

Luxuriant growth of film yeasts may occur at various salt concentrations, and will result in loss of brine acidity. When certain molds accompany this scum growth, the vegetable material may become soft and unusable. Heavy scum yeast and/or mold growth is usually the result of neglect of brined material during the curing and storage period. This is particularly true for stock brined the previous season.

The significance of the presence of the salt tolerant cocci and obligate halophiles is presented in section 47.65

47.32 Finished Pickle Products (Other than Pasteurized)

Fully cured, salt stock vegetables are made into various types of finished pickle products by a series of operations involving leaching out most of the salt, souring with vinegar, and then sweetening with sugar. Preservation of these products is dependent upon sufficient amounts of vinegar alone (for sour pickles), or a combination of vinegar and sugar (for sweet pickles). If the amounts of either ingredient is inadequate, fermentation usually takes place, principally by two groups of organisms, lactic acid forming bacteria and yeasts. Molds and film yeasts may grow on the surface of the liquor chiefly as the result of faulty jar closure.

47.33 Pasteurized Products

Spoilage usually occurs in these products when they are improperly pasteurized and/or improperly acidified so that an equilibrated brine product of pH 3.8 to 4.0 is not achieved. Spoilage is due chiefly to yeasts and/or acid forming bacteria that survive faulty heat treatment, or butyric acid bacteria when the product is not acidified adequately at the outset. Molds and film yeasts are factors in cases of poor jar closure.

47.34 Overnight Dill Pickles (Refrigerated)

The barreled product may be stored for a few days at room temperature and then refrigerated at 36 to 40 F (ca 2 to 5 C). Under such conditions and at equilibrated brine strengths of 10 to 12 salometer (1° salometer = 0.264% salt by weight), microbial growth (chiefly coliforms, gas forming and nongas forming lactics, and fermentative yeasts), and enzymatic activity (pectinolytic and cellulolytic), together with the curing process continues at a slow rate.[16] In a few months, the stored pickles may have lost much of their desired characteristic flavor, texture and color, and also may be bloated because of gaseous fermentation by the principal gas forming microbial groups present (mentioned above). Whether these pickles are made in bulk or in the retail jar, the fact remains that the very nature of the product makes it difficult to maintain good quality pickles for any reasonable length of time. The barreled product reaches the GMP recommended brine pH of 4.5 to 4.6 for low acid food usually before refrigeration or shortly thereafter, and then slowly continues acid development. This recommended condition for brine product pH cannot be assured for the product made in the retail jar because there is no accepted uniform process by the packers wherein the product is acidified at the outset (to equilibrate at pH 4.5 to 4.6) or where it is deliberately incubated for development of natural lactic acid fermentation. Spoilage of this product is caused chiefly by the gas forming microbial groups mentioned earlier. Gas production may be sufficient to reach 15 pounds pressure on the cap. Our sampling of 50 one-quart jars of overnight dills from refrigerated counters or cases of large retail stores located in five geographical areas of the country indicated that every third jar was judged "not acceptable for commercial use." Twenty-five per cent more were placed in the "barely acceptable" to "poor" categories. A followup study on 23 jars in one of the large metropolitan production areas gave better results; jars placed "not acceptable" (½ of the number of the first study), but "barely acceptable" to "poor" ratings were given nearly 40% of the jars. Those jars of pickles placed as "fair," "good" and "excellent" amounted to 17, 13 and 17%, respectively.

47.4 HUMAN DISEASE BACTERIA

There are no authenticated reports to our knowledge of human disease bacteria associated with standard, commercial pickle products prepared under "good manufacturing practices" of acid, salt, and sugar content (and combinations thereof) from brined, salted, and pickled vegetable brinestock—including cucumbers. Even so, certain types of microorganisms that may cause spoilage of the product may, at times, be encountered, such as molds, yeasts, and acid tolerant lactic acid bacteria. These organisms, usually under conditions associated with neglect, may reduce the quality (texture and/or flavor) of the product (prepared in bulk or retail container) and render it unusable. However, these organisms are not considered human pathogens.

Essentially the same pattern of consumer safety applies to fresh pack (pasteurized) pickle products. These have continued to increase in popularity until these items now use about 40% of the annual cucumber crop in the USA. These pickles usually are prepared from raw cucumbers, but may include other vegetables in a mixture; also, vegetables other than cucumbers may be packed, such as various types of peppers, okra, carrots, green beans, green cherry, pear shaped, or regular globe tomatoes, and the like. The process calls for the packed product to be acidifed at the outset with a sufficient amount of food grade organic acid (vinegar, acetic acid, and lactic acid) to result in an equilibrated brine product pH of 4.0 or below (preferably 3.8). Vinegar (10 to 20% strength) is usually the acidulant of choice of industry for cucumber pickle products. The basic pasteurization procedure has been used successfully by industry for over 35 years.[10, 25]

As far as fresh pack (pasteurized) pickle products are concerned, changes in formulation, calling for specifically reduced acidification, or lowering the salt content, or both of these, is probably the most significant and dangerous set of factors to tamper with (assuming that an adequate pasteurization procedure is used). For instance, arbitrarily reducing the vinegar (acid) and salt content of the cover brine of a given product to achieve some abnormally mild flavoring to appeal to some segment of the consuming public, might inadvertently lead to a butyric acid type spoilage problem involving the public health aspect of the product.

47.5 RECOMMENDED METHODS

The methods described here should prove useful to those concerned with, or responsible for, the examination of certain types of manufactured pickles, particularly those products undergoing spoilage as the result of microbial activity. The methods should prove helpful also to research investigators interested in conducting studies on predominating microbial changes occurring in certain brined and salted vegetables during natural fermentation and curing.[26] The methods are also helpful in following the pure culture fermentation of brined cucumbers,[18] green olives,[17] naturally ripe olives,[2] and other vegetables.

47.51 General Procedure

47.511 Collection, storage and preparation of brine samples

Brine or pickle liquor covering vegetable material is required for examination. The size of container to be sampled may range from a small jar of pickles to a 1,000 bushel tank of fermented brine stock. Brine samples from containers, such as tanks and barrels, should be taken for bacteriological analysis as follows:

A suitable length of 3/16″ stainless steel tubing, sealed at one end with lead or solder and perforated with several 1/16″ holes for a distance of 6 to 8 inches from the sealed end, is inserted through an opening between the wooden boards comprising the false head down into the brine toward the middepth of the vegetable material. Withdraw brine through a sanitized, attached piece of rubber tubing into a 12 oz bottle. Fit the receiving bottle with a 2 hole, rubber stopper and 2 short lengths of glass tubing—one for the rubber tubing leading from the stainless steel sampling tube, and the other for a suction bulb to start siphoning action. The length of the steel sampling tube is governed by the depth of the container to be sampled.

Withdraw and discard approximately 24 oz of brine before taking the final sample (about 10 ml) into a sterile, screw cap test tube. If microbial changes during the fermentation are to be followed, start sampling at the time the material is salted or brined, and continue at regular intervals of one to two days during active fermentation. After sampling, wash the whole assembly thoroughly.

For tightly headed barrels such as those used for genuine dills and salted vegetables for nonpickle use, take the sample through the top or side bung.

For smaller containers, such as jars or cans of pickle products, shake thoroughly and take the sample from the center of the material by means of a sterile pipet. Wash the tops of the metal cans with alcohol, flame, and puncture. A beer can opener is useful for puncturing metal tops. If the containers show evidence of gas pressure, carefully release gas by puncturing the sanitized top with a flamed ice pick. Containers under heavy gas pressure may be refrigerated overnight to reduce the gas pressure prior to sampling.

47.512 Storage of samples

Brine samples from actively fermented material should be examined as promptly as possible after collection to prevent changes in the microbial flora. The same is true for samples of packaged pickle products. If it is necessary to ship or store samples, this should be done under the best of refrigerated conditions and the elapsed time from collection to examination should not exceed 12 to 24 hours. When shipment by air is required, samples are collected in sterile 16 × 150 mm test tubes and fitted with plastic screw caps having rubber liners. Pulp and oil liners, or plastic liners such as teflon, may leak due to changes in air pressure.

Brine samples may be preserved for subsequent chemical determinations by the addition of toluene or Merthiolate, 1% aqueous solution, of one to two drops per 10 ml of sample. Collect samples in standard 3 to 4 oz medicine bottles or in 16 × 150 mm test tubes, fitted with screw caps as described above. Shake well to distribute preservative. Caps having pulp-backed vinylite and teflon liners, or pulp backed foil liners should be used; those having cork or composition cork are not satisfactory for prolonged storage.

Samples preserved with the above chemicals are unfit for human consumption and should be so marked.

47.513 Preparation of the sample

Make suitable dilutions of the pickle liquor or brine in the usual manner, except for obligate halophiles. For this group make serial dilutions directly into the recommended liquid medium containing salt.

If poured plates using salt-containing media are desired, the dilution blanks should contain approximately the same salt concentration as the brine sample.

For actively fermenting brines, no specific number of dilutions can be suggested; however, as a guide, such brines, at 5 to 8% salt by weight, may be expected to contain the following populations per ml: acid forming bacteria, 10^6 to 10^9; yeasts, 10^4 to 10^7; obligate halophiles, 10^6 to 10^9; coliforms, 10^6 to 10^8; salt tolerant cocci; 10^6 to 10^7. The expected microbial populations in adequately pasteurized products are normally very low, and composed of resistant spore forming bacteria that remain dormant in the acid liquor. For such products, dilutions of 1 : 10 and 1 : 100 usually suffice. For improperly pasteurized products that are fermenting, the dilutions should cover the estimated range of population suggested for acid-formers and yeasts in fermenting brines. For raw products that are not properly washed, the spore count may reach counts of 1 to 2×10^5/ml even after pasteurization.

47.514 Microscopic examination

Microscopic examination of brine samples for bacteria and yeasts is helpful at times, particularly when carried out in conjunction with plate count observations.

a. Technic for bacteria

Make direct counts for bacteria according to the following procedures:

Place 0.01 ml amounts of brine or liquor on slides, by using a Breed pipet,[5] and spread evenly over a 1 sq cm area; fix with heat.

Stain according to the Kopeloff and Cohen modification of the Gram stain.[32] Count according to the Wang[39] modification of the Breed[5] technic.

Report results as "numbers of different morphological types of gram positive and gram negative bacterial cells per ml of brine."

b. Technic for yeasts

Use the microscopic technic for determining yeast populations in fermenting vegetable brines, and various types of finished pickle products undergoing gaseous spoilage by these organisms, particularly where populations are in excess of 10^4 cells/ml of sample, and where yeast colonies are not required for isolation and study. The use of a vital stain permits differentiation of yeast population into viable and nonviable cells, and increases the usefulness of the direct counting technic.

The counting procedure is essentially the method of Mills[34] as modified by Bell and Etchells[4] for counting yeasts in high salt content brines and in high sugar content liquors:

Add 1 ml of brine or pickle liquor sample to 1 ml of 1 : 5,000 (0.02 per cent) erythrosin stain.

Shake the sample stain mixture to obtain an even suspension.

Using a 3 mm diameter platinum loop, transfer enough of the mixture to the area under the cover glass of an improved Neubauer double-ruled hemacytometer to fill the chamber in one operation.

Allow cells to settle for approximately 5 minutes and count the yeast cells, using a microscope equipped with a 4 mm objective and 15× oculars.

Record cells stained pink as "dead yeast cells," and unstained cells as "live yeast cells."

The number of yeast cells per ml of brine or pickle liquor may be calculated thus:

$$\frac{\text{Number of yeast cells counted} \times \text{dilutions} \times 250{,}000}{\text{Number of large squares counted}} = \text{Numbers per ml}$$

If only one side of the hemacytometer counting chamber is used (25 large squares), the lowest yeast count obtainable is 20,000 per ml, while if both sides are counted (50 large squares), a population as low as 10,000 per ml can be counted.

Report yeast count as "total yeast cells," "live yeast cells," and "dead yeast cells, per ml of sample."

47.515 Titratable acidity and pH

Determinations of titratable acidity and pH of the samples are extremely useful in providing information supplementary to bacteriological analysis.

Determine titratable acidity of a 10 ml sample of the brine or liquor by diluting the sample with 30 to 50 ml of distilled water and titrate with 0.1 N NaOH, using phenolphthalein as the indicator. Report values for brined samples as grams of lactic acid per 100 ml of sample and for finished liquor samples as grams of acetic acid per 100 ml of sample.

For a 10 ml sample, use the following calaculations:

a. ml of 0.1 N alkali used × 0.090 = gm of lactic acid per 100 ml.

b. ml of 0.1 N alkali used × 0.060 = gm of acetic per 100 ml.

When only a small amount of the original sample is available, use a 2 ml amount for titration purposes. Such small samples are not recommended. For the 2 ml sample, multiply the ml of 0.1 N alkali by 5, then by the above number for lactic or acetic acids.

Carry out pH determinations of the samples with a pH meter, checking the instrument frequently with a standard buffer in the pH range of the sample under test.

47.516 Determination of salt content of brine

It is often helpful to know the approximate salt content in performing microbiological examinations of brines. Use a salometer, and test about 200 ml of brine. A chemical test for salt is required for small amounts of sample, or when a higher degree of accuracy is desired than that obtainable by the salometer.

The following method is recommended. Transfer 1 ml of sample to a flask and dilute with 15 to 20 ml of distilled water. Titrate with 0.171 N silver nitrate solution (29.063 gm per liter) using 3 to 5 drops of 0.5% dichlorofluorescein as the indicator. Agitate to keep the precipitate broken up until a light salmon pink color is developed. Report as "gm of sodium chloride per 100 ml of the sample."

When 1 ml of sample is titrated, each ml of silver nitrate solution is equal to 1 gm of sodium chloride per 100 ml.

47.52 Procedure According to Type of Product: Cucumber Pickles and Similar Pickle Products

The three main classes of products under this heading are salt stock vegetables and genuine dills, finished or packaged pickle products made from salt-stock, pasteurized pickles made from fresh stock, and the unheated "overnight" dills (refrigerated). The cucumber is the principal vegetable involved, although substantial amounts of other vegetables, such as onions, peppers, cauliflower, okra, carrots, and green tomatoes, may be used in mixed pickles, relishes, or as individual products.

47.521 Salt stock vegetables and genuine dills

Use the plating technic with differential solid media and decimal dilutions in 0.85% saline diluent. Place decimal dilutions of samples in Petri plates, in duplicate, and fill with medium as follows:

a. Aerobic plate count

Use nutrient agar (Difco) and incubate for three days at 32 C. Overlay the solidified plated samples with about 8 to 10 ml of the same medium to prevent or minimize spreaders.

b. Lactic acid forming bacteria

Use lactobacillus selection medium (BBL), modified carefully to pH 5.6 ± 0.05, plus bromocresol green. Overlay the solidified plates to favor reduced oxygen tension. Lactobacilli colonies appear green to black with a yellow halo.

c. Salt tolerant cocci

Use nutritive caseinate agar, plus 0.1 glucose (Chapter 2) and incubate for three days at 32 C. Count colonies that are grayish white, entire, glistening and of moderate size, and similar colonies that are light orange to yellow in color. Subsurface colonies are lenticular to elliptical in shape. For morphological identification when lactose fermenting yeasts may be present, make stained preparations and examine under the microscope.

d. Coliform bacteria

Use brilliant green lactose bile agar, violet red bile agar or desoxycholate lactose agar (Chapter 2). Incubate for 18 to 24 hrs at 32 C.

e. Yeasts and molds

Use dextrose agar, acidified, and incubate three to five days at 28 to 30 C, or yeast nitrogen base agar plates for estimation of yeasts and molds by the streaking technique (Chapter 2).

f. Film yeasts

For an estimate, pick representative filamentous colonies from the yeast plates into tubes of dextrose broth containing 5 and 10 per cent salt. Incubate three to five days at 32 C, and observe for heavy surface film. Two salt concentrations are suggested for use because some species develop heavier films at the lower salt strength (5 per cent) whereas, with other species, the reverse is true.

g. Obligate halophiles

Use tubes of liver broth plus salt (Chapter 2). Prepare decimal dilutions, seal with sterilized, melted petroleum jelly, and incubate seven days at 32 C. Record positive tubes daily by noting the raising of the petroleum seal due to gas production and the absence of any distinctive odor.

h. Butyric acid forming bacteria

Neutralize the brine sample with an excess of sterile calcium carbonate. Heat a 50 to 100 ml sample in a water bath for 20 minutes at 80 C to kill vegetative cells. Prepare decimal dilutions and inoculate previously heated and cooled tubes of liver broth medium. Seal with melted petroleum jelly and incubate seven days at 32 C. Examine tubes daily for production of gas and a strong butyric acid odor.

47.522 Finished pickle products

Liquor of the sample should be examined for total number of microorganisms, acid forming bacteria, yeasts and molds, film yeasts, and butyric acid forming bacteria, using methods in 47.521.

In undisturbed containers, the surface growth of molds and film yeasts may be obvious. Carefully remove the film after recording the extent of growth, since if shaken up with the sample it will complicate the counts for acid formers and yeasts when the latter groups are present. Examine for coliform bacteria, salt tolerant cocci and halophiles. The test for butyric acid bacteria normally is not required due to acidity of these products.

47.523 Pasteurized types of pickles

There are probably a dozen or more different types of cucumber pickles that fall into this classification (pasteurized), such as various types of fresh dills, fresh sliced cucumber pickles and low acid sweet pickles (from saltstock). Also, many noncucumber products are included: dill tomatoes, sweet and hot peppers, okra, green beans, and fresh vegetable relishes that are prepared from uncured stock.

Examination of the liquor from the products of this class should be made. See section 47.522.

47.524 Overnight dill pickles (refrigerated)

Examination of the liquor from the products of this class should be made as in section 47.522.

47.525 Salted and brined vegetables for nonpickle use

a. Salted vegetables

Most of the vegetables after blanching are preserved according to the dry salt method,[24] using a ratio of 1 : 5 with respect to salt and vegetable weight, and stored in tightly headed wooden casks or metal or plastic containers, preferably at refrigerator temperatures within the range of 1.7 to 4.4 C (35 to 40 F). Vegetables treated in this manner are usually green peas, corn, snap beans, cut okra, small onions, and cut celery. These salted vegetables can be used in the preparation of soups, mixed vegetable products, and strained vegetable products.

b. Brined vegetables

Products such as whole okra, whole celery, and sweet red peppers usually are brined at about 20% salt concentration (equilibrated) rather than by the dry salt method. These are used in the same manner as salted vegetables.

c. Microbiological examination

Brine samples from both types of these vegetables (47.525a and 47.525b) should be examined for aerobic plate count, lactic acid forming bacteria, salt tolerant cocci, coliform bacteria, yeasts and molds, film yeasts, obligate halophiles, and butyric acid forming bacteria by the directions set forth in section 47.521.

47.53 Summary of Procedure

A summary of the bacteriological methods described herein is presented in Table 1. This information is suggested for use as a guide in the examination of certain brined, salted and pickled vegetables and vegetable products.

Some new, some revised, and some different microbiological procedures for cucumber and cucumber product studies have been proposed by Etchells et al.[20] The reader should consult the original reference for details and application of these methods, and for greater detail in specialized instances, consult the related publications.[6, 11–15, 19, 21, 22, 27–31, 36]

47.54 Discussion of the Use of Culture Media and Types of Microorganisms

47.541 Nutritive caseinate agar (Chapter 2) plus 0.1% glucose

Use this medium to save time, effort, and glassware to detect and enumerate several types of bacteria on the same Petri plate. It can be used for enumeration of total bacterial count and salt tolerant cocci, for determining population trends of acid producing bacteria in dill pickles,[31] salt stock,[23] improperly pasteurized fresh cucumber pickles,[22] and in the storage of salted and brined vegetables.[27, 28]

Since this medium contains less agar than usual solid media, no more than 15 ml to insure solidification and prevent dropping of agar when plates are inverted. During hot weather, cool plates prior to inversion and incubation.

Acid forming bacterial colonies show a zone of precipitated casein and a yellow halo in the presence of bromcresol purple. The degree of casein precipitation and color change may vary with the activity and type of acid former.

Surface growth is usually poor. Subsurface colonies are generally elliptical in shape and range in size from 0.5 to 2.5 mm.

Yeasts, other than lactose fermenters, do not grow well on this medium and tend to give a slightly alkaline reaction. In case of doubt, make stained preparations.

While nutritive caseinate agar is not considered a differential medium for salt tolerant coccus forms, the numbers of these organisms in brines of high salt concentration may be estimated. They are indicated by two predominating types of colonies, one grayish white, entire, glistening, and of moderate size, and a similar colony that is yellow to light orange in color. In high salt content, nonacid brines, these organisms are the principal types found on this medium. Due to sensitivity to acid, they are not found usually in active fermentations of the acid type. Deep subsurface colonies may give an acid reaction but, upon prolonged incubation, become alkaline. In highly buffered, salted vegetables, bordering on the range of salt tolerance for acid forming bacteria, care should be exercised that deep colonies of cocci are not recorded as true acid producing bacteria of the lactic group.

47.542 Lactobacillus selection medium (Chapter 2)

This medium should be prepared with 0.0075% bromcresol green dye to aid colony counting, and adjusted carefully to pH 5.6 ± .05 with glacial acetic acid, rather than adding the fixed amount (1.32 ml acetic acid per liter). The modified medium was used successfully[6] for separating relatively low populations of lactic acid bacteria occurring on pickling cucumbers from exceedingly high populations of other microbial groups; thus, the medium is highly selective for the lactobacillus group.

47.543 V-8 medium (Chapter 2)

For determining numbers of lactobacilli, Fabian et al.[30] have shown this medium to give good results. The colonies are green to black with a yellow halo, and develop to a large size in the presence of lactose. The bromcresol green is said to be inhibitory to most of the nonacid formers.

47.544 Brilliant green lactose bile agar (Chapter 2)

This medium is preferred because of the ease of determining the coliform type of colony. Subsurface colonies of the coliform group are deep red against a blue green background. This medium is sensitive to light and should be prepared just prior to use. When this is not convenient, the medium should be stored in the dark. For more complete identification, representative colonies should be streaked on Levine's eosin methylene blue agar.

47.545 Dextrose agar (acidified)

This medium is more inhibitive to the lactic acid types of bacteria and is preferred over malt agar for detecting yeasts in fermenting vegetable brines.[11] Occasionally, yeasts that will not grow on this medium are found in high salt concentrations, 15 to 20 per cent.[21] By reducing the tartaric acid to 3 ml per 100 ml, growth can often be obtained. However, this modification should not be used when the salt concentration is known to be below 15 per cent, since acid forming bacteria will grow enough to make counting of yeasts very difficult.

Mold colonies are distinguished readily from yeasts on this medium, whereas, differentiation of subsurface yeasts and film yeasts present more difficulty. Surface colonies of the common film forming yeasts associated with pickle products and vegetable brines (i.e., species of *Debaryomyces, Endomycopsis, Saccharomyces, Candida* and *Pichia*),[13, 36] are generally dull and very rough as contrasted to the usual round, raised, white, glistening colonies of the fermentative, subsurface yeasts (i.e., species of *Torulopsis, Brettanomyces, Hansenula, Saccharomyces*, and *Torulaspora*).[15, 19, 23] However, even when distinguishing colony characteristics of the two yeast groups exist, they are not considered sufficiently clear cut for separation. Because of

Table 1: Guide to the Bacteriological Examination of Salted, Brined, and Pickled Vegetable Products

Microbial group involved	Culture Medium (Chapter 2)	Classes of products in which microbial group is likely to be present*	Remarks concerning microbial groups
Total count	Plain or dextrose agar	All classes of products 47.111, 47.112, 47.113, 47.114, 47.121 and 47.122	For determination of general microbial populations; in pasteurized products they help to indicate the effectiveness of the treatment.
Acid forming bacteria	LBS agar, modified; V-8 medium	47.111 Fermenting salt stock vegetables and genuine dills 47.112 Finished pickle products 47.113 Pasteurized pickle products 47.114 Overnight dill pickles (Refrigerated)	Acid fermentation; salt-tolerant up to 15 per cent; not likely to be found in brined and salted vegetables above this concentration (47.121 and 47.122).
Salt tolerant cocci	Nutritive caseinate agar	47.12 Brined and salted vegetables for nonpickle use other high salt vegetables without appreciable acidity	No outstanding characteristics of fermentation reported; group salt tolerant but sensitive to acid; can grow at refrigerator temperature (1.7 C) at approximately 10 per cent salt.
Coliform bacteria	Brilliant green lactose bile agar, violet red bile agar, or desoxycholate agar	47.111 Fermenting salt stock vegetables and genuine dills 47.113 Types of pasteurized pickles (not brine cured) 47.114 Overnight dill pickles (refrigerated) 47.121, 47.122 Brined and dry salted vegetables for nonpickle use	Gaseous fermentation; group salt-tolerant but not acid tolerant; most likely absent from finished pickles due to acid content; same is true for brines when appreciable acid is present.

Obligate halophiles	Liver broth plus salt	47.12 Brined and salted vegetables for nonpickle use Other vegetable brines at high salt concentration	Gaseous fermentation; group requires 5 to 15 per cent salt in culture medium and reduced oxygen tension; sensitive to acid; general information or behavior not well known.
Fermentative yeasts, film yeasts and molds	Dextrose agar (acidified), dextrose broth plus salt;** yeast nitrogen base agar for streak plates	All classes of products (47.111, 47.112, 47.113, 47.114; 47.121 and 47.122) for yeasts Molds and film yeasts on liquid surface of products exposed to air and sheltered from sunlight.	Yeasts: gaseous fermentation; acid and salt tolerant; molds and film yeasts: acid and salt tolerant; both groups utilize acid of products and require free oxygen for growth.
Butyric acid group	Liver broth medium without salt	Uncommon in brined and salted vegetables; examination should be made if malodorous fermentation is detected in all classes of products (47.11 and 47.12), particularly 47.114 Overnight dill pickles (refrigerated).	Causes malodorous, gaseous fermentation; not particularly acid or salt tolerant; active fermentations rare in properly brined or salted vegetables.

*Refer to outline for more detailed classification of products listed under 47.11 and 47.12
**For culturing film forming yeasts in general.

this, the procedure outlined under 47.521f should be used. Film yeasts rapidly form a heavy wrinkled surface film at one or both salt concentrations. Certain species, such as *Saccharomyces halomembranis*, form heavier films at 10 per cent salt than at 5 per cent.[11, 13, 21, 36]

47.546 Yeast nitrogen base agar plates for estimation of yeasts and molds by streaking technique[15, 20] (Chapter 2)

Prepare the following in distilled water, equal amounts in separate containers of:

a. 4% agar (Bacto-Difco) + 4% glucose (dextrose)

b. double strength yeast N base broth (Difco Laboratories, Detroit, Michigan) = 1.3% or 1.3 grams of the dehydrated N base powder per 100 ml H_2O.

Sterilize the above containers at 15 lb pressure for 12 minutes. Cool to 50 C, and mix the contents of the two containers, add sterilized tartaric acid (5%) at the rate of 3 ml per 100 ml of mixed agar and N base. This equals 7.5 ml of 5% tartaric acid per 250 ml of media.

Pour plates using 25 to 30 ml of agar per plate; allow to solidify, invert, incubate 24 hours at 30 C, observe for any contaminating colonies, then store inverted in refrigerator until used. Streak the sample, or suitable dilutions of such, on surface with platinum loop (.01 ml capacity). Incubate three to five days at 28 to 32 C, and count.

47.547 Liver broth plus salt (Chapter 2)

This medium has proved satisfactory for detecting obligate halophiles sometimes found in brined and dry salted vegetables. The salt content of the medium should approximate that of the sample. No interference has been encountered by growth of coliforms or yeasts in this medium. This is probably due to the inability of either group to initiate satisfactory early growth in laboratory media even at moderately high salt concentrations in competition with the very fast growing obligate halophiles.

47.548 Liver broth medium (Chapter 2)

This medium has proved useful in detecting saccharolytic, and putrefactive, mesophilic anaerobes. While a positive test is presumptive evidence of mesophilic, spore forming, gas producing anaerobes of the butyric acid forming types, more specific bacteriological tests are required on positive tubes before identification can be made. Also, spore formation in the sample may be negligible due to high acid production in the presence of readily fermentable carbohydrates, and even though previous activity by this group may have been quite high, negative results usually will be obtained in old brines. Positive results in this medium indicate that these types of bacteria were responsible for the malodorous fermentation.

47.6 INTERPRETATION OF DATA

47.61 Salt Stock Vegetables and Genuine Dills

47.611 Significance of observations

The acid fermentation resulting from active growth of lactic acid bacteria is to be expected at brine concentrations below 12 to 15 per cent strength.[23] The acidity developed in the brine, in combination with the salt, results in preservation of salt stock cucumbers, genuine dills, olives, and other brined vegetables.

Yeast counts of viable cells of subsurface species in fermenting cucumber brines may average four to five times that obtained by the plating technic. Clusters of viable cells, no doubt, are responsible for this difference, as each clump of cells forms a single colony on a plate, while the actual number of cells is recorded by the microscopic method.

It should be emphasized that neither microscopic nor plate counting technics give a true picture of the populations of gas forming, subsurface species of yeasts in fermenting brines or pickle samples obtained from containers contaminated with film yeasts originating from luxuriant surface growth. This is applicable to small containers (jars), since large fermenting tanks (1,000 to 6,000 gal capacity) can be sampled in such a way that surface yeasts are not a problem.

47.62 Finished Pickle Products

47.621 Significance of observations

A total count of a few thousand organisms per ml normally is found in unspoiled pickle products. These counts are composed chiefly of resistant, aerobic spore forms that remain inactive in the acid medium of the pickle liquor and tend to decrease during storage. Active yeast fermentation in the product usually is characterized by vigorous gas production which causes the pickle liquor to become highly charged with gas and to possess a tang when tasted. Gas production may be sufficient to blow lids from jars, to break jars, or to distend or burst sealed cans.[16, 25] Also, whole pickles may become "bloaters" (hollow) due to the gaseous fermentation by yeasts and/or gas producing types of acid producing bacteria.[16, 25] The acid content of the liquor may be increased due to growth of acid producing bacteria.

Extensive mold and film yeast growth (on the surface of brine or liquor) usually result in a reduction in acidity of the liquor, and, in advanced stages, the vegetable may be completely softened by such growth.

47.63 Pasteurized Types of Pickles

According to Etchells and Jones,[25] pasteurization to an internal product temperature of 165 F, for 15 minutes, followed by prompt cooling, is required for pickle products that do not contain sufficient amounts of added vinegar and sugar to stop fermentation by certain organisms. Esselen et al[1, 7–9, 33] applied the mathematical method of process calculation, as described by Ball,[3] to the derivation of pasteurization times for pickles. The times and temperatures derived relate to the degree of sterilization or heat units given the product rather than to an internal product temperature per se. Even so, the original pasteurization process described and recommended some 30 or more years ago is still followed by industry.[22, 25] Pasteurization times are based upon procedure wherein the jars are held in a processing tank, or a steam or hot water pasteurizer, at the indicated pasteurization temperature and time. At the end of the pasteurization period, the jars are cooled promptly below 100 F.

A series of experiments were conducted in commercial pickle plants involving the pasteurization of fresh pack dill pickles.[35] Internal product temperatures in the range of 160 to 170 F with an equilibrated acidity of 0.60% acetic acid or greater prevented spoilage by natural fermentation, and produced pickles of good quality. At temperatures less than 160 F, acidities of up to 1.00% acetic acid did not prevent spoilage. Increasingly higher internal product temperatures, from 170 through 200 F, resulted in correspondingly increased amounts of bloater damage to the internal structure of the cucumber. Faster heating rates decreased pickle firmness, particularly for those located in the upper part of the jar. Tightness of pack greatly influenced the heating rate of the fresh pack dill pickles.

The significance of organisms found in these products is essentially that described in section 47.522. Since the acidity is often lower than in finished pickles, and the products may be made from fresh uncured vegetables, it is essential that spoilage types of organisms be detected, and that any improper heat processing be recognized promptly and corrected.

47.64 Overnight Dill Pickles (Refrigerated)

In the pickle industry, the overnight dill is considered a specialty item, and for many years was prepared in bulk, usually by small packers located in or near large metropolitan areas of the country. In recent years, these pickles also have been prepared in quantity directly in consumer size glass containers which are then supposed to be stored, distributed, and retailed under refrigerated conditions (36 to 40 F). Details on the preparation of this product have been described by Schucart[38] and by Etchells et al.[16]

One important characteristic emphasized by Schucart over 30 years ago, for overnight dills made in bulk (barrels), was their "perishability". He also mentioned the low salt and vinegar content of the product and the strong spicing or seasoning, especially with respect to fresh garlic. We can say that, based on our recent studies (unpublished) on the quality of refrigerated overnight dills in glass jars from retail outlets in several metropolitan areas of the country, the same "perishability" characteristic, mentioned so long ago, still exists, plus a high degree of variability both as to the product's generic name (such as Half Sours, Genuine Kosher Dills, Kosher New Dills, Sour Garlic Pickles, Half Sour New Pickles, Fresh Packed Half Sour Pickles, New Half Sours, Home Style New Pickles, Half Sour Kosher New Dills, and the like), as well as regarding the preparation of the product by individual companies, particularly as to their use of acidification, use of preservatives or other chemical additives, use of whole spices, use of spice emulsions, salt content, size and quality of green cucumbers used, type of jar closure used, ratio of cucumbers to brine on a per cent/weight basis, and the presence or absence of proper refrigeration facilities for the product during preparation, distribution and storage, and display at retail outlets.

47.65 Salted and Brined Vegetables for Nonpickle Use

47.651 Significance of observations

In these products gaseous fermentation usually is associated with active development of coliform bacteria, yeasts, and obligate halophiles, all of which can tolerate the high salt concentrations (15% and above) normally employed for preservation. One or more of the above groups may be present. Gas pressure may be sufficient to burst the barrels. The flavor and appearance of the material also may be altered by growth of the above groups.

Numbers of salt tolerant cocci may be found over an extended period in brines, particularly in those containing no appreciable amount of developed acidity. These organisms are extremely salt tolerant, but not acid tolerant. Their fermentation is not gaseous in nature, and no outstanding change in the product has been attributed to their presence, although small amounts of brine acidity may be produced under conditions providing reduced oxygen tension. When numerous colonies showing a decided acid reaction are found on the plates, they should be examined carefully, as it is likely that they will not be acid producing bacteria of the lactic acid group, as might first be suspected, but rather acid producing cocci. This is particularly true in cases where the brine concentration is above 15 per cent salt.

Growth of molds and film yeasts is likely to be a factor when there is air above the brine surface in the container. Casks should be kept filled with brine at all times, irrespective of storage temperature. Unrestricted growth by molds may soften the texture of vegetable material so it is unusable. Heavy scum growth is undesirable, principally from the flavor standpoint and reduction of brine acidity. This may lead to spoilage by salt tolerant organisms that are not acid tolerant.

As mentioned above, refrigerated storage (about 1.7 C) of these brined and salted products is preferred. Under such conditions, and at salt concentrations of 5 per cent and above, microbial activity of the various groups may be restricted greatly. However, at salt concentrations of approximately 10 per cent strength, the cocci may grow rapidly at about 1.7 C (35 F).

47.7 REFERENCES

1. Anderson, E. E., L. F. Ruder, W. B. Esselen, E. A. Nebesky, and M. Labbee. 1951. Pasteurized fresh whole pickles. II. Thermal resistance of microorganisms and peroxidase. Food Technol. **5**:364–368
2. Balatsouras, Georges D. The Chemistry and Technology of Naturally Black Olives. Food and Agriculture Organization of the United Nations, Lecture Series, Athens, Greece. 49 pp
3. Ball, C. O. 1928. Mathematical Solution of Problems of Thermal Processing of Canned Food. University of California Publications, Public Health, **I**, 245 pp
4. Bell, T. A. and J. L. Etchells. 1952. Sugar and acid tolerance of spoilage yeasts from sweet-cucumber pickles. Food Technol., **6**:468–472

5. BREED, R. S. 1911. The determination of the number of bacteria in milk by direct microscopic examination. Zentralblatt für Bakteriologie und Parasitenkunde. II. Abt, **30**:337–340
6. COSTILOW, R. N., J. L. ETCHELLS, and T. E. ANDERSON. 1964. Medium for producing cells of lactic acid bacteria. Applied Microbiol. **12**:539–540
7. ESSELEN, W. B., E. E. ANDERSON, I. S. FAGERSON, and M. LABBEE. 1952. Pasteurized fresh whole pickles. V. Factors influencing pasteurization requirements. The Glass Packer **31**:326–327 and 346–347
8. ESSELEN, W. B., E. E. ANDERSON, L. F. RUDER, and I. J. PFLUG. 1951. Pasteurized fresh whole pickles. I. Pasteurization studies. Food Technol. **5**:279–284
9. ESSELEN, W. B., I. S. FAGERSON, I. J. PFLUG, and E. E. ANDERSON. 1951. Pasteurized fresh whole pickles. III. Heat penetration in fresh pack pickles. The Glass Packer, **31**:175–178
10. ETCHELLS, J. L. 1938. Rate of heat penetration during the pasteurization of cucumber pickles. Fruit Products J. **18**:68–70
11. ETCHELLS, J. L. 1941. Incidence of yeasts in cucumber fermentations. Food Res. **6**:95–104
12. ETCHELLS, J. L. and T. A. BELL. 1950. Classification of yeasts from the fermentation of commercially brined cucumbers. Farlowia **4**:87–112
13. ETCHELLS, J. L. and T. A. BELL. 1950. Film yeasts on commercial cucumber brines. Food Technol. **4**:77–83
14. ETCHELLS, J. L., T. A. BELL, H. P. FLEMING, R. E. KELLING, and R. L. THOMPSON. 1973. Suggested procedure for the controlled fermentation of commercially brined pickling cucumbers—the use of starter cultures and reduction of carbon dioxide accumulation. Pickle Pak Sci. **3**:4–14
15. ETCHELLS, J. L., T. A. BELL, and I. D. JONES. 1953. Morphology and pigmentation of certain yeasts from brines and the cucumber plant. Farlowia **4**:265–304
16. ETCHELLS, J. L., A. F. BORG, and T. A. BELL. 1968. Bloater formation by gas-forming lactic acid-bacteria in cucumber fermentations. Applied Microbiol. **16**:1029–1035
17. ETCHELLS, J. L., A. F. BORG, I. D. KITTEL, T. A. BELL, and H. P. FLEMING. 1966. Pure culture fermentation of green olives. Applied Microbiol. **14**:1027–1041
18. ETCHELLS, J. L., R. N. COSTILOW, T. E. ANDERSON, and T. A. BELL. 1964. Pure culture fermentation of brined cucumbers. Applied Microbiol. **12**:523–535
19. ETCHELLS, J. L., R. N. COSTILOW, and T. A. BELL. 1952. Identification of yeasts from commercial cucumber fermentations in northern brining areas. Farlowia **4**:249–264
20. ETCHELLS, J. L., R. N. COSTILOW, T. A. BELL, and H. A. RUTHERFORD. 1961. Influence of gamma radiation on the microflora of cucumber fruit and blossoms. Applied Microbiol. **9**:145–149
21. ETCHELLS, J. L., F. W. FABIAN, and I. D. JONES. 1945. The *Aerobacter* Fermentation of Cucumbers During Salting. Michigan Agricultural Experiment Station Technical Bulletin No. 2000, 56 pp.
22. ETCHELLS, J. L. and I. D. JONES. 1942. Pasteurization of pickle products. Fruit Products J. **21**:330–332
23. ETCHELLS, J. L. and I. D. JONES. 1943. Bacteriological changes in cucumber fermentation. Food Indust. **15**:54–56
24. ETCHELLS, J. L. and I. D. JONES. 1943. Preservation of Vegetables by Salting or Brining. U.S. Department of Agriculture, Farmers' Bulletin No. 1932, 14 pp.
25. ETCHELLS, J. L. and I. D. JONES. 1944. Procedure for pasteurizing pickle products. The Glass Packer **23**:519–523
26. ETCHELLS, J. L. and I. D. JONES. 1946. Procedure for bacteriological examination of brined, salted and pickled vegetables and vegetable products. A. J. of Pub. Health **36**:1112–1123
27. ETCHELLS, J. L., I. D. JONES, and W. M. LEWIS. 1947. Bacteriological Changes During the Fermentation of Certain Brined and Salted Vegetables. U.S. Department of Agriculture Technical Bulletin No. 947, 64 pp
28. FABIAN, F. W. and H. B. BLUM. 1943. Preserving vegetables by salting. Fruit Products J. **22**:228–236
29. FABIAN, F. W., C. S. BRYAN, and J. L. ETCHELLS. 1932. Experimental Work on Cucumber Fermentation. Michigan Agricultural Experiment Station Technical Bulletin No. 126, 60 pp

30. FABIAN, F. W., R. C. FULDE, and J. E. MERRICK. 1953. A new V-8 medium for determining lactobacilli. Food Res. **18**:280–289
31. JONES, I. D., M. K. VELDHUIS, J. L. ETCHELLS, and O. VEERHOFF. 1940. Chemical and bacteriological changes in dill pickle brines during fermentation. Food Research, **5**:533–547
32. KOPELOFF, N. and P. COHEN. 1928. Further studies on a modification of the Gram stain. Stain Technol. **3**:64–69
33. LABBEE, M., W. B. ESSELEN, and E. E. ANDERSON. 1952. Pasteurized fresh whole pickles. IV. Enzymes and off-flavors in fresh-pack pickles. The Glass Packer **31**:252–253 and 281–282
34. MILLS, D. R. 1941. Differential staining of living and dead yeast cells. Food Res. **6**:361–371
35. MONROE, R. J., J. L. ETCHELLS, J. C. PACILIO, A. F. BORG, D. H. WALLACE, M. P. ROGERS, L. J. TURNEY, and E. S. SCHOENE. 1969. Influence of various acidities and pasteurizing temperatures on the keeping quality of fresh-pack dill pickles. Food Technol. **23**:71–77
36. MRAK, E. M. and L. BONAR. 1939. Film yeasts from pickle brines. Zentralblatt für Bakteriologie und Parasitenkunde. II. Abt. **100**:289–294
37. PEDERSON, C. S. and M. N. ALBURY. 1969. The Sauerkraut Fermentation. New York Agricultural Experiment Station Bulletin No. 824, 84 pp.
38. H. S. SCHUCART. 1942. Practical observations on the manufacture of kosher style dill pickles. Fruit Products J. **21**:206–212
39. SCHU-HSIEN WANG. 1941. A direct smear method for counting microscopic particles in fluid suspension. J. Bact. **42**:297–319

CHAPTER 48

SALAD DRESSINGS

C. P. Kurtzman

48.1 INTRODUCTION

Mayonnaise, cooked starch based dressings resembling mayonnaise, and pourable dressings are the types of salad dressings most commonly marketed. Sufficient cooking to sterilize would destroy the physical integrity of these products; thus preservation usually depends on the vinegar (acetic acid) or lemon juice present. Acetic acid concentration in excess of 1.5% makes a product unpalatable, but a concentration much below this level may permit spoilage. Worrell[28] suggests that the shelf life of properly prepared mayonnaise is between 3 and 6 months.

48.2 NORMAL FLORA

The microorganisms in salad dressings come from the ingredients and from the air. Few species are able to survive the low pH of salad dressings and generally appear in low numbers.[7,16] Bachmann[2] isolated *Bacillus subtilis, B. mesentericus* (*B. pumilis* and *B. subtilis*),[9] micrococci, a diplococcus, and a mold from several types of unspoiled dressings. Fabian and Wethington[7] found no thermophiles, coliforms, or lipolytic bacteria, and only a few yeasts in 103 samples of unspoiled dressings. Some dressings contained a few molds. Of 10 unspoiled dressings examined by Kurtzman et al,[16] nine appeared sterile, and one contained *B. subtilis* and *B. licheniformis* at less than 50 organisms per gram.

48.3 FLORAL CHANGE IN SPOILAGE

Mayonnaise and other salad dressings spoil because of a variety of reasons: separation of the emulsion, oxidation, and hydrolysis of the oils by strictly chemical processes and growth of microorganisms.[8, 11, 24]

Microbiological spoilage is frequently manifest by gas formation that forces out the dressing when the container is opened. Other indicators of spoilage, such as off flavor and changes in color, odor, or texture, may occur. Iszard[12,13,14] was one of the first to report microbiological spoilage of mayonnaise and demonstrated *B. petasites* (*B. megaterium*)[9] as the cause. Spoilage of a "Thousand Island" dressing was caused by *B. vulgatus* (*B. sub-*

IS/A Committee Liaison: John H. Silliker

tilis),[19] and the source of contamination was found to be the pepper and paprika used in the formulation.

Lactobacillus fructivorans, a species considered synonymous with *L. brevis* by some authorities,[4] was first isolated from spoiled salad dressing.[5] Later it proved to be a common spoilage organism[16] which required special isolation media.

Yeasts frequently cause spoilage in a variety of dressings. Fabian and Wethington[6] found a species of *Saccharomyces* in spoiled French dressing and in mayonnaise while Williams and Mrak[27] showed a yeast similar to *Saccharomyces globiformis* caused spoilage of a starch base dressing. Two-thirds of the spoiled dressing samples examined by Kurtzman et al[16] contained *S. bailii*. This yeast species was the only one found. This is significant because the samples came from widely separated areas of the United States.

Appleman et al[1] found a mixture of *B. subtilis* and *Saccharomyces* sp responsible in one instance of mayonnaise spoilage. *S. bailii* and *Lactobacillus plantarum* were both present in high numbers in a blue cheese dressing.[16]

Apparently the microflora causing salad dressings to spoil is quite restricted, and ordinarily consists of a few strains of *Bacillus, Lactobacillus*, and *Saccharomyces*. Occasionally, contaminated ingredients are implicated as the source of spoilage microorganisms, but infrequently cleaned mixing and filling equipment may provide sufficient inoculum for rapid product spoilage. Presumably, such equipment is contaminated initially by airborne cells since the ingredients themselves seldom yield the spoilage microorganisms.

48.4 HUMAN DISEASE MICROORGANISMS

The survival of pathogenic microorganisms in salad dressings is an important consideration. When Wethington and Fabian[25] inoculated mayonnaise and salad dressing with *Salmonella* and *Staphylococcus*, they found that survival time depended upon product pH. At pH 5.0, one strain of *Staphylococcus* survived 168 hours, but at pH 3.2 survival was 30 hours only. The longest survival time for *Salmonella* was 144 hours at pH 5.0, and 6 hours at pH 3.2. Similar data have been reported by other investigators.[10, 15] Vladimirov and Nefedieva[23] reported *Escherichia coli* to survive 1 day only in mayonnaise, and the data of Bachmann[2] suggested survival up to 10 days. From these studies, it is seen that common food poisoning bacteria survive for short periods of time only in salad dressings. However, preparation of dressing at a pH above 5.0 may alter this situation significantly.

48.5 RECOMMENDED METHODS

48.51 Sample Preparation

Stir sample with a sterile glass rod or a sterile metal spatula; place 50 g into a sterile blender jar. Add 450 ml of sterile 0.1% peptone water and blend 2 minutes. Make subsequent dilutions to 10^{-6} with 0.1% peptone water.[16,21]

48.52 Yeasts and Molds

For isolation use YXT agar (Chapter 2) with tetracycline HCl (10 μg/ml) to inhibit bacteria. Incubate plates at 28 C, and examine at 3 and 5 days.[16]

48.53 Aerobic Bacteria

For isolation use plate count agar (PCA) containing 100 μg/ml of cycloheximide for inhibition of fungi. Some yeasts, not likely to be found in salad dressings, are resistant to this level of cycloheximide. Incubate plates at 28 C, and examine at 3 and 5 days.[16]

48.54 Lactobacilli

Fastidious lactobacilli such as *Lactobacillus fructivorans* and *L. brevis* cannot be detected easily on PCA, but they are readily isolated on *Lactobacillus* selective (LBS) agar (Chapter 2). Incubate LBS agar plates at 28 C, and examine at 7 and 14 days.[16]

48.55 Enterobacteria and *Staphylococcus*

These microorganisms may be detected by the methods given in the section on meat and poultry products (Chapter 42).

48.56 Microscopic Observation of Dressings

Yeasts and bacteria are readily stained for microscopic observation by diluting the dressing 1 : 10 with distilled water. To a drop of this dilution add a small drop of 0.5% crystal violet.

48.57 Identification of Spoilage Microorganisms

Yeasts can be identified by the culture techniques of Wickerham[26] and the classification systems found in The Yeasts.[17] Criteria for identification of *Lactobacillus* can be found in Bergey's Manual, 7th ed.,[3] and in papers by Rogosa and Sharpe,[20] Charlton et al,[5] Vaughn et al,[22] and Kurtzman et al.[16] Species of *Bacillus* can be identified on the basis of the scheme presented by Gordon et al.[9]

48.6 INTERPRETATION OF DATA

Sometimes obviously spoiled dressings contain few viable microorganisms (fewer than 10^2 per gram) or none at all. In these instances, the microorganisms died after the nutrients were exhausted or after the accumulation of metabolic byproducts. A direct microscopic examination still may reveal the dead cells.

Gaseous fermentation may not be evident in spoiled dressings until several weeks after manufacture. *Lactobacillus fructivorans* grows slowly, and considerable time is needed for microbial population increase and visible gas buildup. This slow growth is particularly evident on isolation plates; 10 to 14 days may elapse before colonies are observed.[16] It seems likely that this slow growth, as well as the failure to use the proper isolation medium, may account for the relatively few reports of *L. fructivorans* in spoiled salad dressings.

Saccharomyces bailii and certain other haploid species of *Saccharomyces* ferment glucose quickly but may give a delayed fermentation of sucrose.[16,18] This explains the long delay between manufacture and spoilage of products contaminated with these yeasts when sucrose is the sweetener. Kurtzman et al[16] showed 9 out of 13 strains of *S. bailii* from spoiled dressings fermented sucrose vigorously, but fermentation did not begin until 12 to 56 days after inoculation.

Salad dressings average about pH 4 (3.0 to 4.6)[7,16] due to the acetic acid content, and this acidity accounts for the absence of food poisoning microorganisms. Dressings formulated at a significantly higher pH should be examined for the presence of food poisoning microorganisms. Similarly, the mixing of dressings into meat, potato, and similar salads dilutes the acetic acid so that its inhibitory properties are lost.

48.7 REFERENCES

1. Appleman, M. D., E. P. Hess, and S. C. Rittenberg. 1949. An investigation of a mayonnaise spoilage. Food Technol. **3**:201–203
2. Bachmann, F. M. 1928. A bacteriological study of salad dressings. Wisc. Acad. Sci. Arts Lett., Trans. **23**:529–537
3. Breed, R. S., E. G. D. Murray, and N. R. Smith. 1957. Bergey's Manual of Determinative Bacteriology, 7th ed. The Williams and Wilkins Co., Baltimore, Md.
4. Buchanan, R. E., J. G. Holt, and E. F. Lessel, Jr. 1966. *Index Bergeyana*. The Williams and Wilkins Co., Baltimore, Md.
5. Charlton, D. B., M. E. Nelson, and C. H. Werkman. 1934. Physiology of *Lactobacillus fructivorans* sp. nov. isolated from spoiled salad dressing. Iowa State J. Sci. **9**:1–11
6. Fabian, F. W. and M. C. Wethington. 1950. Spoilage in salad and French dressing due to yeasts. Food Res. **15**:135–137
7. Fabian, F. W. and M. C. Wethington. 1950. Bacterial and chemical analyses of mayonnaise, salad dressing, and related products. Food Res. **15**:138–145
8. Frazier, W. C. 1967. Food Microbiology, p. 537. McGraw-Hill, Inc., New York, N.Y.
9. Gordon, R. E., W. C. Haynes, and C. H-N. Pang. 1973. The Genus *Bacillus*. U.S. Dept. Agr. Agr. Handbook No. 427, Washington, D.C.
10. Gram, H. G. 1957. Abtötung von *Salmonellen, Staphylococcus aureus, B. protus*, und *B. alkaligenes* durch mayonnaise. Fleischwirtsch **9**:111–113
11. Gray, D. M. 1927. Bacterial spoilage in mayonnaise, relishes, and spreads. Canning Age **8**:643–644
12. Iszard, M. S. 1927. The value of lactic acid in the preservation of mayonnaise dressing and other dressings. Canning Age **8**:434–436
13. Iszard, M. S. 1927. The value of lactic acid in the preservation of mayonnaise dressing and other products. J. Bacteriol. **13**:57–58
14. Iszard, M. S. 1927. Supplementary report on the use of lactic acid as a preservative in mayonnaise and allied products. Spice Mill **50**:2426–2430
15. Kintner, T. C. and M. Mangel. 1953. Survival of staphylococci and salmonellae experimentally inoculated into salad dressing prepared with dried eggs. Food Res. **18**:6–10
16. Kurtzman, C. P., R. Rogers, and C. W. Hesseltine. 1971. Microbiological spoilage of mayonnaise and salad dressings. Applied Microbiol. **21**:870–874
17. Lodder, J. 1970. The Yeasts, a Taxonomic Study, 2nd ed. North Holland Publishing Co., Amsterdam, Holland
18. Pappagianis, D. and H. J. Phaff. 1956. Delayed fermentation of sucrose by certain haploid species of *Saccharomyces*. Antonie van Leeuwenhock J. Microbiol. Serol. **22**:353–370

19. Pederson, C. S. 1930. Bacterial spoilage of a Thousand Island dressing. J. Bacteriol. **20**:99–106
20. Rogosa, M. and M. E. Sharpe. 1959. An approach to the classification of the lactobacilli. J. Applied Bacteriol. **22**:329–340
21. Straka, R. P. and J. L. Stokes. 1957. Rapid destruction of bacteria in commonly used diluents and its elimination. Applied Microbiol. **5**:21–25
22. Vaughn, R. H., H. C. Douglas, and J. C. M. Fornachon. 1949. The taxonomy of *Lactobacillus hilgardii* and related heterofermentative lactobacilli. Hilgardia **19**:133–139
23. Vladimirov, B. D. and N. P. Nefedieva. 1937. Mayonnaise as a culture medium for microorganisms. Vop. Pitan. **6**:85–96
24. Walker, H. W. and J. C. Ayres. 1970. Yeasts as spoilage organisms, p. 463–527. In A. H. Rose and J. S. Harrison (ed.), The Yeasts, Vol. 3. Academic Press, Inc., New York, N.Y.
25. Wethington, M. C. and F. W. Fabian. 1950. Viability of food-poisoning staphylococci and salmonellae in salad dressing and mayonnaise. Food Res. **15:**125–134
26. Wickerham, L. J. 1951. Taxonomy of yeasts. U.S. Dept. Agr. Tech. Bull. No. **1029:**1–56
27. Williams, O. B. and E. M. Mrak. 1949. An interesting outbreak of yeast spoilage in salad dressing. Fruit Prod. J. **28**:141, 153
28. Worrell, L. 1951. Flavors, spices, condiments, p. 1706–1738. In M. B. Jacobs (ed.), The Chemistry and Technology of Food and Food Products, Vol. 2. Interscience Publishers, New York, N.Y.

CHAPTER 49

CEREAL AND CEREAL PRODUCTS

W. E. Hobbs and V. W. Greene

49.1 INTRODUCTION

Raw cereal grains include wheat, oats, corn, soy beans, durum, rye, rice, and barley. Finished cereal products include flour(s), breakfast cereals, "snacks," corn meal, corn grits, semolina, baked goods, soy protein products, refrigerated doughs, dry mixes (cake, bread, pastry, etc.), and pasta products. The cereal microbiologist is interested also in other products and commodities that ultimately are added to finished "cereal products" (e.g., vitamins, minerals, preservatives, flavorings, colorings, sugar, salt, spices, fats, oils, etc.). Many cereals and cereal products are used in the formulation and manufacture of a variety of other products (e.g., sausages, coldcuts, canned foods, pharmaceuticals, confectioneries, and baby foods).

The three major concerns of the cereal microbiologist are public health, spoilage, and good manufacturing practices. If cereal products are improperly processed, stored, or handled, such classical health hazards as staphylococcal intoxication, salmonellosis, *Escherichia coli* infections, *Clostridium perfringens* poisoning, botulism, mycotoxicosis, or *Bacillus cereus* poisoning may become a problem.

Under improper storage conditions, various "spoilage" microorganisms proliferate on cereal grains and on finished cereal products, rendering them visually undesirable and organoleptically unpalatable. Among these are the "ropy" bacteria, various yeasts and molds, lactic acid bacteria, gas forming bacteria, psychrophiles and psychrotrophs, thermophiles and thermoduric, and "flat sour" organisms. The aerobic plate count, coliform count and yeast and mold count are important as sanitary indices of good sanitation, handling, and storage practices. For further details, see the reviews and published studies on the microbiology of cereals and cereal products.[4, 5, 6, 9, 10, 11, 17, 18]

49.2 FACTORS AFFECTING BIOLOAD*

Cereal products may be divided into the eight general categories described in Table I. The "bioload," as defined here, represents the total mi-

*Bioload refers to the common types and expected ranges of microbial counts present on, or in, various commodities.

IS/A Committee Liaison: Nino F. Insalata

Table 1: Normal Microbiological Profile of Cereal Grains and Cereal Products*

Product Category	Normal Microflora	Quantitative Range	Remarks
I. Raw cereal grains	Molds	10^2-10^4/g	a) Counts representative of "normal" grains in commercial channels; "mildewed" or "musty" or "spoiled" grain would obviously be beyond these ranges.
	Yeasts and yeast-like fungi	10^2-10^4/g	
	Bacteria		
	aerobic plate count	10^2-10^6/g	b) Grains are also routinely tested for salmonellae.
	coliform group	10^2-10^4/g	
	E. coli	10-10^3/g	
	Actinomycetes	10^3-10^6/g	c) Related to amount of soil incorporated in grain sample.
II. Flour(s)*,* cornmeal; corn grits; semolina	Molds	10^2-10^4/g	a) Microbial counts in flour can vary from one storage period to another depending on moisture content and storage conditions. The final observed count is a function of original bioload, proliferation, and die off. Counts frequently decrease during storage; however, increases have also been noted.
	Yeast and yeast like fungi	10-10^2/g	
	Bacteria		
	aerobic plate count	10^2-10^6/g	
	coliform group	0-10/g	
	"rope" spores[9]	0-10^2/g	
			b) Soy flours sometimes contain salmonellae; other flours rarely do.
III. Breakfast cereals and "Snacks"	Molds	0-10^3/g	a) Cereals are additionally tested for *E. coli* and salmonellae.
	Yeasts and yeast like fungi	0-10^2/g	

Table 1: Normal Microbiological Profile of Cereal Grains and Cereal Products*
(*Continued*)

Product Category	Normal Microflora	Quantitative Range	Remarks
			Should be routinely tested for total thermophilic spore formers, flat sour organisms, putrefactive and thermophilic spore formers, and sulfide spoilage organisms.
VII. Pasta products	Bacteria		a) Wide ranges in bioloads of these products reflect difference between egg based and macaroni type products which are all included in "pasta" category. b) Routinely tested for molds and yeasts, salmonellae, and staphylococci.
	aerobic plate count	10^3-10^5/g	
	coliform group	10-10^3/g	
	E. coli	0-10^2/g	
VIII. Dry cereal mixes	Molds	10^2-10^5/g	a) Routinely tested for salmonellae and coagulase positive staphylococci.
	Yeasts and yeast-like fungi	10^2-10^5/g	
	Bacteria		
	aerobic plate count	10^2-10^6/g	
	coliform group	0-10^4/g	
	E. coli	0-10^3/g	

*Table based on "routine" quality control tests normally performed on various items of specified category; data represent industry-wide experience; data presented as "orders of magnitude" for illustrative purposes only.

**Milled rice produced commercially in the southern area of the United States in 1954 was virtually free from internal infection by fungi, and free from internal infection by bacteria, yeast, and actinomycetes.[14] Results from investigations of 1968[7] relative to the conditions that determine the prevalence of individual kinds of fungi in stored rice were similar to those observed in the 1954 study.

Table 1: Normal Microbiological Profile of Cereal Grains and Cereal Products*
(Continued)

Product Category	Normal Microflora	Quantitative Range	Remarks
	Bacteria		b) Snacks are routinely tested for *Salmonella* and coagulase positive staphylococci.
	aerobic plate count	0-10^2/g	
	coliform group	0-10^2/g	
IV. Refrigerated and frozen doughs	Molds	10^2-10^4/g	a) Yeast counts reflect inoculum intentionally added as part of the product formulation-do not represent "contamination."
	Yeast and yeast-like fungi	10^5-10^6/g	
	Bacteria		
	aerobic plate count	10^2-10^6/g	b) Routinely tested for salmonellae.
	coliform group	10-10^2/g	
	E. coli	0-10/g	c) The special case of "buttermilk biscuits" which contain 10^2-10^4 lactic acid producers have been omitted from this compilation.
	psychrotrophs	10-10^3/g	
	coagulase-positive staphylococci	10-10^3/g	
V. Baked goods	Molds	10-10^3/g	a) Routinely tested for salmonellae and coagulase positive staphylococci.
	Yeast and yeast like fungi	10-10^3/g	
	Bacteria		
	aerobic plate count	10-10^3/g	
	coliform group	0-10^2/g	
	E. coli	0-10/g	
VI. Soy protein	Bacteria		a) Quantitative ranges reflect both original contamination and growth during storage of intermediate moisture products.
	aerobic plate count	10^2-10^5/g	
	coliform group	10^2-10^3/g	
	E. coli	0-10^2/g	
	psychrotrophs	10^2-10^4/g	b) Routinely tested for molds and yeasts, salmonellae, and staphylococci.
	Cl. perfringens	0-10^2/g	c) Soy protein products intended for anaerobic storage (e.g., the canning industry)

croflora of the product or raw agricultural commodity. The data in Table I are based on routine quality control tests normally performed on various items of the specified category. The data represent industry-wide experience and are presented for illustrative purposes. These data have meaning to industrial quality control management, and are based on the microbiological procedures described in this compendium.

49.21 Raw Cereal Grains

The microflora of raw cereal grains are found in air, soil, and on animals and plants.[17] The numbers and types of organisms found on any given grain sample depend on such factors as the climate under which the grains are produced (e.g., rainfall, sunlight, temperature, and season), soil in which the plants are grown, biological environment (e.g., birds, insects, and rodents), methods and equipment of harvesting, weather conditions during and immediately following harvesting, conditions and duration of storage (e.g., temperature, bird and rodent droppings, and humidity), and, of course, nutritive composition of the grain. The diversity of these factors gives rise to the variety of organisms found in raw grain samples. Some grains are routinely contaminated with *Cladosporium* molds, while others contain *Aspergillus* and other types.[6, 7] Different varieties of grains often do not differ markedly from each other in respect to microbial populations. Conditions of growth, harvesting, and storage are the factors which control microbial populations of raw grains. Molds, yeasts, and most of the aerobic mesophiles present on raw grains are indigenous to the plants themselves, and grow on plant tissue. External contaminants (coliforms and *E. coli*) may be contributed by birds, man, and rodents which are ecologically associated with cereal grains.

49.22 Flour(s), Corn Meal, Corn Grits, Semolina

Sources of organisms in flour and other milled products originate from the raw materials from which they are manufactured. These factors determine the bioload of the cereal grain and, in turn, influence the microbial profile of the flour. Other sources of potential contamination are transportation facilities, mill unloading devices, conveyors, processing equipment, the milling sequence, and the time that the grain itself is exposed to moisture during the milling process.[9] Grains are tempered, sprayed with water and held in bins for varying periods of time. This procedure may permit microorganisms to proliferate, which will contribute to high microbial counts in commercial flours. Corn meal, corn grits, and some corn flour are traditionally produced by a "dry milling" process that avoids the tempering steps.

Soy flour is manufactured by a different process. The soy beans are moistened, dehulled, flaked, extracted with organic solvent to remove the oil, then "caked" and ground into flour. *Salmonella* have been detected often in soy flour.

49.23 Breakfast Cereals and Snacks

There are three basic breakfast cereal manufacturing processes: the "flaking" process, the "puffing" process, and the "extrusion" process. In each case moisture is introduced in the formulation, thus providing an opportunity for microbial growth. There may be a potential for postheat contamination during enrichment application and "enrobing". These steps add vitamins, minerals, sweeteners, and colorings to the cereal. If the additives are contaminated, or the process is insanitary, the finished product may be contaminated.

49.24 Refrigerated and Frozen Doughs

The sources of organisms in these products may be ingredients in their formulation, including flour, dry milk, eggs, sugar, spices, water, and flavorings. The equipment and environment of manufacture play an important role in the microbiology of the finished product. These commodities leave the manufacturer's supervision in a "raw" state. The final heat treatment (baking) is provided by the consumer. An initial contamination may increase during refrigerated storage in the retail market or the home refrigerator.[12] Lactic acid bacteria may be of special concern in the spoilage of these frozen dough products.[18]

49.25 Baked Goods

The baking process often destroys the microflora of baked goods.[13] Postprocessing contamination may occur during icing, glazing, and filling. If the baked commodity is contaminated after heating and then stored at room temperature, significant undesirable growth may result.

49.26 Soy Protein Products

The ingredients of these products are essentially soy flour and a variety of additives (color, flavoring, vitamins). Each may add to the microbial population of the finished commodity, and some steps of the processes may add contamination. Finished soy protein products range in moisture content from 2% to 64%. The nutritional quality of soy protein and the moisture content of some of these products may be conducive to microbial growth. The range of counts, and the variety of types encountered (Table I), suggest that the problems may be ones of storage and sanitation. Soy protein may be of particular importance if used as an ingredient in retorted, canned products. The organisms present in soy protein which can add to the bioload are thermophilic spore formers, flat sour, putrefactive and thermophilic, putrefactive spore formers, and sulfide spoilage organisms.

49.27 Pasta Products

There are essentially two categories of pasta products (usually manufactured from durum wheat flour): egg based pasta and macaroni type pasta. The former, as its name implies, contains flour, water, enrichment nutrients, and eggs (dried, frozen, or fresh pasteurized). The latter contains only flour, water, and enrichment nutrients. Both products are manufactured in essentially the same fashion, mixed, extruded, shaped, cut, and dried. During manufacture the product is a semisoft, moist, unheated dough in which organisms can proliferate. Also, microorganisms may grow during the slow, low temperature drying process. After drying, microbial activity is terminated. The presence of coagulase positive staphylococci has been a problem in the manufacture of pasta products.[20]

49.28 Dry Cereal Mixes

The manufacture of dry mixes is a dry blending of such ingredients as flour, dried eggs, flavorings, sugar, and dry dairy products. The finished product will reflect the microbial profile of the individual ingredients. There is usually no antimicrobial process involved in the manufacture of dry mixes. Microbial control can be effected by quality control of the ingredients, use of uncontaminated supplies, and maintenance of inhibitory moisture levels in the finished product. The relative degree of success is shown in Table I.

49.3 METHODS

The following summary suggests the routine and special analyses that are employed to determine the microbiological condition of the eight product categories discussed above.

49.31 Routine Analyses

49.311 Mold and yeast determinations (Chapters 4 and 16)

The mold and yeast count is an indication of the sanitary history of the product as well as a prediction of potential future spoilage during storage. Recommended sample size is 50 g diluted 1 : 10 in 450 ml of diluent.

49.312 Aerobic plate count

Sample size: 50 g in 450 ml of diluent. (Chapter 4)

49.313 Coliform organisms and *Escherichia coli* (Chapters 6 and 24)

The finding of coliform bacteria in raw cereal grains and flour does not imply mishandling. *E. coli* in a finished ready-to-eat product may be of public health concern. It may imply recontamination after the product has been processed.

49.314 Staphylococci

Flours intended for pasta product manufacture may be tested routinely for staphylococci. (Chapter 31)

49.315 Salmonellae (Chapters 25 and 26)

49.32 Special Analyses

Performed on certain products under specified circumstances.

49.321 "Rope" Spores

Performed on cereal grains and flours where the prospect of "ropy" dough (from the action of *B. mesentericus*) is of concern.[2]

49.322 Mycotoxins (Chapter 39)

These are toxins elaborated by certain fungi which may grow on moist grain. Epidemiological evidence[15] suggests that humans can be affected by ingestion of these toxins.

49.323 *Staphylococcus* enterotoxin (Chapter 32)

In addition to assaying a cereal product for coagulase positive staphylococci, there are circumstances under which the presence of enterotoxin should be determined. This is desirable when the viable *Staphylococcus* count suggests that there may be some hazard. If the history of the product reveals that growth of staphylococci could have taken place, it is useful to assay the product for toxin.

49.324 *Clostridium perfringens* (Chapter 35)

Clostridium perfringens food poisoning is often associated with meat and poultry. Soy protein products are often designed to substitute for meats. Their physicochemical properties may be considered conducive to the growth of *Cl. perfringens.* These products should be analyzed for the presence of this organism.

49.33 Other Special Tests (Chapters 8–23)

Cereal products may be tested for psychrotrophs, thermophiles, anaerobes, flat sour spores, H_2S producers, lactic acid producers, nitrate-utilizing gas producers, and sulfide spoilage spores. These organisms are of economic importance to the meat packer, canner, dairyman, baby food manufacturer, and others.

49.4 REFERENCES

1. Cereal Laboratory Methods, 7th Edition, American Association of Cereal Chemists. 1962. St. Paul, Minn.
2. Recommended Methods for the Microbiological Examination of Foods. Second Edition. 1966. American Public Health Association, Inc., New York, N.Y.
3. Official Methods of Analyses of the Association of Official Analytical Chemists. Eleventh Edition. 1970. Association of Official Analytical Chemists, Washington, D.C.
4. Bothast, R. J., R. F. Rogers, and C. W. Hesseltine. 1973. Microbial survey of corn in 1970–71. Cereal Science Today, **18**:18–24
5. Bothast, R. J., F. R. Rogers, and C. W. Hesseltine. 1973. Microbiology of Corn and Milled Corn Products. Northern Regional Research Laboratories, Peoria, Ill.
6. Christensen, C. M. 1946. The quantitative determination of molds in flour. Cereal Chemistry, **23**:322–329

7. Christensen, C. M. 1968. Influence of moisture content, temperature, and time of storage upon invasion of rough rice by storage fungi. Phytopathology, **59**:145–148
8. Dack, G. M. 1961. Flour microbiology. Cereal Science Today, **6**:9–10
9. Doty, J. 1961. Bacteria control in the flour milling operation. American Milling Process, **89**:20–21
10. Frazier, W. C. 1967. Food Microbiology. Second Edition. Chapter Thirteen. Contamination, preservation, and spoilage of cereals and cereal products. McGraw Hill Book Company, Inc., New York, N.Y., pp 179–191
11. Hesseltine, C. W. and R. R. Graves. 1966. Microbiological research on wheat and flour. Econ. Bat., **20**:156–168
12. Hesseltine, C. W., R. R. Graves, Ruth Rogers, and H. R. Burmeister. 1969. Aerobic and facultative microflora of fresh and spoiled refrigerated dough products. Applied Microbiol., **18**:848–853
13. Knight, R. A. and E. M. Menlove. 1961. Effect of the bread-baking process on destruction of certain mold spores. J. Sci. Food and Agricul., **12**:653–656
14. Kurata, H., K. Ogasawara, and V. L. Frampton. 1957. Microflora of milled rice. Cereal Chem. **34**:47–55
15. Marth, E. H. 1967. Aflatoxins and other mycotoxins in agricultural products. J. of Milk and Food Technol. **30**:192–198
16. Laboratory Manual for Food Canners and Processors. 1968. Volume 1, pp 105–108. National Canners Association. AVI Publishing Company, Inc. Westport, Conn.
17. Semeniuk, G. 1954. Storage of Cereal Grains and Their Products. AACC Monograph Series, 11, Chapter 111 (Microflora): 77–151
18. Slocum, G. G. 1963. Let's look at some microbiological problems of cereal foods. Cereal Science Today, **8**:313–314
19. Bacteriological Analytical Manual for Foods. 1969. U.S. Food and Drug Administration, Washington, D.C. Superintendent of Documents, U.S. Government Printing Office, Washington, D.C.
20. Lee, W. H., C. L. Staples, and J. C. Olson, Jr., 1975. *Staphylococcus aureus* growth and survival in macaroni dough and the persistence of enterotoxins in the dried products. J. Food Sci., **40**:119–120

CHAPTER 50

CONFECTIONERY PRODUCTS

John S. Hilker

50.1 MICROBIAL FLORA OF CANDY

The microbial flora of confectionery products depends largely on the quality of ingredients and the method of manufacture. The organisms in boiled sweets, for example, are few in number and consist mainly of spore formers which have survived intense heat. Milk chocolate, on the other hand, is made by blending chocolate liquor with sugar, milk solids, and perhaps cocoa butter. Organisms in the ingredients thus will survive and comprise the mixed flora of the final product. Bacteria that produce lactic acid (*Leuconostoc mesenteroides* and species of *Lactobacillus* and *Streptococcus*) and acetic acid (e.g., *Acetobacter* and *Bacillus* species; *Clostridium thermoaceticum*) have been found often in confections along with common yeasts (*Rhodotorula* and *Saccharomyces* species) and molds (*Aspergillus* and *Penicillium*).

The confectionery microbiologist is concerned with the fermentation of cacao beans for flavor development, the control processes, and the environmental conditions which affect preservation characteristics, plant sanitation, and bacteria of significance to public health. Most problems associated with microorganisms result from those deviations in formulation or processing practices which alter the water activity of a confectionery product.

50.11 Microbial Spoilage of Confections

Confectionery products usually are low moisture foods which are not prone to attack by microorganisms in the same way as fish, meats, and milk. Generally, candy is not manufactured or packed under aseptic conditions.

The most important parameter affecting the growth of microorganisms in candy is its water activity (a_w). This characteristic represents the ratio of the vapor pressure of water in a confectionery product to the vapor pressure of free water at the same temperature. Water activity expresses the availability of water in foods for chemical reactions and microbial growth.

Water activity (a_w) values which inhibit the development of various microorganisms are well known. Values less than 0.95 inhibit most pathogenic bacteria, whereas 0.90 prevents growth of most nonpathogens; staphylo-

IS/A Committee Liaison: E. M. Foster

cocci may be an exception, some growing as low as 0.85. Ordinary yeasts are inhibited at 0.88 and most mold growth is stopped at values less than 0.80. Halophilic bacteria will grow at water activity values as low as 0.75, xerophilic molds will tolerate values down to 0.65, and osmophilic yeasts will develop at a water activity as low as 0.61.

The water activity (a_w) of various kinds of candy is shown in Table 1. In general the microbial spoilage of adequately processed candy held under proper storage conditions can be caused only by xerophilic molds and osmophilic yeasts. Pathogenic bacteria do not grow in properly made confections.

Other parameters may affect the biological shelf life of candy. These include the initial microbial load in a finished confection, the product's chemical and physical composition, and the way it is processed, packaged and stored.

Table 1: Water Activity of Various Kinds of Candy

Type of Product	Water Activity (a_w)
Fondant Cream	0.75 to 0.84
Fruit Jellies	0.59 to 0.76
Marzipan	0.65 to 0.70
Turkish Delight	0.60 to 0.70
Marshmallow	0.63 to 0.73
Licorice	0.53 to 0.66
Gums and Pastilles	0.51 to 0.64
Chocolate	0.37 to 0.50
Toffee	Below 0.48
Boiled Sweets	Below 0.30

Available nutrients, the degree of acidity, the redox potential, and the presence of antimicrobial agents influence the growth of microorganisms in candy. Carbohydrates, fats, proteins, minerals, and vitamins in varying concentrations are found in confectionery products. These nutrients are derived from sugars, cocoa, albumin, milk, nuts, oils, and fruit. Antimicrobial constituents consist of various alcohols, essential oils, organic acids, and added preservatives.

Specific properties that determine the ability of various spoilage microorganisms to grow include their tolerance for acid and osmotic pressure, their nutritional demands, and their relative sensitivity to alcohols and preservatives.

These barriers against the growth of microorganisms do not totally eliminate microbial spoilage, however. "Swells," usually accompanied by off odors and off taste can occur in coated creams, chocolate with soft centers, marzipan, fondant, and glacé fruit if the water activity is too high.

Careful attention to recipe formulation, to processing and to storage conditions, accounts for the rarity of microbiological spoilage of candy.

Mold, usually accompanied by off flavors, may occur in fondants, jellies, glacé fruit, and chocolate covered coconut products.

Microbiologically induced rancidity may occur in chocolate coated products containing coconut, marzipan and marzipan like products, chocolate truffles, and wafer cream. It has been reported that this type of rancidity is caused by lipolytic microorganisms which hydrolyse fats prior to use in the confectionery process. The presence of active lipolytic enzymes from nonmicrobial sources also will cause this type of rancidity.

50.2 COATED CREAMS AND CHOCOLATES WITH SOFT CENTERS

Microbial spoilage of coated creams and chocolates with soft centers is evidenced by the breaking or bursting of the confectioner's coating, leakage of the filling, and occasionally by explosion. Spoilage results when gases accumulate beneath the coating following alcoholic, butyric, or lactic fermentations. Alcoholic fermentations are caused by normal and osmophilic strains of yeasts. The butyric fermentations are caused by certain species of *Clostridium*. *Bacillus*, *Leuconostoc*, and *Lactobacillus* species are the causal microorganisms of lactic fermentation.

50.3 MARZIPAN, FONDANTS, GLACÉ FRUIT, AND JELLIES

Microbial spoilage of marzipan, fondants, glacé fruit, and jellies is caused by osmophilic and xerophilic molds. Alcoholic fermentation by yeasts is evidenced by a breaking or splitting of marzipan and fondants, and a "sugaring" or "fermentation" of glacé fruit.

Osmophilic and xerophilic mold spoilage is evident in fondants by holes or colored spots. Jellies and glacé fruit may show colored spots, generalized softening, or visible mold, if circumstances permit fungal development.

Typical yeast spoilage organisms are *Saccharomyces rouxii, Saccharomyces heterogenicus, Saccharomyces mellis, Torulopsis colliculosa, Hansenula anomala,* and *Pichia membranefaciens.* Mold spoilage organisms are *Aspergillus glaucus, Aspergillus niger, Aspergillus sydowi, Penicillium expansum*, and other *Penicillium* species.

The microbial flora causing visible spoilage of confectionery products is shown in Figure 1.

50.31 Bacteria of Significance to Public Health

Historically, confectionery products have an excellent public health record in comparison to reported foodborne illnesses and intoxications of other foods. Only one documented outbreak of food poisoning has been attributable to the ingestion of a confectionery product. This outbreak occurred in the U.S. during 1973 following consumption of chocolate which had been imported from Canada. The outbreak involved 79 people and was diagnosed as salmonellosis. The causal organism was *Salmonella eastbourne*.[4]

Numerous recalls of chocolate confectionery products have occurred in this country because of the presence of salmonellae as a contaminant. However, the Food and Drug Administration, recognizing that there is no way

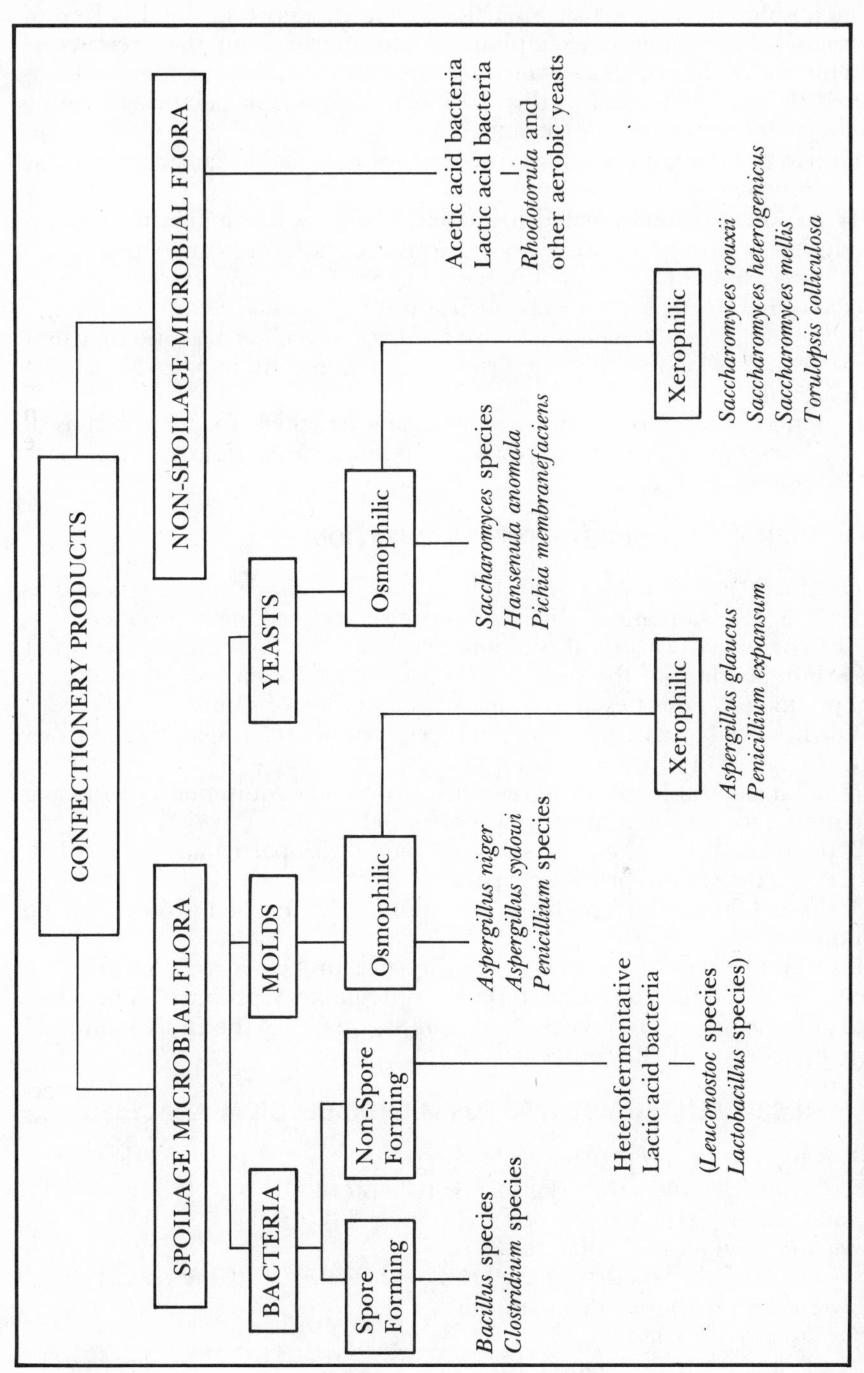

FIGURE 1. Spoilage Microbial Flora of Confectionery Products

to be absolutely certain that a particular lot of nonsterile food is free of *Salmonella*, has specified a sampling and testing plan for the presence of salmonellae in foods.[5] Acceptance or rejection of a "lot" of food is determined by the application of this FDA plan. Regulatory agencies have no tolerance level for these organisms.

Low water activity prevents growth of salmonellae in chocolate but the organisms may gain entry through:

1) the use of contaminated ingredients such as dry milk and milk byproducts, cocoa, egg products, and various vegetable and animal proteins;

2) cross contamination of inprocess material with raw ingredients suspected of harboring salmonellae, such as flour, raw milk, etc.;

3) Wet areas in idle equipment are a source of danger because they may allow growth of pathogenic organisms which then contaminate the next lot of material.

Generally, such food poisoning organisms as enterotoxigenic strains of *Staphylococcus aureus* and *C. perfringens* are not a problem in confectionery products.

50.4 MICROBIAL INDICATORS OF INSANITATION

Enterococci, *Escherichia coli* and other coliforms have been suggested as indicators of insanitation. Their presence in confectionery products cannot be correlated with fecal contamination because they propagate in wet areas independent of the proven presence of fecal contamination. Microbial population densities are used also as an index of sanitation.

The bacterial quality of a confectionery product is a function of any one or a combination of the following factors:

1) the microbial loads of ingredients, processing equipment, processing environments, and manufacturing personnel;

2) the efficiency of the processing operation with particular emphasis directed toward critical processing points;

3) the extent to which postcontamination of the finished product can be minimized.

Indicator microorganisms have no direct or proven correlation with the presence of foodpoisoning bacteria in confectionery products. The presence of these organisms relates to the sanitary quality of the entire manufacturing process and environment.

50.5 RECOMMENDED METHODS FOR MICROBIOLOGICAL ANALYSES

Aerobic Plate Count[2] (Chapter 4)
Escherichia coli and other Coliforms[2] (Chapter 24)
Enterococci (Fecal Streptococci)[2] (Chapter 30)
Staphylococcus aureus[2] (Chapter 31)
Salmonella (Other methods have been described.[1,2,3]) (Chapter 25)
Clostridium perfringens[2] (Chapter 35)
Bacillus cereus[2] (Chapter 33)
Yeasts and Molds[2] (Chapter 16)

The methods indicated above conform to those of the AOAC, and as used by the Food and Drug Administration.

50.6 INTERPRETATION OF DATA

Aerobic plate counts in confectionery products immediately after processing will range from 10^0 to 10^6 per gram. Coliforms, when present, range from 10^0 to 10^2, with *E. coli* generally absent from 1 gram subsamples. Enterococci may be present at levels ranging from 10^0 to 10^4 per gram. Yeasts and molds may vary between 10^0 and 10^3 per gram.

Coagulase (+) *Staphylococcus aureus, C. perfringens, B. cereus*, and salmonellae are rarely present when ingredients and processing conditions are properly controlled.

50.7 REFERENCES

1. Fantasia, L. D., J. P. Schrade, and J. Yager. 1975. Fluorescent Antibody (FA) Method Official First Action, JOAC "Changes in Methods," March, 46.A01
2. Bacteriological Analytical Manual for Foods, 1976. 3rd ed. Food and Drug Administration, Sup't. of Documents, U.S. Gov't. Printing Office, Washington, D.C.
3. Sperber, W. H. and R. H. Deibel. 1969. Accelerated procedure for *Salmonella* detection in dried foods and feeds involving only broth cultures and serological reactions. Applied Microbiol. **17**:533–539
4. Morbidity and Mortality. Vol. 23, No. 10. 1974. Center for Disease Control, U. S. Public Health Service, Atlanta, GA.
5. Olson, J. C., Jr. 1975. Development and present status of FDA *Salmonella* sampling and testing plans. J. Milk and Food Technol. **38**:369–371
6. Official Methods of Analysis, 12th ed. 1975. Assoc. Official Analytical Chemists. Washington, D.C.

CHAPTER 51

NUT MEATS

F. R. Smith and R. E. Arends

51.1 INTRODUCTION

Nut meats are derived from processed nuts which are harvested from trees, shrubs, or plants. They may be subjected to roasting, or consumed in the raw state. Nut meats are sold primarily to the food manufacturing industry, and are incorporated into toppings for baked products, candy, ice cream, cake mixes, etc. Weinzirl[21, 22] found that nut meats can adulterate foodstuffs when he noted that candies containing nut meats had a higher bacterial level than those lacking tree nuts. A further study determined the presence of *Escherichia coli* and the author attributed the contamination to the processing procedures. Thus, nut meats can be a potential source of microbial contamination in food stuffs, and are of public health significance.

The Division of Regulatory Guidance, Bureau of Foods,[7] defines adulterated nuts from the standpoint of microorganisms as those which are moldy or judged to consist wholly or in part of a filthy substance due to the presence of *E. coli*. This adulteration can take place in the field or in processing.

Ostrolenk and Hunter[16] and Ostrolenk and Welch[17] claimed that pecans with unbroken shells were free from *E. coli*, and thus their presence indicates insanitary processing. Marcus and Amling[14] soaked whole pecans, that had no visible breaks or cracks, for 24 hours in a lactose broth solution containing *E. coli*; none of the meats in these nuts became contaminated. However, 24% of sound pecans soaked in water for 48 hours to simulate standing in a rain puddle developed openings along shell suture lines and these did not close completely when the nuts were redried. Understandably, samples from grazed orchards were more heavily contaminated with *E. coli* than those which were not utilized for grazing. Hall[9] feels that since nuts in the unbroken shell and during early maturation do not contain *E. coli*, he must accept the concept of contamination in processing. However, King, Miller, and Eldridge state: "The present findings indicate contamination of nut meats before processing, so their presence on processed nuts does not necessarily indicate poor manufacturing practice."[12]

Their survey, using a statistical sampling plan on almonds as they were received at the processing plant, determined that hard shell varieties had

IS/A Committee Liaison: Joseph C. Olson, Jr.

lower counts than did soft shell varieties, and almonds with complete shells had lower counts than shelled almonds. Almonds harvested onto canvas had lower counts than those knocked onto the ground. Nuts with the least amount of foreign material mixed into the sample had the lowest counts. The genera of bacteria isolated from almonds included *Aerobacter, Escherichia, Streptococcus, Bacillus, Xanthomonas, Achromobacter, Pseudomonas, Micrococcus, Staphylococcus* and *Brevibacterium*.

Meyer and Vaughn state: "The incidence of *E. coli* on meats of commercially hulled black walnuts depended upon the physical condition of the nuts. Apparently tightly sealed ones contained only a few or none, whereas those with visibly separated sutures and spoiled meats yielded the most." In addition, they state: "After husk deterioration, the shell also begins to dehydrate; during this process, the two halves tend to separate, a separation that is not always visible to the eye."[15]

Chipley and Heaton[4] found that the microorganisms associated with commercially shelled pecans are numerous and varied. The genera of bacteria included *Pseudomonas, Corynebacterium, Escherichia, Leuconostoc, Proteus, Aerobacter,* and *Clostridium*. The fungi isolated in this study were *Penicillium, Aspergillus, Fusidium,* and *Tricothecum*. No bacteria and only two fungal microorganisms were found in aseptically shelled pecans. The Committee on Food Protection[5] states that microorganisms persisting in almonds, Brazil nuts, filberts, pecans, and English walnuts derive from (a) damaged or cracked nuts, (b) insect infestation, (c) infected or diseased nuts, (d) contamination during processing.

51.2 PATHOGENIC BACTERIA

The Committee on Food Protection[5] states: "Tree and shrub nuts (almond, black walnut, Brazil nut, butternut, cacao, cashew, chestnut, coconut, cola or kola nut, filbert, hickory nut, lychee, macadamia, pecan, Persian (English) walnut, pili nut, pine nut, pistachio), and ground nuts (groundnut, peanut, water chestnut) often are subject to microbial or fungal contamination, but are seldom, if ever, sources of food poisoning. Occasionally nuts contain *Escherichia coli* and more rarely, *Salmonella*. The presence of *E. coli* in nut meats generally indicates fecal pollution."

Beuchat[3] suspended *Salmonella senftenberg, S. anatum*, and *S. typhimurium* in sterile tap water. In-shell pecans and pecan halves were placed in the water suspension. Survival of the organisms when exposed to thermal treatments normally carried out during the processing of pecans led to the conclusion that these heat treatments are inadequate to destroy salmonellae consistently in highly contaminated nuts. Since organoleptic quality deterioration occurs in pecan nut meats at elevated temperatures, sterilization methods other than thermal treatment may be required for the elimination of viable salmonellae from pecan nuts.

51.3 MOLDS AND MYCOTOXINS

Molds commonly have been found in nuts and on nut meats. Chipley and Heaton[4] note that on germination of the mold spores, the hyphae are capable of penetrating the pecan tissue. The pecan industry attempts to remove in-shell moldy nuts with high velocity air flow. The presence of molds is particularly important to the baking industry. Frequently nut meats are used as a topping without heat treatment and the presence of molds may cause a serious problem due to the loss of product. In addition, the presence of molds can create an injurious health factor due to the toxins that may be produced. This constitutes grounds for regulatory action.

The discovery that moldy peanut meal had caused considerable losses in flocks of turkeys centered attention on possible toxigenic molds. The discovery of *Aspergillus flavus* as the culprit, and the subsequent study of the toxin produced (aflatoxin), has been adequately covered.[2, 19] The authors also outline the action of this toxin on various animal species. In general, the toxins are carcinogenic and cause damage to the liver. They are heat resistant, withstand boiling, and are not destroyed by other simple means. The statement by one author is of interest. "If we conclude that aflatoxin is indeed a health hazard to man, then good husbandry in harvesting and storage of food crops is the only means of reducing or removing aflatoxin from the human environment, since there is no measure which, by its simplicity and effectiveness, is the equivalent of pasteurization and able to act as a second line of defense to clean the contaminated food before people consume it."[2]

Aspergillus flavus was isolated from pecans used in bakery products.[13] In this study of 120 isolates of the *Aspergillus flavus* group, 85 were shown to produce aflatoxin on yeast sucrose medium. In addition, extracts from moldy sections of raw pecans showed aflatoxin-like spots when tested by thin layer chromatography. Schade, McGreevy, King, Mackey, and Fuller[18] detected aflatoxin in almonds.

51.4 PROCESS CONTROL

The literature cited above shows that the in-shell nut meats, if they have broken or cracked shells, can be contaminated with *E. coli*, soil bacteria and molds. The orchard conditions have a noticeable effect on the population of microorganisms on the nut meats. The Committee on Food Protection[5] recognizes the problem involving "accumulators" or pecan brokers. This is a major problem and can lead to contamination with pathogens. After gathering or collecting the nuts, particularly in the case of pecans, many of these collectors have no satisfactory storage place. The nuts may be bagged or stored in containers without protection from rodents or farm animals. Naturally, the product serves as an attractant to mice, rats, and squirrels. Care with in-shell nuts can assist in improving the finished product.

Suggestions have been made that nuts be inspected prior to cracking to remove any with breaks or cracks. It is true that nuts with major breaks can be removed by a sorting table, but those with cracks are difficult to detect. It is our opinion that this is not a workable procedure.

Two areas are particularly critical in the processing of certain nuts. First, the commercial cracker is anxious to prevent shattering of the nut meats in the cracking operation. In order to develop optimum cracking conditions, the in-shell nuts are held under humid conditions to develop the optimum moisture in the nut meat. Usually this requires storage at room temperature for at least 24 hours. Thus, any organisms on the shell or nut meat will increase in numbers. A certain amount of control can be effected by washing the nuts in a chlorine solution (1000 ppm) prior to tempering. A second step is to follow the hydration process by immersing the nuts in water at 90 C. for two minutes. The latter process greatly decreases the numbers of microorganisms, particularly those on the surface of the shells.

The second critical point involves a flotation system to separate shell fragments from nut meat pieces. This process is continuous, and thus, even with changes of water, counts may increase. Our findings are that a minimum dryer temperature of 70 C will reduce this count. However, this temperature also may cause the pecan to be somewhat dark in color.

Gas sterilization or treatment of nut meats as a final step before storage greatly affects the microbiological content. The use of propylene oxide is permitted on processed nut meats (except peanuts) when they are to be further processed into a final food form.[8] The regulations note that the permissible residual content of the sterilizing agent in the nut meat be limited to 300 ppm. The fumigant should be applied in retorts not more than one time and not in excess of four hours duration at a temperature not in excess of 125 F.

Refrigerated storage is utilized to preserve the desired quality. Removal from refrigerated storage, particularly in humid periods, may result in the collection of condensate which will encourage mold growth. Tempering of the stored product to prevent radical temperature change is advisable.

51.5 LABORATORY TESTS

A procedure has been outlined presenting an attribute sampling plan for detection of *Salmonella*.[6] In this procedure, food products were classified according to risk. In addition, the utilization of composite samples is discussed. The latter concept can be of considerable help in giving reasonable assurance of product safety at a lower cost. Other statistical methods of product sampling, involving somewhat more complex interpretation, have been presented by Whitaker and Weiser.[23, 24]

Any sampling plan must be based on a random sample, taken throughout the lot, and is based also on the hypothesis that the defect is distributed randomly throughout the lot. It is important that the sample used incorporates portions taken from various units of the lot. Any sample unit submitted for laboratory testing should contain at least 100 g of product. In the case discussed by the committee on *Salmonella*, a larger sample is recommended.[6] A 25 g analytical unit taken from the 100 g sample unit may be utilized in aerobic plate count, mold count, coliform, and *E. coli* determinations.

Plating for total aerobic counts (Chapter 4) may be made on Standard Methods Agar and for yeasts and molds (Chapter 16) on potato dextrose agar (Chapter 2). The aerobic plates should be incubated at 35 C. for 48 hours and mold plates at 23 C. for five days. The tolerance count levels for both generally are contained in the specifications which are agreed upon by the buyer and the supplier. The decision on permissible levels depends upon the end use of the nut meats.

The procedure for coliform and *E. coli* determination is outlined in Chapter 24. The limiting coliform content is defined by the product specifications. The regulatory requirements for *E. coli* content have been specified as follows: "When *E. coli* MPN of at least 0.36 per gram (IMViC Confirmed) is found in two or more subsamples, if less than 10 subsamples examined; or, in 20% or more of the subsamples if more than 10 subsamples examined,"[7] the product is subject to seizure.

The procedure for the determination of *Salmonella* contamination is given in Chapter 25. Compositing of analytical units is encouraged as a time saving measure.

The determination of aflatoxin in nut meats may follow procedures published in the Journal of the Association of Official Analytical Chemists.[10,11] The first publication follows the FDA procedure and the latter, that of the Peanut Administrative Committee. A total in excess of 20 ppb is unsatisfactory and, at the present, a 15 ppb tolerance is being proposed.[1] A rapid method for measurement is described by Waltking, Bleffert and Kiernan.[20]

51.6 REFERENCES

1. Anonymous. 1975. Compliance program for aflatoxin in peanut products issued by FDA. Food Chem. News. **4**:11–13
2. Barnes, J. M. 1970. Aflatoxin as a health hazard. J. Appl. Bact. **33**:285–298
3. Beuchat, L. R. and E. K. Heaton. 1975. *Salmonella* survival on pecans as influenced by processing and storage conditions. Applied Microbiol. **29**:795–801
4. Chipley, J. R. and E. K. Heaton. 1971. Microbial flora of pecan meat. Applied Microbiol. **22**:252–253
5. Committee on Food Protection. 1975. Food and Nutrition Board. National Research Council. Prevention of Microbial and Parasitic Hazards Associated with Processed Foods. National Academy of Sciences, Washington, D.C.
6. Committee on *Salmonella*. 1969. An evaluation of the *Salmonella* problem. Nat'l. Acad. of Sci. **1683**:147–156, Washington, D.C.
7. Division of Regulatory Guidance, Bureau of Foods 1973. Food and Drug Admin. Admin. Guideline Man. 7412.06
8. Code of Federal Regulations no. 21 CFR. 123.380
9. Hall, Herbert E. 1971. The significance of *Escherichia coli* associated with nut meats. Food Tech. **25**:230–231
10. J. Ass. Offic. Anal. Chem. 1968. "Changes in methods 25. Nuts and nut products," **51**:485–488
11. J. Ass. Offic. Anal. Chem. 1966, 1968. "Changes in methods 25. Nuts and nut products," **49**:229–231; **51**:488–489
12. King, A. Douglas, Jr., Mary Jo Miller, and Linda C. Eldridge. 1970. Almond harvesting, processing and microbial flora. Appl Microbiol **20**:208–214
13. Lillard, H. S., R. T. Hanlin, and D. A. Lillard. 1970. Aflatoxigenic isolates of *Aspergillus flavus* from pecans. Appl Microbiol **19**:128–130
14. Marcus, Karen A. and H. J. Amling. 1973. *Escherichia coli* field contamination of pecan nuts. Appl Microbiol **26**:279–281
15. Meyer, Melvin T. and Reese H. Vaughn. 1969. Incidence of *Escherichia coli* in black walnut meats. Appl Microbiol **18**:925–931
16. Ostrolenk, M. and A. C. Hunter. 1939. Bacteria of the colon-aerogenes group on nut meats. Food Res **4**:453–460
17. Ostrolenk, M. and H. Welch. 1940. Incidence and significance of the colon-aerogenes group on pecan meats. Food Res **6**:117–125

18. Schade, J. E., K. McGreevy, A. D. King, Jr., B. Mackey, and G. Fuller. 1975. Incidence of aflatoxin in California almonds. Appl Microbiol **29**:48–53
19. Spensley, P. C. 1970. Mycotoxins. A menace of moulds. Royal Soc. Hlth. **5**:248–254
20. Waltking, Arthur E., George Bleffert, and Mary Kiernan. 1968. An improved rapid physicochemical assay method for aflatoxin in peanuts and peanut products. J. Am. Oil Chem. Soc. **45**:880–884
21. Weinzirl, J. 1927. Occurrence of the colon group on commercial candies. Am. J. Pub. Hlth. **17**:708–711
22. Weinzirl, J. 1929. Nuts as a possible source of *Escherichia coli* found in candy. Am. J. Hyg. **10**:265–268
23. Whitaker, T. B. and E. H. Weiser. 1969. Theoretical investigations into the accuracy of sampling shelled peanuts for aflatoxin. J. Am. Oil Chem. Soc. **46**:341–384
24. Whitaker, T. B. and E. H. Weiser. 1970. Design and analysis of sampling plans to estimate aflatoxin concentrations in shelled peanuts. J. Am. Oil Chem. Soc. **47**:501–504

CHAPTER 52

CANNED FOODS—TESTS FOR COMMERCIAL STERILITY

James C. Canada, John M. Dryer, and Duane T. Maunder

52.1 INTRODUCTION

When spoilage occurs in a closed container of canned foods, it manifests itself by obvious gas production, swelling of the lid of the container, by a change in the consistency of the product, by a change in the pH of the product, and/or by an increase in the number of microorganisms seen in the microscopic examination of the food. The commercial sterility test is not designed for spoiled food, and therefore when any of the above spoilage criteria are met, the analyst should go directly to Chapter 53 for procedures to determine the cause of spoilage.

Canned foods are those edible foods that have been preserved by heat in hermetically sealed containers. Several definitions have been used for the term "hermetically sealed," but they usually imply a container that excludes the passage of gas or of microorganisms. For the ensuing discussions, an hermetically sealed container is one that prevents the reentry of microbes, thereby preventing spoilage from external sources after the container has been sealed. The canned food should remain unspoiled indefinitely if properly heat processed, and if the seal remains intact. In general, canned foods do not rely on preservatives, or inhibitory agents other than acids, to insure stability. The heat treatment given canned foods is enough to produce commercial sterility, not complete sterility. Absolute sterility is defined as a state completely free of all viable microorganisms. To achieve complete sterility, the heat process required would destroy the product consistency and nutritional values that presently exist in commercially canned foods. The highly heat resistant and nontoxic thermophilic sporeformers (Chapters 20, 21, 22, and 23), under proper handling and storage conditions, remain dormant in commercially canned foods, and present no problems; therefore, the canned foods are called "commercially sterile."

Commercial sterility is defined as:

(a) the absence of microorganisms capable of growing in, and spoiling, the food under normal storage conditions, and

(b) the absence of pathogenic microorganisms capable of proliferating in the food.

Therefore, with this definition in mind, it is permissible for a heat proc-

---

CONTRIBUTOR: Keith Ito

IS/A Committee Liaison: Cleve B. Denny

essed food to contain viable microorganisms. The microorganisms, however, do not increase in numbers, and consequently do not cause physical changes in the food product. Likewise, pathogens are not present, or are incapable of reproducing in the product.

Viable microorganisms normally may be recovered from commercially sterile heat processed foods under three general conditions. (1) The microorganism is an obligate thermophilic sporeforming bacterium and the normal storage temperature is below the thermophilic range. (2) The heat processed food is within the high acid to acid range. Acid intolerant microorganisms may be present but are incapable of growth because of the acidic conditions. (3) Mesophilic or thermophilic sporeformers may be recovered from canned foods which use a combined process of heat and water activity to prevent outgrowth and spoilage. Finding microorganisms in these three cases is normal and the product is considered commercially sterile.

Food processors are able to heat process food to achieve absolute sterility. This is not done for several reasons. Overcooking (excessive time or temperature) encourages the development of off flavors, color and consistency changes, and nutrient losses. "Commercial sterility heat processing" enables the manufacturer to pack a microbially safe, shelf stable food without undue impairment of flavor, color, consistency, and nutrient content. Kautter and Lynt[6] have estimated that in the United States alone, 775 billion containers of food have been produced in the past 45 years, and of these, only four have been associated with botulism leading to death of the consumer. These incidents are, of course, not taken lightly. Manufacturers of foods packed in hermetically sealed containers have a remarkable and enviable record when one considers public health safety and consumer satisfaction.

Food canners have a history of cooperatively sharing technical advances for the improvement of the entire industry. In addition, numerous safeguards are employed (both voluntarily and by regulatory agencies) to assure the adequacy of commercial sterilization cooking procedures. Requirements concerning processing of low acid foods and appropriate records to be kept are found in part 90 and part 128.b of Title 21 of the Code of Federal Regulations.

Sterility testing of canned foods should be conducted on normally appearing canned food by visually examining the container and the product, measuring the pH, and, if necessary, making a microscopic examination to detect large numbers of bacteria. Subculturing rarely is done on a routine basis because it is time consuming, expensive, and has a high risk of laboratory contamination and of faulty interpretation.[2] Nevertheless, procedures for subculturing will be described in this chapter.

52.2 TREATMENT OF SAMPLE

To determine if a certain segment of production meets commercial sterility criteria, the food manufacturer selects test samples representing the production lot and incubates them for 10 days at 30 to 35 C. If conventional retorts are used for sterilization each retort load is examined with at least one sample. If sterilization is accomplished by continuous means, samples are drawn at periodic intervals during the time of pack.

To confirm commercial sterility by collecting containers from the retail shelf, obtain enough samples to validate truly any conclusions drawn from laboratory testing.

After incubation, record the product identity and the manufacturer's code. The container should be examined critically for abnormal conditions such as leakage, swells, flippers, prior opening, etc. (See Chapter 53 for a thorough explanation of container appearances.) If spoilage is evident, use the procedure in Chapter 53. If the container is defective or damaged, the sample is not suitable for commercial sterility testing.

52.3 SPECIAL EQUIPMENT AND SUPPLIES

52.31 Work Areas: Preferred Method

A laminar flow work station meeting ultra clean environment specifications Class 100 is the equipment of preference[4] in sterility evaluations of hermetically sealed containers. This equipment will provide a work environment free of particles 0.3 microns or larger at an efficiency of 99.99% with an air flow of 100 feet per minute. Primary disinfection of the unit can be obtained by a thorough washing of the interior surfaces "excluding the filter screen" with standard iodophores or quaternary ammonium compounds at appropriate concentrations recommended by the manufacturer. After chemical disinfection, the blower should be operated for at least one hour prior to performing analyses within the unit.

Efficiency of the unit should be monitored through the use of opened, uninoculated control media exposed to the work station environment during the entire transfer period. Three controls should be used. There should be one to the right and one to the left of the samples undergoing analysis, and one should be positioned in front of the samples roughly in the middle of the work area.

Sterile air glove boxes also are included under preferred methodology, and should be sterilized and manipulated in accordance with proven aseptic practices. They are not the method of choice however, since the limitations of the gloves make the manipulations involved with opening hermetically sealed containers and subsequently transferring the samples to the culture media somewhat difficult.

52.32 Work Areas: Alternate Method

If the equipment described in the preferred method is not available, the samples may be opened in a room secure from drafts. The counter surface should be scrubbed thoroughly with soap and water and then disinfected with an appropriate bactericidal agent.

52.33 Bacteriological Can Opener

There is a special can opener designed for the aseptic opening of metal containers for bacteriological sampling without distorting the can seams.

(Figure 3 in Chapter 53). Under no circumstances should a common kitchen type can opener be employed, since sample contamination and distortion of the double seam will occur.

52.34 Pipets

Wide bore pipets of appropriate capacity can be used to good advantage when sampling viscous or particulate products. A minimum tip opening of 3 mm is recommended; either glass or disposable plastic is acceptable. All pipets should be cotton plugged to prevent contamination from the user.

Nondisposable pipets used in sterility test work should not be stored in pipet cans, but rather should be wrapped not more than five to a package in heavy Kraft paper. This package should be steam sterilized at 121 C for 20 minutes, then dried in vacuo. Alternately, the packages may be placed in a sterilizing (drying) oven at 160 C to 180 C for two hours.

52.35 pH Meter

Electronic pH meters should be used to determine the pH of the product in question. Accuracy must be within 0.1 pH unit of a known buffer solution. Indicator paper or solution with a suitable range may be used for pH determinations on a large number of homogenous samples. However, any sample showing variation from the normal should be checked on an electronic meter.

52.36 Forceps

Nonserrated forceps[1] at least 8 inches in length should be available to handle large particulates and other nonpipetable products.

52.37 Cork Borer

Cork borers, previously wrapped in Kraft paper and sterilized at 121 C for 30 minutes, provide an excellent means of sampling solid products such as canned meats. Minimum bore is 1/4 inch with a smaller sterile cork borer used as a piston to expel the sampled product into appropriate media.

52.38 Microscope

A suitable bacteriological microscope fitted with oil immersion lens, or a phase system, should be used for direct microscopic examination of the product in question. Microscopic examination of a food product for bacterial contamination should be made at magnifications of not less than 930x.

52.39 Sample Containers

Containers designed for the refrigerated retention of sampled product should be provided. If glass jars are used, openings should be at least 50 mm in diameter to facilitate sample transfer. The jar should be capped with gasketed, threaded, onepiece closures which provide a hermetic seal.

Plastic sample bags previously sterilized with ethylene oxide also are acceptable and are readily available from commercial sources. Screw capped tubes, 1″ × 6″, may be used to retain fluid food samples.

52.310 Culture Tubes

All tubes should have a screw cap closure and be manufactured of borosilicate glass or suitable plastic. If plastic is used, it must be nontoxic and capable of withstanding normal glassware sterilization temperatures. Cotton plugs are not acceptable as closures on either type of tube.

52.311 Petri Dishes

Sterile dishes shall be 100 × 20 or 100 × 15 mm plastic or glass. If plastic is used, it must be nontoxic, and only new, sealed sleeves of disposable plastic dishes shall be used in sterility test work. If glass is used, standard Petri storage cans should be employed for storage of the glassware. Only freshly sterilized containers should be used in sterility test work.

52.4 SPECIAL REAGENTS AND MEDIA

When media is commercially available in the dehydrated form, it should be used in preference to media formulated in individual laboratories.

52.41 Standard Methods Agar (Plate Count agar) (Chapter 2)

52.42 PE-2 Medium (for cultivation of aerobes and anaerobes) (Chapter 2)

52.43 Cooked Meat Medium (for cultivation of anaerobes) (Chapter 2)

52.44 Potato Dextrose Agar (for cultivation of yeasts and molds) (Chapter 2)

52.45 Orange Serum Broth (for cultivation of acid tolerant microorganisms) (Chapter 2)

52.46 Sabouraud Dextrose Agar (for cultivation of yeasts and molds) (Chapter 2)

52.47 APT Broth (for cultivation of *Lactobacillus*) (Chapter 2)

52.48 Acid Products Test Broth (Chapter 2)

52.49 Ziehl-Neelsen's Carbol Fuchsin Stain

52.491 Crystal Violet Stain (0.5 to 1% solution)

52.5 RECOMMENDED CONTROLS

52.51 Glassware

All glassware should be autoclave sterilized at a minimum of 121 C for 20 minutes. Equipment wrapped in Kraft paper should be placed in a sterilizing oven at 170 to 180 C for two hours. Heat sensitive sterilization indicators may be affixed to each autoclave load to identify readily the status of a given unit of equipment. If equipment is to be sterilized well in advance of use, double layer aluminum foil dust covers should be placed over flasks, dilution bottles, etc.

52.52 Media

Media sterility checks should be performed on common liquid media by incubating for 48 hours at appropriate temperatures, and then examining for absence of growth. Solid media require melting if not immediately used. Uninoculated control plates should be prepared and incubated for every lot of media employed in routine test studies.

52.53 Personnel Requirements

Prior to working with samples, hands and face of personnel should be scrubbed thoroughly with an appropriate germicidal hand soap.[3] Personnel should wear clean lab coats, and items of personal clothing such as neckties should be removed or contained. Persons known to have colds, boils, or other similar health problems should not perform these evaluations. Personnel with shoulder length hair and/or beards should wear protective snoods and full head coverings. Full surgical face masks have been suggested by some workers; however, this is not a mandatory requirement of this method, and the use of face masks is optional.

52.6 PROCEDURE

The one reliable test for determining commercial sterility of a container of product is the incubation of that container at an appropriate temperature for a sufficient length of time to allow any significant microorganisms contained therein to grow and manifest their presence.[5] If microorganisms proliferate under the proper conditions imposed, the product is not commercially sterile and should be examined as described in Chapter 53.

52.61 Incubation

Incubation conditions will be governed partially by the purpose of the commercial sterility test.

a. Routine production control:

For low acid products destined for storage at temperatures above about 40 C (hot vending), containers from each sampling period should be incubated at 55 C for 5 to 7 days. For all other low acid products, incubate at 30 to 35 C for ten days, except certain meat products packed under continuous regulatory agency inspection;[7] where faciliteis are limited to 35 C, incubation for ten days may be used. For acid products, incubate at 25 to 30 C for ten days.

b. Examination of a production lot or lots because of suspected non-commercial sterility:

If possible, incubate the entire lot or lots; otherwise, incubate a statistically randomized sample of the lot or of each lot. Use incubation temperatures as in (a) but increase the time to 30 days in the case of mesophiles.

52.62 Examination

Containers may be removed from the incubator whenever outward manifestations of microbial growth appear, i.e., swells, or, with transparent containers, noticeable product change. At the end of the incubation period, some containers should be opened to detect possible flat spoilage.

Weigh each container to the nearest gram and record the weight on the data sheet. Subtract the average tare weight of the empty container from each unit weight to determine the approximate net weight of each sample. (Chapter 53).

If the purpose of the test procedure is to determine merely whether the product is commercially sterile, and provided the analyst is thoroughly familiar with the particular product and with potential types of microbial spoilage of that product, containers may be examined without employing aseptic techniques. Open the containers carefully, note abnormal odors, consistency changes, and frothiness; measure pH electrometrically or colorimetrically; if results are not conclusive at this point, prepare a smear for microscopic examination. The observation of abnormalities in these, compared to containers of normal control product, means that the container (s) of the product is not commercially sterile. In that case, follow the procedures in Chapter 53 on the remaining unopened abnormal product.

If the analyst is unfamiliar with the particular product and its potential spoilage characteristics, it is necessary to use aseptic procedures in opening and examining each container.

52.63 Opening and Examining[3, 5, 8]

The container must be clean. If obvious soil, oil, etc., are present, wash with detergent and water; rinse; wipe dry.

a. Cans

Flood the end without embossed code with a sanitizing agent. Let stand 10 to 15 minutes. Pour off excess liquid; wipe dry with clean, unused towel. Play the flame of a Meeker type burner down onto the end of the can, continuing until visible moisture film evaporates. (If can swells, keep side seam directed away from analyst, and flame cautiously.) Flame a clean bacti-disc cutter, then aseptically cut and remove a disc of metal from the end. (With cans having an "easy open" feature on one end, open the opposite end. Do not disturb the "easy open" end in the event that it is subsequently necessary to examine the score;) subculture.

b. Glass jars

Clean and sanitize closures as described in "Cans." Flame. Open with a flamed beer can opener or with a bacti-disc cutter (avoid unseating or moving the closure with respect to the jar; subsequent seal examination may be desirable); subculture.

Jar with lug type closures should be scribed or otherwise permanently marked on both the cap and the glass. This procedure will allow for the quantitative measurement of the closure security by a qualified container examiner if such an examination is deemed necessary.

If complete removal of the jar closure is necessary, the jar should be cleaned as above, then inverted in a disinfecting solution for an appropriate period of time, e.g., 100 ppm chlorine, for 15 minutes. A sterile cotton pad then is placed on the closure. A flame sterilized bacti-disc cutter point is used to puncture the closure and thus relieve the vacuum. The closure then may be removed with less danger of admitting microbial contaminants over the glass sealing surface.

c. Pouches

Clean and sanitize pouch surfaces with a detergent sanitizing agent. Dry with a sterile, fresh towel. If the product is maneuverable, push it away from one end. With clean lightly flamed scissors, cut the end off just under the pouch seal. (On large pouches a 2 to 3 inch opening may be made by cutting off the corner of the pouch with a diagonal cut.) Open the pouch without touching the cut ends; subculture.

52.64 Subculturing

Immediately after opening the container, transfer 1 to 3 ml of the product with a straight walled pipet (borosilicate glass tubing, 0.7 mm ID × 35 to 40 cm, with fire polished ends is ideal), to appropriate subculture media and incubate; transfer 5 to 15 ml to a sterile tube and refrigerate; this may serve as a sample for subsequent use if necessary. Prepare a smear of the product for microscopic examination; measure the pH of the product; observe the product odor and appearance; record results.

52.65 Subculture Media

The best subculture medium for use in examining spoiled canned foods is a portion of the normal product which has been tubed and sterilized. Its use eliminates a question sometimes raised with the use of artificial laboratory medium concerning whether or not organisms recovered by the laboratory medium actually grow in the product. Liquid products may be tubed without modification. Products consisting of solids and liquids should be blended mechanically, retaining the normal ratio of solids to liquids. Products lacking free moisture should be blended mechanically, adding sterile water as necessary to make a slurry capable of being tubed. Low acid product tubed media should be sterilized in the autoclave (15 minutes at 121 C); acid product tubed media should be sterilized by steaming (30 minutes at 100 C).

Laboratory media (Chapter 2) suitable for canned food examination include the following:

(a) For low acid foods

1. PE-2 Medium, in screw cap tubes, is excellent for the growth of both aerobes and anaerobes encountered in canned food examinations.
2. Cooked Meat medium: for the growth of anaerobes
3. Standard Methods Agar: for the growth of aerobes

(b) For acid foods

1. Orange Serum broth used for bacteria and yeasts encountered in this category of foods. Some lactobacilli do not grow well; therefore, if these organisms are present, use APT broth also.
2. Acid Products Test (APT) broth: a culture broth for acid tolerant microorganisms.
3. Potato Dextrose agar: a plating medium for use when yeasts and molds of the heat resistant types may be present.
4. Sabouraud Dextrose agar: a plating medium for molds and yeasts
5. APT broth: for growth of *Lactobacillus*.

52.66 Subculturing Product Samples

Transfer 1 to 3 ml of product with a straight walled pipet to replicate tubes of subculture media. For products without excess free liquid, transfer portions taken from each of several areas within the continer, e.g., with metal cans, take portions from the center and adjacent to each seam; blend with minimal volume sterile water in sterile mechanical blender. Inoculate subculture medium. For low acid products, inoculate two tubes of sterile product or PE-2 medium for incubation at 30 to 35 C. If cooked meat medium is to be used, inoculate two tubes. It is then necessary to inoculate duplicate plates or tubes of an aerobic medium. Incubate at 30 to 35 C. If normal temperature for storage or handling of the product is higher than 40 C, and if samples have been incubated at 55 C, inoculate duplicate tubes of any medium used for incubation at 55 C; otherwise, do not incubate at 55 C. For acid products, inoculate two tubes of sterile product, or of the laboratory medium or media of choice, and incubate at 25 to 30 C. If molds are suspected, prepare a pour plate and incubate at 30 C. Incubate subcultures for at least five days before declaring them negative.

52.67 Mass Culture Technique; Acid Products Test Broth

This medium has been developed as a culture medium for the selective cultivation of *Lactobacillus*, *Leuconostoc*, and yeasts capable of causing spoilage in acid products. It is an excellent medium for the recovery of minimal contamination and, therefore, the utmost care must be taken by workers to insure that the medium is not contaminated during handling. Since the broth is intended primarily as a mass culture medium, approximately 100 grams of product under test should be inoculated aseptically into 200 ml of sterile medium. Care must be taken to disperse adequately the product throughout the broth.

A minimum of three flasks per sample should be inoculated. An extra retained aseptic sample from each container should be incubated with the flasks. In addition, a retained sample should be held at refrigeration temperatures for use in microscopic comparisons with the incubated product or for repeating the test. Incubation of the cultures should be at 30 C for five days with visual examination for fermentation or biological surface growth daily. The extra retained samples should be incubated for 10 days. At the end of the incubation period all samples should be examined microscopically for evidence of bacterial or yeast contamination.

52.68 Direct Smear for Microscopic Examination

Prepare a wet mount of product liquid (or of surface scrapings of product not having excess free liquid) for examination with a phase microscope, or make a smear for simple staining and examination. Products having a high solids content in liquid portion may be diluted advantageously with approximately equal volumes of water on the slide. A flamed bacteriological loop may be used for making smears. Examine both the product being tested and a normal control product to enable judgment regarding the presence of abnormal levels of microorganisms in the former.

52.69 Confirmation of Positive Laboratory Medium Subcultures

Inherent characteristics of certain food products preclude the germination and outgrowth of some viable bacterial spores present, as well as growth of certain nonsporeformers in meat (Chapter 42). This is particularly true in, but not confined to, acid products. Laboratory media may not exhibit similar bacteriostatic properties, resulting in positive subcultures which may lead to the erroneous conclusion that the product in question is not commercially sterile. The true significance of such positive subcultures, when it is in doubt, may be determined by using them to inoculate the sterile product, either tubed or canned, followed by incubation. Organisms of significance will grow and manifest their presence in the product. Again, use of the sterile product as the primary subculture medium, together with direct microscopic examination, and assessment of product abnormalities (compared to normal control) usually will obviate the necessity of a confirmatory procedure.

52.7 INTERPRETATION OF DATA

Serious consequences can evolve if commercial sterility testing information is based upon faulty laboratory techniques, or if errors are made in evaluating data. The strict observance of the preceding procedures will lessen the chances of designating a product "contaminated," when in actuality, it is "commercially sterile." Likewise, the instances of incorrectly claiming a product to be "commercially sterile," also will be decreased. A similar test has been adopted by the AOAC as official first action.[3]

The development of swells (any degree) may indicate microbial activity. The presence of a swell condition by itself, particularly at the flipper, springer, or soft swell level, however, does not positively reveal noncommercial sterility. Such conditions must be confirmed by demonstrating excessive microorganisms (direct smear), or an abnormal product (pH, consistency, odor, etc.), as compared to the normal product. Some causes of low degree swells (which, of course, cannot be confirmed microbiologically in a commercial sterility test) are overfilling, low filling temperatures, improper vacuum closing procedures, inadequate vacuum mixing (certain comminuted products), incipient spoilage, and chemical swells ("hydrogen springers").[8]

Noncommercial sterility of flat containers is shown if direct smears reveal excessive microorganisms, and if one or more product characteristics are abnormal. Subculture may be necessary to support conclusions.

Duplicate cultures should show comparable biochemical reactions and microbial flora. These in turn should be similar to those in the original product, i.e., predominating flora in direct smears of product will usually appear in subcultures, and biochemical reactions will be comparable.

If only one tube of a duplicate set is positive, adventitious (laboratory) contamination should be suspected. Comparison of the positive culture with the original product, especially in direct smears, often will clarify the situation. If not, new subcultures should be prepared from the retained sample.

A stained smear, or a wet mount examined by phase microscopy which shows only an occasional cell in some fields usually does not suggest a spoiled product. Representative coverage of a smear is necessary to form an accurate evaluation; examination of 10 to 20 fields is necessary to avoid being misled in forming conclusions. Comparison with the normal product is mandatory.

It is sometimes difficult to differentiate bacterial cells from food particles when microscopically examining a food product. On occasion, food grade yeast is an ingredient of the processed food. Check the list of ingredients on the label for yeast. Caution must be taken not to assume that these yeast cells are contaminating microorganisms. Fermented products, e.g., sauerkraut, will show a high normal microbial population by direct smear.

If a recovery medium has a near neutral pH, and the food being tested is an acid food, microbial growth may be recovered. This is normal and expected, and does not indicate the lack of commercial sterility. When testing an acid food, an acid pH recovery medium should be used.

If, when using proper "commercial sterility" testing procedures, microbial growth is recovered from a container of food that displays no evidence of spoilage, the following confirmation program should be carried out:

(a) The bacterial isolate, or isolates, should be grown in pure culture.
(b) An unopened container of food, exhibiting the same manufacturer's code as the one previously tested, should be selected.
(c) Using aseptic techniques, a small puncture hole should be made through the can end or jar closure.
(d) The product should be inoculated (under the surface) with the microbial isolate.
(e) The puncture hole should be flamed to create a vacuum in the headspace, and aseptically sealed with solder or similar material. (It is

sometimes necessary to warm the closed container to 40 or 50 C before puncturing to insure that the seal will be under vacuum.)
(f) The inoculated container should be incubated at 30 to 35 C for 10 days.
(g) The container should be opened and the product examined.

If the previously tested container did contain microorganisms, spoilage indices (or lack thereof) should be identical for both containers. If the first container:

(a) exhibited a normal appearing product, but the analyst recovered growth upon subculture; and
(b) after inoculating the second container with the microbial isolate, finds spoiled product, e.g., gas, separation, etc.;
(c) one should assume that the first container was commercially sterile and that the growth was the result of faulty technique.

Opened containers should be emptied, washed, and dried. Keep them, properly identified, in the event subsequent container examination is deemed desirable.

52.8 REFERENCES

1. Corson, L. M., G. M. Evancho, and D. H. Ashton. 1973. Use of forceps in sterility testing: A possible source of contamination. J. Food Sci. **38:**1267
2. Denny, C. B. 1970. Collaborative study of procedure for determining commercial sterility of low acid canned foods. J. of the AOAC. **53:**713–715
3. Denny, C. B. 1972. Collaborative study of a method for the determination of commercial sterility of low-acid canned foods. J. of the AOAC. **55:**613–616
4. Federal Standard 209a. 1967. U.S. Government Printing Office, #252–522/280. Washington, D.C.
5. Hersom, A. C. and E. D. Hulland. 1975. Canned Foods, An Introduction to Their Microbiology (Baumgartner). 5th Ed. Chemical Publishing Co., Inc. New York, N.Y. 291 pp
6. Kautter, D. A. and R. K. Lynt. 1973. Botulism. Nutr. Rev. **31:**265–271
7. Meat and Poultry Inspection Regulations, Animal and Plant Health Inspection Service, U.S.D.A., Washington, D.C. 1973
8. Schmitt, H. P. 1966. Commercial sterility in canned foods, its meaning and determination. Quart. Bull. Assoc. of Food and Drug Officials of the U.S. **30:**141–151

CHAPTER 53

CANNED FOODS—TESTS FOR CAUSE OF SPOILAGE

D. A. Corlett, Jr.

53.1 INTRODUCTION

The methods and procedures in this chapter are intended for the diagnosis of spoilage in thermally processed foods in hermetically sealed containers. Canned foods exhibiting suspected or actual spoilage conditions are usually submitted for examination because of the abnormal appearance of the container (swelling, leakage, damage, etc.) and/or abnormal condition of the product (off odor, color, or consistency; alleged illness).

The presence of microbiological spoilage is the primary concern in spoilage diagnosis of canned foods; however, the public health significance and the type of microorganisms involved are greatly dependent on the type of food and the circumstances of production and storage. In some instances spoilage, or the appearance of spoilage, does not involve microorganisms and results from physical and/or chemical conditions.

Accordingly, a comprehensive examination procedure is required utilizing ancillary information regarding the history of the defective product, structural integrity of containers, physical and chemical state of the food, and other factors that may be related to the presence of viable microorganisms. These factors, as well as recovery and identification of microbiological groups responsible for spoilage, must be considered to attain accurate and expedient decisions. Failure to utilize all information may lead to an inaccurate diagnosis having obvious economic and/or public health implications.

The methods for spoilage diagnosis are not recommended for tests of commercial sterility of canned foods which appear to be sound. Commercial sterility determination methods are different, and are described in Chapter 52. In addition, the spoilage diagnosis procedures for conventional thermally processed canned foods may not be suitable for examination

---

CONTRIBUTORS: G. R. Bee, H. S. Beerbohm, F. F. Busta, C. B. Denny, D. J. Goeke, Jr., R. R. Graves, G. L. Hays, K. A. Ito, M. B. Jeffrey, M. Johnston, D. A. Kautter, D. E. Lake, H. V. Leininger, R. K. Lynt, A. W. Matthys, D. T. Maunder, W. A. Mercer, C. F. Niven, Jr., R. B. Read, Jr., W. P. Segner, R. V. Speck, F. E. Spencer, H. R. Thompson, P. J. Thompson, R. B. Tompkin, D. J. Wessel, W. V. Yonker.

IS/A Committee Liaison: Cleve B. Denny

of perishable, "pasteurized" canned cured meats, such as refrigerated canned hams. Methods for examination of these products are given in Chapter 42.

53.2 MICROBIOLOGY OF CANNED FOODS

Thermal processing of canned foods is dependent on many factors associated with the type of product, its normal pH, its consistency, the size of the container, and the type of cooker.[10, 33] The most important factor that determines the degree of thermal processing to achieve product stability is pH, because of the inhibitory effect of acidity on survival and outgrowth of indigenous microorganisms.[12, 13] For this reason processed foods are divided into two major pH categories: low acid foods, having a pH above 4.6, and acid foods, having a pH of 4.6 or lower.[39]

53.21 Processing of Low Acid Foods (pH above 4.6)

Low acid foods are thermally processed to produce a condition known as commercial sterility. Commercial sterility is defined as the condition in which all *Clostridium botulinum* spores and all other pathogenic bacteria have been destroyed, as well as the more heat resistant organisms, which, if present, could product spoilage under normal conditions of storage and distribution.[31]

"Commercially sterile" foods, however, are not necessarily completely sterile in the classical sense. Occasionally canned low acid foods may contain low numbers of certain thermophilic spores that will not cause spoilage unless the food is stored at temperatures above 43 C. To completely sterilize a canned product unquestionably would degrade product quality.

53.22 Processing of Acid Foods (pH 4.6 or below)

An acidity of pH 4.6 or less is sufficient to inhibit the growth of *C. botulinum*, and permits the application of a less severe thermal process to the food.

Since there is a significant advantage in quality retention by using "acid cooks", some low acid foods are acidified to pH 4.6 or less.[39] When "controlled acidification" is used, the control of the pH is critical to insure against possible outgrowth of *C. botulinum* in the product. Acid foods, in the higher pH ranges, may contain low numbers of thermophilic sporeformers that do not become a problem unless the food is stored at temperatures above 43 C. Acid foods also may contain dormant mesophilic spores which present no problems.

53.23 Factors Responsible for Spoilage in Canned Foods in General

Microbiological spoilage in canned foods usually is indicated by a swelling of the container. When no abnormal external signs are present, spoilage may be indicated by abnormal odor or appearance of the product. The causes of these general conditions usually are related to one of the following factors:

a. Insufficient processing, permitting survival of mesophilic microorganisms.
b. Inadequate cooling after processing, and/or storage and distribution conditions, permitting growth of various thermophilic microorganisms.
c. Leakage, permitting microorganisms to contaminate the product after processing.

However, other conditions may occur in low acid and acid canned foods which may result in a swelling of the container and abnormal product appearance and thus confuse the investigator. The conditions are:

d. "Incipient spoilage" that occurs before the product and/or ingredient(s) are processed and may be caused by microbial and/or enzymatic action. Incipient microbial action may result in CO_2 production and the presence of excessive numbers of dead microbial cells in the product. Enzymatic action also may result in CO_2 evolution as well as off odors (particularly in vegetables).
e. "Hydrogen swells" that result from chemical action of the food on the container, producing hydrogen gas.[25] When advanced corrosion has occurred, "pinholes" in cans and lids of glass jars may be present. Hydrogen swells are most common in over age merchandise.
f. Nonenzymatic browning (Maillard reaction) which occurs in canned products having high levels of sugar, amino acids and acid. Carbon dioxide may be produced in sufficient amounts to bulge the container, particularly during storage of the product at elevated temperatures. This problem is occasionally encountered in canned fruit concentrates.
g. Product formulation errors and/or mishandling (e.g., freezing of the canned food).

53.24 Microbiological Groups Associated with Spoilage in Low Acid Foods (pH above 4.6)

a. Insufficient processing

Insufficient processing is indicated chiefly by the survival of bacterial spores, particularly those of the *Clostridium* species, and sometimes by *Bacillus* species, which subsequently spoil the product. From the public health standpoint, this is a most serious situation because of the potential development of *C. botulinum* and its toxin.

Generally, anaerobic mesophilic spoilage is associated with putrid odors, and with the presence of rod shaped microorganisms in pure culture with or without typical clostridial sporangia or spores. If both putrid odors and spores are found during examination, the product should be tested for botulinum toxin by the mouse inoculation procedure, regardless of the percent of the pack exhibiting spoilage[37] (Chapter 34).

b. High temperature spoilage (thermophilic spoilage)

Low acid foods may spoil during storage above 43 C from a variety of extremely heat resistant sporeforming thermophilic microorganisms. These bacteria all grow at 55 C, some as high as 70 C. Some are facultative thermophiles, and also may grow at 35 C or lower. Thermophilic bacteria are not pathogenic.

Thermophiles occur naturally in agricultural soils, and their spores frequently are present in low numbers in commercially sterile products. They may occur in substantial numbers if certain ingredients (sugar and starch) used in the product are contaminated with excessive numbers of these organisms, or if growth of thermophiles occurs in poorly controlled preprocessing equipment. Spoilage of this type is more common where the canned product is inadequately cooled or subsequently stored at high temperatures.

It is very important during microbiological examination of low acid foods to avoid confusing thermophilic and mesophilic sporeforming bacterial isolates. They may be confused, because in the vegetative state the cells will grow over a much wider temperature range than when they germinate and grow out from the spore state.[37] For this reason it is essential to produce spores at the temperature of isolation (30 to 35 C or 55 C), heat shock the suspension to destroy vegetative cell forms, and subculture at both 30 to 35 C and at 55 C.[37] The temperature of outgrowth from the spore state indicates whether the isolate is an obligate thermophile (growth only at 55 C), facultative thermophile (growth at both 30 to 35 C and at 55 C), or a true mesophile (growth at 30 to 35 C).

It should be emphasized that heat shocking for the above purposes must be conducted on subcultures. Heat shock of the microorganisms in the food sample may unintentionally destroy sporeformers if they are all in the vegetative cell form, or when acidic conditions prevail in the food, caused by acid producing spoilage organisms.

In low acid foods the most common forms of thermophilic spoilage and the causative microorganisms are categorized as follows:

1. "Flat sour" spoilage is indicated when the container is not swollen, and when the pH of the product is significantly lowered. The causative microorganisms are sporeformers, such as *Bacillus stearothermophilus*, a facultative anaerobe. Spore germination occurs at thermophilic temperatures only, but vegetative cell growth may occur at mesophilic temperatures as well. (Chapter 21)
2. "T. A." spoilage is indicated by swelling, and commonly by the bursting of the container. The condition is caused by obligately thermophilic, sporeforming anaerobes such as *Clostridium thermosaccharolyticum* that produce large quantities of hydrogen and carbon dioxide. The product usually has a "cheesy" odor. Vegetative cells of T.A.'s may grow at temperatures below the thermophilic growth range (Chapter 22).
3. "Sulfide stinker" spoilage is characterized by a flat container in which the contents are darkened and have the odor of "rotten eggs". This type of spoilage is caused by the spore forming an-

aerobic, obligately thermophilic microorganism *Clostridium nigrificans* which produces hydrogen sulfide. No swelling of the container is produced because the hydrogen sulfide is very soluble in the food; however, it does react with any iron present to form black iron sulfide (Chapter 23).

c. Container leakage

Environmental sources such as air, water, or dirty surfaces, contribute to microbiological contamination of processed foods through rough handling of filled containers, or because of defective or damaged containers and closures.[34] Subsequent spoilage results in swelling of the containers when the pathway of the original leak becomes blocked, preventing the escape of gas. If a leak is large enough, leakage of the product may occur. Spoilage in flat containers, with or without vacuum, also may be encountered. This results from contamination of the product with microorganisms that produce little or no gas.

Numerous groups of microorganisms may be found in instances of container leakage including the following: cocci, short and long rods (lactic acid bacteria), yeasts and molds, aerobic sporeformers, mixtures (very common). Generally, but not exclusively, leaker spoilage is characterized by the presence of mixed cultures of cocci, coccoids (oval shaped rods), and rods. All of these organisms exhibit little or no heat resistance.

Aerobic sporeformers are sometimes present in pure or mixed culture as a result of leakage. The presence of sporeformers is becoming more common due to the widespread chlorination (0.5 ppm or more of free residual chlorine) of can cooling water which destroys most vegetative microorganisms.

In addition, postprocessing can-handling equipment may cause seam or closure contamination before the seam compound has had a chance to "set up" and seal the closure. This may occur immediately following thermal processing when containers roll on wet, dirty roll tracks, or receive rough handling in the presence of contaminated surfaces. Rough handling sufficient to cause leakage does not result necessarily in permanent dents in the can; however, permanant dents do not result necessarily in leakage.

The occurrence of anaerobic sporeforming *C. botulinum* must be considered. If, during microbiological examination of the product, there are found large club shaped rods with subterminal spores, particularly in combination with a putrid product odor, the product should be tested for toxicity by the mouse inoculation procedure (Chapter 34).

It is always essential to conduct thorough leakage tests and detailed structural examinations of all containers, although failure to demonstrate potential leakage conditions does not rule out leaker spoilage. This fact cannot be overemphasized when a decision must be made to differentiate spoilage due to leakage or to insufficient processing, especially when the microbial flora found in the product do not necessarily indicate either situation. The original container closure records are available under Section 128 b.8 of Title 21 of the *Code of Federal Regulations* and should be examined if container leakage is suspected.

53.25 Microbiological Groups Associated with Spoilage of Acid Foods (pH 4.6 or Lower)

Microbiological spoilage in acid foods is caused by microorganisms capable of growing at pH 4.6 or lower. *Clostridium botulinum* does not grow in these products (unless abnormal conditions first produce a pH in excess of 4.6).[39]

Generally it is considered necessary to use acidified media to recover acid tolerant microorganisms from acid foods.[1] This encourages the growth of the acid tolerant spoilage organisms, and inhibits the outgrowth of acid sensitive organisms that may have survived the thermal process, and which were inhibited by the normal acidity of the product. Acid sensitive, sporeforming, heat resistant bacteria may be recovered occasionally from acid foods on media having a neutral pH; these have no significance.

After isolation in acid media, it is often convenient to grow acid producing and/or fastidious bacteria on a neutral medium when the presence of these microorganisms is suspected in acid food spoilage.[4] These microorganisms are usually vegetative members of the lactobacilli, streptococci, or pediococci; they may be differentiated from aerobic sporeformers by the catalase test. Lactic acid bacteria are catalase negative (Chapter 15).

Groups of microorganisms that may be associated with spoilage in acid foods are described as follows:

a. Insufficient processing

A variety of acid tolerant sporeforming microorganisms may survive processing. Their survival usually is dependent on excessive preprocessing contamination and/or an attempt to give the product a mild heat process to preserve texture (particularly in fruit products). Improper time temperature treatments may, of course, be responsible. These microorganisms fall into the following groups:

1. "Butyric anaerobes" such as the mesophilic, sporeforming anaerobe *Clostridium pasteurianum* which produces butyric acid, as well as carbon dioxide and hydrogen (Chapter 19).
2. Aciduric "flat sours", particularly *Bacillus coagulans*,[38] in tomato products. This aciduric facultative anaerobic sporeformer may grow at both 30 to 35 C and at 55 C (Chapter 20).
3. Heat resistant molds, particularly in the case of the contamination of juice concentrates and fruits by these fungi prior to processing. The causative microorganism is usually *Byssochlamys fulva*, and related, or similar species which produce very heat resistant ascospores. Spoilage is evidenced by a moldy taste and odor, color fading, the presence of mold mycelia in the product and sometimes by slight swelling of the container lid (Chapter 17).

4. Yeasts and/or asporogenous bacteria, especially in cases of grossly insufficient processing. This type of spoilage may be indistinguishable from leaker spoilage unless the containers are thoroughly examined for leakage and structural defects, and postprocessing handling conditions are known to be satisfactory (Chapter 16).

b. High temperature spoilage (thermophilic spoilage)

Thermophilic spoilage may occur in acid foods, particularly in tomato products, and cultural tests should include "cross temperature" incubation of isolates at 30 to 35 C and at 55 C to differentiate mesophiles from thermophiles.

c. Leakage

Pure or mixed cultures of acid tolerant bacteria, yeast, and mold contaminants are commonly found in leaker spoilage of acid products. The containers may be swollen or flat.

Gas and swelling of the can is commonly produced by bacteria or yeasts, and sometimes by molds. Spoilage in flat cans is caused by bacteria (rods and/or cocci) which do not produce gas. The contents usually show a slight lowering of pH (generally less than 0.2 pH unit) with or without obvious organoleptic changes in the product. Mold spoilage usually is evidenced by the presence of mycelia and fungal spores in flat containers having a leak large enough to permit entrance of oxygen. Sometimes a matt of mycelial growth is present on the surface of the product. An etched ring at the surface level of the liquid inside the freshly opened can often indicates that oxygen has leaked into the can.

53.3 EXAMINATION METHODS FOR DIAGNOSING SPOILAGE CONDITIONS IN CANNED FOODS

The following methods are designed to guide the investigator in conducting a thorough and carefully documented examination. Procedures have been divided into three parts: 53.31. Gathering and Interpretation of Background Information; 53.32. Procedures for Examination of Containers and Contents; and 53.34. Microbiological Culture Procedures. Results obtained by examination of samples according to these procedures may be recorded on the sample data collection forms and compared to the "Keys To Probable Cause Of Spoilage", provided in Section 53.4.

It is recommended that the diagnosis of spoilage in canned foods be conducted by trained microbiologists, or trained technicians under supervision. If difficulties are encountered in the examination of samples or in the interpretation of results, assistance is available from recognized authorities in the field, such as the National Canners Association, the container supplier, the U.S. Food and Drug Administration, or the U.S. Department of Agriculture.

FIGURE 1. Request for Analysis Form

No.____ Date________

Type of Product and Style Pack: ____________________

A. *Information from Point of Collection of Sample(s):* ____________________

1. Source ____________________
2. Number of Samples ____________________

   Sampling Instructions: Please submit 12 suspect units, if available, and 12 units of sound product from a similar code as controls. Submit *all* suspect containers (and an equal number of controls of similar code) if *less* than 12 suspect units are available.
3. Date of Collection ____________________
4. Code(s) ____________________
5. Container Size ____________________

B. *Information Concerning Code Lot of Submitted Sample(s):*

1. Location of Spoiled Goods ____________________
2. Temperature of Storage—Plant:____Warehouse:____Retail:________
3. Number of spoiled and normal containers of each size involved ________
4. Percent spoilage (number spoiled divided by number in lot × 100) ______
5. Packing dates, periods, and lines involved ____________________
6. Date when spoilage first appeared or noted ____________________
7. Are any cans burst, hard swells, soft swells, flippers, or low vacuum, and are any flat cans spoiled (complaints of spoilage in flat cans)? ________
8. Were any irregularities noted during production of product in: ________

   a. Preparation ____________________
   b. Thermal processing ____________________
   c. Cooling temperature ____________________
   d. Chlorination of cooling water ____________________
   e. Sanitation of post-process handling (can lines, labeling area, etc.) ______
   f. Container integrity ____________________
   g. Other (damage, mishandling) ____________________
9. Type of cooker or processor ____________________
10. Process Used—I.T. ________ Time ________ Temp. ________

Please submit appropriate records for any items noted in B

C. *Alleged Illness Complaint Information:*

1. Number of persons involved ____________________
2. Symptoms ____________________
3. Time before onset of symptoms ____________________
4. What other foods and beverages were also ingested ____________________

53.31 Gathering and Interpretation of Background Information

Information regarding the circumstances of production, storage, and the incidence of defect conditions in lots of canned foods suspected of containing spoilage is invaluable for the diagnosis of microbiological conditions.[27] Frequently the cause and nature of spoilage is evident when a thorough review of background information has been made.

A suggested "Request for Analysis" form is provided (Figure 1) for gathering this information. The best way to use the form is to request persons submitting samples to fill out the questionnaire at the time the collected samples are submitted for examination. Some detailed information may not be available initially and may need to be obtained later.

To aid in the utilization of background information, a "Guide for the Interpretation of Background Information" has been provided (53.311).

53.311 Guide for interpretation of background information

Information	*Types of Interpretations*
a. Number of spoiled containers.	i. An isolated can is usually a random leaker, although insufficient processing must be considered. ii. More than one can, especially one per case or more than one per case, may indicate defective containers, rough handling, or insufficient processing.
b. Age of product and storage conditions.	i. Excessive age and/or excessively high temperatures of storage may produce detinning and hydrogen swells. ii. Perforation, due to corrosion or damage of the container, may produce leaking and/or mixed culture leakage spoilage, with swells in some cases (where the perforations are blocked). iii. Thermophilic spoilage may result from high temperature storage, in excess of 43 C.
c. Location of spoilage in stacks; temperature of the warehouse.	i. Spoilage in center of stacks or near ceiling may indicate failure to cool product sufficiently, resulting in thermophilic spoilage. ii. Scattered spoilage may indicate insufficient processing or leakage. iii. Excessive spoilage usually indicates insufficient processing.
d. Processing records, including retort charts, "cook check tags", or other indicator systems for af-	i. Irregular "cooks" may be correlated with spoilage from insufficient processing.

fected lot. (Note: These records must be kept by the canner under Section 128b.8 of Title 21 of the *Code of Federal Regulations*.)	ii. Adequate processing may eliminate insufficient heat treatment spoilage, and indicate leaker or thermophilic spoilage.
e. Daily records, including overcooling conditions, inadequate chlorination and/or contaminated cooling water, and dirty, wet, post-processing equipment.	i. Leaker spoilage may occur with or without defective seams or visible dents.
f. Deviations from normal canning procedures. (Note: Heat processes for each low acid food formulation must be established and kept on record as required in Section 128b.4 of Title 21 of the *Code of Federal Regulations*.)	i. Product preparation spoilage prior to processing may lead to excessive numbers of nonviable microorganisms in product with possible loss of vacuum. ii. Defective seams or rough handling of wet containers after processing may lead to leakage spoilage. iii. Failure to cool cans properly may lead to thermophilic spoilage. iv. Changes in product formulation without reevaluation of process parameters may lead to insufficient processing. v. Inadequate equipment sanitation, existence of dead end piping, etc., may lead to microbial buildups which can overwhelm a proper process.
g. Container inspection records indicating defects.	i. Leakage spoilage is probable.
h. Alleged illness complaint information including symptoms, time before onset of symptoms, and ingestion of other foods which may be more likely causes of food poisoning. Caution must be used in interpreting microbiological findings in food poisoning complaints. The longer a can has been opened, the more it has been handled (or mishandled) before it reaches the laboratory, the less significant the isolation of certain food poisoning organisms will be to the problem, (e.g., coagulase positive staphylococci).	In most cases the information simply does not support the product as the causative agent of microbiological food poisoning where: i. An acid food is involved. ii. Onset of symptoms is immediate. iii. Symptoms may suggest food poisoning by other foods eaten which are recognized to be commonly associated with specific types of relatively non heat resistant microorganisms such as staphylococci, salmonellae, *Clostridium perfringens*, etc.

53.32 Procedures for Examination of Containers and Contents

Conduct all steps in sequence. Record data on the "Spoilage Examination Data Sheet" included for this procedure (Figure 2).

53.321 Sampling

The number of units available for analysis will vary with the circumstances. When the item has been in distribution for a long period of time, or when the item is several years old, few sample units may be available from a consumer complaint. Items with recent codes, or those in plant or wholesale warehouses, are usually available in sufficient quantity.

a. When sufficient samples are available[36]

Select 12 sample units at random, representative of an apparent or suspected spoilage condition, (i.e., swollen containers or reports of abnormal appearance of contents) from each production code involved. Also select 12 control sample units of sound product from a related code (same day, week). Analyze suspect and control sample units.

b. When sufficient samples are not available[36]

If less than 12 are available from each production code involved, analyze all sample units representative of an apparent or suspected spoilage condition. Analyze an equal number of sound control units from a related code. When only one or two sample units are available, it is advisable to request additional units while proceeding with the analysis of those on hand. Avoid testing very hard swells or buckled cans if a sufficient number of softer swells are available. Very hard swells/buckles may burst during examination and/or structural damage to cans may render the containers unsuitable for microleak testing and detailed container examination.

Segregate samples by production code (if more than one code is present). Examine the label of each container for leakage stains and, if present, circle with a marking pen. Make a mark with waterproof ink on the bottom of the label and container to indicate the alignment of the label to the container. Carefully remove labels from containers and attach labels temporarily to the back of the data sheet for reference during external examination of containers.

Consecutively number each container on the side (body) with waterproof ink or copper sulfate solution; take care not to write on the can side seam. Record container numbers, code(s), and can size(s) on the data sheet.

FIGURE 2. Spoilage Examination Data

Date: ________

Identification: ________ Source: ________ Analyst ____ Lab Nos. ____

53.321			53.322	53.323		53.326	53.328	53.329	53.323	53.321
Sample No.	Can Code	Can Size	Can Condition Exterior	Gross Wt. Gms.	Gross Wt. Ozs.	Vacuum or Gas	Odor, Texture Appearance	Microscopic Examination	pH	Can Examination and Microteak Testing

53.322 External examination

Examine the external condition of the containers for swelling of the ends (or lid). Examine containers for structural defects or damage to the following:

a. Cans

1. End seams for evidence of crooked, excessively wide or narrow seams, "cutovers," and "sharp" seams, cracked or ruptured seams, "cable burns" (where the metal is abraded thru the seam), and evidence of leakage.
2. Side seam for possible solder voids, pitting, rusting, or leakage.
3. The embossed code on lid for possible fractures or perforations.
4. The can body for perforations, imperfections (particularly "inclusions" in the tinplate), or leakage.
5. Circle defects with waterproof ink when they are found. Also, mark an "X" on the end of the can exhibiting a defect(s) to show the end which *should not be opened during sampling of the container*. If no defects are found on either can end, save the end with the code (usually the "cannery" end) and mark it with an "X". Never open the lid end of two-piece aluminum or steel cans unless there is evidence of "leaker damage" on the body shell end.
6. Some canners apply a spot or stripe of heat sensitive paint or "thermal sensitive fluid" to the body of the container as a retort check. The paint contains suspended material and has the consistency of "house paint", whereas, the fluid is a solution and has the consistency of coding ink. The choice of color may vary with different canners, and with different types of products. For this reason it may be necessary to confirm the specific color system used on the sample(s).
   Heat sensitive paints and fluids in common use change color during retorting as follows:

Type	*Color Before Retorting*	*Color After Retorting*
Paint	Red	White
	Blue	White
Fluid	Violet	Blue
	Black	Red
	Red	Yellow

A color change indicates only that the product was retorted and does not show that correct process temperature or time was achieved. Therefore, the possibility of insufficient processing cannot be discounted even though the color check conforms to the specified postretort color.

Record observations on data sheet.

b. Glass containers

Examine the following areas for defects or damage:

1. The lid for possible misalignment, damage or leakage.
2. The glass container for "hairline" or small impact cracks, or improper finish.
3. The contents for glass bubbles and abnormal conditions (underfill, digestion, turbidity, etc.).

Record observations on data sheet.

53.323 Weight

Weigh each container to the nearest gram and record the weight on the data sheet. Subtract the average known tare weight of the empty container from each unit weight to determine the approximate net weight of each sample. Then look up the established net weight, as well as the maximum and minimum "permissible" net weights of the product. Following the disposition of contents (53.332), when the empty dried containers are available, weigh them to insure that they fall within the average tare weight for the type of container.

If the net weights of the suspect sample units exceed the net weights of the controls, and exceed the maximum permissible net weight, the suspect containers may be overfilled causing reduced headspace and reduced or zero vacuum. In extreme cases of overfilling, the bulging of the lid gives the external appearance of a swell. Seafoods, due to their size and packing in short-height cans are often overfilled inadvertently, e.g., sardines.

If the suspect units have a lower net weight than controls, and are below the minimum permissible net weight for the product, they may be underfilled, or have leaked. Leakage also may be evidenced by stains or residues on the label and/or on the exterior areas of the container.

53.324 Preparation of area for tests

The area selected for examination should be clean, dust free and located in an area where samples may be taken aseptically for microbiological examinations. A laminar air flow work station is ideal for providing an ultra clean working environment.[22] Analysts should wear clean protective laboratory garments. All sampling utensils and pipets, etc., should be presterilized ready for use. The surface of the workbench used during examination procedures should be washed, sanitized with an effective disinfectant solution, such as 100 ppm chlorine solution, and wiped dry with clean paper towels.

These conditions are essential to minimize incidental microbiological contamination during aseptic sampling, a problem which is most acute when no viable microorganisms are present in the product.[17,18]

53.325 Preparation of samples

Wash all samples in warm soap and water to remove soil and grease. Rinse in clean running water and dry with clean paper towels. Arrange samples in numerical order (order to be opened) on the workbench. Place the end of the can to be opened (unmarked end) in the upward position.

Do Not Open the Container.

53.326 Vacuum measurement or gas analysis

Select a portion of the suspect flat containers for measurement of internal vacuum and/or swollen containers for analyses of headspace gas composition. Also, select a portion of the control sample units for measurement of internal vacuum. Determine the number of units for vacuum or gas test as follows:

Number of Suspect or Contol Containers Available	*Number to select for Vacuum or Gas Tests*
6 to 12	2
2 to 5	1
1	None

When suspect sample units consist of both flat and swollen containers, and only 2 to 5 are available, select a swell for gas analysis. If only one suspect unit is available, save it for aseptic sampling.

a. Vacuum measurement (flat containers)

Use a puncture vacuum gauge with a rubber seal around the puncture needle to obtain the vacuum (in inches of mercury). Take the vacuum through the lid, about 3/4 inches from the edge to avoid clogging the puncture needle. (Cool to room temperature any container which has been removed from incubation before measuring the vacuum.) Record vacuum measurements on data sheet.

b. Gas collection and analysis (swollen containers)

1. Qualitative test for hydrogen

When there is sufficient headspace gas in a swollen can, gas may be collected in an inverted test tube when the lid is punctured. Puncture the lid with a "bacti-disc" opener and hold the inverted tube directly over the point of puncture. Immediately following collection of the gas, place the open end of the tube in a burner flame. The presence of hydrogen is indicated by a loud "pop".

2. Quantitative headspace gas analysis (Hydrogen, Carbon Dioxide, Nitrogen and Oxygen)

Collect a gas sample from a swollen container by use of a rubber gasketed puncture needle connected to a syringe, or to a manometric device having a "gas holding solution."

Analyze the gas sample on a chemical gas analyzer or a "fixed" gas chromatograph.[40]

Analysis of gas from a sound product should yield mostly nitrogen with very small amounts of hydrogen, carbon dioxide or oxygen (generally less than 1%). A large percentage of carbon dioxide usually indicates microbiological growth, or in some products such as vegetables, it may indicate preprocessing product respiration. In high sugar products, carbon dioxide production may be due to the Maillard reaction. A large percentage of hydrogen may indicate can corrosion or detinning (hydrogen swell). A high concentration of both carbon dioxide and hydrogen may be indicative of microbial growth, particularly from thermophilic anaerobes. Record the results on data sheet. *Note:* Sample units used for vacuum measurement of gas collection must not be used for aseptic sampling because of possible contamination from puncture devices.

53.327 Aseptic sampling

For each sample unit, label two sterile large screw cap test tubes (24 mm) with the sample unit number, product lot code, and date of examination. After aseptic sampling by the following procedure, retain each set of food samples obtained from each container for cultural tests.

a. Sterilization of lid

Arrange all containers in numerical order, positioning the end to be opened in the upward condition (the end to be saved should have been marked with an "X").

1. Flat to moderately hard swells

Flood the lid with a disinfecting agent, such as 100 ppm chlorine solution[18] or other equivalent sanitizing agent. (Do not use alcohol because its use does not assure sterility of the lid surface.[20]) Allow to stand 10 to 15 minutes and pour off excess sanitizing agent. Heat with a burner flame over the lid until visible moisture has evaporated. Do not overheat the lid. If a canned sample is under positive pressure, place side seam away from you during the flaming.

2. Hard swells, very hard swells, and buckled cans

It is advisable to chill sample containers having high positive pressure in the refrigerator for several hours prior to lid sterilization. This will minimize possible bursting of the container during flaming of the lid. When thermophilic anaerobes are suspected, do not chill, because chilling often kills the vegetative cells of T.A.'s, and spores are rarely recoverable in thermophilic spoilage.

Note: Since it takes 10 to 15 minutes for disinfection of the lid with the sanitizing agent, flood the lids of the next one or two containers and let them stand while proceeding with opening and sampling. However, once the lid has been flamed, aseptic opening and sampling must be conducted in one continuous operation for each sample unit.

b. Aseptic opening of lid

1. Flat container

Cut a hole slightly off center in the lid using a "Bacti-Disc Cutter" (see Figure 3). A hole 2 inches in diameter is usually sufficient for sampling; however, for a small can or jar, a smaller hole may be necessary. Always leave ½ to ¾ inch of lid remaining in contact with rim for subsequent microleak testing and container examination.

2. Swollen container

(a) Place the container on a large tray in case the contents under pressure spill over during opening.
(b) Invert a large sterilized plastic or metal funnel over can, insert a sterile tipped puncture rod thru funnel, and puncture the lid.
(c) After venting has occurred, remove the funnel and open the container by the procedure described previously for a flat container.

c. Aseptic sampling of contents

Use sterile pipets or utensils to transfer approximately 20 g of container contents to each of two labeled tubes. When pipetting potentially toxic materials always use cotton plugged pipets, or ones that are fitted with a mechanical suction device such as a rubber bulb. The following procedures are suggested for transfer of various types of materials to the tubes:

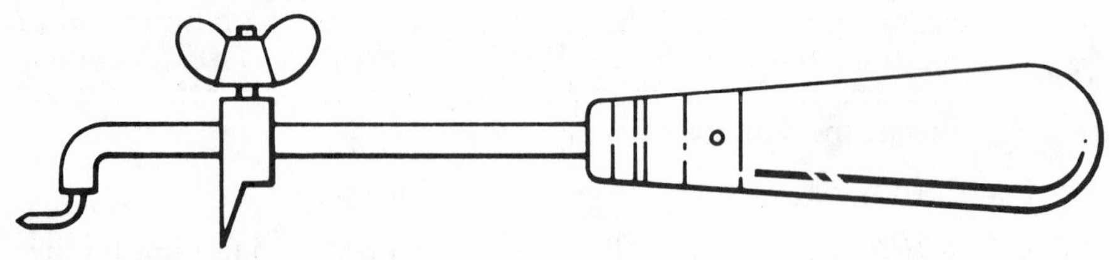

FIGURE 3. "Bacti-Disc Cutter" Can Opener[29]

1. Liquids

   Transfer 20 ml of liquid with sterile 20 ml pipets.

2. Semiliquids, with or without small particular material

   Transfer with sterile 20 ml pipets having large bore tips.

3. Solids in liquids

   (a) Pipet about 10 ml of liquid per tube with a sterile 20 ml pipet having a large bore tip.
   (b) Transfer about 10 g of solid by use of sterile forceps (nonserrated tip type) or spatula.[15] (Do not reuse pipet or forceps without resterilizing in an autoclave.)

4. Semisolid or solid materials

   Transfer about 20 g material with sterile forceps (nonserrated tin type) spatula or spoon.

   Note: All transfer utensils must be wrapped and sterilized prior to use. Alcohol flaming of metal utensils must not be used because it is not an effective means of sterilization.[20]

   Retain all aseptically drawn samples for procedures in 53.33. Store tubes in the refrigerator unless microscopic examination reveals long rods indicating thermophilic anaerobes. Vegetative cells of T.A.'s often die under refrigeration.

53.328 Odor and appearance

Never taste the contents of any container during a spoilage examination because of the potential presence of botulism toxin!

a. Odor

Compare the odor of the suspect sample units to the odor of the controls. This procedure minimizes individual differences in odor perception.

Generally, putrid odors in samples are reason to suspect the presence of mesophilic, anaerobic sporeformers and, if found, the sample unit(s) should be tested for toxin. Fecal odors are generally from coliform bacteria and are indicators of leakage. (For reference to the putrefactive odor characteristic of anaerobic spoilage, smell a laboratory culture of nontoxic Putrefactive Anaerobe #3679 *C. sporogenes* which may be obtained from the National Canners Association.) However, nonproteolytic, nonheat resistant *C. botulinum* strains may be encountered which do not produce putrefactive odors.

Sharp, sour odors are usually caused by an acidic product condition. "Rotten egg" odor is caused by "sulfide stinker" spoilage, usually accompanied by dark discoloration of the product. These odors do not necessarily indicate the presence of toxin.

Determine the odor characteristics of each opened sample unit. The following terms are commonly used in spoilage examinations:

sharp	rotten egg	putrefactive
sour	hydrogen sulfide	putrid
acidic		cheesy
fecal		butyric

b. Appearance

Compare the appearance of the suspect sample units to that of the normal controls. Categorize appearance according to the following characteristics:

1. off color—cooked vs. uncooked;
2. texture—softening, digestion, sliminess;
3. consistency—fluid, viscous, ropiness;
4. clarity—cloudiness in clear liquids, frothiness;
5. foreign materials—particles, mold, etc.;

Record the odor and appearance on the data sheet.

53.329 Microscopic examination

Examine the contents of each suspect and each control sample unit microscopically. With a flamed loop make a smear of the contents from each container on a freshly cleaned microscope slide. Let the smear air dry, flame fix it, stain with crystal violet, and dry. Never Gram stain a product because a Gram stain result depends on the age of the culture. A Gram stain may be made on an 18 to 24 hour culture only.

If a smear will not adhere to the slide during staining, (sometimes encountered with fatty foods) prepare another smear. Air dry and heat fix, rinse with xylol, and let it dry before staining with crystal violet.

If a phase microscope is available, examine a wet mount prepared by transferring a small drop of food slurry to a slide and overlaying it with a coverslip. Endospores appear brightly refractile under phase.

Examine slides at about 1000X under the oil immersion objective, and record observations on the data sheet. Common results and interpretations are:

a. A normal product may exhibit a few microbial cells in every microscopic field examined. The presence of more than a few cells per field, under 1000X magnification, usually indicates an excessive number of microorganisms. (Fermented foods, such as sauerkraut, are an exception; numerous dead cells will be present. Cultural recovery of viable microorganisms in fermented foods is necessary to confirm spoilage of microbial origin.)

b. Mixtures of rods, cocci, coccoids, yeast, and molds usually indicate leakage, but may indicate that the product was not sufficiently thermally processed.

c. Pure cultures of medium-to-long rods with or without detectable spores (directly or indirectly in subculture) may indicate leakage especially if the morphology varies from sample to sample. If subsequent tests indicate that the container is structurally sound, and no postprocessing rough handling of wet cans was observed in the plant, insufficient processing may have occurred.

d. Pure cultures of medium to large rods having sporangia and/or free spores may indicate insufficient processing or leakage. If clostridial rods containing subterminal spores, or only free spores, are present, and/or the odor of the product is putrid or "putrefactive," toxin tests by mouse inoculation should be done immediately.[37]

e. When mixed or pure cultures of cells are observed microscopically, but viable microorganisms are not recovered in subsequent cultural tests, "incipient spoilage" is indicated (except for fermented foods).

Record results on the data sheet.

53.330 pH determination

Determine the pH of each sample unit by insertion of the electrode(s), of a previously referenced pH meter, directly into the product in the container. Record pH on the data sheet.

After measuring the pH of each sample, clean the electrode(s) thoroughly. Use separate electrodes to measure the pH of foods that may be eaten, and foods to be tested for spoilage examinations.

Compare pH measurements of suspect and control samples. In addition, compare results to the normal pH range given for the particular food provided in Section 53.52.

53.331 Disposal of contents and cleaning of container

a. Samples suspected of having toxic spoilage

When one or more sample units are suspected of containing mesophilic, anaerobic spoilage and/or toxin, autoclave the contents after transferring them to another container. Using rubber gloves, rinse the container with 100 ppm chlorine solution and air dry at 35 C. Autoclaving may change the container's lid sealing compound.

b. Low acid foods (pH above 4.6) exhibiting no signs of toxic spoilage

Empty contents into a metal tray and sterilize them in an autoclave before disposal. Use gloves when washing the container.

c. Acid foods (pH 4.6 or lower)

Empty contents into the appropriate disposal container without sterilization.

53.332 Cleaning of container

Wash the interior of the container with warm soap and water, using a bottle brush if necessary to remove food residues from interior seam surfaces. Rinse, soak cans in hot water for 1/2 hour, rinse in hot water, and drain. If available, an automatic dishwater may be used to clean empty cans and bottles.

Prepare cans for container examination by shaking out excess water from the interior of the can, and placing the container, open end up, in an

FIGURE 4. Cultural Tests on Low Acid Food Sample Unit

Sample No: _____ Code: _____ Date Received: _____

Culture Stage	Medium	Number of Tubes or Plates per Test	Incubation Temperature C	Time Period of Incubation Days	Results of Cultural Tests[1]								
					Growth	Acid	Gas[2]	Cell Morphology					Additional Comments
								Rods	Spores	Cocci	Yeast	Mold	
Recovery	GTA (or GTB)	2	30–35 C	4									
		2	55 C	4									
Subculture #1	NAMn	2	30–35 C	10									
		2	55 C	10									

Subculture #2 (Heat shocked 80 C for 10 min.)	NAMn	2	30–35 C	4									
		2	55 C	4									
Recovery	PE-2 (or CMM)	2	30–35 C	10									
		2	55 C	4									
Subculture	*LVA* *(Aerobic)*	2	30–35 C	4									
	LVA *(Anaerobic)*	2	30–35 C	4									

NOTES: (1) 0 = Negative result, e.g., no growth, no acid, etc.
+ = Positive result, e.g., growth, acid, etc.
(2) Note odor of gas or culture in "Additional Comments", particularly from PE-2/CMM and LVA cultures.

air incubator, and drying for 24 hours at 35 C. (Higher temperatures may cause lid sealing compound to seal leaks.) Containers must be dry.

53.333 Container examination

a. Superficial examination

Examine the interior of the can for patches of black or grey discoloration, especially along the side and end seams. Patches of discoloration indicate localized detinning and pitting due to chemical reaction of the can contents with the container, usually producing a hydrogen swell. Detinning may be due to leakage of oxygen into the can, faulty tin plate, abnormally corrosive contents, or excessive age of the product.

In plain (unenameled) cans the tin may be completely corroded from the inner surface of the can. Hydrogen swells are not dependent on concurrent microbiological spoilage, but leaker spoilage may follow when corrosion has caused the container to perforate.

b. Detailed examination

Detailed container examination is necessary when containers have not exhibited obvious defects, especially when leaker spoilage must be clearly differentiated from that caused by insufficient processing.

The detailed examination requires trained personnel and adequate facilities to conduct leakage testing, preferably by the "microleak" detection method,[32] followed by can "tear-down" to evaluate seam and lap construction.[35]

When the expertise for conducting detailed container examinations is not available, it is advisable to seek assistance from the container manufacturer or from the National Canners Association.

Record observations on the data sheet.

53.34 Microbiological Culture Procedures

Microbiological culture procedures are different for the examination of low acid and acid foods. For this reason two separate examination sequences are provided: Section 53.341 for low acid food samples having a normal product pH above 4.6, and Section 53.342 for acid food samples having a normal product pH of 4.6 or lower.

Each section contains a data sheet for recording results during the course of the microbiological examination (Figures 4 and 6). When microbiological tests are completed, compare the results, along with information from 53.31 and 53.32, to the Spoilage Diagnosis Keys provided in 53.4.

Isolates obtained from microbiological tests, along with the reserve sample tube, should always be saved in case further work is necessary.

53.341 Cultural tests for low acid foods having a normal pH above 4.6

a. Description of method[37]

This method employs primary media for the recovery of spoilage microorganisms, and subculture media to further characterize them into spoilage groups. Differentiation into spoilage groups is fundamentally dependent on the growth of isolates under aerobic or anaerobic conditions, growth response at 30 to 35 C and at 55 C, and morphological differentiation of vegetative and sporeforming cultures. Other determinative factors, including acid and gas production, odor of the culture and specific cell morphology (cocci, rods, yeasts, etc.) are also utilized.

The purpose, description, and abbreviated designation for each medium employed is given in the following tables. In several instances a choice of media is given for those in common use. These may be used interchangeably at the option of the investigator.

b. Primary recovery media

Purpose	Primary Recovery Media	Abbreviation
Aerobic recovery media	Glucose tryptone bromcresol purple agar in poured plates.[2]	GTA [(a)]
	Alternate: Glucose tryptone bromcresol purple broth in tubes.[2] (Omit agar)	GTB [(b)]
Anaerobic recovery media	Peptone yeast extract broth containing bromcresol purple indicator and whole peas in tubes.[18]	PE-2 [(c)]
	Alternate: Cooked meat medium, consisting of liver or beef heart infusion broth in tubes.[7, 8]	CMM [(c)]

(a) Growth on GTA plates indicates the presence of aerobic microorganisms. Acid production (color of BCP turns from purple to yellow) may indicate a "flat sour" thermophilic isolate if growth occurs at 55 C. Isolates are characterized morphologically by microscopic examination of colony growth. Prior to inoculation the surface of the agar should be free from excessive moisture.

(b) Growth in GTB tubes is also indicative of aerobic growth, however, anaerobes may grow in the lower portion of the tube. Use of a larger food sample is the chief advantage of GTB.

(c) PE-2 may be used directly. CMM media must be exhausted prior to use by heating to 100 C for 20 minutes to remove oxygen, and rapidly cooled without agitation to the *intended* incubation temperature (30 to 35 C or 55 C). After inoculation the surface of the medium may be layered (stratified) with sterile agar[5] or vaspar[6] to maintain anaerobic conditions. Gas production is evidenced by bubbles in the medium and/or gas under the agar or vaspar seal.

c. Subculture media

Purpose	Subculture Media	Abbreviation
Growth and Sporulation of aerobes.	Nutrient agar containing manganese in poured plates.[26]	NAMn [(d)]
Growth and sporulation of anaerobes (under strict anaerobic conditions).	Liver-veal agar in poured plates.[19]	LVA [(e)]

(d) NAMn supports growth and enhances spore production by aerobic sporeformers[14, 16, 26] and is used primarily to differentiate mesophilic from thermophilic *Bacillus* sp. When rod shaped aerobes in pure culture are isolated on GTA (or GTB) and sporulation is not evident, but there is reason to believe sporeformers are involved in the spoilage, the isolates should be subcultured on NAMn at the temperature of initial isolation. After incubation up to 10 days, if spore production has taken place, the spores are heat shocked to destroy all vegetative cells and cultured again on NAMn at both 30 to 35 C and 55 C. The temperature at which outgrowth occurs from the spore state indicates whether the isolate is an obligate mesophile (growth at 30 to 35 C), an obligate thermophile (growth at 55 C), or a facultative thermophile (growth at 30 to 35 C and at 55 C). Further details are provided on the following cultural examination sequence diagram and in the procedure section.

(c) LVA plates are used to clearly differentiate anaerobes from aerobes. Isolates from PE-2 or CMM are streaked on LVA and incubated aerobically and anaerobically (in anaerobic jar or incubator). The medium supports sporulation of the anaerobic *Clostridium* sp., but it is a good growth medium and under aerobic conditions will grow a great variety of aerobes as well. Prior to inoculation the surface of the agar should be free from moisture.

d. Cultural examination sequence

Figure 5 summarizes the examination sequence for low acid food samples.[37] Review of this sequence prior to utilization of the stepwise testing procedure will aid in determining the extent of the examination. In many cases negative cultural results will preclude the need for further testing. The diagram contains blank lines at various points to indicate that a particular point in the examination has been completed or that further testing is required.

It should be noted that all tests starting with GTA or GTB media are designed to identify groups of aerobic microorganisms. All tests starting with PE-2 or CMM media are intended for identifying the presence of anaerobic microorganisms.

Maximum periods for culture incubation are shown on the diagram; in most cases growth will take place in a much shorter time and continued incubation is unnecessary (exception, NAMn). Detailed test results should be entered on the data sheet provided (Figure 4).

e. Procedure. Recovery methods

1. Select one of the two retained samples (taken aseptically in 53.327 for cultural testing.) Save the other retained sample in the refrigerator in case additional cultural tests are neccessary. For each retained sample selected for cultural tests, assemble the following media (includes duplicate media per test):
   (a) Label four GTA plates (or GTB tubes) with sample number and date. Mark two plates (or tubes) "30 to 35 C" and two "55 C".
   (b) Label four PE-2 (or CCM) tubes with sample number and date. Mark two tubes "30 to 35 C" and two tubes "55 C". CMM tubes must be exhausted by heating at 100 C for 20 minutes and cooled immediately before use.

2. Inoculate each GTA plate by transferring a small quantity of the food, via a flame sterilized loop, to the surface of the plate. Streak in a manner to dilute out the inoculum across the surface of the plate. If GTB tubes are used, aseptically transfer about 1 ml or 1 gm of food into the bottom portion of the tube. Inoculate each tube of PE-2 (or CMM) by aseptically transferring about 1 ml or 1 gm of food into the bottom portion of the tube. Layer the surface of the four inoculated tubes with 1/2 inch of sterile vaspar or agar to provide anaerobic conditions.

3. Incubate media as follows:
   (a) Place GTA (or GTB) and PE-2 (or CMM) media, labeled "30 to 35 C", in an air incubator having a temperature ranging from 30 to 35 C. (This range is provided since different laboratories commonly use 30 C or 35 C as their preferred temperature of incubation.) Incubate GTA (or GTB) up to four days and PE-2 (or CMM) up to 10 days.
   (b) Place GTA plates labeled "55 C" in a 55 C incubator. If GTB tubes are used instead of GTA plates, heat the tubes in a water bath to 55 C before placing in the 55 C incubator. Incubate up to four days. Heat PE-2 (or CMM) tubes labeled "55 C to 55 C" in a water bath before placing them in the 55 C incubator. (Preincubation is necessary to avoid growth and gas production by rapidly growing mesophiles before the tube contents reach thermophilic temperatures in the incubator.) Incubate anaerobic media (PE-2 or CMM) up to 4 days.

4. Record detailed results on the data sheet in spaces provided for the recovery media (GTA/GTB; PE-2/CMM).

FIGURE 5. Cultural Examination Sequence for Low Acid Foods (above pH 4.6)

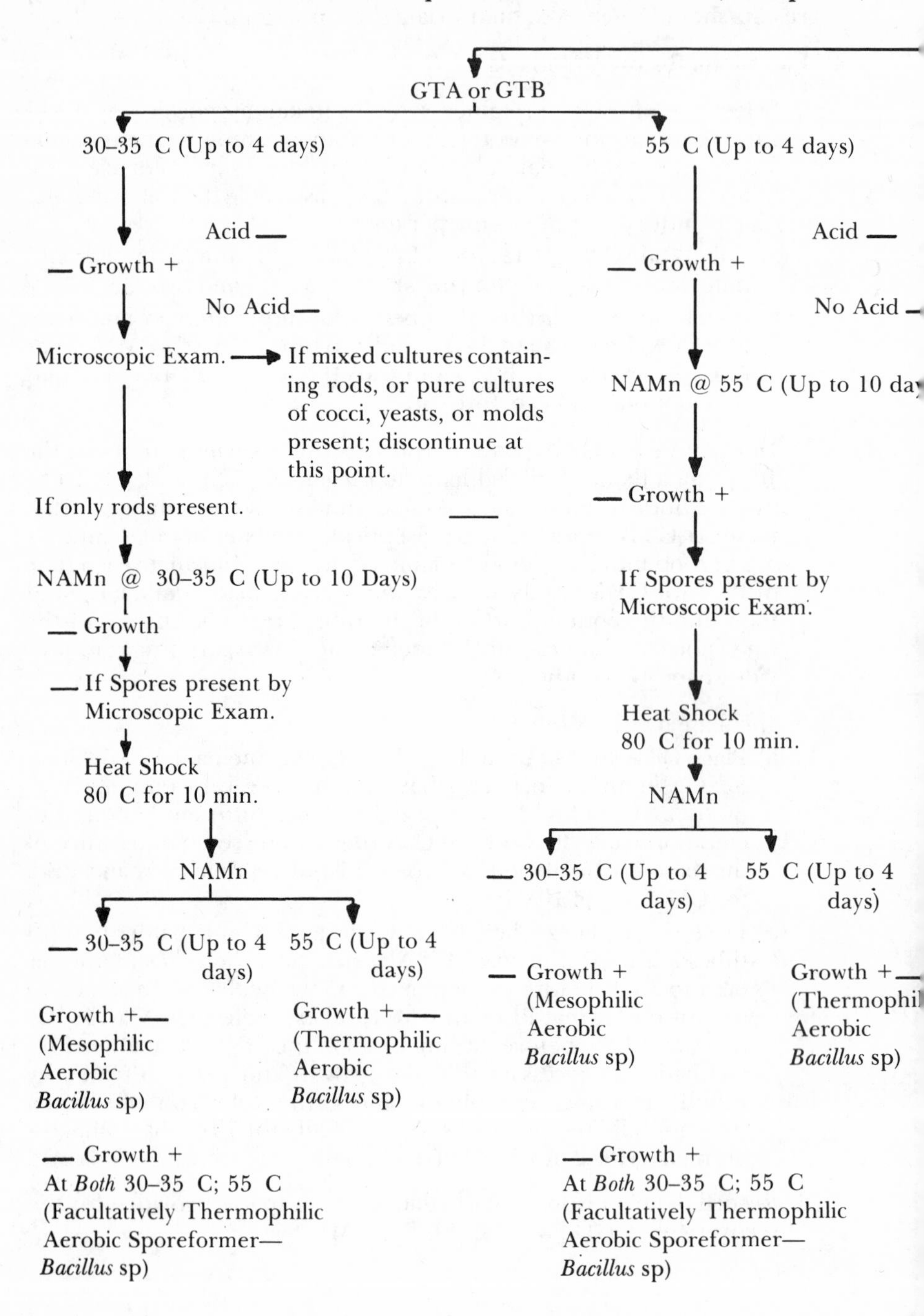

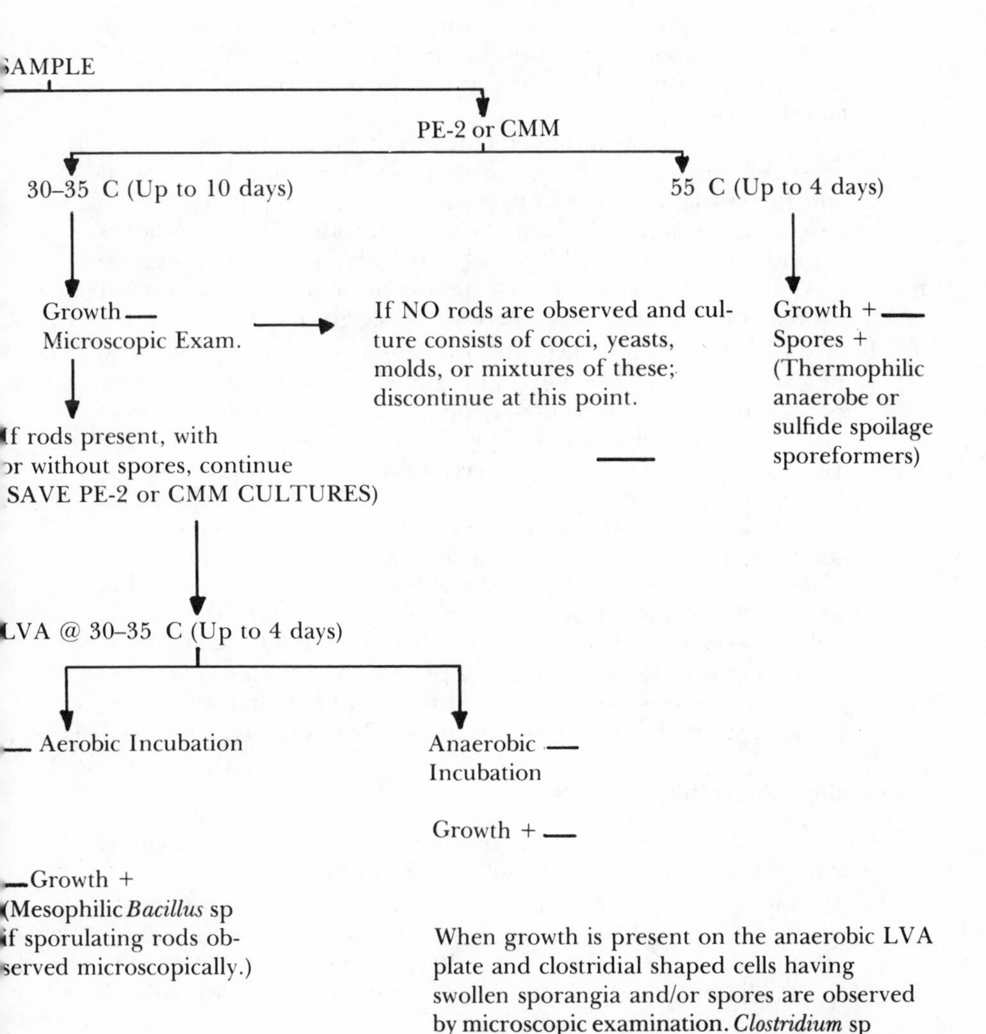

When growth is present on the anaerobic LVA plate and clostridial shaped cells having swollen sporangia and/or spores are observed by microscopic examination. *Clostridium* sp are indicated and the 30–35 C PE-2 or CMM isolation medium should be tested for the presence of botulism toxin by the mouse inoculation procedure.

— Growth + on LVA under aerobic and anaerobic conditions indicates isolation if a mixed culture of an aerobe and an anaerobe, or a pure culture of a facultative anaerobe. However, always test PE-2 or CMM for toxin if growth occurs on the anaerobic plate and microscopic examination reveals a sporeformer.

(a) Growth in GTA (or GTB): Record presence and temperature(s) of growth, acid production (bromcresol purple indicator turns yellow in the presence of acid), and examine growth microscopically under about 1000 × oil immersion (crystal violet stain or phase microscopy). A Gram stain may be made on 18 to 24 hour cultures only.

If growth consists of mixed cultures or pure cultures of cocci, yeasts, or molds, leakage or inadequate processing is indicated and further culturing is unnecessary. Very short (tiny) rods are also an indication of leakage. Generally, rods of various widths are indicative of leakage, but additional testing of them is neccessary. If only rod shaped bacteria are present, with or without spores, continue with subculture sequence given in f1 below.

(b) Growth in PE-2 (or CMM): Record presence and temperature(s) of growth, acid production (in PE-2 medium only), and presence of gas under the seal. Remove the cap from each tube and note odor; replace cap. Examine growth microscopically under about 1000 × under oil immersion (crystal violet stain or phase microscopy). During microscopic examination observe for clostridial morphology, rod shaped cells containing spores, and/or free spores. Spores are brightly refractile.

If rod shaped bacteria, with or without spores, are observed in PE-2 (or CMM) incubated at 30 to 35 C, continue with the subculture sequence given in **f 2** below. Save the PE-2 (or CMM) culture tubes during the subculture procedure, because if anaerobic growth is observed in subculture, the PE-2 or CMM tubes must be tested for the presence of botulism toxin (Chapter 34).

f. Procedure. Subculture methods

1. The presence of mesophilic or thermophilic aerobic sporeforming isolates (*Bacillus* sp.) may be established as follows:

(a) If only rod shaped bacteria were observed on GTA (or GTB) media they should be subcultured on NAMn plates as follows:

(1) Streak culture recovered at 30 to 35 C on GTA (or GTB) media onto two NAMn plates. Incubate the NAMn subculture at 30 to 35 C (only).

(2) Streak bacillus culture recovered at 55 C on GTA (or GTB) media onto two NAMn plates. Incubate the NAMn subculture at 55 C (only).

(3) If growth occurred on GTA (or GTB) media at 30 to 35 C and 55 C, test the 30 to 35 C culture(s) as in (1) and test the 55 C culture(s) as in (2).

Allow NAMn subcultures to incubate up to 10 days to insure sporulation. Verify spore formation by microscopic examination of all cultures before proceeding.

(b) Place two loopfuls of a colony containing spores from each NAMn plate incubated at 30 to 35 C and/or 55 C into separate sterile 16mm glass test tubes containing 1 ml of sterile distilled water. Avoid touching the sides of the tubes with the loop. Swirl tubes gently to disperse culture. Place the tubes in an 80 C water bath for 10 minutes to heat shock spores and to destroy vegetative cells. The level of the spore suspension in the tube should be one inch below the level of the water bath.

(c) Following heat shock of the suspensions, streak a loopful from each tube onto each of four fresh NAMn plates. Incubate two plates at 30 to 35 C and two at 55 C. Incubate up to four days at each temperature and record presence or absence of growth on data sheet. Examine microscopically for spores and note this also.

2. The presence of mesophilic, obligately anaerobic sporeformers (clostridia) may be established as follows:

(a) If rod shaped bacteria occur in PE-2 (or CMM) incubated at 30 to 35 C, streak four LVA plates from each positive PE-2 (or CMM) tube. Inoculate a loopful of culture onto the surface of each LVA plate and streak in a manner to dilute inoculum. Inoculate two of the inoculated LVA plates *aerobically*, in an air incubator 30 to 35 C for up to four days.

Incubate the other two inoculated LVA plates in an anaerobic jar or incubator at 30 to 35 C for up to four days. Insure that the anaerobic jar or incubator is operating correctly by use of a suitable oxidation-reduction indicator.

(b) After two to four day's incubation, examine the aerobic and anaerobic plates for growth. Examine the cultures microscopically and record cell morphology and presence or absence of spores. Note the odor of anaerobic culture and record if it is "putrid", putrefactive, etc.

The following interpretations and/or tests are dependent on the results of subculture growth reactions on LVA plates incubated at 30 to 35 C under aerobic or anaerobic conditions.

(1) When growth occurs only on the aerobic LVA plates, an aerobe has been isolated. No further testing is necessary since anaerobic isolates were not recovered.

(2) When growth occurs only on the anaerobic LVA plates, an anaerobe has been isolated. *Clostridium* species is indicated if microbiological examination reveals the presence of cells having clostridial morphology (cells shaped like "tennis rackets", having terminal spores in swollen sporangia), and/or free spores.

Growth on the anaerobic LVA plate, particularly in the presence of a putrid odor necessitates testing the 30 to 35 C PE-2 or CMM isolation media for the presence of botulism toxin by the mouse inoculation procedure given in Chapter 34. *C. botulinum* may be present.

(3) When growth occurs on both the aerobic and anaerobic LVA plates, this indicates the presence of an aerobe and an anaerobe, or a facultative anaerobe.

Since toxic *C. botulinum* may be present on the anaerobic LVA plate, examine the culture microscopically from the anaerobic LVA plate and conduct the test for botulism toxin as specified in (2).

Growth on the aerobic and anaerobic LVA plates may consist of a pure culture of asporogenous facultatively anaerobic lactobacilli. Microscopically, these organisms appear as slender rods. They will be present in instances of container leakage and are nonpathogenic.

Note: The procedures in **f 2** above are intended to confirm the presence of mesophilic, obligately anaerobic microorganisms indicating the potential presence of *C. botulinum* and the need for conducting the toxin test. However, since mixtures of obligate and/or facultative anaerobes may be recovered, further identification and tests for heat resistance are recommended by methods presented elsewhere in the Compendium. When *C. botulinum* is suspected refer specifically to Chapter 34 for additional test methods.

53.342 Cultural tests for acid or acidified foods having a normal pH of 4.6 or lower

a. Description of method[37]

Aciduric types of microorganisms are the only types capable of growing in an acid product having a pH of 4.6 or lower. For this reason, the subculture media must be selective for spoilage type involved, or approach the normal pH of the acid product to eliminate the possibility of isolating viable organisms dormant in the product, but capable of growing in a neutral medium.[1]

The general media of choice for recovering most acid tolerant microorganisms are orange serum broth acidified to pH 5.5, or thermoacidurans agar plates acidified to pH 5.0. It should be noted, however, that since the pH of these media are somewhat higher than pH 4.6, microorganisms may be recovered occasionally that are normally suppressed in the acid product. For this reason, it is not uncommon to recover spoilage microorganisms from specific acid products in media consisting of the normal product that has been tubed and sterilized. Hence, when the investigator is dealing exclusively with spoilage diagnosis of a specific acid product, it is recommended that recovery of spoilage microorganisms in tubed and sterilized normal product be considered.

Other media are useful to recover specific types or to differentiate microorganisms in acid foods. Potato dextrose agar acidified to pH 3.5 or phytone yeast extract agar with streptomycin and chloramphenicol[11] are suitable for the recovery of yeasts and molds without

Cultural information

Purpose	Medium	Abbreviation	Incubation Temperature
General Recovery	Orange serum broth,[3] pH 5.5, in tubes.[(a)]	OSB	30 to 35 C
	Alternate: Thermoacidurans agar, pH 5.0, plates.[28]	TAA	
B. coagulans. Facultative thermophilic aerobes; aciduric flat sours	Orange serum broth,[3] pH 5.5, in tubes.[(a)]	OSB	55 C
	Thermoacidurans agar,[28] pH 5.0, plates	TAA	30 to 35 C
Butyric anaerobes	Thermoacidurans agar,[28] pH 5.0, deep tubes.	TAD	30 to 35 C
Yeast and molds. Especially the heat resistant mold *Byssochlamys fulva.*	Potato dextrose agar,[4] pH 3.5, plates.	PDA	30 to 35 C
	Alternate: Phytone yeast extract agar with streptomycin and chloramphenicol,[11] plates.	PYE	
Aerobes and acid formers	Glucose tryptone agar with bromcresol purple indicator,[2] plates.	GTA	30 to 35 C
Fastidious lactic acid bacteria.	APT agar,[21] plates.[(b)]	APT	30 to 35 C

(a) If butyric anaerobes are indicated by a butyric odor of the product, or if spoilage by thermophilic anaerobes is indicated by high temperature storage conditions, carbon dioxide and hydrogen, and presence of sporeforming rods, layer the OSB with sterile vaspar or acidified agar to produce anaerobic conditions, or use thermoacidurans agar deep tubes.

(b) The test for catalase consists of flooding growth on this medium with a 3% solution of hydrogen peroxide.[24] Catalase negative isolates belong to the lactic group (lactobacilli, streptococci, pediococci).

the interference of bacteria. Thermoacidurans agar acidified to pH 5.0 is useful for recovery of *B. coagulans* and butyric anaerobes. Unacidified glucose tryptone agar with bromcresol purple indicator may be employed to demonstrate acid producers and unacidified APT agar may be used to support growth of fastidious lactic acid bacteria. The latter two media also support the growth of nonacid tolerant microorganisms and should be used in conjunction with other acidified media.

b. Procedure

1. Select one of the two retained samples (taken aseptically in 53.327) for cultural testing. Save the other sample in the refrigerator in case specific cultural tests become necessary. For each retained sample, selected for cultural tests, assemble the following media (includes duplicate media per test):

   (a) Four OSB tubes (or TAA plates);
   (b) Two TAD deep tubes *if* butyric anaerobe spoilage is suspected;
   (c) Two PDA and PYE agar plates if yeast and/or mold spoilage is suspected;
   Use PDA if spoilage from heat resistant mold (*B. fulva*) is suspected in fruit and fruit products;
   (d) Two GTA plates *if* aerobic acid formers are suspected;
   (e) Two APT plates *if* fastidious lactobacilli are suspected especially in flat cans).
   Label tubes and plates with sample number and date.

2. Inoculate OSB tubes with about 1 gram of food sample. (Record if the medium becomes turbid from the food at time of inoculation.) If TAD plates are used, streak each with a loopful of food as described.
3. Inoculate optional media by transferring a small amount of food to the plate using a flame sterilized loop. Streak across media to dilute out inoculum. Use a pipet for inoculation of TAD tubes; place 1 ml in bottom of melted 47 to 50 C agar, and cool rapidly.
4. Incubate all samples as indicated on the data collection form following. Heat all tubed media intended for 55 C incubation to 55 C in water bath before placing in the incubator.
5. After growth has taken place, make the following observations:

   (a) Presence of acid in or on media containing bromcresol purple indicator (purple to yellow);
   (b) Presence of gas in liquid media (bubbles) or in layered anaerobic media (gas under plug);
   (c) Prepare a crystal violet stain or a wet mount (phase) from colony growth or cell suspension as soon as growth is

apparent. Look for the presence of refractile spores in sporangia and/or free spores, as well as other morphological forms of nonsporeforming microorganisms;

(d) If GTA and/or APT media was used, flood with a 3% solution of hydrogen peroxide. A positive catalase test is indicated by gas bubbles around colony growth; a negative catalase reaction is indicated by the absence of bubbles. Lactic acid bacteria are indicated by negative catalase reaction;

Enter results on worksheet (Figure 6) and compare to diagnostic keys in the following section 53.4.

FIGURE 6. Cultural Test Results for Acid or Acidified Foods Having a pH of 4.6 or Less

Sample No: Code: Date:

Medium	Incubation Temperature(C) (Incubate up to 5 days)	Results								
		Growth	Acid	Gas	Mold	Yeast	Bacterial: Cocci	Rods	Spores	Catalase + or −
OSB[1] or TAA	30–35									
OSB[1] or TAA	55									
	30–35									
TAD	30–35									
PDA or PYE	30–35									
GTA[2] (opt)	30–35									
APT[2] (opt)	30–35									

(1) Note if OSB is layered to produce anaerobic conditions.

(2) Incubation of GTA and/or APT in an anaerobic jar is useful when very poor growth is produced by aerobic incubation procedures. Permit plates to stand exposed to air for a minimum of 30 minutes before testing for catalase.

Table 1: Low Acid Foods—pH Above 4.6[30]
Key to Probable Cause of Spoilage

Condition of Cans	Characteristics of Material in Cans						
	Odor	Appearance	Gas (CO_2 & H_2)	pH	Smear	Cultures	Diagnosis
Swells	Normal to "metal-lic"	Normal to frothy (Cans usually etched or corroded.)	More than 20% H_2	Normal	Negative to occasional organisms	Negative	Hydrogen swells
	Sour	Frothy, possibly ropy brine	Mostly CO_2	Below normal	Pure or mixed cultures of rods, coccoids, cocci or yeasts	Growth, aerobically and/or anaerobically at 30 C, and possibly growth at 55 C	Leakage
	Sour	Frothy; possibly ropy brine; food particles firm with uncooked appearance	Mostly CO_2	Below normal	Pure or mixed cultures of rods, coccoids, cocci or yeasts.	Growth, aerobically and/or anaerobically at 30 C, and possibly growth at 55 C. (If product received high exhaust, only spore formers may be recovered.)	No process given
	Normal to sour-cheesy	Frothy	H_2 and CO_2	Slightly to definitely below normal	Rods med., short to med., long, usually granular; spores seldom seen	Gas anaerobically at 55 C and possibly slowly at 30 C Negative: Thermophilic anaerobes often autosterilize	Inadequate cooling or storage at elevated temps.—thermophilic anaerobes
(Note: Cans are some-times flat)	Normal to cheesy to putrid	Normal to frothy with dis-integration of solid particles	Mostly CO_2; possibly some H_2	Normal to slightly below normal	Rods; possibly spores present	Gas anaerobically at 30 C. Putrid odor	Insufficient processing-mesophilic anaerobes (possibility of *C. botulinum*)

	Slightly off—possibly ammoniacal	Normal to frothy	CO_2	Slightly to definitely below normal	Rods; occasionally spores observed	Growth, aerobically and/or anaerobically with gas at 30 C and possibly at 55 C. Pellicle in aerobic broth tubes. Spores formed on agar and in pellicle	Insufficient processing or leakage—*B. subtilis* type
	Butyric acid	Frothy, large volume gas	H_2 and CO_2	Definitely below normal	Rods, bipolar staining; possibly spores	Gas anaerobically at 30 C Butyric acid odor	Insufficient processing butyric acid anaerobe
No vacuum and/or cans buckled	Normal	Normal	No H_2	Normal to slightly below normal	Negative to moderate number of organisms	Negative	Insufficient vacuum caused by: 1) Incipient spoilage 2) Insufficient exhaust 3) Insufficient blanch 4) Improper retort cooling procedures 5) Over-fill
Flat cans (0 to normal vacuum)	Sour	Normal to cloudy brine		Slightly to definitely below normal	Rods-possibly granular in appearance	Growth without gas at 55 C and possibly at 30 C	Inadequate cooling or storage at elevated temps.-thermophilic flat sours.
	Normal to sour	Normal to cloudy brine; possibly moldy		Slightly to definitely below normal	Pure or mixed cultures of rods, coccoids, cocci or mold	Growth, aerobically and/or anaerobically at 30 C, and possibly at 55 C	Leakage

Table 2: Acid Foods—pH 4.6 or Below
Key to Probable Cause of Spoilage

Condition of Cans	Characteristics of Material in Cans						
	Odor	Appearance	Gas (CO_2 & H_2)	pH	Smear	Cultures	Diagnosis
Swells	Normal to "metallic"	Normal to frothy. (Cans usually etched or corroded)	More than 20% H_2	Normal	Negative to occasional organisms	Negative	Hydrogen swells
	Sour	Frothy; possibly ropy brine	Mostly CO_2	Below normal	Pure or mixed cultures of rods, coccoids, cocci or yeasts	Growth, aerobically and/ or anaerobically at 30 C and possibly at 55 C	Leakage or grossly insufficient processing
	Sour	Frothy; possibly ropy brine; food particles firm	Mostly CO_2	Below normal	Pure or mixed cultures of rods, coccoids, cocci or yeasts	Growth, aerobically and/ or anaerobically at 30 C and possibly at 55 C. (If product received high exhaust, only spore formers may be recovered.)	No process given
	Normal to sour-cheesy	Frothy	H_2 and CO_2	Normal to slightly below normal	Rods, med. short to med. long, usually granular; spores seldom seen	Gas anaerobically at 55 C and possibly slowly at 30 C	Inadequate cooling or storage at elevated temps.-thermophilic anaerobes
	Butyric acid	Frothy; large volume gas	H_2 and CO_2	Below normal	Rods, bipolar staining; possibly spores	Gas anaerobically at 30 C. Butyric acid odor	Insufficient processing-butyric acid anaerobes
	Sour	Frothy	Mostly CO_2	Below normal	Short to long rods	Gas anaerobically; acid and possibly gas aerobically in broth tubes at 30 C. Possible growth at 55 C	Grossly insufficient processing—lactobacili

No vacuum and/or cans buckled	Normal	Normal	No H_2	Normal to slightly below normal	Negative to moderate number of organisms	Negative	Insufficient vacuum caused by: 1) Incipient spoilage 2) Insufficient exhaust 3) Insufficient blanch 4) Improper retort cooling procedure 5) Over-fill
Flat cans (0 to normal vacuum)	Sour to "medici-nal"	Normal		Slightly to definitely below normal	Rods, possibly granular in appearance	Growth without gas at 55 C, and possibly at 30 C. Growth on thermoacidurans agar (pH 5.0)	Insufficient processing-*B. coagulans* (Spoilage of this type usually limited to tomato juice)
	Normal to sour	Normal to cloudy brine; possibly moldy		Slightly to definitely below normal	Pure or mixed cultures of rods, coccoids, cocci, or mold	Growth aerobically and/or anaerobically at 30 C and possibly at 55 C	Leakage, or no process given

53.4 KEYS TO PROBABLE CAUSE OF SPOILAGE

Refer to Tables 1 and 2 to aid in diagnosing a spoilage condition. Each key categorizes possible spoilage conditions for food falling into a given pH range. The normal pH of the product being examined must be used when referring to the charts. Separate keys are provided for low acid and acid food products.

53.5 APPENDIX

53.51 Media and Reagent List

a. Media

Abbreviation	Medium	Reference
1. GTA GTB	Glucose tryptone broth and agar containing bromcresol purple indicator.	(2)
2. PE-2	Peptone yeast extract broth containing bromcresol purple indicator and whole peas (tubes).	(18)
3. CMM	Cooked meat medium in tubes; may be either chopped liver broth or cooked meat medium.	(7, 8)
4. NAMn	Nutrient agar containing manganese (plates).	(26)
5. LVA	Liver veal agar (plates).	(19)
6. OSB	Orange serum broth acidified to pH 5.0 (tubes). (Omit agar from formula given in reference.)	(3)
7. TAA	Thermoacidurans agar (plates).	(28)
8. TAD	Thermoacidurans agar (deep tubes).	(28)
9. PDA	Potato dextrose agar acidified to pH 3.6 (plates).	(4)
10. APT	All purpose medium containing "Tween 80", for cultivation of fastidious lactic acid bacteria.	(21)
11. PYE	Phytone yeast extract agar with streptomycin and chloramphenicol.	(11)

b. Reagents

	Designation	
1.	Crystal violet stain solution.	(9)
2.	Sterile Vaspar for stratification of anaerobic media.	(6)
3.	Sterile agar for stratification of anaerobic media (agar 2%). (Acidity to pH 4.6 for acid media.)	(5)
4.	Hydrogen peroxide solution (3% H_2O_2 in distilled water).	(24)

53.52 Normal pH Ranges of Selected Commercially Canned Foods[23]

Kind of Food	pH Range, approximate	Kind of Food	pH Range, approximate
Apples, whole	3.4–3.5	Figs	4.9–5.0
Apples, juice	3.3–3.5	Frankfurters	6.2
Asparagus, green	5.0–5.8	Fruit cocktail	3.6–4.0
Beans		Gooseberries	2.8–3.1
baked	4.8–5.5	Grapefruit	
green	4.9–5.5	juice	2.9–3.4
lima	5.4–6.3	pulp	3.4
soy	6.0–6.6	sections	3.0–3.5
Beans, with pork	5.1–5.8	Grapes	3.5–4.5
Beef, corned, hash	5.5–6.0	Ham, spiced	6.0–6.3
Beets, whole	4.9–5.8	Hominy, lye	6.9–7.9
Blackberries	3.0–4.2	Huckleberries	2.8–2.9
Blueberries	3.2–3.6	Jam, fruit	3.5–4.0
Boysenberries	3.0–3.3	Jellies, fruit	3.0–3.5
Bread		Lemons	2.2–2.4
white	5.0–6.0	juice	2.2–2.6
date and nut	5.1–5.6	Lime juice	2.2–2.4
Broccoli	5.2–6.0	Loganberries	2.7–3.5
Carrots, chopped	5.3–5.6	Mackerel	5.9–6.2
Carrot juice	5.2–5.8	Milk	
Cheese		cow	6.4–6.8
Parmesan	5.2–5.3	evaporated	5.9–6.3
Roquefort	4.7–4.8	Molasses	5.0–5.4
Cherry juice	3.4–3.6	Mushrooms	6.0–6.5
Chicken	6.2–6.4	Olives, ripe	5.9–7.3
Chicken with noodles	6.2–6.7	Orange juice	3.0–4.0
Chop suey	5.4–5.6	Oysters	6.3–6.7
Cider	2.9–3.3	Peaches	3.4–4.2
Clams	5.9–7.1	Pears (Bartlett)	3.8–4.6
Cod fish	6.0–6.1	Peas	5.6–6.5
Corn		Pickles	
on-the-cob	6.1–6.8	dill	2.6–3.8
cream style	5.9–6.5	sour	3.0–3.5
whole grain		sweet	2.5–3.0
brine packed	5.8–6.5	Pimento	4.3–4.9
vacuum packed	6.0–6.4	Pineapple	
Crab apples, spiced	3.3–3.7	crushed	3.2–4.0
Cranberry		sliced	3.5–4.1
juice	2.5–2.7	juice	3.4–3.7
sauce	2.3	Plums	2.8–3.0
Currant juice	3.0	Potatoes	
Dates	6.2–6.4	white	5.4–5.9
Duck	6.0–6.1	mashed	5.1

Normal pH Ranges of Selected Commercially Canned Foods[23] cont.

Kind of Food	pH Range, approximate	Kind of Food	pH Range, approximate
Potato salad	3.9–4.6	oyster	6.5–6.9
Prune juice	3.7–4.3	pea	5.7–6.2
Pumpkin	5.2–5.5	tomato	4.2–5.2
Raspberries	2.9–3.7	turtle	5.2–5.3
Rhubarb	2.9–3.3	vegetable	4.7–5.6
Salmon	6.1–6.5	Spinach	4.8–5.8
Sardines	5.7–6.6	Squash	5.0–5.3
Sauerkraut	3.1–3.7	Strawberries	3.0–3.9
juice	3.3–3.4	Sweet potatoes	5.3–5.6
Shrimp	6.8–7.0	Tomatoes	4.1–4.4
Soups		juice	3.9–4.4
bean	5.7–5.8	Tuna	5.9–6.1
beef broth	6.0–6.2	Turnip greens	5.4–5.6
chicken noodle	5.5–6.5	Vegetable	
clam chowder	5.6–5.9	juice	3.9–4.3
duck	5.0–5.7	mixed	5.4–5.6
mushroom	6.3–6.7	Vinegar	2.4–3.4
noodle	5.6–5.8	Youngberries	3.0–3.7

53.6 REFERENCES

1. Canned Foods. 1966. *In* Recommended Methods for the Microbiological Examination Of Foods. (J. M. Scharf, Ed.), 2nd Ed., American Public Health Association. New York, N.Y. p. 40
2. Ibid. Appendix A, Culture Media, #21. Dextrose tryptone bromcresol purple agar. p. 170
3. Ibid. Appendix A, Culture Media, #48. Orange serum agar. p. 178
4. Ibid. Appendix A, Culture Media, #51. Potato dextrose agar (acidified). p. 180
5. Ibid. Appendix A, Culture Media, #49. Plain agar for overpouring. p. 179
6. Ibid. Appendix A, Culture Media, #71. Vaspar. p. 186
7. Ibid. Appendix A, Culture Media, #32. Liver broth medium. p. 174
8. Ibid. Appendix A, Culture Media, #41, 42, 43. Meat medium, ground. p. 176–177
9. Ibid. Appendix B, Strains, #B2. Gentian violet stain. p. 189
10. Ball, C. O. and F. C. W. Olson. 1957. Sterilization in Food Technology. McGraw-Hill, New York, N.Y.
11. BBL Manual Of Products And Laboratory Procedures. 1973. Phytone yeast extract agar. Fifth Ed., BBL, Division of Becton, Dickinson and Co., Cockeysville, MD. p. 132.
12. Cameron, E. J. and J. R. Esty. 1940. Comments on the microbiology of spoilage in canned foods, Food Res. **5**:549–557
13. Cameron, E. J. 1940. Report on canned vegetables. J.A.O.A.C. **23**:607–608
14. Charney, J., W. P. Fisher, and C. P. Hegarty. 1951. Manganese as an essential element for sporulation in the genus *Bacillus*. J. Bacteriol. **62**:145–148
15. Corson, L. M., G. M. Evancho, and D. H. Ashton. 1973. Use of forceps in sterility testing: a possible source of contamination. J. Food Sci. **38**:1267
16. Curran, H. R. and F. R. Evans. 1954. The infiuence of iron or manganese upon the formation of spores by mesophilic aerobes in fluid organic media. J. Bacteriol. **67**:489–497
17. Denny, C. B. 1970. Collaborative study of procedure for determining commercial sterility of low-acid canned foods. J.A.O.A.C. **53**:713–715

18. DENNY, C. B. 1972. Collaborative study of a method for the determination of commercial sterility of low-acid canned foods. J.A.O.A.C. **55**:613–616
19. Difco Manual Of Dehydrated Culture Media And Reagents For Microbiological And Clinical Laboratory Procedures. 1962. Liver veal agar—B59. Ninth Ed., Difco Laboratories, Inc., Detroit, 1, MI. pp. 98–99
20. DOYLE, J. E. and R. E. ERNST. 1969. Alcohol flaming—a possible course of contamination in sterility testing. The Amer. J. Clin. Pathol. **51**:507
21. EVANS, J. B. and C. F. NIVEN, JR. 1951. Nutrition of the heterofermatative lactobacilli that cause greening of cured meat products. J. Bacteriol. **62**:599–603
22. Federal Standard 1967. No. 209a, U.S. Government Printing Office 252-522/280.
23. Bacteriological Analytical Manual, 2nd Ed. Food and Drug Administration, Division of Microbiology, 1969. U.S. Department of Health, Education, and Welfare, Washington, D.C. 20204. pH ranges, Section 42.03
24. Ibid. Reagents, hydrogen peroxide solution. Section 14.2.06
25. HARTWELL, R. R. 1951. Certain aspects of internal corrosion in tin plate containers. Adv. in Food Res. **3**:327–383
26. MAUNDER, D. T. 1970. Examination of Canned Foods for Microbial Spoilage. Microbiology, Metal Division Research and Development, Continental Can Co., Inc., 1350 W. 76th St., Chicago, Ill.
27. Diagnosis of Canned Food Spoilage. 1962. Research Information, 62:February. National Canners Association. Washington, D.C.
28. Preparation of media for microorganisms. 1968. In Laboratory Manual For Food Canners And Processors, Vol. 1., Microbiology And Processing. National Canners Association. AVI, Westport, CT. Thermoacidurans agar. p. 20
29. Ibid. Investigating spoilage problems. p. 45
30. Ibid. Investigating spoilage problems. pp. 56–59
31. Ibid. Process calculations. p. 220
32. Construction And Use Of A Vacuum Micro-leak Detector For Metal and Glass Containers. 1972. G. R. BEE, R. A. DECAMP, and C. B. DENNY, eds. National Canners Association. Washington, D.C. Research Laboratory, Washington, D.C.
33. Canned Foods, Principles of Thermal Process Control And Container Closure Evaluation. (A compilation of lectures and study guidelines for use as a textbook in educational programs and technical reference for food industry personnel.) 1975. Edited and Illustrated by staff members of the NCA Research Laboratories. National Canners Association. Distributed by the Food Processors Institute, 1950 Sixth Street, Berkeley, 94710
34. Ibid. Food Container Handling. pp. 45–58
35. Ibid. Container Closure Evaluation. pp. 165–206
36. Canned Food Spoilage Diagnosis Meeting—Session I. January 8, 1974. (Ad hoc Advisory Conference on Methodology for the Canned Foods—Tests for Cause of Spoilage Chapter of the APHA Compendium Of Microbiological Methods.) National Canners Association. C. B. Denny, Convenor. Washington, D.C.
37. Canned Food Spoilage Diagnosis Meeting—Session II. April 30, 1974. (Ad hoc Advisory Conference on Methodology for the Canned Foods—Tests For Cause of Spoilage Chapter of the APHA Compendium Of Microbiological Methods.) National Canners Association. C. B. Denny, Convenor. Washington, D.C.
38. STERN, R. M., C. P. HEGARTY, and O. B. WILLIAMS. 1942. Detection of *Bacillus thermoacidurans* (Berry) in tomato juice and successful cultivation of the organism in the laboratory. Food Res. **7**:186–191
39. TOWNSEND, C. T., L. YEE, and W. A. MERCER. 1954. Inhibition of the growth of *Clostridium botulinum* by acidification. Food Res. **19**:536–542
40. VOSTI, D. C., H. H. HERNANDEZ, and J. B. STRAND. 1961. Analysis of headspace gases in canned foods by gas chromatography. *Food Technol.* **15**:29–31

CHAPTER 54

SOFT DRINKS

Harry E. Korab

54.1 INTRODUCTION

Soft drinks, as the name implies, is a class of nonalcoholic beverages. They may be divided into two subclasses: carbonated soft drinks (sometimes referred to as "soda," "soda pop," etc.) and "still" or noncarbonated soft drinks.[5] Soft drinks, in general, contain 86 to 92% water, 7 to 14% nutritive sweeteners, carbon dioxide (if present), acid, and flavorings. Other ingredients which may be present are listed below.

The Food and Drug Administration's Standard of Identity for carbonated soft drinks[1] defines them as a class of beverages made by absorbing carbon dioxide in potable water. The amount of carbon dioxide used is not less than that which will be absorbed by the beverage at a pressure of one atmosphere and a temperature of 60 F. These drinks contain no alcohol, or only such alcohol (not in excess of 0.5% by weight of the finished beverage) as is contributed by the flavoring ingredient used.

The Standard includes optional ingredients that may be used in carbonated soft drinks, in such proportions as are reasonably required to accomplish their intended effects. These include nutritive sweeteners, flavoring ingredients, natural and artificial color additives, acidifying agents, buffering agents, emulsifying, stabilizing or viscosity producing agents, foaming agents, caffeine, quinine, and chemical preservatives.

"Still" drinks may, in addition to the above list of optional ingredients, contain fruit juice and vitamin C. Should this knowledge be important to the reader, he should check the container label or consult the Food and Drug Administration concerning its latest regulations.

The low pH of acidified beverages, especially carbonated soft drinks, does not permit multiplication, or even survival of pathogens, sanitary indicators, or common mesophiles.[3, 4, 9, 11]

Generally, the pH of beverages falls within these ranges: colas, 2.5 to 2.8; ginger ale, 2.7 to 3.0; lemon-lime, 2.9 to 3.4; orange 3.0 to 3.5. Carbon dioxide dissolves to form carbonic acid, which even without additional edible acids, will depress the pH to 3.3 to 3.8 depending on the degree of supersaturation.[7] Furthermore, carbon dioxide depresses the oxygen tension and thereby inhibits molds and other strict aerobes. Edible acids such as citric, malic, and phosphoric also have specific antimicrobial action.[9, 11]

IS/A Committee Liaison: Nino F. Insalata

54.2 NORMAL FLORA

Mesophilic bacteria, yeasts, and molds may be present in soft drinks in low numbers, without subsequent growth causing a change in flavor or quality.[8] However, unrefrigerated still drinks (noncarbonated) are generally pasteurized prior to, or after, filling the container. In addition, they may contain a chemical preservative. Carbonated soft drinks are not heat processed. Fruit flavored carbonated beverages may contain juices which furnish a good substrate, particularly for yeast growth. As a result, some of these beverages contain up to 0.1 percent benzoic acid for control of growth.[6, 12] The maximum allowed under FDA regulations is 0.1 percent as benzoic acid or sodium benzoate.[2]

The processing of soft drinks relies on the use of ingredients, containers, and processing equipment that carry only a small number of microorganisms. It is not common for the manufacturer to perform daily microbial counts. By using proper sanitizing procedures, he can produce microbiologically acceptable beverages without the need for routine testing. Should spoilage occur, the order of importance of microbiological sources generally will be (1) improperly sanitized syrup and beverage handling and filling equipment; (2) improperly washed and sanitized bottles; (3) sugar, syrups and flavored syrups.

The "Carbonated Beverage Industry Standard" for sugar requires the sugar to contain no more than an average of 10 yeast, 10 mold, and 200 bacteria per 10 grams.[10]

54.3 FLORAL CHANGES IN SPOILAGE

If acidophilic bacteria, and especially carbon dioxide tolerant yeasts, are initially present in a soft drink in sufficient numbers, they may become acclimated to the beverage environment and multiply. A visible sediment or haze will be present. In this case, an "off taste" may be detected. On occasion, certain strains of yeast, even when present in the beverage in low numbers, may multiply and cause visual and organoleptic changes.

Visual haze, particles in suspension and sediments, may be caused by nonmicrobial matter such as impurities in the sugar, plankton, and other aquatic forms in the water used for processing. The investigator should not be misled by visual observation.

Laboratory analysis of microbiologically spoiled beverages performed by the National Soft Drink Association over a period of years indicates that in over 90% of the instances spoilage was due to an excessive number of yeasts. The most frequent have been *Saccharomyces* followed by *Torulopsis*, and *Pichia*. Less frequently, acidophilic bacteria, generally *Lactobacillus, Leuconostoc* or *Acetobacter*, have caused spoilage by forming a haze, sediment, or off flavors. Molds are of little or no problem in carbonated beverages, although some yeasts form clumps which appear somewhat like mold. The investigator should culture the organisms before he renders a final judgment on its morphological characteristics.

54.4 RECOMMENDED METHODS

54.41 Sampling Procedures (Chapter 1)

Select at least two representative bottles or cans of product taken from plant stock or from stock at point of sale.

54.42 Aerobic Plate Count Method (Chapter 4)

a. Open the bottle aseptically and flame the top lightly. Flattop metal cans should be washed, the end wiped with alcohol, flamed, and pierced with a sterile opener.

b. Using a 10 ml pipet having 0.1 ml markings, withdraw a sample of the beverage. If difficulty is experienced with gassing of a carbonated beverage, pour sample into sterile container and agitate to dissipate the gas from the sample before pipetting. The Aerobic Plate Count Method (Chapter 4) described in this text is recommended. Chapter 16 of this text outlines the methods recommended for the detection of yeasts and molds.

54.43 Plate Count Using Membrane Filter Technique (Chapter 4)

The membrane filter technique has application to certain types of beverages. However, a uniform recommended procedure has not been established completely in the soft drink industry. For this reason the investigator who may wish to use this system should obtain the latest recommended equipment, materials, and test procedures from his equipment supplier. To the writer's knowledge the only manufacturers of this system in the U.S. are the Gelman Instrument Company, 600 South Wagner Road, Ann Arbor, Michigan 48106 and the Millipore Corporation, Bedford, Massachusetts 01730.

54.5 INTERPRETATION OF DATA

A soft drink is considered microbiologically spoiled when it is unsightly due to visual turbidity or sediment in a clear or nearly clear beverage, and generally when accompanied by an "off flavor." There is no simple answer as to what constitutes an excessive number of yeasts or bacteria. However, as a rule of thumb, if a beverage contains 50 or more viable yeast cells, or 200 or more bacteria per ml, the organisms may have become acclimated to their environment, and reproduction may occur until the product spoils. Replicate sampling should be performed to determine whether reproduction is continuing. A mold count in excess of 10 per ml indicates a fault in the ingredients or in the processing sanitation.

54.6 REFERENCES

1. Code of Federal Regulations. 1975. Title 21, Food and Drugs. Part 31, Nonalcoholic Beverages. U.S. Govt. Printing Office, Washington, D.C. pp 210–211
2. Code of Federal Regulations. 1975. Title 21, Food and Drugs. Subpart 121.101. *Ibid.* p 316, U.S. Govt. PrintingOffice, Washington, D.C.
3. EAGON, R. G. and C. R. GREEN. 1973. Effect of carbonated beverages on bacteria. Food Res. **22**: (No. 6): 687–688

4. INSALATA, N. F. 1952. Carbon dioxide versus beverage bacteria. Food Engineering. **24**: 84–85, 190
5. KORAB, H. E. 1974. Soft Drinks. (Encyclopaedia Britannica, BENTON, W. and BENTON, H. H. Publishers) Encyclopaedia Britannica, Inc., New York, N.Y.
6. PHILLIPS, G. F. 1971. Imitation Fruit Flavored Beverages and Fruit Juice Bases. Fruit and Vegetable Juice Processing Technology, TRESSLER, D. K. and JOSLYN, M. A., ed. The AVI Publishing Co., Inc., Westport, Conn.
7. QUINN, E. L. and C. L. JONES, 1936. Carbon Dioxide. American Chemical Society Monograph Series No. 72. Van Nostrand Reingold Pub. Corp., New York, N.Y. pp 120–121
8. SHARF, J. M. 1960. The evolution of microbiological standards for bottled carbonated beverages in the U.S. Annales de L'Institut Pasteur de Lille. **11**:117–132
9. SHILLINGLAW, C. A. and LEVINE, M. 1943. Effect of acids and sugar on viability of *Escherichia coli* and *Eberthella typhosa*. Food Res. **8**:404–476
10. Standards and Test Procedures for "Bottlers" Granulated and Liquid Sugar. 1962. National Soft Drink Association, Washington, D.C. pp 1–9
11. WITTER, L. D., J. M. BERRY, and J. F. FOLINAZZO. 1958. The viability of *Escherichia coli* and spoiled yeasts in carbonated beverages. Food Res. **23**:133–143
12. WOODROOF, J. G. and G. F. PHILLIPS, 1974. Beverages Carbonated and Noncarbonated. The AVI Publishing Co., Inc., Westport, Conn.

Index

A

(Numbers following "Table" and "Figure" are *page numbers*.)

B

C

M

N

R

S

XYZ